Imperial College London
2407821490

AF616213

The Selected Papers of
Sir Alan Fersht
Development of Protein Engineering

The
Cambridge Centre
for Protein Engineering
was opened by
The Right Honourable
Margaret Thatcher
OM·FRS·MP
on the first of May
1991

ICP Selected Papers — Vol. 1

THE SELECTED PAPERS OF SIR ALAN FERSHT

Development of Protein Engineering

Editors

Alan R Fersht
University of Cambridge, UK

Qinghua Wang
Baylor College of Medicine, USA

ICP Imperial College Press

Published by

Imperial College Press
57 Shelton Street
Covent Garden
London WC2H 9HE

Distributed by

World Scientific Publishing Co. Pte. Ltd.
5 Toh Tuck Link, Singapore 596224
USA office: 27 Warren Street, Suite 401-402, Hackensack, NJ 07601
UK office: 57 Shelton Street, Covent Garden, London WC2H 9HE

British Library Cataloguing-in-Publication Data
A catalogue record for this book is available from the British Library.

THE SELECTED PAPERS OF SIR ALAN FERSHT
Development of Protein Engineering

ISBN-13 978-1-84816-554-0
ISBN-10 1-84816-554-4

Printed by FuIsland Offset Printing (S) Pte Ltd, Singapore

Preface

I was in the right place at the right time with the right background and ideas to be a pioneer of a new area of science, protein engineering. Protein engineering, the design and construction of novel proteins by mutating their genes, is central to the analysis of the structure and activity of proteins, including their roles in cancer and disease, and is the basis of using designer proteins and antibodies in therapy and biotechnology. Qinghua Wang, my former student, decided that Imperial College Press should publish my selected papers. She went ahead with the planning and, when the Press agreed, informed me! The original plan was that I would choose papers and former colleagues, who would be asked to write about them. I duly selected the papers and started writing notes about each to help the colleagues I had in mind. But, before long, I did all the writing myself, connected everything together, and gave it to Qinghua to edit. But, I did not want this to be just a collection of papers in my honour.

I used various themes to pull the collection of papers into a coherent story. The collection and accompanying dialogue are primarily an autobiographical account of the progress of an individual scientist from raw youth to maturity through a long era that has been particularly exciting for the study of proteins. This personal journey is intertwined with the development of protein engineering. Each paper has been chosen to illustrate what contribution it made at the time, how it contributed to my own scientific development or sprung from it, and what place it had in the development of the field. There are many personal asides about others and myself that give insights into currents of opinion at the time and what makes individual scientists tick. 'Genius is one percent inspiration, ninety-nine percent perspiration,' wrote the great American inventor Thomas Alva Edison. Success in science is somewhat similar. But, in addition, there are essential components of luck and planning. Further, just as Darwinian evolution occurs by natural selection, sometimes adapting features that have evolved for different purposes, so science evolves from the traits of individual scientists. I have traced the roles of planning, serendipity, pre-adaptation and natural selection in my own work.

I have been one of a generation of chemists who have sought to unify chemistry and biology after their disastrous split a century ago. My own work has spanned organic and physical chemistry, biophysics, and molecular and structural biology. My approach is founded on a largely self-taught grounding in fundamental physical-organic chemistry, which I could translate to the study of proteins. In the process of writing this book, I have produced a non-systematic textbook of the evolution of the study of the chemistry of covalent bonds into the analysis of the chemistry of non-covalent bonding, which is the hallmark of biological interactions.

Being a scientist is one of the most fulfilling of careers, not least because the continual interaction with students keeps one young and enthusiastic. This book is dedicated to those students, post-doctorals and colleagues with whom I have enjoyed working so much and who have contributed even more to the work that is associated with me. My former students and post-doctorals have established themselves in Africa, the Americas, Asia, Australia and Europe. I could not want for a better scientific legacy. Perhaps, my writing about some of my own experiences will help and encourage more young scientists to follow their own course through science.

My work would not have been possible without 41 years of continuous, generous and enlightened support from the Medical Research Council of the UK, the MRC. Their support culminated in establishing the MRC Centre for Protein Engineering, which this collection of papers celebrates.

Alan Fersht
Cambridge
August 2010

Acknowledgments

We thank the American Chemical Society for waiving fees and granting us permission to reproduce papers from the *Journal of the American Chemical Society* and *Biochemistry*. We thank *Nature, Proceedings of the National Academy of Sciences USA* and *Journal of Molecular Biology* for allowing reproduction of papers according to the privileges extended to their authors. We thank also Times Newspapers for permission to reproduce the front cover of a *Sunday Times Magazine*.

Contents

Papers Reproduced

List of Colour Photos

Chapter 1

Brief Biographical Introduction

'A [child's] mind is a fire to be kindled, not a vessel to be filled.'

Plutarch

I have always been a scientist, and one who likes to learn unsystematically from experiment, experience and osmosis. When I was seven or eight years old, I dismantled a torch, worked out the electrical connectivity, and then constructed a circuit to light a Christmas crib in my primary school. When I was eleven, I persuaded my father to buy a collection of glassware and chemicals advertised on a notice board in the local newsagent. I quickly worked my way through a 12-volume 1926 encyclopaedia, which provided enough information initially to prepare simple gases and chemicals and then poisons and explosives. The local chemist, Mr. Frank Miller, who had been fined £100 for flying a light plane under Tower Bridge in 1951, was perfectly happy to allow a 12-year-old to gaze over the collection of chemicals in the back of the shop and buy sulphur, phosphorus, mineral acids, potassium ferrocyanide, and if I wanted anything really poisonous my father would have been allowed to sign for me. He was unabashed when, in my ignorance, I pronounced sulphuric acid as sulphurric. Many scientists of my generation entered science from their own curiosity. Sir Alec Jeffreys' entrée into science was dissecting a dead cat he found in a gutter. I was far too squeamish to be a biologist.

My grammar school (11–18 years) put no pressure on the boys to work, and there was little homework. But, this environment is perfect for inquisitive boys to find their own way. An inspiring maths master, 'Tubby' Taylor, instilled into us how to think in mathematics and steer our way through equations without being solely reliant on the algebra and calculus. I loved chemistry and physics because with minimal knowledge one could solve problems based on general principles and logical reasoning. I detested studying geography because I found it difficult to remember, for example, the names and order of towns along the Rhine. Similarly, biology was not for me, as I could not be bothered to learn the Greek names of all the vertebrae etc.

The free time also allowed me to throw myself into playing chess. Chess proved to be a formative intellectual training. I became Essex County Junior Champion and subsequently captain (president) of the Cambridge University Chess Club. My hero was

Emanuel Lasker, World Champion 1894–1921, who had written a chapter in his book *Lasker's Chess Manual* on the importance of 'the plan' in chess. The 'great Emanuel' expounded on how the early chess players tended to have only transient plans and to react to moves rather than have a long-term strategy. That is, they were primarily tacticians, knowing what to do when something has to be done. Wilhelm Steinitz, World Champion 1866–1894, revolutionised chess by the introduction of a coherent strategy — strategy is knowing what to do when there is apparently nothing to be done. Experimental science is like chess, and life in general — a balance of tactics and strategy. In chess, falsehood and bluster are soon exposed, and any success based on trickery is transient. 'In science, truth always wins,' wrote Max Perutz. Unfortunately, it can take much more time to win in science than in chess.

Bobby Fischer, that mad genius of chess, was born a month before me. He was clearly so much better than me that I knew I could never be a great player, especially as I had no desire to crush my opponents mentally, which drove on Fischer. In fact, I did not particularly enjoy beating opponents and just enjoyed playing and the minor successes. Just before going to university, I made a decision to stop taking chess seriously after an inter-county match when analysing some games with Jonathan Penrose, the British champion who led our county of Essex. It just hit me that one could have a very worthwhile career being a good scientist and contributing to the good of the world but to be successful at chess one had to be truly exceptional. My chess coach, Bob (R. G.) Wade was also living in near penury in a bedsitter. I did continue to play throughout my undergraduate days and even became captain of the Cambridge University Chess Club. But, I stopped studying chess. Recently, I came across on the internet excerpts from the school magazine, about my performance at chess and the farewell on leaving school.

> A. R. FERSHT. At School 1954–62. Prefect 1961–62. Captain of Chess 1959–62, and colours. Essex Junior Chess Representative. Captain of Bridge. School Council. Senior Circle. Bulletin Editor. Science Society. Hon. Bursar to Prefects. Physics Lab Assistant. State Scholarship.
>
> One can remember Alan for the warmth of his bubbling, effusive personality in the Prefects' Room. Graced with an extremely pleasant manner and countenance he was the balm and source of encouragement to us all. He will be accepted wherever he goes.
>
> Among his many other gifts we must, of course, record his cool scientific mind. His success in chess (Essex Junior Champion 1959–60, runner-up 1958–1959, 1st reserve for the English Junior Team) and his State Scholarship in Physics and Chemistry are proof of his logical approach to problems requiring careful analysis. He was, too, a great admirer of C. P. Snow in his pursuit of both art and science. Needless to say he held Dr. F. R. Leavis as his arch-enemy.
>
> Must it be said that he had a wealth of background and experience? From 'my ultimate success' as 1st reserve in the Bickersteth Cup Shot, to conducting Ray

Yorke's election campaign with the latest motivational research, he attacked everything he did with vigour and enthusiasm.

That he did not emerge as *eminence grise* of the Prefects (for he was after all their bursar, auditor and accountant, and had an intimate knowledge of character analysis by handwriting) is of great credit to him. His was the greatest opportunity: his too was the selfless acceptance of responsibility.

In whatever life he chooses, we know he will succeed. He carries with him all the good wishes of his innumerable friends at Monoux in taking his place at Gonville and Caius College, Cambridge.

The Monovian 1962

CHESS TEAM CRITICISMS
A. R. FERSHT

Writing a criticism of Alan's chess is extremely difficult, to say the least. Out of his ten inter-school games he has lost only one and drawn only one, and this record surely speaks for itself. As captain of the school team, his own confident play and his encouragement are constant inspirations. His enthusiasm for the game spreads even to the first-formers, and it is largely due to his encouragement of these younger players that the school should have a strong team for the next few years. His interest in school chess is very great, while he plays regularly for West Ham Chess Club. He has gained a position on the British Chess Federation's grading lists, a great achievement. On top of this he is Joint Essex Junior Champion.

R.A.L., Monovian 1961

I hope the same will be said about me as a scientist and head of laboratory.

The overwhelming emphasis in today's schools of constant examinations and assessment — routine project work and low-level exams — would have destroyed me. I loved finding advanced texts and delving into them. The pre-digested pap of the standard school textbooks turned me off. The ancient Greek Plutarch said that 'a [child's] mind is a fire to be kindled, not a vessel to be filled'. The ancient Greeks were so right.

What I wanted to do on leaving school in 1962 was to understand the physical and chemical basis of the world around me. The Natural Sciences Tripos in Cambridge was then the ideal university course, with the students being able to pursue a variety of different subjects. I chose two years of physics, a year each of biochemistry and mathematics, and three years of chemistry, specialising in physical chemistry. But, I had to learn for the first time how to work hard because it took me two years to catch up with the best students from the good private sector schools who were both very bright and had been very well trained. Fortunately, there are no social barriers in science, and science has always been a route for bright children from poor backgrounds or for new immigrants. My grandparents had fled the pogroms in Eastern Europe at the beginning of the 20th century, and I was the first in my family to go to university. Sydney Brenner's father was an illiterate cobbler from Lithuania, probably coming not far from where my grandfather had been born. The background of being able to learn by myself, training my mind with chess and the confidence it gave me, followed by being knocked into shape in Cambridge was a very good preparation to be a research scientist. It prepared me for charting unknown waters, guided by basic principles. I was not a genius, unfortunately, but I had talent in particular areas and the sense of knowing what to do to make the best of my abilities. And, if I saw a crowd running in one direction, I would walk slowly in the opposite.

1.1 1960: Surrounded by chess trophies.

Chess

Element of surprise

SURPRISE is one of the most formidable weapons in a chess player's armoury, and many leading masters place great reliance on preparing new variations which will take their opponents unawares. A classic case is a counter-attack against the Ruy Lopez, which was discovered by the American champion, Frank Marshall. For years Marshall kept his innovation to himself, until he had an opportunity to play it against Capablanca, then recognised as the greatest player of the day. Capablanca, who had probably the finest natural chess talent of all time, threaded his way through the complications and beat off the attack to win.

Since this introduction of Marshall's gambit, his idea has been refined, deeply analysed, refuted, and rehabilitated—the common lot of all popular variations. Nevertheless, the gambit still continues to claim its annual quota of victims, and can be confidently recommended to anyone looking for a way of upsetting the most formidable of all white openings. This week's illustration won the du Mont best game prize in the 1960 London Boys' Championship. Alan Fersht, of Chingford, is a relation of the American grandmaster, Samuel Reshevsky, and has had several tournament successes.

WHITE S. P. Broido	BLACK A. R. Fersht
Ruy Lopez	
1 P-K 4	P-K 4
2 Kt-KB 3	Kt-QB 3
3 B-Kt 5	P-QR 3
4 B-R 4	Kt-B 3
5 Castles	B-K 2
6 R-K 1	P-QKt 4
7 B-Kt 3	Castles
8 P-B 3	P-Q 4
9 P × P	P-K 5

Marshall's original gambit was 9 . . . Kt × P; 10 Kt × P, Kt × Kt; 11 R × Kt, Kt-B 3 (11 . . . P-QB 3 is another popular move here); 12 P-Q 4, Kt-Kt 5; 13 R-K 1, B-Q 3; 14 P-KR 3, Q-R 5; 15 Q-B 3 (if the knight is accepted, Black has a winning attack beginning with 15 . . . Q-R 7 ch), Kt × P. Here Capablanca saw the danger and played 16 R-K 2. If he had accepted the sacrifice by 16 Q × Kt, then Black does not reply 16 . . . B-Kt 6? because of the sudden brilliancy 17 Q × P ch!, R × Q; 18 R-K 8 mate, but plays first 16 . . . B-R 7 ch; 17 K-B 1, and only now 17 . . . B-Kt 6; 18 Q-K 3, B × P!; 19 P × B, Q × P ch; 20 K-Kt 1, Q-R 7 ch; 21 K-B 1, QR-K 1 and wins.

10 P × Kt	P × Kt
11 Q × P	

A dangerous capture, since it enables Black to increase his lead in development by attacking the queen. Preferable is 11 P-Q 4, when a game in the Madrid zonal tournament between Pachman and Neikirch continued 11 . . . B-KKt 5; 12 P-KR 3, B-R 4; 13 P-Kt 4, Kt × P!; 14 Q × P! (if 14 P × Kt, B × P; followed by . . . B-Q 3 and . . . Q-R 5 White's king is exposed to an overwheming attack), Kt-B 3; 15 Q-Kt 2, B-Kt 3; when the weakness of White's king's position is just about balanced by his extra pawn.

11 . . .	B-KKt 5
12 Q-K 3?	

A bad mistake; it is courting trouble for White to place his queen on a file which Black's rook can control in a couple of moves. Correct is 12 Q-Kt 3.

12 . . .	R-K 1
13 P-B 3	

A second blunder, which leaves White's king and queen on the same diagonal as the black bishop, a sure invitation to a combinative attack. He could resist for longer by 13 P-Q 4, B-Q 3; 14 Q-Q 2, B-B 5!; 15 R-K 3, B × R; 16 P × B, Kt-K 5; although White would still suffer from his undeveloped queen's wing.

13 . . .	B-QB 4!
14 Q × B	R × R ch
15 K-B 2	Kt-K 5 ch

A most effective finish, although Black could also win by the simple 15 . . . R × B. If now 16 P × Kt, Q-R 5 ch; 17 P-Kt 3, Q × RP ch; 18 K × R, Q-K 7 mate.

16 K × R	Kt × Q
17 P × B	Q-K 2 ch
18 K-Q 1	Kt-Q 6
19 K-B 2	Q-K 5
White resigns	

LEONARD BARDEN

1.2 1961: "Best game", published in *The Field*.

Chapter 2

The Early Research Years: Influence of William P. Jencks

'Talent does what it can; genius does what it must.'

Edward Bulwer-Lytton

Ph.D. Work: Intramolecular Catalysis as Model for Enzymatic Catalysis (1965–1968)

I graduated with a first class degree in 1965, in the days when only the top 10% were awarded them, and, I believe, I was top in physical chemistry. At one stage, I was going to do my Ph.D. under Sir Cyril Hinshelwood after he had retired from Oxford to Imperial College, on a scholarship from the Gas Council. But, I was warned that he was still following reactions using manometers and not using spectroscopy and direct measurements. Fortunately, I changed my mind and chose to do my Ph.D. research in physical-organic chemistry, which played to my strengths. It is an area where logical analysis is crucial and the highest rigour is demanded. The concepts are relatively simple to grasp for those trained in physical chemistry. Being reasonably numerate is an advantage, but high specialist mathematical skills are not necessary. Consequently, a bright and eager young person can think his or her way through problems. My ambition from my training as a chemist was to solve difficult problems, but most important of all was to have original ideas and do work that was entirely novel and eschew routine studies. My supervisor, Anthony Kirby, left me alone, which was perfect for me and allowed my development into a self-taught kineticist and experimentalist. I delighted in formulating problems for which I could develop kinetic methods of solving. Not being a great mathematician, I had fun working out short cuts to solve kinetic equations by thinking through to the answers, thanks to Tubby Taylor. I also loved designing experiments where the measurements were made in ways to minimize experimental errors. My favourite trick was to measure differences directly or take ratios. These methods should, of course, be part of normal experimental planning. But, it is great fun for a student to work them out for himself. Being allowed to develop with minimum interference is one of the greatest gifts a research supervisor can bestow on a student.

Physical-organic chemists were then trying to understand the huge catalytic power of enzymes by making chemical models. The topic of intramolecular catalysis, where a compound is synthesized in which a catalytic group is close to a substrate radical and its rate of reaction studied, was considered as an excellent model for a reaction in the active site of an enzyme. Aspirin has a carboxylate group as a neighbour of an ester bond. As such, it was a model for the esterolytic activity of enzymes that were then commonly studied. Myron Bender, whom I consider one of my most important indirect mentors via the literature, had published that the reaction mechanism was intramolecular nucleophilic catalysis, whereby the carboxylate attacked the ester to form a highly reactive mixed anhydride. I had never attended classes in physical-organic chemistry, and I learned much of kinetics from Bender's papers. Unfortunately, I found that there was a mistake in Bender's experiments and there was no evidence for the mixed anhydride. My supervisor wanted me to make sulfonate and phosphonate analogues of aspirin. But, I was rotten at the synthetic chemistry and gave up the attempts after about six months of failure when I realised that the proposed mechanisms for the hydrolysis of aspirin were all kinetically equivalent and could not be distinguished between by standard kinetic measurements. Inspired by the work of Victor Gold, I decided the way to distinguish between them was by a combination of Linear Free Energy Relationships (LFERs), trapping experiments and elimination of alternative mechanisms by showing them to be impossible. In my very first paper, I hit on the idea of examining the effects of a substituent in the aromatic ring of aspirin as a measure of the charge development simultaneously on the leaving group and the catalyst. I later discovered that Jaffe had introduced the method some years earlier, but it had rarely been used. To do the experiments, I had to make a whole range of substituted aspirins, but the chemistry was quick and easy.

My very first paper to be published was:

Fersht, A. R. & Kirby, A. J. (1967). Structure and mechanism in intramolecular catalysis. The hydrolysis of substituted aspirins. *J Am Chem Soc* **89**, 4853–7.

2.1 1964: Alan with Tasos Varvoglis, Tony Kirby's other graduate student.

2.2 1969: Operating a stopped-flow spectrometer in Bill Jenck's laboratory in Brandeis.

[Reprinted from the Journal of the American Chemical Society, **89**, 4853 (1967).]

Structure and Mechanism in Intramolecular Catalysis. The Hydrolysis of Substituted Aspirins

A. R. Fersht and A. J. Kirby

Contribution from the University Chemical Laboratory, Cambridge, England.
Received May 3, 1967

Abstract: The rate of hydrolysis of aspirin is notably insensitive to the effects of substituents in the 5 and particularly the 4 position. The 4-methoxy and 4-nitro compounds, for example, are hydrolyzed at the same rate. This is a consequence of the larger but opposite effects of substituents on the interacting carboxyl and ester groups. These effects can be separated using Jaffé's equation, and the ρ values obtained are compared with values for intermolecular catalysis of ester hydrolysis by various mechanisms. They are consistent with a mechanism for intramolecular catalysis in which the ionized carboxyl group acts not as a nucleophile but as a general base.

As part of a detailed investigation of the mechanism of hydrolysis of aspirin,[1] we have studied the effect on the rate of hydrolysis of substituents in the 4 and 5 position. The structure and reactivity of aryl acetates

CO_2H
5
4
$OCOCH_3$

substituted in the *meta* or *para* positions are correlated by the Hammett equation.[2] A second substituent in the *ortho* position affects the reactivity of each one of a series of such esters, but as long as this group is not involved directly in the reaction process their relative reactivity, and thus the ρ value for the reaction, is not affected.[2,3]

If, however, the group in the *ortho* position is involved in bond making or breaking in the transition state, a substituent in the 4 or 5 position will affect the reactivity of both reacting centers. We have attempted to separate the effects of substituents on the carboxylic acid and the ester group in aspirin hydrolysis in the hope that a comparison of the two ρ values thus obtained with those for intermolecular reactions of known mechanism may allow definite conclusions about the mechanism of intramolecular catalysis in the hydrolysis of aspirin.

Experimental Section

Materials. Inorganic salts were of analytical grade, and were used without further purification. Distilled water was further glass distilled twice before use. 5-Bromo- and 5-iodosalicylic acids, as well as the unsubstituted compound, were obtained commercially. The 4-chloro, 4-bromo, and 4-iodo compounds were prepared by the method of Ohta.[4] Published procedures were used also for the preparation of the 5-chloro,[5] 5-methoxy,[6] 4-nitro,[7] and 5-nitro[8] compounds. The 4-methoxy derivative was obtained in 90% yield by diazotizing 4-aminosalicylic acid with amyl nitrite in methanolic HCl, and had mp 158–159.5° (lit.[9] 157°). *Anal.* Calcd for $C_8H_8O_4$: C, 57.14; H, 4.76. Found: C, 57.29; H, 4.82.

Acetylsalicylic Acids. Substituted salicylic acids were acetylated by the method of Ciampa,[5] and recrystallized from ethanol, chloroform, or benzene. The analytical data and melting points are listed in Table I.

Table I. Analytical Data for Substituted Aspirins

Aspirin	Mp,[a] °C	Lit. mp, °C	Calcd, % C	Calcd, % H	Found, % C	Found, % H
4-Cl	133–135	131.5[b]	50.4	3.26	50.1	3.61
4-Br	151–152	. . .	41.7	2.70	41.7	2.92
4-I	157–158	156[c]	35.3	2.29	35.6	2.38
4-NO_2	153–154	155[d]	48.0	3.11	48.1	3.10
4-MeO	123–124	119–121[e]	57.1	4.76	57.4	4.99
5-Cl	147–149	148[f]	50.4	3.26	50.5	3.24
5-Br	155–157	156,[g] 168[f]	41.7	2.70	41.5	2.95
5-I	164–166	166[f]	35.3	2.29	35.2	2.38
5-NO_2	153.5–154.5	. . .	48.0	3.11	47.9	3.12
5-MeO[h]	154–156	. . .	57.14	4.78	57.29	4.82

[a] Uncorrected. Taken on a Kofler block. [b] R. Kuhn and H. R. Hansel, *Chem. Ber.*, **84**, 557 (1951). [c] P. Brenans and C. Post, *Compt. Rend.*, **178**, 1012 (1920). [d] M. Viscontini and J. Pudles, *Helv. Chim. Acta*, **33**, 591 (1950). [e] W. Schulemann and F. Schönhöfer, U. S. Patent 1,588,814; *Chem. Abstr.* **20**, 2563 (1926). [f] Reference 5. [g] P. Brenans and C. Girod, *Compt. Rend.*, **186**, 1128 (1928). The compound was recrystallized to constant melting point from benzene, from ethanol, and from chloroform. [h] The analysis of the acetyl compound was identical with that of the parent salicylic acid, which has the same empirical formula. The melting points are also closely similar. The ultraviolet spectra were sufficiently different, however, to follow the hydrolysis of one to the other.

Kinetic Measurements. The rates of hydrolysis of the substituted aspirins were measured by following the initial rates of release of substituted salicylate anions, at the ultraviolet absorption maxima of the latter (Table II), at 39° and ionic strength 1.0 (KCl). The methods are described in detail in the following paper.

The neutral hydrolysis rate was measured for each ester at five or more different pH values between pH 5.4 and 6.6, in 0.05 *M* phosphate buffers. The observed rates were not independent of buffer concentration, so the buffer constant was obtained for each ester by

(1) A. R. Fersht and A. J. Kirby, *J. Am. Chem. Soc.*, **89**, 4857 (1967).
(2) H. H. Jaffé, *Chem. Rev.*, **53**, 191 (1953).
(3) H. H. Jaffé, *Science*, **118**, 246 (1953).
(4) H. Ohta, *Nippon Kagaku Zasshi*, **78**, 1608 (1957).
(5) G. Ciampa, *Ann. Chim.* (Rome), **54**, 975 (1964).
(6) D. N. Chaudhury, H. I. King, and A. Robertson, *J. Chem. Soc.*, 2220 (1948).
(7) H. Seidel and J. C. Bittner, *Monatsh. Chem.*, **23**, 431 (1902).
(8) H. C. Barary and M. Pianka, *J. Chem. Soc.*, 965 (1946).
(9) M. Gomberg and L. C. Johnson, *J. Am. Chem. Soc.*, **39**, 1687 (1917).

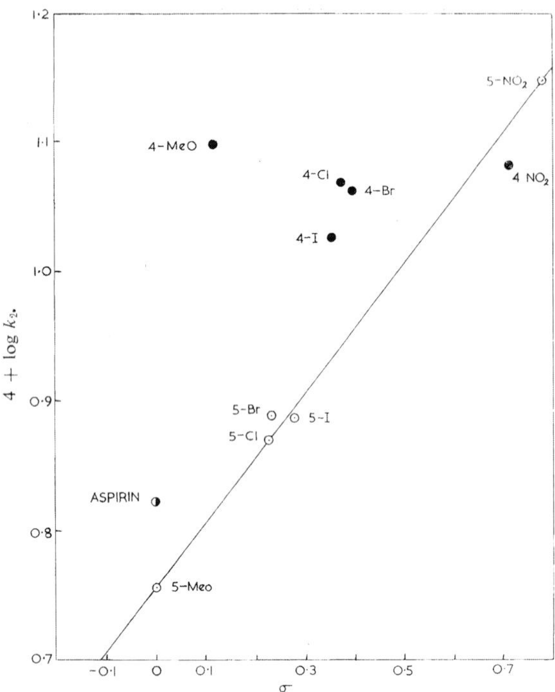

Figure 1. Hammett plot of the hydrolysis data of Table II. σ_m and σ_p values taken from ref 2, except for *p*-MeO. σ_p for the methoxyl group is well known to vary considerably with the reaction concerned.[2] We have used a value of −0.01, obtained by fitting the point for the 5-methoxy compound to the Hammett plot of the phosphate buffer constants from Table II. This value is identical with that obtained by H. van Bekkum, P. E. Verkade, and B. M. Wepster [*Rec. Trav. Chim.*, **78**, 815 (1959)], using data for the ionization of substituted phenols in 8% dioxane–water at 38°. They calculate a value of −0.109 for the same reaction at 25° in water, while a positive value near 0.07 is necessary to account for the rate of hydrolysis of 5-methoxyphenyl salicylate at 59.2° [B. Capon and B. C. Ghosh, *J. Chem. Soc., Phys. Org. Sect.*, 472 (1966)], suggesting that σ_p for the *p*-methoxy group may be temperature dependent.

measuring the hydrolysis rate at three further concentrations of phosphate buffer, up to 0.25 *M* (50% free base). The second-order constant for phosphate catalysis was obtained from the slope of the linear plot of k_{obsd} *vs.* [HPO_4^{2-}].

Results

Pseudo-unimolecular rate constants for the hydrolysis of substituted aspirins are listed in Table II. These values are the means from measurements at five different pH's between 5.4 and 6.6, corrected for phosphate buffer catalysis using the catalytic constants shown. They clearly fall in the pH-independent region, and represent the intramolecularly catalyzed reaction.

The most notable feature of the hydrolysis rates is their very low sensitivity to substituents. Thus the 5-nitro group increases the rate of hydrolysis by a factor of only 2, although *p*-nitrophenyl acetate is hydrolyzed, and reacts with acetate ion, 15 times faster than phenyl acetate.[10] The 4-substituted compounds are all hydrolyzed at very similar rates, the effect of the 4-methoxy group, for example, being identical, within experimental error, with that of the 4-nitro group (Table II).

(10) D. G. Oakenfull, T. Riley, and V. Gold, *Chem. Commun.*, 385 (1966).

Table II. The Hydrolysis of Substituted Aspirins at 39°, Ionic Strength 1.0

Compd	Followed at mμ	$k_{hyd} \times 10^4$,[a] min^{-1}	k_2 for HPO_4^{2-} catalysis, M^{-1} min^{-1} $\times 10^4$ [b]
Aspirin	298.5	6.65 ± 0.01[c]	6.93 ± 0.12[c]
4-Cl	297	11.73 ± 0.07	16.0
4-Br	297	11.54 ± 0.09	16.1
4-I	300	10.62 ± 0.05	14.4
4-NO$_2$	349	12.64 ± 0.09	24.2
4-MeO	291.5	12.52 ± 0.10	10.8
5-Cl	309	7.41 ± 0.06	11.5
5-Br	308	7.75 ± 0.05	10.5
5-I	313	7.70 ± 0.05	12.8
5-NO$_2$	317	14.03 ± 0.09	63.3
5-MeO	318	5.71 ± 0.03	6.74

[a] Mean and standard error from measurements at 5 pH's between 5.4 and 6.6. [b] Buffer constants for substituted aspirins are accurate to ±4%. [c] Data from ref 1.

Discussion

The Hammett plot of the hydrolysis data of Table II for 5-substituted aspirins (using σ_p) gives an acceptable straight line (Figure 1), and a ρ value of 0.50. Such a low value of ρ suggests that the effect of the substituent on the phenolic oxygen atom of the ester is partially offset by an opposite effect on the reactivity of the catalyzing carboxyl group.

The data for 4-substituted compounds, on the other hand, fall apparently randomly on this plot (using σ_m), except that the deviation of the points from the line for the 5-substituted compounds is least for electron-withdrawing 4 substituents (Figure 1). As expected, the simple two-parameter Hammett relation has broken down because substituents have separate, in this case probably opposite, effects on the reactivity of the electrophilic and nucleophilic centers involved in the transition state.

This type of problem has been treated previously by Jaffé;[3,11] who proposed for reactions between side chains of the same ring the equation

$$\log k/k_0 = \sigma_1\rho_1 + \sigma_2\rho_2 \quad (1)$$

where σ_1 and σ_2 are the substituent constants (σ_m and σ_p) of the group in the 4 or 5 position, relative to the reacting groups in the 1 and 2 positions, and ρ_1 and ρ_2 are the reaction constants for the effect of the substituent on the groups in the 1 and 2 positions, respectively. Not surprisingly, this four-parameter equation gives a better fit than the simple Hammett equation for many sets of data, including some where no *ortho* substituent is present. Jaffé considers that such spurious results arise where there is a strong correlation between the values of σ_m and σ_p, and suggests that the usefulness of eq 1 may be limited to series where the correlation coefficient between σ_m and σ_p is less than 0.9.

With this restriction Jaffé[11] obtained two separate ρ values for the effect of substituents on the pK_a's of catechols and of 2-hydroxymethylbenzoic acids. The pK_a's of substituted salicylic acids, however, were not correlated by eq 1 significantly better than by the simple Hammett equation.[11] This has been confirmed recently by a careful study[12] using 17 substituted salicylic

(11) H. H. Jaffé, *J. Am. Chem. Soc.*, **76**, 4261 (1954).
(12) G. E. Dunn and F.-L. Kung, *Can. J. Chem.*, **44**, 1261 (1966).

acids, with the low correlation coefficient of 0.723 between σ_m and σ^- for the substituents.

The application of eq 1 to the hydrolysis of substituted aspirins makes no assumptions about the mechanism of catalysis by the carboxyl groups except that the two groups are affected separately by a given substituent. The hydrolysis of substituted phenyl acetates can be correlated by Hammett's equation whether the rate-determining step is attack by water, hydroxide ion, or a catalyzing carboxylate anion. Also the reactivity of the carboxyl group should follow Bronsted's equation, whether it is acting as a general acid,[13] a general base, or a nucleophile. Also, the pK_a of this group is known to follow a Hammett relation for substituted salicylic acids[12] in which deviations would seem to be more likely, because of intramolecular hydrogen bonding, than for the corresponding acetyl compounds.

Equation 1 can therefore be written in the specific form

$$\log k/k_0 = \rho_{\text{acid}}\sigma_1 + \rho_{\text{phenol}}\sigma_2 \qquad (2)$$

where the ρ's measure the separate effects of the 4 and 5 substituents on the reactivity of the carboxyl[13] and phenol ester groups, σ_1 is σ_p for a 4 substituent, σ_m for a 5 substituent, and σ_2 is σ_m for a 4 substituent and σ_p for a group in the 5 position. Equation 2 can be written as the equation (3) of a straight line.

$$1/\sigma_1(\log k/k_0) = (\sigma_2/\sigma_1)\rho_{\text{phenol}} + \rho_{\text{acid}} \qquad (3)$$

The data of Table II for the hydrolysis of both 4- and 5-substituted aspirins are plotted according to eq 3 in Figure 2. They now give an excellent straight line (with correlation coefficient 0.994), and the slope and intercept, respectively, give values for

$$\rho_{\text{phenol}} = 0.96 \pm 0.04$$

$$\rho_{\text{acid}} = -0.52 \pm 0.03$$

The correlation coefficient between σ_m and σ_p for the substituents used is 0.758, and therefore satisfies Jaffé's criterion discussed above.

Comparison with Intermolecular Catalysis. Three possible mechanisms are discussed in the following paper[1] for intramolecular catalysis of the hydrolysis of aspirin by the carboxyl group. These are (a) intramolecular nucleophilic catalysis by the carboxylate anion; (b) intramolecular general acid catalysis of the attack of hydroxide ion by the undissociated carboxylic acid group; and (c) intramolecular general base catalysis of the attack of a water molecule by the carboxylate anion.

We can make a tentative choice between these three possibilities by comparing the ρ values obtained above with those for similar intermolecular reactions of known mechanism. This procedure is based on the assumption that the reaction constant, ρ, will be the same, or closely similar, for the intramolecular reaction of a given nucleophile with the carbonyl group of a series of substituted phenyl acetates as it is for the corresponding intermolecular reaction, as long as the same mechanism is involved. If it is accepted that the increased rates of

(13) In the case of general acid catalysis by the carboxylic acid group of attack by hydroxide ion, the rate constants correlated by the Bronsted equation, and by eq 2, are the second-order constants for attack by OH^-, rather than the observed pseudo-unimolecular constants. We show in the Appendix that the observed rate constants are also correlated by an equation of the same form as (2), in which the parameter ρ_{acid} is, in fact, a composite constant.

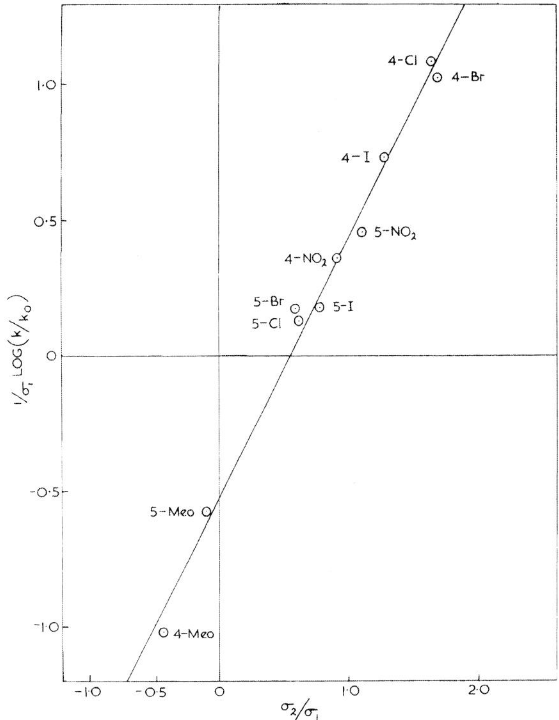

Figure 2. Modified Hammett plot of the hydrolysis data of Table II. σ_m and σ_p values from ref 2, except for *p*-MeO (see caption for Figure 1).

intramolecular reactions are due almost entirely to more favorable entropies of activation this proposition seems unexceptionable. There is, however, not sufficient experimental evidence to establish its generality. The most directly relevant information available comes from a comparison by Bruice and Benkovic[14] of the catalysis of the hydrolysis of substituted phenyl acetates by trimethylamine with the corresponding intramolecular reaction of the ω-dimethylaminobutyrate and -valerate esters. The enthalpies of activation were closely similar for the three reactions, and the higher rates for intramolecular catalysis were due to more favorable entropies of activation. The ρ values were identical (+2.2) for the intramolecular reactions, involving five- and six-membered ring formation, and close to the value for the bimolecular reaction (ρ = 2.5). This was taken as evidence that the same mechanism (rate-determining breakdown of the tetrahedral addition intermediate) is involved in each case.

A direct comparison of ρ values for inter- and intramolecular catalysis can also be made using the data of Table II. For reasons discussed in the following paper[1] we believe that phosphate buffer catalysis of aspirin hydrolysis involves a simple reaction between the ester anion and the phosphate dianion, without catalysis by the carboxyl group. In agreement with this, the bimolecular rate constants for phosphate catalysis of the hydrolysis of both 4- and 5-substituted aspirins (Table II) follow the simple Hammett relation. The ρ value obtained is 0.96, identical with ρ_{phenol} obtained from the hydrolysis data plotted according to eq 3. We have independent evidence[1] that the two reactions involve the

(14) T. C. Bruice and S. J. Benkovic, *J. Am. Chem. Soc.*, **85**, 1 (1963).

same mechanism, so that this equality is further justification for the proposition stated above.

Gaetjens and Morawetz[15] found that the hydrolysis of substituted monophenyl succinates and glutarates, catalyzed by the terminal carboxyl group, is much more sensitive ($\rho = +2.5$) to substituents than the corresponding intermolecular reaction ($\rho = 1.1$[14] or 1.7 (see below)). This difference is taken[14,15] as evidence of a change in the rate-determining step from formation of the tetrahedral addition intermediate, in the bimolecular case, to its breakdown,[14] in the intermolecular reaction. In this second mechanism the bond to the phenol oxygen is broken in the slow step of the reaction, thus accounting for its much greater sensitivity to substitution in the aromatic ring.

The hydrolysis of substituted aspirins ($\rho_{phenol} = 0.96$) is much less sensitive to substitution than these two intramolecular reactions, and we consider that this rules out mechanisms in which the slow step is the breakdown of the tetrahedral addition intermediate. Therefore we limit the discussion to mechanisms involving the addition of a nucleophile to the carbonyl group of the ester in the rate-determining step.

Nucleophilic Catalysis. It is well known that the rates of bimolecular reactions of nucleophiles of similar structure with *p*-nitrophenyl acetate are correlated by the Brønsted equation, with a Bronsted coefficient β near 0.8. This is true in particular for substituted phenols,[16] and corresponds to a ρ value of 1.6–1.7, sharply different from the $\rho_{acid} = 0.52$ observed for substituted aspirin hydrolysis.

ρ_{phenol} for nucleophilic catalysis of the hydrolysis of substituted phenyl acetates by acetate ion can be calculated from the data of Oakenfull, Riley, and Gold,[17] who have separated the contributions of nucleophilic and general base catalysis. The value obtained is 1.7, again much greater than the figure (0.96) for the aspirin reaction.

General Acid Catalysis of the Attack of Hydroxide Ion.[18] In this case the intermolecular reaction has never been identified, although molecules of hydroxylic solvents presumably act as general acids to some extent in the alkaline hydrolysis of esters. ρ for the hydroxide-catalyzed hydrolysis of substituted phenyl acetates is approximately 0.55,[16] and would presumably be lowered still further if more effective general acids than water catalyzed the addition of hydroxide ion to the ester carbonyl group. Since ρ_{phenol} for substituted aspirin hydrolysis (0.96) lies between the values for bimolecular attack by hydroxide and by acetate ion, it seems likely that a nucleophile intermediate in basicity between these two anions is involved in aspirin hydrolysis. An obvious possibility is the partially protonated hydroxide ion which must develop in the general base catalyzed attack of a molecule of water.

(15) E. Gaetjens and H. Morawetz *J. Am. Chem. Soc.*, **82**, 5328 (1960).

(16) (a) T. C. Bruice and R. Lapinski, *ibid.*, **80**, 2265 (1958); (b) T. C. Bruice, T. H. Fife, J. J. Bruno, and N. E. Brandon, *Biochemistry*, **1**, 7 (1962).

(17) D. G. Oakenfull, T. Riley, and V. Gold, *Chem. Commun.*, 385 (1966).

(18) Since intermolecular general acid catalyzed addition of hydroxide ion has not been measured, no Brønsted α is available to compare with ρ_{acid}. It should be noted, however, that since the observed ρ_{acid} is a composite constant for this mechanism,[13] the correct comparison is with $\rho'_{acid} = (\rho_{acid} + \rho_{ionization})$. Since ρ for the ionization of substituted benzoic acids[2] is generally close to 1, α and ρ'_{acid} are probably roughly equal, with a value of about $(-0.52 + 1) = 0.5$.

General Base Catalysis. Since ρ for the ionization of substituted benzoic acids is generally close to 1,[2] ρ_{acid} for the mechanism in which the carboxylate group acts as a general base is numerically close to the Brønsted coefficient, β, and can be compared directly with β values observed for intermolecular reactions.

There are no data available for general base catalysis by substituted benzoate anions of the hydrolysis of a phenyl acetate. Only a small number of general base catalyzed ester hydrolyses have been measured, and the β values observed are for reactions not exactly analogous to aspirin hydrolysis. Some reactions known to be catalyzed by oxyanions acting as general bases are the hydrolysis of ethyl dichloroacetate[19] ($\beta = 0.47$) and of phenyl dichloroacetate[1,20] ($\beta = 0.35$). A similar value (0.30) is observed for the intermolecular catalysis of aspirin hydrolysis by oxyanions, which we consider to be general base catalysis.[1] The observed ρ_{acid} (0.52) is close enough to these figures, with the uncertainties involved, to be considered not inconsistent with the mechanism in which the carboxylate group acts as a general base.

ρ_{phenol} can be compared directly with the value for the general base catalyzed hydrolysis of substituted phenyl acetates by acetate ion. For this reaction ρ can be calculated from the data of Oakenfull, Riley, and Gold[17] as 1.1 ± 0.2. ρ_{phenol} for the hydrolysis of substituted aspirins is 0.96.

Thus in the case of general base catalysis, and in this case only, both ρ_{phenol} and ρ_{acid} are consistent with values observed for the corresponding intermolecular reactions.

Conclusions

Within the limits of the necessary assumptions the separation of the ρ values for the two groups involved in aspirin hydrolysis, and the comparison of these with values observed for intermolecular reactions, does permit an unambiguous conclusion to be drawn: that the most probable mechanism for the hydrolysis of aspirin is that in which the ionized carboxyl group acts as a general base.

The mechanism is discussed in detail in the following paper,[1] in the light of new evidence concerning the hydrolysis of aspirin itself. Consideration of this evidence leads independently to the same conclusion about the hydrolysis mechanism, which is some evidence that the assumptions made in this paper are correct.

Appendix

Equation 2 applies for the observed pseudo-unimolecular rate constants for mechanisms which involve only the aspirin anion, or the anion and one or more molecules of water. For the mechanism involving intramolecular general acid catalyzed attack by hydroxide ion, the equation correlates the second-order constants for the attack of hydroxide

$$\log k_2/k_2^0 = \rho'_{acid}\sigma_1 + \rho_{phenol}\sigma_2$$

k_2 is related to the observed pseudo-unimolecular rate constant k by the expression

$$k_2 = kK_a/K_w$$

(19) W. P. Jencks and J. Carriuolo, *J. Am. Chem. Soc.*, **83**, 1743 (1961).

(20) K. Koehler, R. Skora, and E. H. Cordes, *ibid.*, **88**, 3577 (1966).

where K_a is the dissociation constant of the substituted aspirin. Therefore

$$\log k_2 = \log k + \log K_a - \log K_w$$

$$\log k_2/k_2^0 = \log k/k_0 + \log K_a/K_a^0$$

$$= \log k/k_0 + \rho_{ionization}\sigma_1$$

where $\rho_{ionization}$ is the reaction constant for the ionization of substituted aspirins. Thus

$$\log k/k_0 = (\rho'_{acid} - \rho_{ionization})\sigma_1 + \rho_{phenol}\sigma_2 \quad (4)$$

$$= \rho_{acid}\sigma_1 + \rho_{phenol}\sigma_2 \quad (5)$$

and the observed rate constants are related by an equation of the same form as eq 2. However, the *sign* of the observed ρ_{acid} is not a good criterion of mechanism, and cannot be used to reject the bimolecular mechanism, as explained qualitatively by Capon and Ghosh (see reference in Figure 1 caption).

The importance of the work is that it established intramolecular general-base catalysis, which is now found widespread in protein mechanisms — it was previously thought that the Aspartate side chain would act as a nucleophile. Bill (W. N.) Lipscomb (the 'Colonel') visited Cambridge and asked to meet me. He was very enthusiastic about the work because it provided a basis for the mechanism of carboxypeptidase-A, the enzyme whose structure he had solved. Never underestimate the encouragement and thrill of junior scientists when senior scientists take notice of them and treat them as equals.

I later found that by making appropriate substitutions in the aspirin framework I could change the mechanism to nucleophilic catalysis[1–5], especially by using a second nucleophile to trap the anhydride[4]. Seeing how reaction mechanisms could change with small changes in structure made me realise that there is rarely a unique mechanism for a reaction, and I have always been comfortable with the existence of multiple mechanisms. Some of the bitterest fights in science are between those who believe they have the one true mechanism.

The importance of my Ph.D. work in my development as a scientist was that it allowed me to develop my skills as a kineticist, inculcated in me a lifelong love of linear free energy relationships as a tool to study mechanism, instilled in me the need to analyse objectively all mechanisms that can result in the same kinetics, and that there can be more than one mechanism for the same reaction. I also began to live in the world of the transition state. I could solve complex kinetic equations by visualising their solutions. And, as another prelude for the future, I knew it was essential to be able to make easily a series of substituted reagents to study mechanism in order to probe transition state structure.

Post-Doctoral Work (1968–1969)

I submitted my Ph.D. thesis after two years and eight months as a graduate student. Victor Gold was the external examiner — he left his wife and family to wait in the car park while he examined me, and the viva was less than an hour. I had already been offered a group leader position in the MRC Laboratory of Molecular Biology (LMB) to work on solving the mechanisms of the enzymes whose structures were being solved there by X-ray crystallography. They allowed me to spend a year as a post-doctoral in Brandeis University with Bill (W. P.) Jencks. The stay in Bill's lab was extremely important in my further development. It was the chance to work with a great scientist of the highest integrity. From Bill Jencks, I learned about the importance of non-covalent interaction energies and also the design of simple rapid mixing equipment (a stopped-flow mixing head that could be fitted into a Gilford spectrophotometer), which was my main goal of going to Brandeis.

Bill surprised me one day by telling me that my approach to research was quite different from his. He would choose an area that appeared interesting and would work in it, confident that his skills would solve problems that would emerge, whereas I appeared to formulate a problem and design experiments to solve it before starting, which was true — the training from chess! He also said that he would always do more experiments than the next guy to minimise the chance of errors, a dictum I then followed. It is a myth that science is mainly hypothesis driven, much work is done in an opportunistic manner.

My work with Bill quickly led to the direct observation of the formation of the acetyl-pyridinium ion, a postulated intermediate in pyridine-catalysed acylation reactions[6]. Although a skilled kineticist, I much prefer to demonstrate mechanism by simple direct experiments rather than by indirect subtle methods, as nature always seems to leave unexpected traps — as grandmaster Sergei Tartokower said about chess, 'The mistakes are all there, waiting to be made'. Also, only specialists appreciate very subtle and complex kinetics arguments, and it is depressing giving research seminars that the audience does not understand. The follow-up full paper[7] had some ideas on pre-steady kinetics that have been subsequently invaluable to me. I learned that consecutive first-order reactions give rise to two rate constants (or more complex relaxation constants λ_1 and λ_2 for reversible reactions) whose order cannot be determined without knowing the absolute amplitudes of the signals from the reagent, intermediate and product. This situation arises frequently in protein folding studies, and not many practitioners understand the limitations of the kinetics. The relaxation constants λ_1 and λ_2 are complex functions of the individual rate constants, but I realised the complicating square root terms drop out in the sum of the two, $\lambda_1 + \lambda_2$, and was able to simplify the kinetics and to work out the order of events by determining the dependence on concentrations of added nucleophiles. I rounded off the work by using the principle of microscopic reversibility to study rates of reaction from the reverse direction[8] (related references:[6–10]). On leaving Brandeis, I told Bill I was worried what I would do on return to Cambridge because I had been appointed to work in a new area in which I had no experience. He replied that I should just do good work and it would be appreciated.

I have always been obsessed with 'proof' of mechanism since most evidence is generally only consistent with data and so most evidence is open to interpretation. 'Proof', in its acceptable sense in experimental science, ultimately depends on direct observation. The one-page preliminary communication[6] contains the necessary information to 'prove' that an intermediate is on a reaction pathway: 1) the intermediate was formed fast enough to be on the reaction pathway; 2) it reacted fast enough on the reaction pathway; 3) it could be isolated and characterised. These 'three rules of proof for an intermediate' have been a guiding principle throughout my subsequent mechanistic work on proteins, from chemical mechanisms to protein folding.

[Reprinted from the Journal of the American Chemical Society, **91**, 2125 (1969).]

The Acetylpyridinium Ion Intermediate in Pyridine-Catalyzed Acyl Transfer[1]

Sir:

We wish to report the direct observation of the formation and disappearance of an acetylpyridinium ion intermediate in the course of the pyridine-catalyzed hydrolysis of acetic anhydride in aqueous solution (eq 1). It has been generally believed that this inter-

$$\text{Pyr} + \text{Ac}_2\text{O} \underset{k_{-1}}{\overset{k_1}{\rightleftharpoons}} \text{AcPyr}^+ + \text{AcO}^- \xrightarrow{k_2} \text{Pyr} + \text{AcOH} \qquad (1)$$

mediate is too unstable to permit its accumulation in easily detectable concentrations,[2,3] although the inhibition of the over-all reaction by acetate ion, the rapid reaction rate, and the even faster pyridine-catalyzed exchange of labeled acetate into acetic anhydride provide strong kinetic evidence for such an intermediate.[4] Similarly, inhibition of the pyridine-catalyzed hydrolysis of substituted phenyl acetates by low concentrations of the leaving phenolate ion provides evidence for the same intermediate in these reactions.[5] Estimates of the expected kinetic and thermodynamic stability of the acetylpyridinium ion, based on equilibrium and rate constants for reactions of acetylimidazolium and phosphorylpyridinium ions,[6] led us to search for direct evidence for its formation.

The formation and subsequent disappearance of this intermediate may be followed spectrophotometrically at 280–290 mμ after mixing aqueous solutions of pyridine and acetic anhydride in a stopped-flow apparatus (Figure 1). Increasing concentrations of acetate ion decrease the amount of acetylpyridinium ion formation by increasing the rate of the back reaction (k_{-1}, eq 1). Increasing pyridine concentration was found to increase

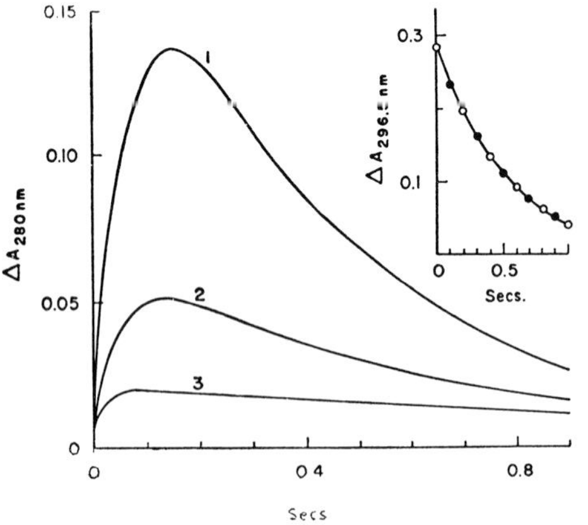

Figure 1. The formation and hydrolysis of acetylpyridinium ion, followed at 280 nm, during the hydrolysis of 2×10^{-4} *M* acetic anhydride catalyzed by pyridine buffer (0.06 *M* free base, pH 5.5) at 25°, ionic strength 1.0, maintained with potassium chloride. Added sodium acetate: curve 1, 4×10^{-3} *M;* curve 2, 10^{-2} *M;* curve 3, 5×10^{-2} *M.* Inset: the disappearance of 10^{-3} *M* (open circles) and 5×10^{-4} *M* (closed circles) *p*-anisidine in the presence of 10^{-4} *M* acetic anhydride and pyridine buffer (0.02 *M* free base, pH 6.5), followed at 296.5 nm, ionic strength 1.0, 25°.

the rate and amount of intermediate formation. Kinetic analysis of these data[7] gave the following approximate values for the rate constants of eq 1 at ionic strength 1.0: k_1, 80 M^{-1} sec^{-1}; k_{-1}, 900 M^{-1} sec^{-1}; k_2, 7.5 sec^{-1}. The value of k_2 is extrapolated to zero pyridine and acetate concentrations to correct for buffer catalysis of the hydrolysis of the intermediate.

The pyridine-catalyzed acetylation of 5×10^{-4} and 10^{-3} *M* anisidine by acetic anhydride follows a pseudo-first-order course with a rate constant which is independent of anisidine concentration (Figure 1, inset). This shows that this acyl transfer reaction occurs through a rate-determining formation of the acetylpyridinium ion intermediate. The second-order rate constants of 83 and 78 M^{-1} sec^{-1} obtained by the use of anisidine and toluidine, respectively, as trapping reagents agree with the value of k_1 obtained from the hydrolysis experiments.

The rate of acetylpyridinium ion hydrolysis is decreased 700-fold to 0.01 sec^{-1} in 9 *M* sodium perchlorate. This large salt effect is similar to that observed with acetylimidazolium ion.[8] The rate constants for the hydrolysis of acetylpyridinium chloride, synthesized at $-60°$,[9] may be determined directly in sodium perchlorate solutions and fall on the same straight line in a plot of log k against salt concentration as those for the hydrolysis of the intermediate formed during the pyridine-catalyzed hydrolysis of acetic anhydride.

The molar extinction coefficient of acetylpyridinium ion at 280 nm in 1 *M* potassium chloride and 4 and 6 *M* sodium perchlorate solutions was estimated to be 3.2×10^3 from the results of kinetic experiments. Difference spectra, corrected for changes in pyridine concentration, gave absorption maxima at 272 nm (ϵ *ca.* 4.3×10^3) and 225 nm (ϵ *ca.* 7×10^3); compare 3-acetylpyridinium chloride, λ_{max} 269 nm (ϵ 3.9×10^3) and 224 nm (ϵ 5.8×10^3) in ethanol.[10]

(1) Publication No. 631 of the Graduate Department of Biochemistry, Brandeis University, Waltham, Mass. Supported by grants from the National Science Foundation and the National Institute of Child Health and Human Development of the National Institutes of Health (HD-1247).
(2) V. Gold and E. G. Jefferson, *J. Chem. Soc.*, 1409 (1953).
(3) D. E. Koshland, Jr., *J. Am. Chem. Soc.*, **74**, 2286 (1952).
(4) A. R. Butler and V. Gold, *J. Chem. Soc.*, 4362 (1961); C. A. Bunton, N. A. Fuller, S. G. Perry, and V. J. Shiner, *Tetrahedron Lett.*, 458 (1961).
(5) W. P. Jencks and M. Gilchrist, *J. Am. Chem. Soc.*, **90**, 2622 (1968).
(6) R. Wolfenden and W. P. Jencks, *ibid.*, **83**, 4390 (1961); W. P. Jencks and M. Gilchrist, *ibid.*, **87**, 3199 (1965); W. P. Jencks, F. Barley, R. Barnett, and M. Gilchrist, *ibid.*, **88**, 4464 (1966).
(7) A. A. Frost and R. G. Pearson, "Kinetics and Mechanism," 2nd ed, John Wiley & Sons, Inc., New York, N. Y., 1961, p 173.
(8) S. Marburg and W. P. Jencks, *J. Am. Chem. Soc.*, **84**, 232 (1962).
(9) A. K. Sheinkman, S. L. Portnova, Yu. N. Sheinker, and A. N. Kost, *Dokl. Akad. Nauk SSSR*, **157**, 1416 (1964).
(10) M. L. Swain, A. Eisner, C. F. Woodward, and B. A. Brice, *J. Am. Chem. Soc.*, **71**, 1341 (1949).

A. R. Fersht, W. P. Jencks
Graduate Department of Biochemistry
Brandeis University, Waltham, Massachusetts 02154
Received February 14, 1969

Chapter 3

Early Independent Work: Enzymatic Catalysis

'You do what you must do and you do it well.'

Bob Dylan

In those days, publishing in the *Journal of the American Chemical Society* (*JACS*) for a chemist had the same kudos as publishing in *Cell* for a biologist today. Within three years (1967–1970), I had published seven full papers and three preliminary communications in *JACS*[1,2,4–8,11–13]. Under normal circumstances, I would have taken a faculty position in a chemistry department as a physical-organic chemist. But, at the age of 26, I returned to Cambridge in 1969 to the MRC Laboratory of Molecular Biology (LMB) as a group leader to apply my skills as a mechanistic chemist to the brand new area of what would be called today structural biology, but then molecular biology. For a chemist to work in a molecular biology laboratory was considered heresy. Feelings were so extreme that a senior chemist colleague confessed to Stephen Benkovic that he refused to work directly on enzymes because it would preclude him from a chair in a UK chemistry department! Lord Todd, who was the professor of organic chemistry at Cambridge when I was a student (and who did support me later on in my career) said in a speech celebrating the centenary of the Royal College of Science that the biggest tragedy in science was the split between chemistry and biology at the beginning of the 20th century. After I was later appointed Professor of Organic Chemistry at Cambridge, a group of German organic chemists told me when I was giving the August von Hofmann lectures in Germany that they were shocked by my appointment!

There was pressure on me to stay in the chemistry department and I was told that LMB had already done its best work and it would be like joining a civil service (government) laboratory. But for me, it was a chance to be independent and work in an exciting new area that was clearly the future. Going to LMB was the best move I could have made. The laboratory was full of inspiring scientists who were all first-rate human beings. I shared a small laboratory with Max Perutz, would have lunch with Cesar Milstein or Aaron Klug or Francis Crick, go down the corridor to sample Fred Sanger's homemade wine in the evening, and sit in lectures with Sydney Brenner in the front row, terrifying visiting speakers with his lightning quick mind and wit. They imbued everyone with the desire to do pioneering research and think big, and provided, with the generous backing of the MRC, the perfect environment for young scientists. Max Perutz was amazing as a director. He read my papers and taught me how to write, and banned me from sitting on any committees as it would distract me from my work.

The serine protease α-chymotrypsin was the then favourite paradigm for mechanistic chemists to study enzymatic catalysis. It is an enzyme of broad specificity that also catalyses the hydrolysis of esters, which can be studied very easily by steady state and pre-steady state kinetics, and could be bought very highly purified in large jars from the Armor Meat Packing Factory. The non-enzymatic catalysis of the hydrolysis of amides and esters in chemical model systems was also being intensely studied by physical-organic chemists. There were a large number of different laboratories all working on the same protein. Using a stopped-flow mixing head that had been designed in Brandeis, I discovered that α-chymotrypsin exists in two conformations in equilibrium, and I was able to measure the equilibrium constant with high precision from the ratio of two spectral changes[14]. A referee claimed that my results were impossibly good, and hence not believable. But, a year or two later I received a grant application from the USA to review in which Rufus Lumry planned to use the 'method of Fersht' for an exhaustive study of the conformational transition, which he then did successfully. The second paper[15] worked out the energetics of formation of a salt bridge, which for many years was used by theoreticians to benchmark simulations. This was my first essay into mapping non-covalent interactions.

3.1 2003: Brian Hartley, Jacques Fastrez and Tony Kirby.

3.2 1969: Jane Sayer, Myriam, Bill and David Jencks, and Alan in Vermont.

J. Mol. Biol. (1972) **64**, 497–509

Conformational Equilibria in α- and δ-Chymotrypsin

The Energetics and Importance of the Salt Bridge

Alan R. Fersht

Medical Research Council Laboratory of Molecular Biology
Hills Road, Cambridge, England

(*Received 30 July 1971*)

Chymotrypsin exists in at least two conformations between pH 2 to 12. One conformation is able, and the other unable, to bind specific substrates and inhibitors. The equilibrium between the two is controlled by a salt bridge between the α-ammonium ion of Ile-16 and the carboxylate ion of Asp-194. Deprotonation of this salt bridge decreases the stability of the active conformation by 2·9 kcal./mole. At 25°C and ionic strength 0·1, 90% of δ-chymotrypsin is in the active conformation between pH 6 and 8. There is a transition, governed by an apparent pK_a of 9·08, as the salt bridge is deprotonated, to a high pH equilibrium where only 12% is in the active form. The active fraction also drops at low pH. The pK_a of Ile-16 when in the salt bridge is 9·96; but when Ile-16 is not constrained the pK_a falls to a normal 7·85. A similar transition is observed in α-chymotrypsin governed by an apparent pK_a of 8·76, and in which the pK_a of unconstrained Ile-16 is 7·94. The enthalpy of ionization of Ile-16 is 9 kcal./mole when unconstrained. A study at 37°C reveals that, at neutral pH, the transition involving the formation of the salt bridge involves an adverse enthalpy term but is favoured by entropy.

The interpretation is based upon knowledge of the enzyme structure derived from X-ray diffraction studies. This affords the first opportunity for measuring the energy of a protein salt bridge.

1. Introduction

The combination of X-ray diffraction and solution studies on α-chymotrypsin has led to the hypothesis that the enzyme may exist in two conformational states (Sigler, Blow, Matthews & Henderson, 1968; Freer, Kraut, Robertus, Wright & Xuong, 1970; Oppenheimer, Labouesse & Hess, 1966; McConn, Fasman & Hess, 1969; Hess, McConn, Ku & McConkey, 1970). One state is able, and the other unable, to bind aromatic substrates and inhibitors. The active conformation predominates at low pH and there is a transition governed by a pK_a of approx. 8·8 to the inactive conformation predominating at high pH. This is based on observations of the pH-dependence of the following parameters: association constants of substrates and inhibitors decrease with increasing pH, the binding falls off governed by a pK_a of approx. 8·8 (Himoe, Parks & Hess, 1967; Bender, Gibian & Whelan, 1966); the catalytic activity, defined by k_c/K_M, falls off at high pH governed by a pK_a of approx 8·8 (Himoe *et al.*, 1967; Bender, Clement, Kezdy & Heck, 1964); the pH dependence of certain physical properties such as optical rotation is also governed by a pK_a of approx. 8·8 (Hess *et al.*, 1970).

Also the binding of substrates or aromatic inhibitors at high pH leads to proton uptake governed by a pK_a of approx. 8·8 (McConn, Ku, Odell, Czerlinski & Hess, 1968; Wedler & Bender, 1969).

In the active conformation Ile-16 is involved in a salt bridge with Asp-194 and the α-ammonium ion must have an unusually high pK_a. In the inactive conformation, the salt bridge is broken, Ile-16 is not constrained and presumably has an unperturbed pK_a. The superposition of the conformational equilibria causes this residue to titrate with a pK_a of 8·8.

As the pH increases above 10 the ability of α-chymotrypsin to bind substrates tends to zero. δ-Chymotrypsin, which differs from α-chymotrypsin in that the sequence Tyr-146-Thr-147-Asn-148-Ala-149 is intact (Neurath, 1957), exhibits a somewhat different behaviour. Dissociation constants of substrates and inhibitors increase at high pH governed by a pK_a of approx. 9 but they plateau to a higher value, constant between pH 10 and 12 (Himoe *et al.*, 1967; Valenzuela & Bender, 1969,1970).

These results are summarized in Figure 1.

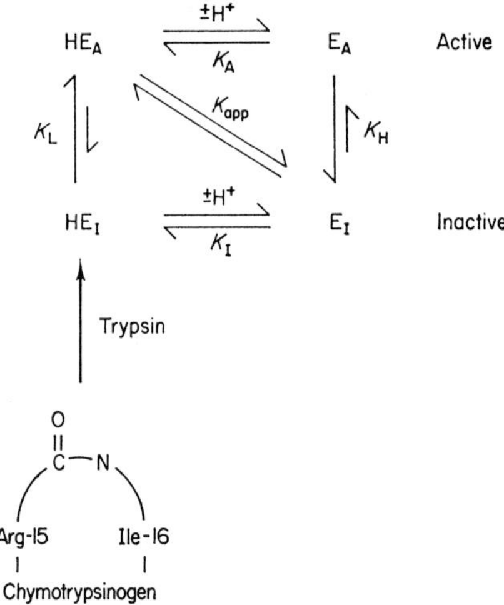

FIG. 1. Scheme for conformational equilibria in chymotrypsin. The observed ionization constant found from titration and binding studies is K_{app}. This is related to the microscopic ionization and equilibrium constants by: $K_A = K_{app}\ (1 + K_L)\ (1 + K_H)^{-1}$ and $K_I = K_{app}\ (1 + 1/K_L)\ (1 + 1/K_H)^{-1}$, where $K_H = E_I/E_A$ and $K_L = HE_I/HE_A$.

We have recently shown that the rate of interconversion of the two conformations in α-chymotrypsin is relatively slow and that the percentage of active conformation may be accurately titrated by mixing with proflavin in the stopped-flow spectrophotometer (see Fig. 2) (Fersht & Requena, 1971).

We have now extended the studies on α-chymotrypsin and have also applied this technique to the δ-enzyme in order to determine the temperature dependence of the conformational equilibria and the stabilization energy of the salt bridge.

2. Experimental Procedure

(a) *Materials*

3 × Crystallized, salt-free, lyophilized α-chymotrypsin (Lot CDI 8LK) and 5 × crystallized chymotrypsinogen A (Lot CGC 8HA) were obtained from Worthington. δ-Chymotrypsin was a salt-free ethanol-precipitate obtained from Sigma, Lot 20C—0150. Proflavin

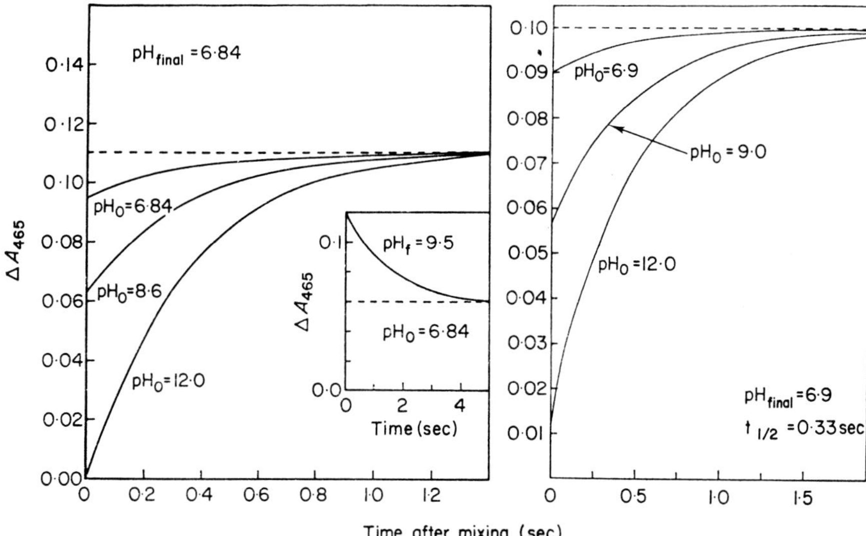

FIG. 2. Formation of chymotrypsin–proflavin complex followed at 465 nm by stopped-flow spectrophotometry. Enzyme incubated in a dilute buffer (25°C, ionic strength 0·1) at pH_0 is mixed with excess proflavin in a concentrated pH 6·84 phosphate buffer to give a final pH of 6·84.

Left-hand side α-chymotrypsin: *inset*; the enzyme initially at pH 6·84 and jumped with proflavin solution to pH 9·5 (Fersht & Requena, 1971). Right-hand side: δ-chymotrypsin: the burst represents the fast association of proflavin with the active conformation. The first-order change is due to the unimolecular conversion of inactive to active and subsequent fast association with proflavin.

hemisulphate was purchased from British Drug Houses and purified as previously described. Protein concentrations were determined spectrophotometrically in 10^{-3} N-HCl at 280 nm using a molar extinction coefficient of 5×10^4 (Dixon & Neurath, 1957). Active site titrations were performed at pH 2·4 using the CBZ-L-tyrosine *p*-nitrophenyl ester method (Kezdy, Clement & Bender, 1964). The α-chymotrypsin contained 94% and the δ-chymotrypsin 65% of the theoretical active sites. The δ-chymotrypsin was shown to contain a negligible concentration of α by end-group analysis for alanine (Gray, 1967) (kindly performed by Dr C. Moore of this laboratory). Some samples of δ-chymotrypsin were purified by affinity chromatography on Agarose–epsilon–amino–caproyl–D–tryptophan methyl ester (obtained from Miles-Yeda) using a slight modification of a known procedure (Cuatrecasas, Wilchek & Anfinsen, 1968): the enzyme was dissolved in 0·05 M-Tris buffer (pH 7·80 at 25°C) at 4°C and chromatographed in the cold. The active enzyme was eluted at 4°C with 0·1 M-acetic acid. This preparation contained 96% of the theoretical number of active sites. A further sample of δ-chymotrypsin was prepared by activating the zymogen (5 mg/ml.) with trypsin (0·14 mg/ml.) for 30 min at 25°C in the 0·05 M-Tris buffer (pH 7·8) and then purifying by affinity chromatography. This preparation was also 96% active. α-Chymotrypsin with 13 of the 15 surface carboxyls blocked with glycine methyl ester was prepared according to Carraway, Spoerl & Koshland (1969).

(b) *Instrumentation and methods*

The stopped-flow spectrophotometer was as previously described except that the dead time had been improved to 4 msec. The driving syringes were set in a hollow brass block through which water circulated from a thermostat. The observation and mixing chambers were thermostatically controlled by the "dual thermo-spacers" of the Gilford 2400 spectrophotometer and the filling syringes were rested on a thermostatically controlled aluminium block. The temperature inside the driving syringes was checked by means of a thermocouple. pH measurements were recorded on a Radiometer pH 26C meter at the same temperature as the kinetic experiments.

Stock solutions of proflavin were prepared daily in deoxygenated water and stored in the dark. Enzyme solutions were prepared daily in 10^{-3} N-HCl and stored on ice.

The dissociation constant for the interaction of proflavin and α-chymotrypsin at 37·0 ± 0·1 deg. C and ionic strength 0·1 was determined by a known method (Glazer, 1965; Brandt, Himoe & Hess, 1967) to be 50 ± 5 μM at pH 6·8. The procedure was checked by first maintaining a constant concentration of proflavin and varying the enzyme concentration and then repeating *vice versa*. A value of 42 ± 4 μM was found for the dissociation constant of proflavin with δ-chymotrypsin at ionic strength 0·1, pH 6·84 and 25·0 ± 0·1 deg. C when the proflavin concentration was held constant and the enzyme varied.

The proportion of chymotrypsin present in the active conformation was determined by the method previously developed (Fersht & Requena, 1971). α-Chymotrypsin (0·5 to 2·0 mg/ml.) was incubated in the stopped-flow spectrophotometer at 37°C and ionic strength 0·1 (added KCl) in a very dilute buffer or potassium hydroxide. This was then mixed on triggering the kinetic run with an equal volume of proflavin (200 μM) in a pH 6·84 phosphate buffer at ionic strength 0·1 (0·05 M). A rapid burst of enzyme–proflavin complex formation, observed at 465 nm, was followed by a slow first-order increase of half-time (40 msec). The ratio of the burst to the total change at 465 nm gives the proportion of enzyme in the active conformation. This procedure was repeated at least 4 times at each pH over a wide pH range. Some autolysis occurred at this temperature as the total change at 465 nm decreased with increasing time. However, the initial burst decreased in a parallel manner and the ratio of the two was time independent. It was not felt desirable to retard the autolysis by the addition of $CaCl_2$ in case this caused specific binding effects (Chervenka, 1959).

The same procedure was used for δ-chymotrypsin at 25°C. δ-Chymotrypsin incubated at pH 10 to 12 still gives an appreciable burst. In order to show that this is not due to a tightly bound impurity at the active site anchoring the enzyme in the active conformation, the experiments were repeated using samples of enzyme purified by affinity chromatography which involves the binding of the enzyme to the resin by means of the active site, displacing any impurity. Identical results were obtained from this, from the sample prepared by activation of the zymogen, and from the commercial sample.

The slow equilibration of the conformational changes ($t_{\frac{1}{2}}$, approx. 300 msec) enabled the following rapid activity assay to be used to check the proflavin assay. A stock solution of carbobenzoxy-L-tyrosine *p*-nitrophenyl ester in acetonitrile was diluted with pH 4·80 acetate buffer (0·2 M) to give a 35 μM solution and immediately placed in the stopped-flow spectrophotometer syringe (this solution rapidly deteriorates). The initial rate of *p*-nitrophenol release on mixing with enzyme (incubated in a dilute pH 6·8 buffer) was monitored over 15 msec at 340 nm. A calibration curve was made using final enzyme concentrations of 9 to 36 μM. 180 μM enzyme was incubated at pH values of 10·2 to 12 and mixed with the nitrophenyl ester at pH 4·8 and monitored in the same way as above. The proportion of active enzyme present was found from the initial rate and the calibration curve, allowing a small correction for the interconversion of the two conformations over the 15 msec duration of the assay and 4 msec dead time of the apparatus.

The detailed analysis of the magnitudes of the initial burst and total enzyme–proflavin complex conformation (observed at 465 nm) has been given previously (Fersht & Requena, 1971). However, as there were small changes in proflavin concentration over the course of the reaction the following equation was used to calculate the proportion of active enzyme (E_A) to total (E_t):

$$\frac{E_A}{E_t} = \frac{\Delta A_o}{\Delta A_t} \times \frac{K_f\,(1 + K)^{-1} + F_o}{F_o} \times \frac{F_f}{F_f + K_f},$$

where ΔA_o is the initial burst at 465 nm and ΔA_t is the total change (extrapolated back to zero time, assuming 4 msec of dead time), F_o is the initial and F_f the final proflavin concentration, K_f is the observed dissociation constant for the enzyme–proflavin complex and K the equilibrium constant E_I/E_A, at pH 6·84.

3. Results

The initial burst contains a contribution from non-specific binding, that is, other than at the active site. Using chymotrypsinogen as a control, it was previously estimated that under the conditions employed here non-specific binding contributes 6% of the total absorbance change to the initial burst. As this figure is significant for δ-chymotrypsin at high pH it was deemed necessary to confirm the proflavin binding results by an activity assay. Fortunately, the relatively slow interconversion of the inactive to active conformation allows the rate assay with carbobenzoxy-L-tyrosine *p*-nitrophenyl ester (described in Experimental Procedure) to be used. Identical results are obtained from both proflavin and activity assays.

The plot of the percentage of α-chymotrypsin against pH at 37°C and ionic strength 0·1 fits well the ionization curve of a base of pK_a 8·76 ± 0·04 for pH values above 7, based on a maximum of 90% and a minimum of 0 (Fig. 3). The proportion of active

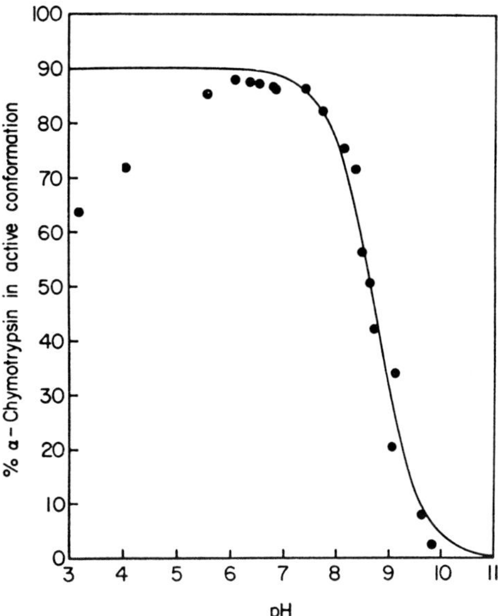

Fig. 3. The percentage of α-chymotrypsin in the active conformation as a function of pH at 37°C and ionic strength 0·1. The solid line is calculated assuming a pK_a of 8·76 and a maximum of 90% active. The decrease at low pH is due to the progressive ionizations of the carboxyl groups.

enzyme reaches a maximum observed value of 88%. Unfortunately, below pH 6 the progressive ionization of the carboxyl groups perturbs the conformational equilibria and the results are difficult to analyse. At high pH the proportion of active conformation tends to zero; the "burst" tends to the 6% attributable to non-specific binding.

The data for δ-chymotrypsin at 25°C and ionic strength 0·1 are presented in Figure 5 and some selected values in Table 1. The proportion of active enzyme at high pH tends to a limiting value of 12·4 ± 0·6% based on the activity assay or 12·4 ± 0·94 based on the proflavin assay; the mean values were adjusted to be exactly equal by using a value of 5·84%, rather than 6%, for the non-specific binding mentioned above. Above pH 8 the plot of the proportion of active conformation against pH fits an ionization curve of pK_a 9·08 ± 0·03 based on limiting values of 12·4 and 94·8%. Again the proportion of active conformation at low pH does not reach this limiting value, the maximum observed proportion being 90·4% at pH 6·84. However, the values for the proportion of active conformation between pH 6 and 12 may be

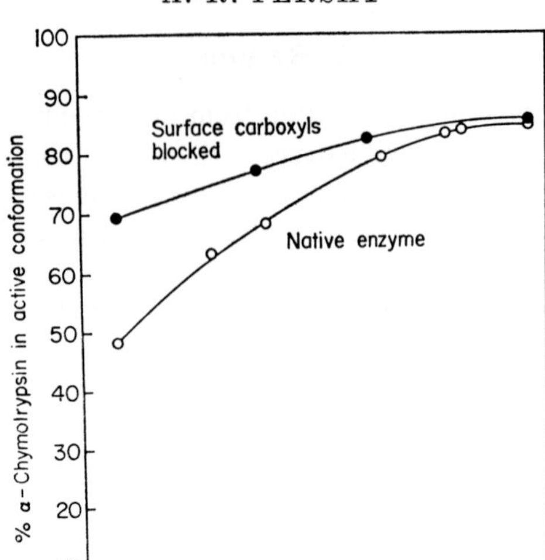

FIG. 4. The effect of carboxyl group ionization on conformational equilibria in α-chymotrypsin at 25°C and ionic strength 0·1. (○) Native α; (●) the enzyme with 13 surface carboxyls blocked by amidation with glycine methyl ester.

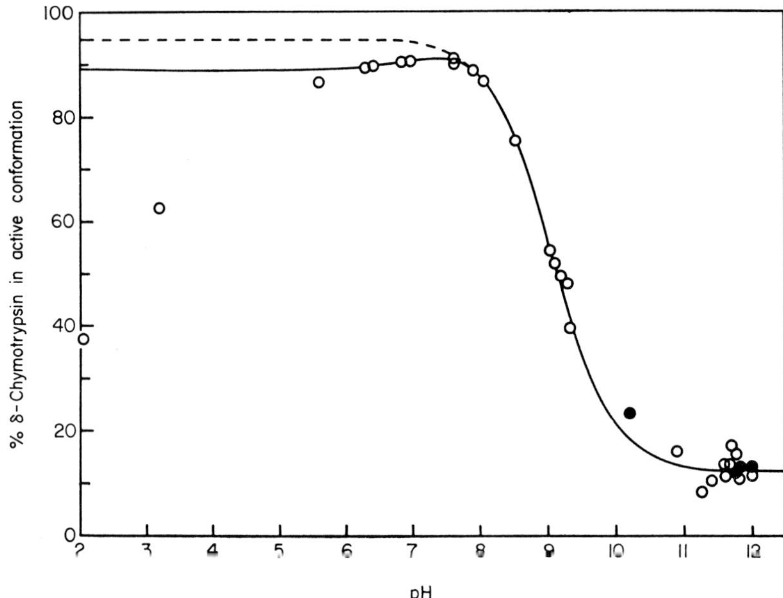

FIG. 5. The percentage of δ-chymotrypsin in the active conformation as a function of pH at 25°C and ionic strength 0·1. The open circles are from the proflavin experiments, closed circles from the rate assays. The solid line is calculated from the equilibrium constants in Fig. 7. The four pK_a values merge into two "apparent" values, an alkaline transition of pK_a 9·08 and a second value of pK_a 7·14. The dotted line is that calculated based on just the 9·08 ionization and a lower pH limit of 94·8% active conformation.

analysed in terms of two ionizations involving three ionic species, H_2E, HE and E. The enzyme H_2E exists as approx. 89% active and approx. 11% inactive. HE is 94·8 ± 0·3% in the active conformation and 5·2% in the inactive. The transition between the two is governed by a pK_a of 7·14. E is 12·4 ± 0·6% active conformation and 87·6 ± 0·6% inactive. The pK_a between HE and E is 9·08. The plot of percentage

TABLE 1

Percentage of δ-chymotrypsin present as active conformation (E_A) at neutral and high pH at 25°C and ionic strength 0·1†

pH	E_A‡ (%)	pH	E_A§ (%)
(a) *Proflavin binding assay*			
6·32	89·24 ± 0·50	11·27	8·48 ± 2·0
6·39	89·48 ± 0·40	11·41	10·33 ± 3·1
6·84	90·40 ± 0·40	11·59	13·26 ± 2·7
6·92	90·06 ± 0·15	11·60	11·88 ±2·0
7·60	90·72 ± 0·15	11·65	13·26 ± 2·0
7·62	90·19 ± 0·40	11·70	17·56 ± 2·5
7·92	88·86 ± 0·44	11·76	15·89 ± 2·9
8·08	86·78 ± 0·65	11·81	11·14 ± 1·9
8·55	72·72 ± 0·50	11·98	11·57 ± 2·4
(b) *Activity assay*			
		11·75	11·92 ± 0·9
		11·80	12·19 ± 0·9
		11·97	12·54 ± 0·9

† Maintained with KCl. ‡ Mean and standard error from at least 5 measurements. The reproducibility is high; a more generous estimate of the absolute error is ± 1%. § Mean and standard error from at least 5 measurements.

TABLE 2

Rate constants for α-chymotrypsin$_{(inactive)} \underset{k_{-1}}{\overset{k_1}{\rightleftharpoons}}$ *α-chymotrypsin*$_{(active)}$ *at ionic strength 0·1*†

pH	25°C k_1 (sec^{-1})	25°C k_{-1} (sec^{-1})	25°C k_1/k_{-1}	37°C k_1 (sec^{-1})	37°C k_{-1} (sec^{-1})	37°C k_1/k_{-1}
5·6	3·1	0·8	3·88	16·3	2·4	5·67
6·3	3·1	0·60	5·17	17·4	1·9	8·21
6·9	3·1	0·59	5·25	16·3	2·0	7·33

† Maintained with KCl.

of active conformation against pH is deceptively smooth between pH 6 and 7·6, and the ionization at pH 7·1 is obscured. However, examination of the rate constants for the interconversion (Table 3) reveals a dramatic change between pH 6·9 and 7·2, although the conformational equilibrium constant is invariant over this range. The pK_a of 7·14 becomes obvious when the data are plotted as in Figure 6. In this plot the percentage of active enzyme between pH 6 and 8·6 is normalized for the 9·08 ionization, that is, the low pH limiting value of active conformation $(E_A)_{max}$, is calculated from the observed value at the pH concerned based on a higher pK_a value of 9·08 and a high pH value of 12·4% active. These results were independent of the enzyme preparation.

The results for α may perhaps be analysed in a similar manner. We have used only the data above pH 7 in the calculations.

The values determined here for the pK_a values of the alkaline transition are in excellent agreement with those determined by Wedler & Bender (1969) and Valenzuela & Bender (1970) from inhibitor binding.

TABLE 3

Rate constants for δ-chymotrypsin$_{(inactive)}$ $\underset{k_{-1}}{\overset{k_1}{\rightleftharpoons}}$ δ-chymotrypsin$_{(active)}$ at 25°C and ionic strength 0·1†

pH	k_1 (sec^{-1})	k_{-1} (sec^{-1})	k_1/k_{-1}‡
5·72	2·2	0·30	6·41
6·43	2·4	0·26	8·35
6·9	2·1	0·20	9·21
7·2	1·6	0·14	10·48
7·84	0·70	0·075	8·33
8·69	0·29	0·87	2·33

† Maintained with KCl.
‡ Measured directly as an equilibrium constant, k_{-1} determined from this and k_1.

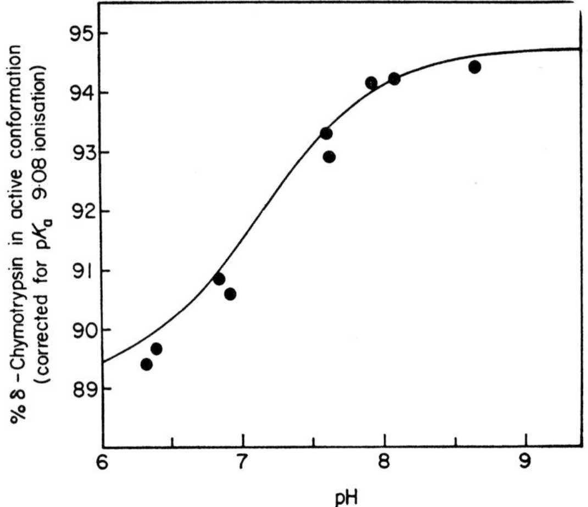

FIG. 6. Illustration of the pK_a 7·14 ionization. The low pH limiting value of HE_A (in Fig. 1) is calculated from the observed proportion of active conformation at the particular pH and assuming one ionization of pK_a 9·08, with E_A = 12·4%. The plot of the calculated values of $HE_{A(low\ pH)}$ against pH fits an ionization curve of pK_a 7·14 and limits of 89 and 94·8% (solid curve). $HE_{A(low\ pH)}$ is calculated from:

$$pH = 9{\cdot}08 + \log (HE_{A(low\ pH)} - E_{A(obs)})/(E_{A(obs)} - 12{\cdot}4).$$

Polymerization of α- and δ-chymotrypsin is negligible at the pH values (>7) used in the calculation of the energy of the salt bridge and the pK_a values of Ile-16 (Krigbaum & Godwin, 1968; Kim & Lumry, 1971). Dimerization is presumably one of the factors influencing the decrease in active conformation at low pH. Kurosky, Graham, Dixon & Hofmann (1971) have shown that Ile-16 is accessible to reaction with nitrous acid at pH 4 and 0°C, confirming that the inactive conformation does occur at low pH.

Blocking 13 of the surface carboxyl groups by amidation with glycine methyl ester inhibits the decrease in the percentage of active conformation at lower pH. It is seen in Figure 4 that above pH 6 the effects of carboxyl group ionizations may be ignored.

The data obtained are of high accuracy as seen in Table 1; the calculations depend only on the measurement of the ratios of two changes. The proflavin jump technique is most accurate for the titration of a large burst and small subsequent change (Fersht & Requena, 1971).

Kim & Lumry (1971) have "pH-jumped" α-chymotrypsin to and from various pH values in the range of 7 to 10 and have obtained the rate and equilibrium constants for the process by stopped-flow fluorimetry. This technique could not be used above pH 10 due to phenolate ion quenching of the fluorescence. Garel & Labouesse (1971) have performed similar experiments at 15°C on δ-chymotrypsin acetylated on tyrosine hydroxyls and all amino groups except Ile-16. The observed rate constant found at each pH by this method is the sum of the pair of rate constants for the forward and reverse reactions at that pH. The individual rate constants and the equilibrium constant are not directly given. The "proflavin-jump" method is far superior as far as analysis of results is concerned. Proflavin is a *specific* titrant for one conformation and the interpretation is less ambiguous. The proportion of enzyme in each conformation is given directly and the conformations characterized by their binding properties. The individual rate constant for the inactive ⟶ active conformational change is isolated when a large excess of proflavin is used and the equilibrium is displaced in favour of the enzyme–proflavin complex. The fluorescence studies, however, provide a useful supplement to the proflavin measurements.

4. Discussion

(a) *pH Dependence of proportion of active conformation*

Below pH 6 both α and δ show a progressive decrease of proportion of active conformation with decreasing pH. This is due to the perturbation of the pK_a values of the carboxyl groups during the conformational change and any dimerization equilibria dependent upon these ionizations. There is evidence, recently adduced from the laboratory of Canady (Royer, Wildnauer, Cuppett & Canady, 1971; Cuppett, Resnick & Canady, 1971), for the existence of at least two salt-dependent forms of α-chymotrypsin at low pH. Kurosky *et al.* (1971) have shown that Ile-16 is accessible at pH 4.

As described in Results, the curve for δ-chymotrypsin is analysed in terms of two ionizations, an apparent alkaline transition of pK_a 9·08 and a secondary perturbation of pK_a 7·14. This perturbation is real. Although the plot of the percentage of active enzyme against pH shows no discontinuity in this region, the observed rate constants for the process change markedly.

At high pH, 12·4% of δ-chymotrypsin remains in the active conformation, with the salt bridge deprotonated. The proportion of active α tends to zero under the same conditions.

Kaplan (1971), using a specific probe for Ile-16 acetylation with labelled acetic anhydride, has shown that this group titrates normally in δ-chymotrypsin with a pK_a of 8·9 until pH 9·8 when there is a discontinuity. A rapid increase in the rate of acetylation of Ile-16 at this pH strongly indicates a further structural change in the enzyme. This does not appear to happen with δ-chymotrypsin; the dipeptide linking Tyr-146 and Ala-149, which is missing in α-chymotrypsin, stabilizes the native conformation.

It is reasonable that a significant fraction of δ-chymotrypsin is found to be in the active conformation at high pH. The previously observed binding of substrates at high pH with pH-independent dissociation constants implies the existence of a pH-independent active conformation. Similarly, the finding that proton uptake on the binding of inhibitors at high pH goes through a maximum then falls above pH 10 also implies the same. (In the case of α-chymotrypsin proton uptake continues at high pH.)

(b) pK_a *Values of Ile-16 (and His-40?)*

The plots of the percentage of active α enzyme as a function of pH above pH 6 fit ionization curves of pK_a 8·76 at both 25°C and 37°C. Using this and the constant for the conformational equilibrium at neutral pH (K_L in Fig. 1) a group in the inactive conformation has a pK_a of 7·94 ± 0·1 at 25°C and 7·68 ± 0·06 at 37°C. The enthalpy of ionization is 9·2 ± 4·2 kcal./mole. These values are in the region expected for an α-amino group, e.g. pK_a 7·8 to 8·2 and ΔH 10 to 13 kcal., are the usual values (Cohn & Edsall, 1943). X-ray diffraction studies suggest that the α-amino group of Ile-16 is the group whose pK_a is perturbed.

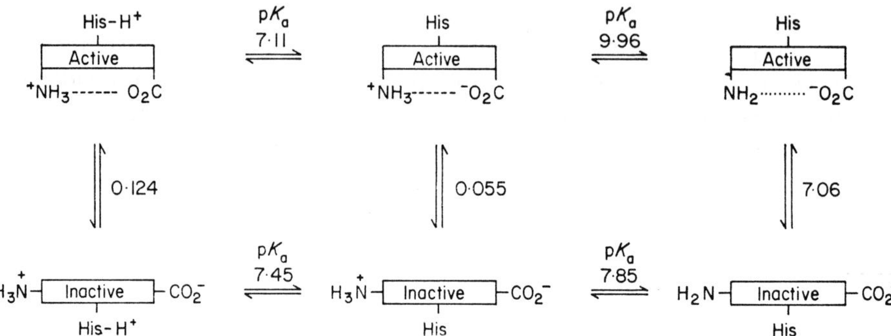

FIG. 7. Summary of derived equilibrium and ionization constants for δ-chymotrypsin at 25°C and ionic strength 0·1. The equilibria on the right-hand side are those due to the ionization of the Ile-16 Asp-194 salt bridge. Those on the left are due to the perturbation of a histidine ionization (His-40?).

A similar value of 7·84 ± 0·05 is calculated for Ile-16 in the inactive conformation of δ at 25°C. As K_H in Figure 1 is measurable in δ it is also possible to measure the pK_a of Ile-16 in the active conformation. The pK_a of Ile-16 in the active conformation, in the salt bridge, is calculated to be 9·96.

Valenzuela & Bender (1970) have determined, under identical conditions, the pK_a of Ile-16 in various enzyme–inhibitor complexes by measuring the proton uptake on binding. These values should be similar to, but not necessarily equal to, the pK_a of Ile-16 in the active conformation. pK_a Values of 9·60 to 10·0 were found.

The salt bridge increases the pK_a of Ile-16 by 2·1 units. There is the possibility of a small conformational rearrangement on the deprotonation of Ile-16, the α-amino group not remaining in juxtaposition with Asp-194, although the enzyme retains the over-all structure of the active conformation. In this case the observed pK_a of 9·96 is an apparent value composed of the microscopic ionization constant and the equilibrium constant for the reorganization process.

Garel & Labouesse (1970,1971) have estimated pK_a values of 7·2 and 10·6 for the inactive and active conformations of acetylated δ-chymotrypsin at 15°C. The latter value is probably correct but 7·2 is in disagreement with our results. This is equivalent to a pK_a of approx. 6·9 at 25°C, a full pH unit below our value. As their results were derived solely from the analysis of pairs of rate constants assuming a single ionization in the pH range 6 to 12 it seems likely that the value of 7·2 represents the second ionization, the perturbation at pH 7·4, noted in the present study.

TABLE 4

Summary of rate and equilibrium constants† for the α-chymotrypsin$_{(active)}$ $\underset{k_{-L}}{\overset{k_L}{\rightleftharpoons}}$ *α-chymotrypsin*$_{(inactive)}$ *transition at low pH ionic strength 0·1*

Temperature (°K)	K_L (k_L/k_{-L})	k_L sec^{-1}	k_{-L} sec^{-1}	pK_{app}
298	0·176	0·546	3·1	8·76 ± 0·06
310	0·111	1·9	17	8·76 ± 0·04

$K_{L\ (298)}$	$\Delta S^o = -27$ eu $\Delta H^o = -\ 7{\cdot}1$ kcal.
$k_{L\ (298)}$	$\Delta S^+ = \ 2$ eu $\Delta H^+ = 18{\cdot}3$ kcal.
$k_{-L\ (298)}$	$\Delta S^+ = 29$ eu $\Delta H^+ = 25{\cdot}4$ kcal.

† Notation as in Fig. 1. The limiting values at low pH by extrapolation of the values above pH 7, i.e. not including the effect of pH on carboxyl group ionization.

This perturbation at pH 7·14 must be that of a histidine. This is presumably His-40, though His-57 is also a possibility. We prefer the former. His-40 is close to Asp-194 in the zymogen (Freer *et al.*, 1970) and is presumably so in the inactive form of the enzyme which is postulated to resemble the zymogen. This is consistent with an increase in pK_a of 7·11 in the active to 7·45 in the inactive.

The pK_a of Ile-16 in the active conformation of α-chymotrypsin has been estimated to be much higher than found here for δ. Limits of >11·5 (McConn *et al.*, 1968; Valenzuela & Bender, 1970) and >10 have been obtained. This may be due to an artifact as there is the possibility of a second structural change at pH 9·8 (Kaplan, 1971).

(c) *Temperature dependence and kinetics*

The proportion of active conformation in α-chymotrypsin increases with increasing temperature. The process is entropy driven and involves an adverse enthalpy change.

TABLE 5

Summary of pK_a values for Ile-16 at ionic strength 0·1, strength of salt bridge

Chymotrypsin	Temp. (°C)	pK_a Ile-16 In salt bridge (active)	pK_a Ile-16 Unconstrained (inactive)	ΔH Ionization kcal.
α−	25	>10	7·94 ± 0·1	9·2 ± 4·2
	37	>10	7·68 ± 0·06	
δ−	25	9·96 ± 0·05	7·85 ± 0·05	

Stabilization energy of Ile-16-Asp-194 salt bridge in δ-chymotrypsin = 2·9 ± 0·1 kcal. mole^{-1}

Approximate activation parameters are given in Table 5. These figures are for the gross changes accompanying the conformational change. The conversion of the inactive to active conformation involves the α-ammonium of Ile-16 shedding its ordered solvent shell to form an unsolvated salt bridge. The main driving force for the total process must be the entropy increase on releasing the ordered water.

Inspection of the activation parameters in Table 5 reveals that the enthalpy of activation of k_{-L}, the rate constant for inactive⟶ active, is approximately the same as that for the total equilibrium. A likely transition state for the process is a situation similar to that of the active conformation but with the salt bridge and other non-covalent bonds stretched.

(d) *Strength of the salt bridge*

It is difficult to isolate the contribution of the salt bridge between Ile-16 and Asp-194 to the total equilibrium constant for the conformational change. The energy of the refolding process obscures the energy of the salt bridge and it is experimentally impossible to determine each from the measurement of *one* equilibrium constant. However, we can define an experimental energy from the measurement of *two* equilibrium constants. The secondary folding interactions are the same at high and neutral pH and so cancel out when the ratio of K_H/K_L (see Fig. 1) is considered. The energy of the salt bridge is given by:

$$\Delta G_{\text{salt bridge}} = -RT\ln(K_H/K_L).$$

The salt bridge between Ile-16 and Asp-194 contributes a stabilization energy of 2·9 kcal. mole^{-1}. If the pH 7 perturbation represents the salt bridge between His-40 and Asp-194 then this has an energy of approximately 0·46 kcal. mole^{-1}.

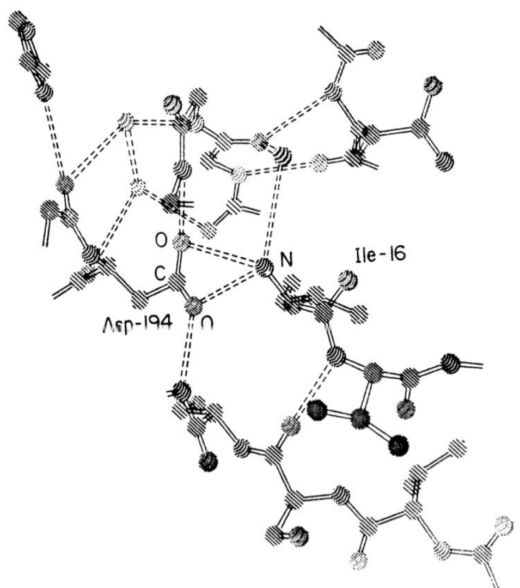

Fig. 8. Environment of the salt bridge between Ile-16: Asp-194. The environment is predominantly hydrophobic except that the β carbon of Asp-194 is exposed to the specificity pocket. Residue 18, just off the right-hand side of the diagram, is at the exterior of the molecule. The chain Ser-189-Ser-190-Cys-191 which, from this viewpoint, lies in front of the salt bridge has been omitted for clarity.

The salt bridges in haemoglobin between the α-carboxylate and α-ammonium ions of the α-chains, which are not so buried as the Ile-16 Asp-194 bridge in chymotrypsin, are presumably weaker (Perutz, 1970).

Scheme I and the conformational importance of Ile-16 have been attacked as being incorrect. This is mainly due to a misunderstanding as regards the equilibrium constants in scheme I. The deprotonation of the salt bridge is not an all-or-none process with 100% inactive enzyme at high pH and 100% active at neutral pH. The equilibria are delicately balanced. The importance of the salt bridge is to displace this balance. In δ-chymotrypsin the 2·9 kcal. energy of the salt bridge is sufficient to change a concentration of 12% active conformation at high pH to 90% in neutral pH. Similarly 84% of α-chymotrypsin in the active conformation at neutral pH changes to less than 6% at high pH. (In this case the percentage at high pH is possibly further decreased by a second structural change.)

I thank Mrs M. Sanders for technical assistance.

REFERENCES

Bender, M. L., Clement, G. E., Kezdy, F. J. & Heck, H. D'A. (1964). *J. Amer. Chem. Soc.* **86,** 3680.

Bender, M. L., Gibian, M. J. & Whelan, D. J. (1966). *Proc. Nat. Acad. Sci., Wash.* **56,** 833.

Brandt, K. G., Himoe, A. & Hess, G. P. (1967). *J. Biol. Chem.* **242,** 3973.

Carraway, K. L., Spoerl, P. & Koshland, Jr, D. E. (1969). *J. Mol. Biol.* **42,** 133.

Chervenka, C. H. (1959). *Biochim. biophys. Acta,* **31,** 85.

Cohn, E. J. & Edsall, J. T. (1943). In *Proteins, Amino Acids and Peptides,* p. 445. New York : Reinhold.

Cuatrecasas, P., Wilchek, M. & Anfinsen, C. B. (1968). *Proc. Nat. Acad. Sci., Wash.* **61,** 636.

Cuppett, C. C., Resnick, H. & Canady, W. J. (1971). *J. Biol. Chem.* **246,** 1135.

Dixon, G. M. & Neurath, H. (1957). *J. Biol. Chem.* **225,** 1049.

Fersht, A. R. & Requena, Y. (1971). *J. Mol. Biol.* **60,** 279.

Freer, S. T., Kraut, J., Robertus, J. D., Wright, H. T. & Xuong, Ng. H. (1970). *Biochemistry,* **9,** 1997.

Garel, J.-R. & Labouesse, B. (1970). *J. Mol. Biol.* **47,** 41.

Garel, J.-R. & Labouesse, B. (1971). *Biochimie,* **53,** 9.

Glazer, A. N. (1965). *Proc. Nat. Acad. Sci., Wash.* **53,** 1238.

Gray, W. R. (1967). *Advanc. Enzymol.* **11,** 139.

Hess, G. P., McConn, J., Ku, E. & McConkey, G. (1970). *Phil. Trans. Roy. Soc.* B**257,** 89.

Himoe, A., Parks, P. C. & Hess, G. P. (1967). *J. Biol. Chem.* **242,** 919.

Kaplan, H. (1971). *Biochem. Biophys. Res. Comm.* **42,** 1042.

Kezdy, F. J., Clement, G. E. & Bender, M. L. (1964). *J. Amer. Chem. Soc.* **86,** 3690.

Kim, Y. D. & Lumry, R. (1971). *J. Amer. Chem. Soc.* **93,** 1003.

Krigbaum, W. R. & Godwin, R. W. (1968). *Biochemistry,* **7,** 3126.

Kurosky, A., Graham, J. E. S., Dixon, J. W. & Hofmann, T. (1971). *Canad. J. Biochem.* **49,** 529.

McConn, J., Fasman, G. D. & Hess, G. P. (1969). *J. Mol. Biol.* **39,** 551.

McConn, J., Ku, E., Odell, C., Czerlinski, G. & Hess, G. P. (1968). *Science,* **161,** 274.

Neurath, H. (1957). *Advanc. Prot. Chem.* **12,** 320.

Oppenheimer, H. L., Labouesse, B. & Hess, G. P. (1966). *J. Biol. Chem.* **241,** 2720.

Perutz, M. F. (1970). *Nature,* **228,** 726.

Royer, G., Wildnauer, R., Cuppett, C. C. & Canady, W. J. (1971). *J. Biol. Chem.* **246,** 1129.

Sigler, P. B., Blow, D. M., Matthews, B. W. & Henderson, R. (1968). *J. Mol. Biol.* **35,** 143.

Valenzuela, P. B. & Bender, M. L. (1969). *Proc. Nat. Acad. Sci., Wash.* **63,** 1214.

Valenzuela, P. B. & Bender, M. L. (1970). *Biochemistry,* **9,** 2440.

Wedler, F. C. & Bender, M. L. (1969). *J. Amer. Chem. Soc.* **86, 3669.**

The hydrolysis of esters catalysed by chymotrypsin goes through an intermediate, the acyl-enzyme, first demonstrated by one of my mentors at LMB, Brian Hartley, who had also introduced the use of *p*-nitrophenol as a chromophore (I once nominated him for a Nobel Prize for his work as he was the unrecognised founder of a whole field in mechanistic chemistry and enzymology based on that chromophore as the alcohol in highly reactive acetate and phosphate esters). The intermediate was easy to prove for activated ester substrates because the acyl-enzyme accumulates in the reaction and can easily be observed. This is not so for amides and peptides. Bill Jencks thought that they were hydrolysed by a different mechanism because the acyl-enzyme, which is a serine ester, is of higher bond energy than a peptide, and there was indirect evidence to support a different mechanism. My first post-doctoral, Jacques Fastrez, and I were able to 'prove' the acyl-enzyme mechanism as being the general mechanism for both esters and peptides in a tour de force of classical enzyme kinetics[16]. Following on from the studies of Myron Bender, George Hess and Freddie (H.) Gutfreund, we could measure directly for activated esters, where the acyl-enzyme accumulates, all the necessary rate constants for its formation and decay in pre-steady state kinetics and show that all data were consistent with steady state kinetics. Then, we extended this mechanism to peptides by studying the reverse reaction, the synthesis of peptides. We measured the rate constants for the formation of peptides by the attack of amines on an acyl-enzyme and, by measuring the equilibrium constant for the hydrolysis of the peptide, we could calculate the compound rate constant (k_{cat}/K_M) for the hydrolysis of the peptide by the acyl-enzyme route. We showed the calculated value of k_{cat}/K_M to be the same as that observed experimentally.

The relationship between k_{cat}/K_M and equilibrium constants was discovered by J. B. S. Haldane, one of my scientific heroes. I had heard him in a BBC interview in about 1962 say that he sat in his office in the biochemistry department in Cambridge doing arithmetic. The biologists left him alone because they thought he was a mathematician, and the mathematicians left him alone because they thought he was a biochemist. I managed to play the same trick between chemists and biologists for a long time.

Jack (J. F.) Kirsch used our acyl-enzyme paper for many years for class work in Berkeley. Like most of my papers, it received one glowing report and a negative one. The then editor of *Biochemistry*, Hans Neurath, had the confidence to ignore negative reports for papers that he believed were good, and he wanted to accept good papers. Unfortunately, nowadays, journals are too conscious of impact factor and their first impulse is to reject. (Related references:[14–25]).

Faraday's advice to a young scientist, the young William Crookes, was: Work. Finish. Publish. To that I would add: Research starts with a question, not an answer. And: Try to disprove your work. Science is littered with mistakes from people so desperate to prove their hypotheses that they choose their own interpretation of their ambiguous data and also select the experiments that support it and ignore or hide the ones that do not. I have always tried to bend my data to see if they would fit the opposite conclusions in case I could be wrong. I love the acyl-enzyme proof paper because it is an example of how careful experiments can distinguish between different mechanisms and provide close to absolute proof of a reaction mechanism.

3.3 1994: Max Perutz at his 80th Birthday meeting which was co-organised by Alan. This photograph, taken by Alan, was one of Max's favourites.

3.4 1994: Alan, Max Perutz, Wayne Hendrickson, Peter Coleman, Don Wiley and John Edsall in the front row at the Royal Institution at Max's 80th Birthday meeting.

[Reprinted from Biochemistry, (1973) 12, 2025.]

Demonstration of the Acyl-Enzyme Mechanism for the Hydrolysis of Peptides and Anilides by Chymotrypsin†

Jacques Fastrez‡ and Alan R. Fersht*

ABSTRACT: The acyl-enzyme mechanism for chymotrypsin was tested by determining the product ratios on the hydrolysis of substrates in the presence of added acceptor nucleophiles which compete effectively with water in the reaction. The product analysis was facilitated by the use of new substrates which could be separated easily from the products by ionophoresis. Direct determination of the ratio of AcPhe (*N*-acetyl-L-phenylalanine) to AcPhe-AlaNH$_2$ produced on the hydrolysis of AcPhe-OMe and AcPhe anilides in the presence of AlaNH$_2$ showed that 1 M AlaNH$_2$ is 44 times more reactive than 55 M water for both substrates. The decrease in V_{max}/K_M for the hydrolysis of AcPhe-AlaNH$_2$ in the presence of AlaNH$_2$ is also accounted for by AlaNH$_2$ being 44 times more reactive. This suggests a common intermediate in the hydrolysis of ester, anilide and peptide substrates. k_{cat}/K_M for the hydrolysis of AcPhe-AlaNH$_2$ was calculated using the above partition ratio, the values of the formation constants of AcPhe-OMe and AcPhe-AlaNH$_2$, the relative reactivities of methanol and water towards the acyl-enzyme derived from ester substrates and the free energy of hydrolysis of AcPhe-OMe. This agrees well with the directly measured value. This is *proof* of the acyl-enzyme mechanism in peptide hydrolysis. Previous attempts to demonstrate a common intermediate have always failed due to an artifact inherent in the approach used. The partitioning of an amide has always been determined indirectly from the decrease in V_{max} on the addition of the leaving group amine (*e.g.*, benzoyltyrosylglycinamide hydrolysis in the presence of added glycinamide). It is pointed out that the addition of certain organic solvents and amines increases V_{max} (and K_M) by a nonspecific effect. This partially compensates the decrease and underestimates the partitioning. In the present study the products from the hydrolysis of tritium-labeled compounds were separated by ionophoresis and assayed directly.

Esters are hydrolyzed by chymotrypsin in a two step process. The first step is the acylation of Ser-195 of the enzyme and the second is the subsequent deacylation (Hartley and Kilby, 1954). The rate-determining step for activated esters of nonspecific substrates and for most esters of specific substrates is deacylation. The acyl-enzyme accumulates in solution and may easily be detected by kinetic or spectral observation. Earlier evidence for the acylation of Ser-195 is summarized by Bender and Kézdy (1964) and Bruice and Benkovic (1966). The ultimate proof for nonspecific substrates is that the structures of the crystalline indolylacryloyl-chymotrypsin (Henderson, 1970) and carbamyl-chymotrypsin (Robillard *et al.*, 1972) have been solved by X-ray diffraction methods. Ser-195 is seen to be acylated.

The hydrolysis of amides is too slow to involve rate-determining deacylation and so acylation of the enzyme is assumed to be the slow step. The acyl-enzyme, if it occurs, is a high-energy intermediate present in low concentration and not easily detectable.

SCHEME I

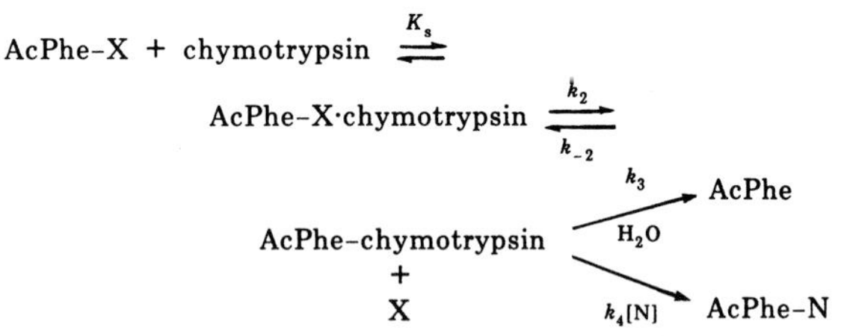

$$\text{AcPhe-X} + \text{chymotrypsin} \overset{K_s}{\rightleftharpoons} \text{AcPhe-X·chymotrypsin} \underset{k_{-2}}{\overset{k_2}{\rightleftharpoons}} \text{AcPhe-chymotrypsin} + \text{X}$$

$$\text{AcPhe-chymotrypsin} \xrightarrow[\text{H}_2\text{O}]{k_3} \text{AcPhe} \qquad \text{AcPhe-chymotrypsin} \xrightarrow{k_4[\text{N}]} \text{AcPhe-N}$$

It should be possible to detect the acyl-enzyme in amide hydrolysis by chemical trapping (see Scheme I). The addition of an acceptor nucleophile, N, will give mixed products, *e.g.*, AcPhe and AcPhe-N. The product ratio AcPhe-N:AcPhe should then be the same for both esters and amides if there is a common intermediate. Experiments along these lines have produced evidence *against* the acyl-enzyme mechanism. For example, Epand (1969) showed that benzoyltyrosine ethyl

† From the MRC Laboratory of Molecular Biology, Hills Road, Cambridge, England. *Received December 6, 1972.*

‡ Chargé de Recherches du Fonds National Belge de la Recherche Scientifique.

SCHEME II

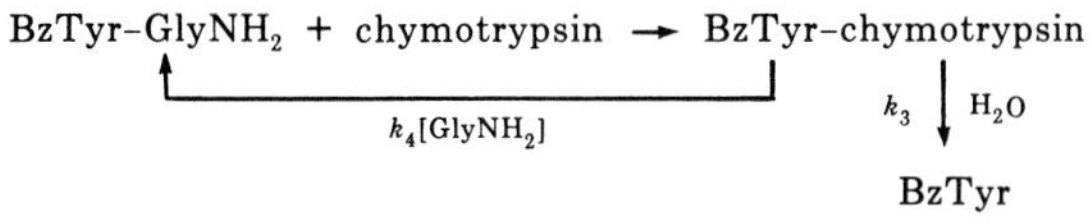

$$\text{BzTyr–GlyNH}_2 + \text{chymotrypsin} \rightarrow \text{BzTyr–chymotrypsin} \xrightarrow[\text{H}_2\text{O}]{k_3} \text{BzTyr}$$

$k_4[\text{GlyNH}_2]$ (BzTyr–chymotrypsin → BzTyr–GlyNH$_2$)

ester and benzoyltyrosylglycinamide apparently do not partition in the same way with $GlyNH_2$. In these experiments the products formed on the hydrolysis of the amide in the presence of the nucleophile were not directly assayed. The partition was found from the inhibition of the hydrolysis of benzoyltyrosylglycinamide by $GlyNH_2$ (Scheme II).

Evidence consistent with the acyl-enzyme mechanism is that the rate of release of *p*-nitroaniline from acetyl-DL-tyrosine-*p*-nitroanilide is unaffected by the addition of 1.6 M hydroxylamine although 50% acetyltyrosinehydroxamic acid is formed (Inagami and Sturtevant, 1964). This indicates that the hydroxylamine is involved after the rate-determining step and that 50% of the reaction must pass through an intermediate subsequent to this step.

We have attempted (a) to devise a system for directly analyzing the products on hydrolysis of substrates by chymotrypsin in the presence of nucleophiles in order to determine the product ratios accurately; (b) to find a sufficiently reactive nucleophile to show that in the hydrolysis of anilides an intermediate occurs after the rate-determining step and more than, say, 90% of the reaction goes through the intermediate; and (c) to show that the rate constants for amide hydrolysis give the correct equilibrium constant for the overall hydrolysis reaction when calculated on the acyl-enzyme scheme.

The following experiments are described. (1) [^{3}H]AcPhe-OMe was hydrolyzed by δ-chymotrypsin in the presence of various nucleophiles ($AlaNH_2$, $GlyNH_2$, and NH_2NH_2). When the AcPhe-OMe had completely reacted, subsequent reactions were quenched and the hydrolysis products, [^{3}H]-AcPhe and [^{3}H]AcPhe-nucleophile, were separated by ionophoresis and assayed. (2) [^{3}H]AcPhe-(*p*-trimethylammonium)-anilide and the (*p*-dimethylamino)anilide were hydrolyzed by δ-chymotrypsin in the presence of various nucleophiles. After about 5% of the substrate had reacted, further reaction was quenched. The products were separated from the substrate by ionophoresis and the [^{3}H]AcPhe and [^{3}H]AcPhe-nucleophile were assayed. In expt 1 and 2 the yields of reaction products and hence the partition ratios were *directly* determined. Rate measurements were not involved. (3) The partition of AcPhe-$AlaNH_2$ between Ala-NH_2 and water was determined *indirectly* by the inhibition of k_{cat}/K_M[1] for the δ-chymotrypsin catalyzed hydrolysis of AcPhe-$AlaNH_2$ by $AlaNH_2$. (4) AcPhe-δ-chymotrypsin was generated *in situ* in the stopped-flow spectrophotometer and its deacylation rate constants were determined in the absence and presence of the various nucleophiles used in the product ratio experiments. The partition of the acyl-enzyme between nucleophiles and water was obtained kinetically and compared with the product ratio expt, 1 and 2. (5) The effect of $AlaNH_2$, $GlyNH_2$, and NH_2NH_2 on the k_{cat}, k_{cat}/K_M, and K_M for the δ-chymotrypsin-catalyzed release of aniline from AcPhe- and AcTyr-anilides was determined. (6) Formation constants for AcPhe-OMe from AcPhe and MeOH and for AcPhe-$AlaNH_2$ from AcPhe and $AlaNH_2$ were measured. From these values and the rate constants for the hydrolysis of AcPhe-OMe and the partition of AcPhe-chymotrypsin between nucleophiles and water k_{cat}/K_M for the hydrolysis of AcPhe-$AlaNH_2$ by δ-chymotrypsin was calculated on the acyl-enzyme scheme and compared with the experimentally measured value.

[1] The parameters k_{cat}, K_M, and V are defined by $V = k_{cat}[E]/(K_M + [S])$, where [E] and [S] are the enzyme and substrate concentrations; μ = ionic strength.

Materials

L-*Alaninamide*. L-Alanine methyl ester was prepared by dissolving 30 g of L-alanine (0.33 mol) in 250 ml of dry methanol containing 42.8 g of thionyl chloride (0.36 mol) which had been prepared in a Dry Ice–isopropyl alcohol bath. After stirring for 15 min, the cold solution stood at room temperature for 1 hr. The solvent was removed below 40°. The ester hydrochloride was recrystallized from methanol after storage under vacuum over sodium hydroxide pellets: mp 108–111° (lit. (Zaoral *et al.*, 1967) mp 109–111°). The ester (30 g) was added at 0° to 250 ml of methanol saturated with ammonia in a pressure bottle. After standing for 2 hr at 0° and 24 hr at room temperature, the solvent was removed by evaporation. The amide was recrystallized from methanol–ether: mp 222–224° and α_D +10.6° (lit. (Beilstein) α_D +10.5°).

Cbz-L-*Phe-AlaNH$_2$* was prepared by adding 2.1 g of Cbz-Phe-*p*-ONph to a mixture of 0.65 g of L-$AlaNH_2 \cdot HCl$ and 0.55 g of triethylamine in 7 ml of dimethylformamide at 0° and leaving overnight in the cold. After precipitation by the addition of 50 ml of H_2O, the compound was washed with 2.5% K_2CO_3, dried, and crystallized from methanol: mp 194–200° (1.5 g).

Acetyl-L-*AlaNH$_2$*. The Cbz derivative (1.5 g) was debenzylated by dissolving in 10 ml of $HBr \cdot CH_3CO_2H$. After 30 min the Phe-$AlaNH_2 \cdot HBr$ was precipitated by ether. The hydrobromide was dissolved in chloroform by the addition of triethylamine and then acetylated with 1 g of acetic anhydride. The white precipitate was recrystallized from methanol (mp 242–244°).

[^{3}H]Acetyl-L-*Phe-AlaNH$_2$*. Cbz-L-$AlaNH_2$ (0.6 g) was debenzylated and the hydrobromide was collected as above. The free base was obtained by dissolving the hydrobromide in 5 ml of H_2O and adding 2 N NaOH until pH 10. After extraction with ethyl acetate the product crystallized on the addition of petroleum ether (bp 40–60°). Phe-$AlaNH_2$ (0.2 g) was dissolved in 2 ml of chloroform containing 25 mCi of [^{3}H]acetic anhydride (500 mCi/mol). After 30 min 0.1 ml of (cold) acetic anhydride was added. The precipitate was recrystallized from methanol: mp 242–244° as above.

[^{3}H]Ac-L-*Phe-OMe*. Phe-OMe·HCl (2.3 g) was dissolved in 15 ml of water and the free base was extracted with 10 ml of ethyl acetate after the addition of 2.5 g of potassium carbonate. Nonradioactive material was prepared by adding 1 g of acetic anhydride to 5 ml of the ethyl acetate solution which had been dried over $CaCl_2$. On addition of petroleum ether at 50° and cooling, the ester crystallized (mp 89–90.5°, lit. (Huang *et al.*, 1952) mp 90–91°). The tritiated material was prepared by adding 25 mCi (25 mCi/mmol) of [^{3}H]acetic anhydride to 2 ml of the dried ethyl acetate extract and after 30 min 0.5 ml of acetic anhydride was added. The ester was crystallized as above: mp 89–90.5°.

Cbz-L-*Phe-(p-dimethylamino)anilide*. *p-N,N*-Dimethylaminoaniline was prepared from the dihydrochloride salt (recrystallized from ethanol–acetone) (2.75 g) by the addition of 27.5 ml of 1 N KOH followed by 30 ml of ethyl acetate. The organic layer was separated, the solvent was removed by evaporation, and the base was dried by the addition of 30 ml of benzene followed by rotary evaporation. Cbz-L-phenylalanine *p*-nitrophenyl ester (3.3 g), dissolved in 50 ml of purified dimethylformamide (previously stirred under vacuum to remove dimethylamine) cooled to 0°, was added to the cooled dimethylaminoaniline and left overnight. Ethyl acetate (125 ml) was added, the solution was filtered, the precipitate was retained, and the solution was extracted several times with potassium carbonate solution when further product precipitated. The combined precipitates were washed with ethyl acetate and potassium carbonate, dried, and then recrystallized from acetone–petroleum ether (mp 187–189°). Further product was obtained by concentrating the ethyl acetate solution and extracting with 1 N HCl. The neutral compound was then precipitated from the aqueous solution by the addition of base to pH 6. The product was recrystallized as above.

L-Phe-(p-dimethylamino)anilide was prepared by debenzylation of the above in HBr–CH_3CO_2H. The ether precipitate was dissolved in water and the free base was liberated by the addition of base to pH 9.5 (mp 92–95°).

Ac-L-Phe-(p-dimethylamino)anilide. The above compound was dissolved in 15 ml of acetonitrile and 1.5 g of acetic anhydride and 1.5 ml of triethylamine were added. The product crystallized and after washing with water was recrystallized from ethyl acetate (mp 239–241°).

AcPhe-(p-trimethylammonium)anilide Iodide. The above (1.3 g) was dissolved in 13 ml of dimethylformamide (previously subjected to evacuation) and 1.5 g of methyl iodide was added. After 3 hr the product was precipitated by the addition of 90 ml of ether, then triturated with hot acetone, and crystallized from ethanol (mp 169–173°).

[^{3}H]Ac-L-Phe-(p-trimethylammonium)anilide iodide was prepared as above except 25 ml (25 Ci/mol) of [^{3}H]acetic anhydride was first added to 120 mg of L-Phe-(*p*-dimethylamino)-anilide and the acetic anhydride and triethylamine after a further 30 min. Methylation was as above.

Ac-L-Phe-p-ONph was synthesized according to Ingles and Knowles (1968).

Satisfactory nuclear magnetic resonance (nmr) spectra and elemental analyses were obtained for the substrates. Optical purity was in all cases >98% as determined by assaying the product release on hydrolysis with chymotrypsin by the kinetic techniques described. Other materials were purified commercial products. The δ-chymotrypsin was obtained from Sigma (lot 20C-0150).

Methods

Product Distribution on the Hydrolysis of Substrates by δ-Chymotrypsin in the Presence of Added Acceptor Nucleophiles

Preparation of Anilide Substrates. Two new anilide substrates were prepared. One was positively charged, and the other could be protonated at low pH so that they could be separated from the negatively charged and neutral products by high-voltage electrophoresis.

$AcPhe^-$ $\quad$ $AcPhe\text{-}AlaNH_2{}^0$ $\quad$ AcPhe—NH—C$_6$H$_4$—$\overset{+}{N}(CH_3)_3{}^+$

$AcPhe\text{-}GlyNH_2$ $\quad$ AcPhe—NH—C$_6$H$_4$—$\overset{+}{N}H(CH_3)_2$

$AcPhe\text{-}NHNH_2$

HYDROLYSIS OF [^{3}H]AcPhe-OMe. To 0.96 ml of a solution containing carbonate buffer, nucleophile ($AlaNH_2$, $GlyNH_2$, or H_2NNH_2) at 25° and ionic strength 1.0 (added KCl) was added

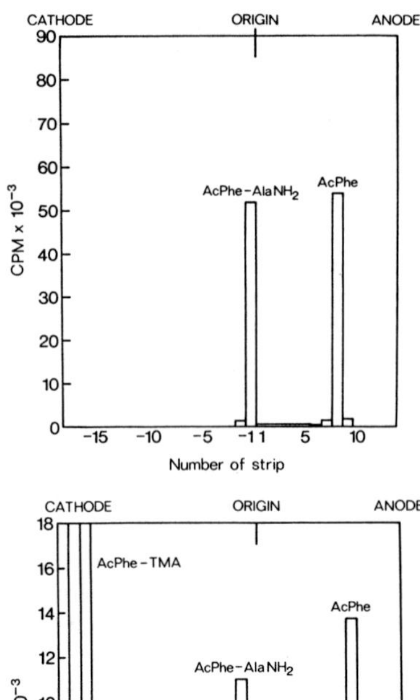

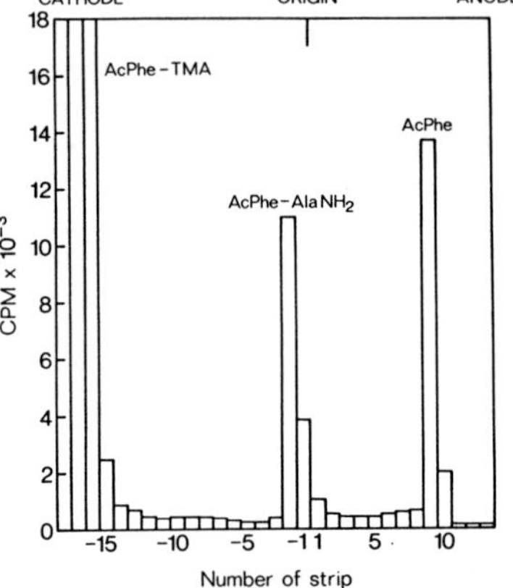

FIGURE 1: Separation of reagents and products by ionophoresis at pH 3.5. (a) Hydrolysis of [^{3}H]AcPhe-OMe in presence of $AlaNH_2$. (b) Hydrolysis of [^{3}H]AcPhe-(*p*-trimethylammonium)anilide in presence of $AlaNH_2$.

10 μl of 1 mM δ-chymotrypsin, followed by 50 μl of 50 mM [^{3}H]AcPheOMe in Me_2SO. The pH was 9.30 ± 0.02. After 30 sec, sufficient time for the substrate to be completely hydrolyzed, the reaction was quenched by the addition of 200 μl of 50% trichloroacetic acid. The solution (10 μl) was spotted on to Whatman No. 3MM paper and the products, AcPhe and AcPhe-nucleophile, were separated by high-voltage electrophoresis. The systems used were either a 1% ammonium carbonate buffer (pH 8.9) or, more generally, pyridine acetate (pH 3.5) at 3 kV with white spirit as coolant or, as a check, pyridine acetate (pH 6.5) at 3 kV using the flat-bed method.

The paper was then cut into about 25 or 30 strips (1 × 3 cm) which were then soaked overnight in vials containing 4 ml of Bray's (1960) solution. The ^{3}H content was then counted on a Nuclear-Chicago Unilux scintillation counter. The histogram thus obtained gave the yield of free acid and acylated nucleophile (see Figure 1). The recovery of products was quantitative.

HYDROLYSIS OF (*p*-TRIMETHYLAMMONIUM)ANILIDE. To 0.99 ml of a solution containing 50 μl of Me_2SO, 5 mM substrate, carbonate buffer, nucleophile, and KCl to give $\mu = 0.95$ and pH 9.30 at 25° was added 10 μl of 1 mM δ-chymotrypsin. After 3.0 min (time for ~5% of the substrate to be hydrolyzed), the reaction was quenched with 200 μl of 50% trichloroacetic acid. The initial products, [^{3}H]AcPheCO_2^- and [^{3}H]AcPhe-nucleophile^0, were separated from the positively charged sub-

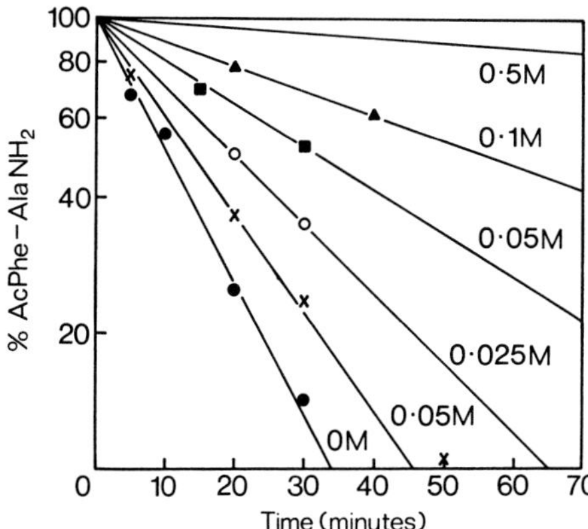

FIGURE 2: Inhibition by $AlaNH_2$ of the hydrolysis of AcPhe-$AlaNH_2$ by δ-chymotrypsin at 25°, μ = 0.95, pH 9.30. AcPhe-$AlaNH_2$ (2.5 mM) is at a concentration well below K_M. The concentration of $AlaNH_2$ is given in M.

strate by high-voltage electrophoresis as above using the pH 3.5 pyridine acetate system. The Whatman No. 3MM paper was cut into strips and counted as before (see Figure 1).

Some experiments were performed with [^{3}H]AcPhe-(*p*-dimethylamino)anilide as substrate. This is protonated under the conditions of the high-voltage electrophoresis and is also easily separated. The results are presented in Table I.

Inhibition of the Hydrolysis of AcPhe-$AlaNH_2$ by $AlaNH_2$.

$$\text{AcPhe-AlaNH}_2 + \text{chymotrypsin} \xrightarrow{k_2/K_S} \text{AcPhe-chymotrypsin} \xrightarrow{k_3} \text{AcPhe}$$
$$(\text{AcPhe-chymotrypsin} \xrightarrow{k_4[\text{AlaNH}_2]} \text{AcPhe-AlaNH}_2)$$

The hydrolysis of [^{3}H]AcPhe-$AlaNH_2$ by δ-chymotrypsin in the presence of varying concentrations of $AlaNH_2$ was determined for [S] $\ll K_M$. Under these conditions the disappearance of substrate should be pseudo first order in [S].

A solution containing 2.5 mM [^{3}H]AcPhe-$AlaNH_2$, pH 9.3 carbonate buffer, and various amounts of $AlaNH_2$ (made up to μ = 1.0) was incubated at 25° and 10 μl of 10^{-3} M δ-chymotrypsin was added. Aliquots of about 200 μl were periodically removed and added to 50 μl of 50% trichloroacetic acid; 10 μl of each quenched mixture was subjected to high-voltage electrophoresis as before and the ratio of [^{3}H]AcPhe:[^{3}H]AcPhe-Ala-NH_2 was determined. The initial slope of the semilogarithmic plot of substrate against time gave V_{max}/K_M. Enzyme autolysis ($t_{1/2}$ ~ 2 hr) caused curvature in the plots for the slower reactions. The results are presented in Table II and Figure 2.

$AlaNH_2$ competes with water for the acyl-enzyme (or any other intermediate covalent or otherwise) and causes noncompetitive inhibition. The partition ratios may be calculated from the inhibition.

If V_{max}/K_M in the presence of nucleophile is designated by the subscript "obsd" and in the absence of nucleophile by subscript "0," then

$$1 - \frac{[V_{max}/K_M]_{obsd}}{[V_{max}/K_M]_0} = \frac{k_4[\text{AlaNH}_2]}{k_3 + k_4[\text{AlaNH}_2]}$$

The results are presented in Tables I and II. (Note: these calculations assume that there is no effective competitive inhibition by the $AlaNH_2$. This is justified later.)

TABLE I: Hydrolysis of AcPhe-OMe, AcPhe-$AlaNH_2$, AcPhe-(*p*-trimethylammonium)anilide (AcPhe-TMA), and AcPhe-(*p*-dimethylamino)anilide (AcPhe-DMA) by δ-Chymotrypsin in the Presence of Acceptor Nucleophiles.[a]

Nucleophile and Concn (M)[b]	% AcPhe-nucleophile Produced on Hydrolysis of: AcPhe-OMe[c]	AcPhe-TMA[c]	AcPhe-Ala-NH_2[d]	AcPhe-DMA[c]	Calcd[e]
$AlaNH_2$					
0.010	24.3	28.4	26.3		28.8
0.025	48.1	47.9	48.4		50.4
0.05	69.0	67.0	67.3	63.2	66.9
0.10	81.8	80.3	81.6		80.14
0.50			97.3		95.3
0.84		>94			97.1
$GlyNH_2$					
0.010		9.0			9.6
0.025		15.8			20.9
0.10	51.8	47.7		46.6	51.4
0.25	75.5	73.9			72.6
0.50	86.0	83.9			84.1
H_2NNH_2					
0.5	51.0	46.3			48.6
0.94	63.8				64.0
$AlaNH_2$[f]					
0.05 (+α-chymotrypsin)	71.3				

[a] 25°, 5% Me_2SO, μ = 0.95. [b] Total ionic forms. [c] From direct radioactivity assays. [d] From inhibition of hydrolysis (k_{cat}/K_M) by $AlaNH_2$. No Me_2SO present. [e] Calculated from mean values of partition (for free base) $AlaNH_2$–H_2O, 44:1; $GlyNH_2$–H_2O, 11.5:1; H_2NNH_2–H_2O, 2.04:1 M^{-1}. [f] Using α-chymotrypsin rather than δ-chymotrypsin.

Kinetic Determination of Partition of AcPhe-Chymotrypsin between Nucleophiles and Water. δ-Chymotrypsin and AcPhe-*p*-ONph were mixed in the stopped-flow spectrophotometer (Gilford 2400 spectrophotometer equipped with rapid mixer, dead time = 4 msec) under conditions that ensured complete acylation of the enzyme and *p*-nitrophenol release in the dead time of the mixer. The subsequent first-order deacylation of the enzyme was followed by monitoring proflavine binding at 465 nm (Bernhard *et al.*, 1966).

One syringe of the mixer contained 5 × 10^{-5} M proflavine, 5 × 10^{-6} M δ-chymotrypsin, pH 9.30 carbonate buffer, and nucleophile, at μ = 1.0. The other syringe contained 5 × 10^{-5} M proflavine, 5 × 10^{-5} M AcPhe-*p*-ONph, and 4% acetonitrile at pH 4.5. After mixing there was a first-order increase in absorbance at 465 nm as the enzyme deacylated. The results are presented in Table III.

Calculation and Measurement of Equilibrium Constants

Partition of AcPhe-Chymotrypsin between Methanol and Water

From Michaelis–Menten Plots. The values of k_{cat} and K_M for the release of *p*-nitrophenolate from AcPhe-*p*-ONph as it is hydrolyzed by δ-chymotrypsin (at pH 9.30, 25°, 5% Me_2SO) in varying concentrations of methanol were found from

TABLE II: Inhibition of AcPhe-AlaNH$_2$ Hydrolysis by AlaNH$_2$.[a]

AlaNH$_2$ (M)[b]	k_{obsd}[c] (sec^{-1} × 10^4)	k_{cat}/K_M[d] (sec^{-1} M^{-1})
0	11.2	132
0.01	8.25	97
0.025	5.78	68
0.05	3.66	43
0.10	2.06	24.2
0.50	0.304	3.6

[a] 25°, μ = 1.0, pH 9.30, 8.5 × 10^{-6} M δ-chymotrypsin, 2.5 mM substrate. [b] Total ionic forms. [c] First-order rate constant for hydrolysis. [d] k_{obsd}/[δ-chymotrypsin].

TABLE III: Deacylation of AcPhe-δ-Chymotrypsin in the Presence of Nucleophiles.[a]

Nucleophile	Concn[b] (M)	k_{obsd}[c] (sec^{-1})	Partition[d] (M^{-1}) (Nucleophile:H$_2$O)	Partition[e] from Product Ratios (M^{-1})
		142 ± 3		
AlaNH$_2$	0.02	232 ± 11	34 ± 4	44
GlyNH$_2$	0.05	215 ± 7	11.1 ± 1	
	0.083	255 ± 11	10.4 ± 1	11.5
H$_2$NNH$_2$	0.2	202 ± 4	2.3 ± 0.2	2.0

[a] Determined directly in stopped-flow spectrophotometer by proflavine displacement, 25°, μ = 1.0. [b] Total ionic forms. [c] Observed first-order rate constant. [d] (k_{obsd} − 142)/([nucleophile] × 142). [e] From the experiments involving ^{3}H-labeled reagents.

stopped-flow experiments monitored at 400 nm. One syringe of the mixer contained enzyme (2.4 μM) in 5% Me$_2$SO, pH 9.30 carbonate buffer, and added KCl at μ = 0.95. The other held 0.95 M KCl, 5% Me$_2$SO, 110 μM AcPhe-*p*-ONph, and varying concentrations of methanol. Tangents were taken at various points along the curve of product against time and a reciprocal plot was obtained to give V_{max} and K_M. In most cases V_{max} only was obtained, by a small extrapolation.

In Scheme I, where X = *p*-nitrophenol and N = MeOH, $k_2 \gg k_3 + k_4$[MeOH], it is easily shown that

$$\frac{d[\text{p-nitrophenol}]}{dt} = \frac{k_3 + k_4[\text{MeOH}][E_0S_0]}{[S_0] + K_s(K_3 + k_4[\text{MeOH}])/k_2}$$

where [S$_0$] = the initial substrate concentration and [E$_0$] the total enzyme concentration.

A plot of V_{max} against [MeOH] should give a slope of k_4[E$_0$] and intercept k_3[E$_0$] (*cf.* Bender *et al.*, 1964a).

From Pseudo-First-Order Kinetics. These experiments were similar to those for the kinetic determination of the partition of AcPhe-chymotrypsin between nucleophiles and water by the proflavine displacement method. The stopped-flow spectrophotometer for these experiments had a path length of 2 mM and dead time of 0.9 msec.

One syringe of the mixer contained 1.45 × 10^{-4} M δ-chymotrypsin, 10^{-4} M proflavine, pH 9.30 carbonate, 5% Me$_2$SO, and KCl to give μ = 0.95. The other syringe contained 3.5 × 10^{-5} M AcPhe-*p*-ONph, 5% Me$_2$SO, 0.95 M KCl, and varying amounts of methanol; 4 msec after mixing there was a good first-order decrease in transmission at 465 nm as the enzyme deacylated. (Reacylation of the enzyme by the methyl ester produced was insignificant.) The results along with those from the zero-order rates are presented in Table IV.

Formation Constants of AcPhe-AlaNH$_2$, AcPhe-GlyNH$_2$, and AcPhe-OMe

Equilibration between AcPhe and AcPhe-AlaNH$_2$ or AcPhe-GlyNH$_2$ at 25° and ionic strength 1.0 was catalyzed by chymotrypsin and the mixtures were assayed as follows.

HYDROLYSIS. To 1 ml of 1 M AlaNH$_2$ (or GlyNH$_2$) at pH 6.80 was added 10 μl of 0.5 M [^{3}H]AcPheOMe in acetonitrile followed by 10 μl of 1 mM α-chymotrypsin. (The partition experiments show that an appreciable fraction of the ester is converted rapidly to the AcPhe-AlaNH$_2$ (or -GlyNH$_2$) which may then hydrolyze.) The mixture was then left for 36 hr to hydrolyze to equilibrium.

SYNTHESIS. A solution (0.5 ml) containing 10 μl of 1 mM δ-chymotrypsin and 10 μl of [^{3}H]AcPhe-OMe (0.5 M in acetonitrile) was incubated for several minutes until hydrolysis was complete; then 0.5 ml of 2 M AlaNH$_2$ (or GlyNH$_2$) at pH 6.8 was added. The mixture was left for 36 hr to synthesize the peptide and equilibrate.

The products were separated by high-voltage electrophoresis and the products were assayed in the scintillation counter.

AcPhe-OMe. Synthesis of AcPhe-OMe from AcPhe (initially 9.8 mM) and hydrolysis of AcPhe-OMe (initially 10.15 mM) catalyzed by 1 N HCl in 4.95 M methanol was followed to equilibrium ($t_{1/2}$ = 4.0 hr) by the alkaline hydroxamate assay. Samples (0.5 ml) were added to 1.0 ml of a freshly prepared mixture of 1.33 M hydroxylamine hydrochloride and 2.67 M NaOH. After 5.0 min 1.5 ml of 20% FeCl$_3$·6H$_2$O in 4 N HCl was added and the absorbance at 540 nm was measured after 12.0 min. A calibration curve was made using

TABLE IV: Hydrolysis and Methanolysis of AcPhe-*p*-nitrophenyl Ester.[a]

[MeOH], M	Michaelis–Menten[b] Kinetics: k_{cat} (sec^{-1})	K_M (μM)	First-Order Kinetics[c] of Deacylation k_3 (sec^{-1})
0	144	3.16	144
0.062			148
0.123			164
0.25	231		249
0.37			260
0.5			264
0.625	330		329
0.75	368		
0.94	416		
1.25	482	12.6	

[a] pH 9.30, μ = 0.95, 5% Me$_2$SO, 25°. [b] From 1/*V* *vs.* 1/[S] plots, [S$_0$] = 55 μM, δ-chymotrypsin = 1.1 μM. [c] δ-chymotrypsin = 9.8 × 10^{-4} M.

TABLE V: Formation Constants for AcPhe-X at 25°, μ = 1.0.

AcPhe-X	pH	Equilibrium Concn (Total Ionic Forms), M AcPhe	AcPhe-X	XH	Direction to Equilibrium of AcPhe-X	[AcPhe-X]/ [AcPhe]$_{total}$-[XH]$_{total}$ (Total Ionic Forms)	[AcPhe-X]/ [AcPhe][XH] (Un-ionized)
AcPhe-AlaNH$_2$	6.80	4.51×10^{-3}	4.85×10^{-4}	0.98	Hydrolysis	0.108	8.81×10^{3}
	6.80	4.50×10^{-3}	5.025×10^{-4}	0.98	Synthesis	0.114	9.25×10^{3}
AcPhe-GlyNH$_2$	6.80	3.99×10^{-3}	1.01×10^{-3}	0.98	Hydrolysis	0.258	2.00×10^{4}
	6.80	3.00×10^{-3}	9.95×10^{-4}	0.98	Synthesis	0.254	1.97×10^{4}
AcPhe-OMe	0	5.38×10^{-3}	4.42×10^{-3}	4.95	Hydrolysis		0.166
	0	5.66×10^{-3}	4.39×10^{-3}	4.95	Synthesis		0.157

standard solutions of AcPhe-OMe. The *N*-acetyl group also slowly hydrolyzed. The end points were extrapolated back to zero time to give a small correction. The results are presented in Table V.

THE pK_a VALUES of AlaNH$_2$, GlyNH$_2$, and AcPhe at 25° and μ = 1.0 were found by titrating 20 mM solutions of the amines (10 ml) and 10 mM of the acid (10 ml) in 1 M KCl with 1 N NaOH from a Burroughs Wellcome Agla syringe. Corrections were made for [H$^+$] in the case of the acid. The pK_a values were as follows: for AlaNH$_2$, 8.24 ± 0.01; for GlyNH$_2$, 8.22 ± 0.01; and for AcPhe, 3.33 ± 0.02.

HYDROLYSIS OF AcPhe-OMe AND AcPhe-AlaNH$_2$. These reactions were carried out in 10-ml volumes in a Radiometer TTT II autotitrator in 0.95 M KCl and 5% Me$_2$SO, using 8×10^{-9} M δ-chymotrypsin with 0.5–3.5 mM AcPhe-OMe, 4×10^{-6} M δ-chymotrypsin and 1.0–8.0 mM AcPhe-AlaNH$_2$.

KINETICS OF ANILIDE HYDROLYSIS IN PRESENCE OF NUCLEOPHILES. The hydrolysis of AcPhe-(*p*-trimethylammonium)-anilide and AcTyr-*p*-chloroanilide were followed using a Gilford 2400 spectrophotometer. The reactions were at 25°, pH 9.30 (carbonate buffer), 5% Me$_2$SO and μ = 0.95, and various concentrations of AlaNH$_2$, GlyNH$_2$, and H$_2$NNH$_2$.

Results

We have used δ-chymotrypsin in preference to α-chymotrypsin since (a) K_M values do not increase as much at alkaline pH because more δ-chymotrypsin remains in the active conformation, (b) the fraction of inactive conformation is lower and this avoids artifacts in presteady-state studies (Fersht, 1972), and (c) we have found that for many substrates K_M is smaller.

The titration of δ-chymotrypsin with Cbz-L-tyrosine *p*-nitrophenyl ester (Kézdy *et al.*, 1964) underestimated the active-site content. A value of 65% was found compared with 88% by the *trans*-cinnamoylimidazole method (Schonbaum *et al.*, 1961). (The δ-chymotrypsin in the paper of Fersht (1972) was also 88% active and not 65% as quoted.) A similar error was found with α-chymotrypsin which had been blocked on the surface carboxyls (Fersht and Sperling, 1973). Again, *trans*-cinnamoylimidazole and *p*-nitrophenyl acetate gave the correct values. The Cbz-tyrosine *p*-nitrophenyl ester method is reliable for α-chymotrypsin.

Product Ratios. We have introduced two new substrates which may be separated easily from their hydrolysis products in order to be able to directly analyze the products of the enzymatic hydrolysis reactions. It is seen in Table I that the product ratios on the simultaneous hydrolysis and acyl-transfer reactions of AcPhe-OMe, AcPhe-(*p*-trimethylammonium)anilide and AcPhe-(*p*-dimethylamino)anilide are identical, within the limits of experimental error, for each set of experimental conditions. Furthermore, the effect of AlaNH$_2$ on the k_{cat}/K_M for the hydrolysis of AcPhe-AlaNH$_2$ is also consistent with the partition ratios. AlaNH$_2$ is 43.6 ± 2.4 times more reactive than water in the reaction of δ-chymotrypsin with AcPhe-OMe, 44.1 ± 1.4 times more reactive for AcPhe-(*p*-trimethylammonium)anilide, and 43.1 ± 2.1 times more reactive for the inhibition of k_{cat}/K_M for AcPhe-AlaNH$_2$ hydrolysis.

The mean values of all the partitioning experiments show that AlaNH$_2$ is 44 times, GlyNH$_2$ 11.5 times and hydrazine 2.0 times more reactive than water toward the acyl-enzyme, AcPhe-chymotrypsin. There is no systematic deviation from the calculated product ratios (Table I, last column) based on these values. There appears to be no saturation of a nucleophile binding site on the *acyl-enzyme*.

Direct measurements, although of somewhat lower accuracy, of the attack of the nucleophiles on the acyl-enzyme generated *in situ* by the acylation of the enzyme by AcPhe-*p*-nitrophenyl ester are consistent with the above (Table III).

It is seen in Table VI that the values of k_{cat} for the release of aniline from AcPhe-*p*-(trimethylammonium)anilide and AcTyr *p* chloroanilide increase for increasing concentrations of AlaNH$_2$. GlyNH$_2$ and H$_2$NNH$_2$ also increase k_{cat}. Dimethylformamide, a secondary amide which is not a nucleophile, also increases k_{cat} for several substrates of chymotrypsin and trypsin (Table VII). AlaNH$_2$ does not cause serious changes in k_{cat}/K_M (Table VI) and it was felt justified to use this parameter in the calculations for the inhibition of AcPhe-AlaNH$_2$ hydrolysis by AlaNH$_2$.

The increase in k_{cat} does not increase linearly with AlaNH$_2$ concentration. This could be due to binding to the *enzyme–substrate complex* with a dissociation constant of about 20 mM.

Reaction of AcPhe-Chymotrypsin with methanol. The attack of methanol on AcPhe-δ-chymotrypsin was initially measured under pseudo-first-order conditions. An excess of enzyme was mixed with the substrate to give rapid acylation and then the deacylation rate constant was measured directly. The results became erratic at higher methanol concentrations due to mixing artifacts in the stopped-flow spectrophotometer. Higher concentrations of methanol were possible under zero-order con-

TABLE VI: Hydrolysis of AcTyr-*p*-chloroanilide and AcPhe-(*p*-trimethylammonium)anilide in the Presence of Nucleophiles.[a]

Substrate	Nucleophile and Concn (M[b])		k_{cat} (sec^{-1})	K_M (mM)	k_{cat}/K_M (sec^{-1} M^{-1})	% Increase in k_{cat}
AcTyr-*p*-chloroanilide[c]	AlaNH$_2$	0	0.0308	0.35	88	0
		0.033	0.0382	0.419	91.1	24
		0.10	0.0405	0.446	90.8	31
		0.5	0.0422	0.45	93.6	37
AcPhe-TMA[d]	AlaNH$_2$	0	0.403	4.3	94.2	0
		0.033	0.504	4.7	107	24
		0.10	0.523	5.37	97.4	29
		0.50	0.557	7.3	76.1	38
	GlyNH$_2$, 0.5		0.642	7.7	81.9	58
	H$_2$NNH$_2$, 0.5		0.543	4.72	115	35

[a] 25°, μ = 0.95, 5% Me$_2$SO, pH 9.30. [b] Total ionic forms. [c] Monitored at 300 nm, $\Delta\epsilon = 1.03 \times 10^3$. [d] Monitored at 295 nm, $\Delta\epsilon = 1.12 \times 10^3$.

ditions of excess substrate over enzyme and at concentrations much higher than K_M. The V_{max} values were converted to k_{cat} values by taking V_{max} in the absence of methanol to be 144 sec^{-1}, as found from the first-order kinetics.

Total hydrolysis of the substrate occurred in less than 800 msec. This is fast compared with the conformational equilibration of the enzyme (Fersht, 1972). The percentage of the enzyme in the active conformation may be calculated from the ratio of the observed V_{max}/[chymotrypsin] (100 sec^{-1}) and the known k_{cat} (144 sec^{-1}) (Fersht, 1972) and is 69% at pH 9.30, 25°, μ = 0.95, 5% Me$_2$SO.

Methanol is 1.9 ± 0.2 times more reactive than water towards AcPhe-chymotrypsin. A similar result was obtained by Bender *et al.* (1964a) for α-chymotrypsin.

Hydrolysis of AcPhe-OMe and AcPhe-AlaNH$_2$. These hydrolyses were performed on the same day using the same stock solutions of enzyme to minimize errors in the ratios of the values of k_{cat}/K_M. AcPhe-AlaNH$_2$ was hydrolyzed at pH 9.30, 5% Me$_2$SO, 25°, and 0.95 M KCl with k_{cat}/K_M = 126 M^{-1} sec^{-1}. Approximate values for k_{cat} and K_M are 5 sec^{-1} and 40 mM. Under the same conditions k_{cat}, K_M, and k_{cat}/K_M for AcPhe-OMe are 135 sec^{-1}, 0.54 mM, and 2.5 × 10^5 M^{-1} sec^{-1}. These data may be tentatively analyzed, using the notation of Scheme I, and assuming k_3 to be 144 sec^{-1}, to give $k_2 \sim 2.2 \times 10^6$ sec^{-1} and $K_S \sim$ 9 mM.

Free-Energy Diagrams. Sufficient data have been derived to plot free energy against reaction coordinate diagrams (Figures 3 and 4).[2] The standard states are 1 M for all reagents except water which is assigned an activity of 1.0. The dissociation constant for AcPhe and chymotrypsin is estimated to be ~10 mM. The rate constants for the association of substrates with chymotrypsin is taken to be ~10^7 sec^{-1} M^{-1} (Hess *et al.*, 1970).

Discussion

Acyl-chymotrypsins have been synthesized and shown to react with amines. Inward and Jencks (1965) isolated the relatively stable furoyl-chymotrypsin and determined its reactivity with a wide range of amines. We prepared AcPhe-chymotrypsin *in situ* and observed its reaction with AlaNH$_2$, GlyNH$_2$, and H$_2$NNH$_2$. It may be inferred from the principle of microscopic reversibility that the reverse reaction, the acylation of chymotrypsin by amides to give the acyl-enzyme, does occur. The question is, then, is acyl-enzyme formation the main route for amide hydrolysis or is it a side reaction?

A method of solving this problem is (a) to demonstrate the acyl-enzyme mechanism for one substrate where it accumulates and then to apply the following criteria to test whether it occurs in the other cases where it does not accumulate. These criteria are (b) "a common acyl portion should generate a common intermediate which should then partition in an identical manner in the presence of added acceptors"; (c) "the rate constants calculated on the acyl-enzyme mechanism must satisfy the overall equilibrium constant for the reaction"; and (d) "the acyl-enzyme occurs after the rate-determining step for the slower reactions and so the addition of nucleophiles should not affect the rate of disappearance of substrate."

In ideal situations, b, c, and d will hold. Criterion c is both

[2] ADDED IN PROOF. The rate constant for the diffusion-controlled association of AcPhe derivatives with δ-chymotrypsin has now been measured (Renard and Fersht, unpublished results). The barrier height is 6.4 kcal M^{-1}, not the estimated value of 8.0 in the figures.

TABLE VII: Effect of Dimethylformamide on k_{cat} for Some Reactions of Chymotrypsin and Trypsin.

Enzyme	Substrate	[DMF], M	k_{cat} (sec^{-1} × 10^2)
α-Chymotrypsin	AcTyr-*p*-chloroanilide[a]	0.32	1.19
		0.65	1.37
	AcTyr-*m*-chloroanilide[a]	0.32	0.93
		0.65	1.11
		1.04	1.36
	FormylPhe-semicarbazide[b]	0	2.88
		0.65	3.25
Trypsin	CbzArg-*p*-toluidide[c]	0.26	0.43
		0.65	0.71
		1.30	0.97

[a] Inagami *et al.* (1969). [b] Fastrez and Fersht (1973). [c] Inagami (1969).

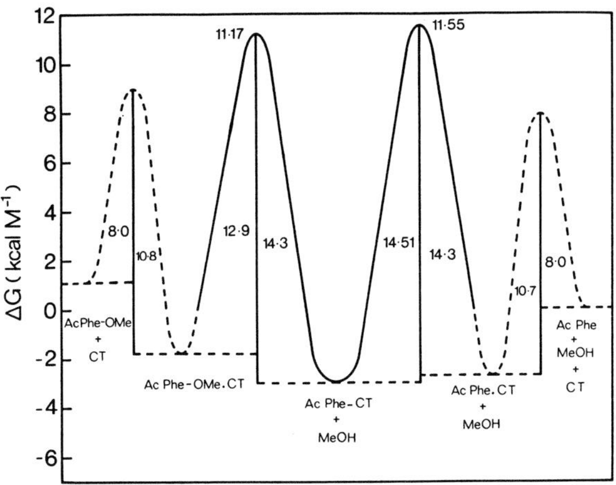

FIGURE 3: Free energy against reaction coordinate for the hydrolysis of AcPhe-OMe by δ-chymotrypsin at pH 9.30, 25°, ionic strength 0.95, 5% Me_2SO. Standard states are 1 M, activity of water is 1. AchPhe is un-ionized.[2]

necessary and sufficient. But artifacts could occur in the measurement of the required individual rate constants. The determination of k_{-2} requires the presence of an added nucleophile and this could perturb the enzyme. Similarly, different enzyme concentrations may be required for each k_2/K_S. Criterion b is not essential. It is possible that the deacylation of the acyl-enzyme may occur while the leaving group is still bound to or slowly diffusing away from the acyl-enzyme and so causing steric or allosteric effects. Criterion d is also not essential. An added nucleophile could act as an allosteric effector or inhibitor and alter the V_{max}.

We shall now apply the above steps, a, b, c, and d.

Acyl-Enzyme Formation. Bender and coworkers have amply demonstrated that AcPhe-OMe and other ester derivatives hydrolyze by acyl-enzyme formation (Bender and Kézdy, 1964). McConn *et al.* (1971) have shown the "burst" of acyl-enzyme formation from AcPhe-OEt and chymotrypsin and the subsequent deacylation by presteady-state kinetics. We have shown that AcPhe-chymotrypsin generated from AcPhe-*p*-ONph deacylates with a rate constant of 144 sec^{-1}, while AcPhe-OMe is hydrolyzed with a k_{cat} value of 135 sec^{-1}. This is consistent with the acyl-enzyme being formed with $k_2 = 2.2 \times 10^3$ sec^{-1} and k_3 144 sec^{-1} (Scheme I). (We shall later argue the uniqueness of the acyl-enzyme mechanism.)

Product Distribution. Direct analysis of the products from the hydrolysis of AcPhe-OMe, AcPhe-(*p*-trimethylammonium)-anilide and, where tested, AcPhe-(*p*-dimethylamino)anilide by δ-chymotrypsin in the presence of $AlaNH_2$, $GlyNH_2$, and hydrazine shows that the esters and anilides partition in an identical manner for each particular nucleophile. The inhibition of k_{cat}/K_M for AcPhe-$AlaNH_2$ by $AlaNH_2$ is consistent with this substrate partitioning identically to the ester and anilides.

The product distribution is consistent with the acyl-enzyme mechanism for esters, anilides, and peptides.

$$\text{AcPhe-OMe} + \text{chymotrypsin} \underset{}{\overset{K_S}{\rightleftharpoons}} \text{AcPhe-OMe·chymotrypsin} \underset{k_{-2}}{\overset{k_2}{\rightleftharpoons}} \text{AcPhe-chymotrypsin} \; (-\text{MeOH})$$

$$\text{AcPhe-chymotrypsin} \underset{k_{-3}}{\overset{k_3}{\rightleftharpoons}} \text{AcPhe·chymotrypsin} \overset{K_P}{\rightleftharpoons} \text{AcPhe} + \text{chymotrypsin}$$

Relationship between Equilibrium Constants and Rate Constants. The Haldane (1930) relationship applied to Scheme III

SCHEME III

$$\text{AcPhe-AlaNH}_2 + \text{chymotrypsin} \overset{K_S}{\rightleftharpoons} \text{AcPhe-AlaNH}_2\text{·chymotrypsin} \underset{k_{-2}}{\overset{k_2}{\rightleftharpoons}} \text{AcPhe-chymotrypsin} + \text{AlaNH}_2$$

$$\text{AcPhe-chymotrypsin} \underset{k_{-3}}{\overset{k_3}{\rightleftharpoons}} \text{AcPhe·chymotrypsin} \overset{K_P}{\rightleftharpoons} \text{AcPhe} + \text{chymotrypsin}$$

gives for the formation constants for AcPhe-OMe (K_{OMe}) and for AcPhe-$AlaNH_2$ (K_{Ala})

$$K_{OMe} = \frac{k_{-3}}{K_P}\frac{k_{-2}}{k_3}\frac{K_S}{k_2} = \frac{[\text{AcPhe-OMe}]}{[\text{AcPhe}][\text{MeOH}]} \quad (1)$$

$$K_{Ala} = \frac{k_{-3}}{K_P}\frac{k_{-2}}{k_3}\frac{K_S}{k_2} = \frac{[\text{AcPhe-AlaNH}_2]}{[\text{AcPhe}][\text{AlaNH}_2]} \quad (2)$$

K_{OMe}, k_3, k_{-2}/k_3, and k_2/K_S ($\equiv k_{cat}/K_M$) have been measured in this study at pH 9.30, etc. Using the un-ionized forms of AcPhe, $AlaNH_2$, etc., and the values in Table VIII we have for AcPhe-OMe (eq 1)

$$0.162 = \frac{k_{-3}}{K_P}\frac{274}{144}\frac{1}{2.5 \times 10^5}$$

hence $k_{-3}/K_P = 2.13 \times 10^4$ M^{-1} sec^{-1}. From this we may calculate K_{Ala} using the observed values of K_S/k_2 k_{-2}, k_3, and the calculated k_{-3}/K_P.

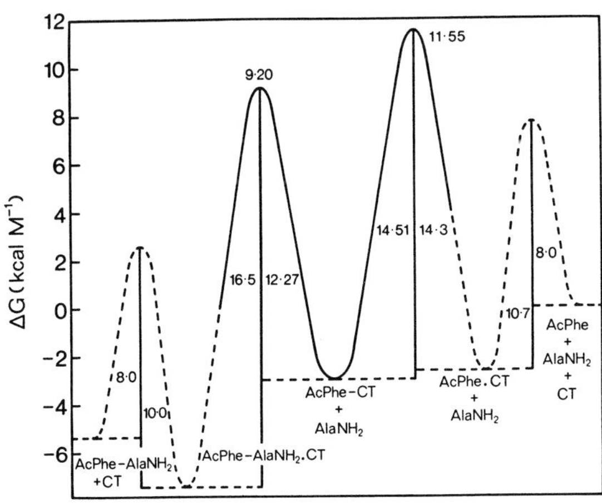

FIGURE 4: Free energy against reaction coordinate for the hydrolysis of AcPhe-$AlaNH_2$ by δ-chymotrypsin at pH 9.30, 25°, ionic strength 0.95, 5% Me_2SO. Standard states are 1 M, activity of water is 1. AcPhe and $AlaNH_2$ are taken as un-ionized. Under normal circumstances both are ionized and the free energy of the products is −6.7 kcal rather than zero as in the figure.[2]

$$K_{\mathrm{Ala}} = 2.13 \times 10^4 \times \frac{44 \times 144}{144} \frac{1}{126} M^{-1} = 7.4 \times 10^3 \, \mathrm{M}^{-1}$$

This is in excellent agreement with the observed value of 9.0×10^3 M^{-1}. This is proof of the acyl-enzyme mechanism being the major pathway for the reaction of chymotrypsin with amides. If the observed value of k_2/K_S for the AcPhe-AlaNH$_2$ hydrolysis were due to another pathway, this would have to be larger than that for the acyl-enzyme mechanism and then the calculated value for K_{Ala} would be much smaller than that observed.

Alternatively, as we know all the rate constants for the formation of AcPhe-AlaNH$_2$ by the acyl-enzyme route (k_{-3}/K_P from the AcPhe-OMe reaction, k_{-2} from the partition of AcPhe-CT between AlaNH$_2$ and H_2O; and k_3 from AcPhe-chymotrypsin deacylation), we can calculate k_2/K_S for the acyl-enzyme route. This is found to be 152 M^{-1} sec^{-1}.

This calculation involves five experimental terms each of an accuracy of $\pm$5–10%. The calculated value is in the range of 152 $\pm$ 50 M^{-1} sec^{-1}. This compares well with the observed value of 126 M^{-1} sec^{-1} for the k_{cat}/K_M ($\equiv k_2/K_S$) for the hydrolysis of AcPhe-AlaNH$_2$.

We have shown so far that the acyl-enzyme pathway is the major route for amide hydrolysis. The next question is whether there is a significant contribution from a second mechanism of direct nucleophilic attack on the noncovalently bound enzyme–substrate complex (Scheme IV). It was hoped to put a

SCHEME IV

$$\text{AcPhe-X} + \text{chymotrypsin} \rightleftharpoons \text{AcPhe-X}\cdot\text{chymotrypsin} \xrightarrow{k_2} \text{AcPhe-chymotrypsin} \begin{cases} \xrightarrow{k_3} H_2O \\ \xrightarrow{k_4 N} \end{cases}$$

$$\text{AcPhe-X}\cdot\text{chymotrypsin} \xrightarrow{H_2O,\ k'_3} \qquad \text{AcPhe-X}\cdot\text{chymotrypsin} \xrightarrow{k'_4 N}$$

maximum limit on the contribution of a competing pathway by repetition of the Inagami and Sturtevant (1964) experiments using a more reactive nucleophile than hydroxylamine. Unfortunately, there are complications in that the values of k_{cat} for the anilide reactions increase slightly on the addition of nucleophiles. The nucleophiles should not be involved in the reaction until after the rate-determining step.

Effect of Nucleophiles on k_{cat} for Anilide Hydrolysis. AlaNH$_2$, GlyNH$_2$, and hydrazine increase k_{cat} for the hydrolysis of anilide substrates (Table VI). We feel that this is not due to the direct chemical participation of the nucleophiles in the rate-determining step of the reaction but is a consequence of alterations in enzyme–substrate interactions. Dimethylformamide, a secondary amide which is not a nucleophile, also causes similar increases in k_{cat} for the reaction of chymotrypsin and trypsin with amides (Table VII).

The direct attack of nucleophiles on the noncovalently bound enzyme–substrate as given in Scheme IV is unlikely for the following reasons. The inhibition of the hydrolysis of AcPhe-AlaNH$_2$ by AlaNH$_2$ can be only after the acyl-enzyme at the k_3 and k_4 steps unless the AlaNH$_2$ binds to the enzyme–substrate complex and sterically inhibits the attack of water (k'_3). If AlaNH$_2$ does not bind the k'_3 route occurs independently of the concentration of AlaNH$_2$. Yet at 0.5 M AlaNH$_2$ the hydrolysis is inhibited by 97.5% (see Table II). This limits k'_3 to being no more than $\sim$2.5% of k_3. The alternative is that nucleophiles such as AlaNH$_2$ may bind to the noncovalent complex and increase the rate by the k'_4 route. This is

TABLE VIII: Rate and Equilibrium Constants Used for Application of (A) the Haldane Equation (Scheme III) and (B) Calculation of Free-Energy Diagrams.

AcPhe-OMe	AcPhe-AlaNH$_2$
Part A	
$k_2/K_S = 2.5 \times 10^5$ M^{-1} sec^{-1}	$k'_2/K'_S = 126$ M^{-1} sec^{-1}
$k_{-2} = 274$ M^{-1} sec^{-1}	$k'_{-2} = 6340$ M^{-1} sec^{-1}
$k_3 = 144$ sec^{-1}	
$k_{-3}/K_P{}^a = 2.13 \times 10^4$ M^{-1} sec^{-1}	
Part B	
$k_2{}^b = 2.2 \times 10^3$ sec^{-1}	$k'_2{}^d = 5$ sec^{-1}
$K_S{}^b = 9 \times 10^{-3}$ M	$K'_S{}^d = 4 \times 10^{-2}$ M
$k_{-3}{}^c = 2 \times 10^2$ sec^{-1}	
$K_P{}^c = 1 \times 10^{-2}$ M	

[a] Calculated from the overall equilibrium constant, etc. (see text). [b] Estimated from k_{cat} and k_3. [c] Estimated by assuming $K_P \sim 10$ mM. [d] Approximate values.

inconsistent with the observed product distribution. If there is a second pathway for the formation of AcPhe-nucleophile then there will be an increase in the partition. For example, k_{cat} for the disappearance of AcPhe-(p-trimethylammonium)anilide is increased by 58% in the presence of 0.5 M GlyNH$_2$. The acyl-enzyme route gives 84% AcPhe-GlyNH$_2$. The total percentage of AcPheGlyNH$_2$ formed by both routes should then be (84 + 58)/158 $\times$ 100%, *i.e.*, 90%. Similarly k_{cat} is increased by 35% in the presence of 0.5 M hydrazine. The product ratio should then increase from 48.6% for the acyl-enzyme mechanism to (48.6 + 35)/135 $\times$ 100%, *i.e.*, 62%. Similar calculations may be made for the partitioning with AlaNH$_2$. These increases would have been easily detected. This alternative is ruled out. (It is also unlikely that the noncovalent enzyme–*ester* complexes are susceptible to attack by external nucleophiles. The enzyme–substrate complex is at a much lower concentration than the acyl-enzyme and a significant rate of attack by water would increase the K_M. The k_{cat} for the hydrolysis of AcPhe-OMe is 135 sec^{-1} and is consistent with a k_3 of 144 sec^{-1}, as found directly from the deacylation of AcPhe-CT, and a k_2 of $\sim 2 \times 10^3$ sec^{-1}. An alternative route would increase k_{cat}. Similarly the direct attack of AlaNH$_2$, AlaNH$_2$, and H_2NNH_2 on AcPhe-CT is consistent with the product distribution on hydrolysis of AcPhe-OMe.)

The increase in k_{cat} on the addition of nucleophiles does not appear to be due to an alternative pathway. This increase could come from the nucleophile binding to the enzyme–substrate complex and either decreasing nonproductive binding of the substrate or causing strain. Both effects would cause an increase in k_{cat} *and* K_M as found.

This stimulation of k_{cat} invalidates the method of determining product ratios using the inhibition of $V_{\max}$ on the addition of leaving group amine. For example, $V_{\max}$ for the hydrolysis of AcPhe-GlyNH$_2$ in the presence of 0.5 M GlyNH$_2$ is decreased by 84% due to reaction of the acyl-enzyme with GlyNH$_2$ but is stimulated by 58% due to the nonspecific effect. The net result is a 75% decrease only. The reactivity of GlyNH$_2$ relative to water is then calculated to be 6:1 instead of 11.5:1. This is the probable cause of the anomaly in Epand's (1969) experiment; 0.10–0.20 M GlyNH$_2$ did not inhibit the hydrolysis of benzoyl-L-tyrosylglycinamide as much as expected from the product ratios measured directly from benzoyltyrosine ethyl ester hydrolysis in the presence of $GlyNH_2$. The failure to observe a leveling off of the rate of the enzymatic synthesis of acetyltyrosinehydroxamate from acetyltyrosine with increasing concentration of hydroxylamine (Caplow and Jencks, 1964) might also be accounted for by a similar effect. In this case the initial formation of AcTyr-ONH_2 followed by the hydroxylamine competing with the enzyme for this could also be a factor.

The observation that greater than 94% of the AcPhe-(*p*-trimethylammonium)anilide reacting with δ-chymotrypsin in the presence of 0.84 M $AlaNH_2$ is converted to AcPhe-$AlaNH_2$ in a step subsequent to the acylation implies that more than 94% of the reaction occurs by the acyl-enzyme mechanism.

In summary, the evidence for the acyl-enzyme mechanism for amides is: (a) the acyl-enzyme generated from ester substrates partitions in an identical manner between nucleophiles and water as the intermediate which occurs in amide hydrolysis; (b) this intermediate occurs after the rate-determining step and accounts for more than 94% of the total reaction; (c) the rate constant for k_2/K_S for the hydrolysis of AcPhe-$AlaNH_2$ is, within experimental error, the same as that calculated on the acyl-enzyme hypothesis using the overall equilibrium constant of the reaction and the experimental values for the rate constants for the synthesis of AcPhe-$AlaNH_2$ from AcPhe and $AlaNH_2$ by the acyl-enzyme route.

The acyl-enzyme mechanism is expected to have an entropic advantage over a reaction which involves the direct attack of water on the enzyme-bound substrate. Jencks and Page (1972) suggest this is a significant effect.

A further advantage is that the nucleophilic -OH of Ser-195 is held rigidly as part of the "charge–relay system" (Blow *et al.*, 1969). The stereochemical orientation is possibly more easily controlled than that of a bound water molecule. Effects such as "strain" require rigid active sites.

Free-Energy Diagrams. The low reactivity of AcPhe-$AlaNH_2$ is due to its thermodynamic stability. The transition state for the acylation of the enzyme by AcPhe-$AlaNH_2$ has a free energy of 9.2 kcal, 2 kcal lower than that involving AcPhe-OMe. But the ground state of the peptide is 6.5 kcal more stable than the ester. The acyl-enzyme is about a kcal more stable than the noncovalent AcPhe-OMe·chymotrypsin complex but is some 4.6 kcal less stable than AcPhe-$AlaNH_2$·chymotrypsin. These results may be compared with those for acetyltryptophan ethyl ester, acetyltryptophanamide and acetyltyrosinehydroxamic acid (Bender *et al.*, 1964b; Rajender *et al.*, 1970; Epand and Wilson, 1964). In the diagram given for AcPhe-$AlaNH_2$ the apparent rate-determining step is deacylation. This is because the reaction conditions are defined for 1 M $AlaNH_2$. At low concentrations of $AlaNH_2$ the rate-determining step is, of course, acylation. Similarly, the equilibrium between the Michaelis complex and the acyl-enzyme is a function of concentration. The rate-determining step in the hydrolysis of AcPhe-OMe changes from deacylation to acylation as the substrate concentration decreases to below the K_M.

References

Bender, M. L., Clement, G. E., Gunter, C. R., and Kézdy, F. J. (1964a), *J. Amer. Chem. Soc. 86*, 3697.

Bender, M. L., and Kézdy, F. J. (1964), *J. Amer. Chem. Soc. 86*, 3704.

Bender, M. L., Kézdy, F. J., and Gunter, C. R. (1964b), *J. Amer. Chem. Soc. 86*, 3714.

Bernhard, S. A., Lee, B. F., and Tashjian, Z. H. (1966), *J. Mol. Biol. 18*, 405.

Blow, D. M., Birktoft, J., and Hartley, B. S. (1969), *Nature (London) 221*, 337.

Bray, G. A. (1960), *Anal. Biochem. 1*, 279.

Bruice, T. C., and Benkovic, S. J. (1966), Bioorganic Mechanisms, New York, N. Y., Benjamin.

Caplow, M., and Jencks, W. P. (1964), *J. Biol. Chem. 239*, 1640.

Epand, R. M. (1969), *Biochem. Biophys. Res. Commun. 37*, 313.

Epand, R. M., and Wilson, I. B. (1964), *J. Biol. Chem. 239*, 4145.

Fastrez, J., and Fersht, A. R. (1973), *Biochemistry 12*, 1067.

Fersht, A. R. (1972), *J. Mol. Biol. 64*, 497.

Fersht, A. R., and Sperling, J. (1973), *J. Mol. Biol. 74*, 137.

Haldane, J. B. S. (1930), Enzymes, London, Longmans, Green & Co.

Hartley, B. S., and Kilby, B. A. (1954), *Biochem. J. 56*, 288.

Henderson, R. (1970), *J. Mol. Biol. 54*, 341.

Hess, G. P., McConn, J., Ku, E., and McConkey, G. (1970), *Phil. Trans. Roy. Soc. London, Sec. B 257*, 89.

Huang, H. T., Foster, R. J., and Niemann, C. (1952), *J. Amer. Chem. Soc. 74*, 105.

Inagami, T. (1969), *J. Biochem. (Tokyo) 66*, 277.

Inagami, T., Patchornik, A., and York, S. S. (1969), *J. Biochem. (Tokyo) 65*, 809.

Inagami, T., and Sturtevant, J. M. (1964), *Biochem. Biophys. Res. Commun. 14*, 69.

Ingles, D. W., and Knowles, J. R. (1968), *Biochem. J. 100*, 561.

Inward, P. W., and Jencks, W. P. (1965), *J. Biol. Chem. 240*, 1986.

Jencks, W. P., and Page, M. I. (1972), *Abstr. 8th Meeting Fed. Eur. Biochem. Soc.*, 216.

Kézdy, F. J., Clement, G. E., Bender, M. L. (1964), *J. Amer. Chem. Soc. 86*, 3690.

McConn, J., Ku, E., Himoe, A., Brandt, K. G., and Hess, G. P. (1971), *J. Biol. Chem. 246*, 2918.

Rajender, S., Han, M., and Lumry, R. (1970), *J. Amer. Chem. Soc. 92*, 1378.

Robillard, G. T., Powers, J. C., and Wilcox, P. E. (1972), *Biochemistry 11*, 1773.

Schonbaum, G. R., Zerner, B., and Bender, M. L. (1961), *J. Biol. Chem. 236*, 2930.

Zaoral, M., Kolc, J., Korenczki, J., Cerneckij, V. P., and Sörm, F. (1967), *Collect. Czech. Chem. Commun. 32*, 843.

Chapter 4

Non-covalent Interactions in Enzyme Catalysis

'This conclusion is surprising, first because it contradicts the widely held belief that strong binding (low K_M) is an important component of enzymic catalysis.'

A. Cornish-Bowden, J. Mol. Biol. 101, 1–9 (1976)

I ended my chymotrypsin phase with a deceptively simple analysis that was a crucial step for my future work[26]. I set out how the use of binding energy could affect catalysis by enzymes, and extended famous ideas of Linus Pauling and J. B. S. Haldane on the role of 'strain' in enzymatic catalysis. The simple equations in this paper relating binding energy, specificity and catalysis were subsequently the basis of my work on the specificity of the binding of amino acids to aminoacyl-tRNA synthetases, analysis of effects of mutagenesis of the tyrosyl-tRNA synthetase on catalysis, and then developing the ideas for protein folding. Surprisingly, the simple relationship for the relative rates of reaction of two substrates A and B competing for catalysis by an enzyme:

$$V_A/V_B = [A](k_{cat}/K_M)_A/[B](k_{cat}/K_M)_B,$$

which is crucial for much of the analysis, was in textbooks only as a special case for when [A] = [B] (e.g. as in the then bible of enzymology, *Enzymes*, by Malcolm Dixon and Edwin C. Webb).

Important conclusions from this study are that it is kinetically unfavourable for intermediates to accumulate during reactions and that for optimal rate, enzymes should have evolved to have values of K_M that are higher than the physiological concentrations of substrates. Those conclusions contradicted the textbooks, which all promoted the idea that enzymes should be saturated with substrates. This paper provoked a lively debate[27,28], and I had quite an argument over the phone with Athel Cornish-Bowden. Athel proved to be extremely magnanimous and later proposed to the Enzyme Nomenclature Commission that k_{cat}/K_M be called the 'Fersht Constant.' It has ended up as the 'specificity' constant, but it should be the 'Haldane Constant' — but there is another Haldane constant from a different Haldane. I have had several scientific arguments in my career, but in nearly all cases have ended up being firm friends with the erstwhile protagonists, who, like me, have all been searching for the truth and not fighting over egos.

On rereading Athel's paper[28] 35 years on, it is clear that there was, in fact, very little difference between us at the time. Athel wrote to me 'it is better to be clear and wrong rather than obscure and right because the former would soon be resolved.' Further, he wrote: 'and no-one could accuse you of writing obscurely.' This paper led on to my activities as a writer of a textbook.

4.1 Circa 1975: Athel Cornish-Bowden.

Proc. R. Soc. Lond. B. **187**, 397–407 (1974)
Printed in Great Britain

Catalysis, binding and enzyme–substrate complementarity

By A. R. Fersht
M.R.C. Laboratory of Molecular Biology, Hills Road, Cambridge CB2 2QH, England

(*Communicated by M. F. Perutz, F.R.S. – Received* 16 *May* 1974)

A simple derivation is given that the catalytic term k_{cat}/K_S† is at a maximum when the structure of the enzyme is complementary to the structure of the substrate in the transition state. In addition, at a constant substrate concentration, [S], the maximum reaction rate is obtained when k_{cat} and K_S are individually high so that K_S is greater than [S]; the overall reaction rate decreases with decreasing k_{cat} and K_S for K_S less than [S]. Two corollaries of this are that intermediates accumulating after the initial Michaelis complex are undesirable and also enzymes whose function is to optimize reaction rates should have evolved to exhibit K_M values above those of accessible substrate concentrations. This could be achieved by an often 'distortionless' strain which consists either of unfavourable interactions in the enzyme substrate complex which are relieved in the transition state or increasingly favourable interactions in the transition state. A possible special role of the backbone NH groups in this context is discussed.

The enzyme need not be complementary to the transition state of the substrate for catalysis to occur.

Introduction

It has long been postulated that the active sites of enzymes have structures which are complementary to the structures of the substrates in their activated complexes or transition states (Haldane 1930; Pauling 1946, 1948). Experimental evidence for this has come from inhibition studies made by using transition state analogues. Compounds thought to resemble the transition states of substrates of proline racemase (Jencks 1966; Cardinale & Abeles 1968), Δ^5-3-ketosteroid isomerase (Wang, Kawahara & Talalay 1963), lysozyme (Secemski, Lehrer & Lienhard 1972)

† k_{cat} and K_M refer to the observed turnover number and Michaelis constant. When K_M is equal to the dissociation constant of the Michaelis complex it is termed K_S. Then, k_{cat} is the first order rate constant for the conversion of the complex to products. $\Delta G_S = RT \ln K_S$. $\Delta G^{\ddagger}$ is the free energy change for ES $\rightarrow$ ES‡. $\Delta G_T^{\ddagger}$ is the free energy change for E + S $\rightarrow$ ES‡. The maximum possible binding energy available is ΔG_b (for the case when there is a perfect fit between enzyme and substrate). ΔG_R is the adverse energy difference between the observed binding energy and ΔG_b. $\Delta G_0^{\ddagger}$ is the inherent activation energy term in $\Delta G^{\ddagger}$ due to the electronic aspects of bond of formation.

and adenosine deaminase (Evans & Wolfenden 1970) bind more tightly than the actual substrates.

In recent review articles (Wolfenden 1972; Lienhard 1973) transition state theory has been applied to enzymes and it has been shown that the *dissociation constants* of transition states and enzymes are always lower than those for the substrates. But this does not imply that enzymes contain structures complementary to those of the transition states of substrates as a dissociation constant is a measure of the free energy *difference* between two states, free and enzyme-bound, and does not give information on the energetics of any of the individual states. In fact, the comparatively low dissociation constants of enzyme-bound transition states probably reflect in the main the inherent instability of transition states in water due to the presence of electrical charges which require solvation and also to the absence of the entropy loss on bringing reagents and catalysts together in solution (Page & Jencks 1971).

In the following analysis I shall demonstrate on simple theoretical grounds the catalytic advantages of enzymes being complementary to transition states; but also emphasize that this is not a necessary criterion for catalysis to occur.

First it should be pointed out that until recently enzyme catalysis was thought to be too efficient to be accounted for by the simple approximation of the substrate and the catalytic groups on the enzyme (Storm & Koshland 1970). A radical advance was made by Page & Jencks (Page & Jencks 1971; Jencks & Page 1972) who presented calculations which indicate that there is a large entropic contribution towards the high rate constants of enzymatic reactions. Multimolecular reactions in solution are slow because the bringing together of catalyst and reagent molecules involves a considerable loss of translational and overall rotational entropies. Each additional molecule in the transition state of the reaction slows down the reaction by a factor of up to 10^8 owing to its losing its independent translational and overall rotational entropies. Enzymatic reactions involve few independent molecules in their transition states. The independent entropy of the substrate is lost on forming the initial enzyme substrate complex and so the chemical steps involve little extra entropy loss. This effective gain in entropy is 'paid for' from the binding energy of the enzyme and substrate. The important conclusion reached by Page & Jencks is that a very large factor in enzymatic rate constants is due to the approximation of the substrate and the catalytic groups of the enzyme. In some cases this factor is of the magnitude expected for enzyme catalysis. Dafforn & Koshland (1973) have reached the same conclusion by using a different approach.

Kinetic and thermodynamic analysis

Whatever the detailed mechanism of catalysis some of the driving energy is derived from the binding energy of the enzyme and substrate. This energy may be distributed between the observed binding and catalytic constants.

Let us analyse the most simple case of enzyme catalysis, Michaelis–Menten kinetics involving just one substrate and one enzyme–substate complex.

$$\mathrm{E}+\mathrm{S} \underset{\Delta G_{\mathrm{S}}}{\overset{K_{\mathrm{S}}}{\rightleftharpoons}} \mathrm{ES} \underset{\Delta G^{\ddagger}}{\overset{k_{\mathrm{cat}}}{\longrightarrow}} \text{products}, \tag{1}$$

$$\mathrm{E}+\mathrm{S} \underset{\Delta G_{T}^{\ddagger}}{\overset{(k_{\mathrm{cat}}/K_{\mathrm{S}})}{\rightleftharpoons}} \mathrm{ES}^{\ddagger}, \tag{2}$$

where K_{S} is the dissociation constant of the enzyme–substrate complex, ES, k_{cat} is the first order rate constant for its decomposition and $\Delta G_{\mathrm{S}} = RT \ln K_{\mathrm{S}}$.

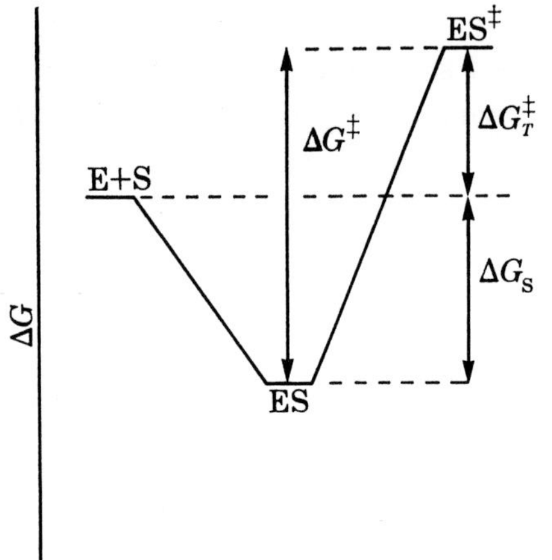

Figure 1. Free energy changes for the simplest Michaelis-Menten kinetic scheme

$$\mathrm{E}+\mathrm{S} \overset{K_{\mathrm{S}}}{\rightleftharpoons} \mathrm{ES} \overset{k_{\mathrm{cat}}}{\longrightarrow} \text{products},$$

where ΔG_{S} is algebraically negative and $\Delta G^{\ddagger}$ and $\Delta G_{T}^{\ddagger}$ positive.

We may consider the enzyme bound transition state, $\mathrm{ES}^{\ddagger}$, as being in equilibrium with the enzyme and substrate separately in solution (equation (2)). The rate constant $k_{\mathrm{cat}}/K_{\mathrm{S}}$ is related to the free energy change in this equilibrium, $\Delta G_{T}^{\ddagger}$, by equation (3)

$$\Delta G_{T}^{\ddagger} = -RT \ln (k_{\mathrm{cat}}/K_{\mathrm{S}}) + RT \ln (kT/h), \tag{3}$$

where k is the Boltzmann and h the Planck constant (Wolfenden 1972). $\Delta G_{T}^{\ddagger}$ is composed of an adverse energy term, $\Delta G^{\ddagger}$, involving the chemical processes of bond making and breaking and an energetically favourable term, ΔG_{S}, due to the realization of the enzyme–substrate binding energy. This is represented in figure 1 where it can be seen that

$$\Delta G_{T}^{\ddagger} = \Delta G^{\ddagger} + \Delta G_{\mathrm{S}}. \tag{4}$$

The binding energy of the substrate and the enzyme will be at a maximum when the enzyme has a structure complementary to that of the substrate. Two extreme cases emerge. The first arises when the enzyme is exactly complementary to the substrate in its original form and the second when it is complementary to the substrate in its transition state structure.

(a) Enzyme complementary to initial substrate

Suppose the maximum amount of intrinsic binding energy available is ΔG_b. In this case it is realized in the initial enzyme substrate complex so that binding will be good; that is K_S, the dissociation constant of the enzyme substrate complex, will be low. But the formation of the transition state will lead to a reduction in binding energy as the substrate geometry changes to give a poorer fit and so will lower k_{cat}. If the adverse energy change is ΔG_R and the free energy of activation due to the chemical bond making and breaking involved in k_{cat} is $\Delta G_0^{\ddagger}$, then the observed free energy of activation for k_{cat} is given by

$$\Delta G^{\ddagger} = \Delta G_0^{\ddagger} + \Delta G_R \tag{5}$$

and

$$\Delta G_S = \Delta G_b. \tag{6}$$

The free energy of activation for k_{cat}/K_S is given by $\Delta G^{\ddagger} + \Delta G_S$, i.e.

$$\Delta G_T^{\ddagger} = \Delta G_0^{\ddagger} + \Delta G_R + \Delta G_b. \tag{7}$$

(b) Enzyme complementary to transition state

This situation was postulated first by Pauling (1946, 1948) but was implied earlier by Haldane (1930). Here the full binding energy ΔG_b is realized in the transition state. There will be an adverse energy term ΔG_R in the initial enzyme substrate complex which will increase K_S but the gain in binding energy as the reaction reaches the transition state will increase k_{cat}.
Thus,

$$\Delta G^{\ddagger} = \Delta G_0^{\ddagger} - \Delta G_R, \tag{8}$$

and

$$\Delta G_S = \Delta G_b + \Delta G_R. \tag{9}$$

Again the free energy of activation for k_{cat}/K_S is given by

$$\Delta G_T^{\ddagger} = \Delta G^{\ddagger} + \Delta G_S,$$

i.e.

$$\Delta G_T^{\ddagger} = \Delta G_0^{\ddagger} + \Delta G_b, \tag{10}$$

and ΔG_R cancels out.

Comparison of equations (7) and (10) shows that k_{cat}/K_S is higher for the enzyme being complementary to the transition state rather than to the initial substrate by a factor of exp $(\Delta G_R/RT)$.

COMPENSATION BETWEEN k_{cat} AND K_S

The value of k_{cat}/K_S in § (b) is independent of the interactions of the enzyme with the initial substrate as the term ΔG_R drops out. If the geometry of the enzyme is subtly varied so that the transition state binding is unaffected but the initial substrate binds more poorly then k_{cat} and K_S will increase individually.

TABLE 1. RELATION BETWEEN RATES OF ENZYME REACTIONS AND k_{cat} AND K_S AT CONSTANT k_{cat}/K_S

($k_{cat}/K_S = 10^3$ s^{-1} mol^{-1} l, [S] $= 10^{-3}$ mol l^{-1})

k_{cat}/s^{-1}	K_S/mol l^{-1}	V†/s^{-1}
10^3	1	0.999
10^2	1×10^{-1}	0.990
10	1×10^{-2}	0.909
1	1×10^{-3}	0.500
1×10^{-1}	1×10^{-4}	0.090
1×10^{-2}	1×10^{-5}	0.009
1×10^{-3}	1×10^{-6}	0.001

† (Moles of substrate/mole of enzyme) × s^{-1}.

Generally a high value of K_S is catalytically advantageous. Tight binding of the substrate is wasteful when the substrate concentration is greater than K_S. Values of reaction rates are calculated in table 1 for various values of k_{cat} and K_S such that k_{cat}/K_S is constant. It is seen that maximal rates are attained for $K_S > 10\times$[S]. Enzymes whose function is to turn over large amounts of substrates rapidly, such as pepsin, lysosyme, chymotrypsin etc., should have evolved to high values of k_{cat} and K_S to maximize the *rate* of reaction. That is, these enzymes would be expected to utilize strain and show specificity in k_{cat}. For optimal catalysis these enzymes should have evolved to give K_M values which are above the usually encountered substrate concentrations.

LACK OF ACCUMULATION OF INTERMEDIATES

A corollary of the maximization of rate by the mutual increasing of k_{cat} and K_M is that no intermediate should accumulate after the initial Michaelis complex. Any intermediate which accumulates lowers the apparent K_M for the reaction causing saturation at lower substrate concentrations. Also as the intermediate accumulates the rate constant for its decomposition will be lower than that for its formation and hence k_{cat} will be lowered. (In general the ratio k_{cat}/K_M will not be affected.)

The observation that the pancreatic trypsin inhibitor–trypsin complex is an otherwise exergonic intermediate which has been highly stabilized (Rühlmann *et al.* 1973) is the converse of the above rule.

Distortionless strain

The evolution of enzymes to bind transition states strongly and substrates weakly to maximize rate is an example of strain (Jencks 1966, 1969) under the general definition of 'the stronger binding of the transition state of the reaction than the substrate'. Although strain may be manifested in some cases by a genuine distortion of the substrate it is likely that the strain will generally be distortionless. This could be due to either the substrate and the enzyme having unfavourable interactions which are relieved in the transition state or the transistion state having additional binding interactions which are not realized in the enzyme substrate complex. In both cases there would be forces that *tend* to distort the substrate towards the transition state. As non-bounded interactions have weak force constants (apart from van der Waals repulsion) and enzymes and substrates are flexible it is difficult to distort by the interaction of the enzyme with the substrate. Rotation about single bonds is possible, such as conformational changes in lysozyme substrates, but the stretching of single bonds or the twisting of double bonds appears to be less likely as this requires strong forces. On the basis of energy refinement calculations, Levitt (1972) has suggested the tentative rule 'small distortions of a substrate conformation that cause large increases in strain energy cannot be caused by binding to the enzyme'. He suggests also that the largest forces that can be exerted are less than 12 kJ mol^{-1} $(10^{-10}\ m)^{-1}$ (3 kcal mol^{-1} $Å^{-1}$) so that to strain a substrate by about 12 kJ (3 kcal), atoms must be moved by 0.1 nm (1 Å). In extreme cases genuine distortion might occur; but in general strain will involve the subtle interplay of favourable and unfavourable interactions.

I feel that a typical example of a strain process is the one proposed for the acylation of chymotrypsin by good polypeptide substrates (Henderson 1970; Fersht, Blow & Fastrez 1973). The carbonyl oxygen of the acyl group which is being transferred to the reactive Ser-195 hydroxyl of the enzyme is weakly hydrogen bonded to the backbone NH of Gly-193. As the reaction proceeds the carbon-oxygen bond length increases as the carbon becomes tetrahedral and the oxygen, bearing a negative charge, moves closer to the NH group forming a stronger hydrogen bond. The initial carbon oxygen double bond is not stretched by the Gly-193, but the tendency to distort is there and the additional binding energy is realized in the transition state. A similar process involving a different system of hydrogen bonds has been proposed for subtilisin (Robertus, Kraut, Alden & Birktoft 1972). It is suggested here that the carbonyl oxygen is not hydrogen bonded in the enzyme substrate complex but flips into the hydrogen bonding position on formation of the tetrahedral intermediate. Strain is also realized in chymotrypsin due to the Ser-195 hydroxyl being forced by the substrate to rotate around the α–β single bond towards the position it takes up in the transition state (Fersht *et al.* 1973).

Strain distributed about enzyme

In the last section the acylation of chymotrypsin by a good polypeptide substrate was described as if the enzyme were rigid. This is too simplistic; Levitt's (1972) calculations suggest that this is not so, the 'distortionless' strain will result in small deformations of the rotational ϕ and ψ bond angles and interatomic distances in the enzyme. In chymotrypsin, rather than the hydrogen bond between

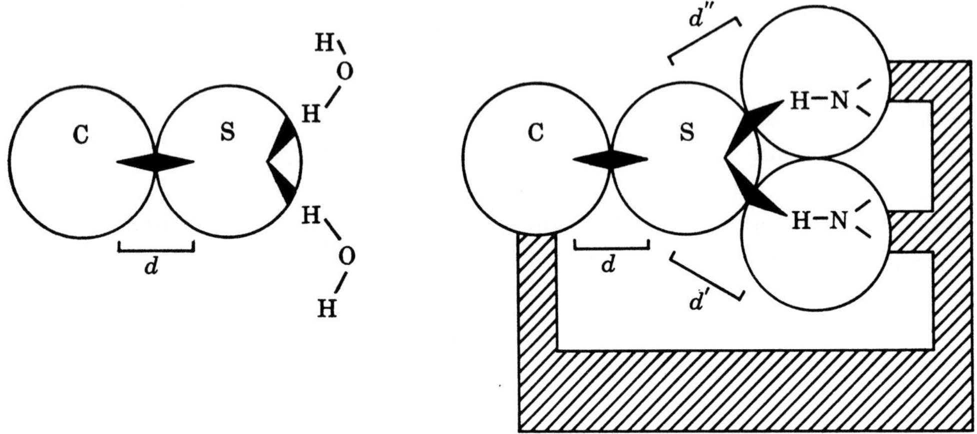

Figure 2. Illustration of the geometric constraints in enzyme catalysis due to catalyst (C) and solvating groups (NH) being part of the enzyme structure. The shaded cones represent the permissible angles for orbital overlap or hydrogen bond formation and *d* the interatomic distance. In the left of the diagram the reaction between the catalyst molecule C and the substrate causes a negative charge to develop which is solvated by water. On the right, the distance between catalyst and hydrogen bond donors is constrained by their being part of the enzyme.

the polypeptide substrate and the backbone NH of Gly-193 not being made in the Michaelis complex the bond will occur at the expense of deforming the enzyme. As the reaction proceeds and the substrate approaches the transition state the enzyme will relax back into its native structure releasing the strain energy. The transition state is stabilized by the same factor, i.e. the energy of one hydrogen bond, as in the 'rigid' description.

Specific solvation of the transition state – a backbone contribution

An important feature to emerge from the X-ray diffraction work on the serine proteases is that the enzyme acts as the solvation shell for the transition state of the reaction (Robertus *et al.* 1972; Fersht *et al.* 1973). This leads to a fundamental difference between simple chemical reactions in solution and enzyme catalysed reactions. In solution the only orientation effects that have to be considered are those between the reagents as the solvent is free to solvate any charges that develop.

In the enzyme reaction on the other hand there is a precise stereochemical relation between the reacting groups and the effective solvating groups which are part of the enzyme. This is illustrated in figure 2.

It is unlikely that enzyme specificity is due to just the requirement for precise angular alinement of orbitals on the substrate and enzyme. Molecular orbital calculations show that there is a considerable latitude of some 10–20° or so in the

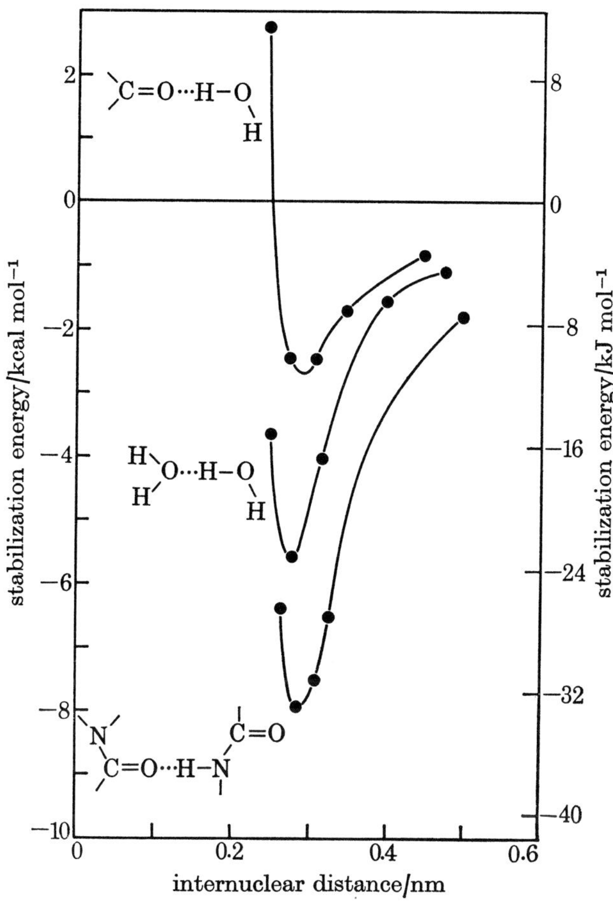

FIGURE 3. Plots of stabilization energy against interatomic distance calculated for various hydrogen bonds.

mutual orientation of the two reagents (Bruice, Brown & Harris 1971). Similarly calculations on hydrogen bonds show that 'bending' of some 15–30° or so from linearity leads to only a small loss of energy (Morokuma & Winick 1970; Morokuma 1971). But *distances* are very important. In figure 3 the strengths of some hydrogen bonds as function of distance have been plotted (Morokuma & Winick 1970; Morokuma 1971; Dreyfus & Pullman 1970). A change of 0.05 nm leads to a significant change in energy.

As discussed in the previous section the rigid positioning of the backbone groups allows a strain contribution to catalysis by making weak hydrogen bonds with the substrate but stronger bonds with the transition state as bond lengths and angles change.

Enzyme not complementary to the transition state

Situations may be conceived in which the maximisation of rate is subordinate to another factor. For example, the sequestering of unstable intermediates, such as acyl adenylates, or the overcoming of unfavourable equilibria between enzyme bound intermediates (Jencks 1969, p. 313) may require that the enzyme structure is closer in complementarity to the intermediates than to the transition states of lowest energy.

However, if the theory of Page & Jencks (1971) is correct, enzymes may exhibit considerable catalysis without utilizing strain. The combination of their theory with a general phenomenon of distortionless strain is attractive. The main driving force of enzyme catalysis in most cases would be entropy leaving strain as a regulatory phenomenon either to maximize rate or exert control.

Specificity

It is often assumed that strain, induced fit and non-productive binding are important in specifivity. For example, if a specific substrate A and a smaller non-specific substrate B compete for the enzyme then the additional binding energy of A may be used to either create strain, compensate for an unfavourable conformational change in the enzyme, or inhibit non-productive binding. The following analysis shows that these phenomena have *no* importance in specificity.

It may be shown from equation (1) that:

$$\left(\frac{\mathrm{d}[A]}{\mathrm{d}t}\right)\Big/\left(\frac{\mathrm{d}[B]}{\mathrm{d}t}\right) = \left(\frac{k_{cat}}{K_S}\right)_A [A] \Big/ \left(\frac{k_{cat}}{K_S}\right)_B [B]. \tag{11}$$

The discrimination between A and B depends on k_{cat}/K_S; i.e. as discussed earlier on the binding of the transition state only and not on the interactions in the enzyme substrate complex. Strain, induced fit and non-productive binding do not affect the ratio k_{cat}/K_S but alter the individual values of k_{cat} and K_S in a compensating manner.

The biological role of strain and induced fit is to maximize rate by increasing k_{cat} at the expense of K_M. No such role can be formulated for the non-productive binding of non-specific substrates; this does not affect the rate of reaction of the specific substrate nor increase specificity. It would appear that non-productive binding is a biologically irrelevant artefact.

The above analyses are based on a single substrate system but the general principles may be extended to multisubstrate systems.

Conclusions

(1) Maximum catalysis (k_{cat}/K_S) occurs when the structure of the enzyme is complementary to that of the transition state (or activated complex) of the substrate. In this situation all the intrinsic binding energy is used to stabilize the unstable intermediate.

(2) Maximum rate (k_{cat}) occurs when the intrinsic binding energy is used for catalysis rather than tight binding. This is brought about by the enzyme evolving to bind the substrate weakly but the transition state strongly. Strain will often just involve the tighter binding of the transition state rather than the actual distortion of the substrate.

(3) Maximum efficiency occurs when the enzyme has evolved to exhibit a K_M value above that of the usual substrate concentrations.

(4) Intermediates accumulating after the initial Michaelis complex may lower the overall reaction rate.

(5) The backbone peptide linkages may make an important contribution to catalysis and specificity by specifically solvating the charges on ionic intermediates.

(6) Specificity is more likely to be due to spatial rather than orientational effects as interatomic interactions are a strong function of distance but a weak function of angle.

(7) Specificity is independent of strain, induced fit and non-productive binding as it is a function of transition state binding only. Non-productive binding is, in most cases, biologically irrelevant.

References

Bruice, T. C., Brown, A. & Harris, D. C. 1971 *Proc. natn. Acad. Sci., U.S.A.* **68**, 658.
Cardinale, G. J. & Abeles, R. H. 1968 *Biochemistry* **7**, 3970.
Dafforn, H. & Koshland, D. E. 1973 *Biochem. biophys. Res. Commun.* **52**, 779.
Dreyfus, M. & Pullman, A. 1970 *Theor. Chim. Acta* **19**, 20.
Evans, B. & Wolfenden, R. 1970 *J. Am. chem. Soc.* **92**, 4751.
Fersht, A. R., Blow, D. M. & Fastrez, J. 1973 *Biochemistry* **12**, 2035.
Haldane, J. B. S. 1930 *Enzymes*, p. 182. Longmans, Green & Co.
Henderson, R. H. 1970 *J. molec. Biol.* **54**, 341.
Jencks, W. P. 1966 In *Current aspects of biochemical energetics* (ed. N. O. Kaplan & E. P. Kennedy), p. 273. New York: Academic Press.
Jencks, W. P. 1969 *Catalysis in chemistry and enzymology*. New York: McGraw-Hill, Inc.
Jencks, W. P. & Page, M. I. 1972 *Federation of European Biochemical Societies 8th Meeting*, vol. 29, 45, Elsevier.
Levitt, M. 1972 Ph.D. Thesis, p. 270, University of Cambridge.
Lienhard, G. E. 1973 *Science, N.Y.* **180**, 149.
Morokuma, K. 1971 *J. chem. Phys.* **55**, 1236.
Morokuma, K. & Winick, J. R. 1970 *J. chem. Phys.* **52**, 1301.
Pauling, L. 1946 *Chem. Engng News* **24**, 1375.
Pauling, L. 1948 *Am. Sci.* **36**, 51.
Page, M. I. & Jencks, W. P. 1971 *Proc. natn. Acad. Sci., U.S.A.* **68**, 1678.

Robertus, J. D., Kraut, J., Alden, R. A. & Birktoft, J. J. 1972 *Biochemistry* **11**, 4293.
Rühlmann, A., Kukla, D., Schwager, P., Bartels, K. & Huber, R. 1973 *J. molec. biol.* **77**, 417.
Secemski, I. I., Lehrer, S. S. & Lienhard, G. E. 1972 *J. biol. chem.* **247**, 4740.
Storm, D. R. & Koshland, D. E. 1970 *Proc. natn. Acad. Sci., U.S.A.* **66**, 445.
Wang, S., Kawahara, F. S. & Talalay, P. 1963 *J. biol. Chem.* **238**, 576.
Wolfenden, R. 1972 *Accts Chem. Res.* **5**, 10.

Chapter 5

Textbook: Enzyme Structure and Mechanism

'When I want to read a good book, I write one.'

Benjamin Disraeli

Enzyme Structure and Mechanism (W. H. Freeman & Co., 1977).
Enzyme Structure and Mechanism 2nd Edn. (W. H. Freeman & Co., 1985).
Structure and Mechanism in Protein Science: A Guide to Enzyme Catalysis and Protein Folding (W. H. Freeman & Co., 1999).

By the mid 1970s, I had spent ten years of my waking hours thinking about chemical and enzymatic catalysis, as well as a lot of subconscious reasoning in my sleep as I often woke in the middle of the night having to write down new ideas that had woken me up. About 1975, I was invited to write a review for *Annual Reviews of Biochemistry*, which I devoted to the role of binding energy in catalysis and specificity. But, the senior editor, P. D. Boyer, took the most unusual step of rejecting it out of hand. I mused on writing a textbook for the new era of the study of proteins built around the rejected review. It would be the first at atomic level descriptions based on X-ray crystallography, emphasize the application of rapid reaction kinetics founded on direct observation, guide students through complex kinetics by teaching them the physical meaning and not just the algebra, and would incorporate some of the results of computer simulation of the flexibility of protein structures by my friend Michael Levitt, who was the wunderkind of the emerging field. Above all, it would be a didactic work that would also contain a collection of novel ideas that would otherwise be dissipated in small papers or reviews. I was also very short of cash. The Labour Science Minister Shirley Williams had written in *The Times* in 1971 'For the scientists the party is over.' Not only were the labs short of funds, but also scientists and academics had been denied a 22% pay increase that every other public sector employee had received because of rampant inflation. I asked Max Perutz if he thought it would be a good idea for me to write the book. He said it was a splendid idea and encouraged me. For five months in 1975–1976, I sat at the end of my bench typing away with two fingers on a worn-out electric machine that had been pensioned off from a typing course at the local technical college, while directing my technician. The final manuscript was a genuine cut and paste job with scissors, glue and Sellotape. Over 30 years later, it is still a standard text, via two revisions.

The first edition was received partly in shock. Some of its reviews had telling points.

> Dr. Fersht is well known for the originality of his researches concerning the kinetics of enzyme catalysis and models thereof. His textbook has a freshness of approach which will be appreciated by both teachers and students of molecular enzymology. For example, the teacher who wishes to persuade students that his lecture course on enzyme kinetics is not just a demonstration of his algebraic virtuosity will find a ready-made lecture or two in chapter 7 demonstrating the use of kinetics in the solution of enzyme mechanisms. Again, students who find kinetic equations somewhat forbidding will be able to impress examiners with their depth of understanding of enzyme catalysis after a close study of chapters 10 and 11 on the theories of enzyme catalysis and specificity. The other three quarters, of the book, equally valuable, ranges from fundamental concepts of chemical catalysis to practical kinetic techniques and from the basic equations of enzyme kinetics to the structures and mechanisms of selected enzymes. This is not an encyclopaedia of molecular enzymology, however, and the author has generally confined himself to those enzymes whose three-dimensional structure is known. Moreover, the reviewer was disappointed there is no description of the contributions made by chemical modification and affinity labelling (apart from four pages in chapter 7) to molecular enzymology. Perhaps it will not cover all the needs of either teacher or student, but it will repay close study by both.
>
> D. T. Elmore, *Endeavour* **2**, 99 (1978).

The first few sentences delighted me: to be acknowledged by an expert for the originality of my work, the freshness of the approach, my guiding students through algebra, and acceptance of the provocative chapters on binding energy made it all worthwhile. But, the negative points — confining to just proteins of known three-dimensional structure and eschewing brute force techniques of chemical modification — were precisely what I was preaching: 'For the protein scientists who ignore three-dimensional structure the party is over.' Even H. Gutfreund, the guru of rapid reaction kinetics, wrote that I had given steady state kinetics short shrift. As useful as such kinetics is for analysing and predicting rates of catalysis, it is not much use for solving mechanisms because all of the evidence from it is indirect, and hence generally ambiguous as to its cause.

The second edition in 1984 was received with uniform praise and enthusiasm. Athel Cornish-Bowden, a grandmaster of steady state kinetics, wrote a wonderful review analysing in detail all the improvements in the extra 100 pages. Later as part of a review of another text he wrote: *'Fersht's book is much stronger on transient-state kinetics, and is more stimulating and even exciting. His is the more likely book to take to read in bed.'*

What more could one want!

The third edition had a major expansion into protein folding. Buzz (R. L.) Baldwin wrote:

> Alan Fersht has revised his classic text *Structure and Mechanism* (Freeman, 1985) and extended it to include applications of the protein engineering method to characterizing transition states in protein folding and enzyme catalysis. His earlier textbook was so good it scared off possible competitors, and its notable features are preserved in the new text. It provides enough information about techniques to make close connections between discussions of mechanism and actual data. It packs an amazing amount of information and viewpoint into a short space without losing readability. It does not just derive equations, it explains their physical meanings and it aims to develop the reader's physical insight. The assumptions behind an approach are laid out clearly and, when a topic is controversial, the author gives his own viewpoint at the end. The pace of research today is such that few leading scientists find time to write textbooks, and the scientific community is surely grateful to Alan Fersht for revising his.
>
> R. L. Baldwin, *Protein Science* **9**, 207 (2000).

Maybe I should thank Shirley Williams after all because she, perhaps, was the final straw that got me to write? Unfortunately, I still squirm when I hear her smug, superior, politician's voice in interviews. Should I also thank P. D. Boyer for rejecting my review on binding energy in catalysis as the initial stimulus? It had been noticed in the 1960s that some multimeric enzymes did not use the active sites on all subunits for catalysis — 'half-of-the-sites activity'. In 1975, I suggested that one reason for this could be that an enzyme could use the binding energy with the substrate at one site to power an energetically unfavourable reaction at another active site via a conformational change[29]. Boyer cited my paper in 1979[30]. Utilization of binding energy at a distant site then became a key feature of the mechanism of synthesis of ATP for which Boyer received part of the Nobel Prize in 1997.

In reality, I wrote the book because I felt I had something to say, and I constructed the book around my themes. I tell my students that when they write a paper, they must have a clear message and that they must construct the writing around a strong storyline. It is like telling a joke: you remember the punchline and lead the listener to it.

In 1965, I visited the USSR as captain of a combined Oxford and Cambridge chess team. Six years later, I presented my first scientific paper at a meeting in Riga. The anti-Semitism there shocked me. I visited the Riga chess club and held my own against the locals. On pointing to a photo on the wall of Mikhail Tal, the brilliant world chess champion, I was told: 'He is not a Latvian, he is a Jew.' A little later, one of my Estonian friends pointed someone out by saying: 'Poor fellow, he can't get a job, he is a Jew.' About the time of writing my book, I became actively involved in helping dissident scientists in the USSR. Under the encouragement of Roger Kornberg, I became secretary of a Cambridge committee for 'refuseniks', almost all of whom were Jews. I visited the Soviet Union and met with several refuseniks, many of whom were eventually able to leave. In the early 1980s, an Estonian friend wanted to work in my lab for a few months. He was asked by the local KGB why he wished to visit me as I 'interfered in their internal affairs'. But, they did agree to his visa. Being in the KGB's records never did me any harm.

5.1 1994: Sydney Brenner, Fred Sanger and John Kendrew at Max Perutz's 80th birthday meeting.

5.2 1994: Cesar Milstein en route to Perutz meeting.

5.3 1994: Gisela Perutz and David Blow at Perutz meeting.

5.4 2009: Aaron Klug.

5.5 1974: Jim Watson and Sydney Brenner (National Academy of Sciences Archives).

Chapter 6

Aminoacyl-tRNA Synthetases: Limits of Specificity and Editing Mechanisms

'He is as English as the donor of the chair.'

Sydney Brenner

I had by 1974 proven that I could solve longstanding problems in classical enzymology and physical-organic chemistry and introduce novel ideas. But, I needed to mature into a scientist with a wider vision, rather than just being a clever problem solver. The opportunity came via Brian Hartley, whose pipe smoke in his tiny office down the corridor was worth bearing for his stimulating, enthusiastic chats. Until the advent of gene cloning, most mechanistic work was on enzymes that were either commercially available or could be easily purified in large amounts. Brian encouraged me to work on the aminoacyl-tRNA synthetases, on which he was collaborating with the Porton Down laboratories to produce proteins in industrial quantities. I tackled those proteins with the skills honed on chymotrypsin, but needed new rapid reaction equipment. I designed a stopped-flow fluorimeter[29,31]. It is still much better in performance and time resolution than any commercially available machine today. It was built around a semi-tangential six-jet mixing chamber that I designed as a sandwich construction because the technician assigned to me in the machine shop was very good at milling surfaces but not at drilling into solids. Years later, I gave the design to Jack Aviv, designer of the Aviv CD spectrometer, in exchange for two good bottles of French wine per annum, which he dutifully provided until the Customs and Excise stopped him. The stability of the fluorimeter was excellent because much of the random jumps and drifts from lamp instability and the 50 cycle noise from the AC mains supply was eliminated by using the same trick as I had used for chemical experiments: the output from a sample of the exciting beam was fed into one socket of a storage oscilloscope and that from the emission beam into the other, and the difference between the two measured directly, cancelling out the artefacts. But, it immediately dawned on me that I also had to relate the easily measurable fluorescence changes to chemical events. This could be done by rapid quenching of reactions and chemical analysis of the components. The usual continuous-flow quenching machines were limited to time ranges of just a few milliseconds or so because the longer time ranges flushed too much protein down the long flow lines. So, I designed a 'pulsed quenched flow' apparatus that would mix two solutions together in a few milliseconds, allow them to incubate and then quench with a third from milliseconds to minutes[32].

What excited me about these enzymes is that they are at the heart of protein synthesis and have to show remarkable specificity for discriminating between the different amino acids. They have clearly evolved to maximise the differences in binding energy between the cognate and competing amino acids to minimise misincorporation. In addition, there was evidence that there could also be active 'editing' or 'proofreading' mechanisms that would remove errors of misacylation of tRNA before incorporation into proteins. My goals were first to put the mechanism of activation of amino acids on a sound basis, then explore the upper limits of protein side-chain interactions that nature is able to evolve, and then work out the basis of proofreading. My tiny group, consisting of just a technician and me, and a brief spell with my second post-doctoral Rod Mulvey, was able to make a significant impact because of the focussed strategy, my in-house designed equipment and the large supply of proteins, purified by Ross Jakes, that allowed us to do a new series of experiments across a whole range of the proteins, rather than work on just one of the 20. I enjoyed every minute of working on them and learned so much.

Linus Pauling had highlighted a fundamental problem in specificity: how can a site tailored to fit the amino acid isoleucine reject valine, which differs only by being shorter by just one methylene group? Valine has to bind in the isoleucine site, and will do so more weakly by just the binding of a methylene group, which was thought from classical experiments on the 'hydrophobic effect' to be worth only about a factor of 3 in association constant. Conversely, a site built to fit valine can reject isloeucine because it is too large to be crammed in. I found that the incremental binding energies of CH_3- and -CH_2- groups are some 50 times higher in a tightly packed hydrophobic pocket than expected from the simple solution measurements, showing the 'hydrophobic effect' in proteins can be dominated by dispersion energies rather than the classical hydrophobic effect made famous by Walter Kauzman[33–38]. We later found the same can be true in protein folding[39,40]. We measured that the repulsion energies of trying to pack a too-large substrate can be higher still[34,37,38].

But, the use of binding energy alone is insufficient to achieve the necessary accuracy. The key to the high specificity for the rejection of valine by the isoleucyl-tRNA synthetase was discovered by Ann Baldwin (the wife of R. L. Baldwin) and Paul Berg in 1966[41]. The transfer of isoleucine to $tRNA^{Ile}$ takes place in two steps. The amino acid is first activated by reaction with ATP to form isoleucyadenylate and then the isoleucine reacts with tRNA. The addition of $tRNA^{Ile}$ to the complex of the synthetase bound to valyladenylate causes its hydrolysis. There is an active proofreading or editing mechanism that corrects errors. I proposed the 'double-sieve' mechanism to explain selection for editing rather than successful transfer[33,34,42] (first expounded in my book). The 'coarse' sieve allows all amino acids isosteric with or smaller than the cognate amino acid into the active site, and rejects the larger ones by steric exclusion. The hydrolytic site is a 'fine' sieve; all activated amino acids smaller than cognate can enter but the cognate is excluded sterically. The best demonstration was with the valyl-tRNA synthetase, which uses a variation of the sieve to edit the isostere threonine[42]. The hydrolytic site would have a polar site for the -OH of threonine which excludes the non-polar CH_3- of valine. The paper is a beautiful example of how a simple rapid quenching experiment with my pulsed-quenched-flow apparatus elucidated a mechanism.

Editing costs energy since a product that requires hydrolysis of ATP is destroyed. I always enjoy deriving simple equations, and I did one for the cost of editing, the cost-selectivity equation that relates the overall specificity of selection of monomers in terms of the specificity at the synthetic and editing steps (Ref. 43 and later editions of my book). Many years later, Paul Schimmel, who has always cited our work, worked out the structural basis of these mechanisms. It was a privilege for me to write a commentary on his first paper 22 years later[44].

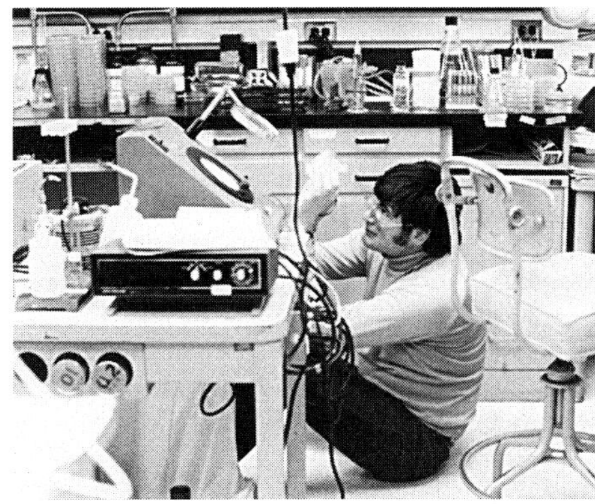

6.1 1978: Counting plaques in Arthur Kornberg's lab.

6.2 1984: Jian-Ping Shi, Alan's first post-doctoral from China.

6.3 2003: Sue Cotterill, Alan's first graduate student, David Lane and Jean Beggs, colleagues at Imperial College, at Alan's 60th Birthday meeting.

6.4 2008: Peter Rigby and David Glover, two of Alan's colleagues at Imperial College at David's 60th Birthday meeting.

6.5 2007: Father Jack Knill-Jones.

6.6 2003: Rabbi Julian Schindler.

[Reprinted from Biochemistry, (1976) **15**, 3342.]

Enzyme Hyperspecificity. Rejection of Threonine by the Valyl-tRNA Synthetase by Misacylation and Hydrolytic Editing†

Alan R. Fersht* and Meredith M. Kaethner

ABSTRACT: Valyl-tRNA synthetase from *Bacillus stearothermophilus* activates threonine and forms a 1:1 complex with threonyl adenylate, but it does not catalyze the net formation of threonyl-$tRNA^{Val}$. On mixing the enzyme-bound threonyl adenylate complex with $tRNA^{Val}$ at pH 7.78 and 25 °C in the quenched flow apparatus it decomposes at a rate constant of 36 s^{-1}. During this process there is a transient formation of Thr-$tRNA^{Val}$ reaching a maximum at 25 ms and rapidly falling to zero after 150 ms. At the peak, 22% of the [^{14}C]threonine from the complex is present as [^{14}C]Thr-tRNA. The reaction may be quenched with phenol and the partially mischarged tRNA isolated. The enzyme catalyzes its hydrolysis with a rate constant of 40 s^{-1}. The data fit a kinetic scheme in which 62% of the threonine from the threonyl adenylate is transferred to the tRNA. This may be compared with the rate constant of 12 s^{-1} at which 84% of the valine is transferred to $tRNA^{Val}$ from the enzyme-bound valyl adenylate, and the rate constant of 0.015 s^{-1} for the subsequent hydrolysis of Val-$tRNA^{Val}$. Inhibition studies indicate a distinct second site for hydrolysis. The translocation of the aminoacyl moiety between the two sites could be mediated by a transfer between the 2′- and 3′-OH groups of the terminal adenosine of the tRNA. The hyperspecificity of the enzyme is based on discriminating between the two competing substrates twice: once against the undesired substrate in the synthetic step, and once against the desired substrate in the destructive step.

During protein biosynthesis the cell distinguishes between certain amino acids with a specificity far greater than that expected from the differences in their structures. For example, as valine is one methylene group smaller than isoleucine it binds to the isoleucyl-tRNA synthetase (IRS), although 100 times more weakly than isoleucine (Flossdorf and Kula, 1973). Yet, it has been found that the overall error rate in protein biosynthesis is less than 1 part in 3000 (Loftfield, 1963; Loftfield and Vanderjagt, 1972). The origin of the increased specificity was suggested by Crick (see Crick, 1975) and found experimentally by (Norris) Baldwin and Berg, to be due to an editing hydrolytic reaction (Norris and Berg, 1964; Baldwin and Berg, 1966). The aminoacylation reaction takes place in two steps (eq 1): the activation of the amino acid by the formation of an aminoacyl adenylate complex followed by the transfer of the aminoacyl moiety to its cognate tRNA (Berg, 1961; Fersht and Kaethner, 1976). Isoleucyl-tRNA synthetase catalyzes the first step with valine and ATP

$$\mathrm{IRS(\pm tRNA)} \xrightarrow[\mathrm{Ile}]{\mathrm{ATP}} \mathrm{IRS \cdot Ile{\sim}AMP\ (\pm tRNA) + PP_i}$$

$$\mathrm{IRS \cdot Ile{\sim}AMP \cdot tRNA \rightarrow IRS + Ile\text{-}tRNA + AMP} \quad (1)$$

but, whereas the addition of $tRNA^{Ile}$ to the isolated IRS·Ile~AMP complex leads to the transfer of about 70% of the isoleucine to the tRNA, the addition of $tRNA^{Ile}$ to the IRS·Val~AMP complex causes its quantitative hydrolysis (Baldwin and Berg, 1966). The net result is that in the presence of $tRNA^{Ile}$ and valine, the IRS acts as an ATP pyrophosphatase, converting ATP to AMP.

The exact mechanism of this editing step is unclear. It has been suggested that the tRNA is first misacylated and the mischarged tRNA specifically hydrolyzed by the enzyme (Yarus, 1972; Eldred and Schimmel, 1972). In support of this it was shown that the aminoacyl-tRNA synthetases may slowly hydrolyze their correctly aminoacylated tRNAs, and in some cases rapidly hydrolyze artificially misacylated tRNAs. However, it has been maintained that, in general, these hydrolysis rates are too low to account for the observed specificities (Bonnet and Ebel, 1974), and that the hydrolysis is nonspecific (Bonnet, 1974; Sourgoutchoff et al., 1974).

We wish to present direct evidence from rapid quenching experiments that the valyl-tRNA synthetase (VRS) from *Bacillus stearothermophilus* discriminates against threonine, which is isosteric with valine, by first mischarging $tRNA^{Val}$ with threonine and then correcting the error by enzymatically hydrolyzing the Thr-$tRNA^{Val}$.

Materials and Methods

The VRS from *B. stearothermophilus* was as previously described (Fersht et al., 1975; Fersht, 1975).

tRNA from *B. stearothermophilus* was obtained from Professor B. S. Hartley, Imperial College of Science and Technology, London. It was initially partially fractionated on benzoylated diethylaminocellulose (eluting NaCl, from 0.4 to 1.0 M, and from 1.0 M to 1.0 M + 20% ethanol, all with 10 mM $MgCl_2$) to obtain $tRNA_1{}^{Val}$ (valyl acceptance = 190 pmol/A_{260}) and $tRNA_2{}^{Val}$ (valyl acceptance = 100 pmol/A_{260}). Purification of $tRNA_1{}^{Val}$ to near homogeneity (valyl acceptance = 1400 pmol/A_{260}) was accomplished by chromatography on DEAE-Sephadex (A-50, eluting with a gradient of 0.375 M NaCl, 8 mM $MgCl_2$–0.525 M NaCl, 16 mM $MgCl_2$; 20 mM Tris-Cl, pH 7.5) followed by the reverse salt gradient procedure with Sepharose 4B (Holmes et al., 1975).

Radioactively labeled amino acids were obtained from The Radiochemical Centre, Amersham, England. The radiochemical purity of the [^{14}C]Thr (batch 16) was quoted at 97–99%. High-voltage electrophoresis at pH 2.1 on Whatman

† From the MRC Laboratory of Molecular Biology, Cambridge CB2 2QH, United Kingdom. *Received December 16, 1975.*

[1] Abbreviations used are: IRS, isoleucyl-tRNA synthetase; VRS, valyl-tRNA synthetase; DEAE, diethylaminoethyl; Tris, 2-amino-2-hydroxymethyl-1,3-propanediol; ATP, adenosine 5′-triphosphate.

AMINO ACID RECOGNITION

No. 4 paper showed that 98% of the radioactivity moved with the threonine and less than 0.5% could be associated with valine.

Experiments were performed in buffers containing 10 mM mercaptoethanol, 0.1 mM phenylmethanesulfonyl fluoride, 10 mM $MgCl_2$, and either 144 mM Tris-Cl (pH 7.78 at 25 °C) or 13 mM Bistris-Cl (pH 5.87 at 25 °C, pH 6.28 at 0 °C). The exchange of [^{32}P]pyrophosphate (2 mM) into ATP (2 mM) at pH 7.78 was measured by the adsorption of ATP on charcoal. The rate of aminoacylation of tRNA was determined by the precipitation of the resultant [^{14}C]Val-tRNA by trichloroacetic acid, collection on nitrocellulose or glass fiber filters. After washing copiously with further trichloroacetic acid (containing valine) and drying, the radioactivity was monitored by scintillation spectrophotometry using a toluene-based scintillant.

The threonine stimulated ATP pyrophosphatase activity was measured from the release of [^{32}P]pyrophosphate from [γ-^{32}P]ATP, the ATP being adsorbed on charcoal and separated by centrifugation (Baldwin and Berg, 1966; Fersht and Kaethner, 1976).

Alkaline phosphatase (*E. coli*) was routinely added where necessary to destroy AMP.

tRNA Stimulated Hydrolysis of VRS·Thr∼[^{32}P]AMP. VRS·Thr∼[^{32}P]AMP was prepared from VRS, threonine, [α-^{32}P]ATP, and inorganic pyrophosphatase, and isolated by gel filtration at 0 °C and pH 6.28 as described for VRS·[^{14}C]Val∼AMP (Fersht, 1975). The rate of hydrolysis of the complex at 0 °C, pH 6.28, was followed by the method of Baldwin and Berg (1966), by monitoring the alkaline phosphate catalyzed release of [^{32}P]orthophosphate from the [^{32}P]AMP that is released. tRNA (20–30 A_{260} units) was added to a solution of the adenylate (200 μl, 4 μM) containing 0.6 unit/ml of alkaline phosphatase.

Transfer of [^{14}C]Thr to tRNA^{Val} at 0 °C. VRS·[^{14}C]-Thr∼AMP was prepared at pH 6.28 from [^{14}C]Thr (232 mCi/mmol), ATP, and inorganic pyrophosphatase, as described for VRS·[^{14}C]Val∼AMP (Fersht, 1975). tRNA was added to a solution of the adenylate (1–4 μM) and alkaline phosphatase (0.6 unit/ml), samples were rapidly quenched with trichloroacetic acid, and the precipitates were collected on nitrocellulose filters.

Preparative Mischarging of tRNA^{Val} with [^{14}C]Thr. tRNA^{Val} (150 μl, 6.4 nmol) was added at 0 °C to a solution of VRS·[^{14}C]Thr∼AMP (1.5 ml, 6.4 nmol) at pH 6.28 and immediately quenched by the addition of a saturated solution of PheOH and chloroform. After centrifugation and precipitation of the tRNA from the aqueous layer (0.2 M sodium acetate, pH 5.5) by the addition of two volumes of ethanol, the tRNA was dissolved in 10 mM $MgCl_2$ (pH 5) and freed from ethanol by gel filtration (10 mM $MgCl_2$, Sephadex G-25), to give 2.6 pmol of tRNA^{Val} that had been 3.8% aminoacylated.

Rapid Quenching Experiments at 25 °C and pH 7.78. (a) Hydrolysis of [^{14}C]Thr-tRNA^{Val}. One syringe of the quenched flow apparatus (Fersht and Jakes, 1975) contained VRS (2 μM) incubated in 288 mM Tris-Cl (pH 7.78), 10 mM $MgCl_2$, 10 mM mercaptoethanol, and 0.1 mM phenylmethanesulfonyl fluoride. The other contained tRNA^{Val} (320 nM) and [^{14}C]Thr-tRNA^{Val} (12 nM) unbuffered in 10 mM $MgCl_2$. The solutions were automatically mixed and quenched with trichloroacetic acid (5%), either 123 or 188 μl being expelled from each syringe.

(b) Hydrolysis of VRS·Thr∼[^{32}P] AMP. One syringe contained tRNA^{Val} (34 μM) in 288 mM Tris-Cl as above, and

TABLE I: Comparison of Pyrophosphate Exchange, Aminoacylation, and ATP Pyrophosphase Activities of VRS.[a]

Reaction	Amino Acid	k_{cat} s^{-1}	K_M mM
Pyrophosphate[b,c] exchange	Val	33	0.03
	Thr	14	7
Pyrophosphate exchange	Val	33	–
in presence of tRNA[b,c,d]	Thr	10	–
Aminoacylation[e]	Val	3	–
ATP pyrophosphatase[c]	Thr	6	7

[a] 25 °C, pH 7.78, 10 mM $MgCl_2$, 2 mM ATP. [b] 2 mM pyrophosphate. [c] 185 nM VRS. [d] 8.6 μM tRNA^{Val}, 37 nM VRS.

the other, VRS·Thr–[^{32}P]AMP (1.86 μM) in 13 mM Bistris-Cl (pH 5.87). The solutions were mixed and quenched with perchloric acid (3.5%) as described above. Aliquots of the effluent were immediately added to Tris-Cl (1.0 M Tris, 0.44 M Tris–HCl, 0 °C) containing alkaline phosphatase (10 units/ml) and quenched with perchloric acid after 20.0 s.

The [^{32}P]orthophosphate released from the [^{32}P]AMP was monitored as described by Baldwin and Berg, 1966.

(c) Transient Formation of [^{14}C]Thr-tRNA. One syringe of the apparatus contained VRS·[^{14}C]Thr∼AMP (1.8 μM) and alkaline phosphatase (0.03 unit/ml) in Bistris-Cl (pH 5.87), the other tRNA^{Val} (34 μM) in Tris-Cl as above. The solutions were mixed and quenched with trichloroacetic acid, and the precipitates collected on nitrocellulose filters.

For b and c the rate of hydrolysis of the VRS·Thr∼AMP complex was monitored periodically by filtering aliquots through nitrocellulose filters (Yarus and Berg, 1970). During the course of the experiments less than 20% hydrolyzed ($t_{1/2}$ = 3 h).

Transfer of [^{14}C]Val to tRNA^{Val}. One syringe of the quenched flow apparatus contained VRS·[^{14}C]Val∼AMP (0.26 μM) in Bistris-Cl, the other tRNA^{Val} (11 μM) in the concentrated Tris-Cl. Aliquots were periodically quenched with trichloroacetic acid.

Hydrolysis of [^{14}C]Val-tRNA^{Val}. [^{14}C]Val-tRNA^{Val} was prepared in situ by adding tRNA^{Val} (0.5 μM) to a solution of VRS·[^{14}C]Val∼AMP (0.11 μM), VRS (4 or 2 μM), and alkaline phosphatase (0.7 unit/ml) at 25 °C, and pH 7.78. Aliquots were quenched at 15-s intervals with trichloroacetic acid.

Results

The VRS from *B. stearothermophilus,* like its counterpart from *E. coli* (Bergmann et al., 1961; Yaniv and Gros, 1969; Owens and Bell, 1970), catalyzes the exchange of [^{32}P]pyrophosphate into ATP in the presence of threonine, as well as valine (Table I), although k_{cat} is some 2.4 times lower and K_m 200 times higher. This exchange is not due to impurities in the threonine. Incubation of the enzyme with ATP, inorganic pyrophosphatase, and a fourfold excess of [^{14}C]Thr leads to the formation and isolation by gel filtration of a VRS·[^{14}C]-Thr∼AMP complex with a stoichiometry of 1.0. Since the [^{14}C]Thr was found to be at least 98% pure, this result cannot be due to a contamination of [^{14}C]Val.

Incubation of the VRS with tRNA, [^{14}C]Thr, ATP, and inorganic pyrophosphatase gives no detectable formation of [^{14}C]Thr-tRNA under a wide variety of conditions, from pH 5.9 to 7.8, and 0 to 75 °C. Instead, the enzyme functions as an ATP pyrophosphatase. The turnover number for this ap-

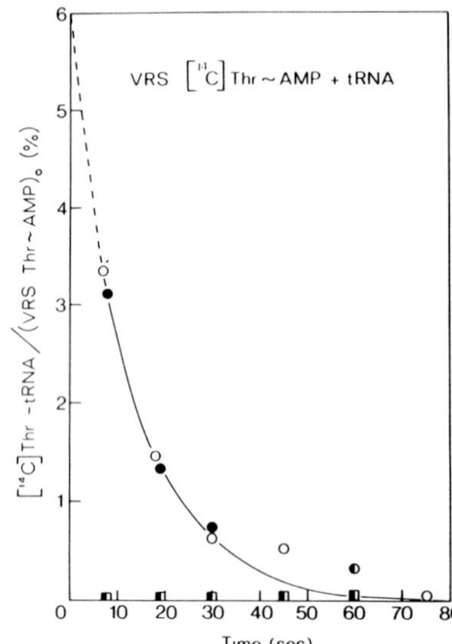

FIGURE 1: Transient formation of Thr-tRNA at 0 °C, pH 6.28. tRNA (10 nmol) was added to a solution of VRS·[^{14}C]Thr~AMP (4 μM, 220 μl) and alkaline phosphatase (0.6 unit/ml) in a buffer containing 10 mM Bistris-Cl, 10 mM $MgCl_2$, and 10 mM mercaptoethanol. (○) tRNA$_1^{Val}$ (0.9 nmol); (●) tRNA$_2^{Val}$ (0.9 nmol); (□) tRNA fractions enriched for either tRNA^{Phe} or tRNA^{Tyr} (tRNA^{Val} < 5 pmol). The ratio of Thr-tRNA to initial concentration of VRS·[^{14}C] Thr~AMP is plotted. The curve is for a rate constant of 0.077 s^{-1}.

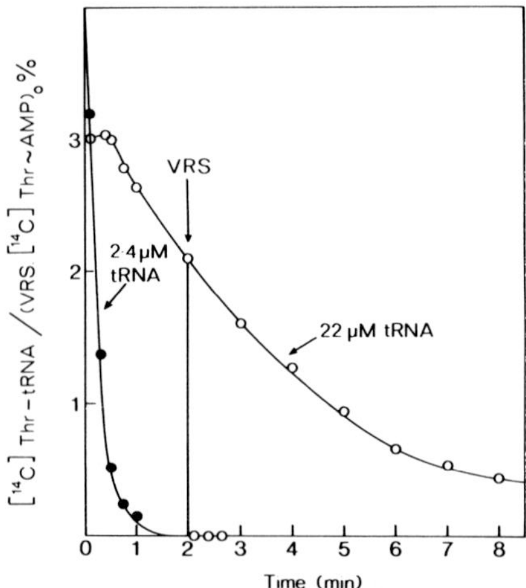

FIGURE 2: Competitive inhibition by tRNA^{Val} of the hydrolysis of Thr-tRNA^{Val} at 0 °C and pH 6.28. VRS·[^{14}C]Thr~AMP (1.7 μM, buffer as for Figure 1) was mixed with either 2.4 μM (●) or 22 μM (○) tRNA^{Val}. In a second experiment with 22 μM tRNA^{Val}, excess VRS was added after 2 min as indicated.

proaches that for the pyrophosphate exchange reaction under similar conditions (Table I).

Reactions of VRS·Thr~AMP with tRNA at 0 °C, pH 6.28. (a) Specificity for tRNA^{Val}. The addition of tRNA enriched in either tRNA$_1^{Val}$ or tRNA$_2^{Val}$ to the VRS·Thr~[^{32}P]AMP complex at 0 °C and pH 6.28 causes the complex to hydrolyze rapidly with a half-life of about 9 s (k = 0.075 s^{-1}). The hydrolysis is enzyme catalyzed, since Thr~AMP is far more stable in the absence of enzyme. The addition of tRNA enriched for tRNA^{Phe} and tRNA^{Tyr} does not stimulate the hydrolysis. Under similar conditions, about 76% of the [^{14}C]Val from the VRS·[^{14}C]Val~AMP complex is transferred to tRNA^{Val}. This behavior is very similar to that observed on the addition of tRNA^{Ile} to the IRS·Val~[^{32}P]AMP complex (Baldwin and Berg, 1966).

(b) Transient Formation of [^{14}C]Thr-tRNA. Addition of fractions of tRNA enriched for tRNA$_1^{Val}$ or tRNA$_2^{Val}$, but not those containing tRNA^{Phe} or tRNA^{Tyr}, to equimolar concentrations of VRS·[^{14}C]Thr~AMP gives a transient formation of [^{14}C]Thr-tRNA^{Val}, extrapolating back to a 5.7% transfer at zero time (Figure 1). Quenching the reaction mixture with phenol within a few seconds of mixing and extracting the tRNA gave a product that had been charged to 3.8% with [^{14}C]Thr.

Addition of excess enzyme (1.9 μM) to this partially misacylated tRNA^{Val} (0.9 μM tRNA^{Val}, 0.034 μM [^{14}C]Thr-tRNA^{Val}) at 0 °C, pH 6.28, deacylated >99% of the tRNA in 7 s.

These results are consistent with Scheme I.

Scheme I

$$\text{VRS·Thr}\sim\text{AMP} \xrightarrow[0.075\ s^{-1}]{\text{tRNA}^{Val}} \text{VRS·Thr-tRNA}^{Val} + \text{AMP}$$

$$\downarrow \sim 1\ s^{-1}$$

$$\text{VRS} + \text{Thr} + \text{tRNA}^{Val}$$

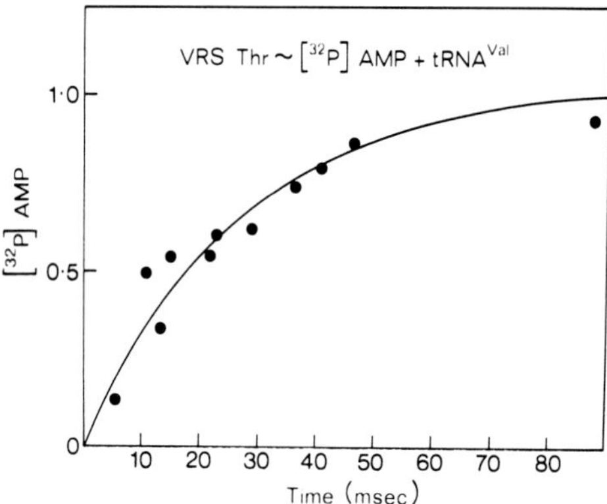

FIGURE 3: tRNA-stimulated decomposition of VRS-Thr~[^{32}P]AMP at 25 °C, pH 7.78, 10 mM $MgCl_2$, and 10 mM mercaptoethanol. Equal volumes of tRNA^{Val} (34 μM) and VRS-Thr-[^{32}P]AMP (1.86 μM) were mixed in the quenched-flow apparatus and the rate of formation of [^{32}P]AMP was measured. The curve is calculated for a rate constant of 36 s^{-1}.

(c) Competitive Inhibition of the Hydrolysis of [^{14}C]Thr-tRNA^{Val} by tRNA^{Val}. Schreier and Schimmel (1972) showed that tRNA^{Ile} competitively inhibits the IRS catalyzed deacylation of Ile-tRNA^{Ile}. A similar situation occurs here (Figure 2). On mixing 22 μM tRNA^{Val} with 1.7 μM VRS·[^{14}C]Thr~AMP, some of the [^{14}C]Thr-tRNA that is formed dissociates from the enzyme and slowly hydrolyzes. On adding an excess of enzyme over tRNA^{Val} so that competitive inhibition is no longer possible, the remaining [^{14}C]Thr-tRNA is completely hydrolyzed within 7 s (Figure 2).

Reactions of VRS·Thr~AMP with tRNA^{Val} and the VRS-Catalyzed Deacylation of Thr-tRNA^{Val} at 25 °C, pH 7.78. The rate constant for the hydrolysis of VRS·Thr~[^{32}P]AMP stimulated by tRNA^{Val} is 36 s^{-1} (Figure 3). The rate constant for the VRS catalyzed hydrolysis of [^{14}C]Thr-tRNA is 40 s^{-1} (Figure 4). The addition of tRNA^{Val} to VRS·[^{14}C]Thr~AMP gives a transient burst of [^{14}C] Thr-tRNA with a peak of 22% of the [^{14}C]Thr transferred at 25 ms, rapidly falling to zero after 150 ms (Figure 5). The time

AMINO ACID RECOGNITION

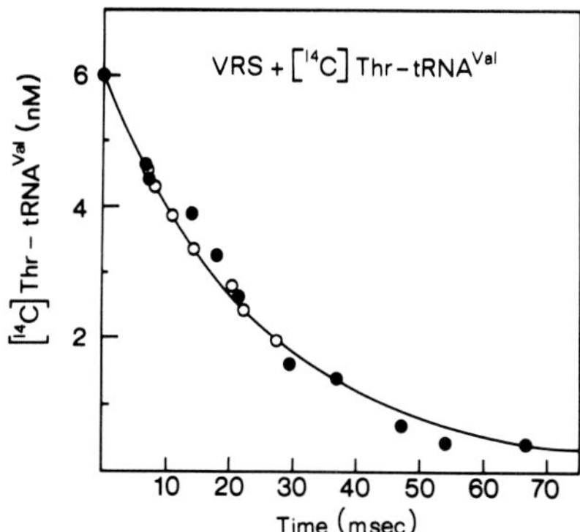

FIGURE 4: VRS-catalyzed hydrolysis of Thr-tRNA^{Val} at 25 °C, pH 7.78, 10 mM $MgCl_2$ and 10 mM mercaptoethanol. (●) Equal volumes of VRS (2 μM) and tRNA ([^{14}C]Thr-tRNA^{Val}, 12 nM; tRNA^{Val}, 320 nM) were mixed in the quenched flow apparatus. (○) Same as in ● but with ATP (4 mM), valine (800 μM), and inorganic pyrophosphatase (1 unit/ml) added to the enzyme to convert the VRS to VRS·Val~AMP. The curve is calculated for a rate constant of 40 s^{-1}.

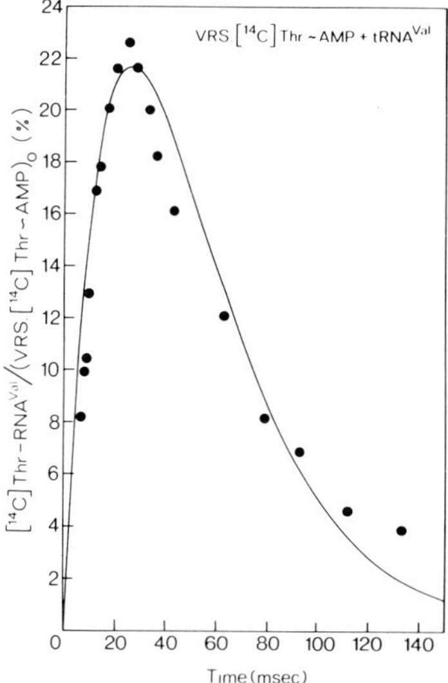

FIGURE 5: The transient formation of ["[^{14}C]Thr-tRNA^{Val} at 25 °C, pH 7.78, 10 mM $MgCl_2$, and 10 mM mercaptoethanol. Equal volumes of tRNA^{Val} (34 μM) and VRS·[^{14}C]Thr~AMP (1.8 μM) were mixed in the quenched-flow apparatus. The curve for the ratio of [^{14}C]Thr-RNA^{Val} to the initial concentration of VRS·[^{14}C]Thr~AMP is calculated for the kinetic scheme involving transfer of 62% of the [^{14}C]Thr from the complex to the tRNA at 36 s^{-1}, followed by the deacylation at 40 s^{-1}.

dependence of the concentration of [^{14}C]Thr-tRNA fits the theoretical curve constructed for Scheme II where 62% of the [^{14}C]Thr is transferred to the tRNA.

Scheme II

$$\text{VRS·[}^{14}\text{C]Thr} \sim \text{AMP·tRNA}^{\text{Val}} \xrightarrow{36\ \text{s}^{-1}} \text{VRS·[}^{14}\text{C]Thr-tRNA} \xrightarrow{40\ \text{s}^{-1}} \text{VRS} + [^{14}\text{C]Thr} + \text{tRNA}$$

These data may be compared with the rate constant of 12 s^{-1} with which 84% of the [^{14}C]Val from VRS·[^{14}C]-Val~AMP is transferred to tRNA (Figure 6), and 1.5×10^{-2} s^{-1} for the VRS-catalyzed hydrolysis of Val-tRNA^{Val}. The reason for these incomplete transfers is unknown (Baldwin and Berg, 1966).

The turnover number for the VRS-catalyzed hydrolysis of Thr-tRNA^{Val} is the highest yet reported (see Bonnet and Ebel, 1974).

The turnover number for the steady-state threonine stimulated ATP pyrophosphatase activity, 6 s^{-1}, is considerably less than the rate constant for the transfer step, 36 s^{-1}. This is possibly due to the formation of threonyl adenylate being partially rate determining, since the k_{cat} for the pyrophosphate exchange reaction is only 10 s^{-1}.

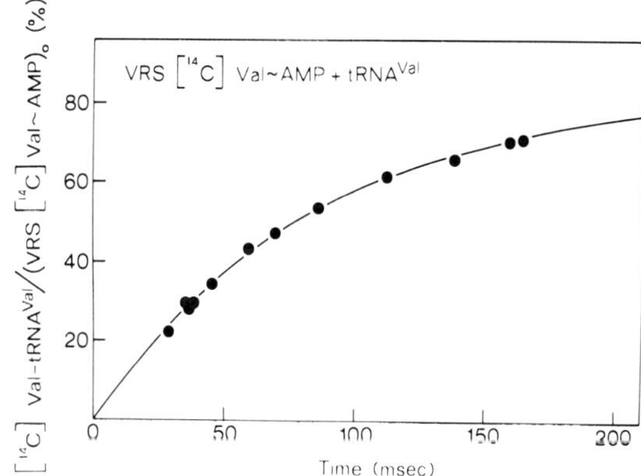

FIGURE 6: Transfer of [^{14}C]Val to tRNA^{Val} at 25 °C, pH 7.78, 10 mM $MgCl_2$, and 10 mM mercaptoethanol. Equal volumes of tRNA^{Val} (11 μM) and VRS·[^{14}C]Val~AMP (0.26 μM) were mixed in the quenched-flow apparatus. The curve is calculated for a rate constant of 11.7 s^{-1} and the transfer of 84% of the [^{14}C]Val.

Discussion

The addition of tRNA^{Val} to the VRS·[^{14}C]Thr~AMP complex leads to a transient production of [^{14}C]Thr-tRNA, reaching a maximum of 22% transfer at 25 ms (Figure 5). The time dependence of the concentration of the Thr-tRNA fits that predicted from the independently measured rate constants for the overall decomposition of the complex (36 s^{-1}) and the VRS-catalyzed hydrolysis of Thr-tRNA^{Val} (40 s^{-1}). The large fraction of misacylated tRNA formed, extrapolating to 62% of the total VRS·[^{14}C]Thr~AMP complex, rules out that this is due to impurities in the reagents. The enzyme prevents the net misacylation of tRNA by first mischarging the tRNA and then correcting the error with "editing" by hydrolysis.

Evidence Concerning the Mechanism. (a) There Are Two Separate Active Sites. The hydrolysis of the misacylated tRNA could be due to either a conformational change causing the acylation site to be converted to a hydrolytic site or the presence of a distinct second site. The latter appears to be the case, since we and others (Schreier and Schimmel, 1972; Yarus, 1972) find that the enzymes continue to hydrolyze aminoacylated tRNA, while the cognate aminoacyl adenylate occupies the acylation site. A second site would appear necessary, since otherwise, in vivo, any misacylated tRNA diffusing from the enzyme would be prevented from returning by the rapid formation of the enzyme bound aminoacyl adenylate.

(b) Deacylation Need Not be Preceded by Dissociation of the Enzyme–tRNA Complex. The following suggests that the deacylation occurs without the tRNA leaving the enzyme. Uncharged tRNA competitively inhibits the hydrolysis of the acylated tRNA (Figure 2). If the charged tRNA left the en-

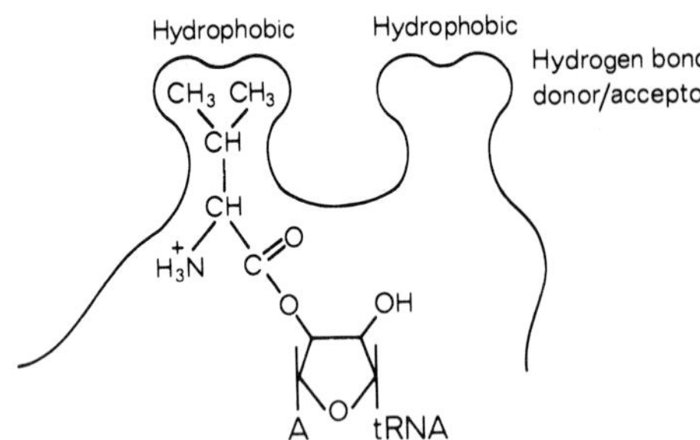

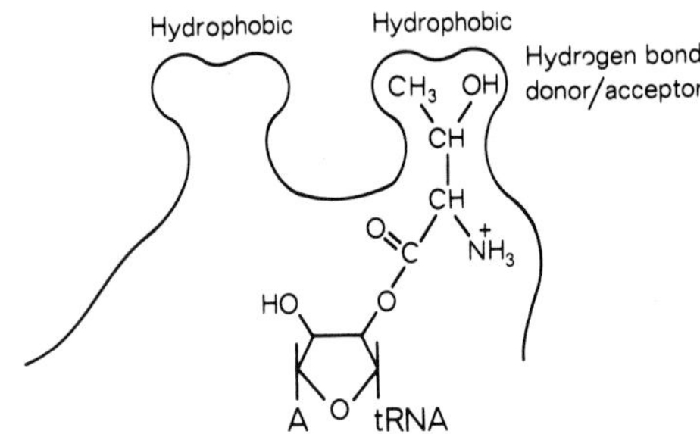

FIGURE 7: Illustration of the specificity mechanism. Top: the hydrophobic acylation site discriminates against threonine. Bottom: the hydrolytic site presumably specifically utilizes the hydroxyl of the threonine for a binding or catalytic effect. The translocation may occur via a 2′ → 3′-OH acyl transfer as illustrated or by a movement of the terminal adenosine.

zyme before hydrolyzing and then returned to a second site, then in the presence of excess uncharged tRNA a large fraction of charged tRNA would build up. However, as seen in Figure 2, only a small fraction of [^{14}C]Thr-tRNA^{Val} "leaks through" and hydrolyzes slowly due to the inhibition. Deacylation is faster than dissociation. (In vivo, since tRNA and the aminoacyl-tRNA synthetases are present in comparable amounts (Yarus and Berg, 1969), the competitive inhibition by uncharged or correctly charged tRNA should be of little consequence.)

The Translocation Step. The translocation of the aminoacyl ester to the hydrolytic site may be simply due to a movement of the CCA terminus of the tRNA. However, an alternative hypothesis may be devised from this study and a recent result from von der Haar and Cramer (1975). They find that on replacing the terminal adenosine of yeast tRNA^{Ile} with 3′-deoxyadenosine, the modified tRNA is still charged by the yeast IRS and ATP. However, it will now also accept valine. Since it is known that aminoacyl groups are rapidly and reversibly transferred between the 2′- and 3′-hydroxyl groups of the ribose ($t_{1/2} \sim 0.2$ ms, Griffin et al., 1966), it is possible that for the IRS and the VRS the acylation site is on the 2′-OH side of the ribose, while the hydrolytic site is on the 3′-OH side so that translocation simply involves the intramolecular migration of the aminoacyl moiety (Figure 7).

Hyperspecificity Depends on Using Structural Differences Twice. Whatever the detailed mechanism of the reaction, the discrimination between valine and threonine depends on the relative rates of acylation and hydrolysis. The enzyme attains its high specificity by using the differences in the structure of the competing substrates *twice:* once by discriminating against the undesired uubstrate in the synthetic step, and once by discriminating against the desired substrate in the destructive step.

References

Baldwin, A. N., and Berg, P. (1966), *J. Biol. Chem. 241,* 839.
Berg, P. (1961), *Annu. Rev. Biochem. 30,* 293.
Bergmann, F. H., Berg, P., and Dieckmann, M. (1961), *J. Biol. Chem. 236,* 1735.
Bonnet, J. (1974), *Biochimie 56,* 541.
Bonnet, J., and Ebel, J. P. (1974), *FEBS Lett. 39,* 259.
Crick, F. H. C. (1975), *Philos. Trans. R. Soc. London, Ser. B: 272,* 193.
Eldred, E. W., and Schimmel, P. R. (1972) *J. Biol. Chem. 247,* 2961.
Fersht, A. R. (1975), *Biochemistry 14,* 5.
Fersht, A. R., Ashford, J. S., Bruton, C. J., Jakes, R., Koch, G. L. E., and Hartley, B. S. (1975), *Biochemistry 14,* 1.
Fersht, A. R., and Jakes, R. (1975), *Biochemistry 14,* 3350.
Fersht, A. R., and Kaethner, M. M. (1976), *Biochemistry 15,* 818.
Flossdorf, J., and Kula, M. R. (1973), *Eur. J. Biochem. 36,* 534.
Griffin, B. E., Jarman, M., Reese, C. B., Sulston, J. E., and Trentham, D. R. (1966), *Biochemistry 5,* 3638.
Holmes, W. E., Hurd, R. E., Reid, B. R., Fimerman, R. A., and Hatfield, G. W. (1975), *Proc. Natl. Acad. Sci. U.S.A. 72,* 1068.
Loftfield, R. B. (1963), *Biochem. J. 89,* 82.
Loftfield, R. B., and Vanderjagt, D. (1972), *Biochem. J. 128,* 1353.
Norris, A., and Berg, P. (1965), *Proc. Natl. Acad. Sci. U.S.A. 52,* 330.
Owens, S. L., and Bell, F. E. (1970), *J. Biol. Chem. 245,* 5515.
Schreier, A. A., and Schimmel, P. R. (1972), *Biochemistry 11,* 1582.
Sourgoutchoff, A., Blanquet, S., Fayat, G., and Waller, J. P. (1974), *Eur. J. Biochem. 46,* 431.
von der Haar, F., and Cramer, F. (1975), *FEBS Lett. 56,* 215.
Yaniv, M., and Gros, F. (1969), *J. Mol. Biol. 44,* 1.
Yarus, M. (1972), *Proc. Natl. Acad. Sci. U.S.A. 69,* 1915.
Yarus, M., and Berg, P. (1969), *J. Mol. Biol. 42,* 171.
Yarus, M., and Berg, P. (1970), *Anal. Biochem. 35,* 450.

Imperial College (1978–1988) via Stanford (1978–1979)

At the age of 34, I was nominated by David Blow and Brian Hartley for the very prestigious Wolfson Research Professorship of the Royal Society, in succession to Dorothy Hodgkin, to be held at Imperial College, where they had moved from LMB. Sydney Brenner, who was on the Royal Society Committee for the selection recounted to me an amusing incident. The professorship was restricted to UK nationals. A rather conservative member of the committee asked whether the name Fersht was English. Lord Todd, the president, jumped in with 'Och, I am a Scot.' Then 'I am of Welsh descent,' and 'Don't look at me, I am half Irish' echoed around the room. Sydney Brenner rounded it off with a characteristically withering remark: 'He is as English as the donor of the chair.' I was offered the chair but dithered about accepting it. My senior colleagues, especially Sydney Brenner, wanted me to stay in Cambridge, but Max Perutz said it was just too good an opportunity to turn down. As usual, he was right: it would give me the opportunity to establish a larger group and do more challenging work. Until then, I had worked with just one technician and an occasional post-doctoral. I was really sad to leave a lab with so many good friends and such support from senior members. Aaron Klug and Sydney Brenner continued to take a strong interest in my career and support me, and both tried hard to recruit me back to Cambridge over the next ten years.

I set up my labs in London and then did a year's sabbatical in Stanford with Arthur Kornberg. Roger Kornberg picked up my wife, Marilyn, two young children, Naomi and Philip, and me at San Francisco airport and told me that his dad had just had a lazy British post-doctoral who worked banking hours and so I should show him what a hardworking Brit could do. In fact, I have always worked regular hours so that I could have time for my family and also refresh myself at weekends. My working practice while still at the bench was to work hard and efficiently from about 8 a.m. to 6 p.m. doing experiments. I would not analyse data during the day any more than necessary for the next experiment. Any time not spent doing practical work during the day I considered a waste of experimental time. I would then take my data home with me and plot graphs in the evening after the children had gone to bed. I would wake up the next morning having planned my day's work the night before, and slept on it. I never did 'trial' experiments to work out conditions but always planned the correct conditions and repeated experiments only for statistical analysis.

I did work hard at Stanford, according to my normal practice, and Arthur communicated a paper on my year's work to *PNAS*[45]. I applied kinetics to the rolling circle replication of bacteriophage øX174 by DNA polymerase III of *Escherichia coli* followed by transfection of the DNA into spheroplasts to analyse its editing mechanism[45]. By varying the concentration of deoxynucleotides, I was able to modulate the fidelity of replication and discover kinetic laws for misincorporation of bases, including the importance of the following (next) nucleotide in fixing errors of mutagenesis. The kinetic assay was based on measuring the number of wild-type and mutant phage produced from the DNA replicated *in vitro*. I jumped up and down with joy, much to the amusement of some students, when I saw the first plaques of viable phage on the lawn of *E. coli.* I had brought with me a pulsed-quench flow machine, but the bacteriology was much more fun! (Related references:[45–54]) My one and only sabbatical year was a very important learning curve. My stay at Stanford taught me by observation how to run a large group and also essential practical skills in molecular genetics.

The post-doctoral who kept my lab running during my year's absence, Julian Schindler, became a rabbi, with an important role in the Chief Rabbi's Office, and my research assistant, Jack Knill-Jones became a vicar. Many years later, another post-doctoral, Douglas Axe, became a prominent figure in the Intelligent Design movement and is Director of the Biologic Institute. At Imperial, I recruited my first graduate student, Sue Cotterill, who is now an established researcher in London. One of the first post-doctorals, Wing Tsui, set up a small biotech company.

My work had come to be recognised internationally. Out of the blue, I was awarded the FEBS Anniversary Prize (for a scientist under 40) at the Federation of European Biochemical Society meeting in 1980 in Jerusalem. Three years later, just before my 40th birthday, I was elected a Fellow of the Royal Society. The citation for election was:

> Distinguished for work on mechanisms of enzyme catalysis, especially by stopped and quenched flow methods. He showed that a slow relaxation of chymotrypsin was not a chemical step on the reaction pathway, but a pH-dependent isomerization between active and inactive forms, and investigated the energetics and equilibria of the transition. He elucidated the leaving-group specificity, leading to a detailed structural interpretation which showed the energetics of 'strain' at the binding site. Another experiment dispelled final doubts about the role of a tetrahedral intermediate. More recently Fersht has studied a more complex group of enzymes, the aminoacyl tRNA synthetases. He demonstrated that their precise specificity depends on consecutive independent recognition steps, and under appropriate conditions he trapped a transiently mischarged aminoacyl tRNA. Fersht has shown how binding energy can be used to enhance either specificity or rate in an enzymatic reaction, leading to a demonstration of thermodynamic limitations on mechanisms of the 'induced fit' type.

The most important legacy from the year at Stanford was that the studies on reverting mutants of øX174 *in vitro* and *in vivo* made me acutely aware of Clyde Hutchison's and Mike Smith's work on site-directed mutagenesis that was based on reverting those same mutants of øX174 using synthetic, mismatched oligodeoxynucleotides. That experience and my general background in structure and mechanism and the desire to understand how nature harnesses non-covalent interactions were the springboard for my subsequent work on protein engineering.

Reprinted from
Proc. Natl. Acad. Sci. USA
Vol. 76, No. 10, pp. 4946–4950, October 1979
Biochemistry

Fidelity of replication of phage ϕX174 DNA by DNA polymerase III holoenzyme: Spontaneous mutation by misincorporation

(editing/proofreading/error frequency/site-directed mutagenesis)

ALAN R. FERSHT*

Department of Biochemistry, Stanford University Medical Center, Stanford, California 94305

Communicated by Arthur Kornberg, July 9, 1979

ABSTRACT DNA from ϕX174 is replicated *in vitro* with a fidelity similar to that found genetically. A mutation of TAG → TGG may be induced, however, by varying the concentrations of deoxynucleoside triphosphates, with a frequency proportional to $[dGTP]^2/[dATP]$. This complex concentration dependence is consistent with the active participation of a proofreading mechanism that hydrolytically excises mismatched base pairs as they are formed. A simple kinetic analysis predicts that the frequency of misincorporation depends on the ratio of incorrect to correct deoxynucleoside triphosphates times the concentration of the *next* triphosphate in the sequence to be added. This suggests that spontaneous mutation by misincorporation depends crucially on the composition of the deoxynucleoside triphosphate pool.

The fidelity of replication of DNA is extremely high. The frequency of spontaneous point mutation in *Escherichia coli* is in the region of 10^{-8}–10^{-10} per base per replication (1), and the frequency in phages is within 2–3 orders of magnitude of this (2, 3). Such accuracy is very difficult to measure *in vitro* by use of synthetic templates and the misincorporation of radioactively labeled deoxynucleoside triphosphates because of trace levels of contaminants. A biological assay has recently been developed to overcome these difficulties and to avoid other artifacts; it uses the rate of reversion of the ϕX174 *am3* mutant during replication of its DNA *in vitro* (4, 5). [The mutant has a G → A transition at position 587 in the Sanger sequence (6), converting the TGG codon for tryptophan to the TAG terminator. The mutation has no effect on phage assembly or DNA replication, but just prevents cell lysis (7).] The initial study involved the partial replication of the viral (+) strand by using DNA polymerase I of *E. coli* and a restriction fragment as a primer (4). The resultant DNA was then transfected into spheroplasts, and the progeny phage or infective centers were assayed for revertants. However, in the absence of added mutagens, no revertants could be detected above the background reversion frequency [$(1.7 \pm 0.6) \times 10^{-6}$], and only a lower limit of 1/7700 could be placed on the accuracy of polymerase I. The second study used the reconstituted T4 replication system to replicate the replicative form (RF I) DNA of ϕX174 to produce progeny double-stranded RF DNA (5). Again, on transfection no revertants were detected above background, placing a lower limit of $5/10^7$ for the accuracy of the T4 polymerase.

In the present study, I apply this procedure to the RF DNA → single-stranded (ss) DNA step in the replication of ϕX174, using the reconstituted enzyme system of DNA polymerase III holoenzyme, *rep* and ssDNA-binding proteins from *E. coli*, and the phage-encoded gene *A* protein. In the presence of ATP and the four deoxynucleoside triphosphates, multiple single-stranded circular copies of the viral (+) strand are made by using the complementary strand of the RF I DNA as template (3, 8, 9). This is the stage *in vivo* at which RF DNA is multiplied to give progeny. The strategy is to vary the concentrations of the deoxynucleoside triphosphates during replication *in vitro* and to examine the concentration dependence of the error rate, if any.

MATERIALS AND METHODS

Materials. Deoxynucleoside triphosphates were obtained from Sigma. Thin-layer chromatography on Polygram polyethyleneimine (PEI) sheets [1 M HCO_2H/0.5 M LiCl or 4 M NaO_2CH (pH 3.4) (10)] showed that dATP, dTTP, and dCTP were greater than 99% triphosphate whereas dGTP contained 6.7% dGDP but no detectable amounts of other triphosphates (<0.1%). Oligo$(dT)_{12}$ and poly(dA) were obtained from Miles.

DNA polymerase III holoenzyme [3.5×10^5 units/mg (11)], ssDNA-binding protein [2.3×10^4 units/mg (12)], gene *A* protein [5.5×10^6 units/mg (13)], *rep* protein [4×10^7 units/mg (14)], and dUTPase [2×10^7 units/mg (15)] were kindly provided by U. Huebscher, R. Fuller, N. Arai, and J. Shlomai of this laboratory. ϕX174 and its DNA were the gifts of J. Kobori. ϕX174 RF I DNA was prepared by a modification of the Hirt procedure (16, 17).

Oligo$(dT)_{12}([^3H]dC)_{0.05}$ and oligo$(dT)_{12}([^3H]dT)_2$ were prepared as described by Brutlag and Kornberg (18).

Methods. ϕX174 RF I DNA (1.4×10^{-2} pmol) was replicated in a reaction mixture (25 μl) containing DNA polymerase III holoenzyme (300 ng, 105 units) ssDNA-binding protein (1.6 μg, 36.8 units), *rep* protein (25 ng, 1000 units), gene *A* protein (9 ng, 50 units), dUTPase (0.1 ng, 200 units), ATP (800 μM), dCTP (40 μM), various concentrations of dGTP, dATP, and $[^3H]$dTTP (900 cpm/pmol), Tris·HCl (55 mM, pH 7.5), dithiothreitol (6.5 mM), glycerol (20% vol/vol), $MgCl_2$ (10.8 mM), NaCl (76 mM), EDTA (0.26 mM), and bovine serum albumin (0.19 mg/ml). (The deoxynucleoside triphosphates were preincubated with the dUTPase for 2 min at room temperature.) After 1 hr at 30°C, duplicate samples were assayed for extent of incorporation of $[^3H]$dTMP by acid precipitation onto nitrocellulose filters while the remainder was extracted with phenol at 65°C (19), precipitated with ethanol and washed. The products were examined by electron microscopy by the Kleinschmidt formamide technique (20).

Spheroplasts from *E. coli* W3350 were prepared by the method of Guthrie and Sinsheimer (19). Transfection and infective center assays were performed as described by Benzinger (21), whereas progeny phage were produced by the method of Pagano and Hutchinson (22). Indicator bacteria were *E. coli*

The publication costs of this article were defrayed in part by page charge payment. This article must therefore be hereby marked "*advertisement*" in accordance with 18 U. S. C. §1734 solely to indicate this fact.

Abbreviations: RF I, covalently closed, circular superhelical (replicative form) ϕX DNA; ssDNA, covalently closed, single-stranded ϕX DNA

* Present address: Department of Chemistry, Imperial College of Science and Technology, London SW7, England.

C for wild type and E. coli CR (su^+) for *am3* (3, 7). Plaques were scored after 3–4 hr of incubation at 37°C. About 10^{-4}–10^{-3} infective centers were formed per DNA particle.

Infective Center Compared with Phage Release Assay. Transfections were performed under conditions where the number of infective centers produced is linear with DNA concentration (19). The number of progeny phage produced, however, is equal to the number of infective centers multiplied by the number of viable phage produced per center (≈100–200 in the present experiments). Thus, when mixtures of wild-type and mutant DNA, which may give different yields of progeny per infection, are analyzed, the infective center assay gives the better measure of the reversion frequency. It is seen later in Table 1, however, that the two procedures agree within a factor of 2.

RESULTS

The Product DNA. The incorporation of [^{3}H]dT into the DNA replicated *in vitro* corresponded on average to about five to eight circles produced per circle of input RF I DNA. The number of infective particles produced, determined by infective center assays using phage extracted samples of ϕX174 ssDNA for calibration, was about 50–60% of this. Examination by electron microscopy of the products formed at the lowest, intermediate, and highest concentrations of dGTP used (Table 1) showed the following compositions, respectively: single-stranded circles, 84, 87, and 85; single-stranded linear molecules, 11, 9, and 9; double-stranded relaxed circles, 4, 4, and 4; and double-stranded linear molecules, 0, 0, and 2. No remaining supercoiled RF I DNA was found. It is seen in Table 1 that the phage used for preparation of the RF I DNA contained about 2×10^{-6} revertants. After transfection of the DNA extracted from this phage, the progeny phage had a slightly increased proportion of revertants, about 4×10^{-6}. Both the RF I DNA and the DNA produced *in vitro* at 8.4 μM dGTP form about 3×10^{-6} revertant phage after transfection. Transfection of the DNA thus produces phage that is phenotypically very similar to the parent.

Increasing the concentration of dGTP relative to dATP to favor the A → G transition, however, caused a dramatic increase in the number of revertants. At 1010 μM dGTP and 10.0 μM dATP, the reversion frequency increased 2000-fold to 1/150, as measured by progeny phage release, and to 1/300, as determined by the number of infective centers. This increase clearly represents mutation *in vitro*. The data appear reliable. The results of the infective center assays paralleled those of the progeny release assays, despite being performed under vastly different ratios of DNA to spheroplasts. The individual assays were also consistent over a 3- to 10-fold range of dilution.

The presence of dGDP as an impurity did not perturb the experiments. Addition of dGDP to 400 μM did not alter the reversion frequency (Table 1).

Dependence of Reversion Frequency on Deoxynucleoside Triphosphate Concentration. The nonlinear relationship between the reversion frequencies and the concentrations of dGTP and dATP may be easily simplified. In Fig. 1, the logarithm of the reversion frequency is plotted against log [dGTP], with [dCTP] and [dTTP] held constant at 40 μM and [dATP] approximately constant at 9.5–12 μM. The slope of +2.0 shows that the frequency varies at $[\mathrm{dGTP}]^2$. A similar plot (Fig. 2) against log [dATP] at 911 μM dGTP has a slope of −1.0, showing that the frequency varies as 1/[dATP]. Combining all the data (at 40 μM dTTP and dCTP), the reversion frequency calculated from the infective center assays is given by:

$$\text{Frequency} = 0.0333\ [\mathrm{dGTP}]^2/[\mathrm{dATP}]. \qquad [1]$$

Calculated from the progeny phase assay, it is:

$$\text{Frequency} = 0.0659\ [\mathrm{dGTP}]^2/[\mathrm{dATP}]. \qquad [2]$$

The fit is very good; the correlation coefficients are, respectively, 0.986 and 0.998.

Revertants Are Mainly Wild Type. Restoration of the lytic function with increasing [dGTP] indicates that the amber codon TAG is removed by the replacement of the T or A by G. The data in Table 1 and Fig. 2 show that dGTP competes with dATP. However, there is no competition between dGTP and

Table 1. Reversion frequencies of ϕX174 DNA synthesized *in vitro*

Concentration, μM			Wild-type revertants/*am3* mutants,* $\times 10^6$	
dGTP	dATP	dTTP	Infective centers	Progeny phage
				2.0 ± 0.2[†]
				3.9 ± 0.2[‡]
				3.0 ± 0.4[§]
8.4	9.5	40	—	2.9 ± 0.5
93	9.5	40	—	54 ± 16
202	12.0	40	156 ± 66	119 ± 5
404	12.0	40	650 ± 100	1110 ± 80
810	12.0	40	1400 ± 200	3280 ± 180
910	12.0	10	1510 ± 330	1930 ± 150
910	12.0	40	2430 ± 470	4670 ± 350
910	12.0	160	3700 ± 800	4480 ± 320
910[¶]	12.0	40	2630 ± 280	4690 ± 980
910	60.0	40	686 ± 72	941 ± 46
910	119.0	40	341 ± 52	564 ± 100
910	238.0	40	121 ± 22	107 ± 13
1010	10.0	40	3380 ± 410	6620 ± 170

* Mean ± SEM from two to five determinations. A total of 18,286 mutants and 12,855 revertants were scored, generally at least 100 for each set.

† Parent phage.

‡ ssDNA from parent phage.

§ RF I DNA template.

¶ 400 μM dGDP was also present.

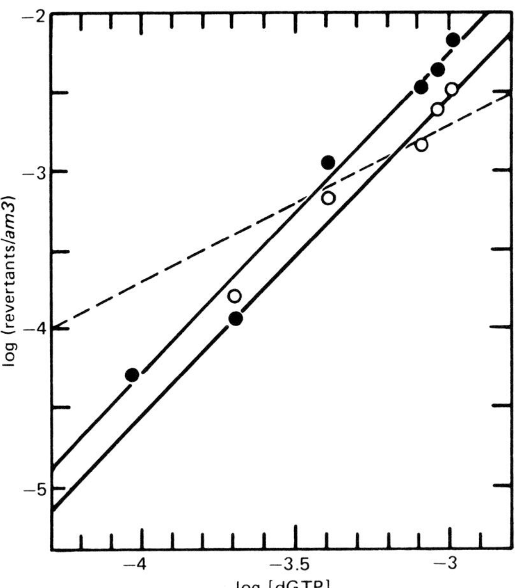

FIG. 1. Dependence of reversion frequency on [dGTP] at approximately constant [dATP] (9.5–12 μM). ●, Progeny phage assays; ○, infective center assays. Slope of solid lines = +2; slope of broken line = +1.

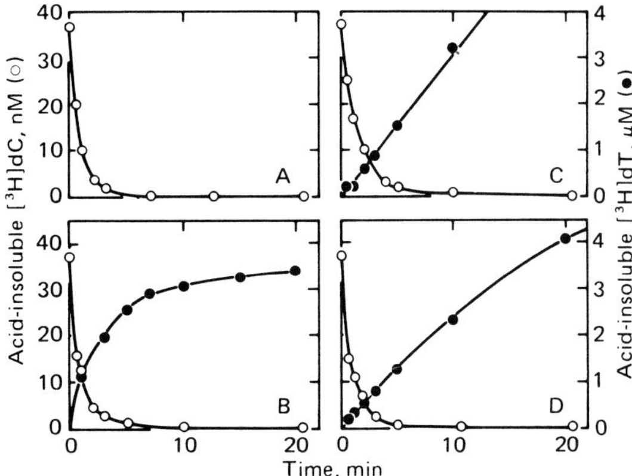

FIG. 3. 3′-5′ Exonuclease activity of DNA polymerase III holoenzyme as a proofreading function. 30°C, 75-μl reaction volumes. DNA polymerase III holoenzyme (600 ng, 200 units), $(dT)_{12}([^3H]dC)_{0.05}$ (665 pmol of nucleotide), Tris·HCl (50 mM, pH 7.5), dithiothreitol (5 mM), glycerol (11%, vol/vol), NaCl (12 mM), and bovine serum albumin (0.15 mg/ml). (*A*) 516 pmol of $(dA)_{600}$ and 0.8 μg of ssDNA-binding protein; (*B*) as *A* but plus 1.5 μmol of dTTP (○) and plus 1.5 μmol of [³H]dTTP (2 × 10⁷ cpm) (●); (*C*) as *B* but plus 6250 pmol of $(dA)_{300}$; (*D*) as *C* but without the DNA-binding protein. In all cases, the 3′-[³H]dC was completely excised in about 10 min under polymerizing conditions.

dTTP; a 16-fold increase in [dTTP] at 911 μM dGTP leads to no decrease in reversion frequency. The codon must therefore be converted to TGG, the wild type.

At the higher concentrations of dGTP, however, the mutation frequencies suggest that multiple mutations may occur. This should not affect the calculations. It is suggested later (Eq. 6) that mutations are strongly biased to changing A-G to G-G. Because there are only about 170 of these pairs in ϕX174 that lead to amino acid changes, the majority of phage at even the highest [dGTP] should have no mutations induced. In support of this, the yield of viable phage was found to be constant over the entire range of concentrations. In any case, the effects of multiple mutation leading to nonviable phage should cancel out in the ratio of revertants to *am3* because both are derived mainly from the progeny.

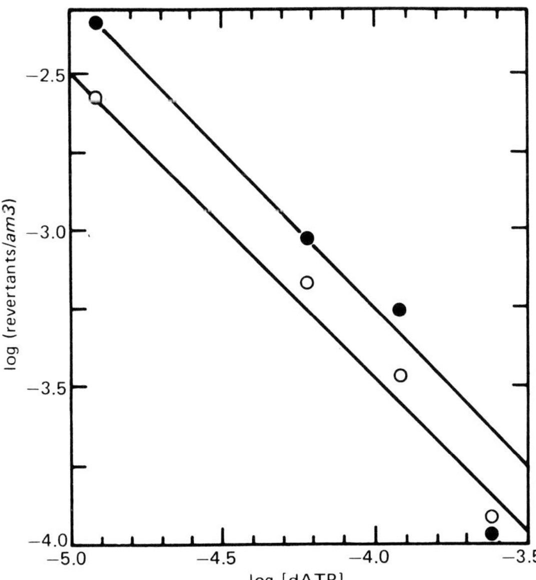

FIG. 2. Dependence of reversion frequency on [dATP] at constant [dGTP] (911 μM). ●, Progeny phage assays; ○, infective center assays. Slope of lines = −1.

3′-5′ Exonuclease Activity of DNA Polymerase III Holoenzyme Is a Proofreading Function. The 3′-5′ exonuclease activity of DNA polymerase I from *E. coli* appears to be a proofreading function for the removal of mismatched base pairs: mismatched bases at the 3′ terminus of synthetic primers are excised before elongation occurs (18). Although it has been shown that polymerase III from *E. coli* has the analogous 3′-5′ exonuclease activity (23, 24), it has not been shown that elongation occurs with the excision of the mismatches. The [³H]dC at the 3′ terminus of $(dT)_{12}([^3H]dC)_{0.05}$ annealed to $(dA)_{600}$ was excised under a variety of polymerizing conditions in the presence of dTTP (Fig. 3). This is not caused by the 5′-3′ activity of the enzyme (23, 24) because, on replacing $(dT)_{12}([^3H]dC)_{0.05}$ by $(dT)_{12}([^3H]dT)_2$, the radioactivity was retained under polymerizing conditions. It is unlikely that the excision was caused by an impurity of polymerase I because the enzyme was prepared from a *polA⁻polB⁻* mutant (11).

DISCUSSION

The replication of ϕX174 RF I DNA to ssDNA by the reconstituted enzyme system has several major advantages as a system in which to study fidelity and spontaneous mutation. (*i*) It uses the DNA polymerase III holoenzyme, believed to be the polymerase responsible for the major amount of replication of the *E. coli* chromosome. (*ii*) Only one strand of the template is copied and only one type of strand is produced, thus reducing complications in analyzing the effects of changing the concentrations of deoxynucleoside triphosphates. (*iii*) Multiple copies of the strand are produced, so that the progeny far outnumber the parent. (*iv*) The synthetic viral strands approximate the infectivity of authentic ϕX DNA.

Reversion Frequencies *In Vitro* Are Consistent with Those Found *In Vivo*. At low concentrations of dGTP, revertants are at background level, but at high concentrations of dGTP to favor the A → G transition, revertants are detected with a frequency of 0.0333 $[dGTP]^2/[dATP]$. Extrapolation of this formula to concentrations of dATP and dGTP found in *E. coli* ([dATP] ≈ 56 μM, [dGTP] ≈ 28 μM, and [dCTP] ≈ [dTTP] ≈ 50 μM; calculated from ref. 25 with a ratio of dry weight to wet weight = 0.25 and water space of *E. coli* = 72% (26)) predicts a reversion frequency of 0.5×10^{-6}, compared with an observed value of 1.7×10^{-6} (4).

Concentration Dependence of Reversion Frequency Is Consistent with a Proofreading Mechanism. If the 3′-5′ exonuclease activity of the polymerase acts during chain elongation as an editing mechanism to remove mismatched base pairs, then the simplest mechanism for misincorporation is the following:

$$\text{(primer)} \underset{\text{dNTP misaddition}}{\overset{k_1}{\longrightarrow}} \text{—N} \underset{\text{dN'TP elongation}}{\overset{k_{el}}{\longrightarrow}} \text{—N—N'} \quad (\text{—N} \xrightarrow{k_{ex}} \text{primer}) \qquad [3]$$

The fraction incorporated is given by the product of the relative rates of addition of incorrect (dNTP) and correct (dN″TP) triphosphates to the chain multiplied by the probability that elongation, rather than excision, occurs. For two substrates competing for a common site, the former term is given by $(k_{cat}/K_M)_N[dNTP]/(k_{cat}/K_M)_{N''}[dN''TP]$ (27, 28), in which k_{cat} and K_M are the usual Michaelis-Menten quantities and N″ refers to dN″TP, etc., and the latter term is given by $k_{el}/(k_{el} + k_{ex})$.

It may be reasoned intuitively that the elongation rate, and hence the proofreading efficiency, must depend to some extent on the concentration of the *next* nucleotide (dN′TP) to be added. For example, at one extreme, when the next nucleotide is absent from the mixture, elongation will not occur and so errors are bound to be excised by the 3′-5′ exonuclease activity. It may be shown (see *Appendix*) that at low concentrations of the following nucleotide, the ratio of incorrect to correct nucleotide misincorporated (ν_{mis}) is given by:

$$\nu_{mis} = \alpha \frac{[dNTP][dN'TP]}{[dN''TP]}. \qquad [4]$$

In the present study, both N and N′ = G and N″ = A. The observed concentration dependence of the reversion frequencies in this study is thus consistent with the 3′-5′ exonuclease activity being a proofreading function. The recent, independent observation that the misincorporation of 2-aminopurine into DNA with T4 polymerases increases with increasing concentrations of triphosphates is also consistent with this proposal (29).

Implications of Concentration Dependence of Error Frequency. If similar kinetic considerations apply *in vivo*, then spontaneous mutagenesis by nucleotide misinsertion depends crucially on the relative concentration of the correct and incorrect deoxynucleoside triphosphates and the absolute concentration of the following nucleotide to be inserted. This predicts that changes in the distribution of deoxynucleoside triphosphate pools will affect mutation rates. It is possible that some mutator genes function by changing the relative concentrations of nucleotides. The same criteria apply to the measurement of error rates *in vitro*. Eq. **4** and those below make predictions that can be tested in other systems and with other codons and that will aid in experimental design.

APPENDIX

Concentration dependence of misincorporation kinetics

To illustrate the intuitive arguments above, consider one of the simplest mechanisms for the elongation/excision step (Eq. **5**). The enzyme, having misinserted N instead of N″, may either

$$\text{E·—N} \underset{k_{-2}}{\overset{k_2}{\rightleftharpoons}} \text{—N E} \underset{k_{-3}}{\overset{k_3[dN'TP]}{\rightleftharpoons}} \text{E·—N·dN'TP} \xrightarrow{k_4} \text{E·—N—N'}; \quad \text{E·—N} \xrightarrow{k_{ex}} \qquad [5]$$

move to the new primer terminus (where binding of dN′TP and elongation occur) or excise N while at the previous terminus. For simplicity, it is assumed that the frequency of excision of correctly paired nucleotides is negligibly small. The ratio of excision to elongation may be calculated either by application of the steady-state assumption or by the method of Cleland (30). By the logic and notation described for Eq. **3**, it may be shown that:

$$\nu_{mis} = \frac{\beta\, k_2 k_3 k_4 [dN'TP]}{k_3 k_4 [dN'TP](k_2 + k_{ex}) + k_{-2} k_{ex}(k_{-3} + k_4)}, \qquad [6]$$

in which $\beta = (k_{cat}/K_M)_N[dNTP]/(k_{cat}/K_M)_{N''}[dN''TP]$.

The term in [dN′TP] follows saturation kinetics: at low [dN′TP], ν_{mis} is linear in [dN′TP] and reduces to Eq. **4**; at high [dN′TP], ν_{mis} becomes independent of [dN′TP] and tends to $\beta\, k_2/(k_2 + k_{ex})$. In the present study, the misincorporation of dGTP is still in the linear region of the curve at 1 mM, whereas the misincorporation of the unnatural 2-aminopurine (which is misinserted at a relatively high frequency) saturates at lower concentrations (29).

Mechanisms more explicit and complex than Eq. **5** (e.g., in refs. 29 and 31) may also be shown to give misincorporation kinetics similar to Eq. **6**.

The work reported in this paper was undertaken during the tenure of an American Cancer Society–Eleanor Roosevelt–International Cancer Fellowship awarded by the International Union Against Cancer. I am deeply indebted to Dr. Arthur Kornberg in whose laboratory this study was performed.

1. Fowler, R. G., Degnen, G. E. & Cox, E. C. (1974) *Mol. Gen. Genet.* **133**, 179–191.
2. Drake, J. W. (1970) *The Molecular Basis of Mutation* (Holden-Day, San Francisco).
3. Denhardt, D. T. & Silver, R. B. (1966) *Virology* **30**, 10–19.
4. Weymouth, L. A. & Loeb, L. A. (1978) *Proc. Natl. Acad. Sci. USA* **75**, 1924–1928.
5. Lin, C. C., Burke, R. L., Hibner, U., Barry, J. & Alberts, B. (1979) *Cold Spring Harbor Symp. Quant. Biol.* **43**, 469–487.
6. Sanger, F., Coulson, A. R., Friedmann, T., Air, G. M., Barrell, B. G., Brown, N. L., Fiddes, J. C., Hutchison, C. A., Slocombe, P. M. & Smith, M. (1978) *J. Mol. Biol.* **125**, 225–246.
7. Hutchison, C. A. & Sinsheimer, R. L. (1966) *J. Mol. Biol.* **18**, 429–447.
8. Eisenberg, S., Scott, J. F. & Kornberg, A. (1976) *Proc. Natl. Acad. Sci. USA* **73**, 3151–3155.
9. Eisenberg, S., Scott, J. F. & Kornberg, A. (1978) in *The Single Stranded DNA Phages*, eds. Denhardt, D. T., Dressler, D. & Ray, D. S. (Cold Spring Harbor Laboratory, Cold Spring Harbor, NY), pp. 287–302.
10. Randerath, K. & Randerath, E. (1964) *J. Chromatog.* **16**, 111–125.

11. McHenry, C. & Kornberg, A. (1977) *J. Biol. Chem.* **252,** 6478–6484.
12. Weiner, J. H., Bertsch, L. L. & Kornberg, A. (1975) *J. Biol. Chem.* **250,** 1972–1980.
13. Scott, J. F., Eisenberg, S., Bertsch, L. L. & Kornberg, A. (1977) *Proc. Natl. Acad. Sci. USA* **74,** 193–197.
14. Scott, J. F. & Kornberg, A. (1978) *J. Biol. Chem.* **253,** 3292–3297.
15. Shlomai, J. & Kornberg, A. (1978) *J. Biol. Chem.* **253,** 3305–3312.
16. Hirt, B. (1967) *J. Mol. Biol.* **26,** 365–369.
17. Eisenberg, S., Harbers, B., Hours, C. & Denhardt, D. T. (1975) *J. Mol. Biol.* **99,** 107–123.
18. Brutlag, D. & Kornberg, A. (1972) *J. Biol. Chem.* **247,** 241–248.
19. Guthrie, G. D. & Sinsheimer, R. L. (1963) *Biochim. Biophys. Acta* **72,** 290–297.
20. Simon, M., Davis, R. & Davidson, N. (1971) *Methods Enzymol.* **21,** 413–428.
21. Benzinger, R. (1978) *Microbiol. Rev.* **42,** 194–236.
22. Pagano, J. S. & Hutchison, C. A. (1971) *Methods Virol.* **5,** 79–123.
23. Livingston, D. & Richardson, C. (1975) *J. Biol. Chem.* **254,** 1748–1753.
24. McHenry, C. S. & Crow, W. (1979) *J. Biol. Chem.* **254,** 1748–1753.
25. Munch-Peterson, A. (1970) *Eur. J. Biochem.* **15,** 191–202.
26. Roberts, R. B., Cowie, D. B., Abelson, P. H., Bollon, E. T. & Britten, R. J. (1957) *Studies of Biosynthesis in Escherichia coli* (Carnegie Inst. Washington, Washington, DC), pp. 58–60.
27. Fersht, A. R. (1974) *Proc. R. Soc. Lond. Ser. B* **187,** 397–407.
28. Fersht, A. R. (1977) *Enzyme Structure and Mechanism* (Freeman, San Francisco), pp. 96–97.
29. Clayton, L. K., Goodman, M. F., Branscomb, E. W. & Galas, D. J. (1979) *J. Biol. Chem.* **254,** 1902–1912.
30. Cleland, W. W. (1974) *Biochemistry* **14,** 3220–3224.
31. Galas, D. J. & Branscomb, E. W. (1978) *J. Mol. Biol.* **124,** 653–687.

Chapter 7

Protein Engineering: A Once in a Lifetime Opportunity

'While site-directed mutagenesis was fun, it was really just the next phase of chemical modification and unlikely to revolutionize understanding of protein folding and enzymology.'

F. M. Richards, March 1985, the first UCLA Symposium on Protein Structure, Folding and Design (Keystone Colorado), reported by P. N. Bryan, Biochem Biophys Acta 1543, 2003–2222 (2000).

Clyde Hutchinson showed in 1971–1972[55,56] that the amino-acid sequence of a protein could be changed experimentally by targeted mutation with a mismatched oligonucleotide, using fragments of wild-type DNA annealing to the gene of an amber mutant from øX174 'marker rescue'. Then later, as first and corresponding author, he and Mike Smith in 1978 demonstrated site-directed mutagenesis using synthetic DNA[57]. But, site-directed mutagenesis was a technique that was languishing because there were precious few ideas of what to do with it. It was mooted that enzymologists could knock out catalytically important groups and confirm reaction mechanisms. But, two of us knew precisely what we wanted to do. Greg Winter, a brilliant young molecular biologist at the LMB, wanted to use site-directed mutagenesis as a tool to replace classical protein chemistry for investigating sequence homologies, and I realised that it was the precise tool I needed to study the whole gamut of how non-covalent interactions from side chains mediated specificity of binding, enzymatic catalysis and protein folding. I saw site-directed mutagenesis as a way to perturb gently the structure of a protein and acquire information just as chemists and physicists had used perturbations over the years. I would be able to analyse proteins using the principles of physical-organic chemistry. We decided to team up and kick off by analysing the tyrosyl-tRNA synthetase, a member of a class of enzyme on which both of us had worked. It was an enzyme of unknown mechanism, with no obvious catalytic groups. David Blow was in the process of solving its structure and we had access to the unpublished co-ordinates. I had all the tools for accurate biophysical study of the enzyme, and Greg had the molecular biology skills and was bubbling with ideas of his own. Within a few years, we had invented a whole series of new methods of analysing proteins that are in wide use today, from scanning of interactions to double mutant cycles, and we made a series of discoveries. We have quite different personalities but quickly became close, lifelong friends.

My strategy was simple. We would make small mutations that just deleted interactions between the enzyme and the substrate. We would then measure quantitatively the effects of the different types of non-covalent interactions on specificity and binding. Crucially, we used the $V_A/V_B = [A](k_{cat}/K_M)_A/[B](k_{cat}/K_M)_B$ equation as it eliminates all the artefacts whereby binding energy is dissipated in 'strain', induced-fit and non-productive binding mechanisms, but redrafted in the form $\Delta\Delta G_{\text{binding}} = -RT\ln((k_{\text{cat}}/K_{\text{M}})_{\text{wild-type}}/(k_{\text{cat}}/K_{\text{M}})_{\text{mutant}})$. Greg went to Canada to Mike Smith's lab to do the initial mutagenesis as he was apprehensive about doing it in Cambridge. Mike was away at the time and Mark Zoller instructed Greg. Greg came down from Cambridge to Imperial College with his precious vial of semi-purified mutated protein. The tyrosyl-tRNA synthetase from *Bacillus stearothermophilus* is a thermostable protein from a thermophile and so we could heat up the extract from its expression in *E. coli* to precipitate most host proteins, and crucially the host tyrosyl-tRNA synthetase so there was no background activity from it. Both of us hovered around Tony Wilkinson (my second student), who fortunately has a very calm temperament, as he made the vital measurements. It was a eureka moment. We had fully prepared in advance to assay the protein, and it worked beautifully: we had changed the activity of the protein. Had we done the obvious experiment and knocked out the activity rather than subtly change it, the result would have been of much less interest, and more open to questions of why it was inactive.

We dashed off the first paper to *Nature*[58], where a referee said that one mutation was not enough and we should make more! Of course, we had already done that.

7.1 2003: Tony Wilkinson and Alex Bullock.

7.2 2003: Greg Winter and Katy Brown.

Redesigning enzyme structure by site-directed mutagenesis: tyrosyl tRNA synthetase and ATP binding

Greg Winter*, Alan R. Fersht†, Anthony J. Wilkinson†, Mark Zoller‡ & Michael Smith‡

*MRC Laboratory of Molecular Biology, Hills Road, Cambridge CB2 2QH, UK
†Department of Chemistry, Imperial College of Science and Technology, London SW7 2AY, UK
‡Department of Biochemistry, Faculty of Medicine, University of British Columbia, 2146 Health Sciences Mall, Vancouver, British Columbia, Canada V6T 1W5

We describe here a general method for systematically replacing amino acids in an enzyme. This allows analysis of their molecular roles in substrate binding or catalysis and could eventually lead to the engineering of new enzymatic activities. The gene encoding the enzyme is first cloned into a vector from which the enzyme is expressed and is then mutated *in vitro* to change a particular nucleotide and hence the amino acid sequence of the enzyme. We have cloned the gene for the tyrosyl tRNA synthetase of *Bacillus stearothermophilus* into a vector derived from the single-stranded bacteriophage M13 to facilitate mutagenesis with mismatched synthetic oligodeoxynucleotide primers. From the recombinant M13 clone we have obtained high levels of the enzyme (~50% of soluble protein) expressed in the *Escherichia coli* host and have converted cysteine (Cys 35) at the enzyme's active site to serine. This leads to a reduction in enzymatic activity that is largely attributable to a lower K_m for ATP.

The tyrosyl tRNA synthetase (TyrTS) catalyses the aminoacylation of $tRNA^{Tyr}$ with tyrosine by a two-step mechanism in which tyrosine is first activated by ATP to form tyrosyl adenylate and then transferred to $tRNA^{Tyr}$ (refs 1, 2). The nucleotide sequence of the gene has been determined from a clone in the plasmid vector pBR322[3] and the enzyme is expressed in *E. coli* from its own promoter to about 10% of total soluble protein[4]. X-ray crystallographic analysis of the tyrosyl enzyme is complete to ~3 Å resolution[5] and the binding sites of tyrosine and ATP have been identified[6,7]. The target for mutagenesis was Cys 35[8], which is also present in the structures of the tyrosyl and methionyl tRNA synthetases of *E. coli*[9]. In the crystallographic model of the tyrosyl enzyme, this residue makes a contact with the 3' hydroxyl of the sugar moiety of tyrosyl adenylate, possibly via a hydrogen bond (D. Blow, personal communication). We decided to replace this cysteine by serine, a close structural analogue, expecting that a conservative change in a binding contact might alter, but not destroy, enzyme activity.

Initially the gene encoding the tyrosyl tRNA synthetase and its promoter were re-cloned from pBR322[4] into the single-stranded bacteriophage M13 to facilitate mutagenesis[10] (Fig. 1). The principle of oligonucleotide-directed mutagenesis relies on the extension of a primer by DNA polymerase (large fragment on a single-stranded circular template[11-24]. The primer, 5' CAAACCCGCT*GTAGAG 3', was complementary to the template except for a single internal mismatch (*) which directs the mutation Cys (TGC) to Ser (AGC). The mismatch is sited in the middle of the primer to protect it from exonuclease action and the precise sequence and length of the primer was chosen to minimize spurious matches elsewhere in the gene or in M13. The newly synthesized strand was sealed with ligase and *E. coli* cells transfected with the cloned circular double-stranded molecule.

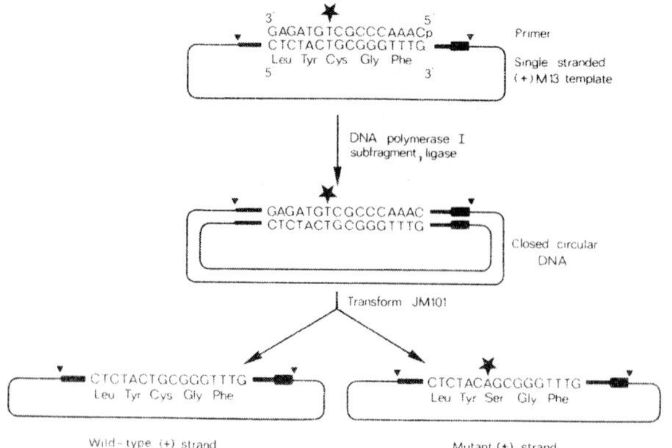

Fig. 1 Scheme for primer-directed *in vitro* mutagenesis of tyrosyl tRNA synthetase. The tyrosyl tRNA synthetase gene was cloned as a 2.5-kilobase(kb) insert into the *Bam*HI site of pBR322 after partial *Sau*3A digestion of *B. stearothermophilus* chromosomal DNA[4]. A 2.5-kB *Sal*I fragment was excized from the TyrTS–pBR322 recombinant plasmid and re-cloned into M13mp93 (J. Messing, unpublished) vector in the orientation indicated. The fragment contains the TyrTS gene (1,257 base pairs, bp) flanked by ~1,000 bp at the 5' end and 296 bp at the 3' end: the 3' portion includes 184 bp from pBR322 corresponding to the section between the *Bam*HI and *Sal*I sites of the plasmid[30]. ▼, The limits of the insert; the thickened line indicates the *B. stearothermophilus* DNA; ■, the portion of pBR322. Mutagenesis was essentially as described elsewhere[10]; 100 pmol of 5' CAAACCCGCTGTAGAG 3' primer (Applied Biosystems) was phosphorylated with 6 units of T4 kinase (BRL) in 30 µl 0.1 M Tris-HCl *p*H 8.0, 10 mM $MgCl_2$, 100 µM ATP, 5 mM dithiothreitol (DTT) at 37 °C for 50 min and then heated at 70 °C for 10 min. Phosphorylated primer (12 pmol) was annealed with 0.5 pmol of TyrTS-M13 template in 10 µl 20 mM Tris-HCl *p*H 7.5, 10 mM $MgCl_2$, 50 mM NaCl, 1 mM DTT at 55 °C for 5 min, then cooled to room temperature. The volume was adjusted to 20 µl 20 mM Tris-HCl *p*H 7.5, 10 mM $MgCl_2$, 25 mM NaCl, 0.5 mM DTT, 0.5 mM dTTP, 0.5 mM dCTP, 0.5 mM gGTP, 0.5 mM ATP, and the primer extended for 5 min with 2 units of DNA polymerase I large fragment (BRL) at limiting dATP (2.5 µM, 10 µCi [α-^{32}P]dATP, NEN) in the presence of 3 units of T4 DNA ligase (BRL). dATP was then increased to 0.5 mM, the reaction incubated at 15 °C for 18 h and then stopped by adding 100 µl 10 mM Tris-HCl *p*H 8.0, 10 mM EDTA. After phenol extraction, an equal volume of 13% polyethylene glycol, 1.6 M NaCl was added to the aqueous phase and the sample was kept on ice for 10 min[31]. DNA was separated from unincorporated [α-^{32}P]dATP label by a 10-min spin in an Eppendorf microfuge and the DNA pellet was dissolved in 200 µl 0.2 M NaOH. Most of the label (80%) had been incorporated into DNA. After 5 min at room temperature the sample was applied to a 5 ml alkaline sucrose gradient (5–20% sucrose, 1 M NaCl, 0.2 M NaOH, 2 mM EDTA) and centrifuged in a Beckman SW50.1 rotor at 37,000 r.p.m., 4 °C for 2 h. Aliquots (6 drops) were collected from the bottom of the tube; the closed circular (cc) DNA (19% incorporated [α-^{32}P]dATP) runs ahead of single-stranded linear and circular molecules. The ccDNA was neutralized with 1 M Tris-citrate, dialysed against 10 mM Tris-HCl *p*H 8.0, 0.1 mM EDTA and used to transform the *E. coli* strain JM101[32] as described elsewhere[33]; 1% of the ccDNA yielded ~500 plaques.

Single-stranded phage was isolated from individual plaques, spotted on to nitrocellulose and the mutants identified by hybridization with 5' ^{32}P-labelled mutagenic primer. At room temperature the primer anneals to both mutant and wild-type DNA but by washing at a higher temperature the primer can be selectively dissociated from wild-type DNA[15,16]. A short wash at $T_D + 2$ °C (T_D is the dissociation temperature calculated by the rule $T_D = 2°$ per dA · dT and 4° per dG · dC nucleotide pair[17]) was sufficient to remove selectively most mismatched primer from wild-type DNA. Of the 36 plaques screened, at least 16 contained mutant phage (Fig. 2). Phage from two of these samples were plaque-purified to ensure the complete segregation of wild-type and mutant molecules, and re-hybridized with primer to identify pure mutant plaques. The required mutation was confirmed by dideoxy sequencing[18] of the pure mutant template with a restriction fragment primer adjacent to the site of mutagenesis (not shown). The directed nature of the mutagenesis should exclude mutations at other sites, therefore we did not sequence the rest of the mutant gene. Neverthe-

Reprinted from Nature, Vol. 299, No. 5885, pp. 756-758, 21 October 1982

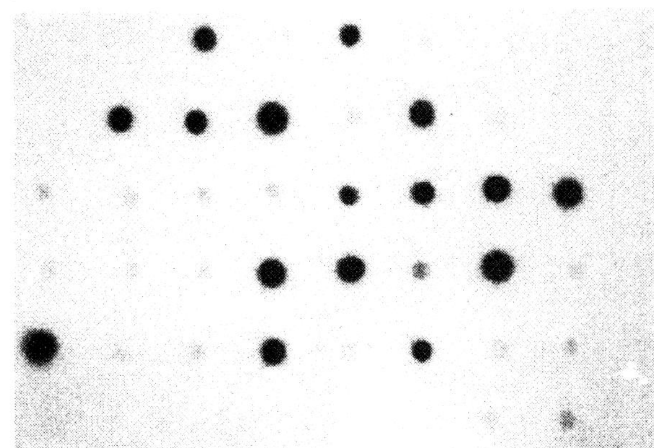

Fig. 2 Screening mutant phage by dot-blot hybridization: 36 plaques were grown in 1-ml cultures and phage prepared as described elsewhere[34]. Phage were dissolved in 50 μl water and for hybridization[35], 1 μl of each was spotted on to nitrocellulose (Schleicher & Schull) and the filter baked at 80 °C for 1 h. After pre-hybridization with 6 × SSC, 10 × Denhardt's solution[36], 0.2% SDS at 67 °C for 1 h, the filter was rinsed with 50 ml of 6 × SSC. The filter was hybridized at room temperature for 1 h by leaving it in an open Petri dish containing 2 ml of 6 × SSC, 10 × Denhardt's solution and 10^6 c.p.m. (Cerenkov) probe. The probe was prepared by treating 20 pmol of mutagenic primer with kinase as described in Fig. 1 legend, except that the following were used: 2 units T4 kinase, no cold ATP and 10 μCi [γ-^{32}P]ATP (>2,000 Ci $mmol^{-1}$; NEN), followed by desalting on a 5-ml column of Sephadex G25F. The filter was rinsed in 50 ml 6 × SSC for 1 min and placed between two sheets of Saranwrap (Dow). Autoradiography for 1 h revealed that each clone had hybridized to the probe. The filter was then washed in pre-warmed 6 × SSC for 4 min at 52 °C (this temperature corresponds to $T_D + 2$ °C). (Previously, we washed the filters at several intermediate temperatures to selectively remove the primer[10], and although the $T_D + 2$ °C rule has worked for several primers we do not know whether it is general.) Autoradiography overnight revealed that 16 of the 36 phage preparations contained mutant; two of these preparations were plaque-purified and six plaques from each were grown up as described above and re-screened to identify pure mutant phage.

Table 1 Activity of TyrTS in cell extracts

Extract of JM101 infected with	TyrTS (nmol of active sites)*	Pyrophosphate exchange rate (nmol s^{-1})	Amino-acylation rate (nmol s^{-1})
M13mp93	<0.02	0.5†	<0.05
M13mp93 + 2.8 nmol of purified TyrTS‡	2.8	7.4	2.7
TyrTS–M13	2.8	7.1§	3.2
TyrTS (Cys→Ser)–M13	2.9	2.1†	1.0

Total enzymatic activities in the extract of 20-ml cultures of infected cells. All kinetic experiments were performed at 25 °C in a standard buffer containing 144 mM Tris-HCl (*p*H 7.8), 10 mM $MgCl_2$, 0.1 mM phenylmethylsulphonyl fluoride and 10 mM 2-mercaptoethanol. The active site titration mixture[23] contained, in addition, 2 mM ATP, 10 μM ^{14}C-Tyr (770 c.p.m. $pmol^{-1}$) and 10 U ml^{-1} inorganic pyrophosphatase. The pyrophosphate exchange[29] mixture contained 2 mM ATP, 2 mM ^{32}P-PP_i and 50 μM tyrosine. The aminoacylation mixture contained 20 μM ^{14}C-Tyr (770 c.p.m. $pmol^{-1}$), 2 mM ATP, 20 μM $tRNA^{Tyr}$ (*B. stearothermophilus*, 200 pmol per A_{260} tyrosine acceptance) and 10 U ml^{-1} inorganic pyrophosphatase. One equivalent of Mg^{2+} was added together with ATP to maintain free Mg^{2+} at 10 mM.

* Determined by active site titration.

† In the absence of added tyrosine there was an exchange rate of 0.45 nmol s^{-1}.

‡ A sample of TyrTS purified from *B. stearothermophilus* was added to the extract of M13mp93-infected cells in the same proportion as produced in the cloned strains. Addition of extract had no effect on active site titration or aminoacylation rate and merely added the background of 0.4 nmol s^{-1} to the pyrophosphate exchange rate.

§ In the absence of added tyrosine there was an exchange rate of 1.4 nmol s^{-1}.

less we showed by kinetic analysis (see below) that extracts from the two independently cloned mutant plaques had the same specific enzymatic activities.

E. coli infected with wild-type or mutant phage was then checked for expression of the enzyme. A sonicate of cells revealed that the tyrosyl tRNA synthetase now constituted >50% of the soluble protein in infected cells (Fig. 3). Such high levels of expression either suggest that the synthetase promoter is strong or reflect the high copy number of M13 replicative form (RF) in the cell. The relative copy number of the vectors might account for the increased expression on re-cloning the gene from pBR322 (20 plasmid copies per cell[19]) into M13 (200 RF copies per cell[20]). Whatever the causes of the high expression in M13, it is most useful and may have wider applications.

Aminoacylation assays on the cell extracts revealed that both mutant and wild-type extracts contained active enzyme, but that the extract from the mutant had a lower activity. The retention of significant activity in the mutant extract indicates that the enzyme does not strictly require a sulphydryl group at this position. Furthermore, as Cys 35 is the only conserved cysteine in the sequences of the tyrosyl tRNA synthetases of *B. stearothermophilus*[3] and *E. coli*[21] it seems that thioacyl intermediates[22] are probably not involved in the mechanism of these enzymes. For a more detailed kinetic analysis we measured the amount of enzyme [E_0] in each crude extract by active site titration involving the formation of tyrosyl adenylate[23]. The specific activity (V_0/[E_0]) of the pyrophosphate exchange and aminoacylation reactions (Table 1) was thus calculated and the results showed that the mutant enzyme was intrinsically less active than the wild type. Further work indicated that on mutation, the value of K_m for ATP in aminoacylation increased from

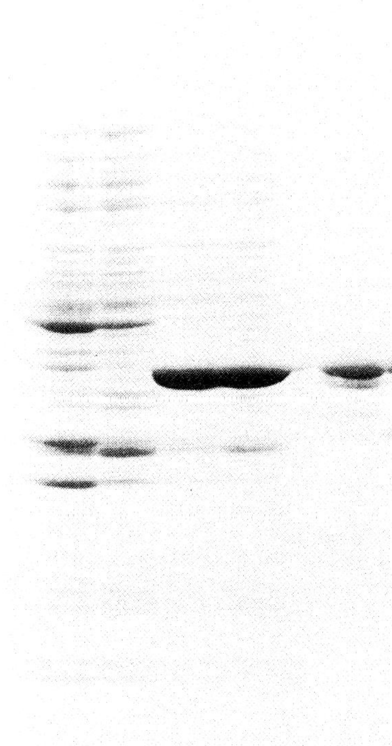

Fig. 3 Expression of tyrosyl tRNA synthetase from M13 in *E. coli*; 1 ml phage stocks in 2 × TY media were prepared from a single plaque as described elsewhere[34] and 50 μl were used to inoculate 20 ml JM101 (D_{550} = 0.05) in 2 × TY. Cells were grown for ~5 h until D_{550} = 1 and then 1 ml of lysate was prepared by freeze-thaw and sonication[4]; 50 μl were analysed on a 10% SDS-polyacrylamide gel[37]. Sonicates of: *a*, JM101 cells; *b*, JM101 infected with M13mp93; *c*, JM101 infected with wild-type TyrTS–M13; *d*, JM101 infected with mutant TyrTS–M13; *e*, TyrTS (*B. stearothermophilus*) marker purified as described in ref. 38.

0.9 to 4.1 mM while V_{max} was reduced (1.4 to 0.9 s^{-1}). The K_m for tyrosine in aminoacylation was, however, unchanged (15 μm).

The fact that the lesion has no effect on tyrosine binding but does affect the binding of ATP is certainly consistent with the location of Cys 35 at the ATP-binding site in the crystallographic model. Why the substitution of serine for cysteine should result in reduced ATP binding is not clear from this experiment alone. Although serine is smaller than cysteine[24,25] and may therefore not make an effective van der Waals' contact with the 3' hydroxyl of the sugar, it can form much stronger hydrogen bonds[26,27]. However, as has been discussed elsewhere[28], such capability need not confer improved binding as the S—H—O and O—H—O hydrogen bonds have slightly different geometries[24,25] and the OH group of Ser 35 could bind water molecules more strongly than does the SH of Cys 35. In fact the serine hydroxyl probably contributes no net binding energy, as the kinetics of a Gly 35 mutant are essentially identical to those of the Ser 35 mutant (unpublished work). Thus, the reduced energy of ATP binding on converting Cys 35 to serine is probably due to the loss of the van der Waals' contact. We therefore envisage that the molecular role of Cys 35 is to make optimal contact with the sugar moiety of the ATP.

Further mutagenesis experiments are planned to delineate the role of individual residues in binding ATP, tyrosine and tRNA and in the catalytic mechanism. The use of M13 vectors for both mutagenesis and expression of the mutant enzyme, and the use of active site titration to measure accurately the amount of enzyme in a crude sonicate of *E. coli*, will facilitate this approach.

We thank Dr D. G. Barker for the TyrTS-pBR322 clone, Dr J. Messing for the vector M13mp93 and Professor D. M. Blow for making available unpublished data and the crystallographic model of TyrTS. M.S. is a career investigator of the MRC of Canada and the work was supported by the MRC of the UK and of Canada.

Received 26 July; accepted 7 September 1982.

1. Loftfield, R. B. *Prog. Nucleic Acids Res.* **12,** 87–128 (1972).
2. Fersht, A. R. & Jakes, R. *Biochemistry* **14,** 3350–3356 (1975).
3. Winter, G., Hartley, B. S., Koch, G. L. E. & Barker, D. G. (in preparation).
4. Barker, G. *Eur. J. Biochem.* **125,** 357–360 (1982).
5. Bhat, T. N., Blow, D. M., Brick, P. & Nyborg, J. *J. molec. Biol.* **158,** 699–709 (1982).
6. Monteilhet, C. & Blow, D. M. *J. molec. Biol.* **122,** 407–417 (1978).
7. Rubin, J. & Blow, D. M. *J. molec. Biol.* **145,** 489–500 (1981).
8. Winter, G. P., Koch, G. L. E., Dell, A. & Hartley, B. S. in *Transfer RNA: Structure Properties and Recognition* (eds Schimmel, P. R., Soll, D. & Abelson, J. N.) 255–265 (Cold Spring Harbor Laboratory, New York, 1979).
9. Barker, D. G. & Winter, G. *FEBS Lett.* **145,** 191–193 (1982).
10. Zoller, M. & Smith, M. *Meth. Enzym.* (in the press).
11. Hutchison, C. A. III *et al. J. biol. Chem.* **253,** 6551–6560 (1978).
12. Smith, M. & Gillam, S. in *Genetic Engineering* Vol. 3 (eds Setlow, J. K. & Hollaender, A.) 1 (Plenum, New York, 1981).
13. Wasylyk, B. *et al. Proc. natn. Acad. Sci. U.S.A.* **77,** 7024–7028 (1980).
14. Miyada, C. G., Soberon, X., Itakura, K. & Wilcox, G. *Gene* **17,** 167–177 (1982).
15. Gillam, S., Waterman, K. & Smith, M. *Nucleic Acids Res.* **2,** 625–634 (1975).
16. Wallace, R. B., Schold, M., Johnson, M. J., Dembek, P. & Itakura, K. *Nucleic Acids Res.* **9,** 3647–3657 (1981).
17. Suggs, S. V., Wallace, R. B., Hirose, T., Kawashima, E. & Itakura, K. *Proc. natn. Acad. Sci. U.S.A.* **78,** 6613–6617 (1982).
18. Sanger, F., Nicklen, S. & Coulson, A. R. *Proc. natn. Acad. Sci. U.S.A.* **74,** 5463–5467 (1977).
19. Clewell, D. B. & Helinski, D. R. *J. Bact.* **110,** 1135–1146 (1972).
20. Hohn, B., Lechner, H. & Marvin, D. A. *J. molec. Biol.* **56,** 143–154 (1971).
21. Barker, D. G., Bruton, C. J. & Winter, G. (in preparation).
22. McElroy, W. D., De Luca, M. & Travis, J. *Science* **157,** 151–160 (1967).
23. Fersht, A. R. *Biochemistry* **14,** 5–12 (1975).
24. Kerr, A. K., Ashmore, J. P. & Koetzle, T. F. *Acta crystallogr.* **B31,** 2022–2026 (1975).
25. Kistenmacher, T. J., Rand, G. A. & Marsh, R. E. *Acta Crystallogr.* **B30,** 2573–2578 (1974).
26. Crampton, M. R. in *Chemistry of the –SH group* (ed. Patai, S.) 379–415 (Wiley-Interscience, New York, 1974).
27. Paul, I. C. in *Chemistry of the –SH group* (ed. Patai, S.) 111–149 (Wiley-Interscience, New York, 1974).
28. Fersht, A. R. & Dingwall, C. *Biochemistry* **18,** 1245–1249 (1979).
29. Heinrikson, R. L. & Hartley, B. S. *Biochem. J.* **105,** 17–24 (1967).
30. Sutcliffe, J. G. *Cold Spring Harb. Symp. quant. Biol.* **43,** 77–90 (1978).
31. Hong, G. F. *Biosci. Rep.* **1,** 243–252 (1981).
32. Messing, J. *Recombinant DNA Tech. Bull.* **2,** 43–48 (1979).
33. Cohen, S. N., Chang, A. C. Y. & Hsu, L. *Proc. natn. Acad. Sci. U.S.A.* **69,** 2110–2114 (1972).
34. Winter, G. & Fields, S. *Nucleic Acids Res.* **8,** 1965–1974 (1980).
35. Grunstein, M. & Hogness, D. S. *Proc. natn. Acad. Sci. U.S.A.* **72,** 3961–3965 (1975).
36. Denhardt, D. T. *Biochem. biophys. Res. Commun.* **23,** 641–646 (1966).
37. Laemmli, U. K. *Nature* **227,** 680–683 (1970).
38. Atkinson, A. *et al. J. appl. Biochem.* **1,** 247–258 (1979).

Greg and I knew immediately that we had done something special: we had made the first site-directed mutagenesis of a residue in a protein of known structure; and we had made the first genetic redesign of a protein. It was only a small step, but we had founded a new area of 'protein engineering'. And, as we both had well-defined strategies, nothing would have held us back. Greg powered forward with making useful novel proteins and founded antibody engineering, and I pushed forward as a protein pathologist, dissecting the structure, activity and folding of proteins. Two papers appeared shortly after our *Nature* paper. One by Jack Richards simply inactivated β-lactamase[59] by mutating the active site nucleophilic serine as did the other paper[60]. The author of the third paper, Irwin Sigal, who also made an inactivating mutation, was one of the victims of Abdul Baset Ali al-Megrahi, being killed six years later in PanAm 103 at Lockerbie when only 35 years old. The work on lactamases fizzled out because it did not have any long-term strategy, other than being a proof of principle.

Tony Wilkinson wrote after seeing a draft of the above:

> 'I will never forget Fersht and Winter poring over the scintillation counter as the print-out emerged on the Cys35 TyrRS mutant.
>
> I did not know about the *Annu Rev Biochem* rejection by Boyer. It makes me wonder, had your review been accepted would I have joined you as a Ph.D. student? Your book was very highly recommended at UCL where Mike Holloway and Bob Rabin taught us enzymology and in my case thoroughly inspired me. Another factor of course for choosing to come to work with you was an inspirational seminar you gave to the Biochemistry Department there, when I was in my third year. It led to a very lively discussion/dispute in which of course you prevailed. I remember the strength of your knowledge and enthusiasm and your eagerness to persuade through reason. This chimes with some of the comments you make about the stimulus of being challenged by counter-argument and debate. It is also inspiring to young scientists to witness intense scientific dialogue. Needless to say I have never regretted my Ph.D. decision and my time at Imperial is filled with happy memories.'

Over the next ten years, we published 16 papers in *Nature*, and I cannot recall a rejection. We had a technological and intellectual lead and, unusually, the editors and referees were also enthusiastic. But, some of the best work was reserved for *Biochemistry* and *Journal of Molecular Biology* (*JMB*). Although these are not very high impact factor journals, both have excellent refereeing and editing and allow the presentation of a full case for ones' work at length and with all the experimental details and data. The subsequent number of citations to the work in those is generally as high as to the *Nature* articles.

Our choice of protein had been almost perfect. The one drawback was that David Blow had no interest in solving the crystal structures of the mutants as he thought point mutants were uninteresting. Eventually, his young American graduate student, Katy Brown, now a faculty member at Imperial College, camped in my lab and solved mutant structures.

Mike Smith was awarded half the Nobel Prize in Chemistry in 1993, for site-directed mutagenesis, the other half to Kary Mullis for the polymerase chain reaction. Clyde Hutchison was elected to the National Academy of Sciences in 1995, and Mike Smith in 1996.

My one and only research paper in *Cell*[61] was unusual for that journal, and very influential. Greg and I introduced double-mutant cycles to detect interactions between residues in proteins. These cycles have been used extensively in studies of catalysis, binding and protein folding, and for monitoring conformational changes in a wide range of processes, including structures of membrane channels.

7.3 1989: Spetses Summer School. Greg Winter, Caroline Winter, Alan and Jim Wells. Alan organised a series of Summer Schools on the Greek Island of Spetses (Spetsai).

7.4 2000: Paul Carter.

Cell, Vol. 38, 835–840, October 1984, Copyright © 1984 by MIT

The Use of Double Mutants to Detect Structural Changes in the Active Site of the Tyrosyl–tRNA Synthetase (Bacillus stearothermophilus)

Paul J. Carter,* Greg Winter,*
Anthony J. Wilkinson,†
and Alan R. Fersht†
* Medical Research Council
Laboratory of Molecular Biology
Hills Road
Cambridge CB2 2QH, England
† Department of Chemistry
Imperial College of Science and Technology
London SW7 2AY, England

Summary

In a previous study, a mutant of tyrosyl–tRNA synthetase in which a threonine residue (Thr51) was converted to proline dramatically improved the affinity of the enzyme for its ATP substrate. How does Pro51 improve the enzyme's affinity for ATP? A priori, Pro51 might interact directly with the ATP, or it might distort the polypeptide backbone and thereby force new or improved contacts elsewhere from the enzyme to ATP. By making mutants of the Pro51 enzyme at two residues that make hydrogen bonds to the ATP substrate, we show that Pro51 greatly improves the strength of one of these contacts. Thus the propagation of a structural change in an enzyme induced by mutation may be detected by the introduction of further mutations.

Introduction

Enzymic catalysis depends on the binding of both the substrate and its transition state to the enzyme. Three-dimensional structures of the active sites of enzymes and of enzyme–substrate complexes can allow us to predict the probable roles of contact residues: for example, the binding of substrate and transition state is often assisted by strong electrostatic interactions and hydrogen bonds with buried polar groups of the enzyme. Site-directed mutagenesis (Hutchison et al., 1978) allows detailed analysis of the roles of such contacts in catalysis, as the side chains at the active site can be systematically replaced. We are using this technique to dissect the structure and activity of the tyrosyl–tRNA synthetase from Bacillus stearothermophilus (Winter et al., 1982). This enzyme catalyzes the aminoacylation of $tRNA^{Tyr}$ in a two-step reaction (Fersht and Jakes, 1975) in which the tyrosine is activated to give enzyme-bound tyrosyl adenylate (reaction 1) and then the tyrosyl moiety is transferred to $tRNA^{Tyr}$ (reaction 2).

$$E + Tyr + ATP \rightleftharpoons E \cdot Tyr\text{-}AMP + PP_i \quad (1)$$

$$E \cdot Tyr\text{-}AMP + tRNA^{Tyr} \rightleftharpoons Tyr\text{-}tRNA^{Tyr} + AMP + E \quad (2)$$

Our studies depend on knowledge of the crystal structure of tyrosyl–tRNA synthetase (Bhat et al., 1982) and of the enzyme-bound tyrosyl adenylate (Rubin and Blow, 1981). This allows us to identify possible hydrogen bonds between enzyme and tyrosyl adenylate (D. M. Blow, personal communication). The adenylate binds in a cleft formed between a strand of β-sheet (residues 31–38) and an α-helix (residues 47–57) such that the side chains of Cys35, His48, and Thr51 may form hydrogen bonds with the ribose (Figure 1). Cys35 and His48 are conserved in the primary sequences of tyrosyl–tRNA synthetase (B. stearothermophilus) and methionyl tRNA synthetase (E. coli) (Barker and Winter, 1982). Since the active sites of these two enzymes are homologous in three dimensions (Zelwer et al., 1982; Bhat et al., 1982), and since Cys35 and His48 occur in corresponding positions in both structures (Blow et al., 1983), these residues are probably important in catalysis and substrate binding. By contrast, Thr51 is not conserved in the methionyl–tRNA synthetase or in the tyrosyl–tRNA synthetase (TyrTS) from E. coli, where it is replaced by glutamic acid and proline respectively. Furthermore, the possible hydrogen bond from Thr51 to ribose in the crystal structure is too long to be strong.

We can measure the importance of Cys35, His48, and Thr51 as substrate contacts by replacing them with residues that do not form hydrogen bonds. For example, converting Cys35 to glycine, thereby completely removing the side chain at residue 35, reduced the affinity of the enzyme for ATP. As the tyrosyl–tRNA synthetase follows Michaelis–Menten kinetics, it is possible to deduce the energy changes in transition state binding from the steady state kinetics (Wilkinson et al., 1983). Thus, by comparing the k_{cat}/K_m values of mutant (Gly35) and wild-type (Cys35) enzymes, we found that the side chain of cysteine contributes 1.2 kcal/mole to transition state binding (Wilkinson et al., 1983). In contrast, the removal of the possible hydrogen bonding donor from Thr51 by converting it to alanine resulted in a small improvement in the affinity of the enzyme (0.4 kcal/mole) for ATP. This is consistent with Thr51's exchanging a good hydrogen bond with water in the free enzyme for a poor one with ATP when it forms the enzyme–substrate complex (Wilkinson et al., 1984). However, on converting Thr51 to Pro (as in the E. coli enzyme) we obtained a massive improvement in the affinity of the enzyme for ATP (Wilkinson et al., 1984). Unlike the other mutations that have been introduced, which are relatively conservative, the proline residue must destabilize the α-helix (residues 47–57).

In principle, the increased affinity for ATP could arise from a new contact between enzyme and substrate or from strengthening an existing contact. We envisaged that the distortion of the α-helix by Pro51 may alter the position of the His48 side chain and so increase its interaction with the ATP (Wilkinson et al., 1984). To test this, we have compared the affinity of the wild-type enzyme TyrTS(His48–Thr51) with the single mutants TyrTS(Gly48) and TyrTS(Pro51) and the double mutant TyrTS(Gly48–Pro51). We show that the effect of Pro51 on substrate affinity relies completely on the presence of the imidazole group at position 48. In addition, by constructing another double mutant, Gly35–Pro51, we show that Pro51 has only

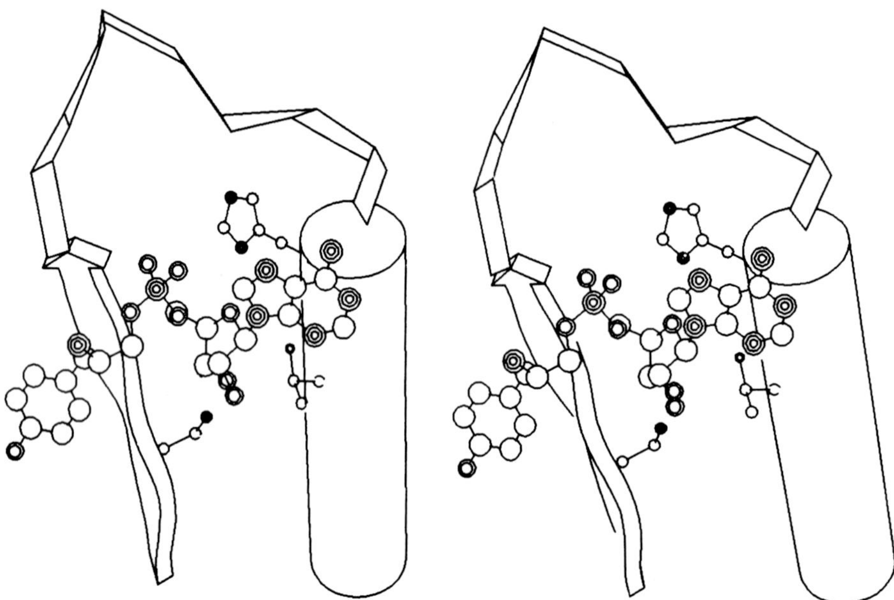

Figure 1. A Stereo Pair of Part of the Active Site of TyrTS (B. stearothermophilus)

Depicted is the cleft formed between the β-sheet, represented by an arrow (residues 31–38), and α-helix, represented by a cylinder (residues 47–57). Cys35 is located in the middle of the β-sheet and interacts with the 3′ OH of the ribose. His48, at the N terminus of the helix, and Thr 51, further down the helix, are H-bonded to the ring oxygen of the ribose.

a small effect upon the interaction of Cys35 and the substrate. Thus a second mutation introduced in the active site of an enzyme can be used to probe the extent of the structural changes introduced by the first mutation.

Results

Kinetics of Amino Acid Activation and tRNA Charging

For variation in [ATP] at saturating levels of tyrosine, both k_{cat} and K_m are affected in the various mutant constructions, as would be expected for mutations in the adenylate binding site (Table 1).

For variation in [tyrosine] in the pyrophosphate exchange reaction note that the levels of ATP (2 mM) are not saturating (except for the Pro51 and Gly35-Pro51 mutants). Consequently the values of k_{cat} for variation in [tyrosine] (Table 2) are generally lower than those for variation in [ATP] (Table 1). However, for the Pro51 and Gly35–Pro51 mutants, which have a high affinity for ATP and for which 2 mM ATP is saturating, the two sets of k_{cat} values are identical.

The K_m values for tyrosine (Table 2) in all the mutants are very similar: as the K_m is a complex term that incorporates rate constants for the chemical steps as well as the dissociation constant for tyrosine, the small differences could arise from the changes at the ATP site. Thus, in the activation of amino acid and tRNA charging, the lesions constructed at the ATP site affect mainly that site.

Table 1. ATP Dependence of Pyrophosphate Exchange and tRNA Charging

Mutant (in italics)	k_{cat} (s^{-1})	K_m (mM)	k_{cat}/K_m (s^{-1} M^{-1})
Cys35 His48 Thr51	7.6 [4.7]	0.9 [2.5]	8,400 [1,860]
Cys35 *Gly48* Thr51	1.9 [2.4]	1.3 [8.7]	1,460 [278]
Gly35 His48 Thr51	2.8 [1.9]	2.6 [6.1]	1,120 [320]
Cys35 His48 *Pro51*	12.0 [1.8]	0.058 [0.019]	208,000 [95,800]
Cys35 *Gly48 Pro51*	2.9 [3.4]	2.1 [6.7]	1,380 [507]
Gly35 His48 *Pro51*	10.3 [2.3]	0.181 [0.051]	56,900 [44,000]
Gly35 Gly48 Thr51	1.32 [1.3]	8.0 [11.6]	177 [110]

All kinetic experiments were performed at 25 ± 0.1°C in a standard buffer containing 144 mM Tris-HCl (100 mM Tris-HCl; 44 mM Tris, pH 7.78), 10 mM $MgCl_2$, 10 mM 2-mercaptoethanol, and 0.1 mM phenylmethanesulfonyl fluoride. [Tyrosine] was 50 μM in pyrophosphate exchange and 100 μM in the charging assays. The figures in brackets correspond to values for tRNA charging. Data taken from this work, from Wilkinson et al., 1983, 1984, and from Winter et al., 1982.

Gly48 Mutant

His48 makes a possible hydrogen bond contact with the ring oxygen of the ribose (Bhat et al., 1982), and mutation to Gly48 reduces k_{cat} and increases the K_m for variation of

Table 2. Tyrosine Dependence of Pyrophosphate Exchange and tRNA Charging

Mutant (in italics)	k_{cat} (s^{-1})	K_m (μM)	k_{cat}/K_m (s^{-1} M^{-1}, $\times 10^{-5}$)
Cys35 His48 Thr51	6.0 [1.2]	2.4 [2.0]	25.0 [6.0]
Cys35 *Gly48* Thr51	1.2 [0.35]	3.2 [3.8]	3.8 [0.9]
Gly35 His48 Thr51	1.0 [0.7]	2.7 [2.0]	3.7 [3.5]
Cys35 His48 *Pro51*	12.4 [0.8]	1.7 [3.1]	73.0 [2.6]
Cys35 *Gly48 Pro51*	1.3 [0.89]	2.9 [2.0]	4.5 [4.5]
Gly35 His48 *Pro51*	10.5 [1.13]	1.21 [2.7]	87.0 [4.2]
Gly35 Gly48 Thr51	0.33 [0.31]	6.15 [3.9]	0.5 [0.8]

Data taken from this work, from Wilkinson et al., 1983, 1984, and from Winter et al., 1982. [ATP] was 2 mM in the pyrophosphate exchange and 10 mM in the charging assays. Brackets: tRNA charging values.

[ATP] (Table 1). By comparing the k_{cat}/K_m terms for the Gly48 mutant with the His48 enzyme, using the equation

$$\Delta G = -RT\ln\left(\frac{k_{cat}/K_m \text{ mut}}{k_{cat}/K_m \text{ wild}}\right)$$

we can deduce (Wilkinson et al., 1983) that the interaction energy of the Gly48 enzyme with the transition state is reduced by 1.2 kcal/mole. This suggests that His48 makes a contact with the ATP worth 1.2 kcal/mole (Figure 2a).

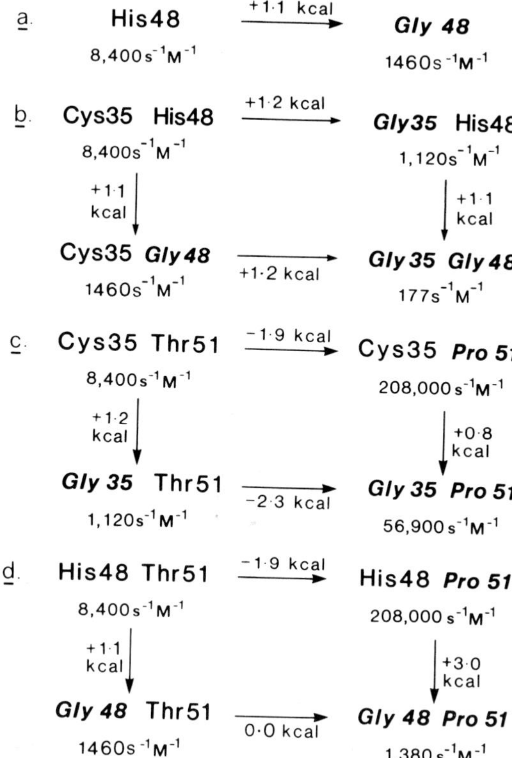

Figure 2. The Energy of Interaction of Each Side Chain with the Transition State (kcal/mole) in the Amino Acid Activation Reaction

Calculations were from k_{cat}/K_m terms (in $s^{-1}M^{-1}$ units in figure), as described in Wilkinson et al., 1983. (See legend to Figure 3.)

Gly35–Gly48 Double Mutant

The kinetics of the purified enzyme (Figure 2b) suggest that the mutations Gly35 and Gly48 are independent: the interaction energy of the sulfhydryl side chain of Cys35 with the transition state is identical within experimental error (1.2 kcal/mole) in the His48 and Gly48 enzymes. Likewise, the interactions of the imidazole side chain of His48 are identical (1.1 kcal/mole) in the Cys35 and Gly35 enzymes. Thus the loss in interaction energy of the Gly35–Gly48 enzyme with transition state (2.3 kcal/mole) is exactly the algebraic sum of the energy changes calculated for each of the single mutants Gly35 and Gly48.

Gly35–Pro51 Double Mutant

The kinetics of the purified enzyme (Table 1) suggest that the mutations Gly35 and Pro51 are relatively independent (Figure 2c); the interaction of the sulfhydryl side chain of Cys35 with the transition state of the substrate is very similar in the Thr51 (1.2 kcal/mole) and the Pro51 (0.8 kcal/mole) enzymes. Likewise, the improved interaction of the transition state with enzyme on introducing Pro51 is very similar in the Cys35 (−1.9 kcal/mole) and Gly35 (−2.3 kcal/mole) enzymes. The improved interaction energy in the Gly35–Pro51 double mutant (−1.1 kcal/mole) is a little more than the value expected (−0.7 kcal/mole) from independent Gly35 (1.2 kcal/mole) and Pro51 (−1.9 kcal/mole) mutants.

Gly48–Pro51 Double Mutant

The kinetics of the purified enzyme (Table 1) are very similar to those of the single mutant Gly48 (Table 1): the improved interaction energy when Thr51 is converted to proline is very much dependent on the presence of an imidazole group at position 48 (Figure 2d). With threonine at position 51 the imidazole group of His48 contributes 1.1 kcal/mole to the binding of the transition state. However, with proline at position 51 the imidazole group contributes 3 kcal/mole, demonstrating that the distortion introduced by Pro51 is propagated to His48 and improves its interaction with the transition state by 1.9 kcal/mole. The loss in interaction energy of the Gly48–Pro51 double mutant (1.1 kcal/mole) contrasts with the improved value expected (−0.8 kcal/mole) from independent Gly48 (1.1 kcal/mole) and Pro51 (−1.9 kcal/mole) mutants.

Discussion

Principle of Using Double Mutants

Site-directed mutagenesis has been used to evaluate the roles of individual side chains in substrate binding and catalysis in cloned enzymes (Winter et al., 1982; Dalbadie-McFarland et al., 1982; Sigal et al., 1982; Villafranca et al., 1983; Wilkinson et al., 1983, 1984). There is an inherent problem in that the extent of structural change introduced by a point mutation is unknown. There may, however, be some clues from kinetics. For example, with a bisubstrate enzyme the lesion introduced at the binding site for one substrate may have little effect on the binding of the other substrate. This may indicate that the structural change is limited to a single site, as appears to be true for the lesions at the ATP site of the tyrosyl-tRNA synthetase, which have little effect on the K_m values for tyrosine. Eventually X-ray diffraction will solve some of the mutant structures: from difference electron-density maps at very high resolution it is possible to detect changes of 0.1–0.2 Å in bond lengths between isomorphous structures (Henderson and Moffat, 1971). As a change of 0.1 Å in a good hydrogen bond with substrate could correspond to 2–3 kcal/mole, it may prove very difficult to explain small differences in rate or affinity by inspection of difference maps. However, gross changes will be readily observed. As an alternative approach to crystallography on mutants we have introduced a second mutation in the active site to help understand how the first mutation affects the substrate affinity. We have applied this technique to a mutant that we would expect to distort the polypeptide backbone (Pro51) and two mutants that should not (Gly35 and Gly48).

If replacement of different side chains induces no structural changes in the enzyme or enzyme–substrate complex their effects will be independent, and the overall change in interaction energy of the enzyme–substrate transition state in double mutants will be the sum of the corresponding terms for the two single mutants. In the thermodynamic terms of Figure 3, the following relations would hold:

$$\Delta G_1 = \Delta G_1' \quad \text{and} \quad \Delta G_2 = \Delta G_2' \qquad (1)$$

For example, Cys35 lies on the side of the active site that is opposite to His48. If the active site were rigid these residues would bind the substrate independently. Removing the Cys35 or the His48 side chains should not disrupt the active site, and this is confirmed by the energetics of the Gly35–Gly48 double mutant (Figure 2b) in which the lesions appear to be independent and equation 1 applies.

If replacements of amino acid side chains were accompanied by extensive structural changes in the enzyme or the enzyme–substrate complex, the change in the interaction energy of the enzyme–substrate transition state calculated for one mutation could vary depending on the presence of the other mutation. Then in Figure 3:

$\Delta G_1 \neq \Delta G_1'$ and $\Delta G_2 \neq \Delta G_2'$

$\Delta G_1 + \Delta G_2' = \Delta G_1' + \Delta G_2$

But

$$\therefore (\Delta G_1 - \Delta G_1') = \Delta G_2 - \Delta G_2' \neq 0 \ldots \qquad (2)$$

Equation 2 may be rearranged to show that the term $|\Delta G_1 - \Delta G_1'|$, which we now define as the coupling energy between the two mutations, measures the energy difference between the double mutant and the two single mutants taken separately. Constructions in which one mutation has no effect on the binding of substrate or transition state at a second site will have zero coupling energy. Constructions in which one mutation alters the binding at a second site for better or worse will have discernible coupling energies.

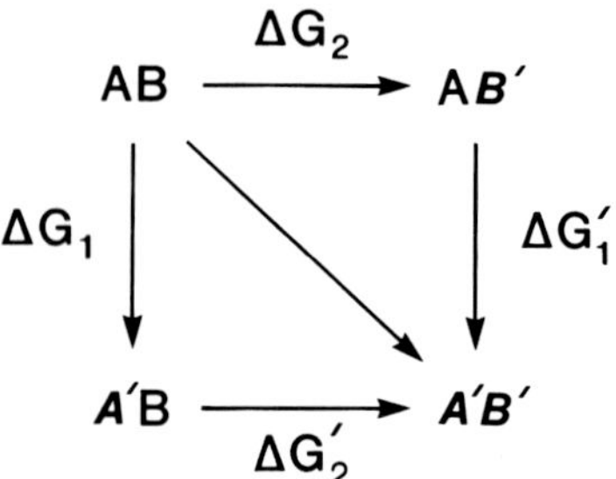

Figure 3. Amino Acid Side Chain A Is Mutated to A′, and B to B′

The changes in interaction energy of enzyme and transition state of the substrate are represented by the ΔG terms. For example, ΔG_1 is the difference in Gibbs-free energy of binding of the substrate (in its transition state structure) to the enzyme forms AB and A′B. $\Delta G_1 = -RT \ln [(k_{cat}/K_m)_{A'B}/(k_{cat}/K_m)_{AB}]$ (Wilkinson et al., 1983).

Since the Gly35 and Gly48 lesions do not seem to elicit structural distortion in the ATP site, these can be used as probes of the possible structural change introduced by Pro51. We find that the Gly35–Pro51 double mutant binds to the transition state a little better (by 0.4 kcal/mole) than would be expected from the two single mutants Gly35 and Pro51 (Figure 2c). Obviously Pro51 does affect the binding of the tyrosyl adenylate transition state to Cys35, reducing the interaction by 0.4 kcal/mole. The main effect of Pro51, however, is via His48. The Pro51 lesion increases the binding of the tyrosyl adenylate transition state to His48 by 1.8 kcal/mole, and thus the Gly48–Pro51 double mutant binds to the transition state much more poorly (by 1.8 kcal/mole) than would be expected from the two single mutants Gly48 and Pro51.

Interaction of His48 with the Substrate

The loss in enzyme activity when His48 is mutated to glycine suggests that the imidazole side chain interacts with the transition state to the extent of 1.2 kcal/mole (Figure 2a). In the crystallographic structure His48 is on the surface of the enzyme, and in the absence of substrate His48 presumably makes a hydrogen bond to a water molecule. The value of 1.2 kcal/mole is probably comprised of two terms: the exchange energy of hydrogen bonds as substrate displaces the bound water molecule and the direct electrostatic energy between the positive charge on the histidine and the negatively charged oxygen atoms on the α-phosphate of ATP. Earlier site-directed mutagenesis experiments on the tyrosyl-tRNA synthetase in which uncharged side chains were removed had suggested exchange energies of 1.2 kcal/mole at Cys35

(Wilkinson et al., 1983), a bond of the correct length, and −0.3 kcal/mole at Thr51, where the bond is too long and hence weak (Wilkinson et al., 1984). It is thus expected that the interaction of His48 with the transition state is not optimal since the electrostatic energy between the positive charge and the negatively charged phosphates should be added to the hydrogen bond energy.

The conversion of Thr51 to proline dramatically improves the affinity of the enzyme for ATP. A priori, this could be caused either by a direct interaction between the pyrrolidine ring of proline and the ribose as in glyceraldehyde phosphate dehydrogenase (Biesecker et al., 1977), or by the distortion of the polypeptide backbone, improving the interaction of a side chain or main chain contact with the substrate. Inspection of the crystallographic model does not suggest a plausible interaction between the pyrrolidine ring and the adenylate. We therefore considered the possibility that the distortion of the polypeptide chain, which must be introduced by proline, might alter the position of His48 and thereby improve its interaction with the substrate. The results show this to be true, although a small part of the improvement presumably corresponds to the loss of the weak hydrogen bond at Thr51 (see Wilkinson et al., 1983). The results also preclude the possibility of a direct interaction of proline and the ribose, as TyrTS(Gly48–Thr51) has an almost identical activity to TyrTS (Gly48–Pro51) in the activation of amino acid. In the Pro51 mutant, His48 now contributes 3 kcal/mole to transition state binding.

How the proline distorts the polypeptide chain so as to alter the position of His48 is not readily predictable. The increased interaction energy might be the result of changes of only a few tenths of an angstrom. Proline can be accommodated within an α-helix and might be expected to kink the helix by up to 20° (Kendrew et al., 1960; Schulz et al., 1974). A detailed inspection of the crystallographic model of TyrTS suggests that the main chain amide of residue 51 could make a hydrogen bond to the main chain carbonyl of either Gly47 (α-helix) or His48 (3_{10} helix). The introduction of Pro51 must disrupt this hydrogen bond, and as a result we see an improved hydrogen bond from His48 and/or electrostatic interaction with the transition state of the substrate.

The procedure of double mutation should be quite general and applicable to other enzymes. Care must be taken, however, in interpretation. For example, there could be a fortuitous cancelling out of the effects of structural mutation. Alternatively, coupling of mutations might arise not through distortion of the structure of the enzyme but through distortion of the substrate or its reorientation in the active site.

Experimental Procedures

Mutant Constructions

The single mutants Cys35 → Gly and Thr51 → Pro were constructed from the TyrTS gene cloned in the phage M13mp93 as described previously (Winter et al., 1982; Wilkinson et al., 1983, 1984) (see Experimental Procedures). The mutant His48 → Gly was constructed using a synthetic mutagenic primer, 5′-GCCAAGCCGCCGATAT-3′, that has a double mismatch with template located in the middle. The frequency of mutants revealed by hybridization with the mutagenic oligonucleotide as probe was 6%. After plaque purification the single strand of one mutant clone was fully sequenced using a family of oligonucleotide sequencing primers staged throughout the TyrTS gene (Wilkinson et al., 1984). The Gly48–Pro51 double mutant was constructed using Pro51 template and the Gly48 mutagenic primer (7% mutants), while the Gly35–Pro51 double mutant was constructed using Gly35 template and the Pro51 mutagenic primer (18% mutants). All mutants were fully sequenced as above.

Oligonucleotide Synthesis

Oligonucleotide primers were synthesized using phosphotriester chemistry on a polydimethylacrylamide kieselguhr support (Gait et al., 1982) and purified by ion exchange HPLC on Partisil 10SAX using a gradient of 1 mM to 200 mM potassium phosphate (pH 6.3) in 60% formamide.

The following procedures are based on Zoller and Smith, 1982.

Phosphorylation of Oligonucleotide

For mutagenesis 100 pmole of mutagenic primer was phosphorylated with 2.5 U T4-polynucleotide kinase (PL Biochemicals) in 20 μl 50 mM Tris-HCl (pH 8.0), 10 mM $MgCl_2$, 5 mM DTT, 1 mM ATP, at 37°C for 45 min and then heated at 70°C for 10 min.

For hybridization screening 15 pmole of mutagenic primer was phosphorylated as above with 30 μCi γ-^{32}P-ATP (3000 Ci/mmole) (Amersham International) as the only source of ATP. The primer was diluted with 3.5 ml 6× SSC (0.9 M NaCl, 90 mM Na citrate, 3 mM EDTA [pH 7.2]) and used directly as a probe.

Synthesis of Covalently Closed Double-Stranded DNA

Ten picomoles of phosphorylated primer was annealed with 0.33 pmole of M13–TyrTS template in 10 μl 10 mM Tris-HCl (pH 8.0), 10 mM $MgCl_2$, by cooling from 80°C to 40°C over about 30 min.

The volume was adjusted to 20 μl 10 mM Tris-HCl (pH 8.0), 10 mM $MgCl_2$, 0.25 mM dATP, 0.25 mM dTTP, 0.25 mM dCTP, 0.25 mM dGTP, 0.25 mM ATP, 5 mM DTT in the presence of 5 μCi α-^{32}P-dATP (Amersham International). The primer was extended with 1 U DNA polymerase I Klenow fragment (BRL) in the presence of 6 U T4 DNA ligase (Biolabs) for 16–20 hr at 12°C. The reaction was stopped by addition of 80 μl 10 mM Tris-HCl (pH 8.0), 10 mM EDTA. An equal volume of 13% PEG 1.6 M NaCl was added and the sample kept on ice for 15 min. DNA was separated from unincorporated α-^{32}P-dATP by a 10 min spin in an Eppendorf microfuge. Of the label, 15%–25% was generally incorporated into the DNA.

Enrichment for Covalently Closed Double-Stranded DNA

The DNA pellet was dissolved in 200 μl 0.2 M NaOH. After 5 min at room temperature the DNA sample was applied to a 5 ml alkaline sucrose gradient (5%–20% sucrose, 1 M NaCl, 0.2 M NaOH, 2 mM EDTA) and centrifuged in a Beckman SW50.1 rotor at 37,000 rpm, 4°C for 2.5 hr. Twenty fractions were collected from the gradient and radioactivity was measured (Cerenkov counts). The closed circular DNA runs ahead of the single-strand linear and circular molecules. The closed circular cDNA fractions were pooled, neutralized with 0.1 vol of 2.6 M NaAc (pH 4.8), and ethanol-precipitated in the presence of 10 μg tRNA carrier using 2.5 vol ethanol. The pellet was taken up in 40 μl water, and small aliquots were used to transfect $CaCl_2$-treated E. coli (Cohen et al., 1972) strain JM101 (Messing, 1979).

Colony Blot-Screening by Oligonucleotide Hybridization

Colony blotting was adapted from Grunstein and Hogness (1975). Two hundred plaques resulting from the transfected closed circular DNA were toothpicked onto L plates and grown up as colonies of M13 TyrTS-infected bacteria for 16–20 hr at 37°C. A nitrocellulose filter (Schleicher and Schüll) was placed on the colony plate for 10 min at room temperature. Filters were transferred to Whatman 3MM paper wetted with 0.5 NaOH (colonies face up) for 3 min at room temperature. Filters were neutralized by placing on Whatman 3MM soaked in 1 M Tris-HCl (pH 7.4) twice for 1 min and then 0.5 M Tris-HCl (pH 7.4), 1.5 M NaCl, for 5 min. The filters were allowed to dry at room temperature and then baked in vacuo for 1 hr at 80°C. The filters were prehybridized for at least 1 hr at 67°C in 100 ml of 10×

Denhardt's solution (Denhardt, 1966), 6× SSC, and 0.2% SDS in a sealable plastic bag.

For hybridization screening prehybridized filters were washed in 50 ml 6× SSC and then hybridized in 3.5 ml probe (see above) in disposable plastic petri dishes for 30 min at room temperature (colonies face down). Hybridized filters were washed three times in 100 ml 6× SSC at room temperature for a total of 2 min and autoradiographed for a few minutes. Good discrimination between wild-type and mutant phages was obtained by a brief wash (1–3 min) at the Wallace temperature for the mutagenic primer (Suggs et al., 1981): $T_m = 4 \times$ (GC pairs) + 2 × (AT pairs) (°C).

Nucleotide Sequencing of Mutant Phages

Phages from a hybridization-positive colony were plaque-purified, and the mutation was verified by dideoxy sequencing (Sanger et al., 1977) using a synthetic oligonucleotide sequencing primer. Mutant genes were then completely sequenced by the dideoxy chain termination method using α-^{35}S-thio dATP and buffer gradient gels (Biggin et al., 1983) with a family of five oligonucleotide primers located at intervals throughout the TyrTS gene (Wilkinson et al., 1984).

Kinetic Assays

The pyrophosphate exchange assay (Calendar and Berg, 1966b), tRNA charging, and active site titration (Wilkinson et al., 1983) were as previously described.

Acknowledgments

We are particularly indebted to Professor D. M. Blow for his advice and the unpublished coordinates of the tyrosyl-tRNA synthetase, and to Dr. A. Lesk for producing Figure 1. We would like to thank Drs. M. F. Perutz, M. S. Neuberger, and A. Lesk for helpful discussions.

The costs of publication of this article were defrayed in part by the payment of page charges. This article must therefore be hereby marked "*advertisement*" in accordance with 18 U.S.C. Section 1734 solely to indicate this fact.

Received May 3, 1984; revised July 23, 1984

References

Barker, D. G., and Winter, G. (1982). Conserved cysteine and histidine residues in the structures of the tyrosyl and methionyl-tRNA synthetases. FEBS Lett. *145,* 191–193.

Bhat, T. N., Blow, D. M., Brick, P., and Nyborg, J. (1982). Tyrosyl-tRNA synthetase forms a mononucleotide-binding fold. J. Mol. Biol. *158,* 699–709.

Biesecker, G., Harris, J. I., Thierry, J. C. Walker, J. E., and Wonacott, A. J. (1977). Sequence and structure of D-glyceraldehyde 3-phosphate dehydrogenase from Bacillus stearothermophilus. Nature *266,* 328–333.

Biggin, M. D., Gibson, T. J., and Hong, G. F. (1983). Buffer gradient gels and ^{35}S label as an aid to rapid DNA sequence determination. Proc. Nat. Acad. Sci. USA *80,* 3963–3965.

Blow, D. M., Bhat, T. N., Metcalfe, A., Risler, J. L., Brunie, S., and Zelwer, C. (1983). Structural homology in the amino-terminal domains of two aminoacyl-tRNA synthetases. J. Mol. Biol. *171,* 571–576.

Calendar, R., and Berg, P. (1966). Purification and physical characterization of tyrosyl ribonucleic acid synthetases from Escherichia coli and Bacillus subtilis. Biochemistry *5,* 1681–1690.

Cohen, S. N., Chang, A. C. Y., and Hsu, L. (1972). Nonchromosomal antibiotic resistance in bacteria: genetic transformation of Escherichia coli by R-factor DNA. Proc. Nat. Acad. Sci. USA *69,* 2110–2114.

Dalbadie-McFarland, G., Cohen, L. W., Riggs, A. D., Morin, C., Itakura, K., and Richards, J. H. (1982). Oligonucleotide-directed mutagenesis as a general and powerful method for studies of protein function. Proc. Nat. Acad. Sci. USA *79,* 6409–6413.

Denhardt, D. T. (1966). A membrane-filter technique for the detection of complementary DNA. Biochem. Biophys. Res. Commun. *23,* 641–646.

Fersht, A. R., and Jakes, R. (1975). Demonstration of two reaction pathways for the aminoacylation of tRNA. Application of the pulsed quenched flow technique. Biochemistry *14,* 3350–3356.

Fersht, A. R., Mulvey, R. S., and Koch, G. L. E. (1975). Ligand binding and enzymic catalysis coupled through subunits: tyrosyl tRNA synthetase. Biochemistry *14,* 13–18.

Gait, M. J., Matthes, H. W. D., Singh, M., Sproat, B. S., and Titmas, R. C. (1982). Rapid synthesis of oligodeoxyribonucleotides. VII. Solid phase synthesis of oligodeoxyribonucleotides by a continuous flow phosphotriester method on a kieselguhr-polyamide support. Nucl. Acids Res. *10,* 6243–6254.

Grunstein, M., and Hogness, D. (1975). Colony hybridization: a method for the isolation of cloned DNAs that contain a specific gene. Proc. Nat. Acad. Sci. USA *72,* 3961–3965.

Henderson, R., and Moffat, J. K. (1971). The difference Fourier technique in protein crystallography: errors and their treatment. Acta Cryst. *B27,* 1414–1420.

Hutchison, C. A., III, Phillips, S., Edgell, M. H., Gillam, S., Jahnke, P., and Smith, M. (1978). Mutagenesis at a specific position in a DNA sequence. J. Biol. Chem. *253,* 6551–6560.

Kendrew, J. C., Dickerson, R. E., Strandberg, B. E., Hart, R. G., Davies, D. R., Phillips, D. C., and Shore, V. C. (1960). Structure of myoglobin. A three-dimensional fourier synthesis at 2 Å resolution. Nature *185,* 422–427.

Messing, J. (1979). Certification of the single-stranded DNA lac-phage multipurpose cloning system. Recombinant DNA. Tech. Bull. *2,* 43–48.

Rubin, J., and Blow, D. M. (1981). Amino acid activation in crystalline tyrosyl-tRNA synthetase from Bacillus stearothermophilus. J. Mol. Biol. *145,* 489–500.

Sanger, F., Nicklen, S., and Coulson, A. R. (1977). DNA sequencing with chain-terminating inhibitors. Proc. Nat. Acad. Sci. USA *74,* 5463–5467.

Schulz, G. E., Elzinga, M., Marx, F., and Schirmer, R. H. (1974). Three-dimensional structure of adenyl kinase. Nature *250,* 120–123.

Sigal, I. S., Harwood, B. G., and Arentzen, R. (!982). Thiol-β-lactamase: replacement of the active-site serine of RTEM β-lactamase by a cysteine residue. Proc. Nat. Acad. Sci. USA *79,* 7157–7160.

Suggs, S. V., Hirose, T., Miyake, T. Kavashima, E. H., Johnson, M. J., Itakura, K., and Wallace, R. B. (1981). Developmental biology using purified genes, D. Brown, ed. (New York: Academic Press).

Villafranca, J. E., Howell, E. E., Voet, D. H., Strobel, M. S., Ogden, R. C., Abelson, J. N., and Kraut, J. (1983). Directed mutagenesis of dihydrofolate reductase. Science *222,* 782–788.

Wilkinson, A. J., Fersht, A. R., Blow, D. M., and Winter, G. (1983). Site-directed mutagenesis as a probe of enzyme structure and catalysis: tyrosyl-tRNA synthetase cysteine-35 to glycine-35 mutation. Biochemistry *22,* 3581–3586.

Wilkinson, A. J., Fersht, A. R., Blow, D. M., Carter, P., and Winter, G. (1984). A large increase in enzyme–substrate affinity by protein engineering. Nature *307,* 187–188.

Winter, G., Fersht, A. R., Wilkinson, A. J. Zoller, M., and Smith, M. (1982). Redesigning enzyme structure by site-directed mutagenesis: tyrosyl tRNA synthetase and ATP binding. Nature *299,* 756–758.

Zelwer, C., Risler, J. L., and Brunie, S. (1982). Crystal structure of Escherichia coli methionyl-tRNA synthetase at 2.5 Å resolution. J. Mol. Biol. *155,* 63–81.

Zoller, M. J., and Smith, M. (1982). Oligonucleotide-directed mutagenesis using M13-derived vectors: an efficient and general procedure for the production of point mutations in any fragment of DNA. Nucl. Acids Res. *10,* 6487–6500.

We introduced residues from the homologous *E. coli* enzyme into the thermophilic protein we had used, and found that we could increase dramatically the apparent affinity of the tRNA-synthetase for its substrate[62]. This increase did not increase activity at the natural concentration in the cell (harking back to my 1974 paper on evolution of K_M[26]). This was a very important finding since it demonstrated that the activities of enzymes could be increased in industrial processes.

7.5 2003: Denise Lowe.

7.6 2003: Danuta Mossakowska.

A large increase in enzyme–substrate affinity by protein engineering

Anthony J. Wilkinson*, Alan R. Fersht*, D. M. Blow†, Paul Carter‡ & Greg Winter‡

* Department of Chemistry and † Department of Physics, Imperial College of Science and Technology, London SW7 2AY, UK
‡ MRC Laboratory of Molecular Biology, Hills Road, Cambridge CB2 2QH, UK

A single point mutation has been engineered in the tyrosyl-tRNA synthetase that improves its affinity (K_M) for its substrate ATP by a factor of 100. In the crystal structure of the tyrosyl tRNA synthetase (of *Bacillus stearothermophilus*), the side-chain hydroxyl of Thr 51 appears to make a weak hydrogen bond with the AMP moiety of the substrate intermediate, tyrosyl adenylate. In the absence of substrate, however, the hydroxyl group should make a strong hydrogen bond with water which would favour dissociation of the enzyme–substrate complex. We have used oligodeoxynucleotide-directed mutagenesis to construct two point mutants at this site: one to remove the hydroxyl group (Thr 51 → Ala 51) and the other, in addition, to distort the local polypeptide backbone (Thr 51 → Pro 51). We report here that both mutants have increased activity (k_{cat}/K_M for ATP) but one mutant (Pro 51) shows a massive 25-fold increase due mainly to a lowered K_M for ATP. This demonstrates dramatically the potential of *in vitro* mutagenesis for improving the affinity of an enzyme for its substrate.

The activities of such bacterial enzymes as β-lactamase[1], 'evolved β-galactosidase'[2,3], amidases[4,5] and ribitol dehydrogenase[6] have been successfully directed towards novel substrates *in vivo* by selecting spontaneous mutations of the enzyme. We are exploring an alternative approach in which we inspect a crystallographic model of the enzyme and attempt to predict the effect of changing a side chain in the active site. We then construct the required mutant by site-directed mutagenesis of the gene, express the enzyme in *Escherichia coli* and measure catalytic activity[7,8].

The tyrosyl-tRNA synthetase (TyrTS) of *B. stearothermophilus* catalyses the aminoacylation of $tRNA^{Tyr}$ in a two-step mechanism in which tyrosine is activated to form enzyme-bound tyrosyl adenylate and then transferred to $tRNA^{Tyr}$. The location of tyrosyl adenylate in the crystal structure indicates that hydrogen bonds are made to the ribose moiety from the side chains of Cys 35, Thr 51 and His 48 (refs 9–11). The Cys 35 interaction has been confirmed by constructing Gly 35 and Ser 35 mutants: these have lowered affinities for ATP resulting either from the complete removal of the hydrogen bond in TyrTS(Gly 35) or from placing the –OH group in TyrTS(Ser 35) slightly too far from the hydrogen bond acceptor on the ATP[7,8]. Interestingly, TyrTS(Ser 35) has a lower affinity for ATP (calculated from k_{cat}/K_M for aminoacylation) than does TyrTS(Gly 35), despite the possibility of a weak hydrogen bond. This probably stems from Ser 35 exchanging a good hydrogen bond with water in the free enzyme for a poor one with ATP on forming the enzyme–substrate complex[8]. As the hydroxyl group of Thr 51 makes a long hydrogen-bonding contact to the ring oxygen (O-1) of the ribose, we predicted that removal of the weak hydrogen bond by constructing an Ala 51 mutant might improve the affinity of the enzyme for ATP.

The Ala 51 mutant was constructed as described previously[7,12] (using a synthetic oligodeoxynucleotide, 5′ AAATGGCGGCCAAGTG 3′) and characterized by dideoxy sequencing[13]. Not only was the Ala 51 lesion confirmed, but also the entire TyrTS gene was sequenced to prove that no other mutations had been introduced inadvertently. For this purpose, a family of five synthetic oligodeoxynucleotide primers staged throughout the TyrTS gene was used (Fig. 1*A*), allowing the complete sequence to be determined on a single salt gradient

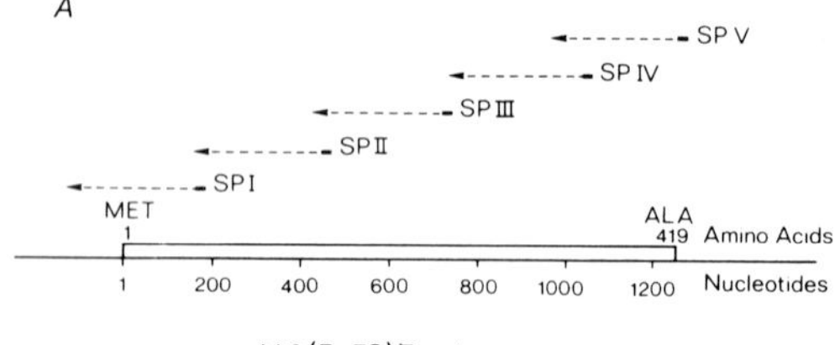

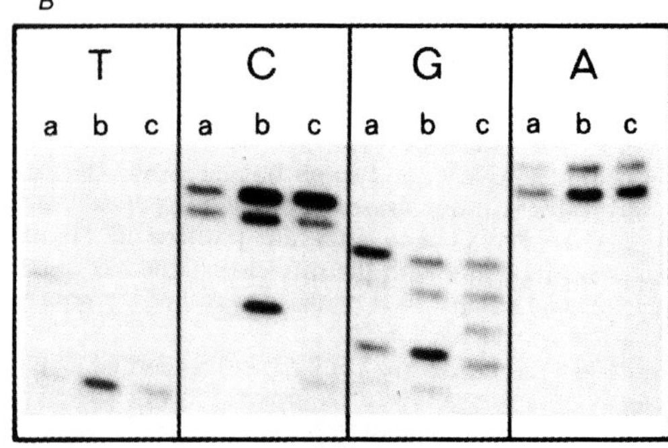

Fig. 1 Sequencing of mutant TyrTS gene. *A*, Oligodeoxynucleotide primers: SPI, 5′ CCCGCCTGCTGGAAGCG 3′; SPII, 5′ TGAACCGACTCTTT 3′; SPIII, 5′ CTCGTACGGCGACGTTT 3′; SPIV, 5′ AAGCGGAACGTCGCCTC 3′; SPV, 5′ TTACTATGCCCAGCGC 3′, complementary to the sequence of the TyrTS gene[15] cloned in M13, were synthesized using phosphotriester chemistry on a polydimethylacrylamide kieselguhr support[16] and purified by ion-exchange HPLC. Templates from the mutants M13TyrTS(Ala 51) and M13TyrTS(Pro 51) were prepared[17] and the genes sequenced by primer extension and the dideoxy chain termination method. As >300 nucleotides from the primer can be read from gradient gels using [α-^{35}S]thio-dATP[14], the complete sequence of each gene may be determined in a single gel. *B*, Sequences of M13TyrTS(Thr 51) (tracks a), M13TyrTS(Ala 51) (tracks b) and M13TyrTS(Pro 51) (tracks c) as determined using the primer SPI. Detection of sequence changes was facilitated by running the T, C, G or A tracks of the three clones in adjacent slots.

gel[14]. By running together the corresponding T, C, G and A tracks of mutant and wild-type clones, sequence changes can be readily detected (Fig. 1*B*).

Mutation of Thr 51 to Ala 51 causes a small change in k_{cat} for both exchange (Table 1) and aminoacylation (Table 2), but decreases K_M for ATP by a factor of two for both reactions. The specificity constant, k_{cat}/K_M, is accordingly increased two-fold, indicating that the enzyme–substrate interaction energy is increased by 0.38 kcal mol^{-1} in the Ala 51 mutant. This is consistent with our prediction that a small increase in enzyme affinity might be engineered by deleting a poor hydrogen bond with the substrate.

Residues 47–61 all have ϕ, ψ angles in the range characteristic of an α-helix (–57°, –48°)[11], but the helix is slightly irregular at the amino-terminus, and the hydrogen bond between Gly 47 (CO) and Thr 51 (NH) may not form. In the homologous *E. coli* enzyme where residue 51 is proline[15], this hydrogen bond could not exist as the nitrogen is locked into the pyrrolidine ring. This must cause a small distortion in the polypeptide backbone around residues 47–51. To examine the effect on the affinity of ATP, we constructed a Pro 51 mutant of the *B. stearothermophilus* enzyme (as above, using the primer 5′ CAAAATCGGGGCCAAGT 3′). The mutation leads to large kinetic changes. In the activation reaction, there is a 15-fold lowering of K_M for ATP and a near doubling of the value of k_{cat}. These combine to give a 25-fold increase in the specificity constant (≡1.9 kcal mol^{-1} of activation energy). In the charging reaction, K_M for ATP is reduced 130-fold for TyrTS(Pro 51), but k_{cat} is also reduced by a factor of 2.6 so that the specificity constant is 50 times higher than that for the wild type (≡2.4 kcal

Reprinted from Nature, Vol. 307, No. 5947, pp. 187-188, January 12 1984

Table 1 Pyrophosphate exchange activity of tyrosyl-tRNA synthetases

Enzyme	$k_{cat}(s^{-1})$	K_M (ATP) (mM)	k_{cat}/K_M (s^{-1} M^{-1})
TyrTS	7.6	0.9	8,400
TyrTS(Ala 51)	8.6	0.54	15,900
TyrTS(Pro 51)	12.0	0.058	208,000

Synthetase exchange activity was measured at 25 °C, *p*H 7.8, in 144 mM Tris-HCl, 10 mM $MgCl_2$, 0.1 mM phenylmethane sulphonyl chloride, 10 mM 2-mercaptoethanol, 2 mM pyrophosphate, 50 μM tyrosine, and 100–250 nM enzyme (assayed by active site titration).

Table 2 Aminoacylation activity of tyrosyl-tRNA synthetases

Enzyme	k_{cat} (s^{-1})	K_M (ATP) (mM)	k_{cat}/K_M (s^{-1} M^{-1})
TyrTS	4.7	2.5	1,860
TyrTS(Ala 51)	4.0	1.2	3,200
TyrTS(Pro 51)	1.8	0.019	95,800

The same buffer was used as for Table 1, plus 100 μM ^{14}C-Tyr, 20 μM $tRNA^{Tyr}$ and 25–50 nM enzyme.

mol^{-1} of activation energy). The detailed stereochemical explanation of this mutation must await further work: for example, it is conceivable that the distortion of the polypeptide backbone introduced by Pro 51 may alter the position of His 48, and increase its interaction with the substrate. However, one factor in the improved affinity must be the removal of the poor hydrogen bond of Thr 51 with ATP.

Would the mutation to TyrTS(Pro 51) confer a selective advantage on the cell, given its improved K_M with ATP? In addition to any change in other properties of the protein that have as yet not been investigated, such as thermostability, there are clear kinetic disadvantages of TyrTS(Pro 51) *in vivo* because of the high cellular ATP concentration. The intracellular concentration of ATP (2–3 mM) is so high that the wild-type and mutant enzymes are nearly saturated with ATP, thus the reaction rate is controlled by k_{cat} and not k_{cat}/K_M (ref. 8). This confers a slight disadvantage on TyrTS(Ala 51) and a greater disadvantage on TyrTS(Pro 51) in the overall charging reaction (Table 2). However, *in vitro*, TyrTS(Pro 51) is far more active at lower ATP concentration (<2 mM). Also, in the tyrosine activation reaction, TyrTS(Pro 51) is far more active at all concentrations of ATP. (Selective pressure is exerted on the overall aminoacylation reaction rather than on the partial reaction of activation.)

Irrespective of the consequences *in vivo*, the dramatic improvement of K_M for ATP *in vitro* has significant implications for industrial enzymology by demonstrating that it is possible to engineer large improvements in enzyme activity by site-directed mutagenesis.

Received 16 September; accepted 4 November 1983.

1. Hall, A. & Knowles, J. R. *Nature* **264**, 803–804 (1976).
2. Hall, B. G. *Genetics* **89**, 453–465 (1978).
3. Hall, B. G. & Zuzel, T. *Proc. natn. Acad. Sci. U.S.A.* **77**, 3529–3533 (1980).
4. Paterson, A. & Clarke, P. H. *J. gen. Microbiol.* **114**, 75–85 (1979).
5. Turberville, C. & Clarke, P. H. *FEMS Microbiol. Lett.* **10**, 87–90 (1981).
6. Wu, T. T., Lin, E. C. C. & Tanaka, S. *J. Bact.* **96**, 447–456 (1968).
7. Winter, G., Fersht, A. R., Wilkinson, A. J., Zoller, M. & Smith, M. *Nature* **299**, 756–758 (1982).
8. Wilkinson, A. J., Fersht, A. R., Blow, D. M. & Winter, G. *Biochemistry* **22**, 3581–3586 (1983).
9. Monteilhet, C. & Blow, D. M. *J. molec. Biol.* **122**, 407–417 (1978).
10. Rubin, J. & Blow, D. M. *J. molec. Biol.* **145**, 489–500 (1981).
11. Bhat, T. N., Blow, D. M., Brick, P. & Nyborg, J. *J. molec. Biol.* **158**, 699–709 (1982).
12. Zoller, M. & Smith, M. *Nucleic Acids Res.* **10**, 6487–6500 (1982).
13. Sanger, F., Nicklen, S. & Coulson, A. R. *Proc. natn. Acad. Sci. U.S.A.* **74**, 5463–5467 (1977).
14. Biggin, M. D., Gibson, T. J. & Hong, G. F. *Proc. natn. Acad. Sci. U.S.A.* **80**, 3963–3965 (1983).
15. Winter, G., Koch, G. L. E., Hartley, B. S. & Barker, D. G. *Eur. J. Biochem.* **132**, 383–387 (1983).
16. Gait, M. J., Matthes, H. W. D., Singh, M., Sproat, B. S. & Titmas, R. C. *Nucleic Acids Res.* **10**, 6243–6254 (1982).
17. Winter, G. & Fields, S. *Nucleic Acids Res.* **8**, 1965–1974 (1980).

An analysis of hydrogen bonding is my most highly cited paper, still getting some 20 citations a year after 25 years. The analysis of how the energetics of hydrogen bonding within proteins is balanced or even dominated by the interactions of the separate components with water is a classic paper[63].

7.7 1989: Eric First.

7.8 1989: George García.

Hydrogen bonding and biological specificity analysed by protein engineering

Alan R. Fersht*, Jian-Ping Shi*, Jack Knill-Jones*, Denise M. Lowe*, Anthony J. Wilkinson*, David M. Blow†, Peter Brick†, Paul Carter‡, Mary M. Y. Waye‡ & Greg Winter‡

* Departments of Chemistry and † Physics, Imperial College of Science and Technology, London SW7 2AY, UK
‡ MRC Laboratory of Molecular Biology, MRC Centre, Hills Road, Cambridge CB2 2QH, UK

The role of complementary hydrogen bonding as a determinant of biological specificity has been examined by protein engineering of the tyrosyl-tRNA synthetase. Deletion of a side chain between enzyme and substrate to leave an unpaired, uncharged hydrogen-bond donor or acceptor weakens binding energy by only 0.5–1.5 kcal mol^{-1}. But the presence of an unpaired and charged donor or acceptor weakens binding by a further ~3 kcal mol^{-1}.

THE hydrogen bond is a ubiquitous feature of biological interactions: it is essential in determining the structure of proteins and nucleic acids; it is a major determinant of specificity in enzyme catalysis and in biological information transfer; and it can influence directly the rate of enzymatic reactions by stabilizing the ionic charges formed in the transition state. Hydrogen bonding in macromolecules and their complexes in aqueous solution is a complex phenomenon because water competes for the hydrogen-bonding sites[1,2]. The calculation of the overall energetics is consequently difficult and there is little knowledge of the energies involved[3]. Here we apply an experimental approach to the problem, using site-directed mutagenesis to produce mutant enzymes that differ in their abilities to form hydrogen bonds with their substrates. The interaction energies can be calculated from kinetic data on the modified enzymes.

Our experimental system is the tyrosyl-transfer RNA synthetase from *Bacillus stearothermophilus*. This enzyme catalyses the aminoacylation of $tRNA^{Tyr}$ in a two-step reaction; activation of the amino acid to form the enzyme-bound tyrosyl adenylate complex followed by transfer to $tRNA^{Tyr}$ (ref. 4).

$$E + Tyr + ATP = E.Tyr\text{-}AMP + PP_i \quad (1)$$

$$E.Tyr\text{-}AMP + tRNA = Tyr\text{-}tRNA + AMP \quad (2)$$

The particular suitability of this enzyme (E) for obtaining accurate kinetic data and its other favourable characteristics for systematic site-directed mutagenesis have been described in detail elsewhere[5]. The three-dimensional structure of the enzyme and its bound aminoacyl adenylate are known from X-ray crystallography[6-9]. Eleven possible hydrogen bonds have been identified that may be formed between the enzyme and substrate, eight of which are from amino-acid side chains that may be mutated (Fig. 1). We are systematically mutating these residues to measure the energetics of their interactions and to test whether hydrogen bonds are actually involved[10]. In all cases, the mutations lead to a smaller side chain that either lacks or has a modified hydrogen-bond donor or acceptor. Our initial data are consistent with hydrogen bonding being an exchange reaction whereby the hydrogen-bond donors and acceptors on the free enzyme and free substrate break their bonds with water on forming the bonds in the enzyme–substrate complex[5,11,12]. Here we analyse the effects of mutation of several different types of hydrogen bonds, including bonds from residues involved in determining the high specificity of the enzyme for tyrosine rather than phenylalanine (Tyr 34), bonds for simple binding of the substrate (Cys 35, His 48, Thr 51, Tyr 169) and a bond closer to the seat of reaction (Gln 195). These bonds may be classified further according to charge; most are between uncharged polar residues, but some involve a charged group on either the enzyme or substrate.

Experimental observations

Our experimental results for the activation of tyrosine are listed in Table 1. The contributions of the side chains to binding were calculated from the values of k_{cat}/K_M as described previously (ref. 5, Table 2). Note that the data refer to the binding of the substrate in the transition state. Such data give, in general, the most reliable measurements of incremental binding energies as the effects of strain, nonproductive binding and induced fit do not affect k_{cat}/K_M (ref. 13). The data for the activation of phenylalanine by the wild-type enzyme and a mutant in the specificity pocket are listed in Table 3. The contributions of side chains to binding energy, discussed in more detail below, may be summarized and classified thus: (1) deletion of a side chain on the enzyme that forms a good hydrogen bond with an uncharged group on the substrate weakens binding energy by only 0.5–1.5 kcal mol^{-1}; (2) deletion of an uncharged side chain on the enzyme that forms a hydrogen bond with a charged group on the substrate weakens binding by ~3.5–4.5 kcal mol^{-1}—only two relevant experiments were performed for this category and so the range could be much wider, but the crucial point is that these energies are considerably higher than in (1); (3) deletion of a group that forms a too-long hydrogen bond actually improves binding.

Effects of removal of side chains

Our measurements on the mutated enzymes give the overall contributions of the side chains to binding. The hydrogen bond donors/acceptors that are removed on mutagenesis are the only portions of the side chains in direct contact with the substrate. But the mutation of an amino acid in a protein from a larger side chain to a smaller may have further structural consequences superimposed on the energetics of the hydrogen bonding. Any fine structure analysis must consider these possibilities and attempt to eliminate artefacts. The obvious experimental procedure is to crystallize the mutants and use protein crystallography to detect structural changes. Such a technique will detect changes of ~0.1 Å or more but will miss a series of smaller changes. (The effects of small relaxations of protein structure on binding are not known.)

Because kinetics measurements are exceptionally sensitive and can detect changes of a fraction of a kilocalorie, we use kinetic methods to increase the reliability of analysis. For example, application of the 'double-mutant' procedure of Carter *et al.*[14] has shown that mutation of residues 35 and 48 does not cause structural changes propagated through the protein[14]. Experiments in which we have mutated side chains by the removal of methyl groups buried in the protein, however, show that these small changes can weaken binding by up to 1 kcal mol^{-1} (A.J.W., A.R.F., P.C. and G.W., unpublished data).

Reprinted from Nature, Vol. 314, No. 6008, pp. 235-238, 21 March 1985

Fig. 1 Hydrogen bonds between the tyrosyl-RNA synthetase and tyrosyl adenylate. MC, main chain.

Table 1 Activation of tyrosine by tyrosyl-tRNA synthetase and its mutants

Enzyme	k_{cat} (s^{-1})	K_M (ATP) (mM)	K_M (Tyr) (μM)	k_{cat}/K_M (ATP) (s^{-1} M^{-1})	k_{cat}/K_M (Tyr) (s^{-1} M^{-1})
Wild type	8.35	1.08	2.23	7,730	3.74×10^6
Tyr→Phe 34	6.86	1.2	4.4	5,720	1.56×10^6
Cys→Gly 35	2.95	2.6	2.7	1,130	1.09×10^6
Cys→Ser 35	2.52	2.4	2.6	1,050	9.65×10^5
His→Asn 48	7.90	1.4	3.8	5,640	2.08×10^6
His→Gly 48	2.00	1.3	3.2	1,540	6.25×10^5
Thr→Ala 51	8.75	0.54	2.0	16,200	4.36×10^6
Gln→Gly 195	0.19	2.5*	100	76*	1.90×10^3
Tyr→Phe 169†	6.05	1.25‡	1,030	4,840‡	5.88×10^3
Δ(321–419)†	7.50	1.08	2.4	6,940	3.12×10^6

Mutant enzymes were prepared, assayed and active-site titrated as described previously[5,14]. Activation was measured by pyrophosphate exchange at 25 °C, *p*H 7.78 in the presence of 144 mM Tris-Cl, 10 mM $MgCl_2$ and 2 mM PP_i. Kinetic constants for variation of ATP were determined in the presence of 0.05 mM tyrosine (except where indicated otherwise) and for tyrosine in the presence of 2 mM ATP. Values of k_{cat} are extrapolated to infinite concentration of tyrosine and ATP.

* 0.3 mM tyrosine.

† Experiments on truncated enzyme with residues 321–419 deleted (the tRNA binding domain)[22].

‡ 1.7 mM tyrosine.

Because of this, we have analysed many different mutants to observe trends. The combination of the extensive and consistent kinetic data (see below) and modelling by molecular graphics suggests that the results so far reflect the direct effects of hydrogen bonding itself and are not dominated by structural change.

Hydrogen-bond strengths *in vacuo*

The stabilization energies of hydrogen bonds *in vacuo*, between compounds X-H and B-Y in equation (3), have been calculated[15]:

$$\text{X-H}\cdots\text{B-Y} = \text{X-H} + \text{B-Y} \qquad (3)$$

(where −H is a hydrogen-bond donor and −B is an acceptor). Representative values for the energies of stabilization for different donors and acceptors are: water/water, $-6.4\ \text{kcal mol}^{-1}$; water/$CH_3SH$, $-3.1\ \text{kcal mol}^{-1}$ for −SH as the acceptor, $-3.2\ \text{kcal mol}^{-1}$ for −SH as the donor; imidazolium/water, $-14\ \text{kcal mol}^{-1}$; acetate/water, $-19.8\ \text{kcal mol}^{-1}$. These values are much greater than those in Table 2 for the overall energies of bonding in aqueous solution.

General model for H bonding in water

Our data fit the formulation of Jencks[16] and Hine[17]:

$$\text{E-H}\cdots\text{OH}_2 + \text{HOH}\cdots\text{B-S} = \text{E-H}\cdots\text{B-S} + \text{HOH}\cdots\text{OH}_2 \qquad (4)$$

For convenience, we assume that the enzyme has a hydrogen-bond donor, −H, and the substrate (S) an acceptor, −B, that pair in the enzyme–substrate complex. In the free enzyme, −H and −B are bound to water molecules. The number and types of hydrogen bonds are conserved in the reaction, which is thus essentially isoenthalpic (within the limits of Hine's[17] equation). For example, as discussed by Wilkinson *et al.*[5], a −SH group is just as effective as a −OH group as a hydrogen-bond donor/acceptor, despite the different absolute strengths of −SH···O and −OH···O bonds in equation (3); if, in equation (4), −H is part of a −SH group, then there is a −SH···O bond on the left-hand side and a −SH···O bond on the right in the enzyme–substrate complex. The contribution of hydrogen bonds to enzyme–substrate binding energy arises from entropy; water hydrogen bonded to the enzyme or substrate has lower entropy than bulk water and the release of hydrogen-bonded water helps drive enzyme–substrate binding[13,16]. Thus, there is a crucial distinction between the absolute energy of a hydrogen bond as in equation (3) and the overall energetics of hydrogen bonding in solution. It is the overall energetics that are important in the binding of enzymes and substrates, and it is the overall energetics that we measure from site-directed mutagenesis.

The removal of one of the hydrogen-bonding groups from equation (4), for example a side chain of the enzyme as in equation (5),

$$\text{E OH}_2 + \text{HOH}\cdots\text{B-S} = [\text{E B-S}] + \text{HOH} + \text{OH}_{2(\text{bulk})} \qquad (5)$$

does not necessarily lead to the loss of the absolute binding energy (enthalpy) of a hydrogen bond as defined by equation (3). Deletion of a hydrophilic side chain means that a water molecule must be located next to a hydrophobic region of the enzyme in the space vacated. This water molecule is at a high energy because it is at an interface and not in bulk water where it can fulfil readily all its hydrogen bonding. There is no hydrogen bond from E to H_2O in the free enzyme and so this partly (or even fully) compensates for the lack of the hydrogen bond from E to −B in the enzyme–substrate complex. When the substrate binds, it displaces the high-energy water molecule (which returns to the bulk solvent), and the hydrophilic hydrogen-bond acceptor of the substrate occupies the 'hydrophobic' site. The overall energetics of the reaction in equation (5) depend on the precise interactions made by the mutant enzyme with the water molecule and the group −B on the substrate. If, for example, the enzyme-bound water molecule in equation (5) is in the same orientation as that in equation (4), then equation (5) may be written as

$$\text{E OH}_2 + \text{HOH}\cdots\text{B-S} = [\text{E B-S}] + \text{HOH}\cdots\text{OH}_2 \qquad (6)$$

There is one hydrogen bond on the right and a similar one on the left of equation (6). This reaction is approximately isoenthalpic when −B is −OH and, to a first approximation, the energetics of equation (6) are the same as those of equation (4). Therefore, deletion of the hydrogen bond in the enzyme–substrate complex does not lead to the overall loss of the absolute strength of a hydrogen bond and may, in certain circumstances, cause no loss of binding energy. The energetics clearly depend on the precise nature of the interactions in each particular example and so will vary from case to case. Before this study, the magnitudes of the energetic changes were unknown. Now we discuss these and show how particular examples may be rationalized and categorized.

Case A. Deletion of a side chain (−H) that interacts with an uncharged group on the substrate. Equation (5) or (6) describes this. The low contribution to specificity of a hydrogen bond between two uncharged residues is exemplified by the mutation Tyr→Phe 34 (Figs 1, 2). This is a very conservative mutation whereby a small group is deleted whose interaction is important in the recognition of tyrosine compared with phenylalanine.

Yet, K_M for tyrosine increases by a factor of only two (Table 1) and the specificity for tyrosine decreases by a factor of only 15 (Table 3). The apparent bond energy of 0.5 kcal mol^{-1} is one of the lowest yet found, perhaps because the tyrosine –OH is a poorer hydrogen-bond acceptor than an aliphatic or water –OH and so the bond from Tyr 34 to the tyrosyl –OH of the substrate is relatively weak. The thiol-containing side chain from Cys 35, which donates to the 3′-OH of the ribose of ATP, contributes 1.1 kcal mol^{-1}. The tyrosyl-tRNA synthetase from *Bacillus caldotenax* differs from the enzyme from *B. stearothermophilus* by only four amino-acid replacements (M.D. Jones, unpublished data). One of these is an asparagine at position 48. The side chain of Asn 48 contributes 0.77 kcal mol^{-1} and that of His 48 contributes slightly more binding energy at 0.96 kcal mol^{-1}.

Note that equation (5) is relatively insensitive to the nature of the side chain deleted (when –B is uncharged). For example, consider the deletion of a charged histidine side chain (–H is imidazolium) as in the mutation His→Gly 48. Then, relative to wild-type enzyme, a strong interaction between imidazolium

Table 2 Comparative binding energies of wild-type and mutant enzymes with substrates

Comparison (Residue and position)		Substrate	ΔG (kcal mol^{-1})
Phe 34	Tyr 34*	Tyr	0.52
Gly 35	Cys 35*	ATP	1.14
Ala 51	Cys 51	ATP	0.47
Gly 48	Asn 48	ATP	0.77
Gly 48	His 48*	ATP	0.96
Ser 35	Cys 35*	ATP	1.18
Phe 169†	Tyr 169*†	Tyr	3.72
Gly 195	Gln 195*	Tyr	4.49
Gly 35	Ser 35	ATP	–0.04
Ala 51	Thr 51*	ATP	–0.44

The first two columns show the residues compared. The apparent contributions (ΔG) of the side chains of different amino acids to the binding energy of the enzyme-transition state complexes were calculated by comparing the ratios of k_{cat}/K_M for activation by wild-type and mutant enzymes as described previously[5], using the equation $\Delta G = -RT \ln \{(k_{cat}/K_M)_{mut}/(k_{cat}/K_M)_{wt}\}$ (subscripts mut and wt refer to mutant and wild-type (or reference) enzymes respectively).
* Wild-type; † truncated enzyme[22].

Table 3 Activation of tyrosine and phenylalanine by tyrosyl-tRNA synthetases

Enzyme	Activation of Tyr k_{cat} (s^{-1})	K_M (μM)	k_{cat}/K_M (s^{-1} M^{-1})	Activation of Phe k_{cat}/K_M (s^{-1} M^{-1})	$(k_{cat}/K_M)_{Tyr}/(k_{cat}/K_M)_{Phe}$ (Relative specificity)
Wild type	5.4	2.2	2.5×10^6	17	1.5×10^5
Tyr→Phe 34	4.4	4.4	1.0×10^6	100	1.0×10^4

Impurities of tyrosine in the phenylalanine were removed by scavenging with tetranitromethane[23]. Conditions as described in Table 1 but data are not extrapolated to infinite concentration of ATP but given for 2 mM ATP. For activation of phenylalanine, the Phe concentration was varied between 5 and 30 mM, well below the K_M value. The specificity of wild-type enzyme is close to that measured previously under different conditions[23].

and the substrate is lost in the mutant enzyme-substrate complex, but this is compensated for by a loss of a strong interaction between imidazolium and water in the free mutant enzyme.

We have analysed several mutants as well as those listed in Table 2. In all cases, the deletion of a side chain that interacts with an uncharged group on the substrate loses between 0.5 and 1.5 kcal mol^{-1} of binding energy.

Case B. Deletion of a side chain that interacts with a charged group on the substrate. This is exemplified by the mutation Tyr→Phe 169 (Fig. 1) whereby the hydrogen bond to the ammonium ion is lost in the enzyme-substrate complex. Equation (5) still describes the situation, but there is now the loss of a strong hydrogen bond between the charged group on the substrate and water when forming the enzyme-substrate complex and 3.7 kcal mol^{-1} of binding energy are lost on Tyr→Phe 169. Gln 195 binds to a carboxylate oxygen of tyrosine in the E.Tyr complex and to the carbonyl oxygen of tyrosyl adenylate (Fig. 1; refs 6, 7, 18). On mutation of Gln→Gly 195 there is a loss of binding energy of 4.5 kcal mol^{-1} in the transition state where the oxygen atom is fractionally charged.

Case C. Deletion of a group on the substrate that interacts with a charged residue on the enzyme, described by the equation.

$$\text{E-H}^+\cdots\text{OH}_2 + \text{HOH S} = [\text{E-H}^+\ \text{S}] + \text{HOH} + \text{OH}_{2(\text{bulk})} \quad (7)$$

This has similar consequences to Case B. A strong interaction in the enzyme-water complex is lost, an example being the 'modification' of tyrosine to phenylalanine. Phenylalanine binds more poorly to the enzyme by ~7 kcal mol^{-1} (Table 3). The phenylalanine either displaces the water molecule that is bound between Asp 176 and Tyr 34 in the enzyme (Fig. 2), leaving two unpaired hydrogen bonds, or, more probably, binds with a water molecule still bound between Asp 176 and Tyr 34. The water is crammed in a space which is too small, so that there are unfavourable interactions with the substrate.

A minimum estimate of the strength of the bond between Asp 176 and the tyrosine –OH can be made from the relative binding of tyrosine and phenylalanine to the mutant Tyr→Phe 34 (5.45 kcal mol^{-1} calculated from the data in Table 3). If phenylalanine displaces the bound water molecule as in Fig. 2, then there will be the loss of the hydrogen bond and also of the dispersion energy between the enzyme and the –OH group. The latter is worth ~1-2 kcal mol^{-1} (ref. 19) so the hydrogen bond is worth 3.5-4.5 kcal mol^{-1}. If it is easier to bind phenylalanine without displacing the water, then the bond must be stronger than this.

Case D. Deletion of a side chain when there are geometrical constraints against hydrogen bonding in an enzyme-substrate complex. The above cases assume that there are no constraints on forming hydrogen bonds in the enzyme-substrate complex. When, as in equation (8), there is poor hydrogen bonding

$$\text{E-H}\cdots\text{OH}_2 + \text{HOH}\cdots\text{B-S} = [\text{E-H B-S}] + \text{HOH}\cdots\text{OH}_2 \quad (8)$$

because the intermolecular distance between –H and –B is too great, there is just one good hydrogen bond on the right-hand side but two on the left of the equation. Deletion of –H may increase the affinity of the enzyme for the substrate because with the mutant enzyme there is one good hydrogen bond on the right but now only one on the left. Mutation of Thr→Ala 51 (ref. 12) is an example of this, as is Ser→Gly 35.

Case E. Deletion of a side chain on the enzyme or a group on the substrate that allows access of water between enzyme and substrate in their complex. It was suggested in Case C that phenylalanine binds to the tyrosyl-tRNA synthetase without displacing the bound water molecule but with unfavourable van der Waals' repulsion. On deletion of a large side chain from the enzyme, there could perhaps be adequate room for a water molecule to occupy comfortably the gap left by the deleted side chain, as in equation (9).

$$\text{E HOH}\cdots\text{OH}_2 + \text{HOH}\cdots\text{B-S} = \text{E HOH}\cdots\text{B-S} + \text{HOH}\cdots\text{OH}_2 \quad (9)$$

A hydrogen-bond inventory shows that the number of hydrogen bonds may be formally conserved, as in Case A, equation (6). The energetics are expected to be similar to Case A for –B being an uncharged side chain. When –B is charged, as in Case B, the retention of solvent water by the charged group in the mutant enzyme-substrate complex will tend to attenuate the effects described in Case B.

Biological specificity

The classical view of biological specificity is that binding energy

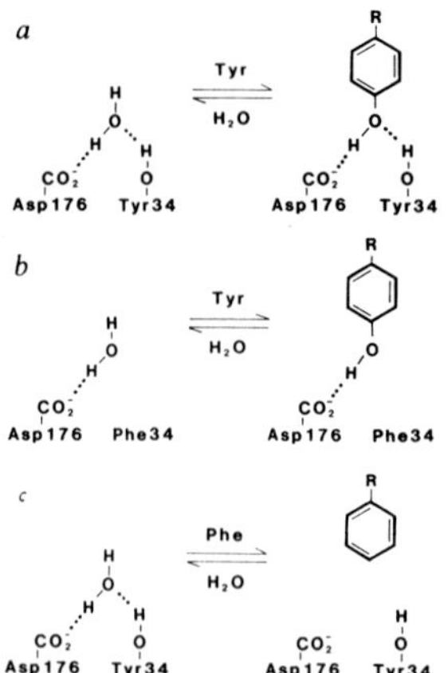

Fig. 2 Illustration of the hydrogen-bond inventory on tyrosine and phenylalanine binding to the tyrosyl-tRNA synthetase. *a*, Binding of wild-type enzyme and tyrosine involves the formation of two hydrogen bonds between its hydroxyl and Asp 176 and Tyr 34, but two equivalent hydrogen bonds of Asp 176 and Tyr 34 with water are broken. *b*, Binding of tyrosine to the Tyr→ Phe 34 mutant involves just the formation of one hydrogen bond with the enzyme, but only one hydrogen bond is broken on displacing the bound water molecule. *c*, The binding of phenylalanine to the enzyme with the concomitant displacement of water would lose a strong hydrogen bond between water and Asp 176 and lead to a partially desolvated charge between the enzyme and substrate. Also, Tyr 34 would be unsolvated. Further, when tyrosine is bound, there are van der Waals' interactions between its phenolic hydroxyl and the enzyme, which are not present when phenylalanine is bound. It is probable that phenylalanine binds to the enzyme with the water molecule remaining bound between Asp 176 and Tyr 34. Note that in *a* and *b* the water molecules released from the enzyme on the binding of tyrosine can be considered to take the place of the tyrosine hydroxyl in solution so that there is no net change in the number of hydrogen bonds in solution.

is provided by dispersion forces and the hydrophobic effect, whilst complementary hydrogen bonds and salt bridges give specificity[16,20]. This study shows that, irrespective of any small structural effects that occur on deletion of side chains, a side chain that forms a hydrogen bond between the enzyme and an uncharged group on a substrate provides only 0.5–1.5 kcal mol^{-1} of binding energy relative to the absence of the side chain. This means that an unpaired uncharged hydrogen bonder provides a factor of only 2.5–15 or so towards specificity. But the absence of a group on the enzyme or a substrate that should form a hydrogen bond with a charged group (as in Cases B and C) affects binding by ~4 kcal mol^{-1} and is worth a factor of 1,000 in specificity. Thus, specificity is caused to some extent by hydrogen bonding but is best mediated by charged residues.

In principle, the same reasoning may be applied to the contribution of hydrogen bonding in base pairing during DNA replication or transcription. There is, however, one important difference when analysing, for example, G·T and A·C mispairing. All the cases discussed here concern the removal of groups so that no unfavourable steric interference is set up between the enzyme and substrate. But, if in base mispairing, G·T takes up the same overall geometry as an A·T base pair and A·C the same as a G·C pair, unfavourable steric and electrostatic interactions will be set up between >NH groups. These effects will compound the loss of hydrogen bonds in the mispairs. In general, unfavourable steric interactions are an important element in specificity, as van der Waals' repulsion energies are such a strong function of interatomic distance. High specificity by steric repulsion is the basis of double-sieve sorting of amino acids by the aminoacyl-tRNA synthetases[21]. Thus, two important components of biological specificity are the avoidance of unsolvated charges and unfavourable steric interactions between enzyme and substrate.

This work was funded by the MRC.

Received 18 December 1984; accepted 12 February 1985.

1. Kauzmann, W. *Adv. Protein Chem.* **14,** 1-63 (1959).
2. Klotz, I. M. & Franzen, J. S. *J. Am. chem. Soc.* **84,** 3461-3466 (1962).
3. Cantor, C. R. & Schimmel, P. R. *Biophysical Chemistry* pt 1, 277 (Freeman, San Francisco, 1980).
4. Fersht, A. R. & Jakes, R. *Biochemistry* **14,** 3350-3356 (1975).
5. Wilkinson, A. J., Fersht, A. R., Blow, D. M. & Winter, G. *Biochemistry* **22,** 3581-3586 (1983).
6. Monteilhet, C. & Blow, D. M. *J. molec. Biol.* **22,** 407-417 (1978).
7. Rubin, J. & Blow, D. M. *J. molec. Biol.* **145,** 489-500 (1981).
8. Bhat, T. N., Blow, D. M., Brick, P. & Nyborg, J. *J. molec. Biol.* **158,** 699-709 (1982).
9. Blow, D. M. & Brick, P. in *Biological Macromolecules and Assemblies* Vol. 2 (eds Jurnak, F. & McPherson, A.) 442-469 (Wiley, New York, 1985).
10. Fersht, A. R. *et al. Angew. Chem.* **23,** 467-473 (1984).
11. Winter, G., Fersht, A. R., Wilkinson, A. J., Zoller, M. & Smith, M. *Nature* **299,** 756-758 (1982).
12. Wilkinson, A. J., Fersht, A. R., Blow, D. M., Carter, P. & Winter, G. *Nature* **307,** 187-188 (1984).
13. Fersht, A. *Enzyme Structure and Mechanism* Ch. 2 (Freeman, New York, 1985).
14. Carter, P. J., Winter, G., Wilkinson, A. J. & Fersht, A. R. *Cell* **38,** 835-840 (1984).
15. Weiner, S. J. *et al. J. Am. chem. Soc.* **106,** 765-784 (1984).
16. Jencks, W. P. *Catalysis in Chemistry and Enzymology* (McGraw-Hill, New York, 1969).
17. Hine, J. *J. Am. chem. Soc.* **94,** 5766-5771 (1972).
18. Monteilhet, C., Blow, D. M. & Brick, P. *J. molec. Biol.* **173,** 477-485 (1984).
19. Fersht, A. R. & Dingwall, C. *Biochemistry* **18,** 1245-1249 (1979).
20. Fersht, A. R. *Trends biochem. Sci.* **9,** 145-147 (1984).
21. Fersht, A. R. & Dingwall, C. *Biochemistry* **18,** 2627-2631 (1979).
22. Waye, M. M. Y., Winter, G., Wilkinson, A. J. & Fersht, A. R. *EMBO J.* **2,** 1827-1829 (1983).
23. Fersht, A. R., Shindler, J. S. & Tsui, W.-C. *Biochemistry* **19,** 5520-5524 (1980).

I had always emphasized that the changes in binding energy observed on mutation are not necessarily the true changes in the interaction energy between the different side chains and the protein. They are, in general, just a quantitative measure of relative binding, that is, specificity. The distinction has invariably been ignored or not understood. So, I later wrote this up as a formal analysis[64], after a simple *Trends in Biochemical Sciences* article[65].

Reprinted from Biochemistry, 1988, *27*, 1577.

Relationships between Apparent Binding Energies Measured in Site-Directed Mutagenesis Experiments and Energetics of Binding and Catalysis†

Alan R. Fersht

Department of Chemistry, Imperial College of Science and Technology, London SW7 2AY, U.K.

Received June 15, 1987; Revised Manuscript Received October 1, 1987

ABSTRACT: The use of binding energy in molecular recognition and enzyme catalysis is currently being probed by experiments on engineered proteins. The interaction energy of an individual side chain with a substrate may be quantified by comparing the binding and rate constants for wild-type enzyme with those for a mutant in which the side chain has been truncated. An apparent binding energy ΔG_{app} is obtained. The physical significance of ΔG_{app} is analyzed with particular reference to hydrogen bonding where one partner in the bond is deleted by mutagenesis. The following conclusions have been drawn for situations where mutagenesis does not unduly perturb the structure of the protein. ΔG_{app} is always a measurement of specificity of binding and catalysis. But, it does not generally measure the incremental binding energy of the hydrogen bond ΔG_{bind}. The discrepancy between ΔG_{app} and ΔG_{bind} is especially large when mutation leaves a charged donor or acceptor unpaired. Here, ΔG_{app} overestimates ΔG_{bind} by possibly several kilocalories per mole. On the other hand, changes in ΔG_{app} ($\Delta\Delta G_{app}$) as a reaction proceeds through its intermediates and transition states are particularly amenable to simple analysis. It is shown that $\Delta\Delta G_{app}$ can measure changes in ΔG_{bind} ($\Delta\Delta G_{bind}$). For example, if there is a change in the energy of an individual bond on going from one state to the next, then $\Delta\Delta G_{app} = \Delta\Delta G_{bind}$. This rule breaks down, however, when the particular step analyzed involves the relevant groups on the enzyme and substrate binding to water in one state but to each other in the next state. In this situation, $\Delta\Delta G_{app}$ does not equal $\Delta\Delta G_{bind}$ and can seriously overestimate it when there is an unpaired charged donor or acceptor.

The importance of binding energy in protein–ligand interactions and enzyme catalysis is now increasingly being explored by site-directed mutagenesis experiments. In particular, the dissection of the structure and activity of the tyrosyl-tRNA synthetase has relied heavily on the measurement of apparent binding energies, which have detected changes in individual interaction energies as the reaction proceeds. The interpretation of the experimental data is, however, not straightforward. It is the purpose of this paper to define the various binding energy terms, to discuss what physical meaning may be attached to experimental data, and to set a rigorous framework for their analysis. The hydrogen bond is the principle interaction analyzed.

Definitions and Basic Equations

Suppose a group Y on a substrate (S) binds to a group X on an enzyme (E). Binding of X and Y is an exchange reaction in which X and Y exchange their interactions with water (w) to interact with each other (eq 1). Similarly, the water

$$\text{E–X·w} + \text{w·Y–S} \rightleftarrows \text{E–X·Y–S} + \text{ww} \qquad (1)$$

solvating X and Y is released to interact with bulk water. The binding energy of X and Y, ΔG_{bind}, may be viewed as a free energy of transfer of X and Y from water to their binding environment. ΔG_{bind} is equivalent to an incremental binding energy in the ES complex. The overall energetics of the interaction depend on the relative individual interaction energies in eq 1 and any changes in entropy and energetics of solvent structure that occur. If the interaction energy of X and w is $G_{X\cdot W}$, of Y and W is $G_{Y\cdot W}$, and of X and Y is $G_{X\cdot Y}$ and the free energy of water entering bulk solvent and any associated changes in the free energy of the solvent is G_W, then

$$\Delta G_{bind} = G_{X\cdot Y} + G_W - G_{X\cdot W} - G_{Y\cdot W} \qquad (2)$$

The value of ΔG_{bind} depends on the fit between X and Y. There is thus a spectrum of values. When there is perfect complementarity between the structures of X and Y and there is no strain or undue loss of entropy on binding, ΔG_{bind} tends to its maximum value, termed the intrinsic binding energy (Jencks, 1981). The value of ΔG_{bind} can also vary throughout a reaction, and changes in ΔG_{bind} ($\Delta\Delta G_{bind}$) are an important component of the energetics of catalysis.

Effects on Binding Energy of Modification of the Group X on the Enzyme. The group X may be modified by site-directed mutagenesis to alter the interactions with Y in two extreme ways (Figure 1, E mutated to E′). First, X may be substituted by another group which can interact with the substrate either favorably or unfavorably. This situation is too difficult to analyze in a general manner as it depends upon the specific substitutions and so will not be pursued. The second is to remove X by deletion, preferably replacing it by a hydrogen atom, so that the interaction between the the group Y and X is simply removed. Deletion can, however, have two extreme results. The first is to leave an empty cavity between the enzyme and substrate. The second is to allow free access of water to the group Y on the substrate and to the mutated region of the enzyme. The two types of deletion have different consequences, which are analyzed below.

We can measure the dissociation constants of the ES and ES′ complexes (K_S and K'_S, respectively) and use these to define the *apparent binding energy* of –XH with the substrate ΔG_{app} (Wells & Fersht, 1986):

$$\Delta G_{app} = RT \ln (K_S/K'_S) \qquad (3)$$

ΔG_{app} can be related in formal manner to the energies of the specific bonds in the complex which have been altered by mutation by setting up a thermodynamic cycle (Figure 2). If the free energy of each complex is denoted by G, then

$$\Delta G_{app} = (G_{E'} - G_E) - (G_{E'S} - G_{ES}) \qquad (4)$$

† This work was supported by the Medical Research Council of the U.K.

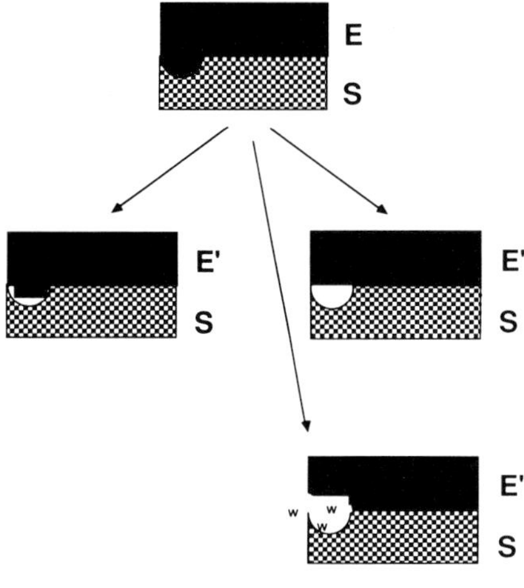

FIGURE 1: Consequences of mutating an enzyme E to E′. (Left) Substitution of a side chain for another that can still interact with the substrate leads to unknown interaction energies. (Right) Deletion of a side chain may leave an empty cavity in the enzyme and deprive the substrate of interactions with water and enzyme. Alternatively, free access of water to the cavity loses approximately the incremental binding energy of the substrate with the mutated group, dependent upon how closely the water in the cavity resembles bulk water.

$$
\begin{array}{ccc}
G_E & K_S & G_{ES} \\
E + S & \rightleftharpoons & ES \\
\updownarrow & & \updownarrow \\
E' + S & \rightleftharpoons & E'S \\
G_{E'} & K'_S & G_{E'S}
\end{array}
$$

FIGURE 2: Thermodynamic cycle relating differences in dissociation constants to differences in free energies of wild-type and mutant enzymes and their enzyme–substrate complexes.

It is seen is seen that ΔG_{app} depends upon the differences in free energy between ES and E′S on the one hand and E and E′ on the other, that is the differences in structure between wild-type and mutant enzymes. Most energy terms cancel out, including those from the covalent modification to the enzyme on mutation. ΔG_{app} thus results from the dissociation energies of the specific bonds between enzyme and substrate that are changed and the energetic changes associated with any structural reorganization of the solvent and enzyme on mutation.

Relationship between ΔG_{app} and ΔG_{bind}. There is a fundamental difference between ΔG_{app} and ΔG_{bind} which is apparent from consideration of the physical processes defining each. ΔG_{bind} relates the binding energy of E and Y with each other compared with their binding energies to water. ΔG_{app} compares the binding energy of Y and E with that of Y and E′. In general, therefore, ΔG_{app} does not equal ΔG_{bind}. The physical meaning of ΔG_{app} is seen from eq 3 to be an experimental measurement of *specificity* of binding since it is derived from relative binding constants.[1]

[1] It is usually more reliable when substrates reacting with enzymes are being analyzed to use an alternative equation from kinetic determinations: $\Delta G_{app} = RT \ln [(k_{cat}/K_M)'/(k_{cat}/K_M)]$, where k_{cat}/K_M and $(k_{cat}/K_M)'$ are the specificity constants for the reactions of E and E′ with S determined from the Michaelis–Menten equation. This avoids complications from nonproductive binding and other phenomena. Since kinetic specificity is defined by the ratio $(k_{cat}/K_M)'/(k_{cat}/K_M)$ (Fersht, 1985), ΔG_{app} is clearly seen to be a direct measure of specificity.

ΔG_{app} is more a measure of ΔG_{bind} for the special case in Figure 1 where deletion allows free access of water to the group Y on the substrate and to the mutated enzyme. This is because mutation removes the interaction between X and Y but allows Y to make its solution interactions; that is, there is no transfer of Y from water when binding to E′. But, even in this example, ΔG_{app} is at best a crude measure of ΔG_{bind} because the properties of water immediately surrounding the enzymes differ somewhat from those of bulk water.

The rest of the analysis concerns ΔG_{app} for deletion mutations where hydrogen bonds are left unpaired. The analysis may be applied to modified substrates binding to a single enzyme by interchanging the symbols E and S where necessary in the equations.

Consequences of Deletion of a Group from a Hydrogen Bond To Leave an Unpaired Donor or Acceptor

Suppose a group –XH in an enzyme interacts with a substrate to form a hydrogen bond (eq 5). The reaction in water

$$\text{S–B}\cdots\text{HOH} + \text{H}_2\text{O}\cdots\text{HX–E} \rightleftharpoons [\text{S–B}\cdots\text{HX–E}] + \text{H}_2\text{O}\cdots\text{HOH} \quad (5)$$

is an exchange reaction in which the donors and acceptors change partners. The overall energetics of the reaction depend on the relative strengths of the individual hydrogen bonds and the entropy changes that occur. Each individual hydrogen bond is characterized by a *hydrogen-bond dissociation energy* which is the depth of the potential energy well of that bond. Representative calculated values of hydrogen bond dissociation energies in vacuo are as follows: $HOH\cdots OH_2$, –6.4 kcal/mol; $H_2O\cdots HSCH_3$, –3.2 kcal/mol; $HOH\cdots S(H)CH_3$, –3.1 kcal/mol; imidazolium/water, –14 kcal/mol, $CH_3CO_2^-\cdots HOH$, –19 kcal/mol (Weiner et al., 1984). Charged groups have greater hydrogen-bond dissociation energies because of the higher electrostatic energies.

ΔG_{bind} is related to the hydrogen-bond dissociation energies by

$$\Delta G_{bind} = G_{EXH\cdot BS} + G_{WW} - G_{EXH\cdot W} - G_{SB\cdot W} + \Delta G_R \quad (6)$$

where $G_{EXH\cdot BS}$ is the hydrogen-bond dissociation energy of $E\text{–}XH\cdots B\text{–}S$, G_{WW} is that of $H_2O\cdots HOH$, etc. ΔG_R is an energy term that contains any entropic or other energetic changes that accompany the reaction, for example, the favorable entropy change accompanying the release of bound water.

Suppose –XH is deleted by mutagenesis to give a mutant E′ so that binding is as in eq 7. The mutation is designed such

$$\text{S–B}\cdots\text{HOH} + \text{H}_2\text{O/E}' \rightleftharpoons [\text{S–B/E}'] + \text{H}_2\text{O}\cdots\text{HOH} \quad (7)$$

that it removes the interaction and does not introduce any steric or other unfavorable or complicating interactions but does not allow access of water to solvate B. To allow for the possibility of the mutation causing energetic changes because of reorganization of the solvent shell of the enzyme or local structure in the enzyme, we add an additional term, ΔG_{reorg}, the "reorganization energy". This contains all the spurious factors arising from the rearrangement of the enzyme and solvent, including any perturbations of the binding of the rest of S–B to the enzyme.

Then, from eq 4

$$\Delta G_{app} = (G_{E'/W} - G_{EXH\cdot W}) - (G_{E'/BS} - G_{EXH\cdot BS}) + \Delta G_{reorg} \quad (8)$$

where $G_{E'/BS}$ is the dissociation energy of the new interaction in the mutant enzyme–substrate complex and $G_{E'/W}$ is the new bond energy between E′ and water. The magnitude of ΔG_{reorg} is unknown and depends on the precise mutations being made. It is the experience of this laboratory from making many series of mutations at many loci that ΔG_{reorg} is probably less than 0.5 kcal/mol for mutations that appear from molecular graphics unlikely to cause gross structural artifacts.

Relationship between ΔG_{app} and ΔG_{bind} for Hydrogen Bonding. $G_{\text{EXH·BS}}$, the hydrogen-bond dissociation energy of E–XH···B–S, is common to eq 6 and 8 and may be eliminated on comparing the two to give

$$\Delta G_{\text{app}} = \Delta G_{\text{bind}} - G_{\text{WW}} + G_{\text{SB·W}} - G_{\text{E'/BS}} + G_{\text{E'/W}} + \Delta G_{\text{reorg}} - \Delta G_{\text{R}} \quad (9)$$

That is, $\Delta G_{\text{app}} \neq \Delta G_{\text{bind}}$ unless the components of the term $(-G_{\text{WW}} + G_{\text{SB·W}} - G_{\text{E'/BS}} + G_{\text{E'/W}} + \Delta G_{\text{reorg}} - \Delta G_{\text{R}})$ cancel out.

Experimental Evidence on Hydrogen-Bond Energetics

(*a*) *Deletion of Hydrogen-Bonding Groups Which Pair with an Uncharged Donor or Acceptor.* A recent compilation of measurements of uncharged hydrogen bonds in enzyme–ligand and nucleotide–nucleotide complexes reveals that both ΔG_{app} and ΔG_{bind} are in the range 0.5–1.8 kcal/mol for bonds where there are no unfavorable steric factors (Fersht et al., 1985; Fersht, 1987). The similarity between ΔG_{app} and ΔG_{bind} may arise for two reasons. First, the term $(-G_{\text{WW}} + G_{\text{SB·W}} - G_{\text{E'/BS}} + G_{\text{E'/W}} + \Delta G_{\text{reorg}} - \Delta G_{\text{R}})$ in eq 9 may cancel out, and rough calculations suggest that this is so. Second, some of the mutations may allow access of water to the remaining donor or acceptor. As explained earlier, ΔG_{app} roughly approximates to ΔG_{bind} under these circumstances. Nevertheless, the similarity between ΔG_{app} and ΔG_{bind} in these cases must be regarded as fortuitous.

(*b*) *Deletion of Hydrogen-Bonding Groups Which Pair with a Charged Donor or Acceptor.* Deletion of a bond to a charged donor/acceptor weakens binding by a much greater amount, some 3–6 kcal/mol for either one or both partners being charged (Lowe et al., 1987). ΔG_{app} may be related to ΔG_{bind} by substituting the hydrogen-bond dissociation energy for a charged bond into eq 9, for example, that for SB^-···HOH ($G_{\text{SB}^-\text{·W}}$), to give eq 10. $G_{\text{SB}^-\text{·W}}$ is by far the most dominant

$$\Delta G_{\text{app}} = \Delta G_{\text{bind}} - G_{\text{WW}} + G_{\text{SB}^-\text{·W}} + G_{\text{E'/BS}} - G_{\text{E'/W}} + \Delta G_{\text{reorg}} - \Delta G_{\text{R}} \quad (10)$$

component in the term $(-G_{\text{WW}} + G_{\text{SB}^-\text{·W}} + G_{\text{E'/BS}} - G_{\text{E'/W}} + \Delta G_{\text{reorg}} - \Delta G_{\text{R}})$. ΔG_{app} thus considerably overestimates ΔG_{bind}. Because $G_{\text{SB}^-\text{·W}}$ is so high, it is likely that there will be compensating interactions of SB^- with water or the protein. In other words, ΔG_{reorg} will be high, and so eq 10 becomes difficult to interpret.

Difference Energy Diagrams and Relationship of $\Delta\Delta G_{\text{app}}$ to $\Delta\Delta G_{\text{bind}}$

Understanding enzyme catalysis requires knowing the interaction energies between the enzyme and substrate throughout the whole course of reaction. Differential binding in reactions of the tyrosyl-tRNA synthetase has been shown by determining ΔG_{app} for the successive intermediates on the reaction pathway (Wells & Fersht, 1985, 1986; Ho & Fersht, 1986; Fersht et al., 1986). Differences in ΔG_{app} along a reaction pathway, $\Delta\Delta G_{\text{app}}$, can be readily interpreted by the following analysis. Suppose there is a bond –XH···B– between EXH and BS and it changes in strength throughout the reaction as it proceeds through the various intermediate and transition-state complexes. We can apply eq 8 to each state. The terms $G_{\text{E'/W}}$ and $G_{\text{EXH·W}}$ are constants in eq 8 for a particular mutation and so their differential with respect to reaction coordinate is zero. If ΔG_{reorg} and $G_{\text{E'/BS}}$ are also constant along the reaction pathway (i.e., $\Delta\Delta G_{\text{reorg}} = 0$ when the structural changes in the solvent and enzyme on mutation do not change during the course of reaction, which are reasonable assumptions for many situations), then

$$\Delta\Delta G_{\text{app}} = \Delta G_{\text{EXH···S}} \quad (11)$$

Or, directly from eq 9 with the same assumption that $\Delta\Delta G_{\text{reorg}} = 0$

$$\Delta\Delta G_{\text{app}} = \Delta\Delta G_{\text{bind}} \quad (12)$$

That is, $\Delta\Delta G_{\text{app}}$ is a direct measure of the changes in bond dissociation energy of –XH···S as it changes as the reaction proceeds through intermediate steps. Thus, site-directed mutagenesis does measure the subtle changes in bond energies as reactions proceed.

An Exception Where $\Delta\Delta G_{\text{app}} \neq \Delta\Delta G_{\text{bind}}$. The rule that $\Delta\Delta G_{\text{app}} = \Delta\Delta G_{\text{bind}}$ when $\Delta\Delta G_{\text{reorg}} = 0$ breaks down when the bond –XH···B– between EXH and BS does not exist at all in one of the enzyme-bound complexes but is formed in a subsequent complex. This could occur when a step in the reaction is accompanied by an isomerization in an ES complex which involves groups on E and S losing bonds to water and forming an intramolecular hydrogen bond as an integral step, as in eq 13. This is now equivalent to the energetic changes as in eq

$$\text{E–XH···OH}_2\text{·S–B···HOH} \rightleftarrows [\text{E–XH···B–S}] + \text{HOH···OH}_2 \quad (13)$$

5 and so the analysis that applied to ΔG_{app} in eq 9 applies to $\Delta\Delta G_{\text{app}}$ (eq 14). For example, just as $\Delta\Delta G_{\text{app}}$ seriously ov-

$$\Delta\Delta G_{\text{app}} = \Delta\Delta G_{\text{bind}} - G_{\text{WW}} + G_{\text{SB·W}} - G_{\text{E'/BS}} + G_{\text{E'/W}} + \Delta G_{\text{reorg}} - \Delta G_{\text{R}} \quad (14)$$

erestimates ΔG_{bind} when a donor to or an acceptor of a charged group is mutated, so $\Delta\Delta G_{\text{app}}$ overestimates $\Delta\Delta G_{\text{bind}}$. For subsequent steps following the isomerization in eq 13, the relationship $\Delta\Delta G_{\text{app}} = \Delta\Delta G_{\text{bind}}$ should again hold.

In general, however, $\Delta\Delta G_{\text{app}}$ is a more reliable quantity than ΔG_{app} for two reasons. First, the unknown quantity on mutagenesis is the value of ΔG_{reorg}, the sum of any of the free energy perturbations in the enzyme and solvent on mutation due to local variation in structure. Although it is likely that ΔG_{reorg} may not be negligible in some cases, ΔG_{reorg} is likely in many cases to alter the energy level of unligated enzyme and all the complexes by the same amount so that $\Delta\Delta G_{\text{reorg}}$ is very small. Second, $\Delta\Delta G_{\text{app}}$ may be measured with high precision because values are calculated from the ratios of first-order rate constants for the interconversion of enzyme-bound species at saturating concentrations of substrates (Wells & Fersht, 1986). These rate constants are, therefore, independent of both the concentration of enzyme and the concentration of substrate and hence problems of purity, active site titer, and dispensing errors.

Illustrative Examples with Tyrosyl-tRNA Synthetase

The tyrosyl-tRNA synthetase catalyzes the formation of enzyme-bound tyrosyl adenylate from tyrosine and ATP (eq 15). Several side chains at the active site have been mutated

$$\text{E} \underset{}{\overset{\text{Tyr}}{\rightleftarrows}} \text{E·Tyr} \overset{\text{ATP}}{\rightleftarrows} \text{E·Tyr·ATP} \rightleftarrows \text{E·Tyr–AMP·PP}_i \rightleftarrows \text{E·Tyr–AMP} + \text{PP}_i \quad (15)$$

to remove hydrogen-bond donating groups, a selection of data being given in Table I. Application of the above theoretical

Table I: Values of ΔG_{app} for Reactions of Tyrosyl-tRNA Synthetase[a]

complex	ΔG_{app} (kcal/mol)				
	Tyr–Phe-34	Cys–Gly-35	His–Gly-48	His–Ala-40	His–Ala-45
E·Tyr	−0.52	0.05	−0.39	0.05	−0.56
E·Tyr·ATP	−0.49	0.08	−0.83	0.17	−0.61
E·[Tyr–ATP]‡	−0.53	−1.26	−1.63	−5.05	−4.11
E·Tyr–AMP·PP$_i$	−0.69	−1.66	−1.62	<−3	<−3
E·Tyr–AMP	−0.88	−1.64	−1.85	0.23	−0.29

[a] Data from Wells and Fersht (1986) and R. J. Leatherbarrow and A. R. Fersht (unpublished results). E·[Tyr–ATP]‡ is the transition state for the reaction.

work enables a deeper analysis of the data than previously given.

There is detailed knowledge about the interactions of Tyr with the enzyme as the crystal structure of the E·Tyr complex has been solved (Brick & Blow, 1987). The structure of the E·ATP complex is as yet unsolved, and the interactions of ATP with the enzyme have been inferred from the crystal structure of the E·Tyr–AMP complex.

Removal of the hydroxyl group of Tyr-34 of the enzyme to give Phe-34 deletes a hydrogen-bond donor to the tyrosine hydroxyl of the substrate. This is reflected in a destabilization of the E·Tyr complex of about 0.5 kcal/mol. Whether or not the value of ΔG_{app} is equal to ΔG_{bind} depends on whether or not there is a water molecule in the cavity in the mutant enzyme which is hydrogen bonded to the substrate tyrosine hydroxyl and how similar the energetics of binding of this water molecule are to those of bulk water. (X-ray crystallographic studies are in progress on the mutant enzyme–tyrosine complex to search for the water.) The subsequent increases in ΔG_{app}, $\Delta\Delta G_{app}$, do give the increases in the hydrogen-bond strength, $\Delta\Delta G_{bind}$, and show that there is increased stabilization of the E·Tyr–AMP complex.

Mutation of Cys → Gly-35 removes a donor from the enzyme to the 3′-OH of the ribose of Tyr–AMP. ΔG_{app} is approximately zero in the E·Tyr·ATP complex, rising to 1.26 kcal/mol in the transition state. As there is as yet no kinetic or direct structural evidence for hydrogen bonding from Cys-35 to the ribose hydroxyl in the E·Tyr·ATP complex, there is the possibility that the change from E·Tyr·ATP to the transition state E·[Tyr–ATP]‡ is an example of eq 13. That is, ATP may bind first without forming a hydrogen bond to Cys-35 (both remaining bonded to water), and there could be a subsequent conformational change to form the bond. If so, the value of $\Delta\Delta G_{app}$ is not necessarily equal to the value of $\Delta\Delta G_{bind}$ for this process. However, the hydrogen bond is clearly present in the transition state, and so the subsequent increases to 1.65 kcal/mol in the E·Tyr–AMP·PPi and E·Tyr AMP complexes do measure the increases in $\Delta\Delta G_{bind}$. His-48 interacts with the nucleotide throughout the reaction, and so the increases in $\Delta\Delta G_{app}$ again measure the increases in $\Delta\Delta G_{bind}$.

There is a binding site for the γ-phosphate or ATP between Thr-40 and His-45 which binds only in the transition state of the reaction (Leatherbarrow et al., 1985). As there is no evidence from the energetics in Table I that hydrogen bonds are present in the E·Tyr·ATP complex, it is possible that this is again an example of eq 13. Indeed, it was suggested that the γ-phosphate swings into the binding site as the transition state is reached, sloughing off hydrogen-bonded water (Leatherbarrow et al., 1985). If so, the value of $\Delta\Delta G_{app}$ for E·Tyr·ATP going to E·[Tyr–ATP]‡ will overestimate the true $\Delta\Delta G_{bind}$, as described above for eq 10 and 14, and so overestimate the extent of transition-state stabilization. The high values of 4–5 kcal/mol support this view. The same reasoning applies to the high values of $\Delta\Delta G_{app}$ observed in the following paper (Fersht et al., 1988). The mutagenesis experiments thus pinpoint crucial charged residues but may overestimate the stabilization energies.

Conclusions

ΔG_{app} is always an experimental measurement of specificity of binding, and so the experimental values for ΔG_{app} may be generally applied to calculating specificities of binding. ΔG_{app} does not in general equal ΔG_{bind} and is at best only a crude approximation of ΔG_{bind} in circumstances where mutation allows access of water to the region where deletion occurs. In particular, ΔG_{app} measured for mutation of partners of charged group can seriously overestimate ΔG_{bind}. Values of $\Delta\Delta G_{app}$ from differences energy diagrams do frequently measure true changes in binding energies as reactions proceed.

References

Brick, P., & Blow, D. M. (1987) *J. Mol. Biol. 194*, 287–297.

Fersht, A. R. (1985) *Enzyme Structure and Mechanism*, 2nd ed., Freeman, New York.

Fersht, A. R. (1987) *Trends Biochem. Sci.* (*Pers. Ed.*) *12*, 301–304.

Fersht, A. R., Shi, J. P., Knill-Jones, J. W., Lowe, D. M., Wilkinson, A. J., Blow, D. M., Brick, P., Carter, P., Waye, M. M. Y., & Winter, G. (1985) *Nature* (*London*) *314*, 235–238.

Fersht, A. R., Wells, T. N. C., & Leatherbarrow, R. J. (1986) *Trends Biochem. Sci.* (*Pers. Ed.*) *11*, 321–325.

Fersht, A. R., Knill-Jones, J. W., Bedouelle, H., & Winter, G. (1988) *Biochemistry* (following paper in this issue).

Ho, C., & Fersht, A. R. (1986) *Biochemistry 25*, 1891–1897.

Jencks, W. P. (1981) *Proc. Natl. Acad. Sci. U.S.A. 78*, 4046–4050.

Leatherbarrow, R. J., Fersht, A. R., & Winter, G. (1985) *Proc. Natl. Acad. Sci. U.S.A. 82*, 7840–7844.

Lowe, D. M., Winter, G., & Fersht, A. R. (1987) *Biochemistry 26*, 6038–6043.

Weiner, S. J., Kollman, P. A., Case, D. A., Singh, U. C., Ghio, C., Alagona, G., Profeta, S., & Weiner, P. (1984) *J. Am. Chem. Soc. 108*, 765–784.

Wells, T. N. C., & Fersht, A. R. (1985) *Nature* (*London*) *316*, 656–659.

Wells, T. N. C., & Fersht, A. R. (1986) *Biochemistry 25*, 1881–1886.

Chapter 8

Catalysis, Specificity and Non-covalent Interactions

'Experiment is the only means of knowledge at our disposal.
The rest is poetry, imagination.'

Max Planck

The next stage of the project was to examine the interplay of binding energy and catalysis by measuring the rate constants for individual steps in reactions using the stopped-flow fluorimeter, which was brought out of retirement. The first results were exciting beyond expectation. My student Tim Wells quickly found that the binding energy between side chains of the protein and the sugar ring of ATP, which was not at the seat of reaction, was used for increasing the rate of reaction rather than giving tighter binding[66]. These were small effects of an order of magnitude, but conceptually very important.

8.1 1985: Imperial College Group. *Top row*: Paul Thomas, Jack Knill-Jones, Tim Wells, Alan, Tammy Gray, Denise Low and Leisha Borgford. *Bottom row*: Robin Leatherbarrow, Thor Borgford, Nigel Brand, Alan Russell, Walter Ward and Hugh Jones.

Hydrogen bonding in enzymatic catalysis analysed by protein engineering

Tim N. C. Wells & Alan R. Fersht

Department of Chemistry, Imperial College of Science and Technology, London SW7 2AY, UK

Two historic publications have suggested that enzymes can utilize the binding energy with their substrates to increase the rate of chemical catalysis—a concept termed strain[1,2]. Specifically, the structure of an enzyme was described as complementary to the structure of the transition state of its substrate rather than to that of the unreacted substrate, so that the substrate would tend to be distorted on binding to the enzyme. The current understanding of strain does not require such distortion. Instead, it is thought that there are groups on the enzyme which interact better with the transition state than with the unaltered substrate, an arrangement known as differential binding of substrates and transition states[3]. The improvement in binding energy caused by going from substrate to transition state lowers the activation energy of the reaction and so increases the catalytic rate. Here we report direct evidence in support of this notion. Rapid reaction kinetics on mutants of a tyrosyl-transfer RNA synthetase produced by protein engineering show that certain hydrogen-bonding side chains some distance from the reacting bonds of the substrate bind to ATP preferentially in the transition state of the reaction rather than in the unreacted state. In this way, the binding energies of the interactions are used primarily to lower the apparent activation energy of the chemical steps rather than to enhance binding.

The hypothesis of better binding in the transition state has been tested using substrates and inhibitors with varying structures. Two lines of supporting evidence have been adduced from classical enzyme kinetics[3]. First, transition-state analogues, which usually mimic the changes in geometry about the bonds that are being broken and made in the reaction, often bind between 10^2 and 10^4 times more strongly than the original substrates. Second, experiments on proteases with broad specificity show that additional groups on a substrate can lead to an increase in k_{cat} rather than stronger binding. These experiments suggest that groups on the substrate that are far from the seat of reaction can alter the catalytic rate. Many of the substrates used in these studies, however, differ greatly from each other or from the specific substrates, and the results may arise simply from the nonproductive binding of the smaller substrates[3].

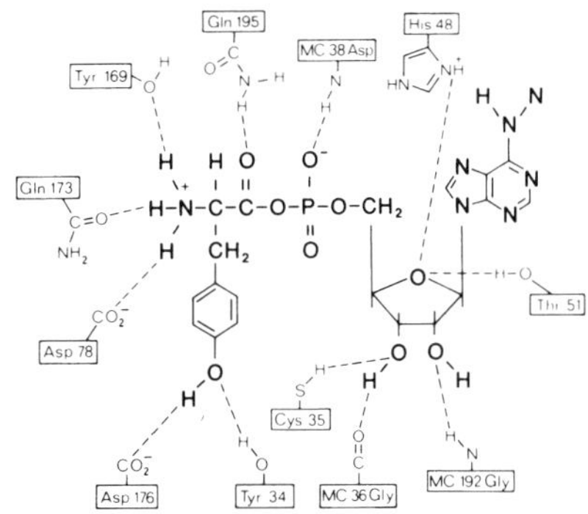

Fig. 1 Side chains of the tyrosyl-tRNA synthetase that form hydrogen bonds with tyrosyl adenylate (courtesy of D.M. Blow and P. Brick).

The advent of protein engineering allows the idea of differential binding of transition states and substrates to be tested directly by varying the structure of the enzyme. An amino-acid side chain that interacts with the substrate can be altered to remove the interaction. If the side chain binds equally well with the substrate in both the ground and transition states, then removal should raise the dissociation constant K_S and not affect k_{cat}. Conversely, if the side chain binds the substrate only in the transition state, then removal should leave K_S unaltered and just lower the value of k_{cat}. Importantly, just one interaction out of many with the substrate may be changed at a time so that artefacts associated with a change of mode of binding of the substrate will be minimized. The contributions of individual interactions to differential binding energies are not known and this approach enables them to be quantified.

Genes of mutant tyrosyl-tRNA synthetases (from *Bacillus stearothermophilus*), which have had the relevant hydrogen-bonding groups removed, have been generated for use in experiments to determine the role of hydrogen bonding in biological specificity[4]. Values of k_{cat} and K_M for tyrosine and ATP for the activation reaction (equation (1)) were measured by pyrophosphate

$$E + Tyr + ATP \rightleftharpoons E\cdot Tyr\cdot ATP \rightleftharpoons E\cdot Tyr{-}AMP + PP \quad (1)$$

exchange. Unfortunately, because of the complex kinetics, the values of k_{cat} and K_M for pyrophosphate exchange are composite

Table 1 Activation of tyrosine by tyrosyl-tRNA synthetases

Enzyme	k_3 (s^{-1})	K'_S (ATP) (mM)	K_S (Tyr) (μM)
Wild type	38	4.7	12
Δ(321-419)*	34	5.2	12
Tyrosine binding-site mutants			
Tyr→Phe 34	35	4.4	29
Tyr→Phe 169†	35	4.6	1,320
ATP binding-site mutants			
Cys→Ser 35	4.7	4.8	8
Cys→Gly 35	4.0	4.5	11
His→Gly 48	9.9	9.9	23‡
Thr→Ala 51	75	4.7	12

Experiments performed at 25 °C at pH 7.78 (144 mM Tris-HCl) in the presence of 10 mM free $MgCl_2$. k_3, Rate constant for the formation of E.Tyr—AMP obtained by stopped-flow fluorimetry[8], with standard error ±5%; K'_S (ATP) dissociation constant of ATP from the E.Tyr.ATP complex, determined as for k_3; K_S(Tyr), dissociation constant of Tyr from the E.Tyr complex obtained from equilibrium dialysis[8].

* Truncated enzyme lacking the tRNA binding domain (residues 320-419)[9].

† Mutation on truncated enzyme with the reference enzyme being the truncated wild-type enzyme.

‡ This value is abnormally high and may be caused by a loss of an electrostatic interaction with His 48.

Table 2 Apparent binding energies of side chains to substrates and to transition states

Side chain	Binding energy with substrate* ($kcal\ mol^{-1}$)	Binding energy with transition state† ($kcal\ mol^{-1}$)
Tyrosine binding site		
Tyr 34 —OH	−0.52	−0.57
Tyr 169 —OH	−2.58	−2.58
ATP binding site		
Cys 35 —SH	−0.03	−1.21
His 48 —imidazole	−0.44	−1.24
Thr 51 —OH	0.00	+0.40

* Calculated from $\Delta G = -RT \ln K_{mut}/K_{wt}$, where K is the relevant dissociation constant of the substrate and mut and wt refer to mutant and wild-type enzymes respectively.

† Calculated from $\Delta G = RT \ln (k_3/K)_{mut}/(k_3/K)_{wt}$.

Reprinted from Nature, Vol. 316, No. 6029, pp. 656-657, 15 August 1985

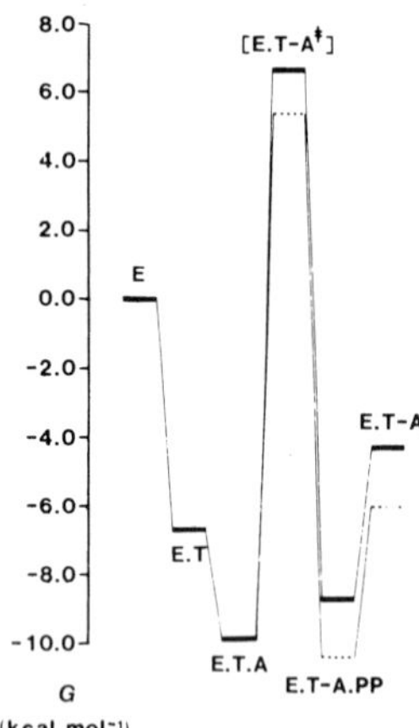

Fig. 2 Gibbs free-energy profiles for the formation of tyrosyl adenylate (E.T-A) by wild-type enzyme (- - - -) and the mutant Cys→Gly 35 (——) (equation (3)). The rate constants for the formation of E.T-A and the transition state E.T−A‡ are taken from Table 1.

$$\mathrm{E} \underset{\pm \mathrm{T}}{\overset{K_S}{\rightleftharpoons}} \mathrm{E \cdot T} \underset{\pm \mathrm{A}}{\overset{K'_S}{\rightleftharpoons}} \mathrm{E \cdot T \cdot A} \underset{k_{-3}}{\overset{k_3}{\rightleftharpoons}} \mathrm{E \cdot T\text{-}A \cdot PP} \underset{\pm \mathrm{PP}}{\overset{K_{PP}}{\rightleftharpoons}} \mathrm{E \cdot T - A} \quad (3)$$

The reverse rate constant k_{-3} ($16\,s^{-1}$ for wild type, $33\,s^{-1}$ for mutant) and dissociation constant for pyrophosphate, K_{PP}(0.61 mM for wild type, 0.63 mM for mutant) were measured by stopped-flow fluorimetry[7]. Free energies were calculated from the standard equations using a standard state of 1 M for tyrosine (T), ATP (A) and pyrophosphate.

values containing rate and equilibrium constants for the pyrophosphorolysis reaction. These data thus gave no conclusive evidence on the interconversion of binding and chemical activation energies. Values for k_3 and the dissociation constants (K_S and K'_S) in equation (2) can, however, be measured

$$\mathrm{E} \underset{K_S}{\overset{\mathrm{Tyr}}{\rightleftharpoons}} \mathrm{E \cdot Tyr} \underset{K'_S}{\overset{\mathrm{ATP}}{\rightleftharpoons}} \mathrm{E \cdot Tyr \cdot ATP} \xrightarrow{k_3} \mathrm{E \cdot Tyr{-}AMP + PP} \quad (2)$$

directly by stopped-flow fluorimetry[5] and equilibrium dialysis (Table 1). From these values, free-energy contributions to binding and catalysis for individual groups can be calculated using the standard equations[3] (Table 2).

The residues that were mutated are all in the binding site of the enzyme, but are not directly involved in bond making and breaking (Fig. 1). Removal of side chains that bind to tyrosine (Tyr→Phe 34 and Tyr→Phe 169) does not significantly alter k_3 but increases the dissociation constant of tyrosine from the E·Tyr·ATP complex (Table 1). The hydroxyl moieties of Tyr 34 and Tyr 169, therefore, can be assumed to interact equally well with tyrosine in its unreacted form and in the transition state of the reaction. The binding energies of the −OH groups of Tyr 34 and Tyr 169 with the substrate and transition state are readily calculated[3] and are seen to be nearly identical (Table 2).

In contrast, removal of side chains that interact with ATP causes significant changes in k_3 but, apart from His→Gly 48, hardly affects the value of K'_S for ATP. Mutation of Cys 35 to Gly 35 or Ser 35 leads to a 10-fold lowering of k_3 with no significant change in K'_S. Changing His→Gly 48 results in a mixture of effects, including a fourfold lowering of k_3 and a twofold increase in K'_S. Mutation of Thr→Ala 51 leads to a twofold increase in k_3 with no discernible effect on K'_S for ATP. This result is entirely consistent with the earlier proposal from structural observations and steady-state kinetic measurements that the bond between Thr 51 and the ribose ring oxygen is long and weak and contributes no binding energy[5,6]. As the value of k_3 is increased on deletion of this bond, it must be even longer and weaker in the transition state. (The presence of an alanine at position 51 gives a more active enzyme, and it is intriguing that the tyrosyl-tRNA synthetase from *Bacillus caldotenax*, which is 99% homologous with that from *Bacillus stearothermophilus*, does have alanine at position 51; ref. 7).

Thus, the side chains of residues 35, 48 and 51 probably have different interaction energies with ATP depending on whether it is in its unreacted form or its transition-state structure. The differences could arise from two effects, or a mixture of the two: (1) an unfavourable interaction in the enzyme–substrate complex that is relieved on formation of the transition state (substrate destabilization); and (2) an interaction that is not properly made in the enzyme–substrate but is realized in the enzyme–transition state complex (transition-state stabilization). We have shown by construction of the full free-energy profile for the formation of tyrosyl adenylate (Fig. 2) that the side chain of Cys 35 does not function purely by substrate destabilization. Removal of the side chain of Cys 35 (Cys→Gly 35) lowers the affinity of the enzyme for tyrosyl adenylate by the full amount expected for the loss of a hydrogen bond[4], showing that it makes favourable interactions with the adenosine moiety of the products as well as with the transition state.

It seems unlikely from model building that the mutation of residues in this study will cause any structural artefact. Further, the double-mutant test has been applied to residues 35 and 48 and no change of structure was detected[6].

We have concluded, therefore, that the interactions of important side chains of the enzyme with tyrosine remain unaltered but that those with ATP are optimized as the enzyme-substrate complex reaches the transition state. Thus we have shown that the binding energy of the enzyme and substrate can be used to enhance the catalytic rate and we have quantified its importance.

This work was supported by the MRC.

Received 30 April; accepted 25 June 1985.

1. Hadane, J. B. S. *Enzymes* 2nd edn (MIT Press, Cambridge, 1965).
2. Pauling, L. *Chem. Engng News* **24**, 1375–1377 (1946).
3. Fersht, A. R. *Enzyme Structure and Mechanism* 2nd edn, Ch. 12 (Freeman, New York, 1985).
4. Fersht, A. R. *et al. Nature* **314**, 235–238 (1985).
5. Wilkinson, A. J., Fersht, A. R., Blow, D. M., Carter, P. & Winter, G. *Nature* **307**, 187–188 (1984).
6. Carter, P., Winter, G., Wilkinson, A. J. & Fersht, A. R. *Cell* **38**, 835–840 (1984).
7. Jones, M. D., Lowe, D. M., Borgford, T. & Fersht, A. R. *EMBO J.* (submitted).
8. Fersht, A. R., Mulvey, R. S. & Koch, G. L. E. *Biochemistry* **14**, 13–18 (1975).
9. Waye, M. M. Y., Winter, G., Wilkinson, A. J. & Fersht, A. R. *EMBO J.* **2**, 1827–1829 (1983).

Then, Robin Leatherbarrow, who was so excited about protein engineering that he commuted initially unpaid to Imperial College from Oxford to join my group as a post-doctoral, found dramatic effects of mutation of side chains that could form hydrogen bonds with the phosphoryl group that was attacked by the nucleophilic carboxylate of the tyrosine[67].

8.2 1985: Tim Wells.

8.3 1985: Alan Russell.

8.4 2003: Mike Sternberg.

Proc. Natl. Acad. Sci. USA
Vol. 82, pp. 7840–7844, December 1985
Biochemistry

Transition-state stabilization in the mechanism of tyrosyl-tRNA synthetase revealed by protein engineering

(enzymology/site-directed mutagenesis)

ROBIN J. LEATHERBARROW†, ALAN R. FERSHT†, AND GREG WINTER‡

†Department of Chemistry, Imperial College of Science and Technology, South Kensington, London SW7 2AY, United Kingdom; and ‡Medical Research Council Laboratory of Molecular Biology, Medical Research Centre, Hills Rd., Cambridge CB2 2QH, United Kingdom

Communicated by William P. Jencks, July 19, 1985

ABSTRACT **The principal catalytic factor in the activation of tyrosine by the tyrosyl-tRNA synthetase is found to be improved binding of ATP in the transition state. The activation reaction involves the attack of the tyrosyl carboxylate on the α-phosphate group of ATP to generate a pentacoordinate transition state. Model building of this complex located a binding site for the γ-phosphate group of ATP, consisting of hydrogen bonds with the side chains of Thr-40 and His-45. Removal of these groups by protein engineering shows that they contribute no binding energy with unreacted ATP but put all of their binding energy into stabilizing the [tyrosine·ATP] transition state [the mutant tyrosyl-tRNA synthetase(Thr-40 → Ala-40; His-45 → Gly-45) has the rate of formation of tyrosyl adenylate lowered by 3.2×10^5 but K_S for ATP is lowered by only a factor of 5]. The side chains of these residues also provide a binding site for pyrophosphate in the reverse reaction. Thus, catalysis is accomplished by stabilization of the transition state by improved binding of a group on the substrate that is distant from the seat of reaction.**

Aminoacyl-tRNA synthetases are responsible for the coupling of an amino acid to its tRNA. For the tyrosyl-tRNA synthetase, as for the majority of aminoacyl-tRNA synthetases, aminoacylation of tRNA proceeds by a two-step process: (*i*) the amino acid is activated by the formation of an enzyme-bound aminoacyl adenylate (Eq. **1**) and (*ii*) there is transfer to tRNA (Eq. **2**) (1) (where E = enzyme).

$$\text{E} + \text{tyrosine} + \text{ATP} \rightarrow \text{E·Tyr–AMP} + \text{PP}_i \quad [1]$$

$$\text{E·Tyr–AMP} + \text{tRNA}^{\text{Tyr}} \rightarrow \text{E} + \text{Tyr-tRNA}^{\text{Tyr}} + \text{AMP} \quad [2]$$

The mechanisms whereby these processes are catalyzed are completely unknown.

The structure of the tyrosyl-tRNA synthetase from *Bacillus stearothermophilus* has been determined to 0.21-nm resolution, and the locations of enzyme-bound tyrosine and tyrosyl adenylate are known (2). The enzyme is a dimer of 2 × 47 kDa. In the crystal structure, however, only the NH_2-terminal 320 amino acids (out of 419) give well-resolved structure in the electron-density map. These residues account fully for the activation reaction, whereas the 99 residues of the COOH terminus are essential for tRNA binding and hence transfer (3). Catalysis and specificity in the tyrosyl-tRNA synthetase from *B. stearothermophilus* have been investigated recently by systematic application of site-directed mutagenesis (4–6).

The amino acid residues of tyrosyl-tRNA synthetase that interact with the bound tyrosyl adenylate are known from the crystallographic studies, but there is no immediate indication of how they catalyze the reaction. The chemical mechanism of the reaction, as shown in Fig. 1, involves the simple attack of the carboxylate group of tyrosine on the α-phosphate group of ATP, resulting in the elimination of magnesium pyrophosphate (7). The classical enzymatic processes of acid–base or covalent catalysis would not seem to be applicable to this reaction as it consists of the attack of a fully ionized good nucleophile on an activated compound with a good leaving group (the magnesium pyrophosphate from ATP).

In this study, we use model building of the probable transition state in the reaction pathway to implicate catalytic interactions with the enzyme. We know the structure of the enzyme-bound tyrosyl adenylate complex, and so we are able to build the transition state for the formation of tyrosyl adenylate by the addition of pyrophosphate to its α-phosphate moiety. To test our predictions, we use site-directed mutagenesis to replace the residues implicated. Kinetic analysis of mutant enzymes then allows quantitation of the interactions involved.

MATERIALS AND METHODS

Construction of Mutant Proteins. The following oligonucleotide primers were synthesized to direct the mutations. Thr-40 → Ala-40: 5′ CCGCCGC*CGGGTCAAAC 3′ (* = mismatched base); His-45 → Gly-45: 5′ GGCCGATA-C*C*CAAACTGT 3′. Site-directed mutagenesis was performed on the tyrosyl-tRNA synthetase gene cloned in the vector M13 mp93 (5, 8). The double-mutant tyrosyl-tRNA synthetase(Thr-40 → Ala-40; His-45 → Gly-45) was constructed by using the His-45 → Gly-45 primer to mutate the tyrosyl-tRNA synthetase(Thr-40 → Ala-40) gene. Mutant enzymes were purified as described by Lowe *et al.* for other mutants (9). All were electrophoretically homogeneous and formed 1 mol of tyrosyl adenylate per mol of enzyme.

Kinetic Procedures. The kinetic constants for the formation of tyrosyl adenylate by tyrosyl-tRNA synthetase were determined in all cases from presteady-state kinetics by direct observation of the formation of enzyme-bound tyrosyl adenylate. Tyrosyl-tRNA synthetase(His-45 → Gly-45) was studied by stopped-flow fluorescence over a period of 0–5 min (10). For tyrosyl-tRNA synthetase(Thr-40 → Ala-40) and tyrosyl-tRNA synthetase(Thr-40 → Ala-40; His-45 → Gly-45), the rate constant for tyrosyl adenylate formation was sufficiently low to allow a single turnover of the enzyme to be followed by manual sampling. The amount of enzyme-bound tyrosyl adenylate was determined at various time intervals by nitrocellulose disc assay [active-site titration (11)]. For tyrosyl-tRNA synthetase(Thr-40 → Ala-40), a complete time course was monitored and the rate constant was determined at various substrate concentrations from the best-fit first-order exponential to the data. For tyrosyl-tRNA synthetase-(Thr-40 → Ala-40; His-45 → Gly-45), the rate constant was determined from the initial rate of the reaction (0–15% completion) and the known enzyme concentration. Michaelis–

The publication costs of this article were defrayed in part by page charge payment. This article must therefore be hereby marked "*advertisement*" in accordance with 18 U.S.C. §1734 solely to indicate this fact.

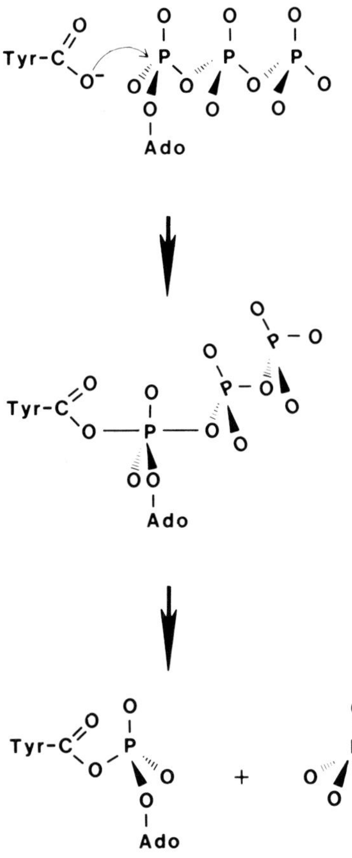

FIG. 1. Chemical mechanism of formation of tyrosyl adenylate from tyrosine and ATP.

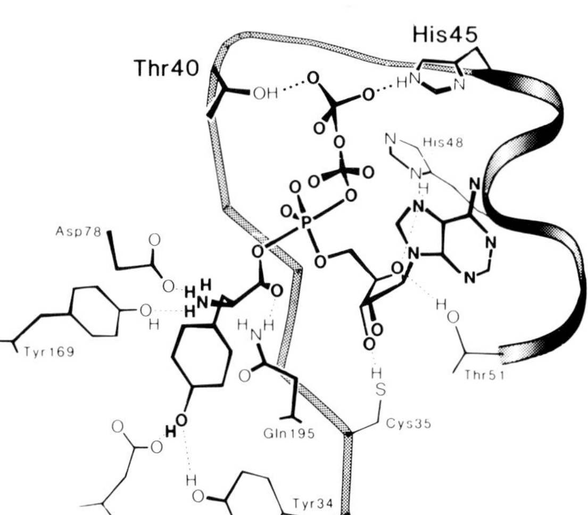

FIG. 2. Model building of the pentacoordinate transition state of the reaction into the crystallographic structure of the tyrosyl-tRNA synthetase. The transition state structure was extrapolated from the known structure of enzyme-bound tyrosyl adenylate. Interactions are shown between the γ-phosphate and the side chains of Thr-40 and His-45.

Menten kinetics were followed by all of the enzymes studied. Values for k_3 and K_S were calculated from the variation of rate constant with substrate concentration. The rates of pyrophosphorolysis of the enzyme-bound tyrosyl adenylates were monitored by nitrocellulose disc filtration on the addition of pyrophosphate to preformed samples of enzyme-bound tyrosyl adenylate. When studying the effect of Mg^{2+} ions, solutions were pretreated, by batchwise adsorption, with Chelex 100 (Bio-Rad).

Computer Graphics. Model building was performed by using the program FRODO using coordinates kindly supplied by P. Brick and D. M. Blow.

RESULTS

Model Building of Reaction Intermediates. Chemical studies have shown that formation of tyrosyl adenylate from tyrosine and ATP results in inversion at the α-phosphorus (7). Since no covalent enzyme-bound intermediates occur, it is implied that the mechanism involves an "in-line" displacement, with the tyrosyl carboxylate acting as a nucleophile and the pyrophosphate being the leaving group (Fig. 1). Such a reaction scheme implies a pentacoordinate transition state. Since enzymic catalysis must involve interactions with the transition state, we adopted the following strategy to delineate the amino acid residues involved. First, starting from the crystallographic structure of the enzyme-bound tyrosyl adenylate, we utilized computer graphics to construct a model of the transition state using a P—O axial bond length of 0.24 nm (12). Finally, by allowing bond rotations around the O—P bonds we looked for interactions with the enzyme that could be implicated in the catalytic mechanism.

From the model building we found that the γ-phosphate group of the intermediate could be placed to interact with the side chains of Thr-40 and His-45 (Fig. 2). These side chains are positioned so that each can make similar polar interactions via the β-OH and NϵH groups, respectively. We should note that His-45 is conserved between various amino acyl-tRNA synthetases (ref. 13; unpublished data), implying an important role for this residue. Although Mg^{2+} is required for the reaction, for simplicity we have not built a Mg^{2+} ion into this model. There is sufficient space for Mg^{2+} to interact with the phosphate groups on the opposite face to that presented to Thr-40/His-45, and we present evidence below that mutation of Thr-40 and His-45 does not affect the binding of magnesium.

Site-Directed Mutagenesis. Although model building suggests a role for the Thr-40 and His-45 side chains, confirmation and quantitation of their involvement require more direct evidence. To provide this, site-directed mutagenesis was used to remove the interacting side chains. The following mutants were constructed and their kinetics were analyzed according to Scheme **1**, which fits the data for catalysis by tyrosyl-tRNA synthetase and all of its mutants. In this scheme, k_{-pp} is found to be fast compared with k_{-3}, and k_{cat} equals k_3 in the presteady state (unpublished data). The second kinetic parameter, K_M, is equal to the substrate dissociation constant, K_S, in the presteady state (i.e., equals k_t'/k_{-t}' or k_a'/k_{-a}' for tyrosine and ATP, respectively).

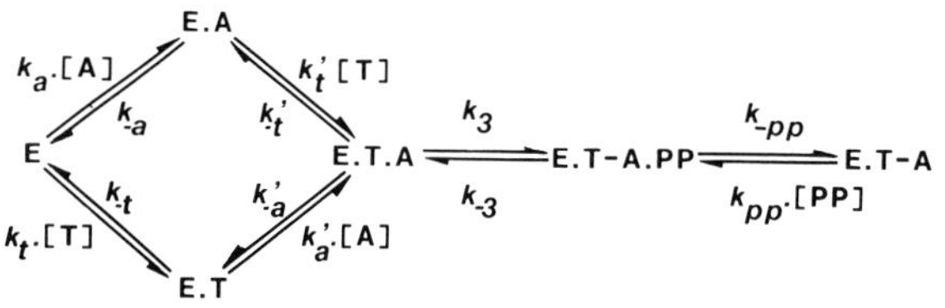

$$E \underset{k_{-a}}{\overset{k_a\cdot[A]}{\rightleftharpoons}} E.A \underset{k_{-t}'}{\overset{k_t'[T]}{\rightleftharpoons}} E.T.A \qquad E \underset{k_{-t}}{\overset{k_t\cdot[T]}{\rightleftharpoons}} E.T \underset{k_{-a}'}{\overset{k_a'\cdot[A]}{\rightleftharpoons}} E.T.A$$

$$E.T.A \underset{k_{-3}}{\overset{k_3}{\rightleftharpoons}} E.T\text{-}A.PP \underset{k_{pp}\cdot[PP]}{\overset{k_{-pp}}{\rightleftharpoons}} E.T\text{-}A$$

SCHEME 1

Tyrosyl-tRNA Synthetase(Thr-40 → Ala-40). In this mutation, the -OH group of the threonine is replaced by an H,

7842 Biochemistry: Leatherbarrow *et al.* *Proc. Natl. Acad. Sci. USA 82 (1985)*

Table 1. Presteady state kinetic parameters for formation of tyrosyl adenylate

Enzyme	k_3,* s^{-1}	K_S for tyrosine, μM	K_S for ATP, mM
Tyrosyl-tRNA synthetase†	38	12	4.7
Tyrosyl-tRNA synthetase(His-45 → Gly-45)	0.16	10	1.2
Tyrosyl-tRNA synthetase(Thr-40 → Ala-40)	0.0055	8.0	3.8
Tyrosyl-tRNA synthetase(Thr-40 → Ala-40; His-45 → Gly-45)	0.00012	4.5	1.1

Experiments were performed at 25°C at pH 7.8 (144 mM Tris·HCl) in the presence of 10 mM $MgCl_2$ (free), 1 unit of inorganic pyrophosphatase per ml, 14 mM 2-mercaptoethanol, and 0.1 mM phenylmethylsulfonyl fluoride under presteady state conditions. k_3 is the forward rate constant for the formation of tyrosyl adenylate (Scheme 1); K_S is the dissociation constant for the substrate (= k'_{-t}/k'_t or k'_{-a}/k'_a for tyrosine and ATP, respectively).
*Extrapolated to infinite substrate concentrations.
†From Wells and Fersht (14). The value of K_S for tyrosine of wild-type enzyme was obtained from equilibrium dialysis and equals k_{-t}/k_t.

removing any hydrogen bonding interaction between residue 40 and the γ-phosphate group of the transition state. The rate constant for formation of tyrosyl adenylate, k_3, by the Ala-40 enzyme is found to be >3 orders of magnitude lower than the native enzyme, whereas values of K_S are hardly altered (Table 1). The time course for a single turnover of the enzyme could be followed by nitrocellulose disc assay over 30–60 min (Fig. 3). In the reverse reaction the rate constant k_{-3} for wild-type enzyme is 14 s^{-1} and K_{pp} (= k_{-pp}/k_{pp}) is 0.7 mM (10). For this mutant, k_{-3}/K_{pp} is 3.6 $s^{-1}\cdot M^{-1}$ and K_{pp} is raised to >8 mM. (Measurements are limited by the solubility of magnesium pyrophosphate.)

Tyrosyl-tRNA synthetase(His-45 → Gly-45). This mutation removes the interacting imidazole side chain. The rate constant for this mutant enzyme is again much less than that of the native enzyme, although it is somewhat higher than for the previous mutant. K_S for tyrosine is unaffected and K_S for ATP is lowered by a factor of 4 (Table 1).

Tyrosyl-tRNA synthetase(Thr-40 → Ala-40; His-45 → Gly-45). As expected, the rate constant for tyrosyl adenylate formation for the double mutant is considerably lower than for either of the previous mutants. The time required for formation of 1 mol of bound tyrosyl adenylate per mol of enzyme is ≈6–8 hr under conditions of saturating concentrations of substrates and is much longer for subsaturating concentrations. The kinetic parameters were therefore obtained from the initial rates of reaction (0–15% completion), monitored by nitrocellulose filter assays over a period of 45 min. k_3 is reduced by 3.2 × 10^5 but values of K_S for ATP and tyrosine are lowered by factors of only 3–4 (Table 1). In the reverse reaction, k_{-3}/K_{pp} is 0.11 $s^{-1}\cdot M^{-1}$ and K_{pp} is >20 mM.

It should be noted that the rate constants for the activation

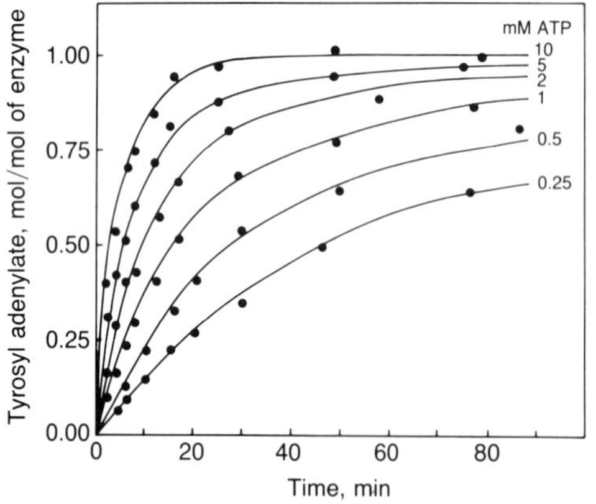

FIG. 3. Time course of formation of enzyme-bound tyrosyl adenylate by tyrosyl-tRNA synthetase(Thr-40 → Ala-40) with varying concentrations of ATP. The concentration of tyrosine was 30 μM.

of tyrosine by the mutants tyrosyl-tRNA synthetase(Thr-40 → Ala-40) and tyrosyl-tRNA synthetase(Thr-40 → Ala-40; His-45 → Gly-45) are so low that they could not be reliably measured by steady-state kinetics. Not only would the rate be difficult to detect but there could be complications that the observed rate may be due to contaminating enzymes or even small amounts of wild-type enzyme generated by infidelities in protein synthesis (misincorporation of amino acids occurs at levels of 10^{-4}–10^{-3}). The measurement of rate constants was possible because presteady-state kinetics was used to detect the rate of formation of 1 mol of tyrosyl adenylate per mol of enzyme. The accumulation of tyrosyl adenylate was in turn allowed because the mutations Thr-40 → Ala-40 and His-45 → Gly-45 reduced only the rate of reaction and not the binding constants of substrates and tyrosyl adenylate.

Effect of Mg^{2+}. It is unlikely that the results of these mutations arise from a change in the Mg^{2+} requirements for the reaction. Removal of magnesium from the reaction mixtures reduces the activity of wild-type enzyme by a factor of >10^3 and abolishes activity of the mutants. Doubling or halving the Mg^{2+} concentration has no effect on the rates of the mutant enzymes, showing that the dissociation constant has not been raised to a high level (results not shown).

DISCUSSION

Mutations at positions 40 and 45 of the tyrosyl-tRNA synthetase have little effect on the dissociation constants of tyrosine, ATP, and tyrosyl adenylate from the enzyme. The binding of pyrophosphate, however, is severely weakened. The most striking feature is the large reduction of the rate constant k_3. Because of the lack of effect on the binding of both substrates and enzyme-bound products, it is unlikely that there is any serious structural change in the enzyme on mutation of these residues. Instead, it is indicated that there is a weakening of binding of pyrophosphate in both the transition state and E·Tyr-AMP·PP complexes. Our model building suggests that Thr-40 and His-45 can form a binding site for the γ-phosphate group of ATP in the transition state and for pyrophosphate in the reverse reaction. The kinetic data show that this binding site contributes no net binding energy for ATP in the enzyme–substrate complex.

Mutation of the γ-phosphate binding site (Thr-40 → Ala-40; His-45 → Gly-45) results in an increase in the energy level of the transition state complex, whereas the energy levels of the enzyme–substrate complex are unaltered (Fig. 4). This results in increased activation energy and consequently lowered k_3. Therefore, these kinetic studies provide direct evidence for stabilization of the transition state as a major factor in the catalytic mechanism of this enzyme.

Magnitude of Catalysis. The tyrosyl-tRNA synthetase has binding sites for tyrosine and ATP that locate the substrates in the correct orientation for reaction. In this way, it converts the nucleophilic attack from an entropically unfavorable

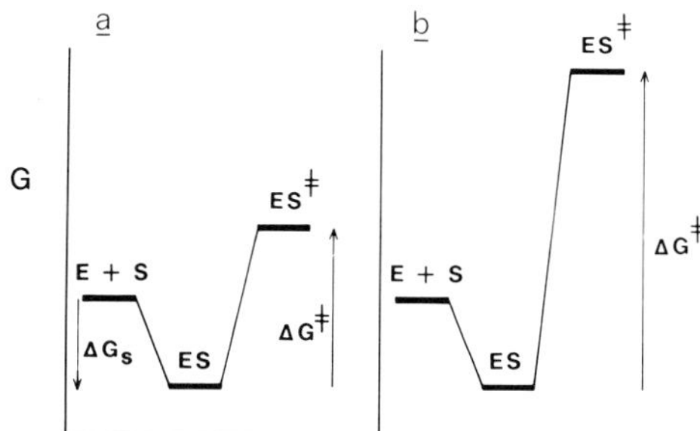

FIG. 4. Schematic Gibbs energy changes for the formation of tyrosyl adenylate by (*i*) tyrosyl-tRNA synthetase and (*ii*) mutant enzymes that have greatly reduced values of k_3 but unaltered K_S. The reaction is $E + S \overset{K_S}{\rightleftharpoons} ES \overset{k_3}{\rightarrow}$ products, where E = enzyme, S = substrate, and ES = enzyme-bound transition state. K_S is proportional to ΔG_S; k_3 is proportional to $\Delta G^\ddagger$ (15). The effect of the mutation is to raise the energy level of the transition state of the reaction.

second-order reaction in solution to a first-order reaction with a concomitant rate enhancement (16). A simple model for the reaction between a carboxylate and a phosphate ester in an enzyme–substrate complex is the cyclization of 2-carboxyphenyl *p*-nitrophenylphosphate (Eq. **3**) (17).

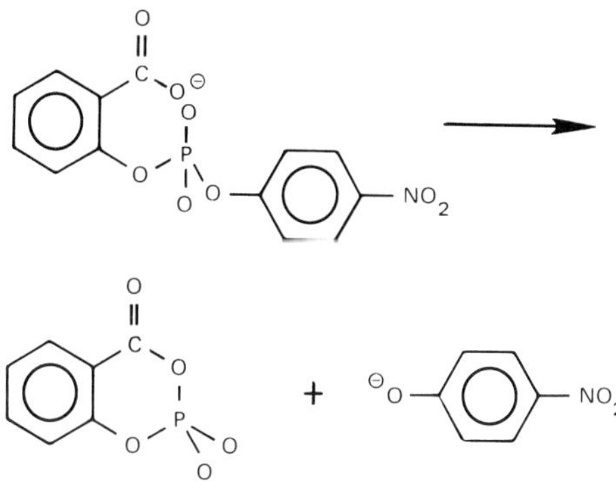

[3]

p-Nitrophenol is an equivalent leaving group to magnesium pyrophosphate because both have similar pK_a values of around 7. The intramolecular cyclization is very rapid compared with its bimolecular counterpart because of the favorable entropy (16) (although there are also possible effects of strain in this compound). The effective concentration of the carboxyl group is estimated to be 6×10^8 M (18), at the upper extreme suggested by Page and Jencks (16). We estimate from the published data at 39°C allowing for a lower pK_a of tyrosine [pK_a of 2 compared with 4 in the model compound (17)] that the equivalent uncatalyzed rate between bound tyrosine and ATP would be *ca.* 10^{-3} s^{-1} at 25°C. This is within the same order of magnitude as that of our least reactive mutant and suggests that the reaction with this enzyme is little more than a simple intramolecular reaction. The wild-type enzyme has a turnover number of 38 s^{-1}, some 4×10^4 times faster than the solution model. Thus, a further factor of 10^4–10^5 or so is required to raise the rate of "uncatalyzed" enzymic reaction, and the transition state stabilization clearly can account for a large fraction of this.

How Does the Enzyme Use the Binding Energies of Thr-40 and His-45 to Stabilize the Transition State and Not the Unreacted Substrate? One possibility, as in Fig. 5, is that the enzyme takes advantage of the large change in bond angles about the α-phosphate group as it goes from four- to five-

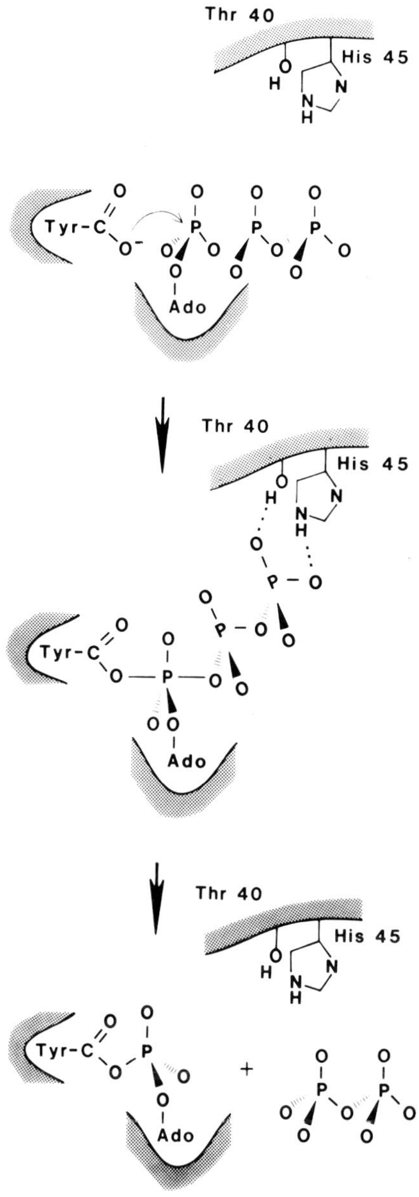

FIG. 5. Schematic mechanism for the catalysis of tyrosyl adenylate formation by tyrosyl-tRNA synthetase. Interactions are made by the side chains of Thr-40 and His-45 to the γ-phosphate group of the transition state.

coordinate. It is possible that the γ-phosphate of substrate ATP does not bind between Thr-40 and His-45 but just remains solvated by water. During the reaction, it swings into its binding site and releases the solvated water. Preliminary data on the activation energies of wild-type and tyrosyl-tRNA synthetase(Thr-40 → Ala-40) support this model because both enzymes have similar enthalpies of activation, but the wild-type enzyme has a more positive entropy of activation (unpublished data). Direct structural evidence on the location of bound ATP would help resolve the mechanistic possibilities. Unfortunately, it has so far proved impossible to solve the crystal structure of the enzyme–ATP complex by x-ray diffraction because of the poor binding (19).

We therefore suggest that the formation of tyrosyl adenylate proceeds by the following mechanism (Fig. 5). (*i*) By providing suitable binding sites for tyrosine and ATP, the enzyme brings the reactive groups together. In this way, binding energy is used to offset the loss of entropy on two molecules coming together (16). (*ii*) By locating the side chains of Thr-40 and His-45 in suitable positions to stabilize the transition state, the enzyme greatly enhances the rate of reaction—loss of these interactions in the tyrosyl-tRNA synthetase(Thr-40 → Ala-40; His-45 → Gly-45) reduces the rate by a factor of >300,000. In this way, binding energy is used to increase catalytic rate.

Use of Binding Energy to Increase Catalytic Rate. It is important to note that interactions made by Thr-40 and His-45 at the γ-phosphate group are distant from the seat of reaction at the α-phosphate. The extent of the rate enhancement provided illustrates dramatically how enzymes can utilize binding energy to increase the rate of chemical catalysis. This concept was originally suggested in 1930 by Haldane, who proposed that appropriate binding sites on an enzyme may be used to help pull apart or push together substrates (20), and was elaborated by Pauling, who proposed that an enzyme should be complementary to, and thus bind preferentially, the transition state rather than the substrate (21). Theoretical considerations (15, 22, 23) show that the rate of reaction is optimized when binding interactions are favored in the enzyme-transition state over the enzyme–substrate complex. This improved binding energy lowers the activation energy of the reaction, increasing the rate. In the tyrosyl-tRNA synthetase, our results show that the side chains of Thr-40 and His-45 make substantially better interactions with the Tyr·ATP transition state than with the enzyme–substrate complex. In addition to the large effects of positions 40 and 45, it has been found that hydrogen bonding groups even further removed from the seat of reaction can cause smaller, but still significant, rate enhancements by preferential binding of the transition state (14).

Previous evidence for transition-state stabilization in enzyme catalysis has come somewhat indirectly from the binding of transition state analogues, from model building, and from kinetic studies on substrate series (23). By the use of protein engineering, we not only provide direct evidence for this type of mechanism in tyrosyl-tRNA synthetase but also identify the groups involved and quantitate the energetics of their involvement.

1. Fersht, A. R. & Jakes, R. (1975) *Biochemistry* **14,** 3350–3356.
2. Blow, D. M. & Brick, P. (1985) in *Biological Macromolecules and Assemblies: Nucleic Acids and Interactive Proteins*, eds. Jurnak, F. & McPherson, A. (Wiley, New York), Vol. 2, pp. 442–469.
3. Waye, M. M. Y., Winter, G., Wilkinson, A. J. & Fersht, A. R. (1983) *EMBO J.* **2,** 1827–1829.
4. Fersht, A. R., Shi, J.-P., Knill-Jones, J., Lowe, D. M., Wilkinson, A. J., Blow, D. M., Brick, P., Carter, P., Waye, M. M. Y. & Winter, G. (1985) *Nature (London)* **314,** 235–238.
5. Winter, G., Fersht, A. R., Wilkinson, A. J., Zoller, M. & Smith, M. (1982) *Nature (London)* **299,** 756–758.
6. Fersht, A. R., Shi, J.-P., Wilkinson, A. J., Blow, D. M., Carter, P., Waye, M. M. Y. & Winter, G. (1984) *Angew. Chem.* **23,** 467–473.
7. Lowe, G. & Tansley, G. (1984) *Tetrahedron Lett.* **40,** 113–117.
8. Carter, P. J., Winter, G., Wilkinson, A. J. & Fersht, A. R. (1984) *Cell* **38,** 835–840.
9. Lowe, D. M., Fersht, A. R., Wilkinson, A. J., Carter, P. & Winter, G. (1985) *Biochemistry* **24,** 5106–5109.
10. Fersht, A. R., Mulvey, R. S. & Koch, G. L. E. (1975) *Biochemistry* **14,** 13–18.
11. Wilkinson, A. J., Fersht, A. R., Blow, D. M. & Winter, G. (1983) *Biochemistry* **22,** 3581–3586.
12. Boyd, D. B. (1969) *J. Am. Chem. Soc.* **91,** 1200–1205.
13. Barker, D. G. & Winter, G. (1982) *FEBS Lett.* **145,** 191–193.
14. Wells, T. N. C. & Fersht, A. R. (1985) *Nature (London)* **316,** 656–657.
15. Jencks, W. P. (1975) *Adv. Enzymol.* **43,** 219–410.
16. Page, M. I. & Jencks, W. P. (1971) *Proc. Natl. Acad. Sci. USA* **68,** 1678–1683.
17. Khan, S. A., Kirby, A. J., Wakselman, M., Horning, D. P. & Lawler, J. M. (1970) *J. Chem. Soc. (B)*, 1182–1187.
18. Kirby, A. J. (1980) *Prog. Phys. Org. Chem.* **17,** 183.
19. Monteilhet, C., Blow, D. M. & Brick, P. (1984) *J. Mol. Biol.* **173,** 477–485.
20. Haldane, J. B. S. (1930) *Enzymes* (Longmans, Green, London).
21. Pauling, L. (1946) *Chem. Eng. News.* **24,** 1375.
22. Fersht, A. R. (1974) *Proc. R. Soc. London Ser. B* **187,** 397–407.
23. Fersht, A. R. (1985) *Enzyme Structure and Mechanism* (Freeman, New York), pp. 311–346.

We had discovered a totally novel reaction: catalysis by the enzyme did not involve any of the classic chemical mechanisms invoking acid, base, or nucleophilic catalysis but the binding energy between the enzyme and substrate drove the reaction because of the tighter binding of the transition state. Such catalysis had long been thought to be a part of enzymatic catalysis, but here it provided all of the catalysis. This mechanism for the reaction has stood the test of time.

By now, we had established experimentally all of the ideas from classical enzymology of how the binding energy of individual side chains could differentially affect the binding of substrates, transition states and products. It had been thought since the speculations of J. B. S. Haldane in about 1930 and then from Linus Pauling in the 1940s and 1950s that the structure of an enzyme should be complementary to that of the structure of the transition state of a reaction. But, we found that a group of side chains had complementarity to the unstable intermediate in the reaction, so as to sequester it for the next step[68].

Reprinted from Biochemistry, 1986, *25*, 1881.

Use of Binding Energy in Catalysis Analyzed by Mutagenesis of the Tyrosyl-tRNA Synthetase†

Tim N. C. Wells and Alan R. Fersht*
Department of Chemistry, Imperial College of Science and Technology, London SW7 2AY, U.K.
Received August 14, 1985

ABSTRACT: The utilization of enzyme–substrate binding energy in catalysis has been investigated by experiments on mutant tyrosyl-tRNA synthetases that have been generated by site-directed mutagenesis. The mutants are poorer enzymes because they lack side chains that form hydrogen bonds with ATP and tyrosine during stages of the reaction. The hydrogen bonds are not directly involved in the chemical processes but are at some distance from the seat of reaction. The free energy profiles for the formation of enzyme-bound tyrosyl adenylate and the equilibria between the substrates and products were determined from a combination of pre-steady-state kinetics and equilibrium binding methods. By comparison of the profile of each mutant with wild-type enzyme, a picture is built up of how the course of reaction is affected by the influence of each side chain on the energies of the complexes of the enzyme with substrates, transition states, and intermediates (tyrosyl adenylate). As the activation reaction proceeds, the apparent binding energies of certain side chains with the tyrosine and nucleotide moieties increase, being weakest in the enzyme–substrate complex, stronger in the transition state, and strongest in the enzyme–intermediate complex. Most marked is the interaction of Cys-35 with the 3′-hydroxyl of the ribose. Removal of the side chain of Cys-35 leads to no change in the dissociation constant of ATP but causes a 10-fold lowering of the catalytic rate constant. It contributes no net apparent binding energy in the E·Tyr·ATP complex and stabilizes the transition state by 1.2 kcal/mol and the E·Tyr–AMP complex by 1.6 kcal/mol. The preferential stabilization of products causes the unfavorable equilibrium constant for the formation of Tyr–AMP and PP_i from Tyr and ATP in solution (3.5×10^{-7}) to be displaced to a value of 2.3 for enzyme-bound reagents. These experiments thus show that (i) the binding energies of side chains remote from the seat of reaction can be used to increase catalytic rates and (ii) the structure of regions of the binding site of an enzyme can be closer in complementarity to an unstable enzyme-bound intermediate than to the transition state for its formation.

The distinctive characteristic of enzyme catalysis compared with solution catalysis is that the enzyme specifically binds its substrate and can use the binding energy to enhance catalytic rate. Pauling (1946) suggested that an enzyme should be complementary in structure to the transition state of the substrate rather than to the substrate itself so that the enzyme would tend to deform the substrate into the transition state. Current ideas on the utilization of binding energy in catalysis support the concept of enzyme–transition-state complementarity but do not demand that the substrate is distorted by the enzyme. The presence of binding sites on the enzyme that form better bonds with the transition state of the substrate than with the unreacted substrate is sufficient to increase the turnover number of the enzyme, k_{cat} (Fersht, 1974, 1985; Jencks, 1975). Evidence has been adduced for the differential binding of transition state and substrate ("transition-state stabilization") from experiments in which the structure of substrates or inhibitors is varied [see Fersht (1985) for review]. But it is now possible to examine directly the catalytic role of binding energy of groups on an enzyme by performing experiments in which the structure of the enzyme is varied by site-directed mutagenesis (Winter et al., 1982; Fersht et al., 1984).

The tyrosyl-tRNA synthetase is particularly suited for such an analysis. Apart from favorable properties in handling and producing accurate kinetic data (Wilkinson, 1983), it is known from the direct solution of the crystal structure of the enzyme-bound tyrosyl adenylate complex combined with mu-

†This work was funded by the MRC of the U.K.

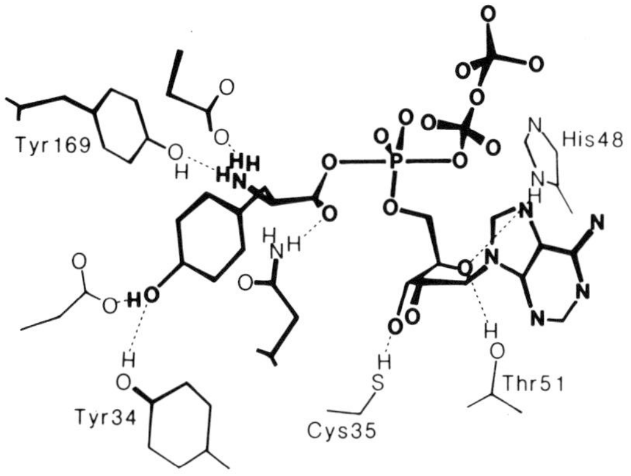

FIGURE 1: Sketch of transition state of tyrosine and ATP during the formation of tyrosyl adenylate illustrating hydrogen bonds made with the enzyme [from model building by extrapolation from the crystal structure of the enzyme–tyrosyl adenylate complex by Leatherbarrow et al (1985)].

tagenesis studies that nine hydrogen bonds are formed between substrates and the side chains of the enzyme (Figure 1). As the potential energy functions of hydrogen bonds are suitable for mediating differential binding effects (Fersht, 1974), analysis of mutants lacking hydrogen-bonding side chains will give the effects of small changes in structure on catalytic activity.

The side chains we have chosen to modify in this study are those that are removed from the seat of reaction and have been shown previously from steady-state kinetics to form good hydrogen bonds with the transition state for the formation of tyrosyl adenylate (Fersht et al., 1985a). We have shown in a preliminary study that they can cause differential binding of substrate and transition state (Wells & Fersht, 1985). We therefore remove interactions that have arisen through evolution and so produce less efficient enzymes. In this study, by determining the free energy profiles of the reaction for wild-type and mutant enzymes, we build up a picture of the effects of adding side chains to the "less-evolved" enzymes. We calculate the contributions of the side chains to the binding energies of the substrates, transition states, and products from direct measurements by pre-steady-state kinetics of the rate and equilibrium constants for the formation of enzyme-bound tyrosyl adenylate and from the measured dissociation constants of the relevant substrates from their complexes with the enzyme.

Experimental Procedures

Materials

Reagents were obtained from Sigma (London) and radiochemicals from Amersham International.

Preparation and Purification of Tyrosyl-tRNA Synthetases. Mutant proteins were prepared with mutant M13mp93 phage constructed as described previously (Carter et al., 1984; Fersht et al., 1985b). An overnight culture of *Escherichia coli* JM101 in 2xTY medium was diluted 100-fold into 500 mL of fresh medium and infected with M13 at high multiplicity. Growth was continued at 37 °C in an orbital shaker, with aeration for 4–5 h. The starter culture was transferred to an MBR bioreactor fermentation system containing 10 L of 2xTY medium. The culture was fermented overnight at 37 °C and pH 6.8 with an aeration rate of 10 L s^{-1}. The cells were harvested by passage of the culture through a Pellicon filtration unit and then by centrifugation of the retentate at 8000 rpm for 20 min. Purification was continued as described previously (Winter et al., 1982), except that the final materials was spplied to a fast protein liquid chromatography (FPLC) Mono Q column (Pharmacia) equilibrated in 20 mM tris(hydroxymethyl)aminomethane hydrochloride (Tris-HCl), pH 7.8, and eluted by using a gradient of 100–300 mM NaCl dissolved in the equilibration buffer. All enzymes were purified to electrophoretic homogeneity and were found to bind 1.0 mol of tyrosyl adenylate/mol of enzyme (Wilkinson et al., 1983).

Methods

All experiments were performed at 25 °C in a standard buffer containing 144 mM Tris-HCl (pH 7.78), 10 mM $MgCl_2$ (free), 14 mM 2-mercaptoethanol, and 0.1 mM phenylmethanesulfonyl fluoride. Additional $MgCl_2$ was added where necessary to compensate for its complexing with ATP and pyrophosphate.

Rate constants for formation of enzyme-bound tyrosyl adenylate and its pyrophosphorolysis were obtained by monitoring protein fluorescence by stopped flow, with excitation at 290 nm, and measuring emission at wavelengths greater than 325 nm after passage through a cutoff filter (Fersht et al., 1975). Typically, the final enzyme concentration was 0.25 μm although higher enzyme concentrations were occasionally used to improve the signal to noise ratio.

Forward Reaction (Tyrosyl Adenylate Formation): ATP Dependence. The forward reactions for all the mutants studied, except TyrTS(Phe-169), were monitored in the presence of saturating concentrations of tyrosine (0.2 mM). One syringe of the stopped-flow fluorometer contained enzyme plus 0.2 mM tyrosine and inorganic pyrophosphatase (2 units mL^{-1}) in the standard buffer. The other syringe contained ATP (the concentration of which was varied over a range from $K_a'/5$ to $5K_a'$) and 0.2 mM tyrosine in standard buffer.

The dissociation constant of tyrosine from the mutant Tyr–Phe-169 is so large that it is not possible to saturate the enzyme with tyrosine because of its low solubility. Rate data for this mutant were determined by saturating the enzyme with ATP (25 mM) and monitoring the reaction under various concentrations of tyrosine.

Reverse Reaction (Pyrophosphorolysis of Tyrosyl Adenylate): Pyrophosphate Dependence. The enzyme-bound tyrosyl adenylate complex was prepared in situ, and the excess ATP, tyrosine, and pyrophosphate were removed by desalting on a Sephadex G 25 column in the standard pH 7.78 buffer [apart from TyrTS(Phe-169) where a dilute pH 6 buffer was used because of the instability of the enzyme-bound tyrosyl adenylate]. The reverse rate constant was then measured by monitoring the fluorescence increase on mixing pyrophosphate with the enzyme–tyrosyl adenylate complex (Fersht et al., 1975). [For TyrTS(Phe-169), the solution of the complex in dilute pH 6 buffer was mixed with 2 × concentrated Tris, pH 7.78, buffer.]

The dissociation constants of tyrosine from mutant enzymes other than TyrTS(Phe-169) were determined by equilibrium dialysis (Fersht et al., 1975). The dissociation constant (K_t) of tyrosine from TyrTS(Phe-169) was determined from the kinetics of pyrophosphate exchange as was K_a', the dissociation constant of ATP from E·Tyr·ATP. It can be shown from standard kinetic theory that the rate constant (k) for the loss of label from [γ-^{32}P]ATP at concentrations $\ll K_a'$ is given by $k = (k_3/K_a')([\mathrm{Tyr}][\mathrm{E}]_0/([\mathrm{Tyr}] + K_t)$, where k_3 is the rate constant for the formation of tyrosyl adenylate, provided [ATP] $\ll$ [PP]. K_t is provided directly from the variation of k with [Tyr], and K_a' may be calculated from the measured values of k_3 from stopped-flow experiments. It was confirmed that the pyrophosphate does not inhibit the enzymes at the concentration used (2 mM). A control experiment on wild-

Scheme I

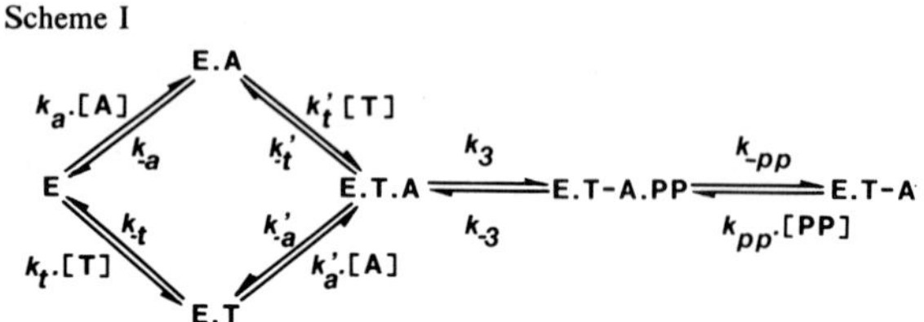

Table I: Rate and Dissociation Constants for the Formation of Enzyme-Bound Tyrosyl Adenylate[a]

enzyme	k_3 (s^{-1})	K_a' (mM)	K_t (μM)
wild type	38	4.7	12
Δ(319–419)[b]	34	5.2	12
tyrosine binding site mutants			
Tyr–Phe-34	35	4.4	29
Tyr–Phe-169[b]	35	4.6	1300
ATP binding site mutants			
Cys–Ser-35	4.7	4.8	8
Cys–Gly-35	4.0	4.5	11
His–Gly-48	9.9	9.9	23

[a] Experiments performed at 25 °C and pH 7.78 (144 nM Tris-HCl, 10 mM $MgCl_2$, and 14 mM 2-mercaptoethanol using stopped-flow fluorescence. Values of k_3 (see Scheme I) are extrapolated to saturating concentrations of ATP and tyrosine. K_a', the dissociation constant of ATP from the E·Tyr·ATP complex, was determined from stopped flow. K_t, the dissociation constant of tyrosine from the E·Tyr complex, was determined by equilibrium dialysis, apart for TyrTS (Phe-169) when kinetics was used. Standard errors are typically ±5% for k_3 and ±10% for dissociation constants. [b] Experiments performed on a truncated enzyme lacking the tRNA binding domain (residues 319–419) (Waye et al., 1983).

type enzyme gave a value of 11.4 μM for K_t from kinetics, compared with 11.6 μM from equilibrium dialysis.

Equilibrium concentrations of enzyme-bound tyrosyl adenylate complexes in mixtures of ^{14}C-labeled tyrosine (1 μM), ATP (400 μM), and pyrophosphate (6–500 μM) were measured by nitrocellulose disk filtration. A solution of wild-type enzyme (100 nM) was incubated at 25 °C in the standard buffer and other reagents were incubated for 1 min; samples (90 μL) were filtered through Schleicher & Schuell 0.45-μm cutoff filters and washed with 3 mL of cold buffer. The enzyme-bound tyrosyl adenylate, which is quantitatively retained by the filter, was assayed by scintillation counting.

Results

Calculation of Energy Levels of Intermediates and Apparent Binding Energies. The activation of tyrosine catalyzed by the tyrosyl-tRNA synthetase is described by Scheme I in which the relevant rate constants are defined ($K_t = k_{-t}/k_t$ = [E][Tyr]/[E·Tyr], $K_a' = k_{-a}'/k_a'$ = [E·Tyr][ATP]/[E·Tyr·ATP], etc.). The dissociation constant of tyrosine from E·Tyr, K_t, was measured directly by equilibrium dialysis [or kinetics for TyrTS(Phe-169)]. K_a' was determined by stopped-flow fluorometric measurements of the rate constants for formation of E·Tyr–AMP on mixing E·Tyr with ATP, which also gave k_3 (Table I). k_{-3} and K_{pp}, the dissociation constant of pyrophosphate from E·Tyr–AMP, were determined by stopped-flow fluorometry of the pyrophosphorolysis reaction on mixing E·Tyr–AMP with pyrophosphate (Table II).

The energy level, G, of each state of the enzyme was calculated from the usual thermodynamic equations by using standard states of 1 M for ATP, Tyr, and pyrophosphate (Table III and Figure 2). Relative to free enzyme (i.e., defining $G_E = 0$)

$$G_{E\cdot Tyr} = RT \ln K_t \quad (1)$$

$$G_{E\cdot Tyr\cdot ATP} = RT \ln (K_t K_a') \quad (2)$$

Table II: Pyrophosphorolysis of Enzyme-Bound Tyrosyl Adenylate Complexes[a]

enzyme	k_{-3} (s^{-1})	K_{pp} (mM)	K_{eq} (k_3/k_{-3})
wild type	16.6	0.61	2.29
Δ(319–419)	15.3	0.68	2.22
tyrosine binding site mutants			
Tyr–Phe-34	22.2	0.45	1.58
Tyr–Phe-169	13.8	0.74	2.52
ATP binding site mutants			
Cys–Gly-35	32.8	0.63	0.12
Cys–Ser-35	31.0	0.44	0.15
His–Gly-48	16.4	0.41	0.60

[a] Conditions as in Table I. Accuracy is ±5% for rate constants and ±10% for dissociation constants (standard error).

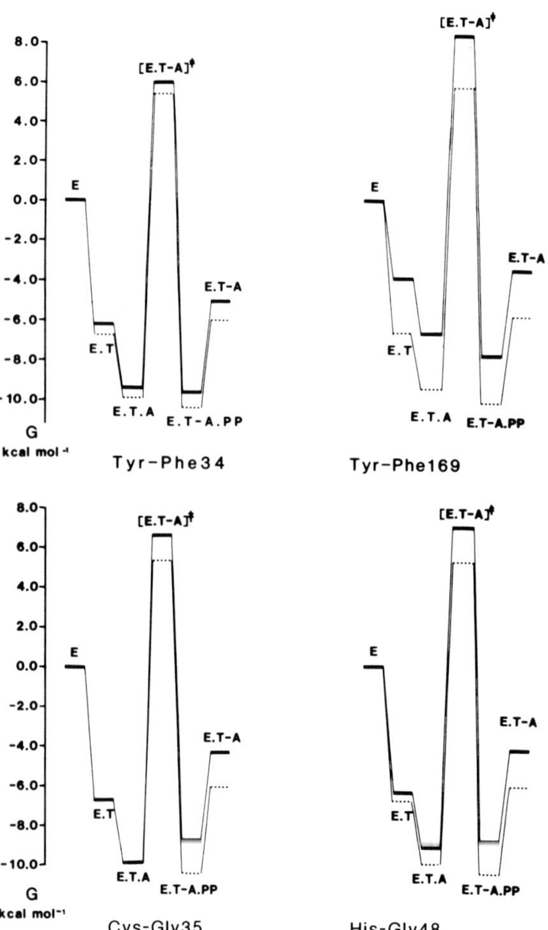

FIGURE 2: Gibbs' free energy profiles for the formation of tyrosyl adenylate and pyrophosphate, as defined in Scheme I, by wild-type (energy levels in broken lines) and mutant (energy levels in heavy lines) tyrosyl-tRNA synthetases, using standard states of 1 M for tyrosine, ATP, and pyrophosphate.

[note that the substrate binding is random ordered (Santi & Pena, 1971) and $K_a'K_t = K_aK_t'$. It is simple to measure $K_t \cdot K_a'$ but difficult to measure the individual components of K_aK_t' since the affinity of the free enzyme for ATP is too low to be measured by equilibrium dialysis and the binding of tyrosine to E·ATP involves highly complex pre-steady-state kinetics (Fersht et al., 1975)]

$$G_{[E\cdot Tyr\cdot ATP]^\ddagger} = RT \ln (k_BT/h) - RT \ln (k_3/K_a'K_t) \quad (3)$$

$$G_{E\cdot Tyr\text{-}AMP\cdot PP_i} = -RT \ln (k_3/k_{-3}K_a'K_t) \quad (4)$$

$$G_{E\cdot Tyr\text{-}AMP} = -RT \ln (k_3K_{pp}/k_{-3}K_a'K_t) \quad (5)$$

Table III: Gibbs Free Energies of Complexes of Wild-Type and Mutant Enzymes[a]

enzyme	kcal mol^{-1}				
	$G_{E\cdot Tyr}$	$G_{E\cdot Tyr\cdot ATP}$	$G_{[E\cdot Tyr\cdot ATP]^‡}$	$G_{E\cdot Tyr-AMP\cdot PP}$	$G_{E\cdot Tyr-AMP}$
wild type	-6.71	-9.89	5.41	-10.38	-5.99
Δ(319–419)[b]	-6.71	-9.82	5.53	-10.31	-5.99
TyrTS(Phe-34)	-6.19	-9.40	5.94	-9.67	-5.11
TyrTS(Phe-169)[b]	-3.93	-7.12	8.23	-7.66	-3.39
TyrTS(Gly-35)	-6.76	-9.97	6.67	-8.72	-4.35
TyrTS(Ser-35)	-6.95	-10.12	6.42	-9.00	-4.43
TyrTS(Gly-48)	-6.32	-9.06	7.04	-8.76	-4.14
standard error	±0.06	±0.08	±0.08	±0.09	±0.10

[a] Standard state = 1 M Tyr, 1 M ATP, 1 M PP_i, and free enzyme (G_E = 0). [b] Truncated enzyme (Waye et al., 1983).

Table IV: Relative Gibbs Free Energies of Complexes of Mutant Enzymes Equal Apparent Binding Energies of Side Chains[a]

enzyme	kcal mol^{-1}				
	$\Delta G_{E\cdot Tyr}$	$\Delta G_{E\cdot Tyr\cdot ATP}$	$\Delta G_{[E\cdot Tyr\cdot ATP]^‡}$	$\Delta G_{E\cdot Tyr-AMP\cdot PP}$	$\Delta G_{E\cdot Tyr-AMP}$
tyrosine binding site mutants[b]					
TyrTS(Phe–34)	0.5	0.5	0.5	0.7	0.9
TyrTS(Phe–169)	2.8	2.7	2.7	2.6	2.6
ATP binding site mutants[c]					
TyrTS(Gly-35)	0.0	0.0	1.2	1.6	1.6
TyrTS(Gly-48)	0.0	0.4	1.2	1.2	1.5
standard error	±0.08	±0.11	±0.11	±0.12	±0.14

[a] The apparent binding energies of the side chains (–OH, –OH, –CH_2SH, and imidazole) are equivalent to the negative value of ΔG. [b] Energies relative to those of wild type. Free enzyme is the standard state (G_E = 0). [c] The E·Tyr complex is the standard state ($G_{E\cdot Tyr}$ = 0) to isolate effects of mutation in ATP site.

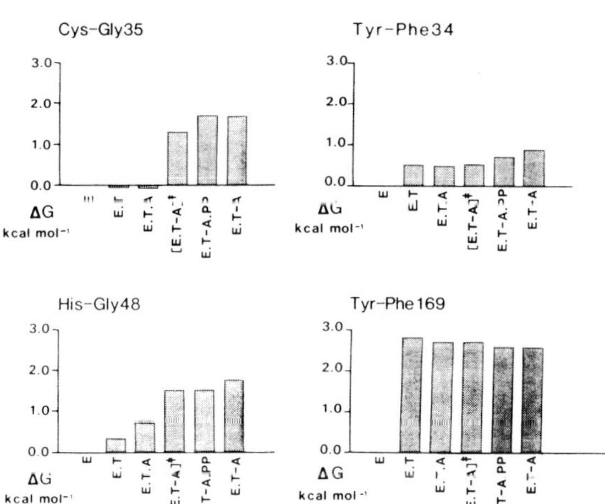

FIGURE 3: Gibbs' free energies of enzyme-bound complexes of mutant enzymes relative to those of wild-type enzyme. These are the global free energy differences, but the negative values of G may be considered as the apparent binding energies of the side chains with the various forms of the substrates during the reaction. Equilibrium concentrations of enzyme-bound tyrosyl adenylate complexes as determined by nitrocellulose disk filtration (100 nM enzyme, 1 μM ^{14}C-labeled tyrosine, 0.4 mM ATP, and indicated concentrations of pyrophosphate in standard buffer; see text). The solid curve is calculated from measured values of K_t, K_a', k_3, k_{-3}, and K_{pp} in Scheme I.

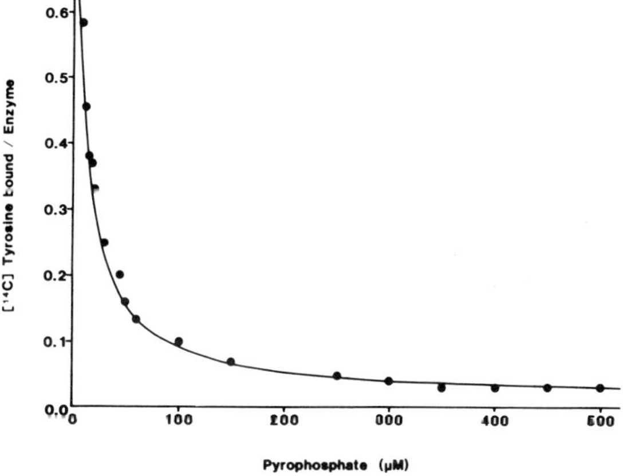

FIGURE 4: Equilibrium concentrations of enzyme-bound tyrosyl adenylate complexes as determined by nitrocellulose disk filtration (100 nM enzyme, 1 μM ^{14}C-labeled tyrosine, 0.4 mM ATP, and indicated concentrations of pyrophosphate in standard buffer; see text). The solid curve is calculated from measured values of K_t, K_a', k_3, k_{-3}, and K_{pp} in Scheme I.

where R is the gas constant, T is the absolute temperature, k_B is Boltzmann's constant, h is Planck's constant, and [E·Tyr·ATP]‡ is the transition state for the formation of tyrosyl adenylate.

The apparent contribution of an individual side chain to the stabilization energy of each state (enzyme-bound complex) can be calculated by subtraction of the energy of the state of the mutant enzyme lacking the relevant side chain from the energy of the same state of wild-type enzyme (Table IV and Figure 3). (The energy differences are, of course, the global properties of the enzymes. But, for analytical purposes, it is often convenient to attribute these to the specifically modified interactions. In many cases, this attribution will be correct. Model building by computer graphics suggests this is so here.) As mutations in the ATP binding site can cause small effects on the dissociation constants of tyrosine, we sometimes use the E·Tyr complex as the standard state (i.e., $G_{E\cdot Tyr}$ = 0) for the ATP-binding site mutants as this isolates the effects of the interactions on the binding of the nucleotide moiety. This does not affect the relative energy levels of the other states or the following discussion.

Detailed Effects of Mutations on Energetics. Gibbs Free Energy Profile of Activation of Tyrosine by Wild-Type Enzyme. The equilibrium constants for the formation of enzyme-bound tyrosyl adenylate complexes were determined, in part, from measurements in the pre steady state of forward and reverse reactions. The validity of this approach was confirmed by measuring the concentrations of these intermediates at equilibrium (Figure 4). The equilibrium concentrations are in good agreement with those calculated from the measured values of K_t, K_a', k_3, k_{-3}, and K_{pp}. The equilibrium constant between the ternary complexes K_3 (=k_3/k_{-3}, = [E·Tyr–AMP·PP$_i$]/[E·Tyr·ATP]) is 2.3.

Removal of Groups Interacting with Tyrosine. Removal of side chains within the tyrosine binding pocket (Tyr–Phe-34 and Tyr–Phe-169) does not significantly alter k_3 or K_a' but increases the dissociation constant for tyrosine. It is thus inferred that the hydroxyl moieties of Tyr-34 and Tyr-169 interact equally well with tyrosine when it is in the unreacted form and when it is in the transition state of the reaction (Figure 3). The value of k_{-3} for TyrTS(Phe-34) is significantly increased so that K_3 is lower for the mutant. Removal of the –OH group destabilizes the E·Tyr–AMP complex more than any other state. Removal of the –OH group from Tyr-169 alters the energy levels of all states equally, within experimental error.

Removal of Groups Interacting with ATP: TyrTS(Cys–Ser-35) and TyrTS(Cys–Gly-35). Mutation of Cys-35 to Gly-35 removes a hydrogen bond to the 3′-OH of the ribose (Figure 1), and mutation to Ser-35 leads to a too-long bond (Winter et al., 1982; Wilkinson et al., 1983). In both cases, this causes a 10-fold lowering of k_3 with no significant changes in K_a' or K_t (Table I). That is, compared with the mutants, the side chain of Cys-35 in wild-type enzyme makes no apparent binding energy contribution in the E·Tyr·ATP complex but does improve the binding in the transition state [E·Tyr·ATP]‡ (Tables III and IV). The pyrophosphorolysis rate constants are increased 2-fold on mutation. The net result is a 15-fold lowering of the equilibrium constant K_3. Removal of the side chain of Cys-35 destabilizes the E·Tyr–AMP and E·Tyr–AMP·PP complexes more than the transition state [E·Tyr·ATP]‡, so the side chain makes an even greater binding energy contribution to the enzyme–product complex.

His–Gly-48. In contrast with the above mutants, the change His–Gly-48 leads to a mixture of effects: a 4-fold lowering of k_3; a 2-fold increase in K_a'; a small change in K_t. The side chain of His-48 causes a net stabilization of the bound adenosine of 0.44 kcal/mol in the E·Tyr·ATP complex, 1.24 kcal/mol in the transition state, and 1.47 kcal/mol in E·Tyr·AMP. The increase in K_t is typical of position 48 mutants and may just reflect the loss of an electrostatic interaction with the carboxylate of the tyrosine (Lowe et al., 1985). It seems unlikely from the double mutant studies of Carter et al. (1984) that the Gly-48 mutation causes any large structural changes within the active site.

Accuracy of Data. Data were collected to a comparable accuracy for all enzymes: the standard error of the mean for dissociation constants is about ±10%, while rate constants are ±5%. It is calculated from theory of errors that the standard error for calculations of $\Delta G_{E \cdot Tyr}$ in Table IV is ±0.08 kcal mol^{-1}. The errors are propagated through the calculations and are calculated to be ±0.11, ±0.11, ±0.12, and ±0.14 kcal mol^{-1} for ΔG in the E·Tyr·ATP, E·[Tyr·ATP]‡, E·Tyr–AMP·PP$_i$, and E·Tyr–AMP complexes, respectively.

Certain of the important conclusions from the data, namely, those concerning the differential contributions of the side chains to the binding energies in the E·Tyr·ATP, E·[Tyr–ATP]‡, and E·Tyr–AMP·PP$_i$ complexes, are based on even higher precision. This is because the relative energies of these complexes may be calculated from just k_3 and k_{-3}, which are measured with an accuracy of ±5%. For example, k_3 for TyrTS(Cys–Ser-35) is lowered by a factor of 9.5 ± 0.07 compared with wild-type enzyme. Thus, relative to wild-type enzyme, the complex E·[Tyr–AMP]‡ in the reaction of mutant is raised by 1.33 ± 0.04 kcal mol^{-1} over the energy of the E·Tyr·ATP complex. Further, as k_{-3} for the mutant is ±5% times higher than for wild-type enzyme, the value of K_3 is (19 ± 2)-fold lower for the mutant. Thus, the energy of the mutant E·Tyr–AMP·PP$_i$ complex relative to the E·Tyr·ATP complex is raised by 1.74 ± 0.06 kcal mol^{-1} compared with wild type. This energy is significantly higher than the change in energy of the transition state. The value of k_{-3} for TyrTS(Tyr–Phe-34) is significantly higher (37%) than that for wild-type enzyme and so indicates that there is a real change in the relative energy levels for that mutant.

Discussion

We have applied a sort of "reverse evolution" to measure the apparent contributions of the binding energies of side chains in the binding site of the tyrosyl-tRNA synthetase to catalysis: removal of the side chains in each case produces a poorer enzyme by either weakening the binding of the substrates or lowering the catalytic rate constant and so represents a backward step in evolution. We are not suggesting that the mutants are the ancestral forms of the present wild-type enzyme, but merely forms of the enzyme that have presumably been tested by evolution and rejected. The rate of spontaneous point mutation is sufficiently high that the mutants must be formed in vivo.

Apparent or Observed Group Binding Energies. The removal of a side chain from an enzyme can have a purely localized effect on binding due to the absence of a specific interaction between the enzyme and substrate or a more global effect due to subtle or gross changes in enzyme conformation. ethe free energy of the unligated enzyme may also be altered on mutation by changes in hydration (Fersht et al., 1985a,b). What is phenomenologically important, however, is the actual effect on catalysis, and this is precisely what we measure in the mutagenesis studies. Where it seems likely from model building and other experiments that the effect on catalysis is caused just by a localized effect, the mutagenesis experiments give empirically the contribution of the side chain to binding energy. This energy contains the contributions from the direct interactions with the substrate that are altered and any additional energy terms due to local changes in structure in the free enzyme and in its complexes. We suggest that the binding energies that are measured from directed mutagenesis be termed *apparent* or *observed* group binding energies, by analogy with "apparent" or "observed" pK_a values. It is stressed that none of the following discussion depends crucially on the effects of mutation being localized.

Differential Stabilization of Substrates, Transition States, and Products. Irrespective of detailed structural changes, we find that removal of side chains can affect the binding energies of substrates, transition states, and products by different amounts. Conversely, the presence of a side chain can contribute to the differential binding of substrates, transition states, and products.

Mutation of the tyrosine binding and ATP binding sites leads to two different types of effects (Figure 3). Removal of the hydroxyl groups of tyrosine residues 34 and 169 that form hydrogen bonds with the substrate tyrosine does not affect the rate constant k_3 for activation but just weakens the binding of tyrosine. Thus, the side chains bind tyrosine equally well in the ground and transition states. Conversely, mutation of Cys-35, which interacts with ATP, has no effect on the dissociation constant of ATP from the enzyme–substrate complex but causes a 10-fold lowering of k_3. Thus, the side chain of Cys-35 contributes no binding energy to the enzyme–substrate complex but binds well to the transition state. Mutation of His-48, which also interacts with ATP, gives a mixture of effects, leading to a 4-fold decrease in k_3 and 2-fold increase in dissociation constant. This side chain thus binds to both the substrate and transition state but binds better to the transition state. In all cases, the side chains contribute most binding energy to the tyrosyl adenylate complex (Figure 3).

Enzyme Complementary to Intermediates. Removal of certain side chains destabilizes the enzyme–intermediate complex more than any other complex. This implies that the structure of those parts of the enzyme that we have modified in this study is closest in complementarity to the structure of the products of the reaction. There is somewhat lower complementarity toward the transition state, and there is least complementarity to the substrates (the predominant effects being on the nucleotide moiety).

It has been suggested that an enzyme could be complementary in structure to enzyme-bound intermediates or products rather than transition states when (i) unfavorable equilibria need to be overcome (Jencks, 1969, p 313) or (ii) a highly reactive intermediate has to be sequestered (Fersht, 1974). Both reasons could be important here. The equilibrium constant for the formation of Tyr–AMP and PP from Tyr and ATP in solution is indeed very unfavorable at 3.5×10^{-7} (T. N. C. Wells, C. K. Ho, and A. R. Fersht, unpublished results). The value for the enzyme-bound species is 2.3 (Table II). The enzyme thus stabilizes E·Tyr–AMP·PP with respect to E·Tyr·ATP by some 9 kcal/mol. The residues mutated in this study account for 3 kcal/mol of this (Table IV). The favorable equilibrium constant for the enzyme-bound reagents means that activation is not inhibited by high concentrations of pyrophosphate and that pyrophosphorolysis of tyrosyl adenylate does not compete with transfer to tRNA. This could well be of phyisological importance since the flux of pyrophosphate produced in protein biosynthesis is so high that pyrophosphate accumulates to a concentration of ~0.5 mM in *E. coli* and some other organisms (Kukko & Heinonen, 1982).

It is also likely that the enzyme–product complementarity is required to slow down the rate of dissociation of Tyr–AMP from the enzyme. The aminoacylation of tRNA in vivo catalyzed by such enzymes as the tyrosyl-tRNA synthetase is probably stepwise: most of the tRNA in *E. coli* is aminoacylated and bound to elonation factor Tu so that the rate-determining step in aminoacylation is the release of discharged tRNA during protein synthesis [summarized by Mulvey & Fersht (1977)]. The enzyme thus exists mainly as the enzyme–tyrosyl adenylate complex in vivo.

Two points must be emphasized. First, enzyme–product or enzyme–intermediate complementarity should only be important for *enzyme-bound* species and not for reactions that require the products or intermediates to diffuse into solution. Second, although the residues modified here are responsible for enzyme–intermediate complementarity, there are residues elsewhere that are responsible for a large factor of enzyme–transition-state complementarity. Leatherbarrow et al. (1985) have located a binding site for the γ-phosphoryl group of ATP that contributes no stabilization energy to the E·Tyr·ATP or E·Tyr–AMP complexes but stabilizes the transition state by some 4–5 kcal/mol. The effect of the residues modified in this study is to fine tune the catalysis.

Mechanistic Interpretation. During the changes in substrate geometry that occur on formation of the transition state and products, the interactions of important side chains with tyrosine remain relatively unaltered, whereas those with ATP are markedly optimized. This suggests that tyrosine moves relatively little with respect to its binding site on the enzyme during the reaction but that there is a significant displacement of the ribose moiety of ATP. The latter could be accomplished either by a movement of the ATP relative to a rigid enzyme, since there are changes in bond lengths and angles during the reaction, or by a conformational change or by a combination of these two processes. It is even possible that there could be a good hydrogen bond between the enzyme and, say, Cys-35 in the E·Tyr·ATP complex but that it causes a distortion of the enzyme or conformational strain in the ATP that is relieved as the reaction proceeds. Resolution of these possibilities will need the direct solution of the structure of the enzyme–ATP complex, which has not proved possible so far by X-ray crystallography (Monteilhet et al., 1984). It is also likely that the position of the ribose moiety in the transition state is closer to its position in the E·Tyr–AMP complex rather than in the E·Tyr·ATP complex since the energy contributions from the side chains are more similar in the transition states and products than in the transition states and substrates.

Registry No. 5′-ATP, 56-65-5; PP_i, 14000-31-8; TyrTS, 9023-45-4; L-Tyr, 60-18-4.

References

Carter, P. J., Winter, G., Wilkinson, A. J., & Fersht, A. R. (1984) *Cell (Cambridge, Mass.) 38*, 835–840.

Fersht, A. R. (1974) *Proc. R. Soc. London, Ser. B B187*, 397–407.

Fersht, A. R. (1985) *Enzyme Structure and Mechanism*, 2nd ed., Chapter 12, W. H. Freeman, Co., San Francisco and Oxford.

Fersht, A. R., Mulvey, R. S., & Koch, G. L. E. (1975) *Biochemistry 14*, 13–18.

Fersht, A. R., Shi, J.-P., Wilkinson, A. J., Blow, D. M., Carter, P., Waye, M. M. Y, & Winter, G. (1984) *Angew. Chem. 23*, 467–473.

Fersht, A. R., Shi, J.-P., Knill-Jones, J., Lowe, D. M., Wilkinson, A. J., Blow, D. M., Brick, P., Waye, M. M. Y., & Winter, G. (1985a) *Nature* (*London*) *314*, 235–238.

Fersht, A. R., Wilkinson, A. J., Carter, P., & Winter, G. (1985b) *Biochemistry 24*, 5858–5861.

Jencks, W. P. (1969) *Catalysis in Chemistry and Enzymology*, McGraw-Hill, New York.

Jencks, W. P. (1975) *Adv. Enzymol. Relat. Areas Mol. Biol. 43*, 219–410.

Kukko, E., & Heinonen, J. (1982) *J. Biochem.* (*Tokyo*) *127*, 347–349.

Leatherbarrow, R. J., Fersht, A. R., & Winter, G. (1985) *Proc. Natl. Acad. Sci. U.S.A. 82*, 7840–7844.

Lowe, D. M., Fersht, A. R., Wilkinson, A. J., Carter, P., & Winter, G. (1985) *Biochemistry 24*, 5106–5109.

Monteilhet, C., Blow, D. M., & Brick, P. (1984) *J. Mol. Biol. 173*, 477–485.

Mulvey, R. S., & Fersht, A. R. (1977) *Biochemistry 16*, 4731–4737.

Pauling, L. (1946) *Chem. Eng. News 24*, 1375–1377.

Waye, M. M. Y., Winter, G., Wilkinson, A. J., & Fersht, A. R. (1983) *EMBO J. 2*, 1827–1829.

Wells, T. N. C., & Fersht, A. R. (1985) *Nature* (*London*) *316*, 656–657.

Wilkinson, A. J., Fersht, A. R., Blow, D. M., & Winter, G. (1983) *Biochemistry 22*, 3581–3586.

Winter, G., Fersht, A. R., Wilkinson, A. J., Zoller, M., & Smith, M. (1982) *Nature* (*London*) *299*, 756–759.

In addition to the work on the tyrosyl-tRNA synthetase, I had a smaller project on the protease subtilisin to explore the nature and magnitude of electrostatic effects in proteins[69]. These studies not only showed quantitatively the importance of long range interactions between single or multiple charges, they also showed the heterogeneity of the dielectric constant of proteins and how there are values greater than that found for pure water[70]. As usual, our data provided the benchmarking for computer simulation[71]. In particular, they validated the use of simple methods for solving the Poisson–Boltzmann equations.

Tailoring the pH dependence of enzyme catalysis using protein engineering

Paul G. Thomas, Alan J. Russell & Alan R. Fersht

Department of Chemistry, Imperial College of Science and Technology, London SW7 2AY, UK

One of the goals of protein engineering is to tailor the pH dependence of enzyme catalysis to optimize activity in industrial processes. Chemical modification studies have shown that the pH dependence of catalysis by serine proteases alters with changes in overall surface charge[1], which suggests that a possible general method of modifying pH dependence is to alter the electrostatic environment of the active site by protein engineering and so change the pK_a values of ionic catalytic groups. Electrostatic effects are of considerable importance in enzyme catalysis and are thought to play a major part in stabilizing charged transition states[2-5]. However, it is extremely difficult to calculate electrostatic effects in proteins because of the microheterogeneity of the dielectric constant. In the present study we show how the change of just one surface charge which is 14-15 Å from the active site of subtilisin has a significant effect on the pH dependence of the enzyme and that the magnitude of the effect changes with ionic strength in a manner qualitatively consistent with electrostatic theory. Such experiments should provide the basis for refining theoretical calculations of electrostatic effects in catalysis.

The subtilisins are a family of extracellular serine proteases (also known as alkaline proteases) produced by species of bacilli prior to sporulation[6]. His 64 at the active site of the enzyme acts as a general base during catalysis, accepting a proton from the nucleophilic residue Ser 221 as it forms a bond with the substrate carbonyl carbon. The enzyme is active at alkaline pH when His 64 is unprotonated, and catalytic activity varies with pH following the ionization of this residue. The mutation we have chosen for our investigation of the magnitudes of electrostatic effects is Asp→Ser 99 in subtilisin from *Bacillus amyloliquefaciens* (BPN' or Novo). Crystallographic data show that this surface aspartate is some 14-15 Å from the imidazole of His 64 (ref. 7). (Note that because of a sequencing error, there is an alanine at position 99 in the early crystallographic models. We built in an aspartate at this position using molecular graphics. The derived distances have been substantiated by a new refined structure (R. Bott, personal communication)—the His 64 C_γ–Asp C_γ distance is 13 Å in the refined map, the same as in our modelling.) Asp 99 is not conserved in highly homologous subtilisins from other organisms: there is a serine at this position in the subtilisins from *Bacillus subtilis* DY[8] and *Bacillus licheniformis*[9] and threonine in subtilisins from *B. subtilis*[10] and *Bacillus amylosaccharíticus*[11]. It is thus expected *a priori* that the mutation Asp→Ser 99 should not have a significant effect on either the structure or catalytic properties of the enzyme other than by electrostatic effects. Removal of the negative charge of Asp 99 should destabilize the low-pH, positively charged form of His 64 and so lower its pK_a from the normal value of 7 by an unknown amount.

To test the predictions and quantify the effects, we have cloned the structural gene encoding subtilisin BPN' from *B. amyloliquefaciens* in the vector pUB110 (refs 12, 13). The structure of pPT30, which harbours the subtilisin gene on a 3.4-kilobase (kb) *Eco*RI fragment, is shown in Fig. 1*a*; its construction will be described elsewhere. The DNA sequence of the structural gene was determined by dideoxy methods[14] using a series of primers (J. Brannigan, unpublished data) and is identical to that published elsewhere[15,16]. Wild-type subtilisin BPN' is expressed and secreted at high levels by *B. subtilis* DB104 transformed with pPT30 (Fig. 1*b*). The strain DB104, given by Dr R. Doi, is deficient in extracellular alkaline and neutral proteases[17]. Subtilisin(Asp→Ser 99), prepared by oligodeoxynucleotide mutagenesis (Fig. 1) is also expressed at similar high levels in DB104.

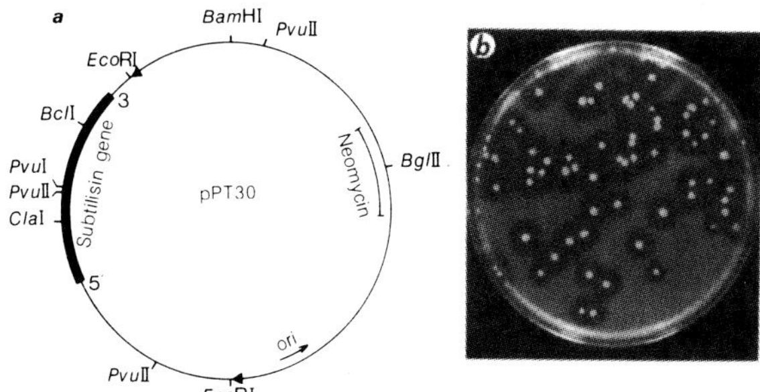

Fig. 1 *a*, Restriction map of the subtilisin expression vector pPT30. The subtilisin gene is contained within a 3.4-kb *Eco*RI fragment (solid block). Sequences derived from pUB110 (4.5 kb) are delineated by the arrowheads. *b*, Secreted subtilisin gives a characteristic clear halo around colonies of *B. subtilis* DB104 harbouring pPT30 after overnight selection on L-agar containing 1% skimmed milk plus 25 μg ml^{-1} kanamycin.

Methods. For mutagenesis, the gene was recloned in M13mp9. M13aprWT was constructed such that the coding strand of the subtilisin gene was ligated to the (+) strand of the viral DNA. M13aprWT was propagated in *Escherichia coli* TG2, a recA$^-$ derivative of JM101 (M. Biggin and T. Gibson, unpublished results) to overcome problems of DNA instability encountered when JM101 was the host. The synthetic primer (5'CGGAACCGC*T*AG-CACG3') contained a double mismatch with the wild-type gene (indicated by asterisks) and was designed to introduce an acceptable Ser codon at position 99. Oligodeoxynucleotide-directed mutagenesis was performed as described elsewhere[20,21], except that the mismatch repair-deficient strain BMH71-18mutL[22] was the recipient in transformations. TG2 provided the lawn to isolate phage from the mutator background. Hybridization screening using the mutagenic primer as a probe showed that 20% of the plaques contained mutant phage. The mutation was confirmed by DNA sequencing after plaque purification. Subtilisin(Asp→Ser 99) was expressed in DB104 on transformation of competent cells[23] with a plasmid constructed by ligating the mutant gene isolated from replicative form DNA into pUB110 in the same orientation as pPT30. Wild-type and mutant enzymes were purified from 36-h culture supernatants grown in L-broth containing kanamycin (25 μg ml^{-1}) and glucose (0.2%). The supernatant was concentrated using a Pellicon membrane filter (PTGC00005, Millipore) before dialysis against 0.01 M phosphate buffer pH 6.2 and purification on a CM-52 cellulose column as described elsewhere (ref. 24).

The activity of subtilisin(Asp→Ser 99) on the synthetic substrate succinyl-L-alanyl-L-alanyl-L-prolyl-L-phenylalanyl p-nitroanilide is similar to that of wild-type enzyme at 25 °C in 0.1 M Tris-HCl buffer, pH 8.6: k_{cat} (the catalytic rate constant) for the wild-type enzyme is 57 s^{-1} and K_m (Michaelis constant) is 0.15 mM, while the corresponding values for the mutants enzyme are 45 s^{-1} and 0.13 mM. The wild-type enzyme has identical activity to that produced from an independently cloned gene (ref. 15 and D. A. Estell and J. A. Wells, personal communication). The pK_a values of the active sites of the free enzymes were determined by kinetics from the pH dependence of k_{cat}/K_m (Fig. 2, Table 1). At ionic strength 0.1, His 64 in the wild-type enzyme has a pK_a of 7.17±0.02; this is decreased to a value of 6.88±0.03 in the mutant, a shift of 0.29±0.04 units. k_{cat}/K_m is decreased by 20% on mutation. Thus, Asp 99 affects both the ionization constant of the active site and also the charged transition state of the reaction.

As a control, it is important to examine the effect of ionic strength on the electrostatic interactions. The presence of high concentrations of ions should mask electrostatic interactions. At high ionic strengths, the observed pK_as should tend to their intrinsic values in the absence of surface charge. According to the classical Debye-Hückel theory and the Poisson-Boltzmann

Reprinted from Nature, Vol. 318, No. 6044, pp. 375-376, 28 November 1985

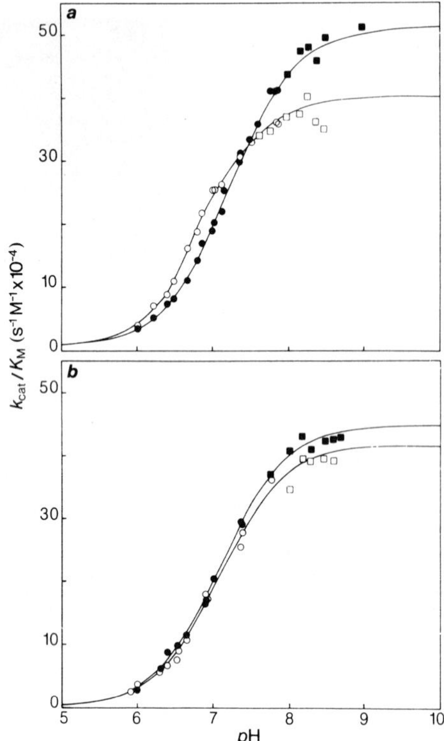

Fig. 2 *p*H dependence of k_{cat}/K_m for the hydrolysis of succinyl-L-alanyl-L-alanyl-L-prolyl-L-phenylalanyl *p*-nitroanilide by wild-type (solid symbols) and mutant Asp→Ser 99 (open symbols) subtilisin at 25 °C, ionic strength 0.1 (*a*) or 1.0 (*b*). Phosphate (circles) and Tris-HCl (squares) buffers of ionic strength 0.1 were used and the ionic strength adjusted by the addition of KCl. The reaction was initiated by adding 15 μl of enzyme (13 μM in 0.01 M phosphate buffer, *p*H 6.2) to 2 ml of buffered substrate (0.02 mM, $\ll K_m$) in a cuvette maintained at 25 °C in a Gilford 2600 spectrophotometer. The increase in absorbance at 412 nm on the release of *p*-nitroaniline was monitored and found to follow, with high accuracy, simple first-order kinetics. The value of k_{cat}/K_m at each *p*H value was determined from the first-order plots exactly as described for the hydrolysis of acetyl-L-tyrosine *p*-acetylanilide by chymotrypsin[25]. The low concentration of substrate used in this method avoids product inhibition. The concentration of each enzyme was determined by active-site titration with *N-trans*-cinnamoyl imidazole[26]. Wild-type enzyme was found to be 91% active and the mutant 96% active compared with concentrations measured by A_{280} ($E_{0.1\%}$ = 1.17). Control experiments showed that the rate of reaction in Tris-HCl buffers was 15% lower than in phosphate buffers. The data for experiments in the presence of Tris-HCl have been corrected for this, but these data were not used in the analysis. The data for the experiments in phosphate buffer were fitted to a theoretical single ionization curve by nonlinear regression (R. J. Leatherbarrow, unpublished). The derived pK_a values are listed in Table 1.

equation, the electrostatic interaction between two ions separated by a distance r is reduced by a factor of $\exp(-r/r_D)$ where r_D, the Debye length, is inversely proportional to the square root of the ionic strength. At ionic strength 1.0, r_D is 3.1 Å and at ionic strength 0.1 it is 9.8 Å. Although the theory is only approximate at high ionic strength, electrostatic effects should be highly masked over a distance of 14–15 Å at ionic strength 1.0. Accordingly, we found (Table 1) that at this ionic strength, both the pK_a values of the active site and the catalytic activities of the wild-type and mutant enzymes converge ($pK_a = 7.11 \pm 0.02$ and 7.09 ± 0.02; $k_{cat}/K_M = (45 \pm 0.6) \times 10^4$ and $(42 \pm 1) \times 10^4$ s^{-1} M^{-1} respectively). Thus, the mutation Asp→Ser 99 has essentially no effect on the catalytic properties of the enzyme when electrostatic effects are masked.

Table 1 The *p*H dependence of hydrolysis by wild-type and mutant subtilisin

Enzyme	pK_a of active site (from kinetics)	Limiting value of k_{cat}/K_m at high *p*H ($s^{-1}M^{-1} \times 10^{-4}$)
Ionic strength 0.1		
Wild type	7.17 ± 0.02	48.9 ± 0.7
Asp→Ser 99	6.88 ± 0.03	40.8 ± 0.9
Ionic strength 1.0		
Wild type	7.11 ± 0.02	44.8 ± 0.6
Asp→Ser 99	7.09 ± 0.02	41.6 ± 0.9

The pK_a values were obtained from the *p*H dependence of k_{cat}/K_m on the hydrolysis of succinyl-L-alanyl-L-alanyl-L-prolyl-L-phenylalanyl *p*-nitroanilide at 25 °C. The *p*H dependence of k_{cat}/K_m gives the pK_a values of the free enzyme and free substrate. In this case, the only ionizing group on the substrate is the succinyl group and its pK_a is below the range of this study. Artefacts which often obscure the interpretation of the *p*H dependence of k_{cat} and K_m individually usually do not affect the *p*H dependence of k_{cat}/K_m (ref. 18). The pK_a of the wild-type enzyme is in excellent agreement with that derived from the hydrolysis of ester substrates in identical conditions (7.15 at ionic strength 0.1; ref. 19). Values are given ±s.e.

We have shown that modification of a single charge in the vicinity of the active site of an enzyme can have a significant effect on the *p*H dependence of the catalytic reaction. Such effects should be larger for charges which are closer to ionic catalytic residues in the enzyme or closer to charges which develop in the transition states. Electrostatic effects will be more pronounced if multiple charges are engineered. It should, therefore, be possible to engineer large effects on both the *p*H dependence and catalytic rate constants by modifying the surface charge.

This work was supported by the SERC of the UK. We thank Mr J. Brannigan for providing sequencing primers and determining the DNA sequence of the wild-type gene. We also thank Dr Bott for pre-publication data on the distance between His 64 and Asp 99, and Dr J. A. Wells for the sequence of the cloned subtilisin gene prior to publication.

Received 20 August; accepted 3 October 1985.

1. Valenzuela, P. & Bender, M. L. *Biochim. biophys. Acta* **250,** 538–548 (1971).
2. Perutz, M. *Science* **201,** 1187–1191 (1978).
3. Warshel, A., Russell, S. T. & Churg, A. K. *Proc. natn. Acad. Sci. U.S.A.* **81,** 4785–4789 (1984).
4. Matthew, J. B., Hanania, G. I. H. & Gurd, F. R. N. *Biochemistry* **21,** 1919–1939 (1979).
5. Warshel, A. *Acc. chem. Res.* **14,** 284–290 (1981).
6. Markland, F. S. & Smith, E. L. in *The Enzymes* Vol. 3 (ed. Boyer, P. D.) 561–608 (Academic, New York, 1971).
7. Kraut, J. in *The Enzymes* Vol. 3 (ed. Boyer, P. D.) 547–560 (Academic, New York, 1971).
8. Nedkov, P., Oberthür, W. & Braunitzer, G. *Hoppe-Seyler's Z. physiol. Chem.* **364,** 1537–1540 (1983).
9. Smith, E. L., Delange, R. J., Evans, W. H., Landon, M. & Markland, F. S. *J. biol. Chem.* **243,** 2184–2191 (1968).
10. Stahl, M. L. & Ferrari, E. *J. Bact.* **158,** 411–418 (1984).
11. Kurhara, M., Markland, F. S. & Smith, E. L. *J. biol. Chem.* **247,** 5619–5631 (1972).
12. Gryczan, T. J., Contente, S. & Dubnau, D. *J. Bact.* **134,** 318–329 (1978).
13. Jalanko, A., Palva, I. & Söderland, H. *Gene* **14,** 325–328 (1981).
14. Sanger, F., Nicklen, S. & Coulson, A. R. *Proc. natn. Acad. Sci. U.S.A.* **74,** 5463–5467 (1977).
15. Wells, J. A., Ferrari, E., Henner, D. J., Estell, D. A. & Chen, E. Y. *Nucleic Acids Res.* **11,** 7911–7925 (1983).
16. Vasantha, N. *et al. J. Bact.* **159,** 811–819 (1984).
17. Kawamura, F. & Doi, R. H. *J. Bact.* **160,** 442–444 (1984).
18. Fersht, A. R. *Enzyme Structure and Mechanism* 2nd edn, Ch. 5 (W. H. Freeman, New York, 1985).
19. Philipp, M. Tsai, I.-H. & Bender, M. L. *Biochemistry* **18,** 3769–3773 (1979).
20. Zoller, M. J. & Smith, M. *Meth. Enzym.* **100,** 468–500 (1983).
21. Winter, G., Fersht, A. R., Wilkinson, A. J., Zoller, M. & Smith, M. *Nature* **299,** 756–758 (1982).
22. Kramer, B., Kramer, W. & Fritz, H. J. *Cell* **38,** 879–887 (1984).
23. Hardy, K. G. in *DNA Cloning, a Practical Approach* (ed. Glover, D. M.) 1–17 (IRL, Oxford, 1985).
24. Estell, D. A., Graycar, T. P. & Wells, J. A. *J. biol. Chem.* **260,** 6518–6521 (1985).
25. Fersht, A. R. & Renard, M. *Biochemistry* **13,** 1416–1426 (1974).
26. Bender, M. L. *et al. J. Am. chem. Soc.* **88,** 5890–5931 (1966).

Rational modification of enzyme catalysis by engineering surface charge

Alan J. Russell & Alan R. Fersht

Department of Chemistry, Imperial College of Science and Technology, London SW7 2AY, UK

Changing the surface charge of subtilisin by site-directed mutagenesis produces enzymes with significantly shifted* p*H-activity profiles, higher catalytic activities and altered specificities. Insight into water as a dielectric, the role of ions in electrostatic shielding and the field effects on catalysis is obtained and suggests rules for tailoring* p*H-activity profiles.

ELECTROSTATIC effects have an important function in enzyme catalysis; many enzyme reactions proceed via charged transition states and so stabilization of charges is a major catalytic factor. The *p*H dependence of enzyme catalysis depends on the ionization of charged groups sensitive to the surface charge of the protein, and electrostatic interactions involved in hydrogen bonds and salt bridges are important in maintaining the structure of proteins. Gaining a quantitative knowledge of electrostatic interactions is difficult, however, because theoretical analysis based on classical macroscopic electrostatics does not easily adapt to the heterogeneous structure of proteins, and experimental data are rare[1-3].

The effect of surface charge on the pK_a values of ionizable side-chains in proteins has been subjected to most analysis, stretching back to the pioneering studies by Linderstrøm-Lang[4] and Kirkwood and colleagues[5]. A variety of theoretical approaches have been applied in recent years, but there are inconsistencies over basic issues such as whether the protein has a high or low dielectric constant[6]. Experimental evidence is required to test theory and provide a rational basis for designing proteins. Earlier experiments on chemical modifications indicated relatively small effects of surface charge on pK_a values of groups in active sites. For example, acetylation of all surface lysine residues of trypsin raises the pK_a of the active-site histidine by only 0.2 units[7] and the succinylation of 14 surface lysine residues of chymotrypsin, giving a change in charge of 28 units, raises the pK_a of the active site by only 1 *p*H unit[8]. In contrast to the earlier studies, we have shown that removal by site-directed mutagenesis of just one surface carboxylate in subtilisin (Asp 99) lowers the pK_a of its active-site histidine (His 64) by 0.3 units at ionic strength 0.1 M, despite the two groups being ~1.3 nm apart[9]. That study, as well as presenting a general method of tailoring the *p*H-dependence of enzyme activity, indicated that it was possible to measure coulombic interactions individually by a combination of protein engineering and kinetics. We now present an extensive survey based on several mutations, both single and double, that give experimental evidence on the magnitudes of electrostatic effects and how they can be used to tailor the properties of enzymes. The results on altering long range electrostatic interactions show: (1) the high electrostatic shielding of surface charges; (2) the importance of counterion binding to charged groups on the protein and how this has obscured earlier studies; (3), the size of electrostatic effects on changes in *p*H-activity profiles; (4), the effects on specificity towards charged substrates; (5), the effects on catalysis.

Subtilisin as model system

The serine protease subtilisin (from *Bacillus amyloliquefaciens*) is a good model system for protein engineering studies: its crystal structure, and that of the highly homologous subtilisin from *Bacillus licheniformis*, has been solved at high resolution, as have their complexes with polypeptide inhibitors[10-12]; it has been cloned and expressed in large quantities from plasmids in *Bacillus subtilis*[9,13]; it may be readily assayed by steady-state kinetics, and absolute rate constants may be obtained after active-site titration. Residue His 64 acts as a general base in catalysis with a pK_a of ~7 (ref. 14). At low *p*H, His 64 is protonated and so the enzyme is inactive.

We have demonstrated that the pK_a of His 64 can be determined with high precision (typically ±0.01-0.02 *p*H units) from kinetics (from the *p*H-dependence of k_{cat}/K_M) so that coulombic interactions may be measured from small shifts in pK_a[15]. The

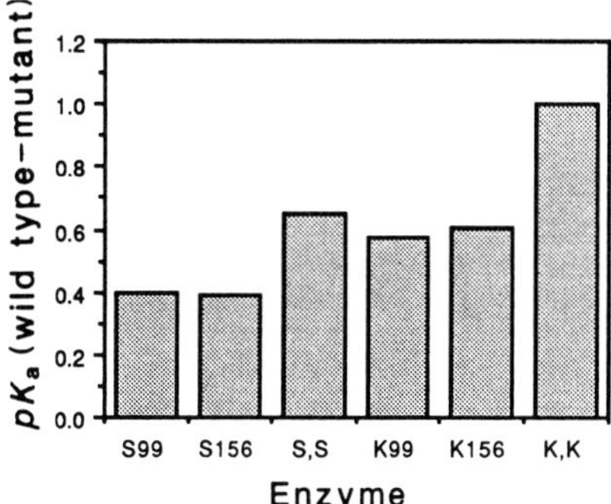

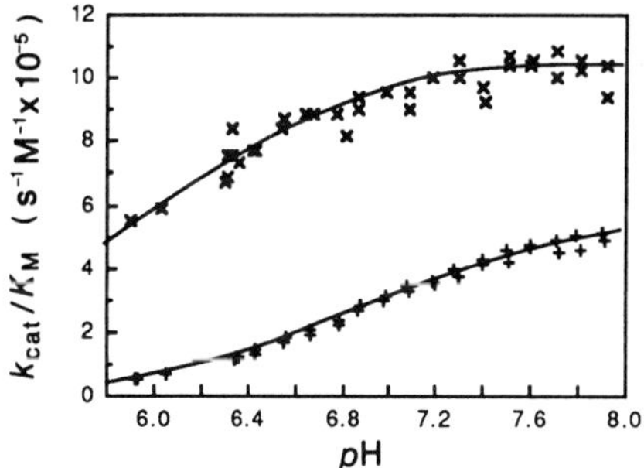

Fig. 1 Top; Shifts in pK_a of His 64 in mutant enzymes relative to wild type at ionic strength 0.001 M (imidazole-imidazolium hydrochloride buffer) and 25 °C. The pK_a of His 64 in wild-type enzyme is 6.91±0.01. Bottom: *p*H dependence of k_{cat}/K_M for hydrolysis of succinyl-Ala-Ala-Pro-Phe *p*-nitroanilide by wild-type (+) and double-mutant Asp→Lys 99; Glu→Lys 156 (×) under the same conditions. Mutagenesis was performed as described previously[9,15]. The mutagenic primers used were as follows: Asp→Ser 99, 5' CGGAACCGCTAGCACCG 3'; Asp→Lys 99, 5' CCGG-AACCTTTAGCACCGA 3'; Glu→Ser 156, 5' GAAGTGCCTGA-GTTACCGG 3'; Glu→Lys 156, 5' AGTGCCTTTGTTACCGG 3'. All mutations were confirmed by DNA sequencing after plaque purification[15]. The mutant proteins were expressed in *Bacillus subtilis* DB104 after recloning into pUB110. Purification and kinetics were also performed as described previously[15]. Imidazole buffers provide better buffering at low ionic strength and span a better *p*H range than the phosphate buffers used previously[15]. The derived pK_as consequently have lower standard errors, typically ±0.01-0.02. The error is higher, however, for the very large shift in the double-lysine mutant (±0.06).

Reprinted from Nature, Vol. 328, No. 6130, pp. 496-500, 6 August 1987

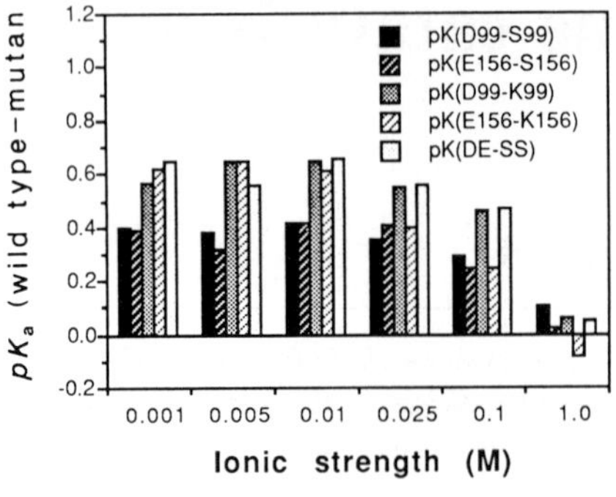

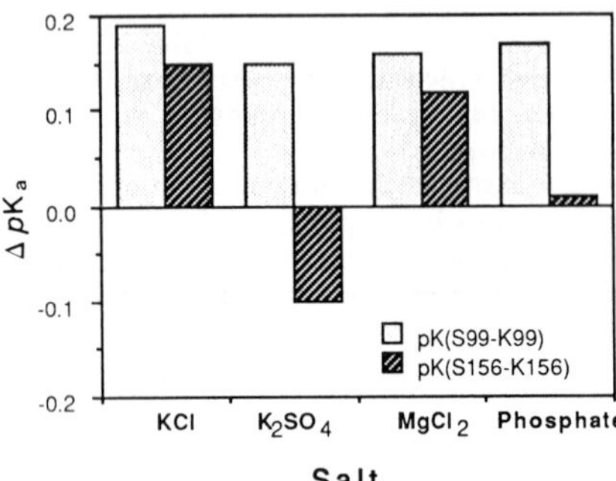

Fig. 2 Effect of ionic strength and nature of buffer on pK_a shift. Top, effect of increasing phosphate buffer from ionic strength 0.005 to 0.1 M followed by addition of KCl to 1.0 M. Bottom, effect of nature of counterion on the pK_a shifts of lysine compared to serine mutants. Ionic strength was maintained at 0.1 M by the electrolyte.

ionization follows a fine titration curve between pH 6 and 8. The data fit precisely a Hill plot of slope 1.00 ± 0.02 under a variety of conditions of ionic strength showing that only a single ionization is being monitored. Two mutants have been made which alter charge in different regions of the protein. One, Glu → Ser 156 removes a negative charge ~1.4-1.5 nm from the imidazole of His 64—the environment between the two is predominantly aqueous. The other, Asp → Ser 99, removes a charge which is separated from His 64 by 1.2-1.3 nm of an environment which is predominantly protein. The conceptual basis of the experiments is that increasing the overall negative charge on the enzyme should raise the pK_a of the active site histidine by stabilizing the protonated form of His 64 whereas increasing the positive charge should lower the pK_a by destabilizing the protonated form of His 64. The pK_a of His 64 in the mutants is lowered by about 0.4 units at infinitesimal ionic strength[15]. Increased changes in charge have now been made by mutating Asp 99 and Glu 156 to lysine residues. The double mutants Lys 99; Lys 156 and Ser 99; Ser 156 were also constructed.

Electrostatic effects cumulative

Electrostatic interactions are masked by concentrated solutions of electrolytes and so electrostatic effects are most apparent at low ionic strength. Plotted in Fig. 1 are changes in pK_a of His 64 for mutants assayed at an ionic strength of 0.001 M, which is effectively zero (experiments in other buffers give the same shifts in pK_a on extrapolation to zero ionic strength). Changing either Asp 99 or Glu 156 to serine lowers the pK_a by about 0.4 units. Changing both simultaneously to give the double mutant with a change of two charge units lowers the pK_a by 0.65 units. Alternatively, making a double charge change by mutating either Asp 99 or Glu 156 to a lysine lowers the pK_a of His 64 by 0.6 units. The quadruple charge change in the double mutant Asp → Lys 99; Glu → Lys 156 gives a shift of 1.0 units, which is so large that it is leaving the region where quantitative analysis of our data is meaningful. The changes in coulombic interactions are thus cumulative. They are approximately additive for the single mutants, allowing for the long span of the side chain of lysine (~0.6 nm). This distance is comparable to the separation of charges in the active site and so there is considerable uncertainty in the distance between the charges: the positively charged ε-NH_3^+ groups of the lysines will tend to stretch away from any positive charge on His 64 and so attenuate the electrostatic effects, especially at low ionic strengths. Indeed, calculations of electrostatic interactions based on the full-extended conformation of lysine suggest that the pK_a-shifts on mutation of Asp or Glu to Ser to Lys are fairly close to being additive (F. Hayes, M. J. E. Sternberg, P. G. Thomas, A. J. R. and A. R. F., unpublished). The changes for the double mutants appear to be less than additive. Additivity is not necessarily expected, however, because of possible changes in microscopic environment on mutation and changes in interactions between charges mediated by orientation of solvent molecules[15].

High effective dielectric constant

The limiting value of ΔpK_a at low ionic strength may be inserted into the standard electrostatic equations to calculate the effective dielectric constant, D, between the group concerned and His 64 ($\Delta pK_a = 244/Dr$, where r is the separation between two unit charges in Å)[15]. In all cases the value is high, independent of whether there is protein or water between the two groups: Asp 99-His 64, $D = 50$; Lys 99-His 64, $D = 80$–100; Glu 156-His 64, $D = 40$; Lys 156-His 64, $D = 60$–80. This is not inconsistent with classical macroscopic electrostatics—the dielectric may be replaced by a surface polarization charge which partly neutralizes the charge it surrounds. Warshel and colleagues have also emphasized this point[3]. They state that, in contrast to common belief, most of the dielectric effect comes from water around the protein and not from the interior of the protein.

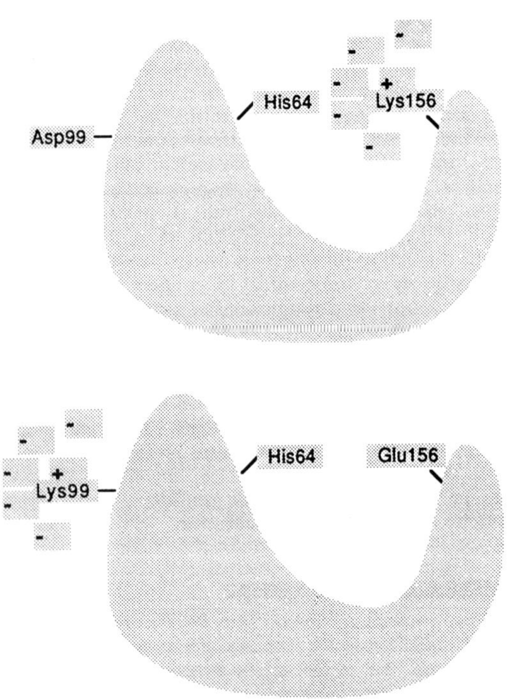

Fig. 3 Schematic representation of ionic atmosphere set up by Lys 99 and Lys 156.

Counterion binding

Basic electrostatic theories, such as the classical Debye-Hückel theory, predict that the charge-charge interactions are masked in a general manner by the field set up by counterions in ionic solutions. Examination of Fig. 2(top) reveals that the difference in pK_a of His 64 between wild-type and mutants Asp→Ser 99 and Glu→Ser 156 does decrease as the ionic strength of a phosphate buffer is increased to 0.1 M, and finally tends to zero when KCl is added to 1.0 M. But careful examination of the figure reveals that there are specific effects superimposed upon the general. The value of ΔpK_a for the mutant Glu→Lys 156 decreases much more rapidly with increasing phosphate concentration than do the other values. We speculated that this results from neutralization of the positive charge on Lys 156 by the binding of phosphate dianion at higher ionic strengths. Support for this hypothesis comes from the shielding effects of other counterions. The values of pK_a were determined in imidazole-imidazolium hydrochloride buffer at ionic strength 0.001 M, maintaining ionic strength at 0.1 M with electrolytes of different charge types: K^+Cl^-; $(K^+)_2SO_4^{2-}$; $Mg^{2+}(Cl^-)_2$. The effects are clearly seen when comparing lysine mutants with serine mutants (Fig. 2, bottom). The difference in pK_a between mutants Ser 99 and Lys 99 is roughly independent of electrolyte. Although there is a difference in pK_a of 0.12–0.15 units between mutants Ser 156 and Lys 156 in the presence of either KCl or $MgCl_2$, in phosphate, which is a mixture of phosphate mono- and di-anions, both mutants at position 156 have the same pK_a. Further, in the presence of the full doubly-charged SO_4^{2-} counterion, there is a reversal of the effect—the Lys 156 mutant has a value of pK_a which is 0.1 units higher than that of Ser 156.

The specific effect of negatively-charged ions is clear on consideration of Fig. 3. The ionic atmosphere surrounding a charge on the surface of a protein is asymmetrical, being in the solvent and away from the protein. Thus, the negative ionic atmosphere surrounding a lysine at position 99 is far removed from His 64 because there is protein between Lys 99 and His 64. But, the negative counterions surrounding Lys 156 are concentrated between Lys 156 and His 64 because there is water between them and protein is on the far side. The positive charge on Lys 156 thus tends to concentrate negative ions in the active site. This is far more noticeable for dianions than monoanions because there is twice the coulombic interaction energy between Lys 156 and a dianion for binding, and the dianion when bound exerts twice as large a coulombic interaction energy with His 64. This observation with Lys 156 is not an isolated result; the double lysine mutant is even more sensitive to salts. Mutation of Lys→Thr 213 gives also a reversal of expected pK_a change in the presence of moderate concentrations of phosphate, and there is a channel of water between residues 213 and 64 (P. Thomas and A.R.F., unpublished—the effective dielectric in this channel is about 150).

The effects of counterions may explain why the earlier experiments using protein chemistry to modify the lysine residues of chymotrypsin and trypsin gave such low changes in pK_a. The kinetic studies were performed in phosphate buffer of ionic strength 0.1 M or greater, and so the very large changes in surface charge would lead to counterions binding to the active site of the enzymes.

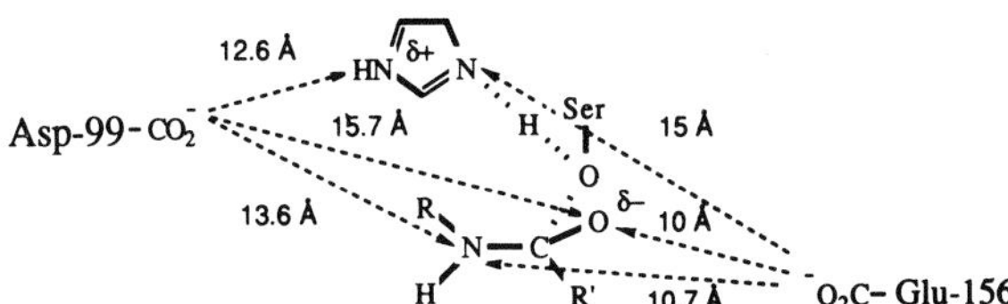

Fig. 4 Stereochemical relationship between charges on Asp 99 and Glu 156 with charges developing on the transition state for the attack of Ser 221 on a peptide substrate. (Crystallographic coordinates kindly provided by Dr M. N. G. James[12]).

Electrostatic effects and kinetic constants

A simple way of analysing and qualitatively predicting electrostatic effects on kinetics is to ascribe charge distributions to atoms in the transition state and calculate their interaction energies with external charges (see ref. 16). The rate-determining step in the hydrolysis of amides by serine proteases is the formation of an acylenzyme with the active site serine[16]. This involves the formation of a tetrahedral intermediate in which an oxyanion is formed. The likely regions where charge will be generated in the transition state are on His 64 of the protein as it acts as a general acid-base catalyst, on the carbonyl oxygen of the substrate as it is partly an oxyanion in the transition state, and, depending whether or not there is rate-determining formation or breakdown of the tetrahedral intermediate, on the N atom of the leaving group[17]. Figure 4 illustrates the charges on the transition state for rate-determining formation of the tetrahedral intermediate and the distances of the charges from Asp 99 and Glu 156 (calculated from the crystal structure of the complex between subtilisin and the barley CI-2 inhibitor[12]). We know from our measurements above that the interaction energy of the negative charge on either Asp 99 or Glu 156 with a full positive charge on His 64 is 0.5–0.6 kcal mol^{-1}. We thus expect that the interaction energies with partial charges in the transition state will be somewhat less than this, unless substrate binding lowers the effective dielectric constant. It is seen in Fig. 4 that Asp 99 is closer to the positive charge on His 64 than the negative charge

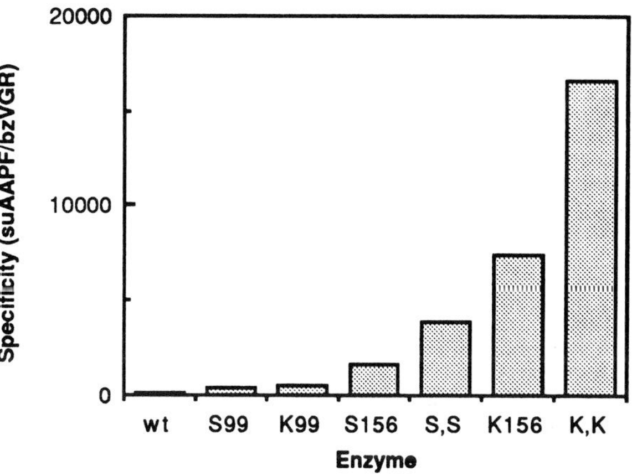

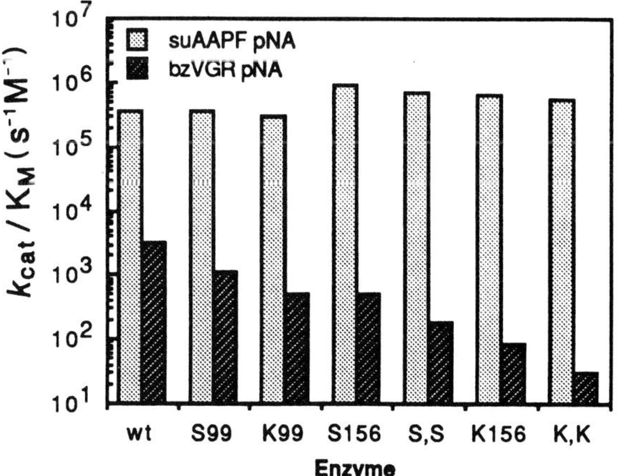

Fig. 5 Specificity of wild-type and mutant enzymes for the substrates succinyl-Ala-Ala-Pro-Phe *p*-nitroanilide and benzoyl-Val-Gly-Arg *p*-nitroanilide at ionic strength 0.01 M (imidazole-imidazolium buffer, *p*H 7.0) at 25 °C. Top: ratio of values of k_{cat}/K_M which gives specificity[16]. Bottom: absolute values of k_{cat}/K_M.

on the oxyanion. Mutation of Asp 99 → Ser → Lys should thus destabilize His 64 more than it stabilizes the oxyanion if there is the same effective dielectric constant for the two interactions. That is, the value of k_{cat} should decrease. This is indeed observed: k_{cat} for the hydrolysis of the substrate succinyl-Ala-Ala-Pro-Phe *p*-nitroanilide (suAAPF *p*NA) changes from 57 to 45 to 30 s^{-1} for these mutations, equivalent to changes in activation energy of 0.14 and 0.24 kcal mol^{-1} respectively. Glu 156, on the other hand, is closer to the oxyanion. Conversely, the value of k_{cat} should increase on mutation of Glu → Ser → Lys. The value of k_{cat} changes from 57 to 55 to 79 s^{-1} after these mutations. There is not the monotonic increase found with mutations of Asp 99, but there is a significant and real increase in catalytic rate constant.

There is a straightforward reason why mutations at residue 99 are well-behaved and those at 156 are not. Residue 99 is not in the substrate-binding site and does not interact sterically with the substrate. Glu 156 is in the site and its side chain is close to the side chain of the substrate (Phe) in the primary site. Thus, electrostatic changes of k_{cat} and K_M on mutation of 156 will be perturbed by direct steric interactions. It is important to note that the steric effects do not perturb the measurement of pK_a by the method employed in this study, the *p*H dependence of k_{cat}/K_M, as this is purely a function of the free enzyme[16].

Effects on K_M can also be qualitatively predicted. The negative charge on the succinyl group of the substrate should ineract more favourably with the enzyme as either or both residues are changed to Ser and Lys. On mutation of Asp 99 → Ser → Lys, there is a monotonic decrease in K_M of 0.17 to 0.13 to 0.10 mM. Mutation of Glu 156 → Ser → Lys is more complicated because it is expected that the long side chain of Lys 156 will interact directly, and unfavourably, with the primary side chain of the substrate (Phe). The K_M changes from 0.17 to 0.05 to 0.13 mM on mutation. The double mutant Asp → Ser 99; Glu → Ser 156 has a low value of 0.03 mM for K_M.

Long range interactions and specificity

Wells *et al.*[17] have shown that the specificity of subtilisin towards charged substrates may be changed where intimate ion pairs are involved. Our data on the effects of mutation on the specificity constant k_{cat}/K_M towards the negatively-charged substrate suAAPF *p*NA and the positively-charged benzoyl-Val-Gly-Arg *p*-nitroanilide (bz VGR *p*NA) show that long-range electrostatic effects also mediate specificity (Fig. 5). The specificity of the wild-type enzyme for negatively-charged substrate compared to positively-charged is 1.2×10^2 and this increases to 1.6×10^4 for the double mutant Asp → Ser 99; Glu → Ser 156 (ratios of values of k_{cat}/K_M). The change is in the direction logically expected from the effects of charges on the enzyme on charges on the substrate.

More active mutant enzymes

We have generated enzymes that are far more active than wild-type under certain conditions. Although the enhanced activities are most apparent at very low ionic strengths where long range electrostatic effects are most significant, low ionic strength is not uncommon. For example, the aqueous component of a washing machine load, which is the destiny of much of the subtilisin that is produced, has an ionic strength close to zero. The activity of the double mutant Asp → Lys 99; Glu → Lys 156 with suAAPF *p*Na is twice that of wild-type subtilisin at high *p*H and ten times greater at *p*H 6 at low ionic strength under k_{cat}/K_M conditions (Fig. 1).

Rules for changing *p*H-activity profiles

Our experiments show that modest, but significant, changes in activity may be made by one or two point mutations. By designing a series of mutations based on this knowledge, larger increases may be engineered. But we have found limits on this procedure because of the effects of counterions in solution. The following rules should be applied for designing changes in *p*H-activity profiles by protein electrical engineering.

(1) Making the surface more negatively charged raises the pK_a values of acidic groups in proteins because they all lose a proton on ionization. Conversely, making the surface more positively charged lowers the pK_a of all acidic groups.

(2) Changes will be maximized at low ionic strengths.

(3) Significant changes will be manifested at ionic strengths as high as 0.1 M if multiply-charged counterions are avoided.

(4) Mutations should be designed so that they do not concentrate counterions in the active site cavity.

The first two rules are clearly implicit in the classical Linderstøm-Lang theory. This study has shown that, contrary to previous experimental data, significant effects may be observed at moderate ionic strengths providing rules (3) and (4) are observed. Changes in rate and binding constants may also be engineered by placing charged groups in the proximity of charges in the transition state or substrate. As most enzymes use general acid-base catalysis and their reactions involve charged transition states, the rational modification of catalytic properties by engineering electrostatics should be generally applicable.

This work was funded by the SERC (Protein Engineering Initiative). A.J.R. has a studentship from the Biotechnology Directorate of the SERC. We thank Dr M. N. G. James for unpublished coordinates of the subtilisin-CI-2 complex.

Received 5 May; accepted 29 June 1987.

1. Rogers, N. K. *Prog. Biophys. molec. Biol.* **48**, 37-66 (1986).
2. Matthew, J. B. *A. Rev. Biophys. Bioengng* **14**, 387-417 (1985).
3. Warshel, A. & Russell, S. T. *Q. Rev. Biophys.* **17**, 283-422 (1984).
4. Linderstrøm-Lang, K. *C. r. Trav. Lab. Carlsberg* **15**, 70-76 (1924).
5. Tanford, C. & Kirkwood, J. G. *J. Am. Chem. Soc.* **79**, 5333-5339 (1957).
6. Warshel, A. *Proc. natn. Acad. Sci. U.S.A.* **81**, 444-448 (1984).
7. Spomer, W. E. & Wootton, J. F. *Biochim. biophys. Acta* **235**, 164-171 (1971).
8. Valenzuela, P. & Bender, M. L. *Biochim. biophys. Acta* **250**, 538-548 (1971).
9. Thomas, P. G., Russell, A. J. & Fersht, A. R. *Nature* **318**, 375-376 (1985).
10. Wright, C. S., Alden, R. A. & Kraut, J. *Nature* **221**, 235-242 (1969).
11. Drenth, J. & Hol, W. G. J. *J. molec Biol.* **28**, 543-551 (1967).
12. McPhalen, C. A. & James, M. N. G. *Biochemistry.* **26**, 261-269 (1987).
13. Wells, J. A., Ferrari, E., Henner, D. J., Estell, D. A. & Chen, E. Y. *Nucleic Acids Res.* **11**, 7911-7925 (1983).
14. Phillip, M., Tsai, I.-H., & Bender, M. L. *Biochemistry* **18**, 3769-3773 (1979).
15. Russell, A. J., Thomas, P. G. & Fersht, A. R. *J. molec. Biol.* **193**, 803-813 (1987).
16. Fersht, A. R. in *Enzyme Structure and Mechanism* (Freeman, New York, 1985).
16. Fastrez, J. & Fersht, A. R. *Biochemistry* **12**, 1067-1074 (1973).
17. Wells, J. A., Powers, D. B., Bott, R. R., Graycar, T. P. & Estell, D. A. *Proc. natn. Acad. Sci. U.S.A.* **84**, 1219-1223 (1987).

Prediction of electrostatic effects of engineering of protein charges

Michael J. E. Sternberg*, Fiona R. F. Hayes*, Alan J. Russell†, Paul G. Thomas† & Alan R. Fersht†

* Department of Crystallography, Birkbeck College, Malet Street, London WC1E 7HX, UK
† Department of Chemistry, Imperial College of Science and Technology, London SW7 2AY, UK

Accurate prediction of electrostatic effects on catalytic activity is an essential component of protein design[1–4]. Site-directed mutagenesis of charged groups in subtilisin of *Bacillus amyloliquefaciens* has provided experimental measurements of electrostatic interactions which may be used to test such theoretical methods. The pK_a of the histidine of the active site has been perturbed by +0.08 to −1.0 units by modifying one or two residues[5–7]. Electrostatic effects in proteins can be modelled by the algorithm of Warwicker and Watson, which uses classical electrostatics and considers both the charge position and the shape of the molecule[8–10]. Here we report that the algorithm can model several pK_a shifts in subtilisin to fair accuracy.

Ab initio modelling of electrostatic effects in proteins requires consideration of the atomic polarizabilities[4] of the heterogeneous protein and the solvent, including both water and counterions. The size of the system makes approaches of this sort difficult and so simplified methods are necessary, such as the use of a macroscopic dielectric (Fig. 1*a*). Simple models use uniform or distance-dependent dielectrics[11,12]. These models ignore the importance of the exact location of the charges as there is far greater solvent screening of surface charges than buried charges. However, these features are incorporated into the Warwicker-Watson algorithm, in which the protein/solvent is divided into cubes, typically of dimension 1 Å and, depending on location, a dielectric (D) is assigned to each cube which corresponds to the protein ($D = 3.5$), the solvent ($D = 80$) or takes an intermediate value for the boundary (Fig. 1*b*). Each charge is spread to the eight corners of the cube. A finite difference solution to Poisson's equation is then obtained. The version of the algorithm was benchmarked[10] by reproducing a previous result that modelled the redox potential shift in cytochrome c_{551} due to the introduction of a charge on a buried propionate group as 87 mV (effective dielectric, D_{eff}, equals 20) compared with experimental values of 65 mV ($D_{eff} = 27$).

We examined whether the Warwicker-Watson algorithm can be used to model the effects of single and double site-specific mutations on the pK_a of the active site His 64 in *Bacillus amyloliquefaciens* subtilisin[5-7] (Table 1). As the algorithm does not include the effect of counterions the theoretical results model experimental values at low ionic concentration. Table 1 shows there is little variation in the results over two to four determinations of the pK_a shift in phosphate and imidazole buffers over a range of 0.001 to 0.01 M.

Coordinates of the crystal structure of *B. amyloliquefaciens* subtilisin BPN' (M. James, personal communication) were used and additional unpublished distances in the mutant enzymes were provided by R. R. Bott. The structures of the mutant

a

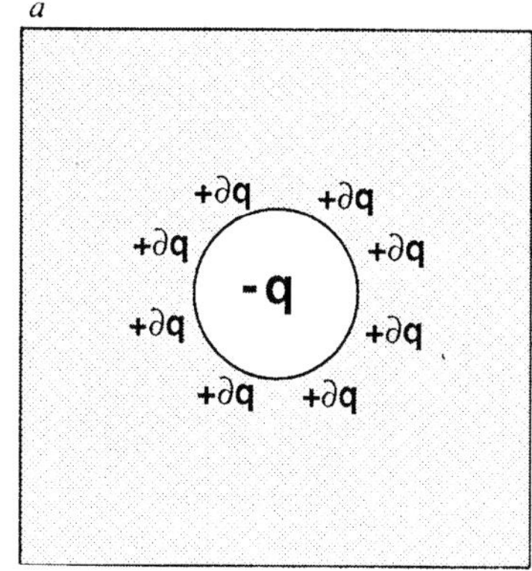

b

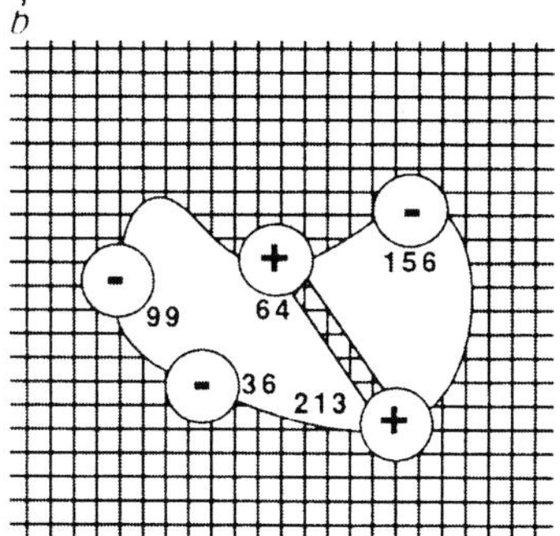

Fig. 1 *a*, The physics of the dielectric effect. A charge q in a continuous medium of dielectric D induces a surface polarization $\Sigma\delta q$ which partially neutralizes the original charge. The effective change inside the dielectrics $-(q - \Sigma\delta q) = -q/D$. *b*, Calculation of electrostatic effects in mutant subtilisins. The Warwicker-Watson algorithm uses a grid and associates dielectrics and charges appropriately. The positions of the charged residues illustrate roughly whether the interaction is mainly through protein or solvent.

Table 1 The difference in pK_a between mutant and wild type in different buffers

Mutant	pK_a shift $I = 0.001$ M imidazole	$I = 0.01$ M imidazole	$I = 0.005$ M P_i	$I = 0.01$ M P_i	Average
Asp → Ser 99	−0.40 ± 0.02	—	−0.38 ± 0.02	−0.42 ± 0.02	−0.40
Glu → Ser 156	−0.39 ± 0.02	—	−0.32 ± 0.02	−0.42 ± 0.02	−0.38
Lys → Thr 213	+0.10 ± 0.02	—	+0.08 ± 0.02	+0.07 ± 0.02	+0.08
Asp → Gln 36	−0.18 ± 0.02†	—	—	—	−0.18
Asp → Lys 99	−0.57 ± 0.02	−0.68 ± 0.02	−0.65 ± 0.03	−0.65 ± 0.04	−0.64
Glu → Lys 156	−0.62 ± 0.02	−0.63 ± 0.02	−0.65 ± 0.04	−0.61 ± 0.03	−0.63
Asp → Ser 99 and Glu → Ser 156	−0.65 ± 0.04	—	−0.56 ± 0.04	−0.66 ± 0.03	−0.62
Asp → Lys 99 and Glu → Lys 156	−1.00 ± 0.06	—	—	—	−1.00

The values of pK_a shift are pK_a (mutant) − pK_a (wild-type) in buffers of different ionic strength (I). Kinetic data was measured and the Lys → Thr 213 and Asp → Gln 36 mutants were prepared and purified as in refs 6 and 7. Primers: 5'-CCCCGTATG*TGTTTCCA (Thr 213) and 5'-TGAGAAGAC*TG*GATACCGC (Gln 36) were used to direct the mutations. †, Value for Asp → Gln 36 is the average of two measurements: −0.17 ± 0.02 and −0.19 ± 0.02. All the substitutions except Asp → Lys 99 and Glu → Lys 156 are observed in different species of subtilisin.

Reprinted from Nature, Vol. 330, No. 6143, pp. 86-88, 5 November 1987

Table 2 Prediction of pK_a shifts due to mutations

Mutant	Mean distance from charge to His 64 nitrogen atoms (Å)	Experimental ΔpK_a	Experimental D_{eff}	Calculated (W&W)* ΔpK_a	Calculated (W&W)* D_{eff}	Calculated ΔpK_a $D=55$	Calculated ΔpK_a D (M&E)†
Asp→Ser 99	12.6	−0.40	48	−0.31	62	−0.35	−0.38
Glu→Ser 156	14.4	−0.38	45	−0.35	48	−0.31	−0.30
Ser→Lys 99	15.0	(−0.25)	65	−0.25	65	−0.30	−0.28
Ser→Lys 156	16.5	(−0.25)	59	−0.30	49	−0.27	−0.24
Lys→Thr 213	17.6 (?)	+0.08	173	+0.19	73	+0.25	+0.21
Asp→Gln 36	15.1	−0.18	90	−0.16	101	−0.29	−0.27
Asp→Lys 99	(13.8)	−0.64	55	−0.56	63	−0.64	−0.64
Gly→Lys 156	(15.5)	−0.63	50	−0.65	48	−0.57	−0.52
Asp→Ser 99 and Glu→Ser 156	(13.5)	−0.63	57	−0.66	55	−0.66	−0.66
Asp→Lys 99 and Glu→Lys 156	(14.7)	−1.00	66	−1.21	55	−1.21	−1.14

Numbers in parentheses denote experimental values calculated from two or more other values rather than directly determined. Distances for mutants were obtained assuming the side chain to be fully extended. Values for the pK_a shifts are the mean values on the two histidine imidazole nitrogens. Distances are the average distance from the side chain nitrogen or oxygens to the two histidine imidazole nitrogens.

* In the Warwicker-Watson (W&W) calculation, the effects of the Asp, Glu or Lys charged side chains were evaluated and other side chains were taken to be uncharged. The first six mutants in the table were calculated directly by the Warwicker-Watson algorithm and the others obtained by addition.

† The values for the pK_a are given using a constant dielectric of 55 or using the distance-dependent equation of ref. 12 (M&E).

enzymes were modelled from the coordinates using the computer graphics program FRODO[13] with side chains being built in the extended conformation. In the algorithm, the effect of charges on the calculated electrostatic potential is additive. Thus the appropriate charge was placed on the side-chain atom and the electrostatic potential at His 64 evaluated. This potential can be related to the shift in pK_a by the Tanford-Roxby[14] equation ($\Delta pK_a = e\Delta\Phi/2.303zKT$, where e is the electronic charge, $\Delta\Phi$ is the change in the electrostatic potential at the titrating site, z the charge at the titrating site (1 for His 64) and T = 298 K). In addition the shifts in pK_a can be expressed[7] as an effective dielectric between the charges using $D_{eff} = 244/\Delta qr\Delta pK_a$, where r is the separation in Å and Δq is the change in charge.

Table 2 shows that for the single mutations, apart from Lys→Thr 213, the calculated pK_a shifts agree with the experimental values to within 0.1 of a unit, better than a factor of 2. Experimentally the pK_a shifts for the double mutants were found to be less than the sum of values for the component single mutations. As the modelling is based on addition of the individual residue changes, poorer agreement for these double mutants was obtained. Assuming additivity of electrostatic effects, the experimental results show a larger pK_a shift for Asp 99 (or Glu 156) to a neutral side chain than for the corresponding mutations of a neutral side chain to Lys 99 (or Lys 156). This is because the extended Lys side chain is farther from the active site histidine than the shorter acidic side chains and this effect is correctly predicted in the modelling.

Classical electrostatic theory shows that in systems with at least two media of different dielectrics, it is possible for the effective dielectric for an interaction to be greater than any individual dielectric[1]. An important feature of the Warwicker-Watson algorithm is that we find it can yield effective dielectrics greater than that of the solvent. Thus for the Asp→Gln 36 mutation, the experimental D_{eff} is 90 and this is modelled in the algorithm with a calculated D_{eff} of 101.

The worst prediction for a single mutation is Lys→Thr 213, where the experimental result is 0.08 compared to the calculated value of 0.23. However, this is a particularly flexible region of the molecule and in crystal structures of several mutant enzymes (R. R. Bott, personal communication) the Nε of Lys 213 is 19 ± 4 Å from His 94. Modelling with a structure that has a separation of 23 Å rather than 17.6 Å and perhaps has the charge more exposed to solvent should yield a better agreement with experiment.

Different algorithms can be applied to model these effects. Independently, Gilson and Honig[15] have shown that the pK_a shifts of the mutations Asp→Ser 99 and Glu→Ser 156 at a range of ionic concentrations may be satisfactorily calculated using a related procedure in the form of a linearized version of the Poisson-Boltzman equation and classical electrostatics. One simpler model, suitable for predicting subsequent mutations, is to use a uniform dielectric of 55, as this is the mean value of the experimental D_{eff} values that are less than 80 (that is, excluding Lys→Thr 213 and Asp→Gln 36). Another method is the distance-dependent dielectric of Mehler and Eichele[12], designed to model electrostatic effects from surface charges, which always uses a D_{eff} less than that of the solvent. Most mutations have an experimental $D_{eff} < 80$ and for these both simpler approaches yield calculated pK_a shifts that agree with experiment as well as the Warwicker Watson algorithm (Table 2). However the limitations of these simpler methods are highlighted in modelling the pK_a shifts for the two mutations with an experimental $D_{eff} > 80$. The simpler methods give poorer agreement with experiment than the results obtained by the Warwicker-Watson algorithm. Although the use of any macroscopic dielectric to model microscopic electrostatic effects is suspect, we conclude that consideration of the protein charge position and shape together with the different dielectric responses of the protein and solvent[8-10,15] provides an approach of modelling that yields good predictions to guide protein engineering.

We thank Dr M. N. G. James for unpublished coordinates of subtilisin, Dr R. R. Bott for distances in mutant subtilisin structures and Dr N. K. Rogers for the computer program of the Warwicker-Watson algorithm. Drs Gilson and Honig provided us with a copy of their paper before publication. This work was funded by the SERC Protein Engineering initiative. A.J.R. receives an SERC Biotechnology Directorate Studentship.

Received 30 June; accepted 17 September 1987.

1. Rogers, N. K. *Prog. Biophys. molec. Biol.* **48,** 37-66 (1986).
2. Tam, S. C. & Williams, R. J. P. *Struct. Bond.* **63,** 103-151 (1985).
3. Matthew, J. B. *et al. CRC crit. Rev. Biochem.* **18,** 91-197 (1985).
4. Warshel, A. & Russell, S. T. *Q. Rev. Biophys.* **17,** 283-422 (1984).
5. Thomas, P. G., Russell, A. J. & Fersht, A. R. *Nature* **318,** 375-376 (1985).
6. Russell, A. J., Thomas, P. G. & Fersht, A. R. *J. molec. Biol.* **193,** 803-813 (1987).
7. Russell, A. J. & Fersht, A. R. *Nature* **328,** 496-500 (1987).
8. Warwicker, J. & Watson, H. C. *J. molec. Biol.* **157,** 671-679 (1982).
9. Rogers, N. K. & Sternberg, M. J. E. *J. molec. Biol.* **174,** 527-542 (1984).
10. Rogers, N. K., Moore, G. R. & Sternberg, M. J. E. *J. molec. Biol.* **182,** 613-616 (1985).
11. Hopfinger, A. J. in *Conformational Properties of Macromolecules* 59-63 (Academic, New York, 1973).
12. Mehler, E. L. & Eichele, G. *Biochemistry* **23,** 3887-3891 (1984).
13. Jones, T. A. *J. appl. Crystallogr.* **11,** 268-272 (1978).
14. Tanford, C. & Roxby, R. *Biochemistry* **11,** 2192-2198 (1972).
15. Gilson, M. K. & Honig, B. H. *Nature* **330,** 84-86 (1987).

After one of my early lectures in the USA, someone reported to me that he had overheard Fred Richards, one of the wise old men of protein chemistry, ask the question 'Why do Alan Fersht's experiments work when no-one else's do?' I thought it was a compliment. Quite the reverse, I subsequently discovered, he simply did not believe them. In a later, very small meeting in Catalonia, he gave a talk based on the premise that the effects of site-directed mutagenesis on protein folding equilibria could not be used to measure the energetics of interactions in proteins because of effects on the denatured state. But, he could do it correctly by measuring the binding of chemically synthesised variants of the S-peptide to RNaseS (a form of RNaseA in which the S-peptide had been cleaved by subtilisin). I gently explained to him why the two approaches were essentially identical and suffered from the same caveats because of effects of mutation on the isolated S-peptide, and so if my work and also that of Nick Pace presented at the meeting were wrong, then so was his. I suppose it is some sort of compliment that ones' ideas are thought to be incorrect as that implies that those ideas are not obvious and trivial. But, such criticism from a 'wise old man', as Fred had been described, did not help.

Chapter 9

Structure–Activity (Rate-Equilibrium or Linear-Free-Energy Relationships) in Mutated Proteins

'In physics, imagination is more important than knowledge.'

Albert Einstein

It had been so exciting using mutagenesis to establish fundamental principles in the use of non-covalent interactions in binding and catalysis, and introducing such techniques as double-mutant cycles and surface (alanine) scanning to map out interactions. But, the excitement was raised to a new pitch when we discovered that the structure of a protein could obey the same type of structure–reactivity relationships that had been used by physical-organic chemists such as Brønsted, Hammett, Leffler and Marcus to analyse the fine details of mechanism. A short paper in *Nature* showed that there could be a linear relationship between the logarithms of the rate constants for engineered mutants and the equilibrium constants for the reaction[72].

9.1 2003: Alison Campbell, Colin Longstaff, Alexei Murzin, Danuta Mossakowska and Katy Brown.

9.2 2003: Alison Campbell and Nigel Brand.

Quantitative analysis of structure–activity relationships in engineered proteins by linear free-energy relationships

Alan R. Fersht, Robin J. Leatherbarrow & Tim N. C. Wells

Department of Chemistry, Imperial College of Science and Technology, London SW7 2AY, UK

Protein engineering is being used increasingly to study the fine details of the structure and activity of enzymes. How can small effects of mutation on activity be reliably quantified and systematized, and artefacts be recognized? A traditional means of doing this in organic chemistry is the use of linear free-energy relationships that link changes in rate constant for a reaction to changes in its equilibrium constant as the structure of the reagents is altered—Brønsted or Hammett plots[1]. We now find that the same type of plot may be applied to enzymatic reactions for variation of the structure of an enzyme with mutation. The activities of many mutant tyrosyl-transfer RNA synthetases fit structure–activity relationships which relate the rate constant for the formation of enzyme-bound tyrosyl adenylate (E.Tyr–AMP) to its equilibrium constant with enzyme-bound tyrosine and ATP (E.Tyr.ATP). This reaction results in an increase in binding energy between certain side chains of the enzyme and the side chain of tyrosine and the ribose ring of ATP. Plots of rate against equilibrium constant for a series of enzymes mutated in the relevant positions indicate that 71% of the binding energy change occurs on formation of the transition state for the chemical reaction and 90% occurs on formation of the E.Tyr–AMP.PP$_i$ complex. Other mutations show a different behaviour which is diagnostic of residues that specifically bind the transition state. Linear free-energy plots show trends and allow exceptions to be readily noted. That the activities of a large number of mutants conform to linear free-energy equations is the best evidence yet that mutation of an enzyme can probe general properties and trends in the relationship between structure and activity.

Mutants of the tyrosyl-tRNA synthetase, which catalyses the formation of tyrosyl adenylate [equation (1)], have been generated by site-directed mutagenesis and studied by kinetics[2-4].

$$\mathrm{E+Tyr+ATP} \rightleftharpoons \mathrm{E.Tyr\text{-}AMP+PP_i} \qquad (1)$$

These mutants have been altered in side chains that make hydrogen bonds with the substrates (Fig. 1). Model building followed by mutagenesis has revealed that there is a major catalytic factor caused by just two side chains that contribute binding energy to the γ-phosphate of ATP only when it is in the transition state (Thr 40 and His 45)[5]. Further, the binding energies of groups far removed from the site of reaction are also used to increase the chemical rate constants for the reaction (Tyr 34, Cys 35, His 48, Thr 51 and Tyr 169)[6]. Subsequent studies have measured the complete free-energy profiles for activation of tyrosine by the mutant enzymes[7,8]. The comparison of these with the profile for wild-type enzyme gives the apparent binding energy of the relevant side chain with the substrates, transition states and intermediates throughout the reaction. These data have been analysed in each individual case to show how the changes in binding energy affect the rate and equilibrium constants for the interconversion of free and enzyme-bound species. Several of the side chains that bind the side chain of tyrosine and the ribose ring of ATP appear to stabilize the enzyme-bound tyrosyl adenylate complex (E.Tyr–AMP) more than the transition state for its formation and more still than the initial enzyme-substrate ternary complex (E.Tyr.ATP)[7,8]. By this type of analysis, we are building up a picture of how each side chain contributes to catalysis.

The question most frequently asked about this approach is:

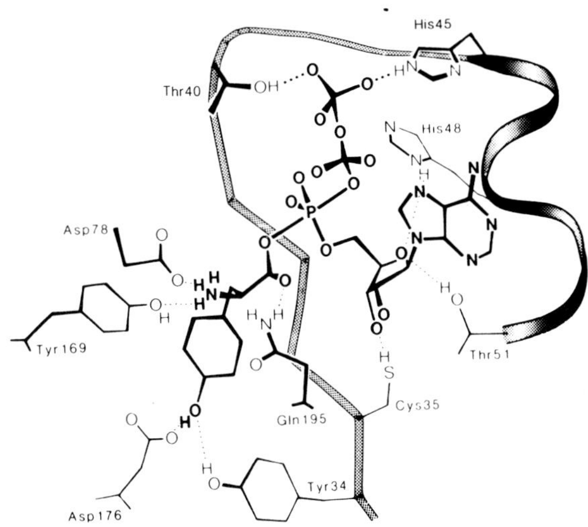

Fig. 1 Side chains of the tyrosyl-tRNA synthetase from *Bacillus stearothermophilus* which form hydrogen bonds with the transition state of ATP and tyrosine during the formation of tyrosyl adenylate (from ref. 5).

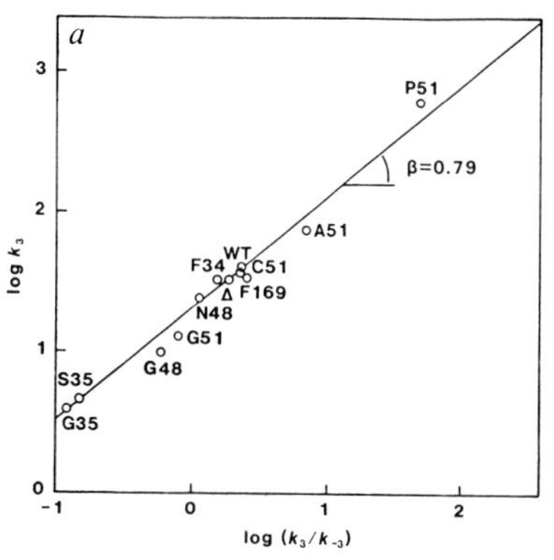

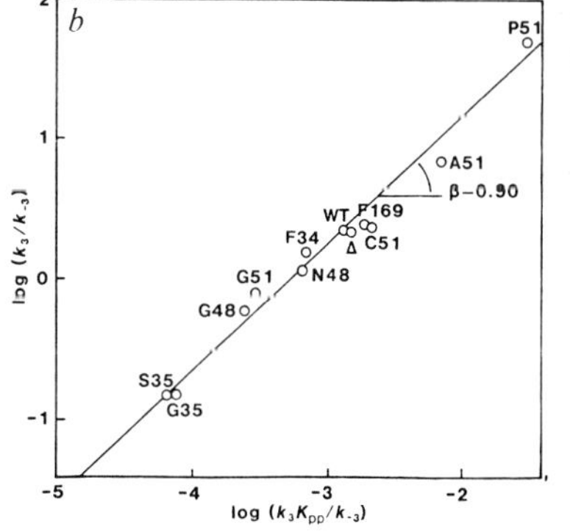

Fig. 2 *a*, Plot of log k_3 against log (k_3/k_{-3}) for equation (5) for mutation of Tyr 34, Cys 35, His 48, Thr 51 and Tyr 169. The mutations are indicated by the one-letter code apart from Δ, which represents the truncated enzyme lacking the tRNA-binding domain[10]. Data are from refs 6 and 7 and T.N.C.W. (unpublished). The plot is equivalent to a plot of $G_{[\mathrm{E.Tyr\text{-}ATP}]} - G_{\mathrm{E.Tyr.ATP}}$ against $G_{\mathrm{E.Tyr\text{-}AMP.PPi}} - G_{\mathrm{E.Tyr.ATP}}$, where G is Gibbs free energy and [E.Tyr.ATP] is the transition state. *b*, Plot of log (k_3/k_{-3}) against log (k_3K_{pp}/k_{-3}). This is equivalent to a plot of $G_{\mathrm{E.Tyr\text{-}AMP.PP_i}} - G_{\mathrm{E.Tyr.ATP}}$ against $G_{\mathrm{E.Tyr\text{-}AMP}} - G_{\mathrm{E.Tyr.ATP}}$. Note that individual variations in rate and equilibrium constants are seen as deviations from the regression line.

Reprinted from Nature, Vol. 322, No. 6076, pp. 284-286, 17 July 1986

How do we know that alteration of a side chain by mutagenesis does not cause additional structural changes in the enzyme which lead to artefactual results? X-ray crystallography of mutants can provide important evidence but may miss a series of small changes or the existence of conformational changes in solution. Further, although crystallography and spectroscopy can solve the structures of the stable states of the enzyme and its substrates, they do not give direct information on the structure which is important for the determination of rate, the transition state. The only way this can be probed experimentally at present is by kinetics. A possible way of both probing the structure of the transition state and detecting artefacts is the use of linear free-energy plots. In these, plots of rate constant against equilibrium constant are made for a reaction in which the structures of the reagents are systematically varied. The plots can give information on how much the transition state resembles starting materials or products. The plots assume that there is the relationship between the equilibrium constant K of a reaction and the rate constant k for the formation of products of the general form:

$$k = AK^{\beta} \qquad (2)$$

where A and β are constants. The plot of log k against log K when equation (2) holds is a straight line of slope β [equation (3)].

$$\log k = \text{constant} + \beta \log K \qquad (3)$$

Since log k is proportional to the free energy of activation of the reaction, $\Delta G^{\ddagger}$, and log K is proportional to the free-energy change for the equilibrium, ΔG_e, equation (3) is equivalent to:

$$\Delta G^{\ddagger} = \text{constant} + \beta \Delta G_e \qquad (4)$$

A qualitative interpretation of equations (2) to (4) is that, all things being equal, a value of β which is close to zero means that the transition state resembles starting materials, and a value of β close to 1 means that the transition state resembles products. Linear free-energy relationships in simple organic reactions generally measure the inductive effects of groups on rate and equilibrium constants. In enzymatic reactions, however, changes in structure remote from the seat of reaction result in changes of binding energy (ref. 9 and A.R.F., unpublished). Can linear free-energy relationships be applied to the changes in rate and equilibrium constants which occur when enzyme structure is varied as in a site-directed mutagenesis experiment? If so, the values of β will, in general, measure the fractions of total changes in binding energy[9].

We have measured all the necessary rate and equilibrium constants defined in equation (5) to construct plots for different stages of the reaction.

$$\mathrm{E} \underset{\mathrm{Tyr}}{\overset{K_t}{\leftrightarrows}} \mathrm{E.Tyr} \underset{\mathrm{ATP}}{\overset{K'_a}{\leftrightarrows}} \mathrm{E.Tyr.ATP} \underset{k_{-3}}{\overset{k_3}{\rightleftharpoons}} \mathrm{E.Tyr\text{-}AMP.PP_i} \underset{\mathrm{PP_i}}{\overset{K_{pp}}{\rightleftharpoons}} \mathrm{E.Tyr\text{-}AMP} \qquad (5)$$

As Fig. 2 shows, a plot of log k_3 against $\log(k_3/k_{-3})$ for mutants in the side chains which bind the side chain of tyrosine and the ribose ring of ATP fits an excellent straight line of slope $\beta = 0.79$. This means that 79% of the binding-energy change on E.Tyr.ATP going to E.Tyr-AMP.PP$_i$ is realized in the transition state for the reaction. Also plotted in Fig. 2 is $\log(k_3/k_{-3})$ against $\log(k_3K_{pp}/k_{-3})$. This is in reality a plot of the change in binding energy on going from E.Tyr.ATP to E.Tyr-AMP.PP$_i$ against the change in going from E.Tyr.ATP to E.Tyr-AMP + PP$_i$. The slope of 0.9 shows that 90% of the change in binding energy on forming E.Tyr-AMP occurs at the stage of formation of E.Tyr-AMP.PP$_i$. Figure 3 summarizes the changes in binding energy. There is a gain in binding energy on going from the ternary complex E.Tyr.ATP to the enzyme-bound tyrosyl adenylate complex, E.Tyr-AMP; 71% of this binding energy is realized in the transition state and 90% in the other intermediate, E.Tyr-AMP.PP$_i$.

The above type of behaviour applies when the residues concerned show a uniformly progressive change in binding energy as the reaction proceeds. A different pattern is expected when

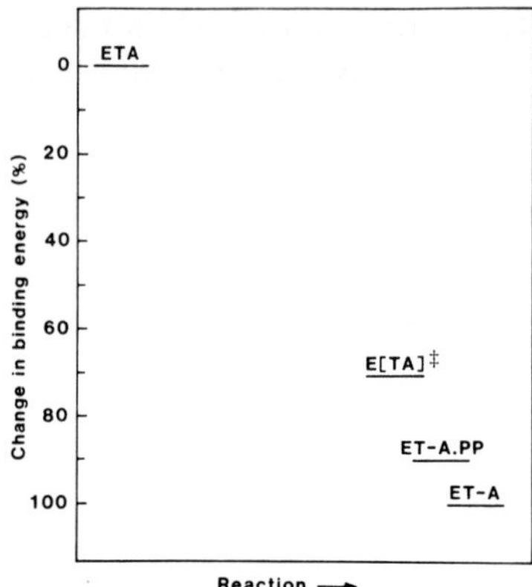

Fig. 3 Summary of the average binding-energy changes calculated for residues Tyr 34, Cys 35, His 48, Thr 51 and Tyr 169 on going from the E.Tyr.ATP complex (ETA) to the transition state (E[TA]‡) to the first intermediate, E.Tyr-AMP.PP$_i$ (ET-A.PP), to the final intermediate, E.Tyr-AMP (ET-A).

the mutation involves residues that bind the transition state of the substrates well but bind the unreacted substrates and products only poorly. Such behaviour should lead to values for β which are much greater than 1. For example, in the extreme case, where a residue binds only at the stage of the transition state, mutation of that residue will not affect the equilibrium constant of the reaction but will affect the rate; that is, β should tend to infinity. This is found quite dramatically with residues Thr 40 and His 45 (Fig. 4), which have already been implicated in such a mode of catalysis[5]. The very high value of β found indicates that they are highly specific for the transition state.

It is quite remarkable that plots in Fig. 2 are so similar to those found for linear free-energy relationships in simple organic reactions. Indeed, the fit to such plots is as good as many examples from physical organic chemistry. Why are the linear free-energy relationships followed so well considering the likelihood of disturbing the structure of the protein? The answer is probably because we have restricted our mutations to the most conservative changes, which have made only small perturbations in the structure.

The demonstration that certain structural changes of the tyrosyl-tRNA synthetase cause changes in rate which may be described satisfactorily by linear free-energy equations is important for the following reasons. First, the equations combine the data from many experiments and so bring order to them. Second,

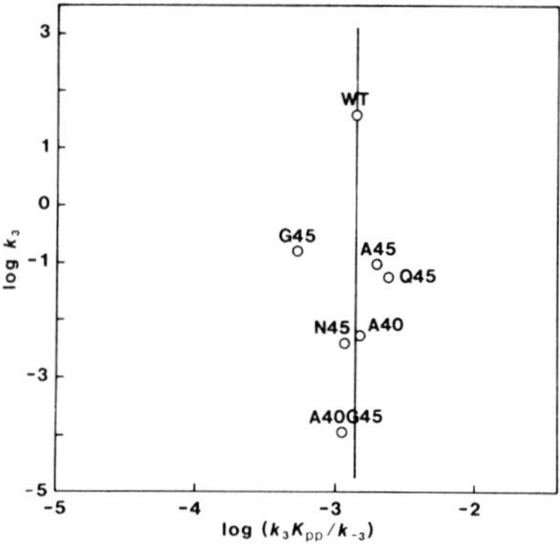

Fig. 4 Plot of log k_3 against log (k_3K_{pp}/k_{-3}) for the mutation of residues Thr 40 and His 45 which are involved in transition-state binding only[5]. This is equivalent to a plot of $G_{[\mathrm{E.Tyr\text{-}ATP}]^{\ddagger}} - G_{\mathrm{E.Tyr.ATP}}$ against $G_{\mathrm{E.Tyr\text{-}AMP}} - G_{\mathrm{E.Tyr.ATP}}$. Data are from ref. 5 and R.J.L. (unpublished).

they quantify aspects of the transition state structure; in particular, they show how much the interactions in the transition state resemble those in the enzyme–substrate and the enzyme–intermediate complexes. Finally, the equations show trends; any mutant that deviates from the trend is immediately apparent.

The last point is particularly important. The results of any one individual mutation can be challenged on the grounds that the mutation might cause a significant change in the protein structure. However, when a whole series of mutational experiments can be fitted to a linear free-energy equation, there is overwhelming evidence that each mutant shows part of a general phenomenon of the relationship between structure and activity. This type of analysis can be applied to any enzymatic reaction where both rate and equilibrium constants may be measured.

Received 6 May; accepted 5 June 1986.

1. Jencks, W. P. *Chem. Rev.* **85,** 511–517 (1985).
2. Winter, G., Fersht, A. R., Wilkinson, A. J., Zoller, M. & Smith, M. *Nature* **299,** 756–758 (1982).
3. Fersht, A. R. *et al. Angew. Chem. int. Edn Engl.* **23,** 467–473 (1984).
4. Fersht, A. R. *et al. Nature* **314,** 235–238 (1985).
5. Leatherbarrow, R. J., Winter, G. & Fersht, A. R. *Proc. natn. Acad. Sci. U.S.A.* **82,** 7840–7844 (1985).
6. Wells, T. N. C. & Fersht, A. R. *Nature* **316,** 656–657 (1985).
7. Wells, T. N. C. & Fersht, A. R. *Biochemistry* **25,** 1881–1886 (1986).
8. Ho, C. K. & Fersht, A. R. *Biochemistry* **25,** 1891–1897 (1986).
9. Fersht, A. R. *Enzyme Structure and Mechanism* 2nd edn, Ch. 12 (Freeman, San Francisco, 1985).
10. Waye, M. M. Y., Winter, G., Wilkinson, A. J. & Fersht, A. R. *EMBO J.* **2,** 1827–1829 (1983).

It was followed up by a definitive one in *Biochemistry* in which we set the ground in terms of mutations to be made and their analysis[73]. We proposed also plotting the progress of realization of the energy of binding as the reaction proceeds to obtain a parameter, which I termed β, in honour of one of my scientific heroes, Brønsted. β is used in acid-base catalysis to measure the extent of proton transfer. Robin Leatherbarrow had suggested using Φ for Fersht, but I though it immodest so to do.

9.3 2003: Nigel Brand, Sue Cotterill and Robin Leatherbarrow.

Reprinted from Biochemistry, 1987, *26*, 6030.

Structure–Activity Relationships in Engineered Proteins: Analysis of Use of Binding Energy by Linear Free Energy Relationships†

Alan R. Fersht,* Robin J. Leatherbarrow, and Tim N. C. Wells
Department of Chemistry, Imperial College of Science and Technology, London SW7 2AY, U.K.
Received January 21, 1987; Revised Manuscript Received April 27, 1987

ABSTRACT: The activity of mutant enzymes can be analyzed quantitatively by structure–activity relationships in a manner analogous to Brønsted or Hammett plots for simple organic reactions. The slopes of such plots, the β values, indicate for the enzymatic reactions the fraction of the overall binding energy used in stabilizing particular complexes. In particular, information can be derived about the interactions between the enzyme and the transition state. The activities of many mutant tyrosyl-tRNA synthetases fit well simple linear free energy relationships. The formation of enzyme-bound tyrosyl adenylate (E·Tyr-AMP) from enzyme-bound tyrosine and ATP (E·Tyr·ATP) results in an increase in binding energy between the enzyme and the side chain of tyrosine and the ribose ring of ATP. Linear free energy plots of enzymes mutated in these positions give the fraction of the binding energy change that occurs on formation of the transition state for the chemical reaction and the various complexes. It is shown that groups that specifically stabilize the transition state of the reaction are characterized by β values $\gg 1$. This is found for residues that bind the γ-phosphate of ATP (Thr-40 and His-45) and have previously been postulated to be involved in transition-state stabilization. The importance of linear free energy plots is that (i) they bring order to the analysis of structure–activity relationships since they allow a large amount of data to be simplified and systematized, (ii) they allow transition-state structure to be inferred from ground-state structure, and (iii), perhaps the most important point at this stage of our knowledge in understanding enzyme structure, they collectively show trends, and exceptions are readily apparent. The observation that the activities of a large number of mutants of the tyrosyl-tRNA synthetase conform to linear free energy equations is the best evidence yet that mutation of the enzyme is probing general properties and trends in the relationship between structure and activity.

The quintessential feature of enzyme catalysis is the use of binding energy to give rate enhancement. It may be shown quite simply by theory (see Appendix) that the increase in rate between an enzyme-catalyzed reaction and its uncatalyzed counterpart in aqueous solution proceeding by the same transition states and intermediates is directly related to the dissociation constants of the substrates, transition states, intermediates, and products from the enzyme. The free energy of transfer of the reagents from water to enzyme is responsible for lowering the free energy of activation and for altering equilibrium constants between enzyme-bound reagents. Thus, to understand enzyme catalysis, one must determine the interaction energies between the enzyme and reagents throughout the complete reaction profile and the energetic changes within the enzyme itself.

Site-directed mutagenesis is eminently suited to studying such structure–activity relationships since the structure of an enzyme may be systematically varied. Unfortunately, protein engineering is an invasive technique since mutagenesis perturbs the very system it studies. The structures of mutant enzymes may be studied by X-ray crystallography or spectroscopy, which may detect changes in structure. But, these techniques can be applied only to stable complexes and cannot be used for examining the important structure for determination of rate, the transition state. Further, X-ray crystallography may not have sufficient resolution for detecting miniscule changes in the structure of an enzyme or its solvation shell that cause significant changes in energetics of catalysis. The only means at present of probing transition-state structure experimentally is the use of kinetics. Accordingly, we are attempting to devise kinetic methods that may be applied in conjunction with X-ray crystallographic data to analyze structure–activity relationships of enzymes and, in particular, how binding energy is utilized. In this paper, we outline the theory of how the use of binding energy may be analyzed by protein engineering, propose criteria for the design of experiments that are open to interpretation by such theory, and apply the theory to reactions of a particular set of mutants of the tyrosyl-tRNA synthetase from *Bacillus stearothermophilus*.

Application of Linear Free Energy Relationships. A traditional way of probing transition-state structure in the reactions of simple organic molecules is the construction of linear

† This work was funded by the Medical Research Council of the U.K.

free energy plots such as the Brønsted and Hammett plots. In these, plots of rate constant against equilibrium constant are made for a reaction in which the structures of the reagents are systematically varied. The plots can give information on how much the transition state resembles starting materials or products. The plots assume that there is the relationship between the equilibrium constant K of a reaction and the rate constant k for the formation of products of the general form

$$k = AK^{\beta} \quad (1)$$

where A and β are constants. The plot of log k against log K when eq 1 holds is a straight line of slope β (eq 2). Since

$$\log k = \text{constant} + \beta \log K \quad (2)$$

log k is proportional to the free energy of activation of the reaction, ΔG^*, and log K is proportional to the free energy change for the equilibrium, ΔG_e, eq 2 is equivalent to

$$\Delta G^* = \text{constant} + \beta \Delta G_e \quad (3)$$

A qualitative interpretation of eq 1–3 is that, all things being equal, a value of β that is close to zero means that the transition state resembles starting materials and a value of β close to 1 means that the transition state resembles products. We have recently shown that linear free energy relationships can be applied to the changes in *binding energy* that occur when enzyme structure is varied in a site-directed mutagenesis experiment (Fersht et al., 1986a).

Linear free energy relationships must not be applied blindly to the kinetics of mutant enzymes since the area is rife with potential artifacts. We thus first classify the effects of a single amino acid substitution to decide which experiments are worthy of extensive analysis.

Classification of Effects of Mutation on Structure. The study of enzyme structure and activity by systematic mutagenesis of the enzyme involves at the very simplest the substitution of just one amino acid for another. More complex experiments may involve multiple substitutions, deletions, or even changing whole domains. But even the substitution of the side chain of an amino acid by another can lead to a variety of effects, some of which may be too complicated for simple analysis. The substitutions with which we are primarily concerned are those that effect the interactions between enzyme and ligand and those that perturb the structure of the enzyme. These are classified as follows.

(*1*) *Nondisruptive Deletion.* A side chain is replaced by another that lacks a group involved in a specific interaction. This is done in such a manner that there are no effects on the structure of the enzyme or any of its complexes with the substrates other than the loss of this interaction. This could be, for example, the removal of an interaction between the enzyme and substrate that does not affect any other interaction between the enzyme and substrate or any interaction within the enzyme itself. These involve almost always the replacement of one side chain by another that is smaller and lacks the functional group under discussion. A typical example is the mutation of Tyr → Phe-34 in the tyrosyl-tRNA synthetase. The mutation just removes a group which, in the free enzyme, makes interactions with solvent only. Nondisruptive deletion may, however, perturb solvation.

(*2*) *Disruptive Deletion.* Replacement of a side chain may lead to a perturbation of structure elsewhere in the protein. One example that we have found in our studies is the removal of the carboxylate of an aspartate residue, which forms part of a hydrogen-bonded network (Lowe et al., 1987).

(*3*) *Conservative Substitution.* A side chain is replaced by one that can substitute in the same interactions. For example, on mutation of His → Asn-48 in the tyrosyl-tRNA synthetase, the amide nitrogen of $-CONH_2$ can on some, but not all, occasions substitute for the δ-NH of histidine in forming a hydrogen bond with the substrate (Lowe et al., 1985).

(*4*) *Semiconservative Substitution.* Some of the function is conserved on replacement. For example, replacement of a histidine by an asparagine when the imidazole ring of the histidine is charged and in a salt bridge. The $-CONH_2$ of the Asn may still form a hydrogen bond but lacks the formal positive charge.

(*5*) *Disruptive Substitution.* Substitution of a large side chain for a small one in a buried close-packed region of a protein may clearly cause severe structural changes because of steric repulsions. There are also substitutions that appear innocuous at first sight but cause severe problems because they add alternative functions. For example, the mutations Asp → Asn and Asn → Asp appear to be extremely conservative since $-CO_2^-$ and $-CONH_2$ are isosteric and both polar. But, whereas both oxygen atoms of $-CO_2^-$ are hydrogen-bond acceptors, as is the oxygen of $-CONH_2$, the $-NH_2$ group is a hydrogen-bond donor.

(*6*) *Nondisruptive Addition.* Bulky groups may be added to the surface of proteins without necessarily causing perturbation of structure. Similarly, the binding cavity of an enzyme may be partly filled so that substrates smaller than the specific substrate can still bind. For example: indole binds to the active site of chymotrypsin and enhances the reaction rate with *p*-nitrophenyl acetate (Foster, 1961; Landis & Berliner, 1980); mutation of residue Gly-166 in subtilisin, which is part of the S_1 subsite for the side chain of a substrate, increases the reactivity toward substrates with small amino acids in the P_1 position (Estell et al., 1986).

The disruptive mutations are likely to be too complicated for simple analysis. The most promising class to analyze is that of nondisruptive removal since, in the ideal case, there is just the removal of a simple interaction with no complications from disruption of structure or the substitution of alternative interactions. Conservative and semiconservative substitutions may also be amenable to analysis. Mutants with nondisruptive additions may perhaps be analyzed where alternative substrates are available. This study concerns primarily nondisruptive deletants, but the principles may be extended to other cases.

Experimental Procedures

The mutant TyrTS(His→Asn-48)[1] was prepared according to Lowe et al. (1985). Kinetic measurements were all performed at 25 °C at pH 7.78 in a buffer containing 144 mM Tris-HCl, 0.14 mM 2-mercaptoethanol, and 10 mM $MgCl_2$ (free). The pre-steady-state kinetic measurements of the rate constants for the activation of tyrosine were determined as before (Wells & Fersht, 1986).

Results

The dissociation constant of tyrosine from the E·Tyr complex with TyrTS(His→Asn-48) (K_t) was found to be 23 μM; the dissociation constant of ATP from the E·Tyr·ATP complex (K'_a), 3.8 mM; the rate constant for the formation of E·Tyr-AMP·PP_i from the E·Tyr·ATP complex (k_3), 27 s^{-1}; the rate constant for the formation of E·Tyr·ATP from E·Tyr-AMP·PP_i (k_{-3}), 23.4 s^{-1}; and the dissociation constant of py-

[1] Abbreviations: TyrTS, tyrosyl-tRNA synthetase; T, Tyr; A, ATP; T-A, Tyr-AMP (tyrosyl adenylate); [T-A]*, transition state for formation of tyrosyl adenylate; ET, enzyme-bound tyrosine; E[T-A]*, enzyme-bound transition state for formation of tyrosyl adenylate; Tris, tris(hydroxymethyl)aminomethane.

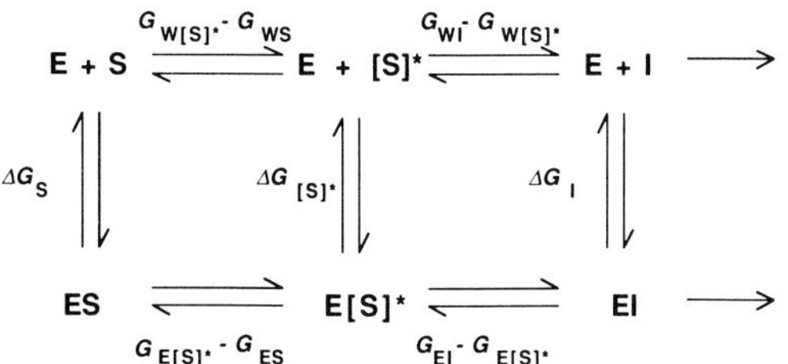

FIGURE 1: Thermodynamic cycles for comparing Gibbs free energy changes for uncatalyzed reaction in solution with its enzyme-catalyzed counterpart. G denotes free energy of reagent (subscript S = substrate, I = intermediate, and [S]* = transition state) in water (subscript W) or bound to enzyme (subscript E). ΔG denotes free energy of transfer of reagent from enzyme to water.

rophosphate from the last complex (K_{pp}), 0.57 mM.

THEORY

Consider a reaction that involves the reversible binding of a substrate S to an enzyme to form a Michaelis complex, ES, followed by the formation of an enzyme-bound transition state, E[S]*, and then an enzyme-bound intermediate, EI (eq 4).

$$E + S \overset{K_S}{\rightleftharpoons} ES \underset{k_{-2}}{\overset{k_2}{\rightleftharpoons}} EI \rightarrow \quad (4)$$

The reaction of the enzyme with its substrates, their transition states, and intermediates may be described by the formal series of thermodynamic cycles in Figure 1. These link the equilibria between the enzyme-bound species and their counterparts in aqueous solution by dissociation constants. From standard transition-state theory and equilibrium thermodynamics

$$RT \ln (k_2/k_{-2}) = G_{ES} - G_{EI} \quad (5)$$

$$RT \ln k_2 = kT/h - (G_{E[S]^*} - G_{ES}) \quad (6)$$

$$RT \ln (k_2/K_S) = kT/h - (G_{E[S]^*} - G_{WE} - G_{WS}) \quad (7)$$

where G is the Gibbs free energy and the subscripts WS and WE refer to the substrate and enzyme in water. It is shown in the Appendix that these equations may be cast entirely in terms of binding energies for analyzing the changes in rate and equilibrium constants that occur when a series of mutated enzymes reacts with the same substrate. That is

$$RT \ln (k_2/k_{-2}) = \Delta G_S - \Delta G_I + \text{constant} \quad (8)$$

$$RT \ln k_2 = \text{constant} - \Delta G_{[S]^*} + \Delta G_S \quad (9)$$

$$RT \ln (k_2/K_S) = \text{constant} - \Delta G_{[S]^*} \quad (10)$$

where ΔG_S, $\Delta G_{[S]^*}$, and ΔG_I are the free energies of binding of the substrate, transition state, and intermediate to the enzymes. The standard equations (e.g., 5, 6, and 7) can be written as usual in terms of the free energies of the relevant states, but it can be considered that the only relevant changes are those in binding energy. Hence, any linear free energy relationships will apply to changes in binding energies. The only assumption is that the structure of the transition state of the substrate does not change on mutation—see Appendix.

Application to Reactions of Tyrosyl-tRNA Synthetase. Mutants of the tyrosyl-tRNA synthetase have been generated that have been altered in side chains that make hydrogen bonds with the substrates (Figure 2) (Winter et al., 1982; Fersht et al., 1984, 1985, 1986b). The first step of the enzymatic reaction is the formation of tyrosyl adenylate (eq 11). Model

$$E + Tyr + ATP \rightleftharpoons E\cdot Tyr\text{-}AMP + PP_i \quad (11)$$

building followed by mutagenesis has revealed that there is a major catalytic factor caused by two side chains that con-

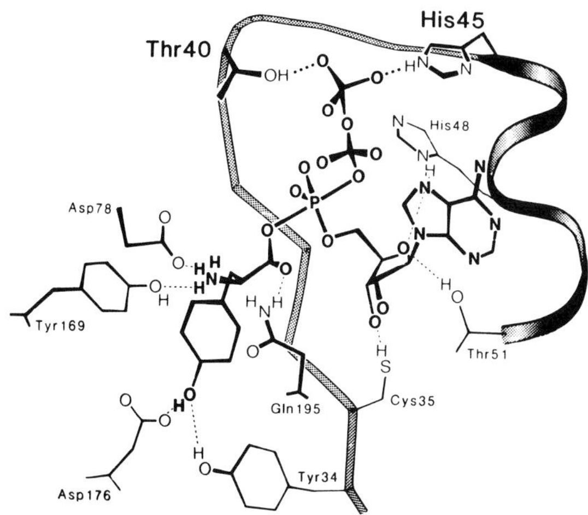

FIGURE 2: Transition state of tyrosine + ATP built into the active site of the tyrosyl-tRNA synthetase [from Leatherbarrow et al. (1985)].

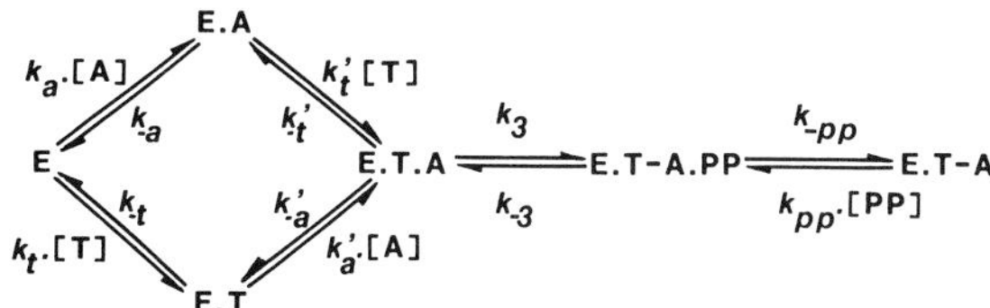

FIGURE 3: Reaction scheme for the activation of tyrosine that defines the relevant rate and dissociation constants.

tribute binding energy to the γ-phosphate of ATP only when it is in the transition state (Leatherbarrow et al., 1985). The binding energies of groups far removed from the seat of reaction are also used to increase the chemical rate constants for the reaction (Wells & Fersht, 1985). Complete free energy profiles for activation of tyrosine by the mutant enzymes have been measured (Fersht et al., 1986a; Wells & Fersht, 1986; Ho & Fersht, 1986). The comparison of these with the profile for wild-type enzyme gives the *apparent* binding energy (Wells & Fersht, 1986) of the relevant side chain with the substrates, transition states, and intermediates throughout the reaction. These data have been analyzed in each individual case to show how the changes in binding energy affect the rate and equilibrium constants for the interconversion of free and enzyme-bound species. Several of the side chains that bind the side chain of tyrosine and the ribose ring of ATP appear to stabilize the enzyme-bound tyrosyl adenylate complex more than the transition state for its formation and more still than the initial enzyme–substrate ternary complex.

We have measured the relevant equilibrium and rate constants for Figure 3, which describes the formation of tyrosyl adenylate by the tyrosyl-tRNA synthetase (Wells & Fersht, 1986; Ho & Fersht, 1986; Leatherbarrow et al., 1985; T. N. C. Wells, R. J. Leatherbarrow, and A. R. Fersht, unpublished results). The free energy terms are defined thus: G_{WE} = free energy of unligated enzyme; G_{ET} = free energy of enzyme–tyrosine complex; G_{ETA} = free energy of enzyme–tyrosine–ATP complex; $G_{E[T\text{-}A]^*}$ = free energy of transition state for the interconversion of the ternary complexes of enzyme-bound tyrosine and ATP and of enzyme-bound tyrosyl adenylate and pyrophosphate; $G_{ET\text{-}A\text{-}PP}$ = free energy of the ternary complex with pyrophosphate and tyrosyl adenylate; $G_{ET\text{-}A}$ = free energy of enzyme-bound tyrosyl adenylate complex.

A series of linear free energy relationships were observed that relate various rate constants with corresponding equilibria. The equations were set up by first writing down the initial,

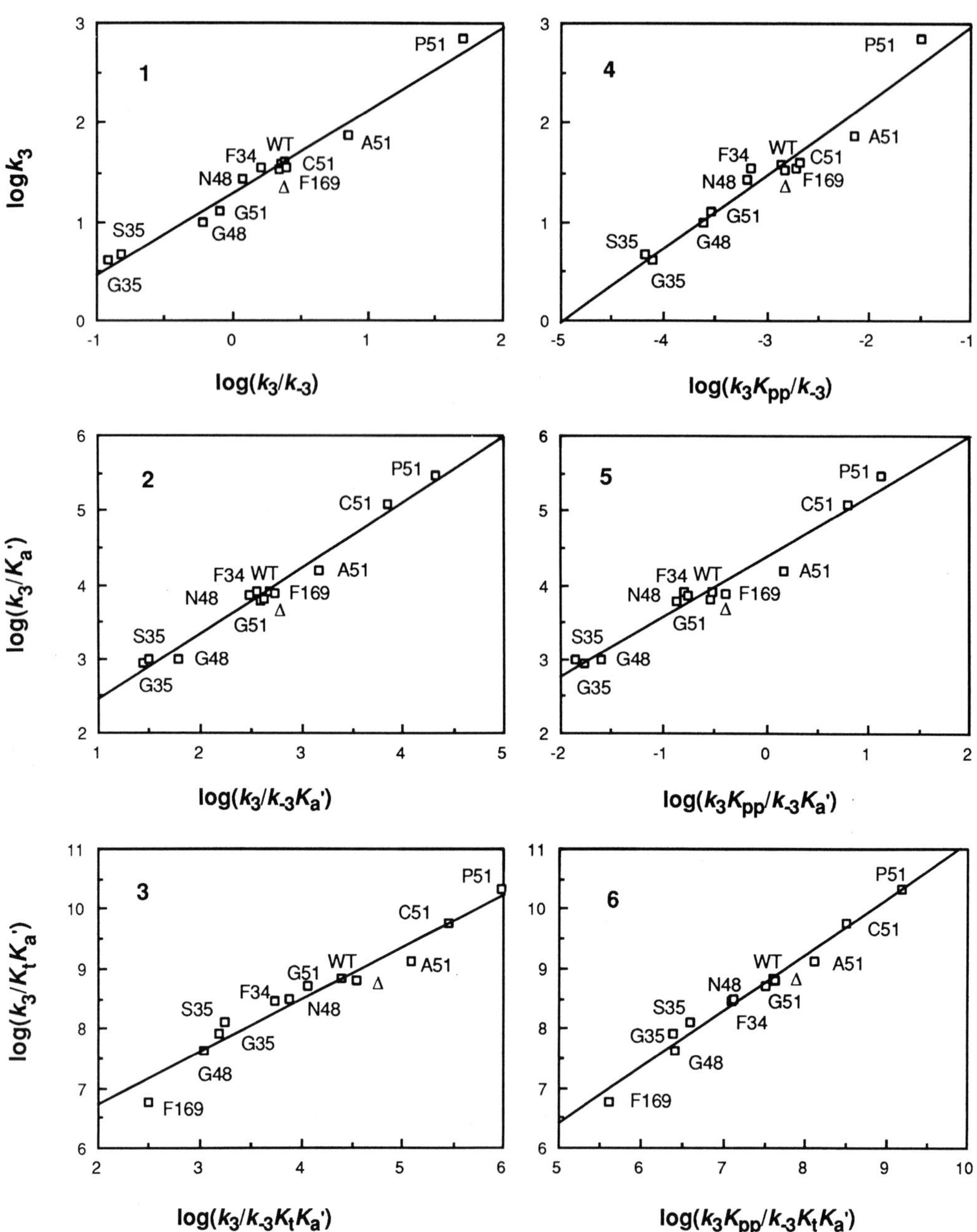

FIGURE 4: Linear free energy plots for different stages of the activation of tyrosine and the equilibria between enzyme-bound complexes.

transition (or intermediate), and final states to be compared. The equilibrium constant relating the initial and final states and the rate and equilibrium constant relating the transition state to the initial state (or, for an intermediate, just the equilibrium constant) were derived. The rate and equilibrium constants were transposed into the free energy terms just by considering the free energies of each state.

(*A*) *Mutations in the Binding Sites for Tyrosine* (*Tyr → Phe-34, Tyr → Phe-169*) *and the Ribose of ATP* (*Cys → Gly-35, Cys → Ser-35, His → Gly-48, His → Asn-48, Thr → Gly-51, Thr → Ala-51, Thr → Cys-51, Thr → Pro-51*). (*Case 1*) *Relationship between k_3 and k_3/k_{-3}*. A relationship of the form

$$k_3 = A(k_3/k_{-3})^{\beta} \qquad (12)$$

is equivalent to

$$G_{E[T\text{-}A]^*} - G_{ETA} = \text{constant} + \beta(G_{ET\text{-}A\cdot PP} - G_{ETA}) \qquad (13)$$

Using the manipulations described in the appendix, eq 13 transforms into

$$\Delta G_{[T\text{-}A]^*} - \Delta G_{TA} = \text{constant} + \beta(\Delta G_{T\text{-}A\cdot PP} - \Delta G_{TA}) \qquad (14)$$

where $\Delta G_{[T\text{-}A]^*}$ represents the free energy of binding of the transition state [T-A]* to the enzyme, ΔG_{TA} represents the sum of the binding energies of tyrosine and ATP, etc. Equation 14 relates the change in binding energy on going from the ETA complex to E[T-A]* (transition state) compared with the total change in binding energy between ET-A·PP (tyrosyl adenylate + PP bound) and ETA. All the states are enzyme-bound and so no terms in K_t, K'_a, or K_{pp} are involved.

We find that a plot of log k_3 against log (k_3/k_{-3}) (Figure 4) gives a good linear free energy relationship of slope 0.79. This suggests that, on average, 79% of the binding energy change on going from ETA to ET-A·PP is realized in the E[T-A] transition state.

(*Case 2*) *Relationship between* k_3/K'_a *and* $k_3/k_{-3}K'_a$. The free energy changes are

$$G_{E[T-A]*} - G_{ET} = \text{constant} + \beta(G_{ET\text{-}A\cdot PP} - G_{ET}) \quad (15)$$

which is equivalent to

$$\Delta G_{[T-A]*} - \Delta G_T = \text{constant} + \beta(\Delta G_{T\text{-}A\cdot PP} - \Delta G_T) \quad (16)$$

Here, the initial state is the enzyme–tyrosine complex (ET). Thus, on formation of ET-A·PP and E[T-A]*, the binding energy of ATP to the ET complex is important and so K'_a enters the equations. β will contain in addition to case 1 the fraction of the total binding energy of ATP to the enzyme in the various states. The logarithmic plot (Figure 4) gives again a good line, but the slope β is now 0.92. That is, 92% of the binding energy change on going from ET to ET-A·PP is realized in E[T-A]*.

(*Case 3*) *Relationship between* $k_3/K_tK'_a$ *and* $k_3/k_{-3}K_tK'_a$. The free energy changes are

$$G_{E[T-A]*} - G_{WE} = \text{constant} + \beta(G_{ET\text{-}A\cdot PP} - G_{WE}) \quad (17)$$

which is equivalent to

$$\Delta G_{[T-A]*} = \text{constant} + \beta(\Delta G_{T\text{-}A\cdot PP}) \quad (18)$$

Here, the initial state is the free enzyme. Thus, formation of E[T-A]* and ET-A·PP from E involves the total binding energies of tyrosine and ATP at the different stages and so K_t and K'_a enter the equations. There is again a satisfactory logarithmic plot (Figure 4). The slope is about 0.95; i.e., 95% of the binding energy change on going from E to ET-A·PP is realized in the E[T-A]* complex.

(*Case 4*) *Relationship between* k_3 *and* k_3K_{pp}/k_{-3}. The energy changes are

$$G_{E[T-A]*} - G_{ETA} = \text{constant} + \beta(G_{ET\text{-}A} - G_{ETA}) \quad (19)$$

which is equivalent to

$$\Delta G_{[T-A]*} - \Delta G_{TA} = \text{constant} + \beta(\Delta G_{T\text{-}A} - \Delta G_{TA}) \quad (20)$$

This is the same as case 1, but the final state is the ET-A complex. The slope of the linear logarithmic plot (Figure 4) is 0.71, slightly less than that for case 1. This indicates that there is a slight increase in binding energy on going from ET-A·PP to ET-A.

(*Case 5*) *Relationship between* k_3/K'_a *and* $k_3K_{pp}/k_{-3}K'_a$. The energy changes are

$$G_{E[T-A]*} - G_{ET} = \text{constant} + \beta(G_{ET\text{-}A} - G_{ET}) \quad (21)$$

which is equivalent to

$$\Delta G_{[T-A]*} - \Delta G_T = \text{constant} + \beta(\Delta G_{T\text{-}A} - \Delta G_T) \quad (22)$$

This is the same as for case 2 except that the final state is now the ET-A complex because the term K_{pp} is involved. Again, β at 0.84 (Figure 4) is slightly lower than that for case 2.

(*Case 6*) *Relationship between* $k_3/K_tK'_a$ *and* $k_3K_{pp}/k_{-3}K_tK'_a$. The energy changes are

$$G_{E[T-A]*} - G_{WE} = \text{constant} + \beta(G_{ET\text{-}A} - G_{WE}) \quad (23)$$

which is equivalent to

$$\Delta G_{[T-A]*} = \text{constant} + \beta(\Delta G_{T\text{-}A}) \quad (24)$$

This is the same as for case 3 except that the final state is ET-A. The value of β (Figure 4) is 0.88, similar to but slightly lower than that for case 3.

Further Plots May Be Constructed with All the Permutations and Combinations of the Rate and Equilibrium Constants. For example, the above results indicate that the binding energy change on going from ETA to ET-A·PP is only about 90% of that on going from ETA to ET-A since the β values are about 10% lower for cases 4–6. This may be shown directly by a plot of log (k_3/k_{-3}) against log (k_3K_{pp}/k_{-3}) (Fersht et al., 1986a). The energy changes are equivalent to

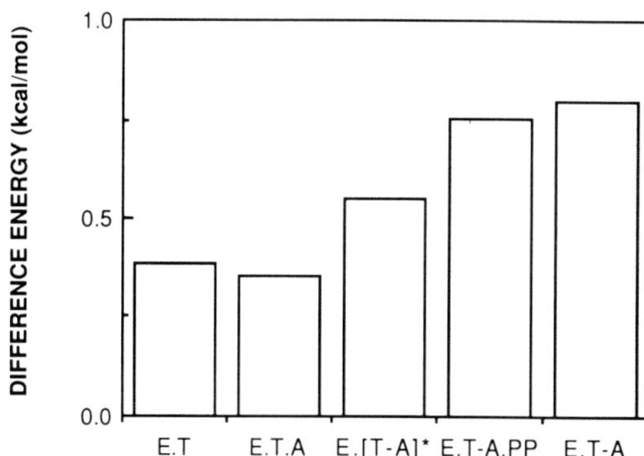

FIGURE 5: Difference energy diagram between the mutant TyrTS-(His→Asn-48) and wild-type enzyme.

$$G_{ET\text{-}A\cdot PP} - G_{ETA} = \text{constant} + \beta(G_{ET\text{-}A} - G_{ETA}) \quad (25)$$

which is equivalent to

$$\Delta G_{T\text{-}A\cdot PP} - \Delta G_{TA} = \text{constant} + \beta(\Delta G_{T\text{-}A} - \Delta G_{TA}) \quad (26)$$

The slope of this plot is seen directly to be 0.9.

Another example is the fraction of the binding energy of ATP used in going from the ET complex to the ETA complex relative to ET going to ET-A. This is found from a plot of log $(1/K'_a)$ against log $(k_3K_{pp}/k_{-3}K'_a)$. This is again a reasonable plot of slope $\beta = 0.12$ (data not shown). The point for Cys-51 deviates from this line, the mutation Thr → Cys-51 being the closest example of a uniform binding energy change (Ho & Fersht, 1986). This deviation is seen again in the following section.

Direct Calculation of β Values from Difference Energy Diagrams. The apparent contributions of the free energies of binding of the various side chains to the binding energies of each state have been calculated previously from the differences in the free energy profiles for the reactions of mutant and wild-type enzymes (Wells & Fersht, 1986; Ho & Fersht, 1986). These may be used to construct individual β values for each mutant. For example, as illustrated in Figure 5, if the apparent contribution of the binding energy of a side chain in the ternary complex ETA is $\Delta\Delta G_{ETA}$, in the transition state is $\Delta\Delta G_{E[T-A]*}$, and in the ET-A complex is $\Delta\Delta G_{ET\text{-}A}$, as defined by Wells and Fersht (1986), then in an example exactly analogous to case 4 (eq 19 and 20)

$$\beta = (\Delta\Delta G_{E[T-A]*} - \Delta\Delta G_{ETA})/(\Delta\Delta G_{ET\text{-}A} - \Delta\Delta G_{ETA}) \quad (27)$$

The values of β derived for individual mutants in Table I are in acceptable agreement with those derived from the slopes of the linear free energy plots. For example, the plot for case 4 (eq 19 and 20) gives a slope of 0.71 whereas the direct calculation gives a range of values averaging 0.61. Similarly, the plot relating to eq 25 and 26 gives a slope of 0.9 whereas the direct calculations give an average of 0.84. Comparison of the individual values in Table I with the values from the linear free energy plots shows how the plots smooth out the individual values for each mutant and shows the overall trend. The small differences between the average values calculated from the individual mutants and the slopes of the linear free energy plots arise from the differences in the methods of statistical analysis. Linear regression tends to weight values

Table I: Direct Calculation of β Values from Difference Energy Plots[a]

mutant	β: $(\Delta\Delta G_{E[T\text{-}A]^{\ddagger}} - \Delta\Delta G_{ETA})/(\Delta\Delta G_{ET\text{-}A} - \Delta\Delta G_{ETA})$	β: $(\Delta\Delta G_{ET\text{-}A\cdot PP} - \Delta\Delta G_{ETA})/(\Delta\Delta G_{ET\text{-}A} - \Delta\Delta G_{ETA})$
Cys → Gly-35	0.78	0.92
Cys → Ser-35	0.69	0.90
His → Gly-48	0.78	0.77
His → Asn-48	0.45	0.87
Thr → Gly-51	0.66	0.66
Thr → Ala-51	0.43	0.70
Thr → Cys-51	0.2	0.91
Thr → Pro-51	~0.89	~0.99

[a] The values of β in the second column are equivalent to eq 19 and 20 and those in the third column to eq 25 and 26.

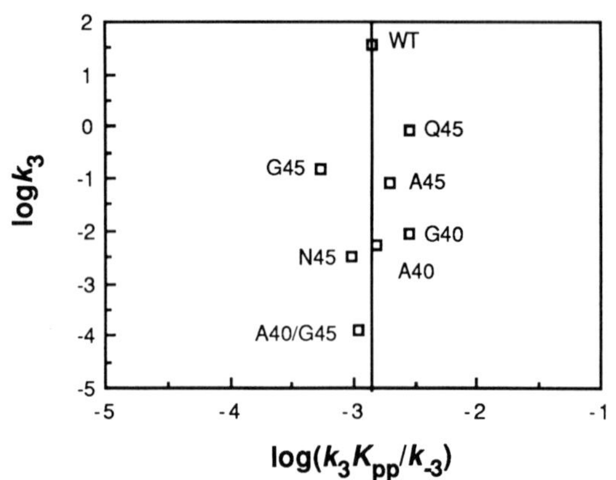

FIGURE 6: Linear free energy plot for mutations in the site for binding the γ-phosphate of ATP in the transition state.

at the extreme ends of the lines, in this case TyrTS(Cys→Gly-35) and TyrTS(Thr→Pro-51), both of which tend to higher values of β in the individual calculations. Further, calculations of individual β values that involve small values of $\Delta\Delta G$ may result in large errors.

(B) Mutations in the Transition-State Binding Site for the γ-Phosphate for ATP (Thr → Ala-40, His → Gly-45, His → Ala-45, His → Asn-45, His → Gln-45). (Case 4) Relationship between k_3 and k_3K_{pp}/k_{-3}. These mutants exhibit a different pattern of behavior from the previous ones (Figure 6). The value of k_{-3} cannot be determined for these mutants because K_{pp} is too high. But the ratio k_{-3}/K_{pp} may be accurately measured and so some of the cases can be analyzed. The slope β in Figure 6 of the logarithmic plot is infinite, that is, $\gg 1$. This is consistent with Thr-40 and His-45 being true examples of transition-state stabilizers—see below.

Discussion

Application of Linear Free Energy Relationship Analysis Requires Careful Design of Mutants and Is Aided by Structural Knowledge. Individual difference energy diagrams are the staple fare for the analysis of energy changes on mutation. The finding of linear free energy relationships correlating a series of mutational experiments is a bonus. We have described in this paper how the structure and activity of a large number of mutant enzymes fit linear free energy relationships with a remarkable consistency—the fit is as good as in many examples from physical organic chemistry. These data comprise *all* of the results of the pre-steady-state kinetics of active site mutants of the tyrosyl-tRNA synthetase that have been published over the past 5 years (Winter et al., 1982; Fersht et al., 1984, 1985; Wells & Fersht, 1986; Ho & Fersht, 1986; Leatherbarrow et al., 1986). But, it must not be considered that these constitute a set of *random* mutants. All the mutants were deliberately and carefully chosen a priori to be in the class that we now term nondisruptive deletions—we have made just small perturbations in the structure to form a highly selected set.

Before embarking on the rigorous analysis, we satisfied ourselves that we were dealing with nondisruptive deletants by a number of criteria. First, the structures of the two mutants at the two extremes of the plots in Figure 4 have been solved by X-ray crystallography [TyrTS(Thr→Pro-51), Brown et al. (1986); TyrTS(Cys→Gly-35), M. Fothergill (unpublished data from this laboratory)], and there are insignificant changes in structure other than the change in side chain and the consequent alteration in solvation. Second, the complete free energy profiles of all the mutants have been determined (Wells & Fersht, 1986; Ho & Fersht, 1986; R. J. Leatherbarrow and A. R. Fersht, unpublished results), and there are no serious perturbations of the energies of the ground-state complexes (E·Tyr, E·Tyr·ATP, E·Tyr-AMP·PP, and E·Tyr-AMP) other than those expected. For example, mutations in the ATP-binding site do not alter the dissociation constant of tyrosine and vice versa, except where there are long-range electrostatic effects. A rare example of the dissociation constant of tyrosine being affected by a mutation in the ATP site is in TyrTS(His→Asn-48) (Figure 5). Here, an electrostatic interaction with the carboxylate of tyrosine is lost on mutation.

Difference energy diagrams are also invaluable for deciding which points to plot on what graphs. For example, it is obvious on examination of Figure 5 that there is a consistent increase in binding energy on going from ETA to E[T-A]* to ET-AA·PP to ET-A. Conversely, in Figure 6, the binding energy changes are seen only in transition-state binding. These clearly represent different phenomena and must be treated separately. If all the experiments were plotted on the same graph without use of chemical knowledge for selection, then the trends in the plots would be obscured. The selection of data for linear free energy plots by use of structural information and accumulated chemical knowledge is entirely precedented by physical organic chemistry. For example, in experiments involving substituted benzene rings, substituents would be segregated into one class consisting of meta and para substituents, and a second class involving ortho substituents because of the known steric effects in the latter. Further, it would be decided whether or not to allow for bond conjugation in choosing the substituent constant for para derivatives. The free energy plots for the mutant enzymes thus fall into two groups, as expected from the difference energy diagrams (Figures 4 and 6).

The first consists of residues that bind to the side chain of tyrosine and the ribose ring of ATP, where the binding energy increases throughout the reaction. Values of β were obtained on the variation of those side chains, which are consistent with the following (Figure 7). On binding ATP to the ET complex, only 12% of the final binding energy of adenosine in the ET-A complex is realized. There is an improvement in binding energy on going from the ternary complex ETA to the transition state E[T-A]* such that 84% of the final binding energy in ET-A is realized. The intermediate ET-A·PP invokes about 91% of the energy. These are, of course, average values for the side chains analyzed.

The second category consists of mutants at positions 40 and 45, which were postulated previously to be involved in binding the transition state only. Mutation of these hardly affects the equilibrium constant between ETA and ET-A and so there is an ill-defined relationship between rate and equilibrium

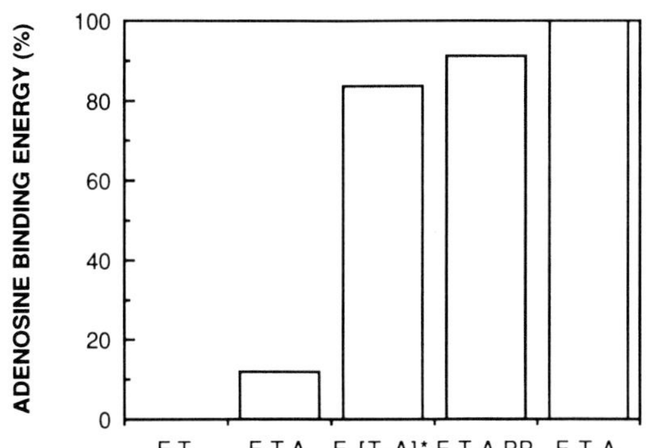

FIGURE 7: Utilization of the binding energy of adenosine throughout the reaction. The initial state is the E·T complex, and the final state is the E·T-A complex, which is taken as the 100% point for realization of the binding energy of adenosine. The relevant interactions are those with Cys-35, His-48, Thr-51, and their mutants. The heights of the bars are 100 times the β value for the relevant plots.

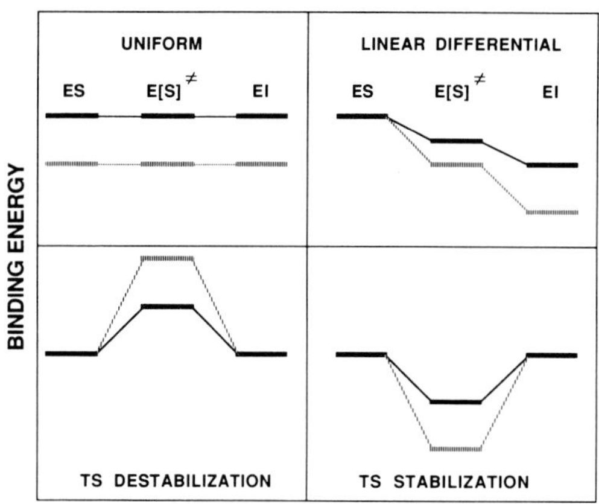

FIGURE 8: Modes of interaction of a side chain of an enzyme with a substrate (S), transition state ($[S]^{\ddagger}$), and intermediate (I) and the effects of mutation. The binding energy of the interaction is plotted as the reaction proceeds. The broken lines represent the interaction energy of the side chain in wild-type enzyme, and the solid lines represent possible energy levels after mutation of the side chain to remove the interaction. In practice, combinations of all effects may be found.

constant that is equivalent to a value of β tending to infinity.

When Should Binding Changes Conform to Linear Free Energy Relationships? Binding energy changes during a reaction may be classified into *uniform*, where a change in structure affects the energy level of each state by the same amount, and *differential*, where each state is altered by different amounts (Albery & Knowles, 1977). These may be further elaborated, as in Figure 8. Suppose the energy levels in differential binding change linearly with reaction coordinate between enzyme-bound substrate, transition state, and product so that the binding energy of the transition state is intermediate between the two ("linear differential"). Suppose, for example, the energy of the EI complex is lowered by a mutation and the ES complex is unaffected. If the linear change in energy is maintained on mutation, the energy level of the E[S]* complex will be lowered by a smaller amount, depending on how much it resembles S or I in its interactions with the enzyme. If it is midway between the two, as in Figure 8, the change in energy will be 50% of that of the change in the EI complex. Thus, linear differential binding will give rise to a linear free energy relationship.

Removal of a group involved in uniform binding will not affect the rate or equilibrium constants of the enzyme-bound states. A particular example of differential binding is when there are interactions that specifically stabilize (or destabilize) the transition state but affect the initial or final states much less, for example, the binding site for the γ-phosphate of ATP. Another example would be electrostatic interactions with charges that are generated in the transition state. As in Figure 8, alteration of these interactions need not alter the binding energies in the ES or EI complex. In the extreme case where the mutated group affects only the transition state, k_2 in eq 4 would alter, but the equilibrium constant k_2/k_{-2} would be unaffected. This generates an infinite value of β, as found in Figure 6. In less examples where the ground states were affected to some extent, high values of β of positive or negative slope would be generated. An observed value of β of greater than 1 is indicative of transition-state binding modes.

Breakdown of Linear Free Energy Relationships. The classes of binding energy changes illustrated in Figure 8 may not be apparent for various reasons even though they are occurring. One reason is that linear free energy relationships are often found to be linear portions of a more extensive nonlinear scheme. Curvature can arise from the nature of the transition state changing with changes in structure of the reagents (Jencks, 1985). The same may be true for binding energy changes. There are, however, more likely sources of problems that obscure the existence of linear free energy relationships. The first stems from the very nature of changes in protein structure on mutation. The existence of a linear free energy relationship arises from the enzyme reacting in a uniform way to mutational change as if structural changes are part of a continuum. That is, the transition state occupies a particular position on the reaction coordinate that responds in a uniform and consistent manner to changes in structure. But a mutation involves a discrete change, and the change could be too large.

Linear free energy relationships also need not occur. There must be many residues that will not fit into a simple pattern: many mutations may cause highly specific effects, each change being unique.

Possible Artifacts in Interpretation of Fit of Mutants to Linear Free Energy Plots. Three phenomena have been encountered that give rise to results that could be misinterpreted (Lowe et al., 1987). The first is a general point about the nature of mutations. There could be effects on mutation that are propagated through the structure, as in disruptive mutations. For example, a mutation in the binding site might cause a small effect on the binding of a substrate but cause a structural change that perturbs another residue which is involved in transition-state stabilization. This would give a deviation from a linear free energy plot and possibly an anomalously high value of β that could be misinterpreted [e.g., TyrTS(Ala-38), TyrTS(Ala-78), etc.]. The second stems from the mathematical analysis. In a plot of, say, log k_3 against log (k_3/k_{-3}), the two variables are not independent. Suppose k_{-3} is measured accurately but k_3 is subject to large error. Then, the error distribution in the log–log plot consists of a line of slope 1.0 passing through the correct value of log k_3. Thus, if the value of β is close to unity, a highly incorrect value of k_3 would not be apparent—as seen for TyrTS(Ala-173) (Lowe et al., 1987). Nevertheless, the fit to the plot requires that the value of A in the Brønsted equation must also be correct. Plots of slope close to unity may well conceal artifactual data.

The third is that the behavior of an individual mutant on a plot may well deviate from the mean. Small deviations are often not apparent on log–log plots because of the compression of data when there are several orders of magnitude of spread.

For example, the difference energy diagram for TyrTS(Phe-169) indicates that there is a uniform binding energy change and not the strikingly progressive increase in binding energy as the reaction procedes. But this is not easily seen above the noise in the log–log plots. The linear free energy plots show the overall trends and smooth out small individual variations. The behavior of any one mutant is best observed from its particular difference energy diagram.

We are confident that the linear free energy plots presented in this study are not artifactual because there is so much consistency between them and the difference energy diagrams: that is, between the general trends and the behavior of individuals.

Conclusions. The demonstration that certain structural changes of the tyrosyl-tRNA synthetase cause changes in rate that may be described satisfactorily by linear free energy equations is important for the following reasons.

(i) The equations combine the data from many experiments and so systematize and bring order to them.

(ii) They collectively delineate aspects of the transition-state structure, in particular by how much the interactions in the transition state resemble those in the enzyme–substrate and the enzyme–intermediate complexes.

(iii) The equations show trends. Any mutant that significantly deviates from the trend is immediately apparent. The accompanying paper (Lowe et al., 1987) contains examples of deviations.

The last point is particularly important. The results of any one individual mutation can be challenged on the grounds that perhaps the mutation propagates a significant change in protein structure beyond the local changes immediate to the mutated side chain. This is difficult to answer experimentally even by X-ray crystallographic studies since they would not detect a series of small changes of the order of 0.1 Å. But, when a whole series of mutational experiments can be fitted to a linear free energy equation, there is overwhelming evidence that each mutant shows part of a general phenomenon of the relationship between structure and activity.

We must end with a strong warning. This type of approach requires a thorough knowledge of the consequences of mutations, in particular from structural data and from difference energy diagrams, before interpretation of results.

Appendix

Demonstration That Free Energy Changes in the Reactions of Mutant Enzymes Are a Consequence of Binding Energy Changes. Wolfenden (1972) showed that the dissociation constant of a transition state from an enzyme may be calculated from the ratio of the rate constants for the enzyme-catalyzed reaction and the uncatalyzed reaction in solution combined with the dissociation constant of the enzyme–substrate complex. The following extends that approach by using it with the treatment of Fersht (1974). In Figure 1, which refers to eq 4, let the free energy of the substrate in water be G_{WS}, the free energy of the enzyme–substrate complex be G_{ES}, and the free energies of the equivalent forms of the transition state be $G_{W[S]^*}$ (in water) and $G_{E[S]^*}$ (enzyme–transition-state complex). The energies of the different states of the intermediate, I, are denoted similarly, and the unligated enzyme has an energy G_{WE}.

The binding energy of the substrate, ΔG_S, is given by

$$\Delta G_S = G_{ES} - G_{WE} - G_{WS} \qquad \text{(A1)}$$

Similarly, the binding energies of the transition state and intermediate are

$$\Delta G_{[S]^*} = G_{E[S]^*} - G_{WE} - G_{W[S]^*} \qquad \text{(A2)}$$

and

$$\Delta G_I = G_{EI} - G_{WE} - G_{WI} \qquad \text{(A3)}$$

From equilibrium thermodynamics

$$RT \ln (k_2/k_{-2}) = G_{ES} - G_{EI} \qquad \text{(A4)}$$

From transition-state theory

$$RT \ln (k_2/K_S) = kT/h - (G_{E[S]^*} - G_{WE} - G_{WS}) \qquad \text{(A5)}$$

$$RT \ln k_2 = kT/h - (G_{E[S]^*} - G_{ES}) \qquad \text{(A6)}$$

(where R is the gas constant, T is the temperature, ad k and h are the Boltzmann and Planck constants).

The energy differences of the enzyme-bound states can be described in terms of binding energies and the energies of the reagents in solution by manipulation of eq A1, A2, and A3 (or by inspection of Figure 1). For example

$$G_{ES} - G_{EI} = \Delta G_S - \Delta G_I + G_{WS} - G_{WI} \qquad \text{(A7)}$$

$$G_{E[S]^*} - G_{ES} = \Delta G_{[S]^*} - \Delta G_S + G_{W[S]^*} - G_{WS} \qquad \text{(A8)}$$

Thus, substitution of eq A7 into A4 gives

$$RT \ln (k_2/k_{-2}) = \Delta G_S - \Delta G_I + G_{WS} - G_{WI} \qquad \text{(A9)}$$

and similarly

$$RT \ln k_2 = kT/h - (\Delta G_{[S]^*} - \Delta G_S + G_{W[S]^*} - G_{WS}) \qquad \text{(A10)}$$

$$RT \ln (k_2/K_S) = kT/h - (\Delta G_{[S]^*} + G_{W[S]^*} - G_{WS}) \qquad \text{(A11)}$$

(Note that $G_{W[S]^*} - G_{WS}$ is the activation energy of the reaction in water that goes through the same transition-state structure of the substrate as on the enzyme.) These equations may be used to construct free energy relationships and calculate binding energies from kinetic data either for the reaction of a series of substrates with a common enzyme or for one substrate with a series of mutant enzymes, the equations being essentially symmetrical. For example, for a series of mutant enzymes, G_{WS}, $G_{W[S]^*}$, and G_{WI} are independent of enzyme, and only the terms involving the enzyme vary. Since kT/h is also constant

$$RT \ln (k_2/k_{-2}) = \Delta G_S - \Delta G_I + \text{constant} \qquad \text{(A12)}$$

$$RT \ln k_2 = \text{constant} - \Delta G_{[S]^*} + \Delta G_S \qquad \text{(A13)}$$

$$RT \ln (k_2/K_S) = \text{constant} - \Delta G_{[S]^*} \qquad \text{(A14)}$$

Equation A12 shows that the equilibrium constant k_2/k_{-2} varies solely as the differences in binding energies of S and I with the enzymes. Equation A13 shows that the rate constant k_2 varies solely according to the differences between the binding energies of the transition state [S] and S with the enzymes. Equation A14 shows that the second-order rate constant k_2/K_S varies solely as the binding energy of the transition state with each enzyme.

Caveats. The above analysis holds for all mutational changes, but artifacts can arise from two causes: (i) the more likely cause is that free energies of binding, ΔG_S, ΔG_I, and $\Delta G_{[S]^*}$, contain the energies of any conformational change in the enzyme induced on binding the substrate, and mutation could cause spurious conformational changes in the enzyme; (ii) the less likely cause is that the assumption that G_{WS}, $G_{W[S]^*}$, and G_{WI} are the same for all mutants could break down if the changes on mutation are so severe that the reaction proceeds by different transition states for different mutants.

Registry No. A, 56-65-5; T, 60-18-4; TyrTS, 9023-45-4; L-Cys, 52-90-4; L-His, 71-00-1; L-Thr, 72-19-5; L-Gly, 56-40-6; L-Ser, 56-45-1;

6038

L-Asn, 70-47-3; L-Ala, 56-41-7; L-Pro, 147-85-3.

REFERENCES

Albery, W. J., & Knowles, J. R. (1977) *Angew. Chem., Int. Ed. Engl. 16*, 285–293.

Brown, K. A., Vrielink, A., & Blow, D. M. B. (1986) *Biochem. Soc. Trans. 14*, 1228–1229.

Estell, D. A., Graycar, T. P., Miller, J. U., Powers, D. B., Burniv, J. P., Ng, P. G., & Wells, J. A. (1985) *Science (Washington, D.C.) 233*, 659–663.

Fersht, A. R., Shi, J. P., Wilkinson, A. J., Blow, D. M., Carter, P., Waye, M. M. Y., & Winter, G. P. (1984) *Angew. Chem., Int. Ed. Engl. 23*, 467–473.

Fersht, A. R., Shi, J. P., Knill-Jones, J., Lowe, D. M., Wilkinson, A. J., Blow, D. M., Brick, P., Carter, P., Waye, M. M. Y., & Winter, G. (1985) *Nature (London) 314*, 235–238.

Fersht, A. R., Leatherbarrow, R. J., & Wells, T. N. C. (1986a) *Nature (London) 322*, 284–286.

Fersht, A. R., Leatherbarrow, R. J., & Wells, T. N. C. (1986b) *Trends Biochem. Sci. (Pers. Ed.) 11*, 321–325.

Foster, R. J. (1961) *J. Biol. Chem. 236*, 2461–2466.

Ho, C. K., & Fersht, A. R. (1986) *Biochemistry 25*, 1891–1897.

Jencks, W. P. (1985) *Chem. Rev. 85*, 511–517.

Landis, B. H., & Berliner, L. J. (1980) *J. Am. Chem. Soc. 102*, 5354–5358.

Leatherbarrow, R. J., Fersht, A. R., & Winter, G. (1985) *Proc. Natl. Acad. Sci. U.S.A. 82*, 7840–7844.

Lowe, D. M., Fersht, A. R., Wilkinson, A. J., Carter, P., & Winter, G. (1985) *Biochemistry 24*, 1506–1509.

Lowe, D. M., Winter, G., & Fersht, A. R. (1987) *Biochemistry* (following paper in this issue).

Wells, T. N. C., & Fersht, A. R. (1986) *Biochemistry 25*, 1881–1886.

Winter, G., Fersht, A. R., Wilkinson, A. J., Zoller, M., & Smith, M. (1982) *Nature (London) 299*, 756–758.

Wolfenden, R. (1972) *Acc. Chem. Res. 5*, 10–18.

But, there are fundamental differences between the simple substitutions that chemists make to study the effects of substituents on the charged transition states of covalent interactions, which are usually electrostatic in nature, and our small changes in the structure of proteins and their affects on binding and catalysis. There was some initial incredulity to the existence of such structure–reactivity relationships. The great Martin Karplus challenged the interpretation of fractional β-values[74] but I pointed out the errors in his assumptions[75], and Arieh Warshel explained clearly the basis of LFERs[76].

The first generation of protein engineers all went on to very successful careers. Robin Leatherbarrow became professor at Imperial College, Tony Wilkinson in York and Alan Russell in Pittsburgh. Tim Wells had a stellar career in biotech and Paul Thomas a happy return to industry. The next generation of Nigel Brand, Colin Longstaff, George A. Garcia, and Johanna Avis also have established groups. Thor Borgford returned to Canada as an academic and has recently founded a biotech company. Walter Ward, Danuta Mossakowska, Denise Lowe, Jorg Eder and Patrick Chène went into successful careers in the pharmaceutical industry and Tony Day and Jim Kellis in biotech. Michael Rheinnecker funds biotech. Eric First flies a variation of the family name as an academic back in the USA.

One day, the President of the Royal Society, Sir Andrew Huxley, casually mentioned to me that the government had decided to give £500,000 each to a handful of scientists to spend as they wished, and the Royal Society had chosen me to be one of the recipients. It was a colossal sum for the mid-1980s. I spent half of it completely re-equipping my lab, but did not know what to with the rest. Imperial College told me to buy a 500 MHz NMR spectrometer, which had just been introduced by Bruker, for use by the chemistry department. A new graduate student Mark Bycroft started his career on it by studying the histidines in subtilisin[77], moved to Cambridge[78] and then solved the solution structure of barnase[79], which was one of the largest structures then solved before the introduction of multi-dimensional NMR. The unexpected grant from the British government, which I would never have received if I had applied for it because I had no background in NMR, was an important landmark in my career because it initiated me into the field. Later on, when I returned to Cambridge, I set up NMR as an important part of the infrastructure of the Centre for Protein Engineering. Mark became an independent group leader in the CPE with his own distinct programme in structural biology.

The recognition of the work as being at the heart of biotechnology also opened the gates to interesting consultancy work. Serving on the scientific advisory board of Merck and Lord Rothschild's Biotechnology Investment fund (BIL) taught me more than I ever contributed. Unfortunately, I could be a little sharp with Victor Rothschild who considered that the company advisers were at his beck and call. He was all set to fire me when in response to a peremptory demand that I should give him advice on what to do with his investments I tersely replied: 'sell the lot, the market is clearly overvalued.' But, a huge financial crash intervened, and I got a rise and the compliment that I had one attribute that he did not, insight. Recently, I warned my Cambridge college on their investment strategy as it was blindingly obvious that there would be a crash. But, our advisors just continued as if markets would rise for ever. I also consulted very early for Genencor and advised David Estell and Jim Wells on their work on subtilisin, not that they needed much advice, but it was great fun.

Greg had realised very early on the importance of protein engineering in making novel proteins, and became a founder of antibody engineering. His first efforts were in grafting human complementarity-determining-region loops on to rodent antibody frameworks to produce therapeutic 'humanised' antibodies that would be better tolerated in humans. These are now used in the clinic. We both joined the advisory board of a small company in Aberdeen, Scotgen, that had licensed Greg's technology from the MRC. This was an entry into the Biotech world and led me into co-founding a number of companies. The most successful was Greg Winter's Cambridge Antibody Technology (CAT), based on his technology for humanising rodent antibodies for therapeutic use. Greg's genius was quickly extended to companies and intellectual property law and he can run rings around most patent lawyers. It did not take much insight to give as much space as possible to Greg in the new Centre for Protein Engineering (see next) to develop his applied work. It paid off in dividends with many hundreds of millions of dollars in royalty income to the MRC, some of which was channelled back to the laboratory. I made sufficient money from CAT, and then another company, Cambridge Discovery Chemistry (founded by Allan Marchington, Ryszard Kobylecki, Steven Ley and I) not to have to worry about financial matters and to be able to fund my passion for collecting early English clocks and chess sets.

My contributions to the study of proteins were clear: the first directed-mutation of a residue in a protein of known structure, accompanied by a clear strategy of how to use the technology, followed by the introduction of new methods and discoveries. But, there was something deeper. Enzymologists and protein chemists had been slow to take the opportunity of recombinant DNA technology. Of the established enzymologists, only Jeremy Knowles, Stephen Benkovic and I leapt in to reap the harvest of the new opportunities. In some ways, we founded the new era of 'Chemical Biology' by showing that the new biology was a respectable and highly productive area for chemists and that there was no need to be intimidated.

Protein engineering had made Greg and me internationally famous. Soon after our first *Nature* paper, I was approached with offers from US and European universities, but my family and I were happy in England, and I also felt that the disruption caused by a move could fritter away the head start I had. Greg received lucrative international prizes and I was awarded academic honours! As early as 1984, I was invited to give the Edsall Lecture at Harvard. In the audience, Frank Westheimer, the father of modern mechanistic enzymology, and Konrad Bloch, who had received a Nobel Prize for his work on enzymes, were in raptures over the work on the tyrosyl-tRNA synthetase. I was elected an Honorary Foreign Member of American Academy of Arts and Sciences in 1988. On 22 November, 1987, my photo appeared on the front cover of *The Sunday*

Times Magazine as one of 'The powers that will be.' That year, I also received an invitation to return to Cambridge as the Herchel Smith Professor of Organic Chemistry. The great organic chemist Albert Eschenmoser was on the election committee, and it was reported to me that he opened the proceedings that there was only one candidate to be considered for the chair, Alan Fersht (I had not applied for it).

Front cover from *The Sunday Times Magazine*.

Chapter 10

MRC Centre for Protein Engineering, 1991–2010

'You fight only the wars that you can win,
and then only the ones you have to.'

Margaret Thatcher

I had dithered about leaving Imperial College because of my involvement with a bid from the University of London for an Interdisciplinary Research Centre (IRC), to be funded by the Science and Engineering Research Council as part of a government initiative. Tom Blundell from Birkbeck, David Colqhoun from University College and I put forward a superb group of people and disciplines. I felt I could not let them down by moving to Cambridge. But, the centre was granted to Oxford on geopolitical considerations as the University of London had been awarded a centre in another discipline. Our failure was a blessing in disguise: I accepted the offer from Cambridge as Herchel Smith Professor of Organic Chemistry from October 1988, and the MRC awarded me a personal unit, the MRC Unit for Protein Function and Design, to be embedded in the chemistry department. I put in a grant proposal to the MRC in the spring of 1988 giving the precise details of our strategy, which was funded after peer review. The MRC, headed by Dai Rees, was in the process of setting up its own IRCs in 1988. Dai asked if I would like to be honorary director of an MRC IRC for Protein Engineering, with Greg as deputy director! The MRC has the tradition of funding work fully and providing the necessary infrastructure, and they did this in spades for us — which we subsequently repaid many times over from our royalty income. For the first time, I had enough space and funding to do everything I wanted. I was able to set up a truly interdisciplinary centre with skilled staff and the necessary equipment in NMR, X-ray crystallography, biophysics, molecular biology and protein chemistry. It was the prototype for many current laboratories and ahead of its time. The IRCs were intended to have a lifetime of five years, but our IRC was renewed and eventually transmuted into a unit: the Centre for Protein Engineering, the CPE.

I always had a small group of very good students and post-doctorals at Imperial College, but now some of the most talented Europeans and a few North Americans beat a path to my unit's door. I now had the ideas, the money, the infrastructure, the proteins and superb colleagues for the next phase. Frankly, we were unstoppable.

Opening of the Centre

Margaret Thatcher, unlike Shirley Williams, was a great supporter of science and really liked scientists — Mrs. T. studied under my predecessor as Wolfson Research Professor, Dorothy Hodgkin, and revered her, though I doubt if the feelings were reciprocated. Mrs. T. invited groups of scientists to give presentations at her official country residence, Chequers. Peter Rigby and I were both invited to one of the first on 1 July, 1984. We were so scared of arriving late driving on an unknown route that we knocked on the door about an hour too early. The housekeeper greeted us by saying that the Prime Minister would have heard us arrive and no doubt would come down to entertain us, which she did. We had a personal guided tour of the gardens and were told how she had won the Falklands War. Harold Macmillan had told her that she had to listen to the generals and work closely with them. She also gave her philosophy: 'You fight only the wars that you can win, and then only the ones you have to.' Useful advice that I follow and that I often tell others embarking on lost causes.

The rest of the afternoon was amusing. Mrs. T. was completely on the ball and would come in with remarks such as: 'When I was in Texas, Brown and Goldstein told me…' Dennis Thatcher lounged around like he was caricatured, bemused with a gin and tonic in hand. Sir Keith Joseph, a senior minister and intellectual force, was ordered around like a schoolboy.

Seven years later, we returned the hospitality by inviting her to open the IRC. Unfortunately, the conservatives had a coup between our sending the invitation and the event, and John Major became Prime Minister. The MRC wanted us to drop Mrs. T. and invite Mr. Major. That type of behaviour is not in Greg's and my character. We had a splendid opening day. Mrs. T. insisted on arriving early in the day so she could listen to the young scientists — she did not want presentations from the senior members. The opening really was memorable, and ended with a dinner attended by Max Perutz, Aaron Klug, Cesar Milstein, Sydney Brenner and other friends.

THE RT. HON. MARGARET THATCHER, O.M., F.R.S., M.P.

2 May 1991

Dear Dr. Fersht,

May I say how very much I enjoyed my visit to the new Centre yesterday and the very interesting programme you had arranged for me.

It was good to meet so many young scientists and hear about their work. I am grateful to them for explaining it so simply to me. You have a very exciting period of discovery ahead.

Thank you too for hosting the evening dinner – and in such historic surroundings. I enjoyed the animated discussion and agree with many of the points that were made. You were most kind in your speech.

After the day at the Centre I came away with renewed faith in our scientific genius and in this generation of young scientists.

With kind regards.

Yours sincerely,

Margaret Thatcher

Professor A. Fersht.

Collage of photographs from the opening day of the CPE 1 May 1991.

Top left: From left to right: Jane Clarke, Andreas Matouschek, Sophie Jackson, Ronnie Loewenthal and Mrs. Thatcher. *Top middle*: Jim Marks, Mrs. Thatcher, Peter Doyle (Astra-Zeneca) and Greg Winter. *Top right*: Max Perutz and Aaron Klug. *Second row left*: Greg Winter, Mrs. Thatcher and Jim Marks. *Second row right*: Dai Rees, Secretary (CEO) of MRC; David Williams, Vice-Chancellor of University of Cambridge; Mrs. Thatcher and Alan Fersht. *Third row left*: Mrs. Thatcher and Alan Fersht. *Third row middle*: Aaron Klug and Max Perutz signing the visitors' book. *Third row right*: Signatures from the book. *Bottom*: Letter from Mrs. Thatcher and plaque engraved by The Kindersley Workshop.

Chapter 11

Protein Folding: Stability

'Often, the greatest difficulty is deciding which of the vast number of possible alterations to make. The prudent investigator, if in doubt, devises elegant genetic procedures to select for those replacements that give the desired effect.'

T. E. Creighton, Science 240, 267 (1988)

The tyrosyl-tRNA synthetase was in many ways the perfect paradigm for developing protein engineering as a general tool for analysing structure and mechanism of enzymes. Its mechanism had been unknown so we were not going over old ground. It relied solely on the utilization of binding energy for specificity and catalysis, and so was the ideal target for mutating side chains. Most importantly, it was so easy to handle. As I have always been interested in uncovering general and universal principles rather than working on specific systems and thought that working on any other enzyme would be an anticlimax after that wonderful enzyme, I turned to a new area for me but one which was perfect for the technology we had developed — protein folding.

The Protein-Folding Problem

I had resolved several years earlier, when still at LMB, to work on the protein-folding problem when we had the right tools available. The 'protein-folding problem' is essentially the prediction of the three-dimensional structure of a protein from its amino acid sequence. An ancillary problem is the prediction of the pathway of protein folding. It was an enigma, posed by Cyrus Levinthal in 1969, that a protein could fold in finite time because of the vast number of conformations that could be explored as a protein folds up (the 'Levinthal Paradox'). Further, solving pathways might be a means of solving structure prediction. But, experimentally, it seemed the most difficult of tasks to measure the pathway of folding at atomic resolution. Quite simply, proteins fold up into their exquisite shapes via a myriad of conformations that were beyond computation, and there was no means of characterising the structures of transition states and transient intermediates.

When I entered the field in the late 1980s, the major groups were working on the folding of classic proteins that had been the well-studied subjects of enzymology and protein chemistry because they were well characterised and readily available in large quantities. The consensus was then that all proteins folded via one or more intermediates and there were two main mechanisms: framework, whereby secondary structure was formed first and then docked to form tertiary structure; and molten globule, whereby there was a hydrophobic collapse to give some native-like tertiary interactions that then directed formation of secondary structure.

Oleg Ptitsyn had founded the Russian school of folding and had introduced the concepts of the molten globule and framework mechanisms. Rainer Jaenicke in Germany and Michel Goldberg in France were performing clever experiments on overall folding. Buzz Baldwin had founded a school of biophysically based protein-folding studies at Stanford and had produced many famous young faculty members in the USA, Europe and Asia. Buzz worked on RNaseA, one of the first proteins to have been solved by X-ray crystallography. But, the disulfide-bridged protein also has a *cis*-peptidyl-prolyl bond and the folding kinetics is dominated by its isomerisation from the predominant *trans* conformation in the denatured state. Chris (C. M.) Dobson at Oxford was working on the folding of the disulfide-bridged hen egg white lysozyme, the first enzyme to have been solved by X-ray crystallography. Tom (T. E.) Creighton was also a major figure and had introduced the trapping of covalent intermediates during the shuffling of disulfide bridges as the tiny protein, bovine pancreatic trypsin inhibitor, folded. S. Walter Englander was pioneering hydrogen-deuterium exchange to study the folding of cytochrome *c*. Disulfide bond formation is a late step in folding and so their studies were characterising late events in folding. David Shortle was pioneering making mutants of a good recombinant protein, *Stapphylococcal* nuclease, to study its denatured state because he reasoned that it held the information to direct the path of folding. Bob (C. R.) Matthews was measuring the folding kinetics of natural mutants of a subunit of tryptophan synthetase. The germs of applying protein engineering to protein folding were in place, but a coherent and new strategy was required. But, what mutations should the experimentalists make?

Tom Creighton in 1988 in a *Perspective in Science* about a protein folding meeting wrote:

> The new technology of protein engineering has made it possible to produce and study experimentally any protein (whether natural, modified, or totally synthetic) and has given a major experimental impetus that was evident in many of the presentations. Often, the greatest difficulty is deciding which of the vast number of possible alterations to make. The prudent investigator, if in doubt, devises elegant genetic procedures to select for those replacements that give the desired effect.[80]

The Rational Plan

To me, that choice of mutation was not the way forward and the choice of such mutations had, in fact, been the stumbling block to progress. The temperature-sensitive mutations found from genetics were so radical that they invariably caused serious destabilization or complex structural changes that did not lead to simple interpretation. A precise chemically based approach was required. Although the protein folding community had heard me talk at a Keystone meeting in 1985 on our strategy for analysing the tyrosyl-tRNA synthetase, 'non-disruptive deletion' mutations and the β approach, few had appeared to appreciate its subtleties, apart from David Goldenberg as a notable exception. My strategy was taken directly from the tRNA-synthetase work: 1) make non-disruptive deletion mutations that delete interactions within the protein with minimal perturbation of structure; 2) measure the change in stability on mutation, which would provide a library of interaction energies; and then 3) measure the changes in rate constants of folding and unfolding to map out the structure of the transition state for folding using LFERs. Basically, I would just 'tickle' the structure of the protein by tiny changes in the structures of amino acids. These changes would gently perturb the energetics of interactions at localised sites, and these would flow through to perturbing the kinetics of folding and unfolding.

Tom Creighton was, in fact, appreciative of our work and later complimented me that I was one of the few honest people in the field. (The protein folding field is, in fact, one of the most honest areas.)

Barnase

It was clear to me that the new generation of technologies needed a recombinant protein that would be small enough for extensive mutagenesis and be amenable to nuclear magnetic resonance (NMR) spectroscopy and computational studies, and not have disulfide bridges or *cis*-peptidyl-prolyl bonds because they over-complicate the kinetics and are a hindrance to studying what initiates folding — the accretion of non-covalent interactions. Having been friendly with Bob (R. S.) Hartley when he was on sabbatical in LMB some ten years before, I knew of his small nuclease, *barnase,* which had the desired properties. So, my group set about cloning and expressing it. On learning that Bob Hartley was doing the same, we agreed with him that we would stop and he would give us his clone as soon as it was available, which he, always the perfect gentleman, did. Ironically, he had offered the protein some years earlier to Buzz Baldwin, who had turned it down to continue working on RNaseA. Had he accepted, my future, and possibly the development of protein folding might have been quite different.

Stage 1. Measuring Interaction Energies in Proteins

Brian Matthews, an old friend from the chymotrypsin days, had realised the importance of solving structures of point mutants. He had been publishing papers on the structures and stability of mutants of T4 lysozyme, based on all 19 amino acid substitutions at temperature-sensitive loci but had not yet used our strategy: we would use our procedure, honed on the tyrosyl-tRNA, of making 'non-disruptive deletion' mutations. We chose side chains that appeared to make important interactions and then pruned back the side chain in the smallest increments, and measured the change in energetics. The very first paper probed the hydrophobic core as Jim Kellis cut back, for example, an isoleucine side chain to valine and then alanine[40]. The loss in free energy of folding for the mutant proteins is 1.0–1.6 kcal/mol per methylene group removed, which is 2–3 times greater than expected from the classical hydrophobic effect. That result was not unexpected to me since it was less than the values I had found a decade earlier when probing the limits of specificity of the aminoacyl-tRNA synthetases (e.g. valine competing with isoleucine for the active site of the isoleucyl-tRNA synthetase), but that work was probably not well known to the protein folding community. As simple as those experiments now seem, it was clearly considered sufficiently important and novel at the time to be enthusiastically accepted for publication in *Nature*, published on 23 June, 1988[40]. It did usher in a new era of making rational mutations for probing protein structure. But, Brian Matthews was hot on our tail, publishing *'Hydrophobic Stabilization in T4 Lysozyme Determined Directly by Multiple Substitutions of Ile-3'* in *Nature* on 4 August, 1988[81].

11.1 1989: MRC Unit of Protein Function and Design, in front of Chemistry Laboratory. *Back row*: Eric First and Alan. *Third row*: Boaz Avron, Javi Sancho, Jim Kellis and Ronny Loewenthal. *Second row*: Dasa Lipovsek, Andreas Matouschek, Derek Parsonage, Liz Meiering and Johanna Avis. *Front row*: Luis Serrano, John Blacker, Sophie Jackson, Pauline Cann and Tammy Gray.

Contribution of hydrophobic interactions to protein stability

James T. Kellis Jr, Kerstin Nyberg, Daša Šali & Alan R. Fersht*

Department of Chemistry, Imperial College of Science and Technology, London SW7 2AZ, UK

A major factor in the folding of proteins is the burying of hydrophobic side chains. A specific example is the packing of α-helices on β-sheets by interdigitation of nonpolar side chains. The contributions of these interactions to the energetics of protein stability may be measured by simple protein engineering experiments. We have used site-directed mutagenesis to truncate hydrophobic side chains at an α-helix/β-sheet interface in the small ribonuclease from *Bacillus amyloliquefaciens* (barnase). The decreases in stability of the mutant proteins were measured by their susceptibility to urea denaturation. Creation of a cavity the size of a $-CH_2-$ group destabilizes the enzyme by 1.1 kcal mol^{-1}, and a cavity the size of three such groups by 4.0 kcal mol^{-1}.

Protein design *de novo* requires a quantitative understanding of the rules for the folding of proteins (see ref. 1 for recent review). One approach is the direct application of computational methods. Important advances are being made as methods improve and computational power increases. A complementary approach is descriptive, surveying known structures and formulating general rules about motifs. An important example is the classification of packing of helices and sheets in proteins[2-7]. There is, however, no direct evidence on the strengths and specificity of the interactions involved and all estimates have been made from model compounds. Direct, simple experimental data on proteins are required in general to test and refine predictive methods.

Significant experimental contributions are now being made using protein engineering (see, for example, refs 8, 9). Work so far has primarily used the selection of temperature-sensitive mutants after random mutagenesis and the construction of large numbers of mutations at the temperature-sensitive loci. We are using an alternative strategy: adopting methods previously employed to dissect the structure and activity of the tyrosyl-tRNA synthetase[10] to provide a quantitative experimental basis to the descriptive approach of refs 2-7. Mutations that remove defined interactions without introducing new or unfavourable interactions, that is, nondisruptive deletions[11] are made on the basis of the crystal structure of the wild-type enzyme. It is especially important to use nondisruptive deletions in analysing protein folding because the free energy of unfolding depends on the difference in free energy between the folded and denatured states and so both should be perturbed as little as possible. Empirical measurements of the energetics of interactions are obtained that may be used directly for experimental design and provide raw data for refinement of computational methods and theory.

The enzyme we have chosen for these studies is barnase, the RNAse from *B. amyloliquefaciens*[12]. This enzyme has almost ideal characteristics as a paradigm for protein folding for the following reasons. It is a monomer of 110 residues of M_r 12,382, one of the smallest known enzymes. The crystal structure has been solved at high resolution[13]; the small size and favourable crystals enabling the resolution to be increased to 1.4 Å. In addition, barnase is small enough for sequence specific two-dimensional NMR assignments to be made and to be tractable to analysis by computational methods. Yet it is large enough to have significant secondary structure. It has, for example, the packing of an α-helix on a β-sheet[13]: the helix formed from residues 6–18 packs against the anti-parallel β-sheet formed from residues 50–55, 70–75, 85–91, 94–101 and 106–108. The protein is composed of one domain and undergoes a reversible thermal or urea-induced denaturation, approximating to a two-state process[14,15]. This enables thermodynamic measurements to be made on the factors that influence protein folding and stability. Equilibrium and kinetic measurements of denaturation may readily be made by difference spectroscopy. Barnase has no disulphide bridges. This reduces structural constraints in the denatured protein and so decreases complications in analysing denaturation. The protein has been cloned and expressed[16].

The mutations chosen in this study are the truncations Ile96 → Val, Ile96 → Ala and Phe7 → Leu. The first two mutations are, in theory, true nondisruptive deletions since simple truncations are involved. Phe → Leu is a conventional conservative mutation, but there is a change in geometry about the γ-carbon (sp^2 in Phe and sp^3 in Leu). The mutations remove hydrophobic inter-

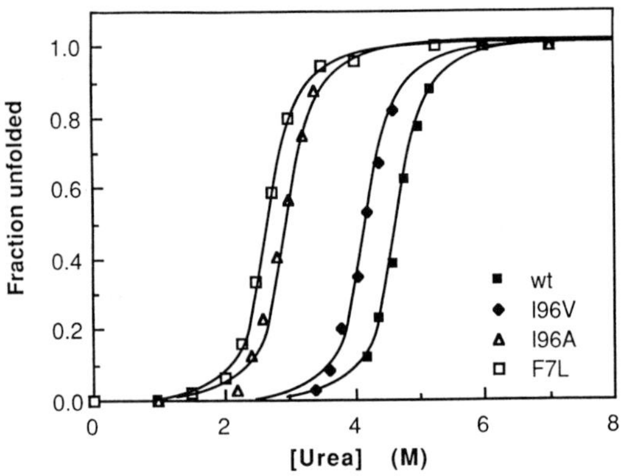

Fig. 1 Urea denaturation of recombinant barnase and engineered mutants. Denaturation was observed spectroscopically using difference spectroscopy at 286 and 270 nm. Experiments were performed at 25 °C in a buffer containing 19.3 mM MES acid and 30.7 mM base (MES = 2-(N-morpholino) ethanesulphonic acid) at ionic strength 0.05 M and *p*H 6.30 (except for Phe → Leu7 which was at *p*H 6.15). The stability of the protein hardly varies between *p*H 6.0 and 6.5 under these conditions. The concentration of barnase was between 9 and 12 μM. There is a decrease of 13.5% in absorbance at 286 nM and an increase of 4.5% at 270 nm on unfolding ($\Delta A_{286} - \Delta A_{270}$) was measured since this both amplifies the signal and minimizes dilution errors as $A_{286} \sim A_{270}$. The solid curves are just for visual aid.

Methods. Wild-type enzyme was secreted from *E. coli* harboring the plasmid pMT410, a derivative of pUC9 (ref. 16 and R. W. Hartley, personal communication). This construction contains the structural gene for barnase fused to the alkaline phosphatase promoter and signal sequence and the gene for barstar (the polypeptide inhibitor of barnase) under the control of its own promoter contained on a 1.4 kilobase (kb) *Eco*RI-*Hin*dIII fragment. For mutagenesis, this fragment was subcloned into M13. The mutated genes were subsequently cloned back into pUC9 for expression. Site-directed mutagenesis was carried out using the double primer method[26]. The mutagenic primers were used together with an M13 universal primer. To improve the yield of mutants, a strain defective in mismatch repair, *E. coli* BMH 71-18 *mut L* (ref. 27) was used in the transfection. Mutant bacteriophages were identified by oligonucleotide hybridization[28] and the mutations verified by dideoxy sequencing of the entire genome. The following primers were used to direct the mutations: 5'TTTTGTAG*AC*CAGCCAG3', Ile → Val96; 5'TTTTGTAAG*C*CAGCCAG3' Ile → Ala96; Phe → Leu7 5'CCCCGTCAAG*CGTGTTG3' (asterisks follow mismatches). Cells were grown in a low phosphate medium[29]. Secreted wild-type and mutant enzymes were purified to homogeneity (D. Mossakowska, K.N. and A.R.F., in preparation).

* To whom correspondence should be addressed.

Reprinted from Nature, Vol. 333, No. 6175, pp. 784-786, 23 June 1988

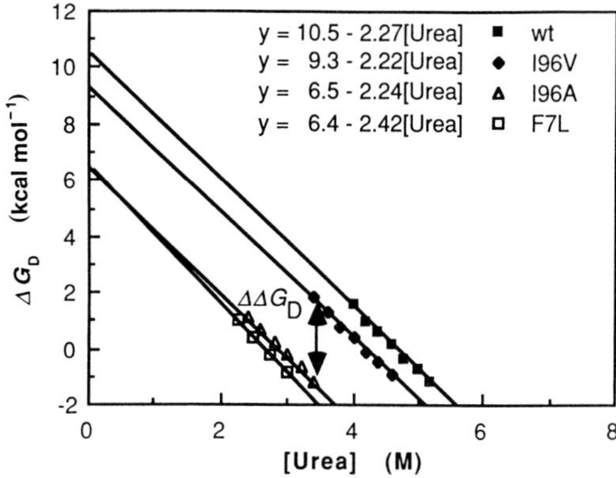

Fig. 2 Linear dependence of ΔG_D on [urea] over experimental ranges allows simple determination of $\Delta G_D^{H_2O}$ and $\Delta\Delta G_D$. The value of $\Delta\Delta G_D$ between, for example, Ile→Val96 and Ile→Ala96, may readily be read off the graph at an intermediate concentration of urea (such as 3.55 M) to be 2.9 kcal mol^{-1}. Even more precise data are obtained when the two denaturation curves overlap as there is no extrapolation and values of ΔG_D for each protein can be made under identical conditions (for example, wild type and Ile→Val96).

actions where the helix composed of residues 6 to 18 docks on to the β-sheet, effectively introducing cavities into the enzyme. The change in stability on mutation is then measured by the change in free energy of the reversible denaturation of the enzyme, determined by UV difference spectroscopy. We have chosen urea-induced denaturation rather than thermally induced denaturation for two reasons. First, it is not a simple matter to apply thermal denaturation for quantitative comparison of different proteins with significantly different values of T_m. The enthalpy of unfolding of proteins varies with temperature and the van't Hoff plots are not linear[17]. Thermodynamic quantities are consequently not easily extrapolated to different temperatures. Urea denaturation, however, may be performed at a single temperature for a whole series of mutants (for example, 25 °C as in Fig. 1) and analysed as below[18]. Second, urea denaturation is thought to give a more completely unfolded protein than does thermal denaturation[17,18].

Wild-type and engineered mutants were expressed in *Escherichia coli*. The yield of mutant enzymes decreases as the mutations become more radical, although specific activity towards hydrolysis of RNA drops by no more than 30%. Mutants Ile96→Ala and Phe7→Leu are expressed at levels of only some 10% of wild type. The decrease in yield presumably reflects a greater susceptibility to proteolytic degradation as the folded structure becomes less stable. Urea denaturation confirms that mutation lowers stability. It can be seen from Fig. 1 that the concentration of urea required to unfold 50% of the protein is 4.63 M for wild-type barnase, 4.19 M for Ile96→Val, 2.90 M for Ile96→Ala and 2.66 M for Phe7→Leu at pH 6.3.

The free energy of unfolding in the absence of urea ($\Delta G_D^{H_2O}$ is conventionally estimated by extrapolating the free energy of unfolding at each individual concentration of urea (ΔG_D) to zero concentration assuming that they are linearly related (equation (1), ref. 18). This procedure gives 10.5 ± 0.2 kcal mol^{-1}

$$\Delta G_D = \Delta G_D^{H_2O} - m[\text{urea}], \quad (1)$$

for $\Delta G_D^{H_2O}$ for wild-type barnase, 9.3 ± 0.2 kcal mol^{-1} for Ile96→Val, 6.5 ± 0.1 kcal mol^{-1} for Ile96→Ala and 6.44 ± 0.04 kcal mol^{-1} for Phe7→Leu. (About 2 parts in 10^5 of the latter two mutants are unfolded in the absence of urea.) The slopes of the plots are quite similar for wild type and mutants (m = 2.27 ± 0.04; 2.22 ± 0.05, 2.24 ± 0.04 and 2.42 ± 0.01, respectively). Wild-type barnase is significantly more stable than T1 ribonuclease ($\Delta G_D^{H_2O}$ = 4 kcal mol^{-1}, ref. 18) and staphylococcal nuclease ($\Delta G_D^{H_2O}$ = 5.5 kcal mol^{-1}, ref. 8), so allowing more radically destabilizing mutations to be made and analysed.

While the above procedure is adequate for estimating the approximate value of $\Delta G_D^{H_2O}$, the extrapolations are over too wide a range of concentrations of urea to measure reliably small differences in stability of mutants (errors in the slope are magnified and the assumption about precise linearity between free energy of unfolding and [urea] is questionable, although the slope m is reasonably constant between 2.5 and 5 M urea). We have used two alternative procedures for measuring differences more directly. The first is the better of the two and makes no assumptions about the dependence of unfolding energy on urea concentration. Wild-type and mutant enzymes of suitably similar stability are denatured in parallel at the same concentrations of urea. Then, at each concentration of urea, the difference in ΔG_D between wild-type and mutant ($=\Delta\Delta G_D = \Delta G_{D(wt)} - \Delta G_{D(mut)}$ is given directly by:

$$\Delta\Delta G_D = RT \ln \frac{\left(\dfrac{[\text{folded}]}{[\text{unfolded}]}\right)_{wt}}{\left(\dfrac{[\text{folded}]}{[\text{unfolded}]}\right)_{mut}} \quad (2)$$

This procedure clearly works well for small differences in stability where denaturation curves overlap and the ratio of [folded]/[unfolded] for wild-type and for mutant can be measured under the same conditions. This cannot be applied to situations where there are large differences in stability between two mutants because the ratio of [folded]/[unfolded] can be measured accurately only in the range 0.1 to 10. Consequently, the ratio will be too extreme for an unstable mutant when the ratio is measurable for wild-type. Eventually, the procedure will be scaled over a series of mutants, each one differing from the next in stability by less than a kcal mol^{-1}. Until then, the second procedure may be applied: a small extrapolation is made to a concentration of urea intermediate between the values required for denaturation of wild type and mutant (or between two mutants, c.f. ref. 19, Fig. 2). This assumes linearity in the dependence of ΔG_D on [urea] only over a small range beyond the experimental data. Values of $\Delta\Delta G_D$ are listed in Table 1. Truncating Ile to Val destabilizes wild type by 1.1 kcal mol^{-1}, and Ile to Ala by 4.0 kcal mol^{-1}. Mutation of Phe to Leu destabilizes by 4.6 kcal mol^{-1}. The experiments thus give direct measure-

Table 1

Mutant	$\Delta\Delta G_D^*$ (in urea)	$\Delta\Delta G_D^{H_2O}$† (at 0 M urea) (kcal mol^{-1})	ΔG_{solv}‡
Ile→Val96	1.1	1.2	−0.12
Ile→Ala96	4.0	4.0	−0.15
Phe→Leu7	4.6	4.1	2.24

* $\Delta G_{D(wt)} - \Delta G_{D(mut)}$ calculated by direct comparison at the same concentrations of urea (wild type and Ile→Val96) or interpolation to an intermediate concentration of urea (Ile→Val96 and Ile→Ala96 at 3.5 M urea, Ile→Ala96 and Phe→Leu7 at 2.8 M urea, see Fig. 2 and text). The similarities in the slopes of the denaturation curves renders the interpolation insensitive to the choice of [urea] within ±1 M.

† $\Delta G_{D(wt)}^{H_2O} - \Delta G_{D(mut)}^{H_2O}$ from extrapolation to [urea] = 0 M (equation (1)). Note that there is expected to be a small discrepancy between this and the previous column because of the different stabilities of the exposed side chains in the denatured state when exposed to water and urea. For complete exposure of the side chain, this is negligible for Ile→Val, +0.15 kcal mol^{-1} for Ile→Ala and is +0.18 kcal mol^{-1} for Phe→Leu at 4 M urea (the larger side chains being more stable in urea)[25].

‡ Data from ref. 23 for the change in free energy of solvation of side chain ($G_{solv(mut)} - G_{solv(wt)}$).

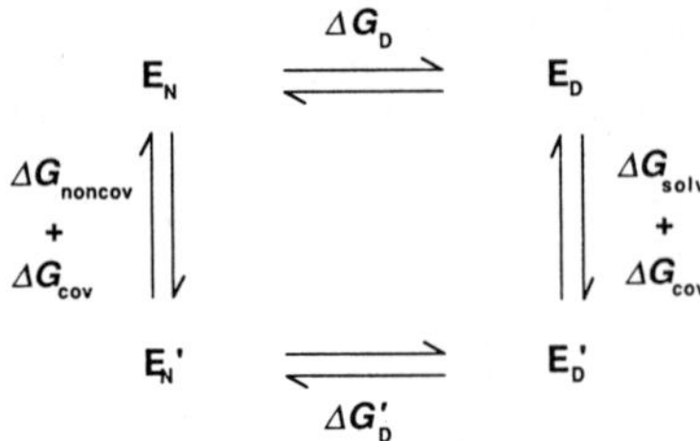

Fig. 3 Thermodynamic cycle relating the unfolding of wild-type (E) and mutant (E′) enzymes. On mutation, there is a change in the free energy of the protein (ΔG_{cov}) because of covalent changes such as the change in chemical bonds on going from Ile to Ala). ΔG_{cov} is the same for both the folded and unfolded states and so cancels out when comparing the right and left hand sides of the cycle. There are also changes in the noncovalent interactions, ΔG_{noncov} $(=G_{E'_N}-G_{E_N})$, on mutation. In the unfolded enzyme, the side chains of the amino acids are more, if not completely exposed to solvent and so $\Delta G_{noncov(D)}$ represents predominantly the changes in the solvation energy of the side chain, ΔG_{solv} $(=G_{E'_D}-G_{E_D})$, on mutation. ΔG_{noncov} for the folded state represents predominantly the loss of interaction energy between the enzyme and those parts of the side chain that are deleted on mutation. ΔG_{noncov} contains, in addition, the energy terms associated with any reorganization of enzyme and associated solvent on mutation. The experimentally determined quantity in our experiments is the difference in unfolding energy of the wild-type and mutant enzymes, $\Delta G_D-\Delta G'_D$. It is seen from the cycle that $\Delta G_D-\Delta G'_D=\Delta G_{noncov}-\Delta G_{solv}$.

ments of the destabilizing effects of cavities in enzymes. This points to a way of stabilizing protein structures: make substitutions to fill in holes that occur in the native structure.

Calculation of the free energy of folding (or unfolding) of proteins *ab initio* involves determining the free energies of the folded and unfolded states (Fig. 3, $\Delta G_D=G_{E_D}-G_{E_N}$, where E_N and E_D are the native and denatured forms of the enzyme). This requires calculating all the noncovalent interaction energies in the folded and unfolded proteins and the solvation energies. One fundamental unknown quantity is the noncovalent interaction energy in the unfolded state. This is probably not a single state but consists of a mixture of many different states of similar energies, in which the exposure of side chains to solvent is unknown and variable. This presents a stumbling block to calculation. It is important to relate the empirical data from the protein engineering experiments, $\Delta\Delta G_D$, $(=\Delta G_D-\Delta G'_D)$ to the energies used in calculations. This is done by a simple thermodynamic cycle (Fig. 3) which considers the native and denatured states for wild-type (E_N and E_D) and mutant (E'_N and E'_D) enzymes. In Fig. 3, the change in the noncovalent interaction energy in the native enzyme on removal of a side chain is defined by $\Delta G_{noncov}=G_{noncov(mut)}-G_{noncov(wt)}$ (where the subscripts refer to the noncovalent interactions in the folded mutant and wild-type enzymes). The equivalent change in noncovalent energy in the denatured state is termed ΔG_{solv} from Fig. 3:

$$\Delta G_D^{H_2O}-\Delta G_D'^{H_2O}=\Delta G_{noncov}-\Delta G_{solv} \quad (3)$$

First, it can be seen from equation (3) that the changes in free energy of unfolding are a function of the interactions in folded and denatured states. Second, equation (3) may be used to gather information about interactions in the denatured state as follows. ΔG_{noncov} may be calculated from the native structures of the wild-type and mutant enzymes using conventional computational methods. Recent developments in computational methods allow the direct calculation of the values of ΔG_{noncov} (and ΔG_{solv}) by the mathematical equivalent of mutating the side chains in slow increments[20-22]. If the side chain in the unfolded state is completely exposed to solvent, then ΔG_{noncov} is simply calculated from equation (3) using relative solvation energies of the amino acid side chains that are available from experiments on model compounds[23,24]. For example, ΔG_{solv} for Ile→Val is −0.12, Ile→Ala is −0.15 and Phe→Leu is 2.24 kcal mol^{-1} (ref. 23). Thus, if the side chain of Phe7 in denatured wild-type enzyme is completely exposed to solvent and so also is Leu7 in the mutant, then ΔG_{noncov} for Phe→Leu would be $4.6+2.2=6.8$ kcal mol^{-1}. Discrepancies between calculated and measured values of ΔG_{noncov} will give clues about side chain interactions in the unfolded states. Further mutational experiments will provide additional direct data on protein stability and more general information on interactions in native and denatured states.

We thank Dr R. W. Hartley for his generous gifts of DNA constructs and encouragement. This work was supported by the MRC and EEC (BAP). D.Š. is supported by funds from The Biochemical Society and IJS.

Received 20 April; accepted 11 May 1988.

1. Jaenicke, R. *Prog. Biophys. molec. Biol.* **49,** 117–237 (1987).
2. Levitt, M. F. & Chothia, C. *Nature* **261,** 552–558 (1976).
3. Janin, J. & Chothia, C. *J. mol. Biol.* **143,** 95–128 (1980).
4. Chothia, C., Levitt, M. & Richardson, D. *J. molec. Biol.* **145,** 215–250 (1981).
5. Chou, K.-C., Nemethy, G., Rumsey, S., Tuttle, R. W. & Scheraga, H. A. *J. molec. Biol.* **186,** 591–609 (1985); **188,** 641–649 (1987).
6. Sternberg, M. J. E. in *Computing in Biological Science* (ed. Geisow, M. J. & Barrett, A. N.) 143–177 (1983).
7. Chothia, C. *Ann. Rev. Biochem.* **53,** 537–572 (1984).
8. Shortle, D. Meeker, A. K. *Proteins* **1,** 81–89 (1986).
9. Albert, T. *et al. Nature* **330,** 41–45 (1987).
10. Fersht, A. R. *Biochemistry* **26,** 8031–8037 (1987).
11. Fersht, A. R., Leatherbarrow, R. J. and Wells, T. N. C. *Biochemistry* **26,** 6030–6038 (1987).
12. Nishimura, S. & Nomura, M. *Biochim. biophys Acta* **30,** 430–433 (1958).
13. Mauguen, Y., Hartley, R. W., Dodson, E. J., Dodson, G. G. Bricogne, G., Chothia, C. & Jack, A. *Nature* **297,** 162–164 (1982).
14. Hartley, R. W. *Biochemistry* **7,** 2401–2408 (1968).
15. Hartley, R. W. *Biochemistry* **8,** 2929–2932 (1969).
16. Paddon, C. J. & Hartley, R. W. *Gene* **53,** 11–19 (1987).
17. Creighton, T. E. *Proteins* (Freeman, New York 1983).
18. Pace, C. N. *Meth. Enzym.* **131,** 266–279 (1986).
19. Cupo, J. F. & Pace, C. N. *Biochemistry* **22,** 2654–2658 (1983).
20. Postma, J. P. M., Berendesen, H. J. C. & Haak, J. R. *Faraday Symp. Chem. Soc.* **17,** 55–67 (1982).
21. Wong C. F. & McCammon, J. A. *J. Am. chem. Soc.* **108,** 3830 (1986).
22. Bash, P. A., Singh, U. C., Langridge, R. & Kollman, P. A. *Science* **236,** 564–568 (1987).
23. Wolfenden, R. L., Andersson, L., Cullis, P. M. & Southgate, C. C. B. *Biochemistry* **20,** 849–855 (1981).
24. Ben Naim, A. & Marcus, Y. *J. chem. Phys.* **81,** 2106–2027.
25. Nozaki, Y. & Tanford, C. *J. biol. Chem.* **238,** 4074–4081 (1963).
26. Zoller, M. J. & Smith, M. *DNA* **3,** 479–488 (1984).
27. Kramer, B., Kramer, V. & Fritz, H.-J. *Cell* **38,** 879–887 (1984).
28. Carter, P. J., Winter, G., Wilkinson, A. J. & Fersht, A. R. *Cell* **38,** 835–840 (1984).
29. Serpersku, E. H., Shortle, D. & Mildvan, A. S. *Biochemistry* **25,** 68–87 (1986).

This paper was followed up in Refs. 39 and 82. There was a further initial rush of publications in *Nature* on the stability measurements[78,83,84], followed by lengthy articles in *Biochemistry* and *JMB*. We measured the energetics of many of the interactions that had been speculated about but not quantified. The next was to probe whether or not the dipole of a helix did have significant interactions with charges, done by a student Dasa Sali (now Lipovsek). We published in *Nature* '*Stabilization of Protein Structure by Interaction of a-Helix Dipole with a Charged Side Chain*' on 8 October, 1988[78], again beating Brian to the post as he published '*Enhanced Protein Thermostability from Designed Mutations that Interact with α-Helix Dipoles*' on 15 December, 1988 in *Nature*[85]. There was another close publication call when we published '*Surface Electrostatic Interactions Contribute Little to Stability of Barnase*' in *JMB* on 5 August, 1991[86] and Brian '*Cumulative Site-Directed Charge-Charge Replacements in Bacteriophage T4 Lysozyme Suggest that Long-Range Electrostatic Interactions Contribute Little to Protein Stability*' in *JMB* on 5 October, 1991[87]. I later learned from one of his then post-docs that we were 'the enemy'. Brian hails from Australia and nothing spurs on an Australian more than a contest with the English. In turn, we 'Poms' love to beat the Aussies in a Test match and win the Ashes. In retrospect it looks like a race, but it was not one at the time!

11.2 1989: Jim Kellis.

11.3 1994: Michael Rossmann and Brian Matthews.

11.4 2003: Liz Meiering and Dasa Sali (Lipovsek).

Stabilization of protein structure by interaction of α-helix dipole with a charged side chain

Daša Šali, Mark Bycroft & Alan R. Fersht*

Department of Chemistry, Imperial College of Science and Technology, London SW7 2AY, UK

The α-helix in proteins has a dipole moment resulting from the alignment of dipoles of the peptide bond which can perturb the pK_as of ionizing groups. One of the two histidine residues (His18) in barnase, the small ribonuclease from *Bacillus amyloliquefaciens*, is located at the negatively charged end (C-terminal) of an α-helix. From NMR titrations of wild-type and engineered mutants we find that the pK_a of His18 is 7.9 in wild-type enzyme, 1.6 units above the value in the urea-denatured enzyme and in model peptides. This implies that there is a favourable interaction between the protonated form of His18 and the α-helix that should stabilize the native structure at neutral pH by 2.1 kcal mol^{-1}. Denaturation at various values of pH of wild-type and muant enzymes engineered at position 18 shows that this is so. The increase in stability of the enzyme as the pH changes from 8.5 to 6.3 is attributable to this interaction, and the pH-stability curve fits pK_a values for His18 in native and urea-denatured enzymes that are consistent with the NMR data.

* To whom correspondence should be addressed. Present address: University Chemical Laboratory, Lensfield Road, Cambridge CB2 1EW, UK.

The stabilization energy of proteins is relatively small; the observed values of some 5–20 kcal mol^{-1} represent the differences between two large numbers—the free energies of the folded and unfolded states—which are closely balanced, so whether or not a protein folds depends crucially on the presence of small interaction energies. Different interactions have been proposed to stabilize enzymes, but accurate experimental data are required to quantify their contributions. We have used barnase as a paradigm for experimental studies on protein stability and folding because of its favourable properties[1]. The enzyme consists of a single polypeptide chain of 110 residues with no disulphide crosslinks and undergoes reversible unfolding[2]. We have used protein engineering to remove (or add) defined interactions that contribute to the stability of the native enzyme, and have then measured the change in free energy of folding from urea-induced denaturation. This provides an empirical measure of changes in stability on change of structure and a data base for refining theoretical calculations.

Long-range electrostatic interactions are important in both catalysis and protein stability, but the magnitudes of charge-charge interactions were until recently difficult to calculate because of a lack of experimental data. Protein engineering has been used to mutate charged surface residues of subtilisin[3]. Another important electrostatic element could be the dipole that results from the aligned peptide units of the α-helix[4–6]: this can participate in electrostatic interactions which affect the binding of charged ligands[6,7], the pK_as of ionizing groups[8] and the stability of the helix itself[9–12]. An anomolously high pK_a of a histidine residue in haemoglobin has been measured by NMR and the perturbation attributed to the field of an α-helix[8]. Negatively charged residues tend to cluster at the positively-charged end of an α-helix and vice versa[9,10]. Changes in T_m of

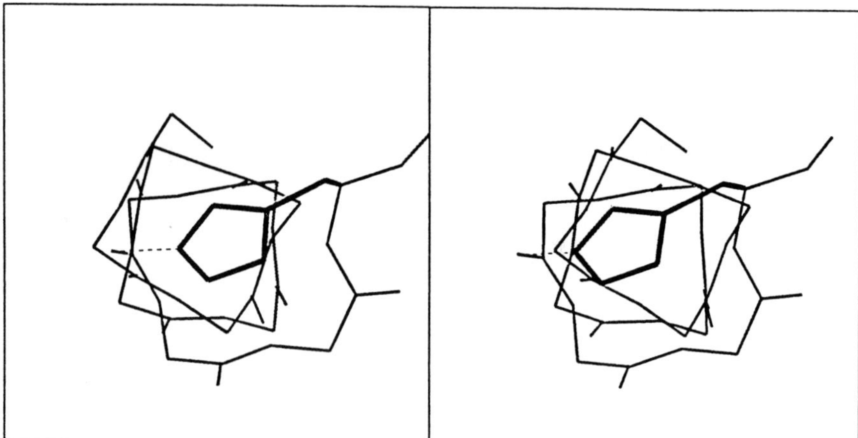

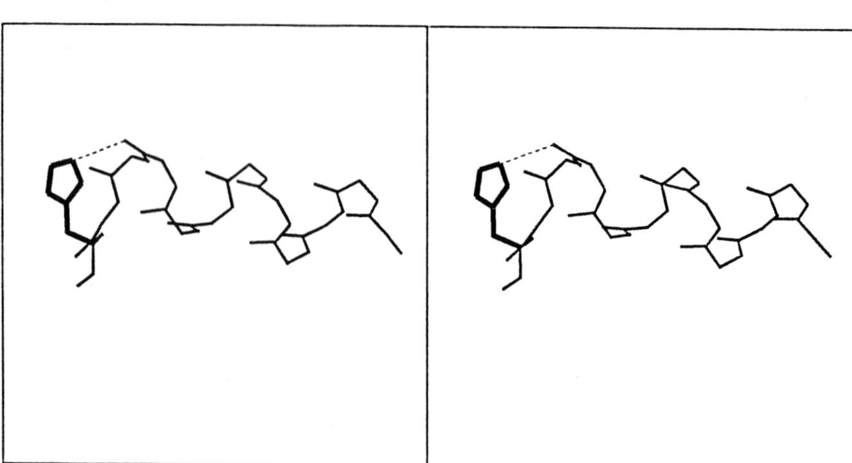

Fig. 1 Two stereo-pairs His18 and its relationship with the α-helix. The hydrogen bond between N_ε(H) of His18 and the backbone carbonyl oxygen of Gln15 is indicated.

Reprinted from Nature, Vol. 335, No. 6192, pp. 740-743, 20 October 1988

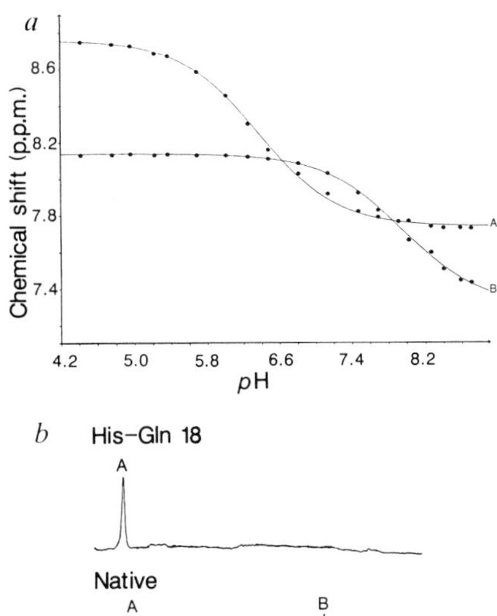

Fig. 2 *a*, Titration curves of C2-H resonances of histidine residues 18 (B) and 102 (A) in native barnase. *b*, Spectra (500 MHz ^{1}H-NMR) of histidine C2-H resonances of native barnase at pH 4.5 and mutant His18 to Gln18 at pH 4.3. NMR titrations were performed on 1 mM protein using a Bruker AM500 spectrometer at 25 ± 1 °C in D_2O, using NaOD and DCl to adjust the pD, with no added salt. Titration of imidazole in D_2O and H_2O showed that the measured pK_a is higher in D_2O by 0.07 units when the pH meter/glass electrode is calibrated with standard buffers in H_2O. The calculated value for the pK_a of His18 in D_2O is 7.93 ± 0.05. This is equivalent to 7.86 ± 0.05 in H_2O. Urea was found to affect the titration of acids and bases in a parallel manner. Standard pH 7.00 phosphate reads 7.30 ± 0.02 in 5M urea and the pK_a of imidazole increases by 0.29 ± 0.02 units in 5 M urea when calibrated against standard buffers in H_2O (Tris and MES (2-(N-morpholino) ethanesulphonic acid) also change their pK_as in an identical manner in urea, see legend to Fig. 4). The measured value of the pK_a of His18 in 5 M urea calibrated against standard buffer in H_2O is 6.59 ± 0.05. This is equivalent to 6.30 ± 0.05 in H_2O. Wild-type and mutant enzyme were prepared as before[1]. The following primer was used to direct the mutation His to Gln18: 5′-GTAGCTTT*TGATATGTC-3′ (where the asterisk follows the mismatch).

small model helices show qualitatively that the helices are stabilized by dipole-charge interactions[11,12]. There is compelling evidence for the importance of electrostatic effects from helix dipoles, but the precise magnitudes are unknown[6]. Here we provide quantitative measurements of the energy of a charge-dipole interaction that stabilizes barnase.

There are two histidine residues in barnase, His18 and His102. His18 is located at the C-terminal end of a helix that extends from residues 6 to 18 (ref. 13) (Fig. 1). The two ring nitrogen atoms of His18 are between 3 and 5.5 Å from the backbone C, N and O atoms of residues 14, 15, 16 and 17 at the end of the α-helix, and the H attached to N_ε makes a hydrogen bond with the backbone carbonyl oxygen of Gln 15. As is usual in proteins, the helix is partially shielded from water, unlike simple model helices. NMR titration (Fig. 2) shows that one histidine ionizes at a nearly normal pK_a of 6.30 ($\pm$0.05), whereas the other has a perturbed pK_a of 7.86 ($\pm$0.05). By analogy with the experiment on haemoglobin[8], we expect the perturbed residue to be His18. We have investigated the possible stabilization of the protein by the helix dipole-charge interaction by measuring the pK_a of each histidine in both native and urea-denatured enzyme by NMR and assigning them using mutants lacking each of the

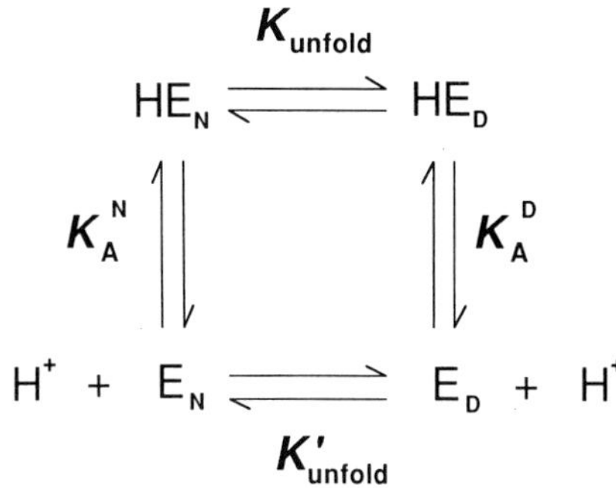

Fig. 3 Thermodynamic cycle relating the acid dissociation constants in the native and denatured forms of the enzyme, K_a^N and K_a^D, with the equilibrium constants for unfolding of the enzyme containing the protonated and unprotonated forms of a residue, K_{unfold} and K'_{unfold}, respectively.

histidine residues. The importance of the electrostatic interaction between the positive charge on the protonated imidazole of His18 and the enzyme was measured by determining the changes in free energy of folding as the imidazole titrates with pH.

Assignment of pK_as and interaction of His18 with enzyme. A mutant having Gln substituted for His at position 18 was constructed and titrated, when the histidine of pK_a 7.86 disappeared and so the high pK_a was assigned to His18. Titration of the proteins in 5 M urea shows that the pK_a of His18 drops to 6.30 ± 0.05, the normal value in small peptides[8], so the pK_a of His18 is, therefore, perturbed by 1.6 units by interaction with the protein. This implies that the protonated form of the imidazole is stabilized by the protein by 2.1 ($\pm$0.1) kcal mol^{-1} relative to the uncharged form.

We infer from the simple thermodynamic cycle in Fig. 3 that the protein is stabilized by 2.1 kcal mol^{-1} on protonation of His18 as follows. If the equilibrium constant for unfolding of the enzyme containing the protonated form of His18 is K_{unfold}, for the unprotonated form is K'_{unfold}, and the ionization constant for His18 in the native enzyme is K_a^N and in the denatured form is K_a^D then

$$K_{unfold}/K'_{unfold} = K_a^N/K_a^D \qquad (1)$$

That is, the equilibrium constant for unfolding changes in parallel with the difference in ionization constants in the native and denatured states. As the pK_a of His102 is 6.3 in both the native and urea-denatured forms, this ionization does not affect the unfolding equilibrium.

Destabilizing the protein by mutation of His18 to Gln18. A crude way of measuring the interaction energy between the protonated His18 and the enzyme is to compare its stability with a mutant in which His18 is substituted by a neutral side chain. But mutation of His18 to Ala18 loses not only the electrostatic interaction, but also the hydrogen bond from the N_ε(H) of the imidazole ring to a main-chain carbonyl oxygen, as well as hydrophobic contacts. The mutant His18 to Ala18 is so destabilized that we were unable to isolate enough for analysis, which we have found before for other mutants where the destabilization is >4-5 kcal mol^{-1}. Mutation of His18 to Gln18 is more appropriate because as shown previously in the tyrosyl-tRNA synthetase, its carboxyamide -NH_2 may form hydrogen bonds equivalent to those of the N_ε of histidine[14] and can also make hydrophobic contacts. Wild-type enzyme is more stable than the His18 to Gln18 mutant at pH 6.3 by 1.6 kcal mol^{-1}. However, it could be argued that the similarity to the expected value for the electrostatic interaction is fortuitous: the dipole from the -$CONH_2$ group could well make a dipole/dipole interaction with the helix, but this is counterbalanced by the loss of internal entropy of rotation in the flexible glutamine side chain on hydrogen bonding back to the main chain. But the mutant His18

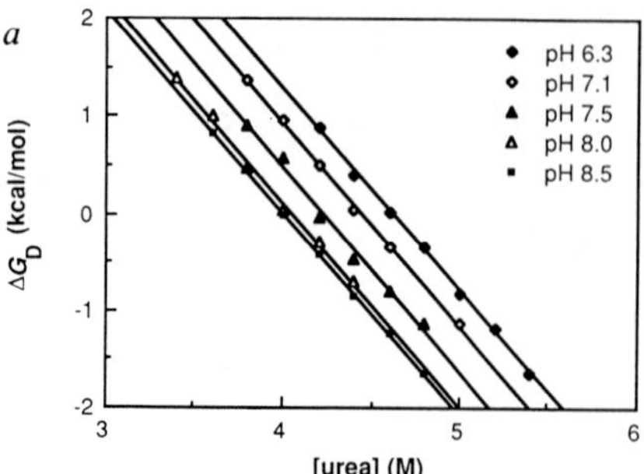

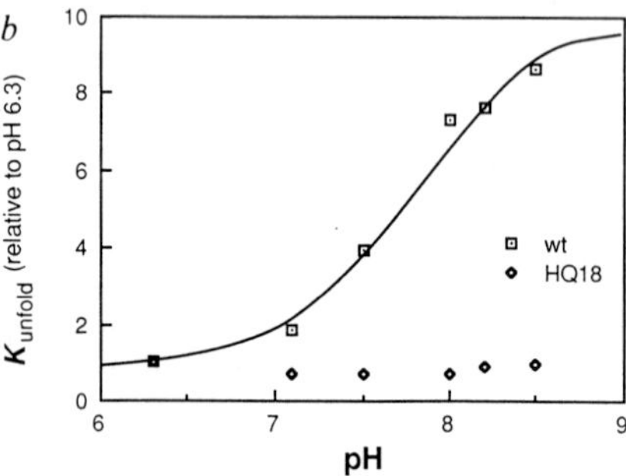

Fig. 4 *a*, Free energy of denaturation of wild-type barnase at different values of *p*H plotted against urea concentration according to the standard equation: $\Delta G_D = \Delta G_D^{H_2O} - m[\text{urea}]$. The data at *p*H 8.2 are omitted because of overcrowding the graph. *b*, Plot of the equilibrium constant for unfolding, K_{unfold}, as a function of *p*H for wild-type enzyme and the mutant His to Gln18. The observed values of K_{unfold} are given relative to the value at *p*H 6.3 for both enzymes. (Note that His18 to Gln18 is less stable than wild type by 1.6 kcal mol^{-1} at *p*H 6.3, but by only 0.3 kcal mol^{-1} at *p*H 8.5.) The data for wild-type-enzyme fit to an ionization curve of pK_a 7.94±0.14 and a ratio of K'_{unfold}/K_{unfold} (as defined in Fig. 3) of 16±4. Denaturation was observed spectrofluorimetrically with excitation at 290 nm and emission at 315 nm. The fluorescence falls by some 70% on denaturation. Experiments were performed at 25 °C in 50 mM MES acid/base buffer at *p*H 6.3 and Tris-HCl at higher values of *p*H (reducing the concentration of buffer tenfold does not affect the results). The concentration of barnase was between 5–10 μg ml^{-1}. The change in free energy of folding as a function of *p*H, $\Delta\Delta G_D$, is calculated very simply and accurately because the denaturation curves overlap. In a manner analogous to that decribed previously[1], $\Delta\Delta G_D$ between two *p*H values pH_x and pH_y, $\Delta\Delta G_D(pH_x - pH_y)$, is calculated from the direct measurement of the fractions of unfolded and folded states of the protein at identical concentrations of urea:

$$\Delta\Delta G_D(pH_x - pH_y) =$$
$$RT \ln\{([\text{folded}]/[\text{unfolded}]pH_y)/([\text{folded}]/[\text{unfolded}]pH_x)\}$$

As there is just a small change in stability from one value of *p*H to the next, the ratios of [folded]/[unfolded] can readily be measured over a range of values of urea concentrations. The values of the pK_as of phosphate (the standard calibration buffer), Tris and MES all increase in 4 M urea by 0.23±0.02 units when calibrated against standard buffers in H_2O. Thus, when calibrated against standard buffers in urea, measured values of *p*H and pK_as are unaffected by the addition of urea. To a good approximation, therefore, urea does not perturb the ionization of groups in the protein. Even if there are small effects (±0.02 units at 4 M urea) on the different types of acids and bases, these are expected to be the same in both native and denatured states and so would not perturb the denaturation equilibrium.

to Gln18 provides a control for the following improved procedure.

Variation of K_{unfold} with *p*H gives electrostatic stabilization energy. The most precise way of using denaturation to determine the interaction energy is to measure the change in stability as the histidine titrates with *p*H. As seen in Fig. 4*a*, addition of urea to wild-type enzyme generates a family of parallel denaturation curves as the enzyme is titrated from *p*H 6.3 to 8.5, whereas the stability of the mutant His18 to Gln18 varies only slightly. A plot of K_{unfold} against *p*H (Fig. 4-b) for wild-type enzyme fits an ionization curve of pK_a 7.94±0.14. The ratio of K'_{unfold}/K_{unfold}, as defined in Fig. 3, is 16±4. Substituting these values into equation (1) gives a value of 6.7±0.2 for the pK_a of His18 in the denatured enzyme. There is a small but systematic variation of the observed unfolding constant for the mutant His18 to Gln18, with *p*H (±20%) that may reflect other ionizations in the protein or environmental effects. To normalize for these in the wild-type enzyme, the ratio of $(K_{unfold})_{wt}/K_{unfold})_{His \to Gln18}$ should be plotted. This gives a slightly poorer fit (not shown), with values of pK_a of 7.7±0.2 for His18 in native enzyme and 6.35±0.25 in the denatured enzyme. The value of K'_{unfold}/K_{unfold} is 22±11. The change in interaction energy of His18 with the protein on ionization is 1.4–2.1 kcal mol^{-1} (that is, it equals $RT\ln(K'_{unfold}/K_{unfold})$). These values are consistent with the NMR data within experimental error.

Data are internally consistent. The two sets of experiments measure the two different pairs of thermodynamic constants in Fig. 3. NMR measures K_a^N and K_a^D whereas denaturation measures K'_{unfold} and K_{unfold}. In addition, the *p*H dependence of denaturation gives K_a^N and K_a^D. The two sets of experiments thus provide a test of the applicability of the simple two-state scheme of Fig. 3 to the denaturation of barnase and the suitability of measuring denaturation. The NMR data are inherently the more accurate, however, and so the thermodynamic constants derived from those should preferably be used.

The electrostatic interaction energy of His18 is predominantly with the α-helix. We have been careful to state so far that the electrostatic interaction energy is between His18 and the (whole) protein because long-range interactions must be made with all the charged groups. But, it does appear that the energy can be attributed predominantly, and probably solely, to the electrostatic interaction of His18 with the peptide dipoles that constitute the α-helix. Firstly it can be seen that positively and negatively charged side chains in the protein are not close to His18 and are reasonably equally distributed from it: the closest charged groups are the ε-NH_2 of Lys19, 9 Å away, and the carboxylate of Asp12, 12 Å away. Indeed, preliminary calculations (S. Wodak, personal communication) indicate that the net field from the charged residues of the protein is close to zero at His18. Secondly, there is direct evidence from the effects of counterions on the pK_a of His18; it is well known from classical measurements that counterions mask electrostatic effects from charges in proteins. Further, recent accurate measurements of effects of mutation of charged groups on the surface of subtilisin show that 1 M KCl almost exactly screens the charge on individual side chains (see ref. 3 for example). The pK_a of His18 increases from 7.86 in 1 mM salt to 7.98 in 1 M KCl, a small increase similar to that found for the titration of imidazole at low and high salt. Thus the high pK_a of His18 does not result from interactions with charged side chains. The lack of dependence of the electrostic interaction on salt suggests that the field of the helix dipole is only weakly shielded by counterions.

We thank Dr R. W. Hartley for his gift of the barnase expression system. This work was supported by the MRC and EEC (B.A.P.).

Received 8 August; accepted 12 September 1988.

1. Kellis, J. T. Jr, Nyberg, K., Sali, D. & Fersht, A. R. *Nature* **333**, 784–786 (1988).
2. Hartley, R. W. *Biochemistry* **7**, 2401–2408 (1968); *Biochemistry* **8**, 2929–2932 (1969).
3. Thomas, P. G., Russell, A. J. & Fersht, A. R. *Nature* **318**, 375–376 (1985); Russell, A. J. & Fersht, A. R. *Nature* **328**, 496–500 (1987).
4. Wada, A. *Adv. Biophys.* **9**, 1–63 (1976).
5. Hol, W. G. H., Van Duijnen, P. T. & Berendsen, H. J. C. *Nature* **273**, 443–446 (1978).
6. Hol, W. G. J. *Prog. Biophys. molec. Biol.* **45**, 149–195 (1985).
7. Warwicker, J. & Watson, H. C. *J. molec. Biol.* **157**, 671–679 (1982).

8. Perutz, M. F., Gronenborn, A. M., Clore, G. M., Fogg, J. H. & Shih, D. T.-b. *J. molec. Biol.* **183**, 491-498 (1985).
9. Blagdon, D. E. & Goodman, M. *Biopolymers* **14**, 241-245 (1975).
10. Wada, S. & Nakamura, H. *Nature* **291**, 757-758 (1981).
11. Shoemaker, K., Kim, P. S., York, E. J., Stewart, J. M. & Baldwin, R. L. *Proc. natn. Acad. Sci. U.S.A.* **82**, 2349-2353 (1985); *Nature* **326**, 563-567 (1987).
12. Mitchinson, C. & Baldwin, R. L. *Proteins* **1**, 23-33 (1986).
13. Mauguen, Y. *et al. Nature* **297**, 162-164 (1982).
14. Lowe, D. M., Fersht, A. R., Wilkinson, A. J., Carter, P. & Winter, G. *Biochemistry* **24**, 1506-5109 (1985).

From then on, we raced ahead, being driven by Luis Serrano. Luis was the perfect postdoctoral for the work, with his amazing ability to dissect mentally a structure and then efficiently perform all the experiments. He came to my lab as a biologist, but then went to EMBL as a fully fledged biophysicist where he took, amongst other successes, the prediction of protein energetics to new levels by writing programs such as AGADIR for predicting helix stability. His first paper on energetics[83] was his beginning of exploring factors that stabilise helices, first quantifying the effects of residues that capped the ends.

11.5 2003: Andreas Matouschek, Mark Bycroft and Luis Serrano.

11.6 2000: Jose Luis Neira.

11.7 2003: Amnon Horovitz, Javier Sancho and Johanna Avis.

11.8 2003: Tony Day (Lars Emrén (Hansen) behind).

Capping and α-helix stability

Luis Serrano & Alan R. Fersht

MRC Unit for Protein Function and Design, Department of Chemistry, University of Cambridge, Lensfield Road, Cambridge CB2 1EW, UK

THE first and last four residues of α-helices differ from the rest by not being able to make the intrehelical hydrogen bonds between the backbone >C=O groups of one turn and the >NH groups of the next[1]. Physico-chemical arguments[1] and statistical analysis[2] suggest that there is a preference for certain residues at the C and N termini (the C- and N-caps)[2] that can fulfil the hydrogen bonding requirements. We have tested this hypothesis by constructing a series of mutations in the two N-caps of barnase (*Bacillus amyloliquefaciens* ribonuclease, positions Thr 6 and Thr 26) and determining the change in their stability. The N-cap is found to stabilize the protein by up to ~2.5 kcal mol^{-1}. The presence of a negative charge of the N-cap adds some 1.6 kcal mol^{-1} of stabilization energy because of the interaction with the macroscopic electrostatic dipole of the helix.

The α-helix is a major component of protein secondary structure. Cogent arguments have been made from simple physical chemistry[1], statistical survey[2] and qualitative experiments on model systems[3,4] to identify the structural features that influence helix stability. Quantitative data on the energetics and importance of these interactions, however, have been unavailable until very recently. The small ribonuclease from *B. amyloliquefaciens*, barnase, has proved to be an excellent system for determining these with high precision from measurements on unfolding of wild-type and mutant proteins[5-8]. Barnase contains two α-helices[9], one from residues 6 to 18 and the other from 26 to 34 (Fig. 1). Both α-helices begin with a threonine residue as N-cap (the first residue in the helix[2], namely Thr 6 and Thr 26, Fig. 1), but the overall composition of the rest of each helix is quite different. The -OH groups of Thr 6 and Thr 26 are positioned so that the oxygen atom may act as a hydrogen bond acceptor from the backbone > NH groups of residues 7 or 27, 8 or 28 and 9 or 29 (N1, N2 and N3 in the Richardson[2] nomenclature). The geometry for hydrogen bonding from the -OH of residues 6 and 26 is good with respect to residues 9 and 29, fair for 8 and 28 and poor for 7 to 27 (Table 1), which is usual with N-caps[2]. The C_γ and C_β of both threonines participate in further interactions in the protein (Table 1). Here we use inferences from recent surveys of helix structure[1,2] to design changes in both barnase helices to give precise information on how hydrogen bonding, and electrostatic[10] and other interactions at the N-terminus of a helix affect its stability.

The most common residues found at the N-cap position of helices are: Ser, Asn, Gly, Asp and Thr[2]. The residues Ala, Leu, Val, Ile, Trp, Arg, Gln and Glu are found rarely[2]. We substituted Thr with the following residues: with Gly, Ala and Val to measure the effect of losing hydrogen bonds from the γ-OH to the > NH groups of the helix and to assess the influence of a different size of side chain; with Ser to measure the loss of hydrophobic interactions from the γ-methyl; with Asn and Gln to determine the effects of substituting the carboxyamide oxygen of each of the γ-OH as a hydrogen bond acceptor, and with Asp and Glu to establish the effects of a charged hydrogen bond acceptor. Two mutants, Thr 26→Asp and Thr 6→Val, could not be expressed, despite numerous attempts using this expression system[11].

Mutation of either Thr 6 or Thr 26 at the N-caps of each helix in barnase gives quite similar changes in stability and consistent trends (Table 2). Considering that the two helices have different compositions and environments, the results are not specific to a particular helix and so presumably report on some general features. A fine-structure analysis can be made by comparing pairs of mutants that differ by just a small change.

Thr→Ser. Mutation of Thr 6→Ser causes the loss of 0.22 kcal mol^{-1} and Thr 26→Ser, 0.56 kcal mol^{-1} of stabilization energy (Table 2). The Cγ of Thr 6 and 26 make van der Waals' contacts with the protein (Table 1), and so this contribution to

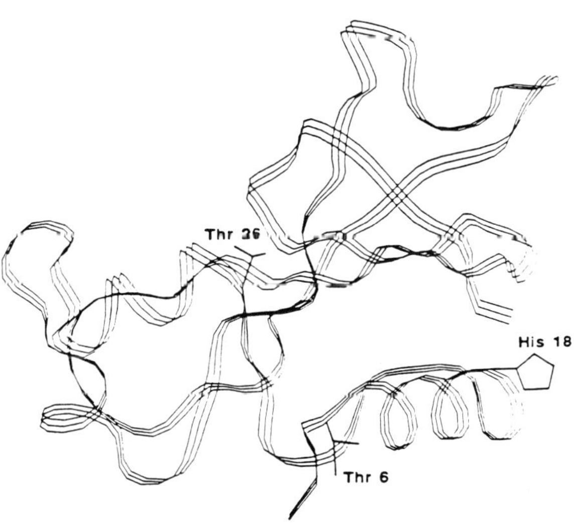

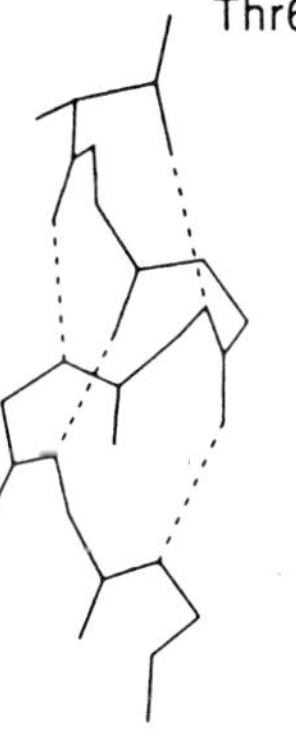

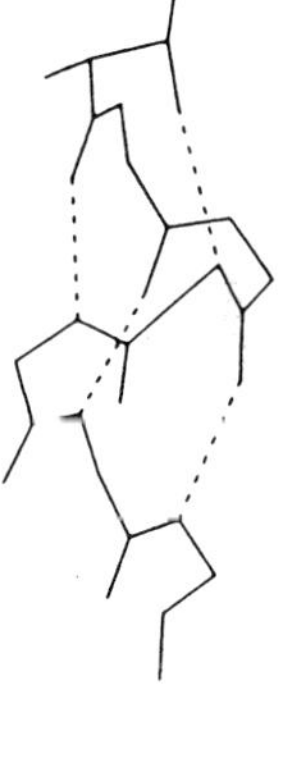

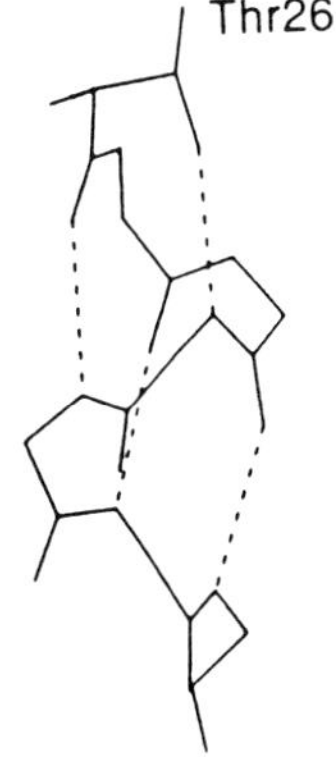

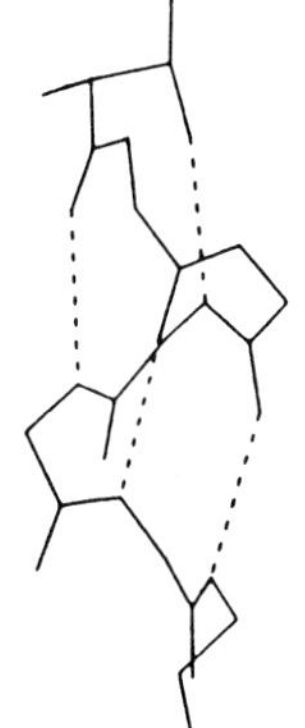

FIG. 1 Top, Sketch of structure of barnase and the residues of interest (produced from coordinates provided by Professor G. Dodson and Dr C. Hill). Thr 6 and Thr 26 act as N-caps[1,2], accepting hydrogen bonds from the backbone group >NHs principally of Gly 9 and Glu 29. The sequences of the two helices used in this work are: Helix 6-18, Thr-Phe-Asp-Gly-Val-Ala-Asp-Tyr-Leu-Glu-Thr-Tyr-His; Helix 26-34, Thr-Lys-Ser-Glu-Ala-Gln-Ala-Leu-Gly. Bottom, Stereo views of the interactions of the γ-OHs of Thr 6 and Thr 26 with main chain residues 7, 8, 9 and 27, 28, 29 respectively. There are good hydrogen bonds with >NHs of 9 and 29 (see text).

Reprinted from Nature, Vol. 342, No. 6247, pp. 296-299, 16th November, 1989

the stabilization is lost. It should be noted that none of these contacts is within the helix. Thus, the advantage of having a threonine rather than a serine at the N-cap seems to depend mainly on the existence of tertiary interactions elsewhere in the structure, and not on interactions within the helix itself. These interactions will vary from protein to protein.

Ser → Gly, Ala or Val. Naïve reasoning would suggest that mutation of Ser → Ala might give an empirical measurement of energy change due to the loss of the hydrogen bonds from the -OH to the backbone >NH groups, and Ala → Gly, a measure of the loss of the van der Waals' energy from the β-methyl group. Mutation of Ser 6 → Ala loses 2.3 kcal mol^{-1} or 1.6 kcal mol^{-1} for replacement of Ser 26 (Table 2). But, mutation of Ser → Gly loses only 1 kcal mol^{-1}. Thus, Gly is more stable than Ala at the N-cap by 0.5–1 kcal mol^{-1}. In general, Ala is energetically preferred to Gly in the helix, except at the N-cap[12], either because of the greater entropic freedom of glycine-containing peptides in the denatured state[13], or because of hydrophobic interactions of the alanine β-methyl group in the helix. Why then is Gly preferred to Ala at the N-cap? We suggest that one contributing factor is that the β-methyl group of Ala sterically interferes with solvation of the >NH groups at the N-terminus of the helix—Gly allows more space for optimal hydrogen bonding by water. Consistent with this notion is that Thr 26 → Val is the most destabilizing mutation, despite the retention of the γ-methyl group with its potential for hydrophobic interactions worth ~0.6 kcal mol^{-1}. The large side chain of Val makes solvation of the >NH groups by water difficult. These results parallel statistical data on the frequency of residues at the N-cap (Table 3). There is evidence elsewhere that an uncharged hydrogen bond contributes some 0.5–1.5 kcal mol of stabilization energy to a complex[14]. The apparent binding energy of the N-cap hydrogen bond from Ser to the >NH groups is in this region.

Thr → Asn, Gln. The previous mutations are relatively easy to interpret in the absence of detailed crystal structures of the mutants because interactions were deleted without introducing new ones, and so the necessary structural information is present in the crystal structure of the wild type. Substitution with Asn, Asp (or Glu or Gln), adds the possibility of further interactions that can obscure simple interpretation. Statistically, Asn is one of the most favourable residues at the N-cap of helices[2]. The carbonyl oxygen of the side chain can act as a hydrogen bond acceptor of the backbone >NH groups. On the other hand, Gln and Glu are rarely found at the N-cap as their side chains cannot make good hydrogen bonds to the backbone >NH groups[2]. Mutation of Thr → Gln in positions 6 to 26 results in destabilizing the protein by 1.9 and 1.7 kcal mol^{-1} respectively. But Thr → Asn also destabilizes both helices by 1.3 kcal mol^{-1}. The reason for this is not obvious, because Asn can be built into the structure with its carbonyl oxygen placed to accept hydrogen bonds from the backbone >NH groups and its own -NH_2 group exposed to water. Perhaps Asn is not inherently as good an N-cap as Ser or Thr, but may be better when the -NH_2 group can hydrogen bond elsewhere in the protein. There should be more protein

TABLE 1 Contact distances (< 5 Å) from Thr 6 and Thr 26 to other residues*

Atom	Neighbouring atom	Distance (Å)
Thr 6		
Oγ	N (Phe 7)	3.63, 3.53, 3.80†
	N (Asp 8)	4.17, 4.06, 4.00†
	N (Gly 9)	3.57, 3.14, 3.05†
Cγ	O$_\delta$ (Asn 5)	3.83
	C (Asn 5)	3.81
	O (Asn 5)	4.20
	O$_\delta$ (Asn 77)	3.98
C$_\beta$	N (Phe 7)	3.05
	N (Asp 8)	4.55
	N (Gly 9)	4.73
	C (Asn 5)	3.64
	O (Asn 5)	4.15
Thr 26		
Oγ	N (Lys 27)	3.37, 3.34, 3.67†
	N (Ser 28)	3.54, 3.38, 3.77†
	N (Glu 29)	3.35, 2.94, 3.23†
	O$_{\delta 1}$ (Asp 54)	4.20
	O$_{\sigma 2}$ (Asp 54)	4.77
Cγ	C (Ile 25)	4.04
	O (Gly 52)	3.52
	C (Gly 53)	4.48
	O (Gly 53)	4.47
	O$_{\delta 1}$ (Asp 54)	4.07
	O$_{\delta 2}$ (Asp 54)	4.66
Cβ	N (Lys 27)	3.08
	N (Ser 28)	4.30
	N (Glu 29)	4.66
	O (Ile 25)	4.34
	C (Ile 25)	3.70
	O (Gly 52)	4.17
	Cγ (Asp 54)	3.88
	O$_{\delta 2}$ (Asp 54)	3.74

* Contact distances from Thr 6 and Thr 26 to other residues were obtained from X-ray coordinates (2 Å resolution), kindly provided by Professor G. Dodson and Dr C. Hill.

† There are three molecules in the unit cell, and these are the individual values for each molecule. The accuracy for bond distances at this resolution is about ±0.2 Å.

TABLE 2 Change in free energy of unfolding ($\Delta\Delta G_U$) on mutation of N-caps (Thr 6 and Thr 26)*

Mutation	Position 6	$\Delta\Delta G_U$ (kcal mol^{-1}) Position 26
Thr → Ser	0.22	0.56
Thr → Val		2.31
Thr → Ala	2.53	2.11
Thr → Gly	1.34	1.58
Thr → Asn	1.27	1.29
Thr → Asp	−0.11	
Thr → Gln	1.87	1.72
Thr → Glu	0.27	0.05
Asp → Asn	1.38	
Glu → Gln	1.60	1.67
Asp → Glu	0.38	
Asn → Gln	0.60	0.43
Ser → Ala	2.31	1.55
Ser → Gly	1.12	1.02
Gly → Ala	1.19	0.53

* $\Delta\Delta G_U$ is the free energy of unfolding of wild-type enzyme minus that of mutant. The free energies of unfolding of wild-type protein and mutants were analysed by urea denaturation monitored by fluorescence[5-7]. This analysis is based on the equation $\Delta G_U = \Delta G_U^{H_2O} - m[\text{Urea}]$, where ΔG_U is the free energy of unfolding in urea solution, $\Delta G_U^{H_2O}$ the free energy of unfolding in water, and m is a constant. We find that there is little difference in calculating $\Delta\Delta G_U$ directly by comparing ΔG_U for wild type and mutant at the same concentration of urea[5] or by using the simpler procedure[7] of $\Delta\Delta G_U = \langle m\rangle(\Delta[U]_{1/2})$, where $\Delta[U]_{1/2}$ is the concentration of urea at which wild type is 50% denatured, minus the concentration of urea at which the mutant is 50% denatured. The average value of m for wild type and mutant is denoted by $\langle m\rangle$. We find that m does not deviate by more than 10% of a mean value which is 2.23 (±0.02 standard error) kcal mol^{-2} for all mutants. $[U]_{1/2}$ never differs by more than 1% (< ±0.04 M) for experiments performed over a period of time and is generally reproducible to ±0.01 M with identical batches of urea. The data are probably reliable to ±0.02 kcal mol^{-1}. As the free energy of unfolding determined by either method is essentially the same, we have used the second method method to obtain the values of the free energy of unfolding of all the proteins.

Proteins were prepared by an improved procedure. Wild-type enzyme was secreted from *E. coli* BL21DE3 pLysS (ref. 16) harbouring the barnase gene construction[12] on the pTZ18u plasmid (Pharmacia). Cells were grown in low-phospate medium[17] and the expression of barnase was induced by the addition of 0.5 mM isopropyl-β-D-thiogalactoside following the conditions described by Studier *et al.*[16]. After further growth for 2.5 h, the protein was purified as before[18]. For site-directed mutagenesis we used the method of Eckstein *et al.*[19]. Mutants were identified either by direct sequencing of single-stranded DNA or after previous screening by the hybridization method[20].

TABLE 3 Comparison of statistical and energetic surveys of N-cap stabilities

Preferred residue at N-cap	Stabilization energy relative to Thr (kcal mol^{-1})*	Frequency at N-cap†	Relative value‡
Asp	−0.1	27	2.1
Thr	0.0	21	1.6
Glu	0.2	5	0.4
Ser	0.4	34	2.3
Asn	1.3	34	3.5
Gly	1.5	33	1.8
Gln	1.8	3	0.4
Ala	2.3	10	0.5
Val	2.3	1	0.1

* Average value at positions 6 and 26.

† Frequency found in survey in ref. 2.

‡ Frequency at N-cap normalized for frequency of occurrence in proteins in general[2].

structure surveys undertaken to look for other factors. Whatever the explanation, Asn is not necessarily a good choice for the N-cap of an helix.

Thr → Asp, Glu. Mutation of Thr 6 → Asp stabilizes the protein by 0.1 kcal mol^{-1}, but Thr 6 → Glu destabilizes it by only 0.3 kcal mol^{-1} and Thr 26 → Glu destabilizes it by less than 0.1 kcal mol^{-1}, despite the low frequency of Glu at the N-cap[2]. The α-helix possesses a macroscopic dipole with its positive pole at the N-terminus[10]. Recent studies have suggested values of about 1.4–2.1 kcal mol^{-1} for the interaction energy of a single charge with the dipole of a helix[6,15]. A charge–dipole electrostatic interaction is the probable reason for the stability of the mutants with Asp and Glu. Comparison of Asp with Asn and Glu with Gln gives another estimate of the importance of the charge–dipole interaction: Asp 6 → Asn gives 1.4 kcal mol^{-1}; Glu 6 → Gln, 1.6 kcal mol^{-1}; and Glu 26 → Gln, 1.6 kcal mol^{-1}. These results are strikingly similar to those of the earlier studies, despite there being an unfavourable electrostatic interaction at position 6 with the negative charge on Asp 8. As so many different substitutions with different hydrogen bonding patterns and at different ends of the helix give similar values, the estimate of 1.4–2.1 kcal mol^{-1} for the helix-charge interaction is valid.

There is a very rough parallel between the energetics of N-cap formation and statistical surveys of the frequencies found in practice (Table 3). Notable exceptions, however, are Asn and Glu. Our data indicate that Glu is quite acceptable at the N-cap, being only marginally less stable than Thr. Asn is statistically over-represented, given the data in this study.

In conclusion, the N-cap is important to stability and substitution of selected residues can cause an energy change of over 2 kcal mol^{-1}. It is clear, however, that when designing a helix, the choice of residue depends on the context of the overall tertiary structure—that is, on interactions with residues removed in sequence but stereochemically close to the helix. Even so, it seems that some residues are particularly favourable. For example, Ser is inferior to Thr only when hydrophobic interactions are favourable for the γ-methyl group. For barnase, the Ser mutants are only 0.2 to 0.6 kcal mol^{-1} less stable, although the methyl group can be worth up to 1.5 kcal mol^{-1} when packing is particularly good[5,7]. In other circumstances, steric exclusion of the γ-methyl group would prohibit the use of Thr. Asp is clearly the best for an exposed N-terminus that has no adjacent negative charges. □

Received 23 June; accepted 6 October 1989.

1. Presta, L. G. & Rose, G. D. *Science* **240,** 1632–1641 (1988).
2. Richardson, J. S. & Richardson, D. C. *Science* **240,** 1648–1652 (1988).
3. Kim, P. S. & Baldwin, R. L. *Nature* **307,** 329–334 (1984).
4. Shoemaker, K., Kim, P. S., York, E. J., Stewart, J. M. & Baldwin, R. L. *Proc. natn. Acad. Sci. U.S.A.* **82,** 2349–2353 (1985); *Nature* **326,** 563–567 (1987).
5. Kellis, J. T. Jr, Nyberg, K., Sali, D. & Fersht, A. R. *Nature* **333,** 784–786 (1988).
6. Sali, D., Bycroft, M. & Fersht, A. R. *Nature* **335,** 741–743 (1988).
7. Kellis, J. T. Jr, Nyberg, K. & Fersht, A. R. *Biochemistry* **28,** 4914–4922 (1989).
8. Matouschek, A., Kellis, J. T., Serrano, L. & Fersht, A. R. *Nature* **340,** 122–126 (1989).
9. Mauguen, Y. *et al. Nature* **297,** 162–164 (1982).
10. Hol, W. G. J. *Prog. Biophys. molec. Biol.* **45,** 149–195 (1985).
11. Paddon, C. J. & Hartley, R. W. *Gene* **53,** 11–19 (1987).
12. Strehlow, G. H. & Baldwin, L. R. *Biochemistry* **28,** 2130–2133 (1989).
13. Matthews, B. W., Nicholson, H. & Becktel, W. J. *Proc. natn. Acad. Sci. U.S.A.* **84,** 6663–6667 (1987).
14. Fersht, A. R. *Trends Biochem. Sci.* **12,** 301–304 (1987).
15. Nicholson, H., Becktel, W. J. & Matthews, B. W. *Nature* **336,** 651–656 (1988).
16. Studier, F. W., Rosenberg, A. H. & Dunn, J. J. *Meth. Enzym.* (in the press).
17. Serpesu, E. H., Shortle, D. & Mildvan, A. S. *Biochemistry* **25,** 68–87 (1986).
18. Mossakowska, D. N., Nyberg, K. & Fersht, A. R. *Biochemistry* **28,** 3843–3850 (1989).
19. Sayers, J. R. & Eckstein, F. In *Genetic Engineering: Principles and Methods* Vol. 10 (ed. Setlow, J. K.) 109 (Plenum, New York and London, 1988).
20. Carter, P. J., Winter, G., Wilkinson, A. J. & Fersht, A. R. *Cell* **38,** 835–840 (1984).

ACKNOWLEDGEMENTS. We thank Professor Guy Dodson and Dr Chris Hill for providing coordinates from unpublished high resolution data on barnase. L.S. is an EMBO fellow.

The next paper[84] (followed up in[88]) was particularly interesting because not only did Luis show that alanine stabilises helices by far more than expected but also that there was a linear dependence on the burial of solvent accessible surface area. Amnon Horovitz measured the relative contributions of different amino acid side chains to stability in the middle regions of α-helices[89]. It was thought for a very long time that the reason why glycine destabilises α-helices by about a kcal/mol is because of the conformational freedom (high entropy) of unstructured glycine-containing peptides[90]. Only recently did Valerie Daggett and I show that alanine has the same stability as glycine in a fully denatured peptide as the conformational entropy advantage of glycine is balanced by alanine's better solvent interactions[91].

11.9 2006: Celebrations after Honorary Degree at Hebrew University. *Top row*: Amnon Horovitz, Gideon Schreiber, Tammy Gray, Sandra Kostyn, Alan, Stephen Kostyn, Ronnie Loewenthal, Ruth and Yossi Sperling. *Bottom row*: Dorit and Assaf Friedler, Yael Schreiber and Limor Loewenthal.

LETTERS TO NATURE

Effect of alanine versus glycine in α-helices on protein stability

Luis Serrano, Jose-Luis Neira, Javier Sancho & Alan R. Fersht

MRC Unit for Protein Function and Design, Cambridge IRC for Protein Engineering, MRC Centre, Hills Road, Cambridge CB2 2QH, UK

THE rational design of proteins requires knowledge of the helix-forming propensities (s-values) of the different amino acids. There is, however, considerable controversy about the relative values for alanine and glycine[1–12]. We find from experiments on mutants of barnase that the relative effect of Ala versus Gly on helix stability depends crucially on the position in the helix (whether they are at the ends (caps) or are internal) and the context (the influence of their neighbours). Glycine is greatly preferred at the N and C caps. At internal positions, Ala stabilizes the helix relative to Gly by 0.4 to 2 kcal mol^{-1}. The variation results from a combination of burial of hydrophobic surface on folding and interference with hydrogen bonding of the protein with solvent. There is a good empirical correlation between the relative stabilizing effects of Ala and Gly with the total change in solvent-accessible hydrophobic surface area of the folded protein on mutation of Gly to Ala. It is not valid to assign to each amino acid a unique s-value that is generally applicable to all positions in all helices in all proteins.

Several groups have examined the helix-forming propensities of various amino acids by the host-guest method[1,2] or by making mutations in momomeric[3–7,10] or dimeric helix-forming peptides[8] and more complex systems[11]. There are considerable discrepancies between measurements of the relative helix-forming propensities (helix propagation parameter, s)[10] of alanine and glycine from copolymer host-guest experiments, analysed by the Zimm-Bragg formulation ($s_{Ala} = 1.07$, $s_{Gly} = 0.62$, equivalent to Ala stabilizing a helix by 0.3 kcal mol^{-1} relative to Gly)[13] and the more complex Lifson-Roig[14] theory ($s_{Ala} = 1.56$, $s_{Gly} = 0.015$, equivalent to 2.8 kcal mol^{-1} stabilization by Ala relative to Gly)[10]. (The change in stability of a protein on mutation of Ala to Gly at a particular position, $\Delta\Delta G_{unf}$, is related to the helix-forming propensities by $\Delta\Delta G_{unf} = -RT \ln(s_{Ala}/s_{Gly})$.) There can be problems in using model peptides because they are poorly characterized in solution. They are generally less than 100% helical and it is likely that the ends fray more than the middle[8,10,11]. Different results are obtained when applying multistate or two-state models[6,10]. Experiments on proteins are more amenable to interpretation because the parent structures are well defined and the helices are held intact by the bulk of the protein. The derived experimental data may also be analysed by a simple two-state model.

Barnase is suitable model protein to study the effect of single mutations on protein stability[15–21]. It has two regular α-helices (residues 6–18 and 26–34) which contain different residues and are partly exposed to solvent. Our strategy consists in selecting those residues that are on the solvent-exposed face of the helices and do not make extensive contacts with the rest of the protein. These residues are mutated to Ala and Gly, and the changes in protein stability ($\Delta\Delta G_{unf}$) measured with respect to the Ala mutants. In the first helix, we have selected the N-cap residue (Thr 6), Asp 8 and Asp 12, Thr 16, Tyr 17 and the C-cap residue (His 18). In the second helix, we have chosen the N-cap residue (Thr 26), Lys 27, Ser 28, Glu 29, Gln 31 and Ala 32, all of which make very few or no contacts outside the α-helix, and the C cap (Gly 34). The analysis of Ala→Gly mutations at the N cap has already been published[15]. The analysis of all the mutants is shown in Table 1. There is a clear-cut difference between the effects if the different mutations at the N cap, C cap and at internal locations. Gly is favoured over Ala at both caps (−0.5 or −0.9 kcal mol^{-1} at the N cap; −1.2 or −3.1 kcal mol^{-1} at the C cap). At all the other positions, Ala is favoured over Gly but there is considerable disparity in the differences in free energy of unfolding between Ala and Gly at the different internal positions.

The positional dependence of the relative helix-stabilizing propensities of Ala and Gly in the helices derived from small peptides in solution has been attributed to fraying at the ends[10]. The difference in α-helix propensity between Ala and Gly in these helices decreases monotonically towards the N or C caps but it never results in a higher stability for a peptide having Gly instead of Ala[10]. The results on barnase show clearly that there is a genuine positional and context dependence for the relative helix-stabilizing propensities of Ala and Gly that does not result from fraying. The data may be readily rationalized by simple physical principles that relate to solvent exposure (Table 2). First, the N and C Caps have exposed NH and CO groups, respectively, that require hydrogen-bond formation with either groups on the enzyme or by solvent[22,23]. Bulky substituents at the N (ref. 15) or C caps that cannot satisfy the hydrogen-bond requirements of the first and last four residues of helices inhibit the solvation. Second, there is burial of hydrophobic surfaces on the formation of the helix, and this should be a stabilizing factor. It has been shown for small molecules that there is a linear relationship between free energies of transfer of hydrocarbon side chains from water to hydrophobic solvents and the change in accessible surface area[24–27]. There is thus a physical reason for attempting to correlate hydrophobic effects with changes in solvent-accessible hydrophobic area. We find that plotting $\Delta\Delta G_{unf}$ against ΔA_{HP}, the change in solvent-accessible hydrophobic area in the folded protein on mutation of Gly→Ala (Table 2), gives a good correlation (Fig. 1):

$$\Delta\Delta G_{unf} = 1.85 - 0.055\Delta A_{HP} \quad \text{kcal mol}^{-1} \qquad (1)$$

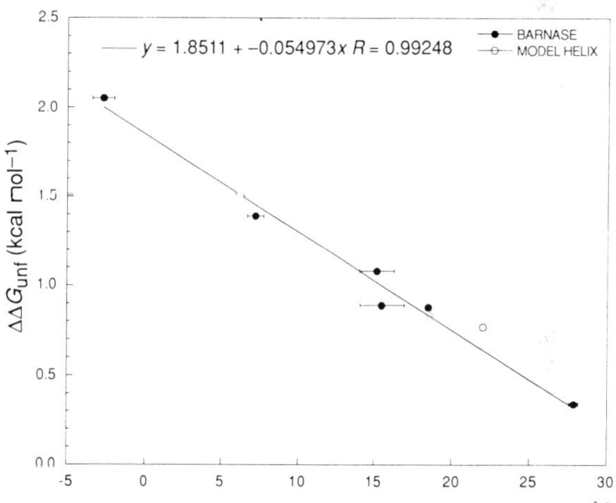

FIG. 1 Correlation between total changes in solvent-accessible hydrophobic surface area on mutation of Ala to Gly in the folded protein and the effect on protein stability. ○, Value for mutation of Gly to Ala found by O'Neil and DeGrado[8] using surface-accessible areas calculated by modelling their peptide using Quanta and the Polygen suite of programs. The mutation of Gly→Ala is calculated to increase the solvent-accessible hydrophobic surface area by 22 Å^2, which is close to that calculated for a model polyalanine helix (21 Å^2). The error bars represent the spread from averaging of data for the three molecules of barnase in the unit cell. The linear correlation holds only for the folded state. A simple thermodynamic analysis suggests that the correlation should be between $\Delta\Delta G_{unf}$ and the change in surface area of the folded state on mutation minus the change in that for the unfolded state. Presumably, the energy difference between Ala and Gly in the unfolded state of the protein is roughly independent of sequence and so is a constant term in the equation.

LETTERS TO NATURE

The details of the calculation are given in Table 2; ΔA_{HP} = (total solvent-accessible hydrophobic surface area of mutant containing Ala − total solvent-accessible hydrophobic surface area of mutant containing Gly)$_{\text{folded protein}}$. The data for the N and C caps are excluded from this plot, as are also those for other positions where mutation of Gly→Ala buries exposed hydrophilic surfaces th.˙ require intermolecular hydrogen bonding (exposed backbone NH group at the N termini and the CO at the C termini plus surrounding side chains and main chain atoms; Table 2). (There is also a problem at the C cap of position 34 because Gly 34 in the native protein has ϕ- and ψ-angles that are energetically unfavourable for Ala and larger amino acids.) Note that the value of 55 cal mol^{-1} Å^{-2} relating changes in energy to changes in surface area is some 2 times higher than found[27] from transfer free energies. This value is, however, consistent with extensive experimental data from other hydrophobic mutations[28,29] and similar to new theoretical estimates[30].

We have also searched for correlations with changes in solvent-accessible hydrophilic area on mutation, although there is not precedence for this. Multiple linear regression for the variation of $\Delta\Delta G_{unf}$ with changes in both areas shows no significant dependence on the change in solvent-accessible hydrophilic surface area for the changes in internal positions in the helix. There is a correlation, however, if the only hydrophilic surfaces that are considered are those that have hydrogen-bond donors or acceptors that should make hydrogen bonds with solvent. Multiple linear regression for the variation of $\Delta\Delta G_{unf}$ with changes in solvent-accessible hydrophobic surface area and with decrease in solvent accessible area of such groups (ΔA_{HB}; Table 2) gives the correlation:

$$\Delta\Delta G_{unf} = 1.68 - 0.041\Delta A_{HP} - 0.19\Delta A_{HB} \quad \text{kcal mol}^{-1} \qquad (2)$$

(where the standard errors for the three terms are ±0.24, ±0.013 and ±0.03, respectively). The simpler equation (1) fits the internal residues slightly better, but there is a fair fit of observed values and those calculated by equation (2) (Fig. 2).

The absolute values for the helix-forming propensities of the different amino acids result from a variety of factors that change between the folded and unfolded states. For example, the backbone around glycine has a greater rotational freedom than for any other residue in the unfolded state, which favours unfolding[31]. It is clear, however, that the relative helix-forming propensities of Ala and Gly are strongly context-dependent, and one factor is changes in solvent exposure. When the Ala that is

TABLE 1 Stability of barnase and mutants with Ala and Gly residues at different positions in the major α-helices

Position and residue in wild type	Change in energy on substitution of Ala→Gly ($\Delta\Delta G_{unf}$) in kcal mol^{-1}
N cap (Thr 6)	−0.93
N cap (Thr 26)	−0.50
N+1 (Lys 27)	0.76
N+2 (Asp 8)	0.34
N+2 (Ser 28)	0.86
N+3 (Glu 29)	0.56
M (Asp 12)	0.88
C−3 (Gln 31)	1.08
C−2 (Thr 16)	1.39
C−2 (Ala 32*)	0.91
C−1 (Tyr 17)	2.05
C cap (His 18)	−1.15
C cap (Gly 34)	−3.12

The free energies of unfolding of wild-type protein and mutants were analysed by urea denaturation monitored by fluorescence[11,17]. The data for a single mutant are reliable to ±0.02 kcal mol^{-1}. The change in stability between mutant proteins with Ala or Gly at a certain position is obtained by subtracting the difference in stability between wild-type and an Ala mutant, from the difference in stability of wild-type and a Gly mutant. (−, Gly more stable than Ala.) Mutants and Proteins were prepared as previously described[11,17].

* Data from A. Horovitz and J. Matthews (unpublished results).

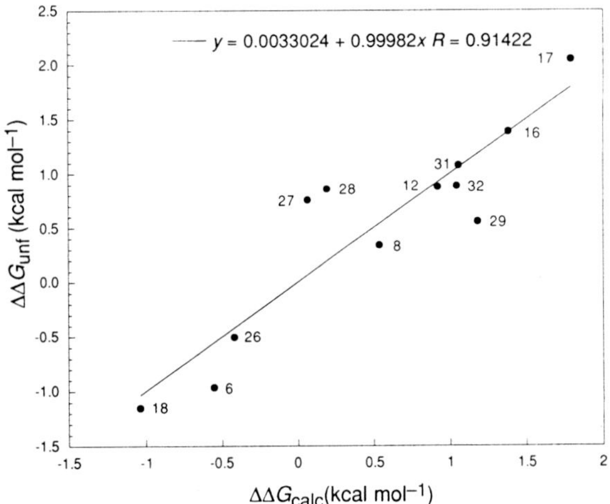

FIG. 2 Comparison of observed changes in free energy of unfolding on mutation of Ala to Gly and those calculated from equation (2) ($\Delta\Delta G_{unf} = 168 - 0.041\Delta A_{HP} - 0.19\Delta A_{HB}$) for the different positions in the helices.

TABLE 2 Changes in solvent-accessible area in folded protein on mutation of Ala→Gly at different positions in the two helices of barnase

	Increase in total solvent-accessible area		Decrease in solvent-accessible area of hydrophilic groups that require solvation†
Position	Hydrophobic* (ΔA_{HP}) (Å^2)	Hydrophilic* (Å^2)	(ΔA_{HB}) (Å^2)
6 (N cap)	22.4	−12.4	6.8
8	27.8	−13.5	0
12	18.5	−9.8	0
16	7.2	−4.1	0
17	−2.7	−3.7	0
18 (C cap)	15.8	−14.4	10.7
26 (N cap)	3.1	−10.0	10.2
27	7.3	−11.0	6.8
28	27.2	−8.0	1.9
29	3.7	−4.2	1.8
31	15.2	−6.6	0
32	15.5	−6.4	0

* The figures tabulated are the differences in solvent-exposed hydrophobic surface area in the folded protein having Ala instead of Gly. These were calculated using the program of Shrake and Rupley[33], as implemented by Miller *et al.*[34]. The Ala or Gly mutants were constructed by deleting the surplus side-chain atoms from the coordinate file of barnase. Similar results were obtained using the program Quanta (Polygen) with which the target side chains were mutated to Ala or Gly using the protein design package of the problem. The exposed hydrophobic or hydrophilic area of the whole protein was calculated from the surface generated by the locus of the centre of a sphere of radius 1.4 Å rolling over the exposed van der Waals surface. The calculated solvent-accessible hydrophobic or hydrophilic area for the mutant containing Gly was subtracted from that calculated for the mutant containing Ala to give the value ΔA_{Hp} or the hydrophilic equivalent.

† The changes in solvent-accessible area of non-intramolecularly hydrogen-bonded groups and hydrophilic groups of the protein, on mutation of Ala to Gly, were calculated for the folded protein using the program of Miller *et al.*[34]. These are mainly for changes in the exposure of the backbone NH groups at the N termini of the helices, and the CO at the C termini, with some contributions from neighbouring side chains and backbone.

mutated is screened from solvent so that there is little change in solvent exposure on mutation to Gly, the helix is stabilized by some 2 kcal mol^{-1} by Ala. At the other extreme, where the Ala is highly exposed to solvent, it stabilizes the helix by only some 0.4 kcal mol^{-1}, relative to Gly. We have applied equation (1) to a model polyalanine helix ($\phi = -65°$ and $\psi = -41°$). The predicted change in free energy of unfolding on mutation of Ala to Gly in an internal position of the α-helix is 0.7 kcal mol^{-1}. This is close to the experimental value found by O'Neil and DeGrado[8] (0.77 kcal mol^{-1}) for an internal position in a helical peptide that has a change in solvent-accessible hydrophobic surface on mutation of Ala$\rightarrow$Gly similar to that in a model polyalanine helix (Fig. 1) and at the extreme of the experiments by Kemp *et al.*[11] (0.17 to 0.77 kcal mol^{-1}). The peptide used by O'Neil and DeGrado[8] is unlikely to fray, whereas that of Kemp *et al.*[11] is complicated by end-fraying. The positional dependence for the relative helix-forming propensities of Ala versus Gly may account for the high ratio of s values estimated[10] from the Lifson–Roig theory, which gives a value of 2.8 kcal mol^{-1} for the relative stability of Ala versus Gly in a polyalanine model helix instead of 0.7 kcal mol^{-1}. It was assumed in the calculations that s is independent of position in the model substituted-polyalanine helix and, implicitly, that the variation in apparent propensity is caused solely by fraying at the ends[10].

Our empirical correlations show the importance of hydrophobic effects in protein folding and how they can influence helix-forming and, presumably, other structure-forming propensities of side chains, as suggested by Richards and Richmond[32], and also the importance of solvent accessibility of hydrogen-bonding groups[22]. It is not valid to assign to each amino acid a unique s value that is generally applicable to all positions in all helices in proteins. □

Received 18 October 1991; accepted 8 January 1992.

1. Sueki *et al.* *Macromolecules* **17,** 148-155 (1984).
2. Scheraga, H. A. *Proc. natn. Acad. Sci. U.S.A.* **82,** 5585-5587 (1985).
3. Altmann, K. H., Wojcik, J., Vasquez, M. & Scheraga, H. A. *Biopolymers* **30,** 107-120 (1990).
4. Padmanabhan, S., Marqusee, S., Ridgeway, T., Laue, T. M. & Baldwin, R. L. *Nature* **344,** 268-270 (1990).
5. Merutka, G., Lipton, W., Shalongo, W., Park, S. H. & Stellwagen, E. *Biochemistry* **29,** 7511-7515 (1990).
6. Lyu, P. C., Lip, M. I., Marky, L. A. & Kallenbach, N. R. *Science* **250,** 669-673 (1990).
7. Strehlow, K. G. & Baldwin, R. L. *Biochemistry* **28,** 2130-2133 (1989).
8. O'Neil, K. T. & DeGrado, W. F. *Science* **250,** 646-651 (1990).
9. Marqusee, S., Robbins, V. H. & Baldwin, R. L. *Proc. natn. Acad. Sci. U.S.A.* **86,** 5286-5290 (1989).
10. Chakrabartty, A., Schellman, J. A. & Baldwin, R. L. *Nature* **351,** 586-588 (1991).
11. Kemp, D. S., Boyd, J. G. & Muendel, C. C. *Nature* **352,** 451-454.
12. Lesk, A. M. *Nature* **352,** 379-380.
13. Altman, K. H., Wojcik, J., Vasques, M. & Scheraga, H. A. *Biopolymers* **30,** 107-120 (1990).
14. Lifson, S. & Roig, A. J. *J. chem. Phys.* **34,** 1963-1974 (1961).
15. Serrano, L. & Fersht, A.R. *Nature* **342,** 296-299 (1989).
16. Sali, D., Bycroft, M. & Fersht, A. R. *Nature* **335,** 563-567 (1988).
17. Kellis, J. T. Jr, Nyberg, K., Sali, D. & Fersht, A. R. *Nature* **333,** 784-786 (1988).
18. Kellis, J. T. Jr, Nyberg, K. & Fersht, A. R. *Biochemistry* **28,** 4914-4922 (1989).
19. Horovitz, A., Serrano, L., Avron, B., Bycroft, M. & Fersht, A. R. *J. molec. Biol.* **216,** 1031-1044 (1990).
20. Horovitz, A., Serrano, L. & Fersht, A. R. *J. molec. Biol.* **219,** 5-9 (1991).
21. Serrano, L., Bycroft, M. & Fersht, A. R. *J. molec. Biol.* **218,** 465-475 (1991).
22. Presta, L. G. & Rose, G. D. *Science* **240,** 2632-1641 (1988).
23. Richardson, J. S. & Richardson, D. C. *Science* **240,** 1648-1652 (1988).
24. Hermann, R. B. *J. phys. Chem.* **76,** 2754-2759 (1972).
25. Harris, M. J., Higuchi, T. & Rytting, J. *J. phys. Chem.* **77,** 2694-2702 (1972).
26. Reynolds, J. A., Gilbert, D. B. & Tanford, C. *Proc. natn. Acad. Sci.* U.S.A. **71,** 2925-2927 (1974).
27. Chothia, C. *Nature* **248,** 338-339 (1974).
28. Kellis, J. T. Jr, Nyberg, K. & Fersht, A. R. *Biochemistry* **28,** 4914-4922 (1989).
29. Serrano, L., Kellis, J. T., Cann, P., Matouschek, A. & Fersht, A. R. *J. molec. Biol.* (in the press).
30. Sharp, K. A., Nicholls, A., Friedman, R. & Honig, B. *Biochemistry* **30,** 9686-9697 (1991).
31. Matthews, B. W., Nicholson, H. & Becktel, W. J. *Proc. natn. Acad. Sci. U.S.A.* **84,** 6663-6667 (1987).
32. Richards, F. M. & Richmond, T. *Ciba Fdn Symp.* **60,** 23-37 (1978).
33. Shrake, A. & Rupley, J. A. *J. molec. Biol.* **79,** 351-371 (1973).
34. Miller, S., Janin, J., Lesk, A. M. & Chothia, C. *J. molec. Biol.* **196,** 641-656 (1987).

ACKNOWLEDGEMENTS. We thank C. Chothia for discussions and K. Henrick and A. Cameron for coordinates of barnase. L.S. holds a European Community fellowship.

Double-mutant cycles were brought back with a vengeance[86,92–99]. They are precision tools for measuring the interaction energies between two groups that can be mutated separately and together. We measured the interactions between surface electrostatic charges[86,93,100], and aromatic-aromatic[94] and histidine-aromatic side-chain interactions[95]. Later on, we probed buried salt bridges[97,101]. The keenest exponent of double-mutant cycles was Amnon Horovitz who had been using a similar formalism to analyse co-operativity in allosteric changes during his Ph.D. in Jerusalem. Amnon has subsequently extended their use. Amnon was the first of my Israeli post-doctorals, all of whom have gone back to become professors, Amnon, Gideon Schreiber and Dan Tawfik at the Weizmann Institute and Assaf Friedler at the Hebrew University. Boaz Avron became the top administrator in the Weizmann. One Israeli turned down an offer of a place in my lab just as Gideon was leaving because he thought that no lab could keep up a 100% success rate in obtaining positions in Israel! Ronny Loewenthal, an MD who became my research student, has a senior medical position in Israel. Amnon's first paper sets out formally our strategy[92]. Gideon is also an ace exponent of double-mutant cycles, so the Weizmann Institute is a real centre of excellence in cycles.

11.10 2006: Honorary Degree at Hebrew University with Tom Friedman and Aaron Ciechanover.

J. Mol. Biol. (1990) 214, 613–617

COMMUNICATIONS

Strategy for Analysing the Co-operativity of Intramolecular Interactions in Peptides and Proteins

Amnon Horovitz and Alan R. Fersht

MRC Unit for Protein Function and Design
Department of Chemistry, University of Cambridge
Lensfield Road, Cambridge CB2 1EW, U.K.

(Received 5 March 1990; accepted 12 April 1990)

Double mutant cycles enable the measurement of pairwise interactions in proteins. This method is extended for mutations at any number of positions in the protein. This provides a way for determining the context dependence of pairwise interactions on other neighbouring residues.

In order to study experimentally the interaction of a particular amino acid residue in a protein with other residues in that protein, it has become common practice to mutate the residue of interest in a manner considered to be non-disruptive (Fersht *et al.*, 1987). Because proteins are highly co-operative structures, mutation of one residue may lead to disruption of more than one interaction. Further, in many cases, interactions of several amino acid residues are coupled to one another and may not be reduced simply to a sum of pairwise interactions. Results of single mutation experiments are, for these reasons, easily prone to misinterpretation. This problem may be at least partially overcome by invoking double mutant cycles (see below; Ackers & Smith, 1985), first applied to protein engineering by Carter *et al.* (1984) and further analysed by Horovitz (1987). It is possible to determine from such cycles whether an interaction between two residues exists and in some cases even to quantify it. Examples for the application of this approach include a demonstration that the amino acid residues at positions 175 and 211 in the α subunit of tryptophan synthase interact with each other (Hurle *et al.*, 1986) and, more recently, that a long-range electrostatic interaction exists between residues at positions 28 and 139 of dihydrofolate reductase (Perry *et al.*, 1989). It is not possible to tell using this approach whether the observed interaction is dependent on other neighbouring residues. A general method for analysing the interaction between any number of residues ($N \geq 2$) is given here and some of its main features are described.

An interaction between two residues may be detected by constructing a double mutant cycle as follows:

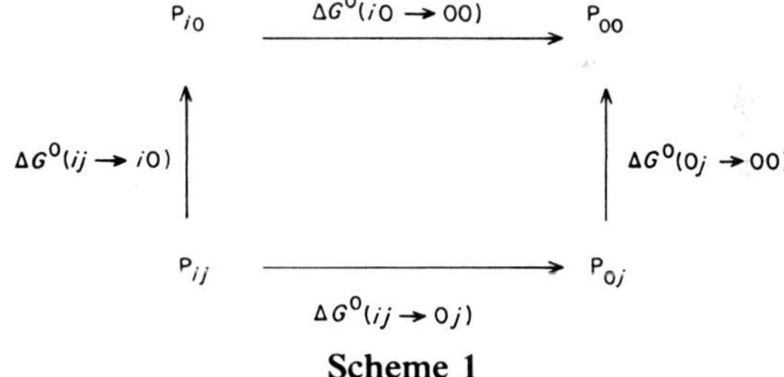

Scheme 1

where P stands for protein, i and j for particular amino acid residues in that protein and 0 for their replacement by another given amino acid. $\Delta G^0(ij \rightarrow 0j)$ and $\Delta G^0(i0 \rightarrow 00)$ stand for the free energy changes upon mutation of residue i when leaving residue j unchanged and when residue j has already been mutated, respectively. $\Delta G^0(ij \rightarrow i0)$ and $\Delta G^0(0j \rightarrow 00)$ are similarly defined for mutation of residue j. If the change in free energy due to mutation of residue i is independent of residue j, then $\Delta G^0(ij \rightarrow 0j) = \Delta G^0(i0 \rightarrow 00)$. It follows from the symmetry of the cycle that also $\Delta G^0(ij \rightarrow i0) = \Delta G^0(0j \rightarrow 00)$, i.e. the effect of mutation of residue j is independent of residue i. When the effects of the two mutations are not independent of each other, then $\Delta G^0(ij \rightarrow 0j) \neq \Delta G^0(i0 \rightarrow 00)$ and $\Delta G^0(ij \rightarrow i0) \neq \Delta G^0(0j \rightarrow 00)$. This indicates that the two residues are coupled. The free energy of coupling between residues i and j, $\Delta G_I(ij)$, is given by:

$$\Delta G_I(ij) = \Delta G^0(ij \rightarrow 0j) - \Delta G^0(i0 \rightarrow 00)$$
$$= \Delta G^0(ij \rightarrow i0) - \Delta G^0(0j \rightarrow 00). \quad (1)$$

The term coupling is used here, since the interaction between the two residues may not be direct, but rather propagated through the molecule indirectly. The free energy changes in the above cycle can, in some instances, be accurately determined by measuring the unfolding free energies of the proteins that comprise it. Scheme 1 is now expanded to include the unfolding reactions of the proteins in it as follows:

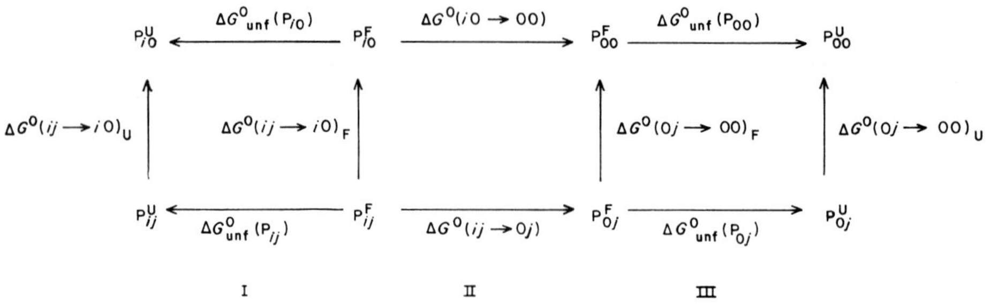

Scheme 2

where U and F stand for the unfolded and folded states of the proteins, respectively; $\Delta G^0_{unf}(P_{ij})$, $\Delta G^0_{unf}(P_{0j})$, $\Delta G^0_{unf}(P_{i0})$ and $\Delta G^0_{unf}(P_{00})$ for the unfolding free energies of the respective proteins. From the principle of free energy conservation, it follows that for cycles I and III in Scheme 2 one has:

$$\Delta G^0(ij \rightarrow i0)_F = \Delta G^0(ij \rightarrow i0)_U + \Delta G^0_{unf}(P_{ij}) - \Delta G^0_{unf}(P_{i0}). \quad (2)$$

$$\Delta G^0(0j \rightarrow 00)_F = \Delta G^0(0j \rightarrow 00)_U + \Delta G^0_{unf}(P_{0j}) - \Delta G^0_{unf}(P_{00}). \quad (3)$$

Combining equations (1), (2) and (3) one obtains:

$$\Delta G_I(ij)_F = \Delta G^0(ij \rightarrow 0j)_F - \Delta G^0(i0 \rightarrow 00)_F$$
$$= (\Delta G^0_{unf}(P_{ij}) - \Delta G^0_{unf}(P_{i0}))$$
$$- (\Delta G^0_{unf}(P_{0j}) - \Delta G^0_{unf}(P_{00}))$$
$$+ (\Delta G^0(ij \rightarrow i0)_U - \Delta G^0(0j \rightarrow 00)_U). \quad (4)$$

The coupling free energy between residues i and j in the unfolded state of the proteins, $\Delta G_I(ij)_U$, is given by: $\Delta G_I(ij)_U = \Delta G^0(ij \rightarrow i0)_U - \Delta G^0(0j \rightarrow 00)_U$. If these residues are far apart in the unfolded state and do not interact, then $\Delta G^0(ij \rightarrow i0)_U = \Delta G^0(0j \rightarrow 00)_U$. The same would be expected for a completely random fluctuating coil if the residues are not far apart. In both cases, equation (4) is then reduced to:

$$\Delta G_I(ij)_F = (\Delta G^0_{unf}(P_{ij}) - \Delta G^0_{unf}(P_{i0}))$$
$$- (\Delta G^0_{unf}(P_{0j}) - \Delta G^0_{unf}(P_{00})). \quad (5)$$

If residues i and j do interact in the unfolded state, then the differences in unfolding free energies are a measure for the difference in the interaction energy between the folded and unfolded states as given by the following equation:

$$\Delta G_I(ij)_F - \Delta G_I(ij)_U = (\Delta G^0_{unf}(P_{ij}) - \Delta G^0_{unf}(P_{i0}))$$
$$- (\Delta G^0_{unf}(P_{0j}) - \Delta G^0_{unf}(P_{00})). \quad (6)$$

The above cycle provides no information as to the possible dependence of an interaction on a third residue, k. In order to determine if the interaction between residues i and j is dependent on a third residue k, it is necessary to construct a triple mutant box as follows:

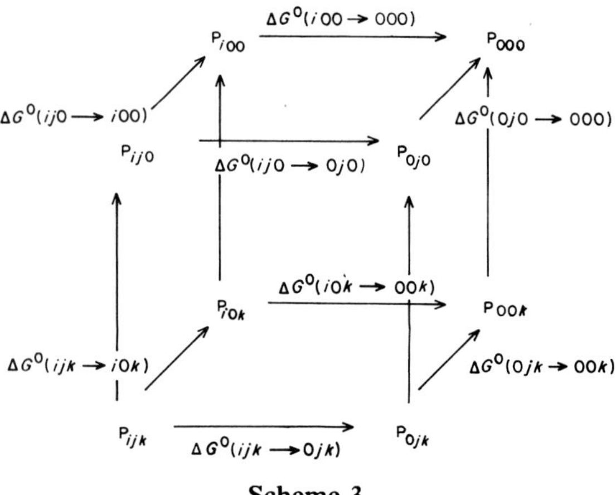

Scheme 3

where k stands for a third amino acid residue, and all other notations are as before. The bottom cycle of the above cube, formed by the proteins P_{ijk}, P_{i0k}, P_{0jk} and P_{00k}, corresponds to the cycle shown by Scheme 1. The difference between them is that in the bottom cycle of the cube it is explicit that residue k has not been mutated, and not implicit as in Scheme 1. The coupling energy between residues i

and j when residue k has not been replaced, $\Delta G_I(ij, k)$, is given by:

$$\Delta G_I(ij, k) = \Delta G^0(ijk \rightarrow i0k) - \Delta G^0(0jk \rightarrow 00k)$$
$$= \Delta G^0(ijk \rightarrow 0jk) - \Delta G^0(i0k \rightarrow 00k). \quad (7)$$

The coupling energy that may be measured by the top cycle of the cube shown in Scheme 3, formed by the proteins P_{ij0}, P_{i00}, P_{0j0} and P_{000}, is for the interaction between residues i and j when residue k has been replaced. This coupling energy, $\Delta G_I(ij, 0)$, is given by:

$$\Delta G_I(ij, 0) = \Delta G^0(ij0 \rightarrow i00) - \Delta G^0(0j0 \rightarrow 000)$$
$$= \Delta G^0(ij0 \rightarrow 0j0) - \Delta G^0(i00 \rightarrow 000). \quad (8)$$

The difference between the coupling energies measured by the top and bottom cycles of the cube, $\Delta\Delta G_I(ijk)$, is a measure of the effect of residue k on the interaction between residues i and j. This difference is:

$$\Delta\Delta G_I(ijk) = \Delta G_I(ij, k) - \Delta G_I(ij, 0)$$
$$= (\Delta G^0(ijk \rightarrow i0k) - \Delta G^0(0jk \rightarrow 00k))$$
$$- (\Delta G^0(ij0 \rightarrow i00) - \Delta G^0(0j0 \rightarrow 000)). \quad (9)$$

Because of the symmetry of the cube, equivalent expressions are obtained for the effect of residue i on the interaction between residues j and k, and for the effect of residue j on the interaction between residues i and k. It is important to note that the information obtained from a cube is qualitatively different from that obtained from a cycle, since $\Delta\Delta G_I(ijk)$ is not a more accurate measure of the coupling energy between two residues but, as stated above, it is just a measure of the effect of a third residue (e.g. k) on the interaction between two other residues (e.g. i and j).

When two residues (i and j) are targeted and mutated, the construct required for detecting an interaction between those residues, i.e. the double mutant cycle, has a dimension of two. The coupling energy, $\Delta G_I(ij)$, is the difference (eqn (1)) between free energies corresponding to parallel one-dimensional elements of the two-dimensional construct. When three residues (i, j and k) are replaced, the construct required for detecting an effect of one residue on the interaction between the other two, i.e. the cube, is three-dimensional. The measure of that effect, $\Delta\Delta G_I(ijk)$, is the difference (eqn (9)) between coupling energies corresponding to parallel two-dimensional elements of the three-dimensional construct. The information gained from the cycle and the cube, although qualitatively different, is derived in both cases by comparing the free energy terms corresponding to parallel elements of a dimension one less than that of the construct itself. We now claim that it is general that the information to be gained from any N-dimensional mutant construct will come from comparing the free energy terms corresponding to "parallel" elements of a dimension one less than that of the construct itself. Since what is meant by parallel elements is not clear for dimensions higher than three, a more rigorous definition is required. The number of mutants in any N-dimensional construct is 2^N where N is the dimension. On going from an N-dimensional mutant construct to an $(N+1)$-dimensional mutant construct, the number of mutants is doubled (Eigen *et al.*, 1988). The two free energy terms to be compared in any $(N+1)$-dimensional mutant construct correspond respectively to the 2^N mutants present also in the N-dimensional construct and to the remaining 2^N mutants present in the $(N+1)$-dimensional construct but not in the N-dimensional construct.

The above general principle is now applied to the less intuitive case of a four-dimensional mutant construct. A scheme of such a construct is given as follows:

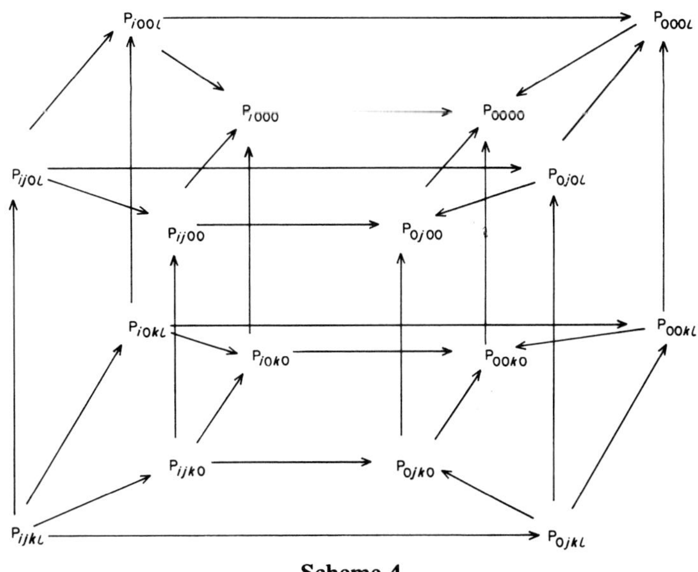

Scheme 4

where l stands for a fourth amino acid residue and all other notations are as before. It may be seen that this construct consists of two cubes, one of which (the outer one), corresponds to the previous three-dimensional construct (Scheme 3) and the other (the inner one), contains the remaining "new" mutants. The information to be gained from this four-dimensional construct, $\Delta\Delta\Delta G_I(ijkl)$, should come from comparing the two free energy terms corresponding to these two three-dimensional constructs. The first free energy term, which is for the previous three-dimensional construct (see Scheme 3 and eqn (9)) is:

$$\Delta\Delta G_I(ijk, l) = \Delta G_I(ij, kl) - \Delta G_I(ij, 0l). \quad (10)$$

In equation (10) it is explicitly stated that residue l has not been replaced. This equation is otherwise equivalent to equation (9). The second free energy term, which is for the second cube in Scheme 4, is:

$$\Delta\Delta G_I(ijk, 0) = \Delta G_I(ij, k0) - \Delta G_I(ij, 00). \quad (11)$$

In equation (11) it is explicit that residue l has been mutated. The information to be gained from a four-dimensional construct, $\Delta\Delta\Delta G_I(ijkl)$, is given by:

$$\begin{aligned}\Delta\Delta\Delta G_I(ijkl) &= \Delta\Delta G_I(ijk, l) - \Delta\Delta G_I(ijk, 0) \\ &= (\Delta G_I(ij, kl) - \Delta G_I(ij, 0l)) \\ &\quad - (\Delta G_I(ij, k0) - \Delta G_I(ij, 00)). \quad (12)\end{aligned}$$

So far, no physical meaning has been attributed to $\Delta\Delta\Delta G_I(ijkl)$. Equation (12) has a formal resemblance to equation (5) for a double mutant cycle, which suggests that $\Delta\Delta\Delta G_I(ijkl)$, which is for a four-dimensional structure, also corresponds to some kind of double mutant cycle. This cycle is given as follows:

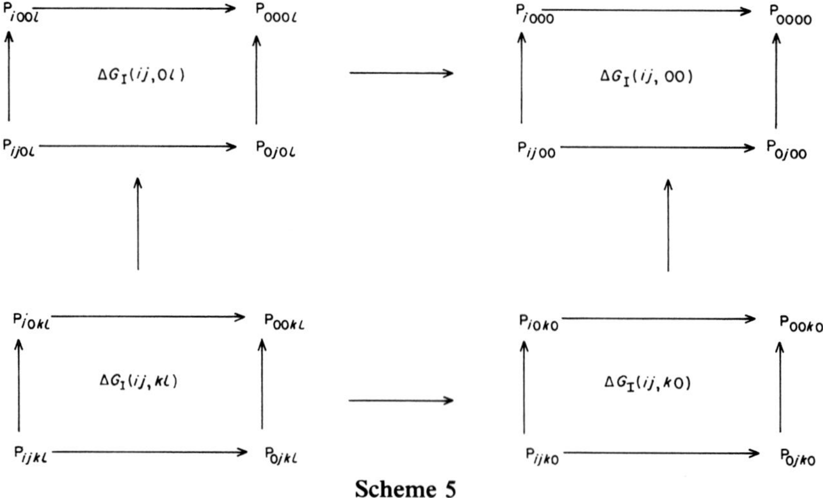

Scheme 5

Scheme 5 shows a cycle that consists of four sub-cycles. Each of the four sub-cycles is a double mutant cycle for residues i and j. The overall cycle shown in Scheme 5 is a double mutant cycle for residues k and l. $\Delta\Delta\Delta G_I(ijkl)$ is, therefore, a measure of the effect of the interaction between residues k and l on the interaction between residues i and j. In other words, it is a measure of the coupling between two interactions. Owing to the symmetry of the four-dimensional construct, one may measure from it the effect of the interaction between any two of the four residues i, j, k and l on the interaction between the other two residues.

By applying the principles so far used, this treatment may be easily extended for dimensions higher than four. The general rule is that any N-dimensional mutant structure may be represented by a hierarchy of structures of dimensions that add up to N. Thus, a four-dimensional mutant construct was represented by Scheme 5 as a two-dimensional cycle that consisted of sub-cycles also of dimension two. The analysis of a five-dimensional mutant construct will be based on the difference between two free energy terms corresponding to two four-dimensional mutant constructs. This difference in free energy terms corresponds to a cube with a cycle at each corner, or to a cycle with a cube at each corner. The dimensions of the cube and its component cycles or, alternatively, the dimensions of the cycle and its component cubes, add up to five, which is the dimension of the structure being represented. From this construct, it is possible to measure the effect of a fifth residue on the coupling between the interactions of two pairs of residues.

We conclude by giving some concrete examples for possible applications. Very common structural motifs in peptides and proteins are the γ and β-turns (Rose *et al.*, 1985). These consist of three and four residues, respectively, and are usually stabilized by intraturn hydrogen bonds or salt bridges between residues i and $i+2$ in γ-turns and residues i and $i+3$ in β-turns. In the case of γ-turns, for example, by measuring the strength of these interactions using double mutant cycles for positions i and $i+2$, the effect of the residue at position $i+1$ on this interaction would not be revealed. In this case a triple mutant box needs to be constructed. A typical γ-turn might consist, for example, of lysine, proline and aspartic acid residues at positions i, $i+1$ and $i+2$ of the turn, with a salt bridge between the lysine and aspartic acid residues. By constructing a double mutant cycle for the residues in the ion-pair only, the contribution of the proline residue to the strength of the salt bridge would not be determined. A triple mutant box would allow us to compare this interaction both in the presence of the proline residue and when it is mutated to glycine, for example, and thus to determine the contribution of the conformational constraint due to the proline residue to the overall stability of the γ-turn. In the case of β-turns, in order to determine the importance of the residues at positions $i+1$ and $i+2$ on the interaction between residues i and $i+3$, it would be necessary to make a four-dimensional mutant construct. In α-helices one often finds salt bridges between residues i and $i+3$ or between residues i and $i+4$ (Marqusee & Baldwin, 1987). Here again, in order to ascertain the importance of the residues that do not actually take part in forming the salt bridge, it would be necessary to make either four or five-dimensional mutant constructs.

The approach described here is particularly suitable for the analysis of synthetic peptides, since a large number of variants may be generated with relative ease. The thermal dependence of the helix coil transition of two sets of polyalanine peptides with glutamic acid and lysine residues positioned at alternating $i+4$ and $i+1$ spacings was recently analysed (Merutka & Stellwagen, 1990). The two sets correspond to triple mutant boxes given as follows:

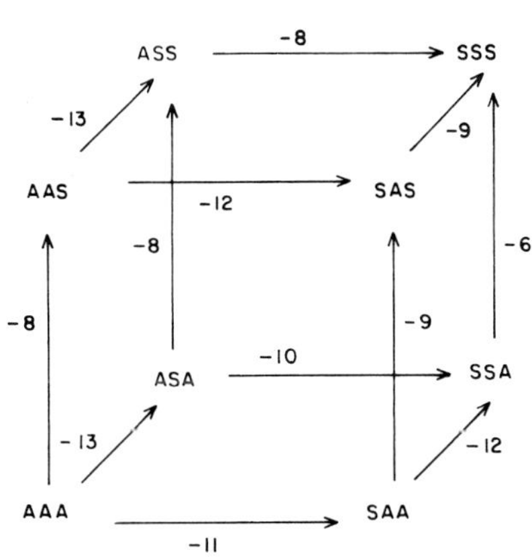

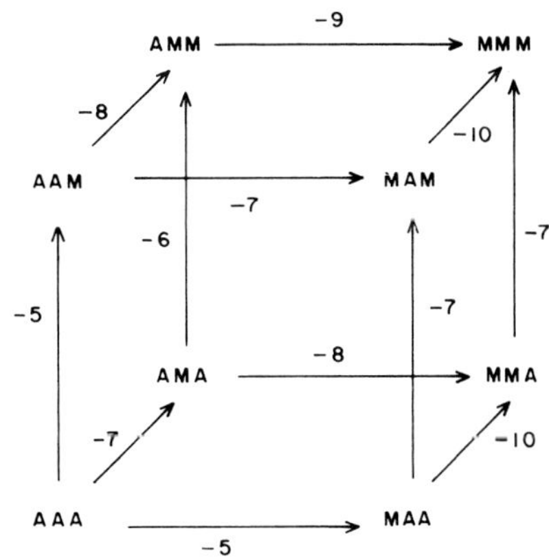

Scheme 6

where S and M stand for serine and methionine residues that were introduced at three $i+2$ positions in the peptide sequence, the numbers are the differences in melting temperatures of the peptides (Δt_m, (°C)). Small changes in t_m are directly proportional to changes in free energy if the entropy change is constant (Becktel & Schellman, 1987). The effects of the amino acid replacements on the helix coil transition were reported to be additive and position independent. However, by applying the approach described here it may be shown, for example, that although the contributions to helix stability of serine residues introduced at only two positions in the peptide are indeed additive, they become non-additive in the presence of a serine at a third position. This effect is not seen when methionine residues, considered to be helix stabilizing, are introduced at the same positions in the peptide. In sequence-based secondary structure prediction methods, the helical tendency of an amino acid, for example, is determined by its frequency in α-helices relative to its overall frequency in proteins. These methods do not take into account the effects of neighbouring amino acid residues on the helical tendency of an amino acid mainly because of the need for much larger data bases. By applying the approach described here, it should be possible to obtain direct experimental evidence for the relative importance of neighbouring amino acid residues in determining the propensities of amino acid residues to form different secondary structures.

References

Ackers, G. K. & Smith, F. R. (1985). *Annu. Rev. Biochem.* **54**, 597–629.

Becktel, W. J. & Schellman, J. A. (1987). *Biopolymers*, **26**, 1859–1877.

Carter, P. J., Winter, G., Wilkinson, A. J. & Fersht, A. R. (1984). *Cell*, **38**, 835–840.

Eigen, M., Oswatitsch-Winkler, R. & Dress, A. (1988). *Proc. Nat. Acad. Sci., U.S.A.* **85**, 5913–5917.

Fersht, A. R., Leatherbarrow, R. J. & Wells, T. N. C. (1987). *Biochemistry*, **26**, 6030–6038.

Horovitz, A. (1987). *J. Mol. Biol.* **196**, 733–735.

Hurle, M. R., Tweedy, N. B. & Matthews, C. R. (1986). *Biochemistry*, **25**, 6356–6360.

Marqusee, S. & Baldwin, R. L. (1987). *Proc. Nat. Acad. Sci., U.S.A.* **84**, 8898–8902.

Merutka, G. & Stellwagen, E. (1990). *Biochemistry*, **29**, 894–898.

Perry, K. M., Onuffer, J. J., Gittelman, M. S., Barmat, L. & Matthews, C. R. (1989). *Biochemistry*, **28**, 7961–7968.

Rose, G. D., Gierasch, L. M. & Smith, J. A. (1985). *Advan. Protein Chem.* **37**, 1–97.

Edited by S. Brenner

One result that excited me as an enzymologist was that Liz Meiering found an inverse relationship between stability and activity on mutation of residues in the active site of barnase, showing that binding sites and catalytic residues are a source of strain[102].

11.11 1989: Liz Meiering.

J. Mol. Biol. (1992) **225**, 585–589

Effect of Active Site Residues in Barnase on Activity and Stability

Elizabeth M. Meiering, Luis Serrano and Alan R. Fersht

MRC Unit for Protein Function and Design
Cambridge IRC for Protein Engineering
University Chemical Laboratory
Lensfield Road, Cambridge CB2 1EW, U.K.

(Received 4 October 1991; accepted 30 January 1992)

We have mutated residues in the active site of the ribonuclease, barnase, in order to determine their effects on both enzyme activity and protein stability. Mutation of several of the positively charged residues that interact with the negatively charged RNA substrate (Lys27→Ala, Arg59→Ala and His102→Ala) causes large decreases in activity. This is accompanied, however, by an increase in stability. There is presumably electrostatic strain in the active site where positively charged side-chains are clustered. Mutation of several residues that make hydrogen bonds (Ser57→Ala, Asn58→Asp and Tyr103→Phe) causes smaller decreases in activity, but increases or has no effect on stability. Deletion of hydrogen bonding groups elsewhere in proteins has been found previously to decrease stability by 0·5 to 1·5 kcal mol^{-1}. Conversely, we find that two mutations (Asp54→Asn and Gln104→Ala) decrease stability and increase activity. Another mutation (Glu73→Ala) decreases both activity and stability. It is clear that many residues in the active site do not contribute to stability and that for some, but not all, of the residues there is a compromise between activity and stability. This suggests that certain types of local instability may be necessary for substrate binding and catalysis by barnase. This has implications for the understanding of enzyme activity and the design of enzymes.

Keywords: mutagenesis; protein stability; activity; enzyme; protein folding

How do residues that are involved in catalysis and substrate binding affect protein stability? Although detailed studies of the contributions of individual residues to activity (Fersht, 1987; Knowles, 1991) or to stability (Shortle, 1989; Alber, 1989; Matthews, 1991) have been carried out for various proteins, the relationship between function and stability remains poorly understood. Barnase, a small ribonuclease from *Bacillus amyloliquefaciens*, has been used to study protein stability (Kellis *et al.*, 1988) and activity (Mossakowska *et al.*, 1989; Day *et al.*, 1992). In this paper, we study the effects of individual residues in the active site of barnase on activity and stability.

Barnase is a member of a family of guanine-preferential microbial endoribonucleases (Hill *et al.*, 1983). Both the crystal structure and the solution structure of barnase have been determined (Maugen *et al.*, 1982; C. Hill, personal communication; Bycroft *et al.*, 1991). The crystal structure of the homologous enzyme, binase, from *Bacillus intermedius*, complexed with 3′-guanosine monophosphate has also been solved (Sevcik *et al.*, 1990; Pavlovsky *et al.*, 1988). Barnase consists of a single polypeptide chain in which two N-terminal α-helices pack onto a C-terminal five-stranded antiparallel β-sheet. The active site is located in a shallow groove on the surface of the protein (Fig. 1). At the side of the active site, a conserved loop (residues 56 to 63) between the first and second strands of the β-sheet forms a binding site for the guanine base (Sevcik *et al.*, 1990). His102 and Glu73 act as general acid–base catalytic groups during cleavage of RNA (Mossakowska *et al.*, 1989).

Most of the residues in barnase that have been studied to date make interactions that stabilize the protein. For example, hydrogen bonds have been found to stabilize barnase by some 0·5 to 1·5 kcal mol^{-1} (1 cal = 4·184 J) (Serrano & Fersht, 1989; Matouschek *et al.*, 1989), $-CH_2-$ groups involved in hydrophobic interactions in the protein core by up to 2 kcal mol^{-1} (Kellis *et al.*, 1988, 1989), and surface salt bridges by up to 1 kcal mol^{-1} (Serrano *et al.*, 1990; Horovitz *et al.*, 1990; Sali *et al.*, 1991). Here, we mutate active site residues (Fig. 1) and find a different pattern of results: (1) the basic (i.e. positively charged) residues destabilize the protein; (2) the side-chains that make hydrogen bonds and

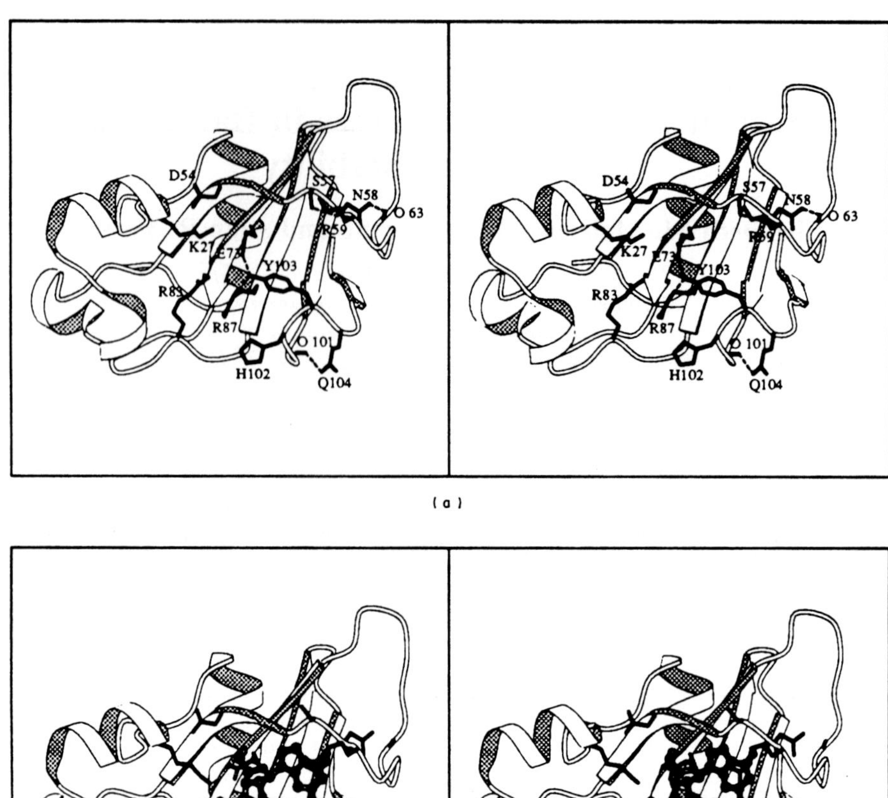

Figure 1. (a) Stereo overview of the active site of barnase (using the program *Molscript* by Dr P. Kraulis; co-ordinates courtesy of Dr C. Hill and Professor G. Dodson). Charged residues and hydrogen bonding residues that have been mutated are shown, as are residues Arg83 and Arg87 (R in 1-letter code). The end of the side-chain of Arg59 is not shown because it is not defined in the crystal structure. (b) Stereo view of the reaction product, 3′-guanosine monophosphate, bound to the active site of binase (co-ordinates courtesy of Drs A. G. Pavlovsky and M. Ya. Karpeisky). Binase, an RNase from *B. intermedius*, is identical to barnase for all but 17 residues. All residues illustrated are the same as for barnase except Gln104 (Q in 1-letter code), which is an alanine in binase.

van der Waals' contacts destabilize the protein or contribute little to stability; (3) the negatively charged side-chains stabilize the protein. All the mutations decrease enzyme activity, as measured by k_{cat}/K_m for hydrolysis of guanylyl(3′-5′)uridine 3′-monophosphate (GpUp), except for two mutations that increase activity and decrease stability.

Mutations of positively charged residues

Positively charged residues are clustered in the active site of barnase (Fig. 1). The $-NH_3^+$ of Lys27 interacts with the reactive phosphate group (Meiering *et al.*, 1991) and is important in stabilizing the transition state of dinucleotide hydrolysis (Mossakowska *et al.*, 1989). This side-chain points into the active site and makes several van der Waals' contacts with the rest of the protein. Mutation of Lys27 to Ala greatly decreases the activity of the enzyme, mainly due to a decrease in k_{cat}. The resultant protein is more stable than wild-type barnase by 0·36 kcal mol^{-1} (Table 1). Arg59 is located in the guanine binding loop and interacts with the guanine base in the crystal structure of binase complexed with 3′ guanosine monophosphate (Fig. 1(b)). The side-chain of this residue is exposed to solvent and makes few intramolecular contacts. Truncation of Arg59 to Ala stabilizes the protein by 0·64 kcal mol^{-1} and causes a considerable decrease in activity due to both an increase in K_m and a decrease in k_{cat} (Table 1). The side-chain of the

Table 1

Activity and stability of barnase mutants

Enzyme	$\Delta\Delta G_u$† (kcal mol^{-1})	Activity‡ k_{cat} (s^{-1})	K_m ($\times 10^6$ M^{-1})	k_{cat}/K_m ($\times 10^6$ s^{-1} M^{-1})	Relative k_{cat}/K_m
Wild-type	0·00	55·4	17·5	3·17	1·00
Arg59→Ala	0·64	33·2	67·2	0·479	0·15
Lys27→Ala	0·36	0·616	138	0·00446	0·0014
His102→Ala	−0·24	n.d.§	n.d.	n.d.	<0·0001
Asn58→Asp	0·40	47·5	133	0·356	0·11
Ser57→Ala	0·15	48·8	18·0	2·71	0·85
Tyr103→Phe	0·00	54·2	20·0	2·72	0·85
Gln104→Ala	−0·16	63·5	18·2	3·49	1·10
Asp54→Asn	−2·67	33·7	9·70	3·47	1·09
Glu73→Ala	−2·30	n.d.	n.d.	n.d.	0·02¶

† Change in stability on mutation, relative to wild-type barnase. Positive values indicate an increase in stability compared with wild-type and negative values indicate a decrease in stability relative to wild-type. Free energies of unfolding in 50 mM-Mes at pH 6·3 and 25 °C were determined by urea denaturation monitored by fluorescence as described (Kellis *et al.*, 1989). The values of $\Delta\Delta G_u$ are reliable to $\pm 0{\cdot}03$ kcal mol^{-1}. Mutants (Arg59→Ala, Asn58→Asp, Ser57→Ala, Tyr103→Phe, Gln104→Ala and Asp54→Asn) were made and expressed as described by Serrano & Fersht (1989) and Mossakowska *et al.* (1989), respectively. Preparation of the other mutants has been reported (Mossakowska *et al.*, 1989).

‡ Activity measurements were made using the substrate guanylyl(3′-5′)uridine 3′-monophosphate (GpUp) in 100 mM-sodium acetate/acetic acid buffer (pH 5·8), containing 50 μg molecular biology grade BSA (Sigma)/ml, at 25 °C by monitoring the increase in absorbance at 275 nm on hydrolysis of GpUp to Gp and Up. Active site titration is not available for barnase. The activity of wild-type barnase is independent of batch within ±3%. The reproducibility of activity, the high stability against irreversible denaturation, the monophasic denaturation curves for reversible equilibrium unfolding, and the monophasic kinetic traces for unfolding of barnase suggest that it is 100% active.

§ Not determined.

¶ k_{cat} and K_m could not be determined for Glu73→Ala because of high levels of substrate inhibition for this mutant. The rate of hydrolysis by Glu73→Ala relative to that by wild-type, corrected for enzyme concentration, is given for 50 μM-GpUp.

catalytic residue, His102, makes many van der Waals' contacts with residues 84 to 86 and 101. The mutant His102→Ala is 0·24 kcal mol^{-1} less stable than wild-type and has no detectable activity. However, the pK_a of His102 in wild-type barnase is 6·3 (Sali *et al.*, 1988) and so this residue is approximately 50% protonated under the conditions where the stability was measured. At pH 4·5, where the histidine is fully protonated in wild-type, His102→Ala is 0·55 kcal mol^{-1} more stable than wild type. Thus, the positive charge of His102 also destabilizes the protein. All the positively charged residues studied here are important for activity and destabilize the protein.

Mutations of hydrogen bonding residues

Residues in the active site that make hydrogen bonds were also mutated (Fig. 1(a)). Asn58 is involved in forming the structure of the guanine binding loop by hydrogen bonding between its side-chain and the main chain of residues 62 and 63 (Sevcik *et al.*, 1990; C. Hill, personal communication). Mutation of Asn58 to Asp deletes a hydrogen bond between its side-chain NH and the backbone >CO of residue 63 but, instead of destabilizing the protein as expected, it increases the stability by 0·40 kcal mol^{-1} (Table 1). This mutation causes a decrease in k_{cat} and an increase in K_m so that overall there is a substantial decrease in activity. Ser57 is also in the guanine-binding loop. The γ-OH of Ser57 makes a long hydrogen bond to the backbone >NH of residue 59. Mutation of Ser57 to Ala deletes the hydrogen bond, but stabilizes the protein by 0·15 kcal mol^{-1}. k_{cat} and consequently k_{cat}/K_m are decreased slightly. Tyr103 and Gln104 are in the loop formed by residues 101 to 105 between the fourth and fifth strands of the β-sheet. The –OH of Tyr103 makes hydrogen bonds to the carboxyl of Glu73 and to the N$^\varepsilon$ of Arg87. Mutation of Tyr103 to Phe deletes these two hydrogen bonds, but the mutant protein has the same stability as wild-type and is slightly less active than wild-type because K_m is increased (Table 1). The side-chain >NH of Gln104 makes a hydrogen bond across the loop to the backbone >CO of residue 101. The rest of the side-chain makes very few contacts with the rest of the protein. In contrast to the other mutations, truncation of the side-chain of Gln104 to Ala decreases the stability by 0·16 kcal mol^{-1}. But, due to an increase in k_{cat}, the mutant enzyme is slightly more active than wild-type (Table 1). In general, the hydrogen bonds studied here either destabilize the protein or contribute little to stability and have moderate affects on activity.

Mutations of negatively charged residue

The side-chain of Asp54 is close (~6 to 7 Å) to the positively charged residues Lys27, Arg72 and

Arg83, and makes a hydrogen bond with the backbone >NH of residue 27. Mutation of Asp54 to Asn destabilizes the enzyme by 3 kcal mol^{-1}, probably largely due to the loss of favourable electrostatic interactions, but the mutant is again more active than wild-type (Table 1). The increase in k_{cat}/K_m is mainly due to tighter binding of substrate (i.e. decreased K_m). The catalytic residue, Glu73, makes a strong electrostatic interaction with Arg87, as well as interacting with Arg83 and Lys27. It makes a few van der Waals' contacts and a hydrogen bond to the –OH of Tyr103. The mutant Glu73→Ala is different from the other mutants in that it is both 2·3 kcal mol^{-1} less stable than wild-type and has greatly decreased activity.

The effects of the individual residues in the active site of barnase on enzyme activity and protein stability are complex. Residues in an active site may be involved in catalysis and substrate binding, or stabilizing the structure of the active site, or both. The negatively charged side-chains studied here are examples of residues that contribute to stability. Glu73 is also essential for catalysis, however, while Asp54 seems to have a small adverse effect on the activity of wild-type by weakening substrate binding.

A significant number of the mutations studied here cause a decrease in activity with an associated increase or no change in protein stability. These mutations can be grouped into two types: positively charged side-chains and hydrogen bonding side-chains. The positively charged side-chains, Lys27, Arg59 and His102, are necessary for catalysis and binding of the negatively charged RNA substrate. Mutation of these residues to alanine stabilizes the protein. This suggests that the clustering of like-charged residues in the active site (Fig. 1) causes electrostatic strain, which is destabilizing. Similarly, the *apo* form of calbindin D_{9k} is destabilized by a group of negatively charged residues in a calcium binding site (Akke & Forsén, 1990). Electrostatic strain is likely to be found locally in many proteins where clusters of like-charged residues or aligned dipoles are used for binding charged species such as metal ions, substrates, cofactors or other ligands.

Mutation of the hydrogen bonding residues, Asn58, Ser57 and Tyr103, causes relatively smaller decreases in activity than mutation of the positively charged residues, and either stabilizes the protein or has no effect on stability. Hydrogen bonds have been found previously to stabilize proteins by approximately 0·5 to 1·5 kcal mol^{-1} (Fersht *et al.*, 1985; Alber *et al.*, 1987; Serrano & Fersht, 1989; Matouschek *et al.*, 1989). The data suggest that the hydrogen bonds studied here may be intrinsically weak, or that the stabilizing contribution of the hydrogen bonds is counterbalanced by destabilizing interactions. These results also imply that there may be some strain in the active site.

In addition to the mutations that decrease activity and increase stability, two mutations have the opposite effect: Asp54→Asn and Gln104→Ala have increased activity and decreased stability relative to wild-type barnase. Thus, there is an inverse relationship between stability and activity for many of the mutations studied here. This relationship does not hold for all the mutations, since Glu73 contributes to both stability and activity, nor is the relationship quantitative, since some residues make much larger contributions to activity than stability and *vice versa*. The data do suggest, however, that certain types of local strain or instability may be inevitable in active sites and may, in fact, be necessary for substrate binding and catalysis.

The presence of destabilizing or weak interactions may allow the active site to be flexible, and so facilitate the binding of substrate, conformational changes during catalysis, and release of product. Further, active sites have evolved to make intermolecular interactions with substrates and transition states, and so groups on the enzyme must be available for hydrogen bonding and hydrophobic interactions. This may be at the expense of intramolecular interactions and burying hydrophobic surfaces, and so lower the stability of the free enzyme. Local instability or strain may occur generally in proteins in regions that interact with substrates, cofactors, allosteric effectors, proteins and other ligands.

E.M.M. is supported by the Natural Sciences and Engineering Research Council of Canada. L.S. is an EC postdoctoral fellow.

References

Akke, M. & Forsén, S. (1990). Protein stability and electrostatic interactions between solvent exposed charged side chains. *Proteins Struct. Funct. Genet.* **8**, 23–29.

Alber, T. (1989). Mutational effects on protein stability. *Annu. Rev. Biochem.* **58**, 765–798.

Alber, T., Dao-pin, S., Wilson, K., Wozniak, J. A., Cook, S. P. & Matthews, B. W. (1987). Contributions of hydrogen bonds of Thr157 to the thermodynamic stability of phage T4 lysozyme. *Nature (London)*, **330**, 41–46.

Bycroft, M., Ludvingen, S., Fersht, A. R. & Poulsen, F. M. (1991). Determination of the solution conformation of barnase using NMR spectroscopy. *Biochemistry*, **30**, 8697–8701.

Day, A. G., Parsonage, D., Ebel, S., Brown, T. & Fersht, A. R. (1992). Barnase has subsites that give rise to large rate enhancements. *Biochemistry*, in the press.

Fersht, A. R. (1987). Dissection of the structure and activity of the tyrosyl-tRNA synthetase by site-directed mutagenesis. *Biochemistry*, **26**, 8031–8037.

Fersht, A. R., Shi, J. P., Knill-Jones, J., Lowe, D. M., Wilkinson, A. J., Blow, D. M., Brick, P., Carter, P., Waye, M. M. Y. & Winter, G. (1985). Hydrogen bonding and biological specificity analysed by protein engineering. *Nature (London)*, **314**, 235–238.

Hill, C., Dodson, G., Heinemann, U., Saenger, W., Mitsui, Y., Nakamura, K., Borisov, S., Tischenko, G., Polyakov, K. & Pavlovsky, S. (1983). The structural and sequence homology of a family of microbial ribonucleases. *Trends Biochem. Sci.* **8**, 364–369.

Horovitz, A., Serrano, L., Avron, B., Bycroft, M. & Fersht, A. R. (1990). Strength and co-operativity of contributions of surface salt bridges to protein stability. *J. Mol. Biol.* **216**, 1031–1044.

Kellis, J. T., Jr, Nyberg, K., Sali, D. & Fersht, A. R. (1988). Contribution of hydrophobic interactions to protein stability. *Nature (London)*, **333**, 784–786.

Kellis, J. T., Jr, Nyberg, K. & Fersht, A. R. (1989). Energetics of complementary side chain packing in a protein hydrophobic core. *Biochemistry*, **28**, 4914–4922.

Knowles, J. R. (1991). Enzyme catalysis: not different, just better. *Nature (London)*, **350**, 121–124.

Matouschek, A., Kellis, J. T., Jr, Serrano, O. & Fersht, A. R. (1989). Mapping the transition state and pathway of protein folding by protein engineering. *Nature (London)*, **340**, 122–126.

Matthews, B. W. (1991). Mutational analysis of protein stability. *Curr. Opin. Struct. Biol.* **1**, 17–21.

Mauguen, Y., Hartley, R. W., Dodson, E. J., Dodson, G. G., Bricogne, G., Chothia, C. & Jack, A. (1982). Molecular structure of a new family of ribonucleases. *Nature (London)*, **297**, 162–164.

Meiering, E. M., Bycroft, M. & Fersht, A. R. (1991). Characterization of phosphate binding in the active site of barnase by site-directed mutagenesis and NMR. *Biochemistry*, **30**, 11348–11356.

Mossakowska, D. E., Nyberg, K. & Fersht, A. R. (1989). Kinetic characterization of the recombinant ribonuclease from *Bacillus amyloliquefaciens* (barnase) and investigation of key residues in catalysis by site-directed mutagenesis. *Biochemistry*, **28**, 3843–3850.

Pavlovsky, A. G., Borisova, S. N., Strokopytov, B. V., Sanishvili, R. G., Vagin, A. A. & Chepurnova, N. K. (1988). In *Metabolism and Enzymology of Nucleic Acids Including Gene Manipulations*, pp. 217–221, Plenum Press, New York.

Sali, D., Bycroft, M. & Fersht, A. R. (1988). Stabilization of protein structure by interaction of α-helix dipole with a charged side chain. *Nature (London)*, **335**, 740–743.

Sali, D., Bycroft, M. & Fersht, A. R. (1991). Surface electrostatic interactions contribute little to stability of barnase. *J. Mol. Biol.* **220**, 779–788.

Serrano, L. & Fersht, A. R. (1989). Capping and α-helix stability. *Nature (London)*, **342**, 296–299.

Serrano, L., Horovitz, A., Avron, B., Bycroft, M. & Fersht, A. R. (1990). Estimating the contribution of engineered surface electrostatic interactions to protein stability by using double-mutant cycles. *Biochemistry*, **29**, 9343–9352.

Sevcik, J., Sanishvili, R. G., Pavlovsky, A. G. & Polyakov, K. M. (1990). Comparison of active sites of some microbial ribonucleases: structural basis for guanylic specificity. *Trends Biochem. Sci.* **15**, 158–162.

Shortle, D. (1989). Probing the determinants of protein folding and stability with amino acid substitutions. *J. Biol. Chem.* **264**, 5315–5318.

Edited by A. Klug

It used to be thought that surface residues of proteins were not important in stability because they are the least conserved. Luis Serrano shattered that illusion by systematically mutating barnase into its near identical sibling, binase[103]. Further, the energetic changes on mutation are additive. The punchline at the end of the abstract:

> These results suggest a simple way to improve the stability of proteins: choose two homologous proteins that have high similarity, mutate individually all of the residues that differ between the two, and combine the mutations that increase the stability in a multiple mutant.

was subsequently used by us to make a stable variant of the tumour suppressor p53, which will be seen later to be the key to our structural studies.

11.12 1993: Members of CPE. *Top*: Jonathan Blackburn, Laurent Jespers, Sam Williams, Graham Cook, Olga Perisic, Paul Dear, Ahuva Nissim, Cyrus Chothia, Steve Russell, Reinhard Grisshammer and Tim Green. *Third row*: Christine Strachan, Roger Williams, Gerald Walter, Jonathan Cox, John Doorbar, Jim Embleton, Tim Buss, Moira Tong, Jane Clarke, Hennie Hoogenboom, Peter Waterhouse, A Visitor, Kim Henrick and Ed Nerou. *Second row*: Tim Bonnert, Yahouda Harpaz, Gideon Schreiber, Andreas Matouschek, Martin Long, Joanne Stubbs, Miri Hirshberg, Yu Wai Chen, Jesus Sanz, Ian Tomlinson and Chris Johnson. *Front row*: Paul Barker, Greg Winter, Ronny Loewenthal, Alain Tissot, Mikael Oliveberg, Stephane Vuilleumier and Pete Jones.

J. Mol. Biol. (1993) **233**, 305–312

Step-wise Mutation of Barnase to Binase

A Procedure for Engineering Increased Stability of Proteins and an Experimental Analysis of the Evolution of Protein Stability

Luis Serrano†, Anthony G. Day and Alan R. Fersht

MRC Unit for Protein Function and Design
Cambridge Centre for Protein Engineering
MRC Centre, Hills Road, Cambridge CB2 2QH, U.K.

(Received 9 March 1993; accepted 26 May 1993)

We have chosen two members of the microbial RNase family, barnase and binase, which have 85% identity (17 substitutions and 1 deletion) and almost identical three-dimensional structure, to study the evolution of protein stability. The 17 residues that differ are scattered throughout the molecule. Each of the 17, differing residues has been mutated independently and the effect on protein stability analysed. Each point mutation has an effect on protein stability that ranges from $+1{\cdot}1$ to $-1{\cdot}1$ kcal mol^{-1}. These changes in energy are additive. There is no clear correlation between the type of mutation and the effect on protein stability. A multiple mutant having six of the single mutations that increase the stability of barnase is 3·3 kcal mol^{-1} more stable than wild type and has the same activity. There could be selective pressure to maintain proteins at a certain stability and, consequently, mutations that decrease stability tend to be counterbalanced by stabilizing mutations. Alternatively, there could simply be pressure to maintain stability above a certain level, and any further increases in stability need not be maintained during evolution. These results suggest a simple way to improve the stability of proteins: choose two homologous proteins that have high similarity, mutate individually all of the residues that differ between the two, and combine the mutations that increase the stability in a multiple mutant.

Keywords: protein stability; protein activity; co-operativity; protein evolution; barnase

1. Introduction

The enhancement of protein stability by rational design is one of the great goals of protein engineering. Two approaches have been followed to alter the stability of proteins. First, the analysis by protein engineering of the contribution of the different interactions that take place in a protein has resulted in some general rules about possible ways to increase the stability of a protein (reviewed by Fersht & Serrano, 1993). Second, the comparison between homologous proteins from thermophiles and mesophiles has provided some insight into the reasons why related proteins performing the same function could have very different stabilities (Argos *et al.*, 1979; Imanaka *et al.*, 1986; Vihinen, 1987; Menedez-Arias & Argos, 1989; Mrabet *et al.*, 1992). A complete analysis of the differences between homologous thermo-stable and normal proteins is, however, difficult, since in general they have diverged significantly. Consequently, it is very likely that there will be slight conformational rearrangements and co-operative effects from those residues that are different (Imanaka *et al.*, 1987). It is important to examine proteins that are at the beginning of divergence, since their structures are still almost identical and the effect of single differences can be easily analysed.

Barnase, a small ribonuclease from *Bacillus amyloliquefaciens*, has proved to be a good paradigm for thermodynamic studies. This protein consists of a single polypeptide chain of 110 amino acid residues. Its three-dimensional structure has been determined by X-ray crystallography (Mauguen *et al.*, 1982; Baudet & Janin, 1991) and by NMR in solution (Bycroft *et al.*, 1991). Barnase has a close relative, binase, whose sequence is identical for all but 18 residues (Hill *et al.*, 1983; Pavlovsky *et al.*, 1983, 1989) and whose three-dimensional structure

† Present address: EMBL Biostructures and Biocomputing Programme, Meyerhofstrasse 1, Heidelberg, Germany.

is known (Pavlovsky *et al.*, 1983, 1989). Barnase undergoes reversible denaturation, mediated by denaturants or heat, which closely follows a two-state process (Hartley, 1975; Kellis *et al.*, 1988). Changes in the free energy of unfolding can be accurately measured (Kellis *et al.*, 1988, 1989; Serrano *et al.*, 1990; Horovitz *et al.*, 1992). In this study, we have independently mutated each one of the 17 residues† that differ between the two proteins and have determined the changes in stability. We have also made multiple mutants in order to look at the additivity of changes in stability and have measured the activity of the mutant enzymes.

2. Experimental Procedures

(a) *Materials*

The substrates, GpUp and RNA type XI from bakers' yeast were obtained from Sigma. The buffer used in the denaturation experiments was 2-[*N*-morpholino]ethanesulphonic acid (MES) from Sigma. The MES stock was a 1·00 M solution containing 387 mM acid form and 613 mM sodium salt, which gives a pH of 6·3 at 25°C on dilution to 50 mM. Urea was enzyme grade from Bethesda Research Laboratories, MD, U.S.A. Radiochemicals were from Amersham International, and SP-Trisacryl was obtained from IBF, Villeneuve La Garenne, France. *Escherichia coli* BL21DE3 pLysS was a generous gift from Dr F. W. Studier. Plasmid pTZ18U and the helper phage M13KO7 were obtained from Pharmacia, Sweden. Wild-type barnase in the plasmid pUC19 (Paddon & Hartley, 1987), generously donated by Dr Hartley, was recloned into the pTZ18U plasmid (Serrano & Fersht, 1989; Serrano *et al.*, 1990). The X-ray co-ordinates for barnase were provided by Dr K. Henrick, Professor G. Dodson and A. Cameron. Wild-type binase from *Bacillus intermedius*, was generously provided by Professor M. Ya. Karpeisky. Binase was also expressed from a synthetic gene which was constructed as described in section (b), below.

(b) *Mutagenesis*

Single-stranded DNA was obtained from the modified pTZ18U, harboured in *E. coli* TG2 after infection with the helper phage M13KO7 (Pharmacia), following the conditions described by Pharmacia. Site-directed mutagenesis was carried out using the method of Eckstein (Sayers *et al.*, 1988) and the kit supplied by Amersham (using 1/5 of the recommended amounts of DNA and kit). Mutants were identified by directly sequencing the single-stranded DNA. The efficiencies of mutation ranged from 50 to 99%. The mutations Thr16 → Arg, His18 → Lys, Ile55 → Val and Leu89 → Val have been previously described (Serrano *et al.*, 1990, 1992; Loewenthal *et al.*, 1992). The synthetic binase gene was constructed by repeated inverse polymerase chain reaction (PCR) mutagenesis on the barnase plasmid using 6 pairs of mutagenic primers (Clackson *et al.*, 1991).

† There are 17 amino acid substitutions and a deletion, Gln2 → Δ, on going from barnase to binase. The Gln2 → Δ deletion lies in a region of the protein that is not visible in the crystal structures and that appears disordered in the NMR solution structure (Bycroft *et al.*, 1991). This deletion was not investigated in this work and was not incorporated in the synthetic binase gene that was constructed.

(c) *Expression and purification of barnase and synthetic binase*

These were performed as described previously (Serrano *et al.*, 1990).

(d) *Urea denaturation*

The reversible unfolding was monitored by fluorescence spectroscopy with excitation at 290 nm and emission at 315 nm, as described previously (Horovitz *et al.*, 1992).

(e) *Solvent-accessible area calculations and atom distances*

The solvent-accessible area calculations and the distances between atoms were calculated from the surface generated by the locus of the centre of a sphere of radius 1·4 Å rolling over the exposed van der Waals' surface using the program WHAT IF (Vriend, 1990).

(f) *Activity measurements*

These were performed as described previously for the dinucleotide substrate GpUp (Day *et al.*, 1992) and for RNA type XI from bakers' yeast (Mossakowska *et al.*, 1989).

3. Results

(a) *Description of the mutants*

The majority of the substitutions between barnase and binase are conservative, with seven exceptions (Gln15 → Ile, Thr16 → Arg, Gln31 → Ser, Gly65 → Ser, Lys66 → Ala, Thr79 → Val and Gln104 → Ala) and only two involve completely buried residues (Ile88 → Leu and Leu89 → Val)

Table 1
Solvent accessibility of the residues that differ between barnase and binase

Position	Barnase residue accessibility (%)	Binase residue accessibility (%)
15	31	35
16	31	57
18	46	49
19	73	41
29	40	42
31	47	37
44	64	64
55	46	48
62	55	50
65	59	70
66	55	39
79	63	62
85	20	27
88	0	0
89	0	0
104	62	34
108	47	47

The percentage of the accessibilities of the residues in the folded state with respect to the unfolded state in a Gly-X-Gly tripeptide (Miller *et al.*, 1987) are indicated.

Table 2

Hydrogen bonds made by the side-chains of those residues that are different between barnase and binase

Residue	Donor	Acceptor	Distance (Å)
Gln15 (barnase)	$N^{\varepsilon 2}$	O Trp94	2·5
Thr16 (barnase)	$O^{\gamma 1}$	O Asp12	3·0
Arg19 (binase)	$N^{\eta 2}$	OH Tyr13	2·6
Glu29 (barnase)	$O^{\varepsilon 1}$	N Thr26	2·8
Ser31 (binase)	O^{γ}	O Ser28	2·8
Lys62 (barnase)	N^{ζ}	O Gln104	2·9
Arg62 (binase)	$N^{\eta 1}$	O Ala104	2·9
Gln104 (barnase)	$N^{\varepsilon 2}$	O Asp101	3·1
Lys108 (barnase)	N^{ζ}	O Arg110	2·9
Arg108 (binase)	$N^{\eta 1}$	O Arg110	2·5

(Table 1). In the first α-helix, there are three mutations: Gln15 → Ile, Thr16 → Arg and His18 → Lys. Mutation of Gln15 to Ile results in the deletion of a hydrogen bond between the side-chain CO group of Gln15 and the main-chain NH group of Trp94 (Table 2) and an increase in the hydrophobic area buried inside the protein (33 $Å^2$). The mutation Thr16 to Arg has been previously analysed in detail (Serrano *et al.*, 1990); the increase in stability upon mutation is due to better van der Waals' contacts with Tyr17 (0·2 kcal mol^{-1}) and to a favourable electrostatic interaction with Asp12. The His18 to Lys mutation has been described previously (Loewenthal *et al.*, 1992); the change in stability, which is pH-dependent, is due to a favourable interaction of the charged His residue with Trp94, and is less than 1 kcal mol^{-1} at pH 7·0 (Sali, 1990). Lys19 is located on the loop connecting the first and second α-helices of barnase. It is highly exposed to the solvent in barnase (Table 1) and makes an electrostatic interaction with Asp22. Arg19 in binase makes a hydrogen bond with the OH of Tyr13 (Table 2). There are two mutations in the second α-helix of barnase. The side-chain of Glu29 in barnase makes a hydrogen bond with the main-chain NH group of Ser28 (Table 2) that is not made by Gln29 in binase. Gln31 is highly exposed in barnase and the equivalent residue in binase (Ser31) makes a hydrogen bond with the main-chain CO group of Ser28. In the small one-turn α-helix of barnase, there is one mutation, Asp44 to Glu. The side-chain of Asp44 is ill-defined in the crystal structure of barnase as is that of Glu44 in binase. Ile55 in the first β-strand is highly exposed to the solvent (Table 1), and the equivalent Val residue in binase makes similar van der Waals' contacts (Serrano *et al.*, 1992). Lys62, Gly65 and Lys66 are located on the guanosine-binding loop (Sevcik *et al.*, 1990). The side-chain of Lys62 makes a hydrogen bond with the side-chain CO group of Gln104. In binase, the side-chain of Arg62 makes a hydrogen bond with the main-chain CO of Ala104. Gly65 and Lys66 are highly exposed in barnase as well as in binase (Table 1). Thr79 is located in the loop connecting the second and third β-strands of barnase. Its side-chain does not make a hydrogen bond and it is highly exposed to the solvent, as is Val79 in binase (Table 1). There are three mutations in the third β-strand of barnase: Ser85 → Ala, Ile88 → Leu and Leu89 → Val. Ser85 is the first residue of the third β-strand of barnase and it is highly exposed to the solvent, as is Ala85 in binase (Table 1). The side-chain of Ile88 is completely buried and is the central residue of the main hydrophobic core of barnase (Serrano *et al.*, 1992), as is Leu88 in binase (Table 1). Leu89 is the central residue of the active-site hydrophobic core of barnase (Serrano *et al.*, 1992) and is completely buried as is Val89 in binase (Table 1). Gln104 is located in the loop between the fourth and fifth β-strands in barnase. The side-chain $O^{\varepsilon 1}$ of Gln104 makes a hydrogen bond with the side-chain N^{ζ} of Lys62 in barnase and is highly exposed (Table 1). The equivalent residue Ala104 in binase is also exposed to the solvent. Lys108 is in the C-terminal region of the protein. The side-chain of Lys108 in barnase makes a hydrogen bond with the CO main-chain group of Arg110, as does the side-chain $N^{\eta 1}$ group of Arg108 in binase.

(b) *Thermodynamic analysis*

The thermodynamic analysis, by urea denaturation, of the 17 mutants is shown in Table 3. With only one exception (His18 → Lys), the effect on the free energy of denaturation at pH 6·3 of all the single mutations is less than ±1 kcal mol^{-1}. Some of the mutations that increase the stability of the protein were selected, a triple and a multiple mutant were made and their effect on protein stability was determined (Table 3). The stability of binase was also determined (Table 3).

(c) *Activity*

It has been noted that there is an inverse correlation between stability and activity for mutations of some residues in the active site of barnase (Meiering *et al.*, 1992). There are no significant correlations found here for residues that are not involved in binding and catalysis (Table 4). However, we found that binase is 25-fold more active than barnase with respect to the substrate GpUp (Table 4). We also found that binase is fivefold less active than barnase with respect to the substrate RNA type XI from bakers' yeast (Table 4). The rates observed with RNA as substrate were identical at 1·5, 2 and 3 mg/ml RNA, indicating that V_{max} is measured at these concentrations. We find that rates vary with different batches of RNA substrate and comparisons between enzymes can be made only if the same batch of RNA is used for all measurements, as was done here.

4. Discussion

(a) *Analysis of changes in stability on mutation*

There is no specific localization of the differences between barnase and binase to any particular region of the protein. The 17 amino acid substitutions are

Table 3

Changes in the free energies of unfolding of wild-type barnase and the different mutants at the two regular α-helices of barnase

Mutant	m^{a} (kcal mol^{-2})	$\Delta G_{u}^{H_2O}$ [b] (kcal mol^{-1})	$[Urea]_{50\%}$ [c] [m]	$\Delta\Delta G_{u}^{[urea]}{}_{50\%}$ [d] (kcal mol^{-1})
Q15I	2·02	10·2	5·06	−0·96
T16R	1·98	9·6	4·84	−0·53
H18K	2·00	7·9	3·96	1·19
K19R	2·08	9·7	4·68	−0·21
E29Q	1·88	8·6	4·56	0·01
Q31S	2·05	9·1	4·44	0·25
D44E	1·84	8·5	4·61	−0·08
I55V	1·93	8·5	4·42	0·29
K62R	2·04	8·8	4·32	0·48
G65S	2·03	9·8	4·83	−0·51
K66A	2·09	9·8	4·70	−0·25
T79V	1·91	9·0	4·72	−0·29
S85A	2·02	9·1	4·51	0·12
I88L	1·98	8·8	4·42	0·28
L89V	1·92	8·5	4·43	0·27
Q104A	1·98	8·8	4·46	0·21
K108R	1·91	9·6	5·05	−0·93
Q15I/T16R/K19R	2·00	10·8	5·42	−1·66
Multiple[e]	1·86	11·8	6·34	−3·44
Binase[f]	2·02	10·2	5·04	−0·93
Barnase	1·92	8·8	4·58	0·00

The free energies of unfolding of wild-type protein and mutants were analysed by urea denaturation monitored by fluorescence.

[a] m is the slope of the linear denaturation plot, $-d\Delta G_{u}/d[urea]$.

[b] ΔG^{H_2O} was determined by urea denaturation of the proteins and extrapolation of the data to zero (denaturant).

[c] $[Urea]_{50\%}$ is the concentration of urea at which 50% of the protein is unfolded.

[d] $\Delta\Delta G_{u}^{[urea]}{}_{50\%}$ was determined by subtracting $^{[Urea]}{}_{50\%}$ for each mutant from that for wild type, and multiplying the difference by the average slope $\langle m \rangle$ using sloping baseline fitting ($\langle m \rangle = 1{\cdot}95$) (Horovitz *et al.*, 1992). Note that the values of $\Delta\Delta G_{u}^{[urea]}{}_{50\%}$ are fairly precise measurements, whereas the values of $\Delta G_{u}^{H_2O}$ are relatively imprecise due to a long extrapolation of data (see Kellis *et al.*, 1989; Serrano *et al.*, 1992, for a discussion of accuracy).

[e] The multiple mutant has the following mutations: Q15I, T16R, K19R, G65S, K66A and K108R.

[f] Binase was wild type and was a generous gift from Professor M. Ya. Karpeisky.

scattered throughout the protein; six of them are in the α-helices, four in the β-strands and six in loops and turns. Two of the differences occur at the central positions of the two hydrophobic cores of the protein, four of the differences are close to the residues involved in the binding of the guanosine base in the guanosine binding loop, and one is close to the catalytic His102 residue (Sevcik *et al.*, 1990). Seven of the mutations are not conservative: Gln15 → Ile, Thr16 → Arg, Gln31 → Ser, Gly65 → Ser, Lys66 → Ala, Thr79 → Val and Gln104 → Ala, and there are three Lys to Arg mutations (19, 62 and 108). All except three of the mutations (Gln15 → Ile, His18 → Lys and

Table 4

Activity of the mutants

Protein	V_{m} (RNA)[a] ($\Delta A_{298{\cdot}5}$ s^{-1} M^{-1})	K_{m}(GpUp)[b] (μM)	k_{cat}(GpUp)[b] (s^{-1})	k_{cat}/K_{m}(GpUp)[b] (s^{-1} M^{-1})
Barnase	$(1{\cdot}50 \pm 0{\cdot}10) \times 10^{5}$	$21{\cdot}7 \pm 1{\cdot}2$	$58{\cdot}5 \pm 1{\cdot}5$	$(2{\cdot}67 \pm 0{\cdot}28) \times 10^{6}$
Q15I/T16R/K19R	—	$19{\cdot}0 \pm 0{\cdot}2$	$54{\cdot}9 \pm 0{\cdot}3$	$(2{\cdot}90 \pm 0{\cdot}06) \times 10^{6}$
Multiple[c]	—	$26{\cdot}5 \pm 1{\cdot}9$	$69{\cdot}9 \pm 1{\cdot}5$	$(2{\cdot}64 \pm 0{\cdot}20) \times 10^{6}$
W-T binase[d]	—	$5{\cdot}5 \pm 1{\cdot}3$	403 ± 45	$(73 \pm 19) \times 10^{6}$
Synthetic binase[d]	$(2{\cdot}78 \pm 0{\cdot}06) \times 10^{4}$	$6{\cdot}8 \pm 0{\cdot}8$	437 ± 11	$(64{\cdot}3 \pm 7{\cdot}7) \times 10^{6}$

All experiments were performed at 25°C.

[a] Activity with bakers' yeast RNA, type XI, was determined in 100 mM Tris·HCl (pH 8·5) containing 100 μg/ml RNase-free BSA.

[b] Activity with GpUp was determined in 100 mM sodium acetate (pH 5·8), containing 100 μg/ml RNase-free BSA.

[c] The multiple mutant has the following mutations: Q15I, T16R, K19R, G65S, K66A and K108R.

[d] Synthetic binase has a Gln residue at position 2 that is not present in wild-type binase. The concentration of protein was determined spectrophotometrically at 280 nm based on the absorption coefficient for barnase, where $\varepsilon_{0{\cdot}1\%} = 2{\cdot}2$ (Loewenthal *et al.*, 1991). The kinetic analyses were performed as previously described (Day *et al.*, 1992; Mossakowska *et al.*, 1989).

Table 5
Additivity of changes in stabilization energies of the mutants

Protein	$\Delta\Delta G_{wt\ barnase}$ (kcal mol^{-1})[a]	$\Sigma\Delta\Delta G_{wt}$ (kcal · mol^{-1})[b]
Binase	−0·9	−0·7
Q15I/T16R/K19R	−1·7	−1·7
Multiple[c]	−3·4	−3·4

[a] $\Delta\Delta G_{wt\ barnase}$, the change in stabilization energy, was calculated as indicated for Table 1.
[b] $\Sigma\Delta\Delta G_{wt}$ was calculated by summing the effects of changes in stability of barnase of the single mutations.
[c] The multiple mutant has the following mutations: Q15I; T16R; K19R, G65S, K66A and K108R.

Lys108 → Arg) have an effect on protein stability that ranges from −0·5 to 0·5 kcal mol^{-1}.

(i) *The mutations that hardly affect the stability of barnase*

These are Glu29 → Gln, Asp44 → Glu and Ser85 → Ala. The three are conservative mutations of solvent-exposed residues. The mutation of Glu29 to Gln results in a decrease in stability because of breaking a hydrogen bond with the NH main-chain group of residue Thr26 and an increase in stability because of the elimination of an electrostatic repulsion between the side-chains of Asp54 and Glu29 (Sancho *et al.*, 1992) and an increase in the entropic freedom of the side-chain of Gln29 compared with that of Glu29. The destabilizing effect is also partly counterbalanced by the increase in solvent exposure of the main-chain NH group of Thr26 (from 4 to 29%). Mutation of Asp44 to Glu does not result in any significant change in the interactions of the side-chain of Asp44 with the rest of the molecule since, in both cases, the side-chains are highly mobile and do not interact with any residue in particular. Finally, the Ser85 → Ala mutation has a destabilizing effect due to a decrease in the hydrophobic surface (6 Å^2) and a stabilizing effect because the OH side-chain group of Ser85 is partly buried and screened from solvent without being hydrogen bonded within the protein (solvent exposure = 18 Å^2).

(ii) *The mutations that increase the stability of barnase*

These are Gln15 → Ile, Thr16 → Arg, Lys19 → Arg, Gly65 → Ser, Lys66 → Ala, Thr79 → Val and Lys108 → Arg. It is interesting that none of these changes takes place in the β-strands. Five of the changes occur in loops and in all the cases solvent-exposed residues are involved. Two of the stabilizing mutations increase the hydrophobic surface buried (Gln15 → Ile and Thr79 → Val). Mutation of Gln15 to Ile breaks one internal hydrogen bond, leaving a hydrogen bond donor group without acceptor. This could result in a decrease of protein stability about 1·5 kcal mol^{-1} (Fersht, 1987). The concomitant increase in the hydrophobic surface buried of 33 Å^2 should result in a net increase in the stabilization of the protein (Matsumura *et al.*, 1988). Mutation of Thr79 to Val increases the hydrophobic surface buried by around 3 Å^2, and there is no effect due to the mutation of the side-chain OH group to a methylene group, since the OH is almost 100% exposed to the solvent. The mutation Gly65 → Ser decreases the entropy of the unfolded state, and the stabilization found is in the range previously predicted for this type of mutation (Matthews *et al.*, 1987). One of the mutations, Thr16 → Arg, introduces a favourable electrostatic interaction with Asp12 and makes better van der Waals' contacts with Tyr17 (Serrano *et al.*, 1990). Three of the mutations (Lys19 → Arg, Lys108 → Arg and Lys66 → Ala) are of the type predicted, from comparing thermo-stable and normal proteins, to decrease flexibility and increase thermo-stability (Menendez-Arias & Argos, 1989).

(iii) *The mutations that decrease the stability of barnase*

These are His18 → Lys, Gln31 → Ser, Ile55 → Val, Ile88 → Leu, Leu89 → Val and Gln104 → Ala. The decrease in stability on mutation of His18 → Lys results from losing a favourable interaction of the charged ring of the His residue with Trp94 (Loewenthal *et al.*, 1992). Mutation of Gln31 to Ser results in the formation of a hydrogen bond between the side-chain OH group of Ser31 and the main-chain CO of Ser28. There is a destabilizing effect from a decrease in the hydrophobic surface buried (17 Å^2) and a stabilizing effect from the NH_2 side-chain group of Gln31 being partly buried without being hydrogen bonded to the protein (solvent-accessible area of 37 Å^2). Mutation of Ile55 to Val decreases the hydrophobic surface buried by 4·6 Å^2. Mutation of Ile88 to Leu maintains the hydrophobic surface buried but changes the packing arrangement slightly. Mutation of Leu89 to Val decreases the hydrophobic surface buried and changes the packing arrangement of the surrounding residues.

(b) *Co-operativity of the effects of mutation*

The folding of a protein is highly co-operative. The effect on protein stability of many of the interactions between two or more residues may also be co-operative, and so breaking a particular interaction may weaken a further set of interactions (Horovitz *et al.*, 1991). There seems to be no co-operativity connected with the 17 substitutions between barnase and binase, since the sum of the effects of the single mutations on the stability of the protein is very close to the value obtained for the difference in stability between the two proteins. Moreover, the stability of a triple or sextuple mutant is similar to the sum of the effects of the single mutations (Table 4). This is particularly interesting in the case of mutations of residues that are close to each other and in the same element of secondary structure (Gln15 → Ile and Thr16 → Arg in the first α-helix; Glu29 → Gln and Gln31 → Ser in the second α-helix; Gly65 → Ser and Lys66 → Ala in the guanosine binding loop and Ser85 → Ala, Ile88 → Leu and Leu89 → Val in the third β-strand). Co-operativity is, perhaps, expected because of stabilisation or destabilisation of secondary structure elements. The effect of the mutations located in the same secondary structure element has, in general, the same algebraic sign and, at least for the first α-helix and guanosine binding loop, they are additive. This suggests that the stabilizing or destabilizing effect of a single mutation in a secondary structural element does not affect the strength of the interactions made with the rest of the protein by another residue in the same secondary structural element.

In a similar study on subtilisin BPN′, using a different approach, it was found that small independent changes in the stability of the protein were additive and resulted from very subtle changes in the structure of the protein. As in barnase to binase, the increase in stability is due to several different effects, ranging from better van der Waals' contacts to improved hydrogen bonding (Pantoliano *et al.*, 1989). Also there are two close substitutions (positions 217 and 218) that do not exhibit co-operativity. Not all of the interactions found in a protein are co-operative, and probably one of the more important lines of research on protein stability in the future will be the elucidation of which interactions are co-operative and which are not.

(c) *Protein activity*

It is suggested that at mesophilic temperatures thermo-stable proteins are less flexible (Vihinen, 1987). Flexibility is essential for the regulation and function of several proteins. Consequently, thermo-stable proteins should have an optimum activity at higher temperatures than for the equivalent mesophilic proteins. A protein engineering study on barnase indicates that there is an inverse correlation between stability and activity for some of the interactions that take place in the active site (Meiering *et al.*, 1992). The analysis of the activity of our more stable multiple mutants indicates that there need not be an inverse relationship between activity and stability for mutations of residues that are not in the active site and thus it is possible to have more stable proteins with the same activity at the same temperature as wild-type protein.

The one odd result is that the activity of binase towards the dinucleotide, GpUp, is 25 times higher than that of barnase (at pH 5·8) and the stable multiple mutants, and that the activity of binase towards RNA is some five times lower than that of barnase (at pH 8·5). This is true for both wild-type binase and binase expressed from an artificially constructed gene that does not have the deletion, Gln2 → Δ, found in wild-type. There are five differences between barnase and binase that are located close or in the guanosine binding loop (Ile55 → Val, Lys62 → Arg, Gly65 → Ser, Lys66 → Ala and Leu89 → Val; Sevcik *et al.*, 1990), which could account for this difference, either independently or co-operatively. The pH dependence of the activity of barnase towards dinucleotides is sharply bell-shaped (Mossakowska *et al.*, 1989). A complete analysis of the pH dependence of activities of mutants will be required for more detailed studies. This, and the construction of further hybrids, is the subject of ongoing investigation in this laboratory.

(d) *Protein evolution*

One of the central questions in protein evolution is how the changes in the amino acid sequence of a protein result in changes in stability and structure and, therefore, in its function. It is generally assumed that small changes in the sequence of a protein should not have large effects on its structure or stability and that changes on the surface of a protein should have a smaller impact than those in buried regions. Consequently, and according to the neutral theory of evolution, the great majority of mutant substitutions will be caused by random fixation through sampling drift of selectively neutral mutants (Kimura, 1991). It follows from this theory, that unless there is a selective pressure for changing the stability of a protein, the stability of two related proteins should be similar and the effect of single or multiple mutations along the evolutionary pathway should not be bigger than the difference in stability between the two proteins (neutral corridor) (Malcom *et al.*, 1990). Under selective pressure to increase the stability of an enzyme, there will be an accumulation of small stabilizing mutations that will result in an overall increase of the stability of the protein. This is in agreement with the analysis of thermo-stable enzymes that has indicated that many small changes over the entire polypeptide chain are responsible for protein thermo-stabilization (Argos *et al.*, 1979).

Our results do not prove that the evolution from barnase to binase, or *vice versa*, has followed a neutral corridor, since we do not have other proteins intermediate between the two. However, the observation that the effect on protein stability of every one of the mutants is not higher or lower than the difference in stability between barnase or binase, and that the effects are additive, could be interpreted as there being an evolutionary pressure to maintain the stability of these proteins within a narrow range of $\pm 10\%$ of their total stability. These results would then be in good agreement with the neutral theory of evolution. Alternatively, there could be simply selective pressure to maintain stability above a certain threshold level. There will then be no further pressure to increase stability or to retain any mutations that do increase stability. Our results do, however, show that it is possible to increase significantly the stability of a protein by the addition of stabilizing mutations.

5. Conclusions

This study suggests a method to improve the stability of proteins without any prior knowledge about their structure or the principles that govern protein stability. Choose two highly related proteins, make all the individual mutants between the two, determine the changes in stability, select the mutations that increase stability and then construct the multiple mutants that contain the favourable mutations. This is most likely to work with proteins that have not diverged too much, so minimizing the number of mutations to be made and the likelihood of many co-operative changes in structure or stability, as has been found in the case of highly divergent proteins (< 45% homology; Imanaka *et al.*, 1986).

L.S. was supported by an EC fellowship.

References

Argos, P., Rossmann, M. G., Grau, U. M., Zuber, H., Frank, G. & Tratschin, J. D. (1979). Thermal stability and protein structure. *Biochemistry*, **18**, 5698–5703.

Baudet, S. & Janin, J. (1991). Crystal structure of a barnase d(GpC) complex at 1·9 Å resolution. *J. Mol. Biol.* **214**, 123–132.

Bycroft, M., Ludvigsen, S., Fersht, A. R. & Poulsen, F. M. (1991). Determination of the three dimensional solution structure of barnase using nuclear magnetic resonance spectroscopy. *Biochemistry*, **30**, 8697–8701.

Clackson, T., Güssow, D. & Jones, P. T. (1991). *PCR: A Practical Approach* (McPherson, M. J., Quirke, P. & Taylor, G. R., eds), pp. 187–214, IRL Press, Oxford.

Day, A. G., Parsonage, D., Ebel, S., Brown, T. & Fersht, A. R. (1992). Barnase has subsites that give rise to large rate enhancements. *Biochemistry*, **31**, 6390–6395

Fersht, A. R. (1987). The hydrogen bond in molecular recognition. *Trends Biochem. Sci.* **12**, 301–304.

Fersht, A. R. & Serrano, L. (1993). Principles of protein stability determined from protein engineering experiments. *Curr. Opin. Struct. Biol.* **3**, 75–83.

Hartley, R. W. (1975). A two state conformational transition of the extracellular ribonuclease of *Bacillus amyloliquefaciens* (barnase) induced by sodium dodecyl sulfate. *Biochemistry*, **14**, 2367–2370.

Hill, C., Dodson, G., Heinemann, U., Saenger, W., Mitsui, Y., Nakamura, K., Borisov, S., Tischenko, G., Polyakov, K. & Pavlovsky, S. (1983). The structural and sequence homology of a family of microbial ribonucleases. *Trends Biochem. Sci.* **8**, 364–369.

Horovitz, A., Serrano, L., Avron, A., Bycroft, M. & Fersht, A. R. (1991). Strength and co-operativity of contributions of surface salt-bridges to protein stability. *J. Mol. Biol.* **216**, 1031–1044.

Horovitz, A., Matthews, J. & Fersht, A. R. (1992). α-Helix stability in proteins. I. Factors that influence stability at internal positions. *J. Mol. Biol.* **227**, 560–568.

Imanaka, T., Shibazaki, M. & Takagi, M. (1986). A new way of enhancing the thermostability of proteases. *Nature (London)*, **324**, 695–697.

Kellis, J. T., Jr, Nyberg, K., Sali, D. & Fersht, A. R. (1988). Contribution of hydrophobic interactions to protein stability. *Nature (London)*, **333**, 784–786.

Kellis, J. T., Jr, Nyberg, K. & Fersht, A. R. (1989) Energetics of complementary side-chain packing in a protein hydrophobic core. *Biochemistry*, **28**, 4914–4922.

Kimura, M. (1991). Recent development of the neutral theory viewed from the Wrightian tradition of theoretical population genetics. *Proc. Nat. Acad. Sci., U.S.A.* **88**, 5969–5973.

Loewenthal, R., Sancho, J. & Fersht, A. R. (1991). Fluorescence spectrum of barnase: Contributions of three tryptophan residues and a histidine-related pH-dependence. *Biochemistry*, **30**, 6775–6779.

Loewenthal, R., Sancho, J. & Fersht, A. R. (1992). Histidine–aromatic interactions in barnase: Elevation of histidine pK_a and contribution to protein stability. *J. Mol. Biol.* **224**, 759–770.

Malcom, B. A., Wilson, K. P., Matthews, B. W., Kirsch, J. F. & Wilson, A. C. (1990). Ancestral lysozymes reconstructed, neutrality tested, and thermostability linked to hydrocarbon packing. *Nature (London)*, **345**, 86–89.

Matsumura, M., Becktel, W. J. & Matthews, B. W. (1988). Hydrophobic stabilization in T4 lysozyme determined directly by multiple substitutions of Ile3. *Nature (London)*, **334**, 406–410.

Matthews, B. W., Nicholson, H. & Becktel, W. J. (1987). Enhanced protein thermostability from site-directed mutations that decrease the entropy of unfolding. *Proc. Nat. Acad. Sci., U.S.A.* **84**, 6663–6667.

Mauguen, Y., Hartley, R. W., Dodson, E. J., Dodson, G. G., Bricogne, G., Chothia, C. & Jack, A. (1982). Molecular structure of a new family of ribonucleases. *Nature (London)*, **297**, 162–164.

Meiering, E. M., Serrano, L. & Fersht, A. R. (1992). Effect of active site residues in barnase on activity and stability. *J. Mol. Biol.* **225**, 585–589.

Menendez-Arias, L. & Argos, P. (1989). Engineering protein thermal stability: Sequence statistics point to residue substitutions in alpha-helices. *J. Mol. Biol.* **206**, 397–406.

Miller, S., Janin, J., Lesk, A. M. & Chothia, C. (1987). Interior and surface of monomeric proteins. *J. Mol. Biol.* **196**, 641–656.

Mossakowska, D. E., Nyberg, K. & Fersht, A. R. (1989). Kinetic characterization of the recombinant ribonuclease from *Baccillus amyloliquefaciens* (barnase) and investigation of key residues in catalysis by site-directed mutagenesis. *Biochemistry*, **28**, 3843–3850.

Mrabet, M. N., Vandenbroeck, A., Vandenbrande, I., Stanssens, P., Laroche, Y., Lambeir, A. M., Matthijssens, G., Jenkins, J., Chiadmi, M. & Vantilbeurgh, H. (1992). Arginine residues as stabilising elements in proteins. *Biochemistry*, **31**, 2239–2253.

Paddon, C. J. & Hartley, R. W. (1987). Expression of *Bacillus amyloliquefaciens* extracellular ribonuclease (barnase) in *Escherichia coli* following an inactivating mutation. *Gene*, **53**, 11–19.

Pantoliano, M. W., Whitlow, M., Wood, J. F., Dodd, W., Hardman, K. D., Rollence, M. L. & Bryan, P. N. (1989). Large increases in general stability of subtilisin BPN′ through incremental changes in the free energy of unfolding. *Biochemistry*, **28**, 7205–7213.

Pavlovsky, A. G., Vagin, A. A., Vainstein, B. K., Chepurnova, N. K. & Karpeisky, M. Ya. (1983). Three-dimensional structure of ribonuclease from *B. intermedius* 79 at 3·2 Å resolution. *FEBS Letters*, **162**, 167–170.

Pavlovsky, A. G., Sanishvili, R. G., Strkopytov, B. V., Vainshtein, B. K., Vagin, A. A., Chepurnova, N. K. & Borisova, S. N. (1989). The structure of 2 crystal modifications of *Bacillus intermedius* ribonuclease. *Kristallografiya*, **34**, 137–142.

Sali, D. (1990). Interactions contributing to barnase stability. Ph D thesis, Cambridge University.

Sancho, J., Serrano, L. & Ferscht, A. R. (1992). Histidine residues at the N-termini of α-helices: perturbed pK_as and protein stability. *Biochemistry*, **31**, 2253–2258.

Sayers, J. R., Schmidt, W. & Eckstein, F. (1988). 5′–3′ Exonucleases in phosphorothioate-based oligonucleotide directed mutagenesis. *Nucl. Acids Res.* **16**, 791–802.

Serrano, L. & Fersht, A. R. (1989). Capping and α-helix stability. *Nature (London)*, **342**, 296–299.

Serrano, L., Horovitz, A., Avron, B., Bycroft, M. & Fersht, A. R. (1990). Estimating the contribution of engineered surface electrostatic interactions to protein stability using double-mutant cycles. *Biochemistry*, **29**, 9343–9352.

Serrano, L., Kellis, J. T., Cann, P., Matouschek, A. & Fersht, A. R. (1992). The folding of an enzyme. II. Substructure of barnase and the contribution of different interactions to protein stability. *J. Mol. Biol.* **224**, 783–804.

Sevcik, J., Sanishvili, R. G., Pavlovsky, A. G. & Polyakov, K. M. (1990). Comparison of active sites of some microbial ribonucleases: structural basis for guanylic specificity. *Trends Biochem. Sci.* **15**, 158–162.

Vihinen, M. (1987). Relationship of protein flexibility to thermostability. *Protein Eng.* **1**, 477–480.

Vriend, G. (1990). WHAT IF: A molecular modeling and drug design program. *J. Mol. Graph.* **8**, 52–56.

Edited by J. Karn

Chapter 12

Protein Folding: Pathways and Φ-Value Analysis

'For these reasons, it is unlikely that the unfolding approach will provide new insight into the still enigmatic process of protein folding.'

J. Buchner and T. Kiefhaber, Nature 343, 601–602 (1990)

Stage 2. Using Changes in Energetics to Analyse Transition States

We were now ready to map out the transition state of folding using structure–activity relationships by combining kinetic and equilibrium measurements. The refolding of barnase is somewhat complicated because there is a folding intermediate, and so I chose to look at the unfolding of the protein by mixing an aqueous solution of the protein with concentrated solutions of denaturant urea. Remember that my work on leaving Brandeis mapped out the transition states for very slow acyl-transfer reactions by looking at the reverse reaction, and using the principle of detailed balance or microscopic reversibility to work out the forward reaction. Protein unfolding kinetics has the advantage that it starts from the best-defined state on the pathway — the native state. To do the kinetics for protein unfolding, I designed a hand-pushed rapid-mixing device that had a long narrow delay tube with long dead time of some 13 ms before reaching the observation cell. By that time, the dense urea solutions had completely mixed with the aqueous buffer and the tube prevented the diffusion of further reagents into the cell. Liz Meiering introduced me at a Gordon Research Conference by relating her amusement at seeing me energetically urging Andreas Matouschek to 'push harder' while we were testing the mixer and estimating the flow rate from the distance the water jet hit the floor of the lab after being projected from a bench. Andreas' subsequent work led to Φ-value analysis: the transfer to protein folding of the LFERs that I had adapted from classical physical-organic chemistry to analyse catalysis and binding in the tyrosyl-tRNA synthetase.

I had recently reported[64] that the analysis of the energetics of mutagenesis differs from that of covalent changes and so the term β, which we had used for the LFERs for the tyosyl-tRNA synthetase, is not strictly analogous to the β used for classical analysis. So, I threw modesty to the wind and used Φ to denote the fraction of binding energy formed in the transition state. Our paper[104], although only five pages long, contains all the essence of Φ-value analysis, including the choice of mutations, the effects of the different types of mutation on the denatured state, the use of thermodynamic cycles, and the interpretation of Φ-values for the different types of mutation. Very importantly, we studied the unfolding reaction of barnase rather than the folding because the latter has complications.

Just as Tony Wilkinson had been the lucky student to be at the very beginning of protein engineering and Tim Wells at the beginning of mechanistic studies, Andreas Matouschek was the lucky one to be there at the beginning of our folding studies. 'Lucky' is in Louis Pasteur's sense of 'chance favouring the well-prepared mind,' which means in practice of having the drive, intelligence and skills to recognise a golden opportunity and grab it with both hands. Andreas was the first of a succession of German Ph.D. students to study with me. Also, several German students did their diploma work in my lab, the most successful being Patrick Cramer, who has had a stunning career and was elected to the German Academy of Sciences Leopoldina in 2009.

Andreas's first paper was a landmark in the study of protein folding[104]. We had been able to reconstruct the transition state for the unfolding of barnase at near-atomic level detail. In terms of the reverse reaction, the final rate-determining transition state for the folding of barnase, we had worked out to what extent the various helices, strands of β-sheet and hydrophobic core had been formed.

12.1 2006: Luis Serrano, Andreas Matouschek and Valerie Daggett at Bader Award Symposium.

12.2 1991: Nadia el-Masry and Ronny Loewenthal.

12.3 1992: Patrick Chene, Marco Moracci and Michael Rheinnecker.

12.4 1991: Joerg Eder.

Mapping the transition state and pathway of protein folding by protein engineering

Andreas Matouschek, James T. Kellis Jr, Luis Serrano & Alan R. Fersht*

MRC Unit for Protein Function and Design, Department of Chemistry, University of Cambridge, Lensfield Road, Cambridge CB2 1EW, UK

In the transition state for unfolding of barnase, the hydrophobic core between the major α-helix and β-sheet is somewhat weakened, the C terminus of the major helix is largely intact but its N terminus is exposed and a major loop has been invaded by solvent.

Two fundamental problems to be solved in the mechanism of protein folding are the stability of the folded state and the pathway of folding—in classical terms, the thermodynamics and kinetics. Just as pathways in classical enzyme kinetics have been most rigorously characterized when stable intermediates are present, so progress in understanding protein folding has been most impressive when intermediates have been detected; for example, by trapping transient covalent linkages between cysteines[1] or transient secondary structure by isotope exchange and NMR[2,3]. Recently, kinetic measurements have been made on the folding of mutant proteins that give information on the roles of specific residues in the processes[4-7]. We now present a general approach for examining the role of non-covalent interactions in protein folding based on kinetic and thermodynamic methods developed for studying catalysis by the tyrosyl-transferRNA synthetase using protein engineering[8,9]. The overall strategy is: (1) establish and measure the interactions that stabilize different parts of the molecule from experiments on the equilibrium stability of mutant proteins; (2) use kinetic measurements of protein folding and unfolding to determine what fraction of the stabilization energy measured in 1 is used to stabilize the transition state for folding. This procedure has the potential of discovering the sequence of folding of different parts of the molecule and the interactions in the transition state. Each mutation engineered into the protein acts as a reporter group for the progress of folding at the appropriate position. Crucial to this approach is the choice of mutations. First, as described below, these are strategically located around the molecule to report back on the behaviour of different types of structural elements in the folded enzyme. Second, each mutation was designed to remove a small interaction that stabilizes a discrete region of structure—complex mutations that affect multiple interactions or introduce adverse steric effects were carefully avoided. The effects of more radical and random mutations are often too complicated to interpret. We also show how the changes in kinetics and equilibria of unfolding of mutant enzymes are related and how the relationship may be analysed.

The enzyme studied is barnase, the RNase from *Bacillus amyloliquefaciens* which consists of a single polypeptide chain with no crosslinks. Its virtues for measuring equilibrium interaction energies have been discussed elsewhere[10-12]. In particular, changes in the free energy of folding on mutation can be accurately measured with little intrusion from artefact[12]. It is also a useful system for studying the role of non-covalent interactions in the kinetics of folding. First, there are no disulphide linkages to constrain the unfolded state. Second, the three prolines in barnase, unlike those in RNase A for example, are all *trans*[13].

* To whom correspondence should be addressed.

The rate-determining step in folding of proteins that contain *cis* prolines in the native structure is sometimes the slow isomerization of proline from the dominant *trans* conformation in the denatured protein[14]. Only a small fraction of the denatured barnase exists with *cis* prolines and proline-independent rate constants for unfolding and refolding are readily measured by stopped-flow fluorescence spectroscopy (unpublished data). Here, we are concerned with just measurements of unfolding.

Transition state

The transition state of a single-step chemical reaction is defined simply as the structure at the highest point on the free-energy profile. For a reaction with a series of intermediates, the reaction of each intermediate involves a separate transition state. The accepted convention is that the highest energy transition state is referred to as the transition state for the overall reaction. The transition state for protein unfolding/folding must be an extreme example of a reaction with many intermediates. All physical and spectroscopic evidence on protein structure indicates that individual parts of proteins breathe and reversibly unfold locally at relatively rapid rates. As the structure of the whole of the molecule changes during unfolding and there will be local minima for many of the breathing and local unfolding motions, the highly crenulated free-energy profile for unfolding of Fig. 1 is expected. (This does not contravene the two-state assumption in equilibrium studies because all of the intermediates are at high energy with respect to folded and unfolded states.) The transition state for unfolding is likely to be fuzzy in some parts because the mobile parts of the protein can rapidly interconvert between different conformations of similar energies.

Definitions of rate-determining transition states can become complicated when there are intermediates along the reaction pathway which are more stable than the initial state. This can happen during protein folding[2,3]. The rate-determining transition state in the reaction is then the highest energy transition

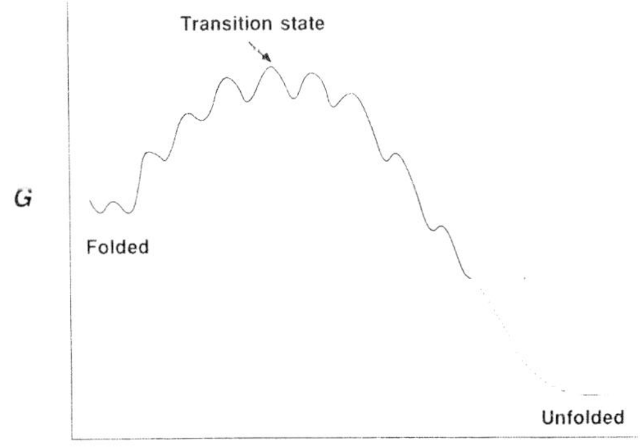

FIG. 1 Illustration of the free-energy profile and transition state for protein unfolding/folding. There are likely to be many small minima during the progress to the highest energy transition state (that is the 'transition state' for the overall reaction) because of the accessibility of different energy states for localized transitions.

Reprinted from Nature, Vol. 340, No. 6229, pp. 122-126, 13 July 1989 state after the stable intermediate. In this case, the kinetics provide information on the energy difference between the transition state and the stable intermediate. To avoid complications in this study from stable intermediates, we have studied the unfolding reaction because we know the structure of the initial and most stable state—that of the native enzyme.

Mutations

Mutations have been prepared that probe various parts of the folded protein (Fig. 2). The hydrophobic core of the protein is formed by the packing of the major α-helix (residues 6–18) on the antiparallel β-sheet. These interactions can be perturbed by changing Leu 14, Ile 88 and Ile 96 to other amino-acid residues by mutation[10,12]. The protonated imadazole of His 18 interacts with the dipole of the major α-helix at the C cap[11] and also makes a hydrogen bond with the backbone NH of Gln 15. Mutation of His 18 is thus a probe of helix formation. The hydroxyl of Thr 6 stabilizes the N terminus of the major helix by acting as a hydrogen-bond acceptor from the backbone >NH groups of Phe 7, Asp 8 and Gly 9 (L.S. and A.R.F.; manuscript in preparation). The hydroxyl of Thr 26 does the same for residues 27, 28 and 29 of the minor α-helix (residues 26–33) (data not shown). New probes constructed here are Tyr→Phe 78 that disrupts two hydrogen bonds across a loop of the enzyme to the backbone NH and CO of Gly 81, and Thr→Ser 16 on the solvent-exposed surface of the major helix, close to the C terminus of the helix. In all cases, the residues were mutated by making non-disruptive deletions[15] that remove small interactions by making minimal changes in structure that are unlikely to cause serious steric effects. This is even more important in studying the effects of mutation on folding than on catalysis because it is likely that the transition state for folding (Fig. 1) is more easily changed than that for a chemical reaction.

Urea denaturation

Urea denaturation was used in the kinetic and equilibrium measurements. The basic ideas used are those described by Tanford[16]. Proteins unfold in urea because side chains and polypeptide backbone are more soluble in urea than in water and unfolding leads to an increase in the exposure of buried structural elements. It is found experimentally that the free energy of transfer of side chains and peptide backbone to urea solutions is approximately linearly proportional to the

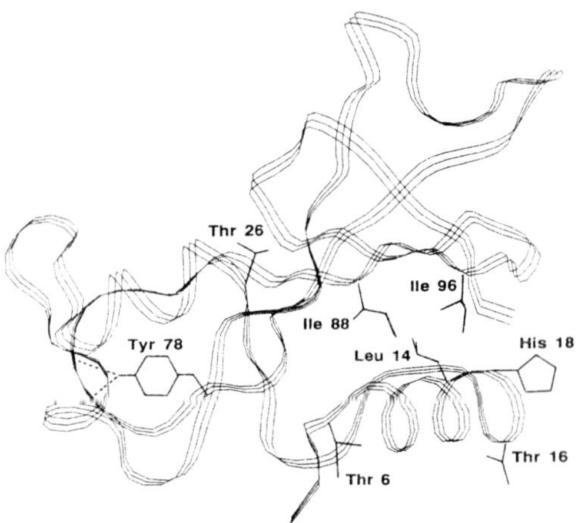

FIG. 2 Sketch of structure of barnase and the residues of interest (produced from coordinates kindly provided by Professor G. Dodson and Dr C. Hill). Thr 6 and Thr 26 act as N caps[20,21], accepting hydrogen bonds from the backbone NH groups of principally Gly 9 and Glu 29. The hydroxyl group of Tyr 78 forms hydrogen bonds with the backbone NH and CO groups of Gly 81. The protonated imidazole of His 18 makes an electrostatic interaction with the α-helix dipole, a hydrogen bond with the backbone NH of Gln 15 and hydrophobic interactions with Trp 94 (ref. 11). Ile 88, Ile 96 and Leu 14 form part of the hydrophobic core. Mutant enzymes were prepared as described by Serrano and Fersht (manuscript in preparation) and in refs 10–12. The mutation Tyr→Phe 78 was directed by the primer: 5'-ATAATTGAA*ATGTAGTC, and Thr→Ser 16 by 5'-TAGAAGTCA*GTATAGTA, where an asterisk follows the mismatch.

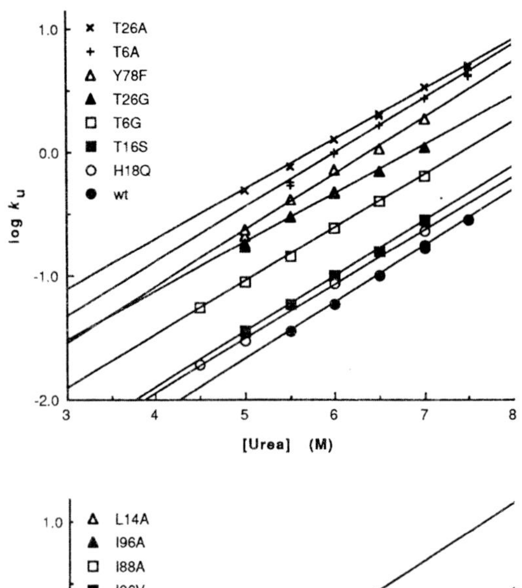

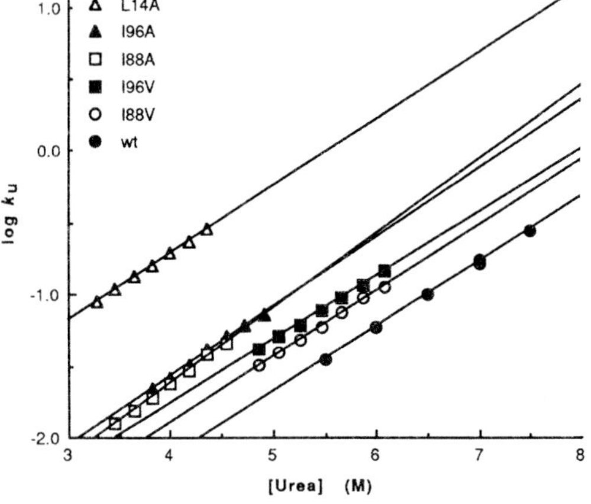

FIG. 3 Plots of log k_u versus [Urea] for wild-type and mutant barnases. Rate constants measured in units of s^{-1}. All experiments were performed in 50 mM 2-(*N*-morpholino)ethane sulphonic acid buffer (pH 6.3) and at 25 ± 0.1 °C, monitoring the unfolding by the change in intrinsic fluorescence of the protein (λ_{ex} = 290 nm, λ_{em} = 315 nm). Top; unfolding of wild-type enzyme and mutants altered in the helices and loop. Reactions were followed with a Perkin Elmer MPF-44B fluorescence spectrophotometer equipped with a rapid mixing head. The mixing device contained a T-jet mixing chamber followed by a 30-ms delay loop ensuring complete mixing of the solutions before observation. A cylindrical quartz cuvette of 3-mm diameter was used as the observation cell. The solutions were driven through the mixing chamber manually from two Hamilton syringes of different volumes to allow various mixing ratios. Data were acquired with a Tandon Target Microcomputer, a DT2801 Data Translation board and the Bio-kine data acquisition software package by Bio-logic and analysed using the program Enzfitter by R. J. Leatherbarrow. Unfolding was initiated by diluting the aqueous protein solution (~0.2 mg ml^{-1}) into urea solutions of different concentrations at a ratio of 1:10. Final concentrations of urea in the observation cell were between 4 and 8 M. Bottom, unfolding of mutants altered in the hydrophobic core. Reactions were followed after manual dilution of barnase (80 μl) into 0.8 ml of buffered urea solution using a Perkin Elmer LS50 fluorescence spectrophotometer (apart from L14A which was monitored as above). Wild-type data from the top are included for reference. The slopes of all lines are within ±5% of 0.45 except for those of Thr→Ala 26 and Thr→Gly 26 which are 10% lower and Ile→Ala 88 which is 16% higher.

concentration of urea ([Urea])[16]. This is consistent with the experimental observation that the free energy of unfolding (ΔG_U) is linearly proportional to [Urea]:

$$\Delta G_U = \Delta G_U^{H_2O} - m[\text{Urea}] \tag{1}$$

where $\Delta G_U^{H_2O}$ is the free energy of unfolding in water and m is a constant that is proportional to the increase in degree of exposure of the protein on denaturation[16]. Rate constants for unfolding and folding also change with [Urea][16]. For example, the rate constant for unfolding, k_u, is often found to increase with increasing [Urea] according to equation 2 where $k_u^{H_2O}$ is the rate constant in H_2O (ref. 16).

$$\log k_u = \log k_u^{H_2O} + m_{k_u}[\text{Urea}] \tag{2}$$

In terms of activation energies for the reaction[16]:

$$\Delta G_U^{\ddagger} = (\Delta G_U^{H_2O})^{\ddagger} - m_u^{\ddagger}[\text{Urea}] \tag{3}$$

The transition state is intermediate between the folded and unfolded states in the degree of exposure to solvent. The fractional increase in exposure of the protein in the transition state for unfolding relative to that on complete unfolding is given by $m_u^{\ddagger}/m$. This ratio has been used in the past to estimate the average fractional increase in exposure in the transition state[16].

The classical behaviour is followed by the mutants in this study (Fig. 3). In all cases, plots of log k_u versus [Urea], at concentrations where the enzymes are greater than 95% unfolded at equilibrium, give excellent straight lines which are close to being parallel. The lowered stability of the mutants is reflected in higher rates of unfolding. The average slope (m_{k_u}) is 0.448 which is equivalent to a value of 0.61 for $m_u^{\ddagger}$. Compared with a value of m for complete unfolding of 2.2, the fractional increase in solvent exposure of the enzyme on going to the transition state is 0.28.

Kinetics and equilibria

The kinetic data show that the mutant enzymes have lower activation energies of unfolding than does wild type. The change in activation energy is defined by $\Delta\Delta G_U^{\ddagger} = \Delta G_U^{\ddagger} - \Delta G_U'^{\ddagger}$, where $\Delta G_U^{\ddagger}$ is the activation energy for the unfolding of wild-type enzyme and $\Delta G_U'^{\ddagger}$ is that for mutant. Experimentally, $\Delta\Delta G_U^{\ddagger} = -RT \ln (k_u/k_u')$. Important new information about the transition state is available by comparing changes in activation energies from kinetics with overall changes in equilibrium energies. The change in free energy of unfolding is defined by $\Delta\Delta G_U = \Delta G_U - \Delta G_U'$, where ΔG_U is the free energy of unfolding of wild-type enzyme and $\Delta G_U'$ is that of mutant. Experimentally, $\Delta\Delta G_U = -RT \ln (K_U/K_U')$, where K_U is the equilibrium constant for unfolding. It is tempting to take the ratio of $\Delta\Delta G_U^{\ddagger}/\Delta\Delta G_U$ and equate it directly to β-values in physical-organic chemistry (ref. 17, chapter 2) whereby β is a measure of the extent of bond breaking in the transition state ($\beta = 0$ for no bond breaking, and $\beta = 1$ for complete bond breaking). This cannot be assumed here because here there is an additional term in the equations that is required to account for changes in solvation energy on mutation[18]. A full analysis is given in Fig. 4, where $\Delta\Delta G_U^{\ddagger}$ and $\Delta\Delta G_U$ are related by a pair of thermodynamic cycles, and in equation 4 (compare with ref. 17 chapters 2 and 13):

$$\Delta\Delta G_U^{\ddagger}/\Delta\Delta G_U = (\Delta G_F - \Delta G_{\ddagger})/(\Delta G_F - \Delta G_{solv}) \tag{4}$$

where, as defined in Fig. 4, ΔG_F is the difference in the non-covalent interaction energy between the folded states of wild-type and mutant enzymes, $\Delta G_{\ddagger}$ between the transition states, and ΔG_{solv} the unfolded states.

ΔG_F, $\Delta G_{\ddagger}$ and ΔG_{solv} are, of course, global energy terms that refer to the energies of the whole protein. It can be useful, but not essential, for interpretation to attribute ΔG_F and $\Delta G_{\ddagger}$ mainly to the changes at the bonds disrupted on mutation. This is justifiable when small changes are made that do not cause a significant reorganization of the protein[18]. Similarly, for a fully denatured protein, ΔG_{solv} is the difference in solvation energy between wild-type side chain and the mutant. The intrusion of ΔG_{solv} in equation 4 is the complicating factor that prevents a simple interpretation. It is possible, however, to find circumstances where $\Delta\Delta G_U^{\ddagger}/\Delta\Delta G_U$ may be simply interpreted. (To emphaseize the difference from β-values, $\Delta\Delta G_U^{\ddagger}/\Delta\Delta G_U$ is abbreviated to ϕ.) Three examples follow:

(1) *The target region is exposed to solvent in the transition state to the same extent as in the unfolded state.* Under these circumstances, $\Delta G_{\ddagger} = \Delta G_{solv}$. By inspection of the cycles, $\Delta\Delta G_U = \Delta\Delta G_U^{\ddagger}$, therefore:

$$\Delta\Delta G_U^{\ddagger}/\Delta\Delta G_U = \phi = 1 \tag{5}$$

(2) *The interaction energies in the transition states and folded states are very similar.* If $\Delta G_F = \Delta G_{\ddagger}$, then $\phi = 0$. If $\Delta G_F \approx \Delta G_{\ddagger}$, and $\Delta G_{\ddagger} \gg \Delta G_{solv}$, ϕ is small.

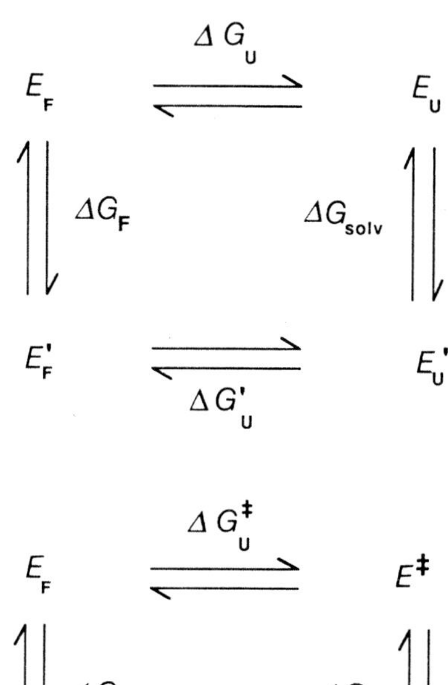

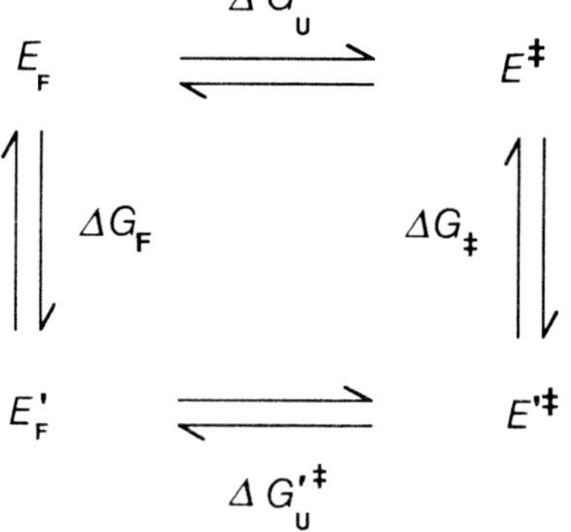

FIG. 4 Thermodynamic cycles that relate free-energy changes in unfolding or wild-type (E) and mutant (E′) enzymes. Top, changes in equilibrium unfolding. The difference between the non-covalent energy of folded wild-type and mutant enzymes is defined by $\Delta G_F = G_{E'_F} - G_{E_F}$. Similarly, the difference in the interaction energies between the unfolded proteins and solvents is given by: $\Delta G_{solv} = G_{E'_U} - G_{E_U}$ (compare with ref. 9). The difference in free energy of unfolding $\Delta\Delta G_U = \Delta G_U - \Delta G_U'$, where $\Delta G_U = G_{E_U} - G_{E_F}$. Bottom, the analogous cycle for the formation of the transition states E‡ and E′‡. By transition state theory (see ref. 17); the chemical activation energy, $\Delta G_U^{\ddagger} = G_E^{\ddagger} - G_{E_F}$. The difference in free energy of activation $\Delta\Delta G_U^{\ddagger} = \Delta G_U^{\ddagger} - \Delta G_U'^{\ddagger}$. The difference between the non-covalent energy of the transition states of wild-type and mutant enzymes is defined by $\Delta G_{\ddagger} = G_{E'^{\ddagger}} - G_{E^{\ddagger}}$. $\Delta G_{\ddagger}$ and ΔG_F may be related by comparing the two cycles as follows. From the top cycle, $\Delta G_F + \Delta G_U' = \Delta G_{solv} + \Delta G_U$; and from the bottom, $\Delta G_{\ddagger} + \Delta G_U^{\ddagger} = \Delta G_F + \Delta G_{U\ddagger}'$. Therefore, $\Delta\Delta G_U = \Delta G_F - \Delta G_{solv}$, and $\Delta\Delta G_U^{\ddagger} = \Delta G_F - \Delta G_{\ddagger}$. The ratio of $\Delta\Delta G_U^{\ddagger}/\Delta\Delta G_U$ is thus equal to $(\Delta G_F - \Delta G_{\ddagger})/(\Delta G_F - \Delta G_{solv})$. Note that the definition of the Brønsted β for unfolding is $\beta = (\Delta G_F - \Delta G_{\dagger})/\Delta G_F$. ($\beta$ for folding $= \Delta G_{\ddagger}/\Delta G_F$.) If $\Delta G_{solv} = 0$, then $\Delta\Delta G_U^{\ddagger}/\Delta\Delta G_U = \beta$ for unfolding.

TABLE 1 Changes in activation ($\Delta\Delta G_U^{\ddagger}$) and free energies of unfolding ($\Delta\Delta G_U$) on mutation of barnase

Mutation	Function of position	$\Delta\Delta G_U$ kcal mol^{-1}	$\Delta\Delta G_U^{\ddagger}$ (at 4 M urea) kcal mol^{-1}	$\Delta\Delta G_U^{\ddagger}$ (in H_2O) kcal mol^{-1}	$\Delta\Delta G_U^{\ddagger}/\Delta\Delta G_U$ (at 4 M urea)	$\Delta\Delta G_U^{\ddagger}/\Delta\Delta G_U$ (in H_2O)
Thr→Gly 6	N cap*	1.34	0.86	1.01	0.64	0.76
Thr→Ala 6	N cap*	2.23	1.66	1.76	0.74	0.79
Thr→Gly 26	N cap*	1.58	1.33	1.67	0.84	1.06
Thr→Ala 26	N cap*	2.14	1.91	2.19	0.89	1.02
His→Gln 18	C cap—charge/helix dipole	1.60	0.23	0.36	0.14	0.22
Thr→Ser 16	Hydrophobic on helix surface	1.87	0.28	0.37	0.15	0.20
Tyr→Phe 78	Bridges loop	1.50	1.39	1.41	0.92	0.94
Leu→Ala 14	Hydrophobic core	4.80	1.91	1.86	0.40	0.39
Ile→Val 88	Hydrophobic core	1.49	0.30	0.28	0.20	0.19
Ile→Ala 88	Hydrophobic core	4.46	0.68	0.33	0.15	0.07
Ile→Val 96	Hydrophobic core	0.98	0.48	0.55	0.49	0.56
Ile→Ala 96	Hydrophobic core	3.52	0.74	0.60	0.21	0.17

All data at 25 °C and pH 6.3. Values of $\Delta\Delta G_U$ are taken from refs 11, 12 and unpublished data (L.S. and A.R.F; manuscript in preparation), apart from those for Thr→Ser 16 and Tyr→Phe 78 which were measured as described in ref. 12.* The nomenclature of ref. 21 for residue that occurs at the N terminus of α-helix and can make hydrogen bonds with the backbone NH groups not involved in intrahelical backbone hydrogen bonding[20].

(3) $\Delta G_{solv}=0$. Mutation of a hydrophilic side chain to a hydrophobic side chain, for example Ser→Ala or Asp→Ala, leads to very large changes in ΔG_{solv} as the hydrophilic groups have large solvation energies, whereas the hydrocarbon groups do not. Conversely, $\Delta G_{solv}\approx 0$ for small changes in aliphatic hydrophobics (for example, Ile→Ala or Val have values of ΔG_{solv} in water for fully exposed side chains of only −0.21 and −0.16 kcal per mol respectively)[19]. Then for mutations such as Ile→Ala or Val and the side in chains in the unfolded structure are exposed to water, ΔG_{solv} can be ignored when ΔG_F is significant and:

$$\Delta\Delta G_U^{\ddagger}/\Delta\Delta G_U = \phi = (\Delta G_F - \Delta G_{\ddagger})/\Delta G_F \quad (6)$$

Relationship between β and ϕ. The above examples are simple to interpret because β and ϕ are, in fact, identical for those circumstances. In case 3, $\Delta G_{solv}=0$, which is the condition for $\phi=\beta$ under all conditions (see legend to Fig. 4). In case 1 where there is complete bond breaking in the transition state, $\beta=1$ by definition, and ϕ is seen to be equal to 1 (equation 5). In case 2 where there is no bond breaking in the transition state, $\beta=0$ by definition, and also $\phi=0$. But, in cases other than 1 or 2 where $\Delta G_{solv}\neq 0$, $\phi\neq\beta$. There will be an increasing discrepancy between ϕ and β in cases 1 and 2 as ϕ differs further from 0 and 1, and will be particularly large where hydrophilic groups are mutated to hydrophobic. Another discrepancy where $\phi\neq\beta$ arises when there is residual structure in the unfolded protein because ΔG_{solv} will then contain the energetics of those structural interactions. Barnase was chosen as there is minimal likelihood of residual structure in its unfolded states, and this assumption is supported by the finding that there are identical free-energy changes of folding measured by thermal-, urea- and guanidinium hydrochloride-induced denaturation[12]. Given this reasoning and assumptions, we can now describe the transition state for unfolding at the positions where mutations have been made.

Unfolding

The ratios $\Delta\Delta G_U^{\ddagger}/\Delta\Delta G_U$ (that is, ϕ) are listed in Table 1. There are only minor differences between the ratios observed at 4 M urea and those extrapolated to 0 M urea, indicating that there are no artefacts from the binding of urea to specific groups of the protein. The interpretations divide into the three classes above:

(1) *Altered region is exposed to solvent in the transition state,* $\Delta\Delta G_U^{\ddagger}/\Delta\Delta G_U\approx 1$. The side chains of the threonines in the N caps form hydrogen bonds with the backbone NH groups at the beginning of the helices. The values of ϕ are close to unity for the mutation Thr 26→Ala or Gly, and the values for Thr 6 mutations are very high. Either the helices are unwound in the transition state in these regions or just the N termini are exposed. (As seen below, it is likely that the helices still exist).

Alteration of Tyr→Phe 78 also has $\phi\approx 1$. In global energy terms this means that the portion of the loop is as exposed to solvent in the transtion state as is the unfolded state. In specific terms, the hydrogen bonds between the tyrosyl hydroxyl and Gly 81 that stablize the folded structure are broken in the transition state and hydrogen bonds are made with water.

It could be argued that the ϕ-values of 1 arise not because the loop has become invaded with solvent but because hydrogen bonds in the loop or N cap are stretched in the transition state and weakened by an energy which is, fortuitously, the same as that observed for complete unfolding. If this is so then there are two states available to the loop which are of equal energy—strained or exposed to solvent. Given the free-energy profile of Fig. 1, it would be expected that both states would be equally occupied in the transition state as they are of the same energy. Thus, a ϕ-value of 1 in this context implies that at least 50% of the loop is exposed to solvent in the transition state.

(2) *Transition state has largely the folded structure,* $\phi\approx 0$. His 18 interacts with the electrostatic dipole at the C terminus of the major α-helix (residues 6–18). The ϕ-value for mutation of His→Gln 18 is only ~0.2. This indicates that the interaction is largely intact in the transition state. For this to occur, the helix must still be present in the transition state. Further evidence for the integrity of the C terminus of the helix is given below.

(3) *Hydrophobic interactions,* $\Delta G_{solv}\approx 0$. The methyl group of Thr 16 makes hydrophobic interactions with the aromatic ring of Tyr 17 in the C terminus of the helix which are lost on unfolding. The ϕ-value of 0.15–0.2 means, according to equation 6, that 80–85% of the non-covalent interaction energy is maintained in the transition state. This is additional evidence that the helix is largely intact in the transition state.

Changing the hydrophobic residues in the core gives ϕ-values in the range 0.3±0.2. Thus some 70% of the interaction energy remains in the transition state. The core is weakened, but still has significant energy.

In conclusion, the transition state for unfolding has significant secondary and compact structure. The major α-helix (residues 6–18) is largely intact, the hydrophobic core is weakened but still important, and the major α-helix is still docked on the central β-sheet. One loop is considerably exposed to solvent in the transition state as are the N termini of both helices.

Folding pathway

The transition state for folding is, by the principle of microscopic reversibility, the same as that for unfolding under identical conditions and for a reversible process. We have thus shown that the structure of the transition state for folding is like the structure of the folded state, with much of the secondary structure having been formed and with significant interactions in the hydrophobic core. These results are consistent with recent NMR evidence that during folding an early framework intermediate forms in which there is significant secondary structure[2,3] and the proposal that the transition state occurs late on the folding pathway[1]. The initial results from our protein engineering study have begun to map out the energetics of these processes and to show new details such as a major loop not being formed in the transition state or the lack of hydrogen bonds from N caps to the exposed backbone NH groups of the helices. These studies now point to new questions that can be answered by studying further mutations. For example, are all the loops exposed to solvent in the transition state and are they the last elements of structure to be formed? □

Received 20 April; accepted 7 June 1989.

1. Creighton, T. E. *Proc. natn. Acad. Sci. U.S.A.* **85,** 5082-5086 (1988).
2. Udgaonkar, J. B. & Baldwin, R. L. *Nature* **335,** 694-699 (1988).
3. Roder, H., Elöve, G. A. & Englander, S. W. *Nature* **335,** 700-704 (1988).
4. Beasty, A. M. *et al. Biochemistry* **25,** 2965-2974 (1989).
5. Garvey, E. P. & Matthews, C. R. *Biochemistry* **28,** 2083-2093 (1989).
6. Chen, B.-I., Baase, W. A. & Schellman, J. A. *Biochemistry* **28,** 691-699 (1989).
7. Goldenberg, D. P., Frieden, R. W., Haack, J. A. & Morrison, T. B. *Nature* **338,** 127-132 (1989).
8. Winter, G., Fersht, A. R., Wilkinson, A. J., Zoller, M. & Smith, M. *Nature* **299,** 756-758 (1982).
9. Fersht, A. R. *Biochemistry* **26,** 8031-8037 (1987).
10. Kellis, J. T., Jr, Nyberg, K., Sali, D. & Fersht, A. R. *Nature* **333,** 784-786 (1988).
11. Sali, D., Bycroft, M. & Fersht, A. R. *Nature* **335,** 741-743 (1988).
12. Kellis, J. T., Jr, Nyberg, K. & Fersht, A. R. *Biochemistry* **28,** 4914-4922 (1989).
13. Mauguen, Y. *et al. Nature* **297,** 162-164 (1982).
14. Schmid, F. X. & Baldwin, R. L. *Proc. natn. Acad. Sci. U.S.A.* **75,** 4764-4768 (1978).
15. Fersht, A. R., Leatherbarrow, R. J. & Wells, T. N. C. *Biochemistry* **26,** 6030-6038 (1987).
16. Tanford, C. *Adv. Protein Chem.* **24,** 1-95 (1970).
17. Fersht, A. R. *Enzyme Structure and Mechanism* 2nd edn (Freeman, New York, 1985).
18. Fersht, A. R. *Biochemistry* **27,** 1577-1580 (1988).
19. Wolfenden, R., Andersson, L., Cullis, P. M. & Southgate, C. C. B. *Biochemistry* **20,** 849-855 (1981).
20. Presta, L. G. & Rose, G. D. *Science* **240,** 1632-1641 (1988).
21. Richardson, J. S. & Richardson, D. C. *Science* **240,** 1648-1652 (1988).

ACKNOWLEDGEMENTS. J.T.K. is a recipient of a NATO postdoctoral fellowship. L.S. is a recipient of an EMBO postdoctoral fellowship.

Like most breakthroughs, there was initial scepticism. Two very talented young German students, who quickly became my good friends, attacked the paper because of the use of thermodynamic cycles and doing protein unfolding and microscopic reversibility. The rebuttal is quite clear[105].

12.5 2009: Alan, Thomas Kiefhaber and Johannes Buchner at Manchot Award dinner.

12.6 1993: Taken by surprise at a carefully kept secret 50th birthday party at Caius College. Derek Elliott serving champagne. Photos by Sophie Jackson.

12.7 1993: Laura Itzhaki and Alan Jasanoff.

12.8 1993: Daniel Otzen, Doug Axe and Sarah Perrett.

Folding pathway enigma

SIR—Fersht and co-workers recently published an interesting approach to studying the transition state and pathway of protein unfolding[1]. They propose a new way to interpret the kinetics of urea-induced unfolding of wild-type and mutant forms of barnase. They find significant differences in the importance of specific protein regions on the energy level of the transition state for unfolding. Their interpretation is that some regions are already exposed to solvent whereas others, such as the N-terminal helix, still show native-like structure in the transition state.

These experiments provide additional insight into the mechanisms of urea-induced protein unfolding, but they do not necessarily provide information on the transition state and pathway of refolding of proteins as suggested by the authors.

Their data analysis is based on the assumption that barnase exists only in two distinct conformational states: native and denatured. The validity of such a two-state model could only be proven by calorimetric measurements which, to our knowledge, have not yet been carried out.

Also, the thermodynamic cycles proposed in the article (see Fig. 4, ref. 1) are purely hypothetical, as the authors have to assume an equilibrium between wild-type and mutant protein which is obviously out of range of experimental handling.

The main shortcoming, however, is the use of the unfolding kinetic data to elucidate the transition state and pathway of protein folding. This is only legitimate under identical conditions, where the principle of microscopic reversibility is valid, that is, under strongly denaturing conditions. The authors' comparison of their results with recent NMR studies on early refolding steps[2,3] is misleading because these studies were carried out under "strongly native conditions". Under these conditions (as Fersht and co-workers agree) the folding pathway is essentially different from the mechanism of unfolding because partially folded intermediates are formed rapidly in the refolding of many proteins[4-6].

For these reasons, it is unlikely that the unfolding approach will provide new insight into the still enigmatic process of protein folding.

JOHANNES BUCHNER

Institut für Biophysik und Physikalische Biochemie, Universität Regensburg, Universitätstrasse 31, D-8400 Regensburg, FRG

THOMAS KIEFHABER

Laboratorium für Biochemie, Universität Bayreuth, Postfach 10 12 51, D-8580 Bayreuth, FRG

FERSHT *ET AL.* REPLY—Any novel analysis will, and should, attract intense scrutiny. All of the points that have been raised have precedents, but we welcome the opportunity to discuss them. The analysis of a forward pathway by studying the reverse, for example, is well-established in physical chemistry and enzymology[7] and in protein folding[8].

First, the use of hypothetical thermodynamic cycles is perfectly valid and is frequently employed in thermodynamic analysis; for example, the classic Born-Haber cycle described in most elementary physical chemistry texts. This is allowed by the first law of thermodynamics, which states that energy changes depend only on initial and final states and not on the pathway between them. Cycles equivalent to those we use[1] are also used by the theoreticians for their free-energy calculations (see, for example, ref. 9).

Second, the data analysis is not based on the assumption that barnase exists in only two distinct conformational states. For the equilibrium unfolding cycle, the analysis is based simply on the energy difference between initial (folded) and final (unfolded) states and the first law of thermodynamics. Just as the law allows us to add hypothetical intermediates, it allows real ones to be ignored. For the kinetic cycle, as explained in the article[1], transition-state theory when applied to the unfolding kinetics avoids the problems of intermediates accumulating, because this happens before the rate determining step only on the refolding pathway. Incidentally, microcalorimetry is not the only way of detecting folding intermediates. Kinetics can detect intermediates that are not observed by equilibrium microcalorimetry and, from measurements of both folding and unfolding rates, either show that a two-state transition is obeyed or that an intermediate accumulates (ref. 10, and our unpublished work). We did indeed, detect an intermediate on the refolding pathway of barnase by kinetic methods and H-D exchange (see refs 2, 3), but its presence does not affect our analysis of unfolding as it occurs after the rate-determining step in the unfolding direction and accumulates only slightly in the transition region (unpublished data).

Third, we stated clearly that microscopic reversibility applies under identical reaction conditions and for a reversible process (ref. 1; see also ref. 7, p. 89-90). It holds not just for "strongly denaturing conditions" but for all concentrations of urea, because folded and unfolded protein always exist in equilibrium even though the equilibrium constant may be far from unity. At any one of these concentrations, the transition states for folding and unfolding are the same. The question is whether or not the transition state changes with change of denaturant. Creighton[11] has summarized the evidence that the nature of the transition state for folding-unfolding does not change with denaturant concentration — the rate of unfolding changes uniformly with changing conditions but refolding changes non-uniformly because the nature of the unfolded state changes[8]. (This is a principal reason why unfolding kinetics is easier to analyse.)

This laboratory has espoused the need for determining the full free-energy pathways of reactions by measuring rate constants in both the forward and reverse directions (see studies on the tyrosyl-tRNA synthetase). We have measured the rate constants for refolding of barnase and many of its mutants and combined these with the unfolding data to characterize the structure of the folding intermediate and the transition states by extension of our protein engineering analysis (manuscript in preparation). The combined data give new information on the folding intermediate but no more information on the transition state than did the unfolding data alone.

Creighton has stated: "Instead of searching for nucleation sites in unfolded proteins, it might be more relevant to search for unfolding nucleation events in the native conformation" (ref. 11). For this reason, and for the necessity of constructing free-energy profiles, it is an essential element of folding studies to measure unfolding rate data. Furthermore, because refolding kinetics is fraught with complications arising from the presence of folding intermediates, *cis-trans* isomerization of prolines, and chemical processes that occur when the denatured protein is incubated for many minutes, unfolding kinetics is simpler to analyse and far less prone to artefact. Unfolding studies by themselves can be used to define the last transition state on the refolding pathway. Unfolding studies might also be more relevant than refolding for relating to phenomena *in vivo*, because the pathway of folding *in vivo* depends on the sequential formation of proteins during translation and may be aided by chaperones, whereas unfolding does not depend on these factors. Far from the unfolding approach being unlikely to provide new insight into protein folding, it should be clear from the above that unfolding studies are an essential element in the understanding of folding, and are just as important as refolding studies.

A.R. FERSHT
J.T. KELLIS JR
A.T.E.L. MATOUSCHEK
L. SERRANO

MRC Unit for Protein Function and Design, University Chemical Laboratory, Lensfield Road, Cambridge CB2 1EW, UK

Reprinted from Nature, Vol. 343, No. 6259, pp. 601–602, 15th February, 1990

1. Matouschek, M., Kellis, J.T., Serrano, L. & Fersht, A.R. *Nature* **340**, 122–126 (1989).
2. Udgoankar, J.B. & Baldwin, R.L. *Nature* **335**, 694–699 (1988).
3. Roder, H., Elove, G.A. & Englander, S.W. *Nature* **335**, 700–704 (1988).
4. Schmid, F.X. & Baldwin, R.L. *J. molec. Biol.* **135**, 199–215 (1979).
5. Kim, P.S. & Baldwin, R.L. *A. Rev. Biochem.* **51**, 459–489 (1982).
6. Kuwajima, K. *Proteins* **6**, 87–103 (1989).
7. Fersht, A.R. *Enzyme Structure and Mechanism*, 89–90, 205–206, 237, 427 (Freeman, New York, 1985).
8. Segawa, S.-I. & Sugihara, M. *Biopolymers* **23**, 2475–2488, 2489–2498 (1984).
9. Rao, S.N., Singh, U.C., Bash, P.A. & Kollman, P.A. *Nature* **328**, 551–554 (1987).
10. Kuwajima, K., Mitani, M. & Sugai, S. *J. molec. Biol.* **206**, 547–561 (1989).
11. Creighton, T.E. *Proc. natn. Acad. Sci. U.S.A.* **85**, 5082–5086 (1988).

Although our approach is theoretically sound, we subsequently demonstrated by Φ-value analysis for several other small proteins that they do fold and unfold via the same transition state, and theoreticians have generated transition states computationally that partition equally to reagents and products. Thomas Kiefhaber continues to make good-natured criticisms of my methods, and I still continue with the rebuttals. He, with the support of Johannes Buchner, recently got me appointed Manchot Visiting Professor at their university in Munich (November 2009).

About a year after this paper, I gave a talk at Stanford. Buzz Baldwin introduced me by saying: 'Alan has an advantage over the rest of us — he has a game plan.' Of course, it was the chess training again.

We followed up by examining the refolding reaction[106]. One conventional approach is to plot the logarithms of the rate constant for folding and unfolding against concentrations of denaturant when, according to earlier work by Charles Tanford, they should be straight lines. But these had never been found for the refolding reaction. Sure enough, the unfolding limb of barnase was straight, but the refolding was curved downwards, tending to sigmoidal. From my training in studying the hydrolysis of aspirin for my doctorate, I knew immediately that this was prima facie evidence for a change in rate determining step for folding with change in denaturant concentration, and hence proof of an intermediate. My interpretation is now called 'rollover'.

12.9 1993: Mike Lubienski, Mark Proctor, Ben Davis and Jacqui Matthews.

12.10 1993: Chris Johnson and Mikael Oliveberg.

Transient folding intermediates characterized by protein engineering

Andreas Matouschek, James T. Kellis Jr, Luis Serrano, Mark Bycroft & Alan R. Fersht*

MRC Unit for Protein Function and Design, Department of Chemistry, University of Cambridge, Lensfield Road, Cambridge CB2 1EW, UK

Kinetic experiments on engineered mutants of barnase detect an intermediate on the folding pathway and allow the mapping of the tertiary interactions of the side chains and their energetics. Many of the interactions present in the final folded state tend to be either fully formed or not formed at all in the intermediate or subsequent transition state for folding, but the hydrophobic core becomes progressively consolidated. These methods in combination with NMR provide extensive structural characterization of the folding intermediate and the sequence of events in the folding pathway.

THE pathway of protein folding is dominated by noncovalent interactions that determine structure and stability. There is growing evidence that the refolding of proteins *in vitro* proceeds through intermediates that can accumulate transiently[1-5]. Unlike intermediates in covalent chemistry, which can frequently be isolated and characterized because of the relatively stable chemical bonds, the intermediates in protein folding, aside from -S-S- linkages, differ just in noncovalent bonding. The ephemeral nature of such bonds causes severe problems in the characterization of folding intermediates and requires novel methods for their investigation[6-9]. A recent high-resolution approach uses NMR, which can detect backbone NH groups that undergo hydrogen exchange slowly with solvent because they are hydrogen bonded within α-helices, β-sheets or other structural elements. Rapid quenching experiments on hydrogen exchange during refolding can detect secondary structure that is formed faster than the final folded structure. For example, regions of β-sheet and α-helix respectively, are formed rapidly in the refolding of RNase A (ref. 6) and cytochrome *c* (ref. 7). Earlier work also shows intermediate states. Spectroscopy and light and X-ray scattering measurements on α-lactalbumin have shown the existence of a state, termed the molten globule, which is stable under mildly denaturing conditions (acid pH, moderate concentrations of denaturants or high temperatures[1-5]). This state has significant secondary structure and is compact, but has a fluctuating hydrophobic core. It is difficult to study by NMR because of broad linewidths and extensive overlapping of peaks but methods have been developed to confirm some of its characteristics[8]. An intermediate state similar to or identical with the molten globule, has been suggested as a general occurrence on folding pathways[10,11].

We have applied protein engineering to analyse the unfolding of barnase, a small RNase from *Bacillus amyloliquefaciens*, and have shown how this approach can characterize the noncovalent structural changes in the transition state, which is inaccessible to the usual spectroscopic methods[12]. Mutations were made that remove individual interactions stabilizing different regions of the folded structure. Each mutation acts as a specific probe for structural changes during unfolding. We show here how kinetics can detect a transiently formed intermediate in the refolding pathway and present the full theory of how the protein engineering procedure can characterize structural properties of the intermediate that cannot be detected by NMR-hydrogen-exchange, such as events in the hydrophobic core or in hydrogen bonds that are in fast exchange, in addition to giving indirect evidence on secondary structure. The procedure also gives the energetics of noncovalent interactions within these conformations. The kinetic studies are complemented and reinforced by a separate study in which NMR-H-exchange experiments identify further secondary structure in the intermediate[13].

* To whom correspondence should be addressed.

Kinetics detects folding intermediate

From experiment, the logarithm of the equilibrium constant for the unfolding of small proteins in urea (K_u) is proportional to the urea concentration[14]:

$$\log K_u = \log K_u^{H_2O} + m_e[\text{urea}] \quad (1)$$

where $k_u^{H_2O}$ is the equilibrium constant for unfolding in water

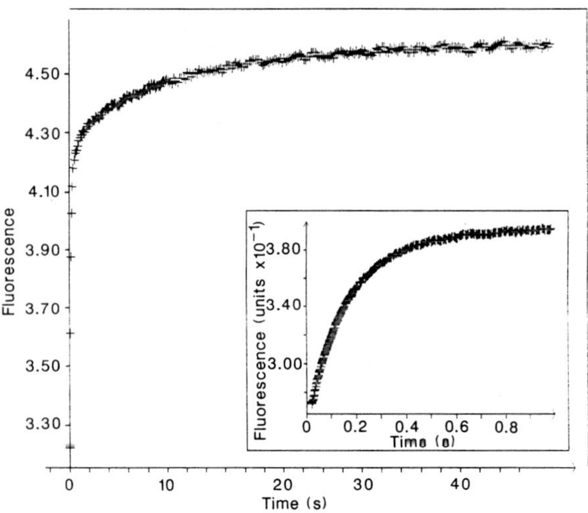

FIG. 1 Refolding of wild-type barnase monitored by increase in tryptophan fluorescence, excitation at 290 nm and emission at 315 nm. Refolding was initiated by mixing protein denatured in 7 M urea, 50 mM MES buffer, pH 6.3 with 10 volumes of 50 mM MES, pH 6.3 at 25 °C to give 0.636 M urea, using a Perkin-Elmer spectrofluorimeter MPF 44B fitted with a stopped-flow mixer[12]. The time scale for the slower phase is 0 to 50 s. Inset, fast phase from 30–950 ms. The fast phase has a rate constant of 6.5 s^{-1} and amplitude 0.131 units whereas the slow phase increases at 0.094 s^{-1} with amplitude 0.032 units. The rate constant for the slow phase varies only within a range of ±20% for different mutants under the same conditions. The thickness of the trace does not represent the noise in the signal but is an artefact of the symbols used by the plotter. The signal-to-noise ratio is better than 50:1. The amplitude of the slow phase is that expected if ~5% of each peptidyl prolyl linkage is *cis*. A value of 0.044 to 0.053 has been recently reported for the equilibrium constant for *cis/trans* Lys 116–Pro 117 in unfolded staphylococcal nuclease[24,25].

Reprinted from Nature, Vol. 346, No. 6283, pp. 440–445, 2nd August, 1990

and m_e is a constant that is proportional to the increase in degree of exposure to solvent of the protein on denaturation. The rate constant for unfolding, k_u, is generally found to increase with increasing urea concentration according to equation (2) where $k_u^{H_2O}$ is the rate constant in H_2O (refs. 5,12–15).

$$\log k_u = \log k_u^{H_2O} + m_{ku}[\text{urea}] \quad (2)$$

Equations (1) and (2) imply that the rate constant for folding, k_F, of a protein that involves the reversible transition between just two states, E_F and E_U, (equation (3))

$$E_F \underset{k_F}{\overset{k_u}{\rightleftharpoons}} E_U \quad (3)$$

must follow the rate law

$$\log k_F = \log k_F^{H_2O} - (m_e - m_{ku})[\text{urea}] \quad (4)$$

(As $\log k_F^{H_2O} = \log (k_u^{H_2O}/K_u^{H_2O})$, the value of $\log k_F$ may be calculated simply from the unfolding rate constants and equilibrium data.)

The kinetics of protein folding is often complicated by the slow and frequently rate-determining isomerization of proline residues[16]. Barnase has three proline residues, which are all *trans* in the native structure. Earlier work suggests that some 5–10% of each of the prolines in the denatured protein should exist in the *cis* conformation and that the conformations should interconvert with a half-life of tens of seconds and an activation energy of about 15 kcal mol^{-1} (ref. 17). We find that when barnase that has been incubated in 6–8 M urea is diluted with 10 volumes of water there is a biphasic regain of folded structure with the two kinetic phases well resolved (Fig. 1). The first phase, 83% of the amplitude measured by stopped-flow fluorescence spectroscopy[12], occurs with a half-life about 50 ms at low urea concentration that increases with increasing urea. The slow phase, 17% of the amplitude, has a half-life of 10–20 s, changes only slowly with urea concentration and has an activation energy of about 13 kcal mol^{-1}. These are just the characteristics expected for the conversion of ~5–10% each of three proline residues in the *cis* conformation in the unfolded enzyme, to *trans* in the folded. (This interpretation has subsequently been confirmed directly by showing that the slow phase is efficiently catalysed by peptidyl prolyl isomerase, L. Meiering, unpublished results). The fast and main phase is the folding of the fraction of unfolded protein that has all its prolines in the *trans* conformation in solution. Measurements of regain of activity on renaturation demonstrate that the fast phase corresponds to the formation of active enzyme. The following discussion refers just to the fast phase.

Figure 2 shows that whereas the simple rate law of equation (1) describes the unfolding of wild-type barnase, $\log k_F$ does not follow equation (4). The value of $\log k_F$ at low urea concentration deviates by many orders of magnitude from that predicted by the kinetics of a two-state equation. This is classic evidence for the existence of a metastable intermediate on the refolding pathway: the intermediate accumulates on a fast time scale and the observed rate constant is that for the reaction of the intermediate and not for its formation. The deviation from ideality is not caused by the equilibrium and rate equations (1) and (2) breaking down at lower concentrations of urea. The evidence for this is: $\log k_u$ can be measured with suitable mutants under more acidic conditions down to 0.5 M urea and is linear with urea concentration over the relevant concentration range (Fig. 2); $K_u^{H_2O}$ measured from the linear extrapolation of

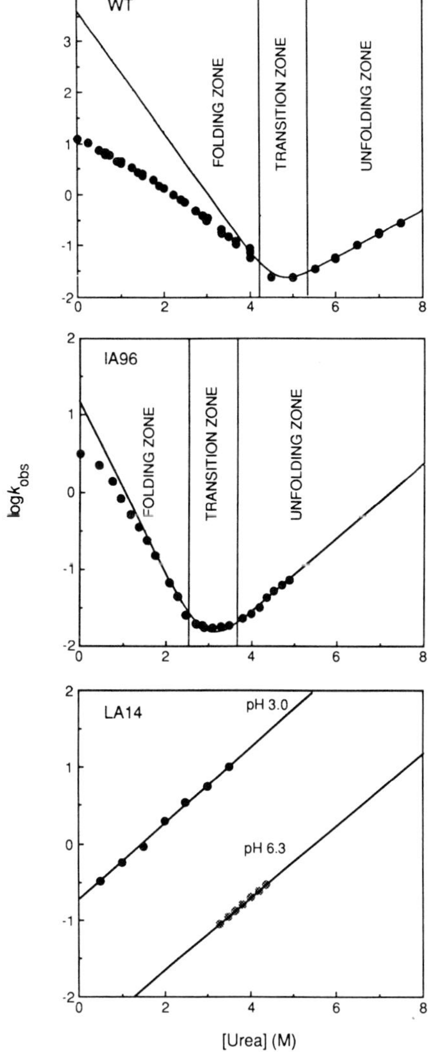

FIG. 2 Urea dependence of rate constants for denaturation and renaturation of wild-type and mutant enzymes. Rate constants are measured in units of s^{-1} at 25 °C, pH 6.3 and 50 mM MES buffer. Upper panel, $\log k_{obs}$ for folding of wild-type enzyme. The experiments at low concentrations of urea consist of solutions of enzyme in high concentrations of urea diluted 11-fold into refolding buffer in the stopped-flow fluorimeter[12]. In this region, k_{obs} is the rate constant for refolding. (The point at 0 M urea was obtained from the renaturation of acid-denatured enzyme, after a pH jump from low pH.) At high urea concentration, folded enzyme in aqueous buffer was mixed with 10 volumes of urea solution, to observe the kinetics of unfolding. In the transition region, both folded and unfolded protein exist at measurable concentrations. The rate constants in this region may be measured by mixing either unfolded or folded protein with urea and monitoring the progress to equilibrium ($k_{obs} = k_u + k_F$ for a simple two-state transition when both forward and reverse rate constants are significant)[26]. The solid curve is that calculated for a simple two-state system (equation (3)) and $\log k_{obs} = \log (k_u + k_F)$ (ref. 26). The rate constant k_u was calculated from the experimental values for the unfolding kinetics, which conform to equation (2) with $k_u^{H_2O} = 1.1 \times 10^{-4}$ s^{-1} and $m_{ku} = 0.463$ M^{-1}. k_F was calculated by inserting the value of k_u into equation (1) using the values of $K_u^{H_2O}$ and m_e from equilibrium unfolding measurements (The two-state approximation holds in the transition region). Middle panel, $\log k_{obs}$ for the mutant IA96 (Ile→Ala). The two-state approximation holds nicely down to less than 2 M urea and well below the transition region. The solid curve was calculated as above using $k_u^{H_2O} = 3.2 \times 10^{-4}$ s^{-1} and $m_{ku} = 0.479$ M^{-1}. Bottom panel, $\log k_u$ measured at low concentrations of urea for the destabilized mutant LA14 (Leu→Ala). The stability of barnase decreases at low pH (ref. 27). At pH 3.0 LA14 is >95% unfolded at 0.5 M urea. LA14 was dialysed against distilled water at pH 7 and mixed in the stopped-flow fluorimeter, as above, with 10 volumes of 110 mM sodium formate, pH 3.0, containing various concentrations of urea. The data for pH 6.3 are given for comparison.

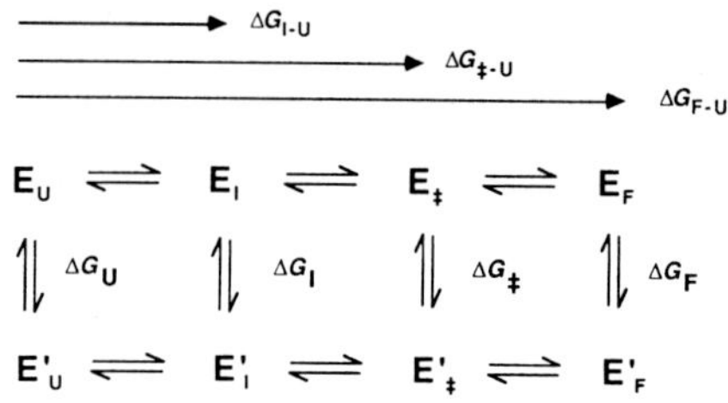

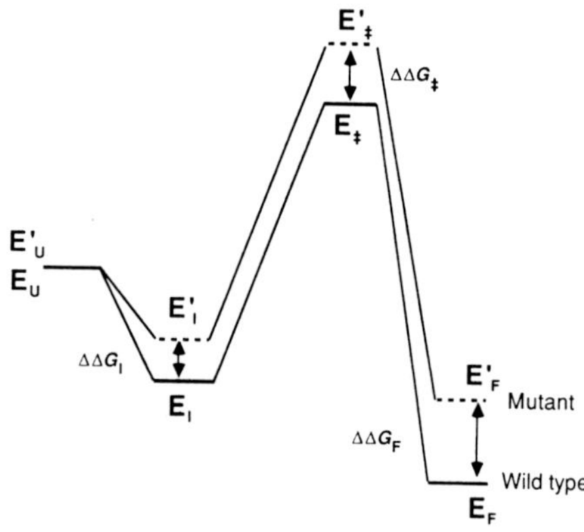

FIG. 3 Thermodynamic analysis. Top, Cycles relating properties of mutant and wild-type enzymes. The energy difference between E_F and E_U, $\Delta G_{F\text{-}U}$ $(=G_{E_F}-G_{E_U})$, is measured from equilibrium unfolding. The energy difference between E_F and E_I, $\Delta G_{F\text{-}I}$ $(=G_{E_F}-G_{E_I})$, is calculated from the ratio k_u/k_{-u}. $\Delta G_{I\text{-}U}$ $(=G_{E_I}-G_{E_U})$ is calculated from $\Delta G_{F\text{-}U}-\Delta G_{F\text{-}I}$. The free-energy of the transition state, $\Delta G_{\ddagger\text{-}F}$ $(=G_{E_\ddagger}-G_{E_F})$ can be calculated from transition state theory[26] and the value of k_u, the rate constant of unfolding[12]. $\Delta G_{\ddagger\text{-}U}$ $(=G_{E_\ddagger}-G_{E_U})$ is calculated by adding $\Delta G_{\ddagger\text{-}F}$ and $\Delta G_{F\text{-}U}$. There may be uncertainties in calculating the absolute values of $\Delta G_{\ddagger\text{-}F}$ by transition state theory but the errors cancel out when measuring $\Delta\Delta G_\ddagger$, the difference in $\Delta G_{\ddagger\text{-}F}$ between wild-type and mutant enzymes, as only ratios of rate constants are used. For wild-type enzyme, the following values are calculated in H_2O at 25 °C and pH 6.3: $\Delta G_{I\text{-}U}=3.2$ kcal mol^{-1}; $\Delta G_{\ddagger\text{-}U}=9.6$ kcal mol^{-1}; and $\Delta G_{F\text{-}U}=-10.2$ kcal mol^{-1}. In the vertical virtual equilibria, ΔG_U is the difference in noncovalent energy between unfolded wild-type and mutant enzymes, $\Delta G_I=G_{E_I}-G_{E'_I}$ is the difference between the folding intermediates, $\Delta G_\ddagger$ is the difference between the transition states and ΔG_F between the folded states. Bottom, Calculation of difference energies. The difference energies are defined as: $\Delta\Delta G_F=\Delta G_{F\text{-}U}-\Delta G'_{F\text{-}U}$; $\Delta\Delta G_\ddagger=\Delta G_{\ddagger\text{-}U}-\Delta G'_{\ddagger\text{-}U}$; $\Delta\Delta G_I=\Delta G_{I\text{-}U}-\Delta G'_{I\text{-}U}$ where the prime denotes mutant. In practice, all energies are calculated as difference energies and relative to the folded state ($\Delta\Delta G_{F\text{-}U}$, $\Delta\Delta G_{F\text{-}I}$, $\Delta\Delta G_{F\text{-}\ddagger}$) and then converted to the difference energies relative to the unfolded state by subtracting them from $\Delta\Delta G_{U\text{-}F}$. The difference energies are the experimentally determined values. These may be related to the true differences in noncovalent energy between wild-type and mutant proteins by extending theory developed previously[12,21] (see text). The difference energies may be measured as above with high precision. $\Delta\Delta G_F$ measured from equilibrium denaturation is reproducible to ±0.03 kcal-mol^{-1}. The value of $k_{-u}^{H_2O}$ is measured to better than ±5% (standard error). The precision of values of $k_u^{H_2O}$ vary from a standard error of ±14% for wild-type from the extrapolation to 0 M urea of experiments conducted at concentrations above 5.5 M, to ±2% for the shorter extrapolation of unfolding data for a destabilized mutant such as LA14 (see Fig. 2). The consequent errors in $\Delta\Delta G_\ddagger$ and $\Delta\Delta G_I$ are estimated from the method of calculation to be less than ±0.14 kcal mol^{-1}. The differences between $\Delta\Delta G_\ddagger$, $\Delta\Delta G_I$ and $\Delta\Delta G_F$ are even more precisely known because $\Delta\Delta G_\ddagger$ and $\Delta\Delta G_I$ are measured relative to $\Delta\Delta G_F$. The error in $\Delta\Delta G_\ddagger-\Delta\Delta G_F$ and in $\Delta\Delta G_\ddagger-\Delta\Delta G_F$ and is less than ±0.11 kcal mol^{-1}.

measurements of log K_u at high urea concentration agrees with that measured by temperature- or guanidinium chloride-induced denaturation[18]; the folding kinetics of some less stable mutants follow the two-state equation down to much lower concentrations of urea because the intermediate is destabilized and accumulates to a much smaller extent (for example, mutant IA96 in Fig. 2).

It has been suggested that the deviation in the rate constants for refolding is not a consequence of an intermediate accumulating but is caused by the *cis → trans* isomerization of one of the proline residues becoming rate determining at low urea. This is unlikely for the following reasons. First, the amplitude of the fast phase (83%) is so high that it would require the *cis* conformation to be the major form in the unfolded enzyme. Second, the rate constant of 13 s^{-1} is far too high for a *cis–trans* isomerization of a proline in a denatured protein. Third, the high amplitude accounts for most of the reaction and so there would still be a discrepancy between the equilibrium constant for unfolding calculated from the ratio of rate constants and that from calculated equilibrium studies.

The NMR study[13] shows directly that on the folding pathway there is an intermediate with considerable formation of secondary structure. The refolding of wild-type enzyme is a multiphasic reaction of which we measure only the slowest steps in the stopped-flow experiments. The faster steps are lost in the dead time of the stopped-flow spectrophotometer, which is 30 ms because of problems in mixing urea with water. The slowest step is the isomerization of prolines in the denatured state. This step has the same rate constant for all the mutants studied. The following analysis of the kinetic consequences of the folding intermediate on the refolding pathway applies to the phase that is observed in the stopped-flow time range before the *cis–trans* isomerization.

Free energy profile for folding

The role of binding energy of many groups at the active site of the tyrosyl-transfer RNA synthetase was elucidated by comparing the free-energy profiles of wild-type and mutant enzymes throughout the reaction[19]. What is crucial is not so much the free-energy profile itself but the differences in energy levels between mutant and wild-type enzymes. The 'difference energy' diagrams readily give a qualitative picture of the importance of each side chain throughout the reaction[19,20] and the results can be interpreted quantitatively[21]. The same approach may be applied to study the energetics of interactions of side chains during protein folding.

First, the free-energy profile of the folding pathway of the wild-type protein must be (partly) constructed. The kinetics of folding from 30 ms fit, with high precision, a single exponential trace with a slow tail for proline isomerization[13]. There may be many intermediates on the reaction pathway. Some of these occur before the formation of the one observed, E_I, but are lost in the dead time of the apparatus because they are formed and react too fast. Others will occur between E_I and E_F but are after the rate-determining step for folding (see discussion in ref. 12). The observed rate constant, k_{-u}, is simply that for E_I transforming to E_F (equation (5)). The observed rate constant for unfolding, k_u (equation (2)), is the rate constant for the formation of E_I from E_F, which is the rate-determining step for unfolding.

$$E_U \rightleftarrows \cdots \rightleftarrows E_I \underset{k_u}{\overset{k_{-u}}{\rightleftarrows}} E_F \tag{5}$$

(If 'E_I' is a mixture of intermediates, then k_{-u} is a weighted mixture of rate constants, and reports back on the weighted average of the properties of the mixture.)

The schemes for analysing the kinetics and equilibria are given in Fig. 3. The analysis is in terms of difference energy diagrams or apparent energy differences. The free-energy profile for folding of wild-type enzyme is calculated from the measured free energy of unfolding and the rate constants $k_u^{H_2O}$ and $k_{-u}^{H_2O}$, the values in the absence of denaturant, to give the energies of the folded state, transition state for unfolding and the intermediate relative to the unfolded state. The unfolded state is assigned a standard energy of 0. The same is done for a mutant. An accurate procedure for calculating the differences in the energy levels is given in the legend to Fig. 3.

Difference energy diagrams

In Fig. 4 are plotted the difference energy diagrams for a variety of mutants, each chosen as before[12], to remove a small interaction that stabilizes a specific part of the structure. The crucial step in the interpretation is to relate the experimentally measured difference energies ($\Delta\Delta G_I$, $\Delta\Delta G_\ddagger$ and $\Delta\Delta G_F$ in Fig. 3) to the true differences in noncovalent energy between each of the states. The true differences are defined as:

$$\Delta G_U = G_{E_U} - G_{E'_U}, \qquad \Delta G_I = G_{E_I} - G_{E'_I},$$
$$\Delta G_\ddagger = G_{E_\ddagger} - G_{E'_\ddagger} \quad \text{and} \quad \Delta G_F = G_{E_F} - G_{E'_F}$$

(where G_{E_U} is the free energy of unfolded wild-type enzyme, $G_{E'_U}$, that of unfolded mutant, and so on). Applying the first law of thermodynamics to the cycle in Fig. 3 gives:

$$\Delta\Delta G_F = \Delta G_F - \Delta G_U \qquad (6)$$

$$\Delta\Delta G_\ddagger = \Delta G_\ddagger - \Delta G_U \qquad (7)$$

$$\Delta\Delta G_I = \Delta G_I - \Delta G_U \qquad (8)$$

The apparent energies ('$\Delta\Delta G$') differ from the true by a constant term for each mutant, ΔG_U, the difference in free energy between unfolded mutant and wild-type enzymes. An important corollary of equations (6)–(8) is that the change in the experimentally determined apparent energies on going from one state to another in the folding profile is equal to the difference in true energies. For example, from equations (6) and (7):

$$\Delta\Delta G_F - \Delta\Delta G_\ddagger = \Delta G_F - \Delta G_\ddagger \qquad (9)$$

and from equations (7) and (8):

$$\Delta\Delta G_\ddagger - \Delta\Delta G_I = \Delta G_\ddagger - \Delta G_I \qquad (10)$$

Changes in apparent energies ($\delta\Delta\Delta G_{app}$) are always equal to changes in true stabilization energies ($\delta\Delta G_{true}$) along a profile, that is, $\delta\Delta\Delta G_{app} = \delta\Delta G_{true}$. Equations (6)–(10) hold for all mutations in all circumstances. But they apply to the global energies of the different states of wild-type and mutant enzymes. The crucial question in the interpretation of these equations is, therefore, whether the changes in energy relate just to the local changes in bond energies around the site of mutation. It is possible that $\delta\Delta\Delta G_{app}$ has a significant contribution from longer range global changes that vary from state to state between wild type and mutant. This can be analysed by dividing the change in noncovalent energy on mutation of a side chain into two terms: one, ΔG_{local}, arising from changes where the mutated side chain was previously in contact with other residues; and a second, ΔG_{reorg}, the reorganization energy[21], arising from changes in the energy of the protein elsewhere, in both conformational energy and solvation. Thus, $\delta\Delta\Delta G_{app} = \delta\Delta G_{local} + \delta\Delta G_{reorg}$. If $\delta\Delta G_{reorg} = 0$, then, $\delta\Delta\Delta G_{app}$ reports back on just the local events during folding. If, $\delta\Delta G_{reorg} \neq 0$, then $\delta\Delta\Delta G_{app}$ is uninterpretable in a simple manner.

The two major assumptions in this study are: the mutations are such that ΔG_{reorg} and $\delta\Delta G_{reorg}$ are small and do not obscure the results; and the mutations do not perturb the pathway of folding. This achieved by using two categories of mutation. The first is deletion where a portion of a side chain is removed without introducing a new function (such as Ile→Val, Tyr→Phe). If there are no changes in protein structure on mutation other than removal of part of the side chain, the mutation is said to be nondisruptive[20] and $\delta\Delta G_{reorg}$ is zero. Under these conditions, $\delta\Delta\Delta G_{app} = \delta\Delta G_{local}$. Further, where there is an empty cavity in the mutant enzyme at the site of deletion, $\delta\Delta G_{local}$ is a direct measure of the interaction energy of the side chain in the wild-type enzyme with the contiguous residues. Mutations that cause an increase in the size of a buried side chain or add new contacts are avoided as they are likely to lead to disruptions of structure and the high probability that $\delta\Delta G_{reorg} \neq 0$.

The second category is mutation of surface residues. Here, solvent can move to accommodate changes. Further, for changes at the surface where there is free access of water to the mutated sites, $\Delta\Delta G_{app}$ approximates to the true binding energy of the group that is deleted[21]. We prefer to make mutations that give relatively small changes in stabilization energy as not only are they likely to cause insignificant values of ΔG_{reorg} and $\delta\Delta G_{reorg}$, they are also the least likely to perturb the pathway of folding.

We find for many of the mutations analysed in this study that there is a greatly simplifying feature: the change in stability of the intermediate or transition state on mutation is either close to zero or close to 100% of that of the change in stability of the folded enzyme. These are the changes expected if the interactions are, respectively, either not formed at all or fully formed in the intermediate or transition state.

The difference energies are calculated with high precision. For each mutant relative to wild type, the error in $\Delta\Delta G_F$ is less than ±0.03 kcal mol^{-1}, and in $\Delta\Delta G_\ddagger$ and $\Delta\Delta G_I$ less than ±0.14 kcal mol^{-1} (see legend to Fig. 3). Relative to $\Delta\Delta G_F$, $\Delta\Delta G_\ddagger$ and $\Delta\Delta G_I$ are known to an accuracy of better than ±0.11 kcal-mol^{-1} as the error in $\Delta\Delta G_F$ affects the three energy terms equally. Problems in interpreting difference energies thus do not arise from uncertainties in the accuracy of the measurements.

Analysis of mutants

YF78, TA6, TA26–probing a loop and the N termini of helices. Tyr 78 stabilizes a major loop by its -OH group forming hydrogen bonds with the NH and C=O of Gly 81 in the folded state (Fig. 5). Mutation of Tyr→Phe (YF78) is a deletion that should not cause significant structural changes. It was shown previously[12] that in this mutant all the stabilization energy from the hydrogen bonds is lost in the transition state for unfolding. All the energy is lost in the intermediate as well: $\delta\Delta\Delta G_{app}$ is almost zero between E_I and $E_\ddagger$. The -OH of Thr 26 acts as the N-cap of the second helix, forming hydrogen bonds to the NH groups of residues 27–29 (ref. 22). Mutation of the N-cap to Ala (TA26) is a good mutation to analyse as it is simply an alteration of a surface residue that is replaced by solvent water. Again, it was shown previously by mutating TA26 that all the energy of the N-cap is lost in the transition state for unfolding and the same is seen to happen in the intermediate. Thr 6 forms the N-cap of the first helix. In the mutant TA6 (Thr→Ala) 80% of the difference energy is lost in the transition state for unfolding, and the same amount of energy is lost in the intermediate as in the transition state.

HQ18, TS16–probes of the major α-helix. The charge on the protonated form of His 18 makes a coulombic interaction with the dipole of the α-helix from residues 6–18, and an imidazole NH makes a hydrogen bond with the backbone CO of Gln 15. The interaction energy is thus a probe of the integrity of the C terminus of the helix. In the mutant HQ18(His→Gln), most of this interaction energy is maintained in the transition state[12]. It is maintained equally in the intermediate. A second residue that probes the C terminus is Thr 16. The γ-methyl of Thr 16 makes a very strong hydrophobic interaction with the aromatic ring of Tyr 17, the next residue in the helix in the folded structure. In the mutant TS16 (Thr→Ser) this is maintained in the transition state for unfolding[12], and is seen now to be equally maintained in the intermediate, adding further evidence for early helix formation. These mutations are most suitable as they alter just surface side chains that are exposed. Under these circumstances, values of $\Delta\Delta G_{app}$ are close to the true binding energies[21].

LA14, IA88, IA96, IV88, IV96–probes of the hydrophobic core. The principal α-helix (residues 6–18) packs onto the antiparallel β-sheet by the interdigitation of hydrophobic side chains, of which Leu 14, Ile 88 and Ile 96 are representative, to form a hydrophobic core. Mutation of these differs from the previous

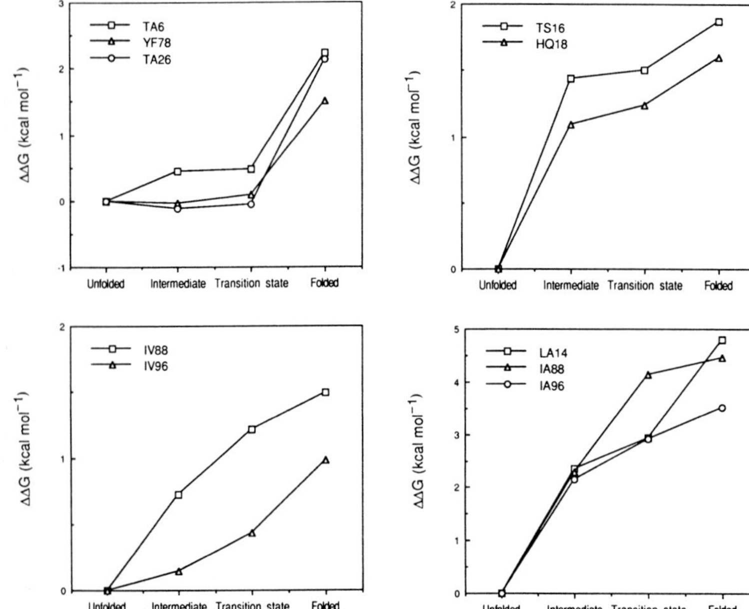

FIG. 4 Difference energy diagrams. The difference energies, $\Delta\Delta G_{app}$, calculated according to Fig. 3 are plotted for the mutants. The following rate constants and equilibrium data have been used.
Wild type: $k_u^{H_2O} = 1.1 \times 10^{-4}\ s^{-1}$, $k_{-u}^{H_2O} = 13\ s^{-1}$;
TA6: $k_u^{H_2O} = 2.2 \times 10^{-3}\ s^{-1}$, $k_{-u}^{H_2O} = 15\ s^{-1}$;
TA26: $k_u^{H_2O} = 4.6 \times 10^{-3}\ s^{-1}$, $k_{-u}^{H_2O} = 14\ s^{-1}$;
YF78: $k_u^{H_2O} = 1.2 \times 10^{-3}\ s^{-1}$, $k_{-u}^{H_2O} = 13\ s^{-1}$;
TS16: $k_u^{H_2O} = 2.1 \times 10^{-4}\ s^{-1}$, $k_{-u}^{H_2O} = 14\ s^{-1}$;
HQ18: $k_u^{H_2O} = 2.1 \times 10^{-4}\ s^{-1}$, $k_{-u}^{H_2O} = 12\ s^{-1}$;
LA14: $k_u^{H_2O} = 2.7 \times 10^{-3}\ s^{-1}$, $k_{-u}^{H_2O} = 5.6\ s^{-1}$;
IV88: $k_u^{H_2O} = 1.8 \times 10^{-4}\ s^{-1}$, $k_{-u}^{H_2O} = 6.5\ s^{-1}$;
IA88: $k_u^{H_2O} = 2.0 \times 10^{-4}\ s^{-1}$, $k_{-u}^{H_2O} = 0.7\ s^{-1}$;
IV96: $k_u^{H_2O} = 2.9 \times 10^{-4}\ s^{-1}$, $k_{-u}^{H_2O} = 9.2\ s^{-1}$;
IA96: $k_u^{H_2O} = 3.2 \times 10^{-4}\ s^{-1}$, $k_{-u}^{H_2O} = 4.1\ s^{-1}$.
Identical values for $k_{-u}^{H_2O}$ are obtained from either extrapolation of the rate constants for refolding at low urea concentrations to 0 M or directly from renaturation by a pH jump from pH 1.5 to 6.3 in the absence of urea. $k_u^{H_2O}$ is obtained by linear extrapolation (see data in ref. 12). A referee has noted that the values of $\Delta\Delta G_F$ for just the interactions measured here account for most of the observed free-energy of folding of the protein ($\Delta G_{F-U} = -10.2$ kcal mol^{-1}). This is not anomalous because, as defined in Fig. 3, $\Delta G_{F-U} = G_F - G_U$. G_F and G_U are both much larger than ΔG_{F-U}, possibly being in the region of thousands of kcal mol^{-1}. Thus, the values of $\Delta\Delta G_F$ are <1% of G_F. Whether or not a protein folds depends on the small difference between G_F and G_U. Just a few mutations can tip the balance so that the unfolded form is favoured at equilibrium.

examples as there are obvious progressive changes in the difference energies as the protein folds (Fig. 4). Mutation of these apolar side chains should lead to small values of ΔG_U (ref. 12) and so changes in difference energies most probably do reflect the true energy changes. Mutations of isoleucines to valines (IV88, IV96) are better probes than mutation to alanine (IA88, IA96) as the smaller the mutational change the less the perturbation of structure. Ile→Val is almost an ideal mutation whereas Ile→Ala leaves a larger cavity. Surrounding residues could relax into the cavity with some probability of a significant reorganization energy term, or there could be ingress of solvent. For IV96 and IV88, there is a clear progression of $\Delta\Delta G_{app}$ as the reaction proceeds, and the same is true for the larger mutations. There are individual variations in the energetics but, in all cases, the stabilization of the intermediate is about midway between that of the unfolded and folded states whereas the transition state has up to 80–90% of the final stabilization. The individual variation might reflect the differences in binding of the different methylene groups in each side chain or could be a consequence of ΔG_{reorg} terms. Nevertheless, it is clear that the hydrophobic core becomes consolidated as folding proceeds.

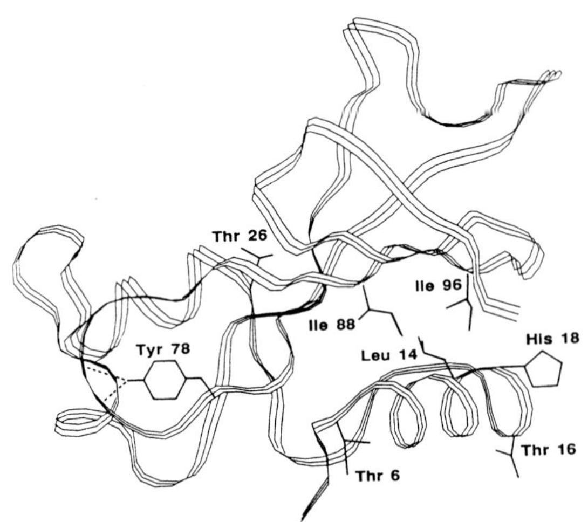

FIG. 5 Sketch of barnase illustrating the target residues.

Reliability of structural conclusions

The deductions about changes in structure from changes in energetics are, of course, indirect. Aside from the assumptions that $\delta\Delta G_{reorg}$ is small and the mutations do not perturb the pathway of folding, what other problems could there be to obscure the interpretations? The results can be divided into two extreme classes to illustrate these: those where interactions are maintained on changing state and those where interactions are lost. Maintenance of an interaction energy does not prove that the particular interaction itself is maintained; it could be that a different interaction of equivalent energy is formed in the new state. Total loss of an interaction energy in an intermediate, however, is conclusive evidence for the loss of an energetically significant interaction as there is not the ambiguity of an alternative interaction. The evidence for loss of the N-caps during the unfolding of barnase is, thus, unambiguous. Evidence for maintenance of the interactions at the C terminal of the major helix is very good. Two separate mutations, based on different properties of the helix, give consistent results, as do NMR experiments on hydrogen exchange[13]. Results on the hydrophobic core are less conclusive; the changes in interaction are intermediate between the two classes. The core is unambiguously weakened in the transition state and intermediate, but it could be changed even more than indicated by the mutational data as there could be alternative hydrophobic interactions taking place.

Intermediate and folding events

Protein engineering on Thr 16 and His 18 shows that the C-terminal portion of the α-helix extending from residues 6–18 is formed in the intermediate. N-capping of the two larger helices occurs after the rate-determining transition state for folding as

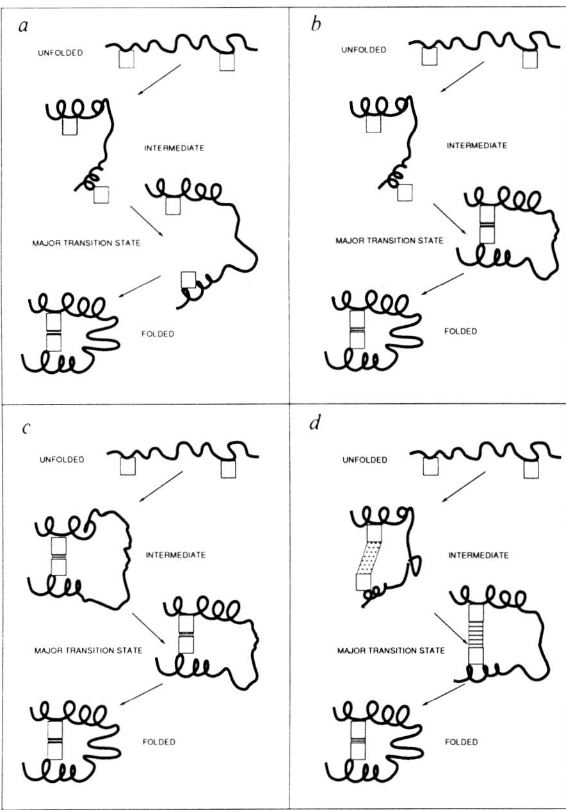

FIG. 6 Possible steps in the folding of barnase. *a*, Two residues do not interact in the unfolded, intermediate or transition states but form after the rate-determining step. Mutation of these does not affect the kinetics of folding but alters the unfolding rate constant. Examples of this are the N-caps and Tyr 78. *b*, The residues interact in the transition state and folded state. No examples yet found. *c*, Interactions occur from the intermediate onwards. Examples approximating to this are His 18 and Thr 16 interacting with residues in the major α-helix. *d*, Progressive bond formation, as with the hydrophobic core.

does the closing of the loop containing Tyr 78. The hydrophobic core between the 6–18 helix and the antiparallel β-sheet consolidates as folding proceeds, implying that the antiparallel β-sheet must be present to some extent. Much of the structural detail of the intermediate is the same as that deduced earlier for the transition state[12], the most striking difference between the intermediate and transition state discovered so far being the weaker hydrophobic core. The weak hydrophobic interactions in the intermediate may be due either to the core being fluid and weakly packed or to the intermediate being composed of an equilibrium mixture of many states, some of which are well packed and others not. In either case, formation of the final folded structure from the intermediate requires the precise docking of elements of the preformed secondary structure to give a closely packed core and intact loops. Interestingly, the interactions studied here, apart from those in the core, are either close to being fully formed or not formed at all in the transition state or intermediate (Fig. 6). NMR-H-exchange[13] confirms the existence of the secondary structure and shows that the C termini of all three α-helices are formed in the intermediate as are many of the interactions in the sheet.

Earlier work on stable molten globule intermediates showed general evidence for secondary structure and qualitative evidence for a weakened hydrophobic core[1–9]. This description fits the intermediate found on the folding pathway of barnase. But NMR studies on the molten globule state of α-lactalbumin, which has extensive secondary structure and a condensed hydrophobic core, show only a few NH groups that are significantly protected against exchange[2,9] whereas the NH groups in barnase are extensively protected. The protein engineering methods suggest changes that can be made to stabilize the intermediate in barnase so that it can be studied directly by NMR. Certain mutations, such as those in the loops and N-caps, destabilize the folded structure but not the intermediate. Judicious choice of a combination of these mutations should destabilize the native folded state to such an extent that the intermediate should become the predominant compact state at low concentrations of urea and so may be studied directly. Kinetic studies on further mutants should enable us to map most of the structure of the intermediate as well as that of the transition state[12].

Our results show the strengths and limitations of the NMR and protein engineering methods, which are in many ways complementary. NMR gives detailed information about the extent and timing of formation of secondary structure. Protein engineering gives detailed information about the extent, timing and strengths of formation of tertiary interactions, and indirect evidence about secondary structure. Protein engineering is more general as most side chains can be readily modified, and it can be applied to proteins which are too large to be examined by NMR. Kinetics is the only method that can be used to study transition states. Both methods are indirect, as interactions are deduced rather than observed directly. But the consistency of results between the two methods implies that they are giving correct information.

One approach to predicting tertiary protein structure is to predict the repeating secondary structures, such as α-helices and β-sheets, and then dock them (see ref. 23). Our results suggest that docking of preformed elements is indeed part of the rate-determining process in folding and provides encouragement for that theoretical approach. □

Received 14 December 1989; accepted 18 June 1990.

1. Dolgikh, D. A. *et al. FEBS Lett.* **136**, 311–315 (1981).
2. Dolgikh, D. A. *et al. Eur. Biophys. J.* **13**, 109–121 (1985).
3. Dolgikh, D. A., Kolomiets, A. P., Bolotina, I. A. & Ptitsyn, O. B. *FEBS Lett.* **165**, 88–92 (1984).
4. Gast, H., Zirver, D., Welfe, H., Bychkova, V. E. & Ptitsyn, O. B. *Int. J. Biol. Macromol.* **8**, 231–236 (1986).
5. Kuwajima, K., Mitani, M. & Sugai, S. *J. molec. Biol.* **206**, 547–561 (1989).
6. Udgaonkar, J. B. & Baldwin, R. L. *Nature* **335**, 694–699 (1988).
7. Roder, H., Elöve, G. A. & Englander, S. W. *Nature* **335**, 700–704 (1988).
8. Baum, J., Dobson, C. M., Evans, P. A. & Hanley, C. *Biochemistry* **28**, 7–13 (1989).
9. Dobson, C. M. & Evans, P. A. *Nature* **335**, 666–667 (1988).
10. Ptitsyn, O. B. *J. Protein Chem.* **6**, 277–293 (1987).
11. Ptitsyn, O. B., Pain, R. H., Semisotnov, G. V., Zerovnik, E. & Razgulyaev. O.I. *FEBS Lett.* **262**, 20–24 (1990).
12. Matouschek, A., Kellis, J. T., Serrano, L. & Fersht, A. R. *Nature* **340**, 124–126 (1989).
13. Bycroft, M., Matouschek, A., Kellis, J. T., Serrano, L. & Fersht, A. R. *Nature* **346**, 488–490 (1990).
14. Tanford, C. *Adv. Protein Chem.* **24**, 1–95 (1969).
15. Segawa, S.-I. & Sugihara, M. *Biopolymers* **23**, 2475–2488, 2489–2498 (1984).
16. Schmid, F. X., Grafl, R., Wrba, A. & Beintema, J. J. *Proc. natn. Acad. Sci. U.S.A.* **83**, 872–876 (1986).
17. Brandts, J. F., Halvorson, H. R. & Brennan, M. *Biochemistry* **14**, 4953–4963 (1975).
18. Kellis, J. T., Jr., Nyberg, K. & Fersht, A. R. *Biochemistry* **28**, 4914–4924 (1989).
19. Wells, T. N. C. & Fersht, A. R. *Biochemistry* **25**, 1881–1886 (1986).
20. Fersht, A. R., Leatherbarrow, R. J. & Wells, T. N. C. *Biochemistry* **26**, 6030–6038 (1987).
21. Fersht, A. R. *Biochemistry* **27**, 1577–1580 (1988).
22. Serrano, L. & Fersht, A. R. *Nature* **342**, 296–299 (1989).
23. Cohen, F. E., Richmond, T. J. & Richards, F. M. *J. molec. Biol.* **132**, 275–278 (1979).
24. Alexandrescu, A. T., Hinck, P. A. & Markley, J. L. *Biochemistry* **29**, 4516–4525 (1990).
25. Evans, P. A., Kautz, R. A., Fox, R. O. & Dobson, C. M. *Biochemistry* **28**, 362–370 (1989).
26. Fersht, A. R. *Enzyme Structure and Mechanism 2nd edn* (*W. H. Freeman & Co., New York*, 1985).
27. Hartley, R. W. *Biochemistry* **8**, 2929–2932 (1969).

ACKNOWLEDGEMENTS J.T.K. and L.S. are EMBO fellows. A.M. was supported by the European Community. We thank A.H. Caldecoat and T.P. Meehan for constructing the stopped-flow equipment.

The folding reaction is very complicated and has at least two intermediates. The rollover, as shown many years later, comes primarily from a change of the ratio of partitioning of one intermediate from reactants to products with changing concentrations of denaturant[107,108].

Two years later, 5 April, 1992, we published 10 papers back to back in *JMB* delineating the folding pathway of barnase, comprising 127 pages[82,95,109–116]. There were some complaints to the editor of *JMB*, but those papers have accumulated over 2000 citations. One paper sets out in detail the analysis[112].

12.11 1993: Jane Clarke and Andrea Hownslow.

12.12 1991: Jesus Sanz, Tim Hubbard and Joo Tan.

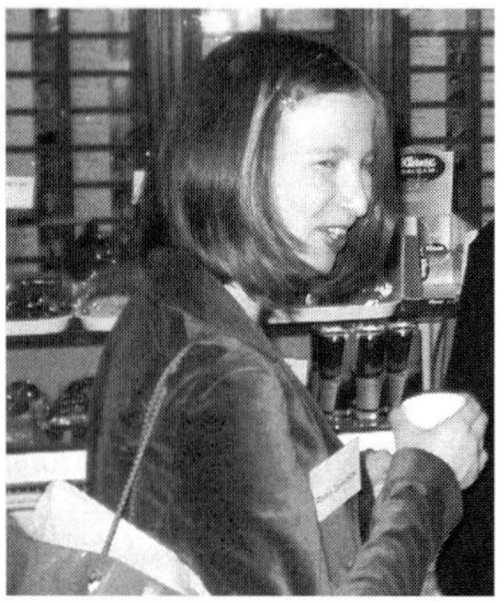

12.13 2003: Rivka Isaacson.

12.14 1993. Cyrus Chothia and Yahouda Harpaz.

J. Mol. Biol. (1992) **224**, 771–782

The Folding of an Enzyme

I. Theory of Protein Engineering Analysis of Stability and Pathway of Protein Folding

Alan R. Fersht, Andreas Matouschek and Luis Serrano

MRC Unit for Protein Function and Design
Cambridge IRC for Protein Engineering
Department of Chemistry
University of Cambridge, Lensfield Road, Cambridge CB2 1EW, U.K.

(Received 23 September 1991; accepted 5 December 1991)

The theory, assumptions and limitations are outlined for a simple protein engineering approach to the problem of the stability and pathway of protein folding. It is a general procedure for analysing structure–activity relationships in non-covalent bonding, including enzyme catalysis, that relates experimentally accessible data to changes in non-covalent bonding. Kinetic and equilibrium measurements on the unfolding and refolding of mutant proteins can be used to map the formation of structure in transition states and folding intermediates. For example, the ratio of the changes in the activation energy of unfolding and the free energy of unfolding on mutation is measured to give a parameter ϕ. There are two extreme values of ϕ that are often found in practice and may be interpreted in a simple manner. A value of $\phi = 0$ implies that the structure at the site of mutation is as folded in the transition state as it is in the folded state. Conversely, $\phi = 1$ shows that the structure at the site of mutation is as unfolded in the transition state as it is in the unfolded structure. Fractional values of ϕ are more difficult to interpret and require a more sophisticated approach. The most suitable mutations involve truncation of side-chains to remove moieties that preferably make few interactions with the rest of the protein and do not pair with buried charges. Fractional values of ϕ found for this type of mutation may imply that there is partial non-covalent bond formation or a mixture of states. The major assumptions of the method are: (1) mutation does not alter the pathway of folding; (2) mutation does not significantly change the structure of the folded state; (3) mutation does not perturb the structure of the unfolded state; and (4) the target groups do not make new interactions with new partners during the course of reaction energy. Assumptions (2) and (3) are not necessarily essential for the simple cases of $\phi = 0$ or 1, the most common values, since effects of disruption of structure can cancel out. Assumption (4) may be checked by the double-mutant cycle procedure, which may be analysed to isolate the effects of just a pair of interactions against a complicated background. This analysis provides the formal basis of the accompanying studies on the stability and pathway of folding of barnase, where it is seen that the theory holds very well in practice.

Keywords: protein folding; protein engineering; catalysis; enzyme; non-covalent bond

1. Introduction

Biology is dominated by the chemistry of the non-covalent bond. All of molecular recognition, from the binding of ligands by proteins to the folding and assembly of macromolecules, is determined primarily by non-covalent interactions. Enzyme catalysis is simple solution catalysis modulated by differential non-covalent binding of substrates, transition states, intermediates and products. Although we know the nature of the forces and interactions involved and can calculate some with high precision in *vacuo*, their magnitudes and importance in aqueous solution have been more difficult to assess. The physical heterogeneity of solutions of proteins and nucleic acids at the atomic level renders the calculation of the electrostatic components of interactions most difficult. Water competes with other molecules for binding, and so it is frequently the difference between binding of

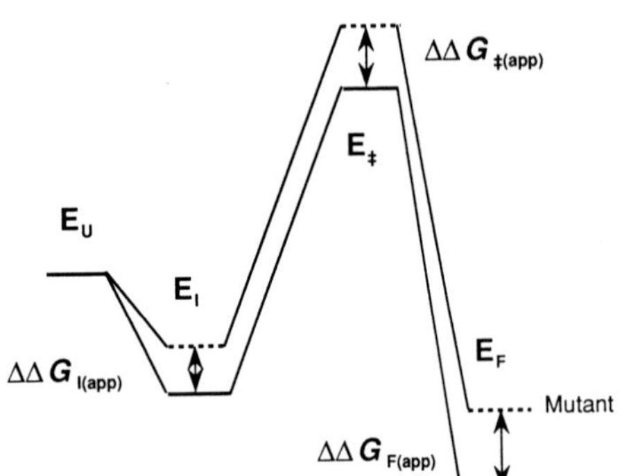

Figure 1. Free energy profiles for the folding of wild-type and mutant enzymes according to the scheme $E_U \rightleftharpoons E_I \rightarrow E_F$ *via* a transition state $E_‡$. The step $E_U \rightleftharpoons E_I$ is a pre-equilibrium. The unfolded enzymes are arbitrarily set as the standard states. The apparent energy differences between wild-type and mutant energy levels differ from the true, therefore, by a term equal to the difference in energy between unfolded wild-type and mutant (see eqn (1)).

enzyme to water and another ligand that has to be calculated. The introduction of protein engineering in the early 1980s has enabled many of the interactions to be studied directly by experiments on purposely modified proteins.

We have developed a simple procedure for studying non-covalent interactions involving proteins using protein engineering and the following strategy. (1) Identify, from examination of the structure of a protein, an interaction of a side-chain that appears to be of importance in binding and/or kinetics. (2) Modify that interaction by changing the side-chain in a structurally sensible manner by site-directed mutagenesis (usually by deleting a small part of the side-chain). (3) Determine, from equilibrium and kinetic measurements, the free energy profiles for the reaction of mutant and wild-type proteins (Fig. 1). (4) Construct a *difference energy diagram*: that is, plot the difference in energy levels between the complexes of wild-type and mutant (Winter *et al.*, 1982; Wells & Fersht, 1986; Fersht *et al.*, 1987; Fersht, 1988; Matouschek *et al.*, 1989, 1990).

This procedure gives an empirical measurement of the magnitude and importance of a particular interaction and how it affects kinetics, equilibria and mechanism. Each mutation acts as a specific probe of the extent of formation of structure in the immediate locality. Many of the basic procedures were devised for studying binding and catalysis in the reactions of the tyrosyl-tRNA synthetase (Winter *et al.*, 1982; Fersht, 1987*b*). One of the important features in the enzyme–substrate complexes of the tyrosyl-tRNA synthetase is that many of the side-chains of the enzyme each make just a single non-covalent bond with the substrate. This simplifies the analysis of the interactions by mutagenesis procedures, since only one relevant bond is broken on mutation. In protein folding, however, many interactions are often broken on mutation, and each may affect stability. The analysis is correspondingly more difficult and complex.

Some of the theory has been presented in preliminary form, and it has been shown possible to map the structures of transition states and intermediates in the folding of barnase (Matouschek *et al.*, 1989, 1990). We have now extended the procedure to provide a detailed description of those structures in the accompanying series of studies. We discuss first the methods for the study of single mutations, the limitations and the assumptions. We then discuss multi-mutational procedures that simplify the problems that arise on changing several bonds on mutation using thermodynamic cycles. The cycles are extended to analyse co-operative interactions.

2. Theoretical Basis for Relating Experimental Data on Mutant Proteins to Changes in Energy of Protein Folding

(a) *The measured quantities: apparent stabilization energies and difference energy diagrams*

The rate and equilibrium constants along the folding (or reaction) pathway for any individual protein are, in principle, directly measurable by experiment. These measurements may be converted into free energy changes that are used to construct the (partial) free energy profiles for the reactions of wild-type and mutant proteins (Fig. 1). The free energies of ground states are calculated from equilibrium thermodynamics using the standard equation $\Delta G = -RT\ln K$. The free energies of transition states may be calculated from transition-state theory using the standard equation, $k = (k_B T/h)\exp(-\Delta G‡/RT)$, where k is the rate constant, k_B the Boltzmann, h the Planck constant and $\Delta G‡$ the free energy of activation or the difference in energy between the transition and ground states. There may be some objections to the validity of this equation for calculation of absolute values of $\Delta G‡$ but, in practice, we use it quantitatively only to calculate differences in $\Delta G‡$. Under these circumstances, complicating additional factors cancel out and the approach is valid†. The free energy profiles

† The approach is not tied to transition state theory but may be derived from any rate theory that involves an activated complex. The activation energy term in the rate equation for kinetics in enzymic reactions may be partitioned into two components: that of the chemical step involving the energetics of bond making and breaking; and the change in the non-covalent interactions. The chemical step requires a quantum mechanical description but the non-covalent energy terms may be described by classical thermodynamics. It is assumed when comparing the reactions of a wild-type enzyme and a mutant that the quantum mechanical component is the same for both and thus cancels out in the ratios. The differences involve just the classical component. In protein folding, the total system may be described by classical thermodynamics.

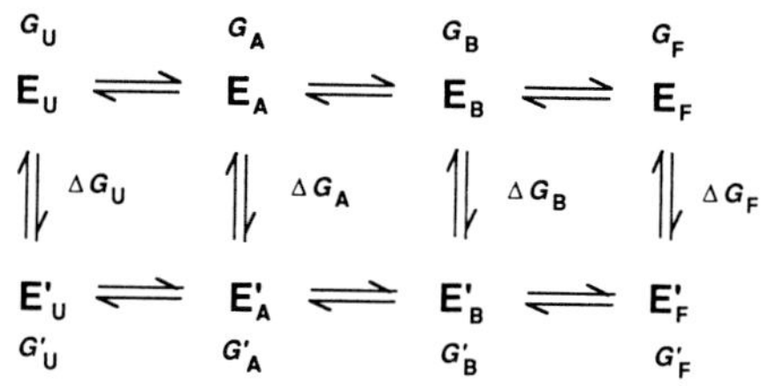

$$\begin{array}{ccccccc} G_U & & G_A & & G_B & & G_F \\ E_U & \rightleftharpoons & E_A & \rightleftharpoons & E_B & \rightleftharpoons & E_F \\ \updownarrow \Delta G_U & & \updownarrow \Delta G_A & & \updownarrow \Delta G_B & & \updownarrow \Delta G_F \\ E'_U & \rightleftharpoons & E'_A & \rightleftharpoons & E'_B & \rightleftharpoons & E'_F \\ G'_U & & G'_A & & G'_B & & G'_F \end{array}$$

Figure 2. Thermodynamic cycles relating states of wild-type and mutant enzymes. E_A and E_B can be any two states during folding, including transition states. The "horizontal" equilibria may be directly measured. The "vertical" equilibria are virtual, but may be calculated.

for wild-type and mutant are calculated, and the difference in energy levels between wild-type and mutant plotted to construct the difference energy diagram (Wells & Fersht, 1986).

It should be noted that all measurements are made using one state of the protein as the reference state. In practice, the reference state in protein folding is generally the unfolded state, E_U. (In enzyme catalysis, the reference state is the unligated enzyme.) This does not affect the relative energies of complexes along the pathway of an individual enzyme but it does introduce a complicating factor when comparing two different enzymes (e.g. wild-type and a mutant), since the two unfolded species are arbitrarily given the same energy (Fig. 1). Because of this, the energy differences in Figure 1 are denoted by the subscript app (= apparent). The relationship of the measured energies to the true differences in energies between wild-type and mutant proteins is analysed by constructing a series of thermodynamic cycles as in Figure 2. These relate the states that occur during the folding of a wild-type enzyme, E†, from its unfolded state E_U with the analogous states in the folding of a mutant, E′. The mutant protein differs from wild-type because a side-chain has been changed. Each state of the mutant protein differs in energy from wild-type by the energy of the covalent change plus changes in the non-covalent interactions. The difference in covalent energy is assumed to be constant throughout the cycles and cancels out in the appropriate manipulations. The changes in free energy along a reaction pathway may be measured experimentally as described above. For example, consider any state E_A on the pathway. The energy of E_A relative to E_U may be measured for wild-type and that of E'_A relative to E'_U for mutant protein. The difference in energy between states E_A and E'_A, relative to the unfolded states E_U and E'_U, $\Delta\Delta G_{A-U}$, is defined by the experimentally measurable quantities $(G_A - G_U)$ and $(G'_A - G'_U)$ thus:

$$\Delta\Delta G_{A-U} = (G_A - G_U) - (G'_A - G'_U). \quad (1)$$

Historically, we have used the generic term *difference energy* to describe the quantity $\Delta\Delta G_{A-U}$ (Wells & Fersht, 1986). In the particular case of a mutation that deletes a group that makes an interaction, we describe $\Delta\Delta G_{A-U}$ as the *apparent binding energy* or *apparent stabilization energy* of that group (in Fig. 1, $\Delta\Delta G_{F(app)} = \Delta\Delta G_{F-U}$, etc.).

(b) *Relationship between apparent binding energies and true non-covalent energy differences between wild-type and mutant proteins*

The apparent energies may be related to the true energy differences between E_A and E'_A ($G_A - G'_U = \Delta G_A$) and between E_U and E'_U ($G_U - G'_U = \Delta G_U$) by rearranging equation (1):

$$\Delta\Delta G_{A-U} = (G_A - G'_A) - (G_U - G'_U). \quad (2)$$

That is:

$$\Delta\Delta G_{A-U} = \Delta G_A - \Delta G_U. \quad (3)$$

The same treatment applies to the changes between any states, e.g.:

$$\Delta\Delta G_{A-B} = \Delta G_A - \Delta G_B. \quad (4)$$

That is, the measured energy differences, $\Delta\Delta G_{A-B}$, are always equal to the true differences in non-covalent energies between two sets of states. A corollary of equation (4) is that for any states L, M and N: $\Delta\Delta G_{L-M} = \Delta\Delta G_{L-N} - \Delta\Delta G_{M-N}$ etc.

(c) *Relationship between global values of* ΔΔG *and individual interaction energies*

Consider a group X on the protein that interacts with a group Y in the folded protein, and X is mutated to Z. For simplicity, consider that X, Y and Z in the unfolded enzyme interact with water (eqns (5) and (6)):

$$E_U(X \ldots mH_2O + nH_2O \ldots Y) \rightleftharpoons E_F(X \ldots Y, [pH_2O]) + (m+n-p)H_2O \quad (5)$$

$$E'_U(Z \ldots m'H_2O + n'H_2O \ldots Y) \rightleftharpoons E'_F(Z \ldots Y, [p'H_2O]) + (m'+n'-p')H_2O. \quad (6)$$

The energies to be considered are divided up into local terms, involving the direct interactions at the site of mutation, and long-range effects involving interactions elsewhere in the protein. For example, in calculating the value of ΔG for the interaction between X and Y in wild-type compared with Z and Y in mutant, the direct interactions are: $G_{F(X \ldots Y)}$, the interaction energy between X and Y in the folded protein; $G_{F(X \ldots E)}$, the interaction energy between X and the rest of the folded protein; $G'_{F(Z \ldots Y)}$, that between Z and Y in the folded mutant protein; $G_{U(X \ldots H_2O)}$, the interaction (solvation) energy between X and water in the unfolded enzyme; $G_{U(Y \ldots H_2O)}$, that between Y and water in the unfolded protein; $G'_{U(Z \ldots H_2O)}$, that between Z and water in unfolded mutant; and so on. In addition, if mutation leaves a cavity in the protein, the solvation energy of Z and Y in the folded mutant protein must be considered, $G'_{F(Z \ldots H_2O)}$ and

† Abbreviations used: E, wild-type enzyme; S, substrate; n.m.r., nuclear magnetic resonance.

$G'_{F(Y...H_2O)}$, as well as any increase in the solvation of the protein on mutation, $G'_{F(E...\Delta H_2O)}$. If the site of mutation is accessible to water in the folded wild-type and mutant protein, then the solvation energies of X and Y ($G_{F(X...H_2O)}$ and $G_{F(Y...H_2O)}$) must be included. Longer-range effects on the energy of the protein on mutation of $X \rightarrow Z$ are lumped together in a term, the reorganization energy; G_{Ureorg} for the unfolded enzyme; and G_{Freorg} for the folded state (G_{reorg} = change in non-covalent free energy of the enzyme on mutation other than energy terms directly involving X, Y, Z and areas of E that are exposed directly to solvent on mutation). Since $\Delta\Delta G_{F-U} = (G_F - G'_F) - (G_U - G'_U)$:

$$\begin{aligned}\Delta\Delta G_{F-U} = {} & G_{F(X...Y)} + G_{F(X...E)} + G_{F(Y...E)} \\ & + G_{F(X...H_2O)} + G_{F(Y...H_2O)} \\ & - G'_{F(Z...Y)} - G'_{F(Z...E)} - G'_{F(Y...E)} \\ & - G'_{F(Z...H_2O)} - G'_{F(Y...H_2O)} \\ & - G'_{F(E...\Delta H_2O)} - G_{U(X...H_2O)} \\ & - G_{U(Y...H_2O)} - G_{U(E...H_2O)} \\ & + G'_{U(Z...H_2O)} + G'_{U(Y...H_2O)} \\ & + G'_{U(E...H_2O)} \\ & + G_{U(reorg)} - G_{F(reorg)}. \end{aligned} \quad (7)$$

There is not a simple analysis of equation (7) when $G_{Ureorg} \neq G_{Freorg}$; that is when $\Delta G_{reorg} (= G_{Freorg} - G_{Ureorg}) \neq 0$. ΔG_{reorg} is likely to be significant when substitutions are made that add steric bulk, new functional groups are introduced that can make new interactions, buried charged groups are removed that disrupt electrostatic pairing, or large groups are deleted. These are termed disruptive mutations. Non-disruptive mutations, that is where $\Delta G_{reorg} = 0$, are most likely to occur when small deletions are made. If the mutation is non-disruptive, $G_{F(Y...E)} = G'_{F(Y...E)}$, $G_{U(Y...H_2O)} = G'_{U(Y...H_2O)}$ etc. and so equation (7) simplifies to:

$$\begin{aligned}\Delta\Delta G_{F-U} = {} & G_{F(X...Y)} + G_{F(X...E)} - G'_{F(Z...Y)} \\ & - G'_{F(Z...E)} + G_{F(X...H_2O)} \\ & + G_{F(Y...H_2O)} - G'_{F(Z...H_2O)} \\ & - G'_{F(Y...H_2O)} - G'_{F(E...\Delta H_2O)} \\ & - G_{U(X...H_2O)} + G'_{U(Z...H_2O)}. \end{aligned} \quad (8)$$

The structural effects of mutations can be checked by structural studies such as protein crystallography or high field nuclear magnetic resonance (n.m.r.). Non-disruptive deletions are the easiest to analyse. Some simple cases follow.

Case (1). Non-disruptive deletions without access of water to site of mutation. A buried group X is deleted without causing any disruption of the enzyme, and water does not occupy the space of the deleted moiety ($\Delta G_{reorg} = 0$, Z = H, empty cavity replaces X). There is no interaction to replace that of X with Y and so $G'_{F(Z...Y)} = 0$. $G'_{F(E...\Delta H_2O)} = 0$ as there is no change in solvation around the site of mutation. The value of $G'_{U(Y...H_2O)}$ is assumed to equal that of $G_{U(Y...H_2O)}$. Equation (8) reduces to:

$$\begin{aligned}\Delta\Delta G_{F-U} = {} & G_{F(X...Y)} + G_{F(X...E)} \\ & - G_{U(X...H_2O)} + G'_{U(H...H_2O)}. \end{aligned} \quad (9)$$

This is not the incremental binding energy of X and Y, which is conventionally defined as $\Delta\Delta G_{binding} = G_{X...Y} - G_{X...H_2O} - G_{Y...H_2O}$. Equation (9) holds for any intermediate state, I, during the folding process. That is:

$$\begin{aligned}\Delta\Delta G_{I-U} = {} & G_{I(X...Y)} + G_{I(X...E)} \\ & - G_{U(X...H_2O)} + G'_{U(H...H_2O)}, \end{aligned} \quad (10)$$

where $G_{I(X...Y)}$ is the energy of the interaction of X and Y in that state. Note that the terms $G_{U(X...H_2O)}$ and $G'_{U(H...H_2O)}$ are the values for the unfolded state and so are constant as the reaction proceeds from one intermediate state to another. This means that changes in the energy of the bond between X and Y are simply given by:

$$\delta\Delta\Delta G = \delta(G_{F(X...Y)} + G_{F(X...E)}). \quad (11)$$

Thus, the changes in apparent binding energy for non-disruptive deletions report back on the changes in energy between X and Y, and X with E, in the wild-type protein during folding (Fersht, 1988).

Case (2). Non-disruptive deletions with unrestricted access of water to site of mutation. A surface group X is deleted without causing any disruption of the enzyme and water has free access to Y and E ($\Delta G_{reorg} = 0$, Z = H, water replaces X). There is no interaction other than solvation to replace that of X with Y. Equation (8) reduces to:

$$\begin{aligned}\Delta\Delta G_{F-U} = {} & (G_{F(X...Y)} + G_{F(X...E)}) \\ & + (G_{F(X...H_2O)} - G_{U(X...H_2O)}) \\ & + (G_{F(Y...H_2O)} \\ & - G'_{F(Y...H_2O)} - G'_{F(E...\Delta H_2O)}) \\ & + (G'_{U(H...H_2O)} - G'_{F(H...H_2O)}). \end{aligned} \quad (12)$$

Equation (12) may be readily interpreted for certain limiting examples.

(1) *X is a hydrophobic side-chain that is surrounded by other hydrophobic groups on the enzyme.* ($G_{F(X...Y)} + G_{F(X...E)}$) is the sum of the van der Waals' interactions that are made by X. The remaining terms in equation (12) represent the loss of solvation energy of X and the hydrophobic pocket in the enzyme with which it interacts: ($G_{F(X...H_2O)} - G_{U(X...H_2O)}$) is the loss of solvation energy of X; ($G_{F(Y...H_2O)} - G'_{F(Y...H_2O)} - G'_{F(E...\Delta H_2O)}$) that of the pocket; and ($G'_{U(H...H_2O)} - G'_{F(H...H_2O)}$) = 0. Thus, $\Delta\Delta G_{F-U}$ is the true incremental binding energy of the hydrophobic interaction (= Σ van der Waals' − ΣΔsolvation energies).

(2) *X interacts only with Y and not with the rest of E.* X is a surface group that interacts just with solvent and with Y. $G_{F(X...E)}$ is zero, as are both $G'_{F(E...\Delta H_2O)}$ and ($G'_{U(H...H_2O)} - G'_{F(H...H_2O)}$). Equation (12) reduces to:

$$\begin{aligned}\Delta\Delta G_{F-U} = {} & G_{F(X...Y)} + (G_{F(X...H_2O)} \\ & - G_{U(X...H_2O)}) + (G_{F(Y...H_2O)} - G'_{F(Y...H_2O)}). \end{aligned} \quad (13)$$

The term ($G_{F(X...H_2O)} - G_{U(X...H_2O)}$) is equal to the change in solvation energy of X on binding to Y, and ($G_{F(Y...H_2O)} - G'_{F(Y...H_2O)}$) is equal to the change in solvation energy of Y on binding to X. Equation (13) is thus identical to that for the incremental binding energy of X and Y (cf. Fersht, 1987*a*, 1988).

3. Probing the Structure of Transition States and Intermediates using Kinetics

(a) *Brønsted equation*

One of the most powerful techniques in physical organic chemistry for probing the structure of transition states during a reaction is the use of structure–activity relationships. This consists of preparing a series of substituted reagents, measuring the change in equilibrium constant on the change in structure, and then measuring the changes in the rate constants for the reaction. It is frequently found that the Brønsted equation applies:

$$\log k = \text{constant} + \beta \log K, \quad (14)$$

where k is the rate constant and K the equilibrium constant for the reaction. β, the Brønsted coefficient, is a measure of the similarity of the structure of the transition state to that of the starting materials or products. An equivalent formulation is:

$$\Delta G\ddagger = \text{constant} + \beta \Delta G_{eq}, \quad (15)$$

where $\Delta G\ddagger$ is the activation energy and ΔG_{eq} the free energy change at equilibrium. A value of $\beta = 0$ means that there is no bond breaking of the relevant interaction between the initial and transition states. Complete bond breaking is characterized by $\beta = 1$. This relationship applies to the some of the non-covalent interactions between the tyrosyl-tRNA synthetase and its substrates during the formation of tyrosyl adenylate.

The Brønsted equation when applied to just two compounds may be written in the simplified form: $\beta = \Delta\Delta G\ddagger/\Delta\Delta G_{eq}$. A version of this equation has been applied to analyse protein folding. The following strategy is used. (1) Mutant proteins are made in which groups that contribute to stability are deleted or changed. (2) Each of the mutations is used as a probe of structure during the pathway of protein folding or unfolding. This is done by making kinetic measurements on the folding of wild-type and mutant enzymes and relating the changes in activation energy and of energy levels of intermediates with the changes in equilibrium energies of folding. This is equivalent to applying a series of two-point Brønsted plots around the protein.

This approach is more readily applied to the process of protein unfolding than refolding (Fig. 3). This is because: (1) the rate constant for the unfolding of many small proteins is a simple monophasic exponential process; and (2) the initial structure is that of the folded protein, which is the best-characterized structure. Refolding, conversely, is frequently multiphasic and proceeds from an unknown structure. The rate constants for the unfolding of wild-type and a mutant protein give $\Delta\Delta G_{\ddagger-F}(= (G_\ddagger - G_F) - (G'_\ddagger - G'_F))$, where ‡ denotes the transition state. Equilibrium unfolding data give $\Delta\Delta G_{U\text{-}F}$. The ratio of the two is termed ϕ. This is similar to, but can differ significantly from, the Brønsted β:

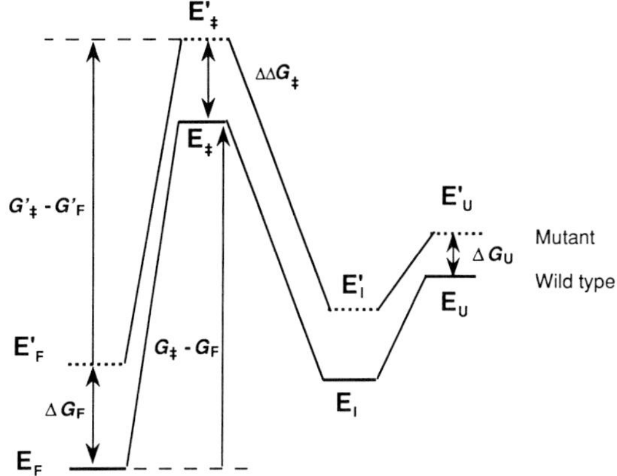

Figure 3. Free energy profile for the unfolding of a protein *via* 1 major transition state and intermediate.

$$\phi_{unf} = \Delta\Delta G_{\ddagger-F}/\Delta\Delta G_{U-F}. \quad (16)$$

Substitution of equation (4) gives:

$$\phi_{unf} = (\Delta G_\ddagger - \Delta G_F)/(\Delta G_U - \Delta G_F). \quad (17)$$

The Brønsted coefficient β generally deals with just one bond being made or broken. However, ϕ differs from β because many different energy terms are involved in the overall energetics. One problem is the term ΔG_U that appears in the denominator. These terms are discussed below. Values of ϕ may be calculated for all states on the reaction pathway by the construction of difference energy diagrams as in Figure 1. Values of ϕ for unfolding and refolding are related simply by the equation: $\phi_{unf} = 1 - \phi_{refold}$.

(b) *Simple interpretations of ϕ_{unf} ($\phi = 1$ or 0)*

It must be emphasized in the following discussion that we are not assuming that there is a Brønsted or linear free-energy relationship. ϕ is just related to the ratio of the binding energy in one state relative to that in a reference state. There are two situations where ϕ may be interpreted in a straightforward manner with few assumptions (Matouschek *et al.*, 1989, 1990). These are found, fortunately, to occur frequently in practice (Fig. 4).

Case (1). $\phi_{unf} = 1$, $\Delta G_\ddagger = \Delta G_U$. This situation occurs when the region of the protein containing the mutation is as completely unfolded in the transition state as in the unfolded enzyme for both wild-type and mutant proteins, and the other regions are unaffected by the mutation.

Case (2). $\phi_{unf} = 0$, $\Delta G_\ddagger = \Delta G_F$. This situation occurs when the region of the protein containing the mutation is as completely folded in

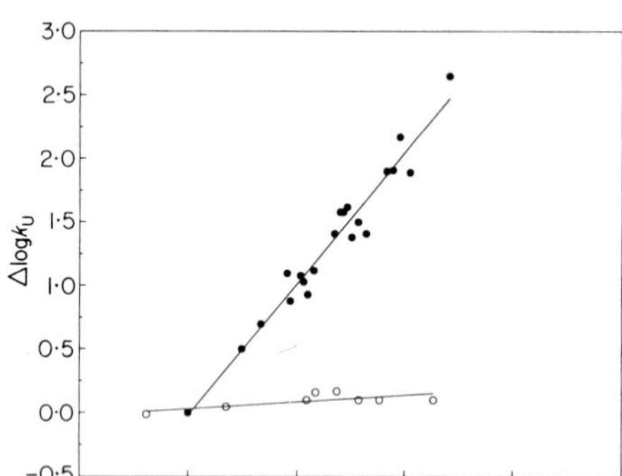

Figure 4. Brønsted plot for the change on mutation of the logarithm of the rate constant for unfolding of barnase against the change in the logarithm of the equilibrium constant for unfolding. Data are from paper III in this series (Serrano *et al.*, 1992). Of 38 mutations studied, 20 fall on the line β $(=\phi)=1$, and 8 on the line β $(=\phi)=0$. (The slopes and standard errors of the linear regression lines are $1{\cdot}04\pm0{\cdot}06$ and $0{\cdot}05\pm0{\cdot}04$.)

the transition state as in the folded enzyme for both wild-type and mutant proteins and the other regions are unaffected by the mutation. There could, however, be an alternative situation that gives rise to $\phi=0$. This is when the group targeted for mutation makes a new interaction in the transition state that, fortuitously, has the same energy as in the folded state.

ϕ is equal to the classical Brønsted β coefficient for the above two examples. The interpretation is greatly simplified because it is at a coarse level. The value of ΔG_U need not be known. The value of ΔG_{reorg} is unimportant if it is the same in ground and transition states. It does not matter if the group that is mutated makes multiple interactions because they are either all made or all broken.

(c) *Fractional values of* ϕ

(i) *General situation*

Values of ϕ other than zero or unity can be very difficult to interpret. The interpretation has to be at the level of individual interactions rather than the more global descriptions allowed above. The value of ΔG_U has to be known, as well as that of ΔG_{reorg}. All the energy terms in equation (7) must be considered. As before, mutations that substitute one group for another are very difficult to analyse. Even if ΔG_{reorg} is insignificant on mutation of X to Z, it is the change in the value of $G_{F(X\ldots Y)}-G'_{F(Z\ldots Y)}$ that is measured rather than just that of $G_{F(X\ldots Y)}$. One of the simplest cases of non-disruptive deletions may, however, possibly be interpreted.

(ii) *Non-disruptive deletions without access of water to site of mutation*

Making the same assumptions as before for equation (9), and for the situation where X and Y are not exposed to water in the transition state, equation (17) gives:

$$\phi_{unf}=\frac{G_{\ddagger(X\ldots Y)}+G_{\ddagger(X\ldots E)}-G_{F(X\ldots Y)}-G_{F(X\ldots E)}}{G_{U(X\ldots H_2O)}-G'_{U(H\ldots H_2O)}-G_{F(X\ldots Y)}-G_{F(X\ldots E)}}. \quad (18)$$

The quantitative interpretation of ϕ requires knowing quantitatively $G_{U(X\ldots H_2O)}-G'_{U(H\ldots H_2O)}$. This can be very large when X is hydrophilic and is deleted to leave H attached to a hydrophobic side-chain. The value of $G_{U(X\ldots H_2O)}-G'_{U(H\ldots H_2O)}$ on changing of an isoleucine to a valine, however, is very small, and the value for Ile → Ala is not large. For those hydrophobic → hydrophobic mutations:

$$\phi_{unf}\approx\frac{G_{\ddagger(X\ldots Y)}+G_{\ddagger(X\ldots E)}-G_{F(X\ldots Y)}-G_{F(X\ldots E)}}{-G_{F(X\ldots Y)}-G_{F(X\ldots E)}} \quad (19)$$

$$\approx 1-\frac{G_{\ddagger(X\ldots Y)}+G_{\ddagger(X\ldots E)}}{G_{F(X\ldots Y)}+G_{F(X\ldots E)}}. \quad (20)$$

This is the fractional loss of van der Waals' interaction energies.

The treatment applies more accurately to the mutation for Ile → Val than for Ile → Ala. Mutation of Ile → Val leaves only a small cavity and only a few interactions broken. Ile → Ala gives a large cavity. This is more likely to allow side-chains to move into the cavity with significant values of ΔG_{reorg} or the ingress of water. Further, many interactions may be broken and so equation (19) has to be applied to each individual interaction. The observed value of ϕ is the average for all these interactions.

Non-disruptive deletions with access of water to site of mutation are very complicated because of the terms involving solvation of the mutated site in the folded mutant enzyme. These can vary throughout the course of the reaction.

The value of ϕ differs from the classical β value for the above cases as ϕ differs from zero and unity. The further the deviation from zero or unity, the greater the deviation of ϕ from β. The amount of deviation depends on the size of $G_{U(X\ldots H_2O)}-G'_{U(H\ldots H_2O)}$ relative to $G_{F(X\ldots Y)}$ in equation (19). Fractional values of ϕ are especially difficult to interpret where mutations are made that are not deletions but new interactions are introduced. Fractional values indicate that there is partial formation of structure but there is not necessarily a linear relationship between ϕ and the extent of formation of structure. It is also possible that a mixture of different states, some with the structure fully formed and others with the elements of structure completely unfolded, is responsible for a fractional value of ϕ.

(d) *Fine structure analysis of* ϕ *for multiple mutations of a single side-chain*

A family of mutations may be made from a single side-chain to measure the interactions of individual

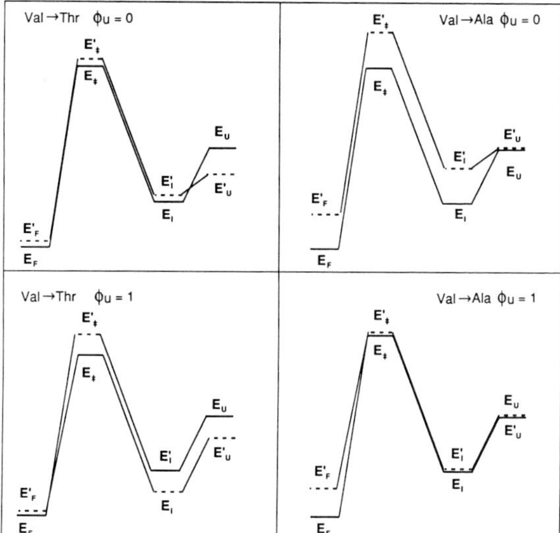

Figure 5. Free energy profiles for the mutation of a (buried) valine to a threonine and alanine. The 2 extreme cases of $\phi = 0$ and 1·0. The destabilizing effect of threonine stems primarily from stabilizing the unfolded state. The destabilizing effect of alanine stems primarily from destabilizing the folded state.

moieties. For example, Thr51 of the tyrosyl-tRNA synthetase has been mutated to Ser, Ala, Gly (Fersht *et al.*, 1985). Comparing the changes in energetics on Thr → Ser51 gives the role of the γ-methyl group of Thr; Ser → Ala51, the role of the γ-OH; and Ala → Gly51, the β-methylene. The same may be done with ϕ values. The two mutations, Ile → Val and Ile → Ala, generate a third comparison, Val → Ala, so that the ϕ values of the δ-methyl of Ile and γ-methylenes may be separated.

(c) *Val/Thr, hydrophobic/hydrophilic switch*

A useful set of probes for structure analysis is the substitution of valine by threonine and by alanine (Fig. 5). Valine and threonine are isosteric and, to a first approximation, make van der Waals' interactions of similar strengths. Substitution of Thr for a buried Val destabilizes a protein primarily by stabilizing the unfolded state of the mutant Val → Thr. Conversely, mutation of a buried valine by alanine destabilizes a protein primarily by the loss of van der Waals' interactions in the folded state. Experiments on the series Val → Thr → Ala thus probe the same region of structure by entirely different structural considerations and double-check results.

The effects of mutation on the stability of the unfolded state may be estimated by using the extended unfolded state in the gas phase as the standard state, and the free energies of hydration of amino acid side-chains from the gas phase (Wolfenden *et al.*, 1981). The hydration energies of the side-chains of Ala and Val are very similar at 1·94 and 1·99 kcal mol^{-1}, respectively (1 cal = 4·184 J). Thus, unless there is compact structure in the unfolded state in solution, the free energies of transfer of an unfolded protein containing Val or Ala from the gas phase to water are practically identical and so the proteins have the same free energy in solution. The hydration energy of Thr is −4·88 kcal mol^{-1} so that mutation of Val → Thr stabilizes the unfolded state in solution by 6·87 kcal mol^{-1}.

4. Assumptions of the Method

There are several assumptions and caveats in the previous derivations that are required for a simple analysis of the kinetic and equilibrium data (Matouschek *et al.*, 1989, 1990).

(1) *Mutation does not alter the pathway of folding* ($G_{\ddagger reorg}$ *and* G_{Ireorg} *are insignificant*). The free energy of folding of barnase, a typical small enzyme is only 9 kcal mol^{-1}. The free energy of activation of unfolding of barnase is about 20 kcal mol^{-1}. Mutations can be made that alter each of these by 4 to 5 kcal mol^{-1}. Clearly a large change in energy on mutation could change the folding kinetics by altering the energy of the transition state to such an extent that either there is a switch to an alternative parallel pathway or there is a change in rate-determining step in a sequential pathway.

(2) *Mutation does not significantly change the structure of the folded state* (G_{Freorg} *insignificant*).

(3) *Mutation does not perturb the structure of the unfolded state* (G_{Ureorg} *is insignificant*).

Simple and analysable results may arise even when assumptions (2) and (3) break down. If, for example, the structure of the folded state is perturbed on mutation but the transition state is perturbed to the same extent (e.g. $G_{\mathrm{Freorg}} = G_{\ddagger\mathrm{reorg}}$), then the perturbation may cancel out, especially for $\phi = 0$ or 1. The same is true for effects on the unfolded states. We find in the following papers many examples of $\phi = 0$ or 1, even for mutations that do appear to be disruptive. The theory is quite robust for examples of $\phi = 0$ or 1.

(4) *The target groups do not make new interactions with new partners during the course of reaction (perils of identifying interactions from energies)*. There are two possibilities. The first is that during the course of folding of wild-type protein the target residue (X) makes a significant new interaction with a new partner. This will not be apparent by examining the energetics. The second possibility is that the mutant residue (Z) makes an interaction not found in the wild-type, which is considered by assumption (1).

Assumptions (1) to (4) are at their most reasonable when small non-disruptive deletion mutations are made, e.g. Ile → Val, Thr → Ser, Ala → Gly, Tyr → Phe, etc. These change the energetics by only small amounts (~ 1 to 2 kcal mol^{-1}), do not introduce radical new functional groups that are likely to make new interactions, and so minimize the likelihood of a gross change in structure of the transition and ground states. The simple theory of equations (19) and (20) applies best to the case when just a single bond is being made or broken. Application to a reaction where several non-covalent bonds change gives a weighted average change for all bonds. The fewer interactions made or broken, the better will the theory apply.

(5) *Refolding proceeds by the reverse of the route for unfolding*. This applies principally to the procedure of inferring information about refolding from the unfolding pathway. It is known from the principle of microscopic reversibility that the folding and unfolding pathways must be the reverse of each other under the same reaction conditions at equilibrium. The question has been raised of whether or not the unfolding in concentrated solutions of denaturant may be extrapolated to the refolding pathway at low concentrations of denaturant (Buchner & Kiefhaber, 1990). This is a valid extrapolation, and has been discussed in detail (Fersht *et al.*, 1990). Practical complications can arise from slow events, however, such as proline isomerizations and so analysis is simplified by considering only the more rapid phases in refolding in which proline isomerization is not rate determining. The logarithm of the unfolding rate constant for barnase and its mutants is linear with urea concentration from strongly denaturing conditions of 8 to 9 M-urea to low concentrations (0·5 M) that are usually renaturing conditions (Matouschek *et al.*, 1990). The unfolding data may be linearly extrapolated to 0 M-urea and so refolding and unfolding may be compared in the absence of denaturant. Further, the plots for wild-type and mutants are close to parallel, so the changes in energies on mutation found by extrapolation to 0 M-urea are very similar to those measured at higher concentrations (Matouschek *et al.*, 1989). Direct evidence for unfolding and refolding pathways being the reverse of each other has come from studies on the barley CI-2 inhibitor, which refolds without an intermediate being kinetically important (Jackson & Fersht, 1991). The ratios of the first-order rate constants for unfolding and unfolding measured in separate experiments in guanidinium chloride give the same equilibrium constants as measured directly by equilibrium denaturation over the whole range of concentrations of guanidinium chloride. This straightforward demonstration cannot be applied in a simple manner to the unfolding and refolding of wild-type barnase from 0 to 8 M-urea, since a folding intermediate accumulates and refolding cannot be described by a single exponential at low concentrations of urea. At higher concentrations of urea, however, the intermediate is less stable than the unfolded state and becomes kinetically unimportant. The criterion that the ratio of measured forward and reverse rate constants is the same as the measured equilibrium constant does hold in the transition region for folding and unfolding at 4 to 5 M-urea. Further, it has proven possible to destabilize the refolding intermediate by mutation so that it does not accumulate except at very low concentrations of urea that are well below the folding/unfolding transition. The ratios of the unfolding and refolding rate constants for such mutants accurately give the measured equilibrium constants until very low concentrations of urea are used, and so show directly that refolding and unfolding are the reverse of each other (Matouschek *et al.*, 1990).

5. Double-mutant Cycles

(a) *General theory*

Many of the drawbacks on analysing the energetics of single mutations may be overcome by using double-mutant cycles (Serrano *et al.*, 1990, 1991; Horovitz *et al.*, 1990; Horovitz & Fersht, 1990), a technique that was developed for analysing enzyme–substrate interactions in the tyrosyl-tRNA synthetase (Carter *et al.*, 1984). The two residues that are presumed to interact, X and Y, are mutated singly and pairwise to generate the family of mutants, E−XY (wild-type), E−X, E−Y and E (Fig. 6). The changes in free energy for the mutations are denoted by: $\Delta G_{\text{E-XY}\rightarrow\text{E-X}}$, $\Delta G_{\text{E-Y}\rightarrow\text{E}}$, $\Delta G_{\text{E-XY}\rightarrow\text{E-Y}}$ and $\Delta G_{\text{E-X}\rightarrow\text{E}}$. A formal analysis of the energy changes has been given elsewhere (Serrano *et al.*, 1990, 1991; Horovitz *et al.*, 1990) but may be briefly summarized. Mutation of E−XY to E−X results in the loss of the interaction energy of X with Y, the interaction energy of Y with the rest of the protein and solvent, changes the solvation energy of X and of the rest of the protein. There may also be a significant value of G_{reorg}. For a single mutation, it is difficult to separate the contribution from the breaking of the X . . . Y bond from that of the interaction of Y with the rest of the protein. However, making the same mutation on a mutant that lacks X, i.e. E−Y → E, loses the interaction energy of Y with the rest of the protein and solvent, and may cause some G_{reorg}. All things being equal, the interaction energy of Y with the rest of the protein will be very similar in E−XY and E−Y, and the value of G_{reorg} will tend to be similar for both mutations. These terms will then cancel out in $\Delta G_{\text{E-XY}\rightarrow\text{E-X}} - \Delta G_{\text{E-Y}\rightarrow\text{E}}$. An experimental energy

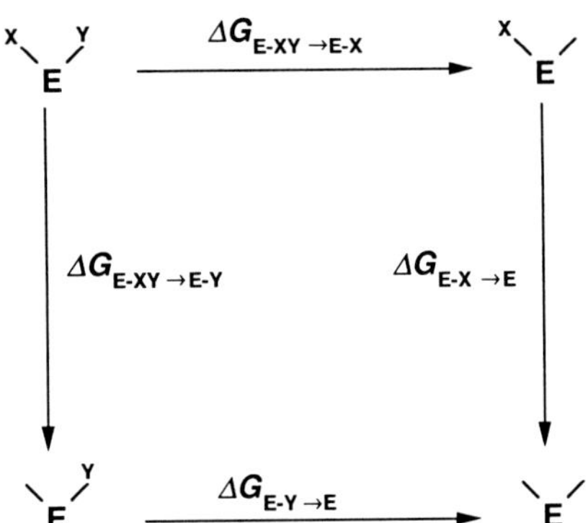

Figure 6. Double-mutant cycle.

value is obtained, $\Delta\Delta G_{int}$ (= the coupling energy), defined by:

$$\Delta\Delta G_{int} = \Delta G_{E-XY \rightarrow E-X} - \Delta G_{E-Y \rightarrow E} = \Delta G_{E-XY \rightarrow E-Y} - \Delta G_{E-X \rightarrow E}. \quad (21)$$

$\Delta\Delta G_{int}$ is equal to the energy of the wild-type protein minus the free energies of the two single mutants, plus the free energy of the double mutant. That is:

$$\Delta\Delta G_{int} = G_{E-XY} - (G_{E-X} + G_{E-Y}) + G_{E}. \quad (22)$$

(b) *Non-disruptive deletions*

Formally, for the deletions of X to H and Y to H indicated in Fig. 6:

$$\begin{aligned}\Delta\Delta G_{int} = {} & [G_{X\ldots Y} + G_{E\ldots Y} + G_{E\ldots X} \\ & + G_{Y\ldots H_2O} + G_{X\ldots H_2O}]_{E-XY} \\ & - [G_{E\ldots X} + G_{X\ldots H_2O} + G_{H\ldots H_2O} \\ & + G_{(E\ldots \Delta H_2O)} + G_{reorg}]_{E-X} \\ & - [G_{E\ldots Y} + G_{Y\ldots H_2O} + G_{H\ldots H_2O} \\ & + G_{(E\ldots \Delta H_2O)} + G_{reorg}]_{E-Y} \\ & + [G_{H\ldots H_2O} + G_{H\ldots H_2O} \\ & + G_{(E\ldots \Delta H_2O)} + G_{reorg}]_{E},\end{aligned} \quad (23)$$

where $G_{X\ldots Y}$ is the non-covalent interaction energy between X and Y, $G_{E\ldots Y}$ is the non-covalent interaction energy between Y and the enzyme other than X, and $G_{E\ldots X}$ is the equivalent term for X. $G_{Y\ldots H_2O}$ and $G_{X\ldots H_2O}$ are the solvation energies of Y and X, and $G_{H\ldots H_2O}$ is the solvation energy of the H that replaces X or Y. $G_{(E\ldots \Delta H_2O)}$ is any additional increase in solvation energy of the enzyme on mutation. The mutational form of the enzyme is denoted by the subscripts E−XY, E−X, etc. Equation (23) simplifies for non-disruptive deletions, since the value of the individual energy terms is independent of the subscript E-XY, E-X, etc. Two classes emerge.

Case (1). Non-disruptive deletions without access of water to site of mutation. If the unwanted interactions do cancel out and X is not further exposed to solvent on the mutation of Y and *vice versa*, then:

$$\Delta\Delta G_{int} = G_{X\ldots Y}. \quad (24)$$

That is, $\Delta\Delta G_{int}$ directly measures the interaction energy between X and Y.

Case (2). Non-disruptive deletions with unrestricted access of water to site of mutation. If the unwanted interactions do cancel out and X is exposed to solvent on the mutation of Y and *vice versa*, and $[G_{(E\ldots \Delta H_2O)}]_{E} = [G_{(E\ldots \Delta H_2O)}]_{E-X} + [G_{(E\ldots \Delta H_2O)}]_{E-Y}$, then:

$$\begin{aligned}\Delta\Delta G_{int} = {} & G_{X\ldots Y} + ([G_{X\ldots H_2O}]_{E-XY} \\ & - [G_{X\ldots H_2O}]_{E-X}) \\ & + ([G_{Y\ldots H_2O}]_{E-XY} - [G_{Y\ldots H_2O}]_{E-Y}).\end{aligned} \quad (25)$$

Equation (25) is very similar to equation (12) and represents the incremental binding energy of X and Y. That is: $\Delta\Delta G_{int} = G_{X\ldots Y} - \Delta G_{X\ldots H_2O} - \Delta G_{Y\ldots H_2O}$, where $\Delta G_{X\ldots H_2O}$ is the interaction energy of X with the water that replaces Y on the mutation of E−XY to E−X, and $\Delta G'_{Y\ldots H_2O}$ is the interaction energy of Y with the water that replaces X on the mutation of E−XY to E−Y.

The strength of the double-mutant approach is that, in the ideal situation, the interaction energies of X and Y with the rest of the protein cancel out to leave just the interaction energies between X and Y. Further, the cycle tolerates non-zero values of ΔG_{reorg} terms as long as these can cancel out. Even if those interaction and reorganization energies do not cancel out exactly, the errors will be reduced to second-order effects. A further advantage is that the cycles can resolve the dilemma of whether a ϕ-value of zero on a single mutation is caused by the retention of the interaction or that the residue acquires a new interaction of identical energy. This is because both partners in the interaction are mutated and the interaction between the pair is measured directly.

(c) *Reference states in double mutant cycles*

Values of $\Delta\Delta G_{int}$ measured from free energies of unfolding, $\Delta\Delta\Delta G_{int(unf)}$, are (cf. eqn (22)):

$$\begin{aligned}\Delta\Delta\Delta G_{int(unf)} = {} & (G_{U(E-XY)} - G_{F(E-XY)}) \\ & - (G_{U(E-X)} - G_{F(E-X)}) \\ & - (G_{U(E-Y)} - G_{F(E-Y)}) \\ & + (G_{U(E)} - G_{F(E)}).\end{aligned} \quad (26)$$

On collecting together the terms in G_U and G_F, equation (26) is rearranged to:

$$\Delta\Delta\Delta G_{int(unf)} = \Delta\Delta G_{Uint} - \Delta\Delta G_{Fint}, \quad (27)$$

where $\Delta\Delta G_{Uint}$ and $\Delta\Delta G_{Fint}$ are the values of $\Delta\Delta G_{int}$ for the folded and unfolded states, respectively. In general, the value of $\Delta\Delta G_{Uint}$ is zero for side-chains of residues that are not adjacent in the unfolded state and so $\Delta\Delta\Delta G_{int(unf)} = -\Delta\Delta G_{Fint}$.

Complications can arise, however, if measurements are made on two states in which both have non-zero values of $\Delta\Delta G_{int}$. An example is the measurement of the rate constants for protein unfolding in which the reference state is the folded protein to measure the co-operativity of interactions in the transition state. The co-operativity in the free energies of activation for the unfolding rate constants, $\Delta\Delta\Delta G_{int(\ddagger)}$, is given by:

$$\Delta\Delta\Delta G_{int(\ddagger)} = \Delta\Delta G_{\ddagger int} - \Delta\Delta G_{Fint}, \quad (28)$$

where $\Delta\Delta G_{\ddagger int}$ is the true interaction energy in the transition state. It is seen from equation (28) that if there is exactly the same co-operativity in the transition and folded states, i.e. $\Delta\Delta G_{\ddagger int} = \Delta\Delta G_{Fint}$, then there will be no measured co-operativity in the rate constants for unfolding. It is safer, therefore, to use the unfolded state as the reference state in all cases.

6. Interpretation of $\Delta\Delta G_{int}$ as a Probe of Structure during Protein Folding

A very important feature of the double-mutant method is that $\Delta\Delta G_{Uint}$ is generally zero; that is, there are not co-operative interactions in the unfolded state. This is unlike the analysis of single mutations where values of ϕ are complicated by the term ΔG_U, the difference in energy between the denatured states of mutant and wild-type enzymes. The definition of ϕ for single mutations contains this term (cf. eqn (17)). A simpler definition of ϕ for $\Delta\Delta G_{int}$, ϕ_{int}, may be formulated by:

$$\phi_{int} = \Delta\Delta G_{Zint}/\Delta\Delta G_{Fint}, \quad (29)$$

where $\Delta\Delta G_{Zint}$ is the value in a particular state, E_Z, of the protein during folding, where $\Delta\Delta G_{Uint}$ is assumed to be zero. As before, there are two simple cases that may be interpreted in a straightforward manner. If $\phi_{int} = 0$, then the interaction energy is lost and so the two residues X and Y are no longer in contact in E_Z or they are separated to the extent that there is no net interaction energy. If $\phi_{int} = 1$, then X and Y are in full contact. Fractional values of ϕ_{int} are a function of all the energy terms in equation (23), and can only be interpreted subject to evidence about the individual energy terms. A fractional value of ϕ_{int} does, however, indicate that the two residues are in contact to some extent.

For *non-disruptive deletions without access of water to site of mutation*, fractional values of ϕ are directly interpretable since, from equation (24), $\Delta\Delta G_{Zint} = G_{Z(X\ldots Y)}$, where $G_{Z(X\ldots Y)}$ is the interaction energy between X and Y in state Z. Thus, for example, from equation (29), ϕ_{int} for unfolding $= G_{\ddagger(X\ldots Y)}/G_{F(X\ldots Y)}$. ϕ is thus a direct measure of the fractional change in non-covalent energy between X and Y. As in previous examples, ϕ_{int} gives an average value for all the interactions between X and Y.

(a) *Reliability of* ϕ_{int} versus ϕ

ϕ_{int} is a far more reliable indicator of structural changes than is ϕ. Measurement of ϕ_{int} tends to isolate the contributions of the interactions between the two target residues. The effects of changes in interactions between X and Y and the rest of the protein tend to cancel out and G_{reorg} is reduced to second-order effects. The major assumption in interpreting $\Delta\Delta G_{int}$ is that the co-operativity thus measured results from a direct interaction between the two residues concerned and not an indirect effect mediated by changes in other residues. However, examination of the structure of the folded protein should provide the necessary evidence. Interpretation of ϕ is at its most reliable when the moiety of the side-chain that is removed by mutagenesis makes only a few contacts with the rest of the protein. It is found in practice, in the accompanying papers, that ϕ is often very reliable.

(b) *COSMIC method (Horovitz* et al., *1991)*

One further aspect of the double-mutant cycle method is that a series of them may be combined. For example, if residue A interacts with B, which in turn interacts with C, which interacts with D etc., then a series of interconnecting cycles may be made to analyse the chain of interactions in folding. This technique, termed COSMIC (combination of sequential mutant interacting cycles) has been used to leapfrog along the major α-helix of barnase.

(c) ΔG_{int} *and higher-order co-operativity in proteins*

Some of the interactions in proteins are co-operative or synergistic; that is, the energy of the whole is greater than the sum of the parts. $\Delta\Delta G_{int}$ has been shown to be a measure of co-operativity (Horovitz & Fersht, 1992). $\Delta\Delta G_{int}$, as defined in equation (22), directly measures the excess interaction energy of having two side-chains simultaneously present as compared with each of them separately. This may be extended to higher dimensions by defining and analysing $\Delta\Delta G_{int}$ for multiple mutations of multiple interacting side-chains (Horovitz & Fersht, 1992).

7. Catalysis

It has been shown experimentally from protein engineering studies on the tyrosyl-tRNA synthetase that linear free-energy relationships occur between changes in rate constants and changes in equilibrium constants (Fersht *et al.*, 1987; Fersht & Wells, 1991). These result from the binding energy component that modulates binding and catalysis. In view of the complexities in analysing binding energy changes in folding, it is worthwhile showing how much simpler is the analysis of enzyme–substrate interactions because one can frequently modify just one bond between E and S. We shall restrict the analysis to this situation.

Dissociation and rate constants for the reaction of substrates and products may be measured by conventional means for wild-type and mutant enzymes. Wild-type and mutants may be compared by thermodynamic cycles similar to those for folding (Fig. 7). For any state, e.g. an ES complex, an apparent binding energy relative to the free enzyme, may be derived, e.g. $\Delta\Delta G_{ES-E}$, which is the change in free energy of binding of S on mutation:

$$\Delta\Delta G_{ES-E} = (G_{ES} - G_E - G_S) - (G'_{ES} - G'_E - G_S) \quad (30)$$

$$= (G_{ES} - G_E) - (G'_{ES} - G'_E). \quad (31)$$

Enzyme-bound states, e.g. ES and EP complexes, may also be compared:

$$\Delta\Delta G_{EP-ES} = (G_{EP} - G_{ES}) - (G'_{EP} - G'_{ES}) \quad (32)$$

or

$$\Delta\Delta G_{EP-ES} = (G_{EP} - G'_{EP}) - (G_{ES} - G'_{ES}). \quad (33)$$

A ϕ value for catalysis may be defined. For example, for ES → ES‡ → EP:

$$\phi = \Delta\Delta G_{ES\ddagger-ES}/\Delta\Delta G_{EP-ES}. \quad (34)$$

We now show for two simple cases that ϕ does equal true changes in bond energies without the

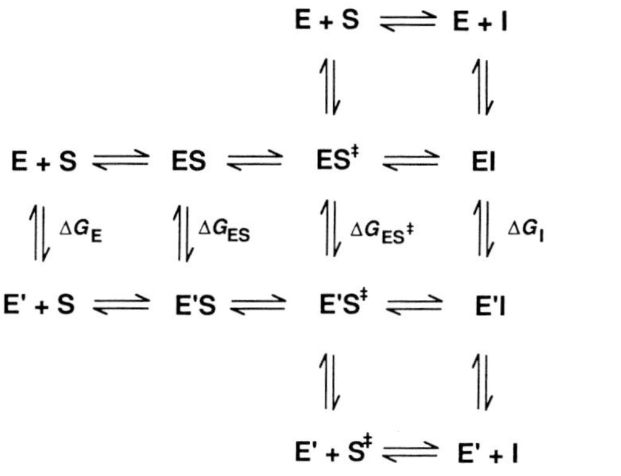

$$
\begin{array}{ccccccc}
 & & & & \mathrm{E+S} & \rightleftharpoons & \mathrm{E+I} \\
 & & & & \updownarrow & & \updownarrow \\
\mathrm{E+S} & \rightleftharpoons & \mathrm{ES} & \rightleftharpoons & \mathrm{ES^{\ddagger}} & \rightleftharpoons & \mathrm{EI} \\
\updownarrow \Delta G_E & & \updownarrow \Delta G_{ES} & & \updownarrow \Delta G_{ES^{\ddagger}} & & \updownarrow \Delta G_I \\
\mathrm{E'+S} & \rightleftharpoons & \mathrm{E'S} & \rightleftharpoons & \mathrm{E'S^{\ddagger}} & \rightleftharpoons & \mathrm{E'I} \\
 & & & & \updownarrow & & \updownarrow \\
 & & & & \mathrm{E'+S^{\ddagger}} & \rightleftharpoons & \mathrm{E'+I}
\end{array}
$$

Figure 7. Thermodynamic cycles for states of wild-type and mutant enzymes during catalysis, including I transition state and I intermediate. Product release is not shown. The "horizontal" equilibria may be directly measured. The "vertical" equilibria are virtual, but may be calculated.

complicating factors found for the mutations that affect folding.

Case (1). Non-disruptive deletions without access of water to site of mutation (water does not take place of mutated side-chain in E'S complex). Consider the enzyme having a group X that interacts with group Y on the substrate. X is mutated to H, water does not enter the cavity in the complex, and there is no reorganization energy terms in the ES complexes. All solvation terms are assumed to be the same, other than in the direct interaction of X with the water molecule that is displaced by Y. The interactions are simplified to:

$$\mathrm{E{-}X \ldots H_2O + S{-}Y \ldots H_2O \rightleftharpoons [E{-}X \ldots Y{-}S] + 2H_2O} \quad (35)$$

$$\mathrm{E{-}H \ldots H_2O + S{-}Y \ldots H_2O \rightleftharpoons [E{-}H \ldots Y{-}S] + 2H_2O.} \quad (36)$$

Similar equations may be written for the ES‡ and EP complexes. Thus, using the notation $G_{ES(X \ldots Y)}$ = energy of the interaction of X and Y in the ES complex etc:

$$\Delta\Delta G_{EP-ES} = G_{EP(X \ldots Y)} - G_{ES(X \ldots Y)} \quad (37)$$

and

$$\Delta\Delta G_{ES\ddagger-ES} = G_{ES\ddagger(X \ldots Y)} - G_{ES(X \ldots Y)}. \quad (38)$$

Thus, for $\phi = \Delta\Delta G_{ES\ddagger-ES}/\Delta\Delta G_{EP-ES}$,

$$\phi = (G_{ES\ddagger(X \ldots Y)} - G_{ES(X \ldots Y)})/(G_{EP(X \ldots Y)} - G_{ES(X \ldots Y)}) \quad (39)$$

and so ϕ is equal to the true ratio of the change in bond energies of X and Y on progression to the ES‡ and EP complexes from the ES.

Case (2). Non-disruptive deletions with access of water to site of mutation (water takes place of mutated side-chain in E'S complex). The assumptions are as for above except that water enters the cavity produced on mutation:

$$\mathrm{E{-}X \ldots H_2O + S{-}Y \ldots H_2O \rightleftharpoons [E{-}X \ldots Y{-}S] + 2H_2O} \quad (40)$$

$$\mathrm{E{-}H \ldots H_2O + S{-}Y \ldots H_2O \rightleftharpoons [E{-}H \ldots \mathit{n}H_2O \ldots Y{-}S] + 2{-}\mathit{n}H_2O.} \quad (41)$$

Thus:

$$\Delta\Delta G_{EP-ES} = G_{EP(X \ldots Y)} - G'_{EP(H \ldots H_2O)} - G'_{EP(Y \ldots H_2O)} - (G_{ES(X \ldots Y)} - G'_{ES(H \ldots H_2O)} - G'_{ES(Y \ldots H_2O)}). \quad (42)$$

It is not unlikely that $G'_{EP(H \ldots H_2O)} \approx G'_{ES(H \ldots H_2O)}$ and $G'_{EP(Y \ldots H_2O)} \approx G'_{ES(Y \ldots H_2O)}$. Thus, $\Delta\Delta G_{EP-ES} \approx G_{EP(X \ldots Y)} - G_{ES(X \ldots Y)}$, as in equation (37). A similar set of equations may be derived for the ES‡ complex. Again, with a ϕ-value defined as $\phi = \Delta\Delta G_{ES\ddagger-ES}/\Delta\Delta G_{EP-ES}$:

$$\phi \approx (G_{ES\ddagger(X \ldots Y)} - G_{ES(X \ldots Y)})/(G_{EP(X \ldots Y)} - G_{ES(X \ldots Y)}). \quad (43)$$

ϕ is approximately equal to the true ratio of the change in bond energies of X and Y on progression to the ES‡ and EP complexes from the ES.

There is an important difference between linear free-energy relationships derived from binding energy changes in catalysis and those from conventional covalent chemistry. In covalent chemistry, the parent compound is modified by substituents that are removed from the seat of reaction. All substituents exert their affect on just the specific chemical bonds that undergo the chemical reaction, often by a field effect. In enzyme catalysis, the substrate makes many non-covalent bonds with the enzyme, and these may change as the reaction proceeds. Mutation of the side-chains in these interactions affects different parts of the substrate. There is no *a priori* reason why these should all change to the same extent on going from ES → ES‡ → EP. The binding energy changes constitute an ensemble of different ϕ values. Indeed, this was found for the tyrosyl tRNA synthetase (Fersht *et al.*, 1987). The ensemble of ϕ values may be grouped into sets that have the same individual values of ϕ. A linear free energy relationship means simply that there is a set of individual interactions that happen to have the same values of ϕ. A common value of ϕ for a set of interactions suggests that there is a co-ordinate movement of the enzyme and substrate relative to each other at the positions of those interactions (Fersht & Wells, 1991).

8. Summary

(1) Measurements of equilibrium and rate constants in the folding of wild-type and mutant proteins can be analysed to give the true changes in the global non-covalent energies on mutation.

(2) The simplest mutations to analyse are those involving deletions of groups that make defined contributions to binding, rather than mutations

that give rise to substitutions that can make alternative interactions.

(3) Measurements on small deletion mutations should report back on the energetics of the target group in wild-type enzyme.

(4) The measured changes in stabilization energies are not, in general, equal to the incremental binding energies of the groups that are deleted. Where, however, there is open access of solvent to the site of mutation with the same solvation energies as in bulk solution, the measured change in stabilization energy on mutation can equal the true incremental binding energy of the deleted group.

(5) Changes in the energies of intermediates and transition states may be compared with the overall change in stability on mutation. A series of ϕ values may be obtained that gives information on the extent of formation of structure at the point of mutation. The two extreme values of $\phi_{\text{unf}} = 0$ and $\phi_{\text{unf}} = 1$ may be readily interpreted as indicating complete retention and complete breaking of structure, respectively. Fractional ϕ values indicate partial formation of structure but in general, there is not a linear relationship between ϕ and extent of structure formation.

(6) Double-mutant cycles extend the range of mutational analysis. They are used to isolate the energetics of pairwise interactions. Since both partners in the interactions are mutated, the double-mutant cycles detect unambiguously whether new interactions are made or the original pairwise interactions are maintained for $\phi = 0$. Complicating effects from reorganization energies may cancel out in the cycle analysis. Even where they do not cancel, their effects should be reduced. The analysis of ϕ values for double mutant cycles is simplified, since values of interaction energies in the unfolded state are frequently zero.

References

Buchner, H. & Kiefhaber, T. (1990). Protein folding enigma. *Nature (London)*, **343**, 601.

Carter, P. J., Winter, G., Wilkinson, A. J. & Fersht, A. R. (1984). The use of double mutants to detect structural changes in the active site of the tyrosyl-tRNA synthetase (*Bacillus stearothermophilus*). *Cell*, **38**, 835–840.

Fersht, A. R. (1987*a*). The hydrogen bond in molecular recognition. *Trends Biochem. Sci.* **12**, 301–304.

Fersht, A. R. (1987*b*). Dissection of the structure and activity of the tyrosyl-tRNA synthetase by site-directed mutagenesis. *Biochemistry*, **26**, 8031–8037.

Fersht, A. R. (1988). Relationships between apparent binding energies measured in site-directed mutagenesis experiments and energetics of binding and catalysis. *Biochemistry*, **27**, 1577–1580.

Fersht, A. R. & Wells T. N. C (1991). Linear free energy relationships in enzyme binding interactions studied by protein engineering. *Protein Eng.* **4**, 229–231.

Fersht, A. R., Wilkinson, A. J., Carter, P. & Winter, G. (1985). Fine structure–activity analysis of mutations at position 51 of tyrosyl-tRNA synthetase. *Biochemistry*, **23**, 5858–5861.

Fersht, A. R., Leatherbarrow, R. & Wells, T. N. C. (1987). Structure–activity relationships in engineered proteins: analysis of use of binding energy by linear free energy relationships. *Biochemistry*, **26**, 6030–6038.

Fersht, A. R., Kellis, J. T., Jr, Matouschek, A. & Serrano, L. (1990). Protein folding pathway (reply). *Nature (London)*, **343**, 601.

Horovitz, A. & Fersht, A. R. (1990). Strategy for analysing the co-operativity of intramolecular interactions in peptides and proteins. *J. Mol. Biol.* **214**, 613–617.

Horovitz, A. & Fersht, A. R. (1992). Co-operative interactions in protein folding. *J. Mol. Biol.* **224**, 733–740.

Horovitz, A., Serrano, L., Avron, B., Bycroft, M. & Fersht, A. R. (1990). Strength and co-operativity of contributions of surface salt bridges to protein stability. *J. Mol. Biol.* **216**, 1031–1044.

Horovitz, A., Serrano, L. & Fersht, A. R. (1991). COSMIC analysis of the major α-helix of barnase during folding. *J. Mol. Biol.* **219**, 5–9.

Jackson, S. E. & Fersht, A. R. (1991). Folding of chymotrypsin inhibitor 2 1: evidence for a two-state transition. *Biochemistry*, **30**, 10,428–10,435.

Matouschek, A., Kellis, J. T., Jr, Serrano, L. & Fersht, A. R. (1989). Mapping the transition state and pathway of protein folding by protein engineering. *Nature (London)*, **340**, 122–126.

Matouschek, A., Kellis, J. T., Jr, Serrano, L., Bycroft, M. & Fersht, A. R. (1990). Characterizing transient folding intermediates by protein engineering. *Nature (London)*, **346**, 440–445.

Serrano, L., Horovitz, A., Avron, B., Bycroft, M. & Fersht, A. R. (1990). Estimating the contribution of engineered surface electrostatic interactions to protein stability by using double-mutant cycles. *Biochemistry*, **29**, 9343–9352.

Serrano, L., Bycroft, M. & Fersht, A. R. (1991). Aromatic-aromatic interactions and protein stability: investigation by double-mutant cycles. *J. Mol. Biol.* **218**, 465–475.

Serrano, L., Matouschek, A. & Fersht, A. R. (1992). The folding of an enzyme. III. Structure of the transition state for unfolding of barnase analysed by a protein engineering procedure. *J. Mol. Biol.* **224**, 805–818.

Wells, T. N. C. & Fersht, A. R. (1986). Use of binding energy in catalysis analysed by mutagenesis of the tyrosyl-tRNA synthetase. *Biochemistry*, **25**, 1881–1886.

Winter, G., Fersht, A. R., Wilkinson, A. J., Zoller, M. & Smith, M. (1982). Redesigning enzyme structure by site-directed mutagenesis: tyrosyl-tRNA synthetase and ATP binding. *Nature (London)*, **299**, 756–758.

Wolfenden, R., Andersson, L., Cullis, P. M. & Southgate, C. C. B. (1981). Affinities of amino acid side chains for solvent water. *Biochemistry*, **20**, 649–855.

Edited by A. Klug

The work was summarised by me at the FEBS meeting in Stockholm in 1993, where I received the Datta Medal, and my Datta Lecture was published in FEBS letters[117]. Also in 1993, as a 50th birthday present, I was elected to the National Academy of Sciences of the USA as a Foreign Associate. I had always wanted to be a member in order to publish in their journal, but had never, even in my wildest dreams, had even considered it would happen. It did, and it changed my choice of journal. It was so fast and easy to publish in *Proc. Natl. Acad. Sci. USA* that I sent papers there that I would normally have entrusted to the slow pace and increasingly belligerent refereeing of *Nature*.

Those two awards were completely unexpected as I did not know I had been nominated. The unexpected awards give far more pleasure than the ones for which you know you have been nominated.

Φ-value analysis had completely revolutionised the study of protein folding. For the first time, we could get near atomic resolution details of the transition state for protein folding. In some examples, we could infer the structure of elements directly. In others, we provided the raw data to benchmark simulation. I suggested that Φ-values could be used by theoreticians in the same way as NOEs are used by NMR spectroscopists as constraints for solving structures[118].

Most theoreticians were delighted with Φ-value analysis and especially the data and mechanism for chymotrypsin inhibitor 2 (see next section). But a tiny rump of theoreticians queried the results until they gradually accepted their authenticity. But, for a number of years, I and my former students and post-doctorals were the only practitioners of Φ-value analysis. It was either steadfastly ignored or criticised, and it took ten years of battle to get the technique universally accepted. John Ladbury, at the 50th anniversary meeting of *JMB* on 1 October, 2009, told me that he was warned in the early 1990s that I held views on proteins that were not generally accepted. I knew that it had finally been accepted when some of those who had first ignored and then criticised Φ-analysis claimed they had invented it (comments in referees reports!). Fifteen years after the first publication as applied to protein folding, I published a critique that rehearsed the original paper[104], which stands still unflawed today with all of the basic concepts and how they should be applied[119].

Φ-Value analysis and the nature of protein-folding transition states

Alan R. Fersht* and Satoshi Sato

Medical Research Council Centre for Protein Engineering, Hills Road, Cambridge CB2 2QH, United Kingdom

Contributed by Alan R. Fersht, April 16, 2004

Φ values are used to map structures of protein-folding transition states from changes in free energies of denaturation ($\Delta\Delta G_{D\text{-}N}$) and activation on mutation. A recent reappraisal proposed that Φ values for $\Delta\Delta G_{D\text{-}N} < 1.7$ kcal/mol are artifactual. On discarding such derived Φ values from published studies, the authors concluded that there are no high Φ values in diffuse transition states, which are consequently uniformly diffuse with no evidence for nucleation. However, values of $\Delta\Delta G_{D\text{-}N} > 1.7$ kcal/mol are often found for large side chains that make dispersed tertiary interactions, especially in hydrophobic cores that are in the process of being formed in the transition state. Conversely, specific local interactions that probe secondary structure tend to have $\Delta\Delta G_{D\text{-}N} \approx 0.5$–2 kcal/mol. Discarding Φ values from lower-energy changes discards the crucial information about local interactions and makes transition states appear uniformly diffuse by overemphasizing the dispersed tertiary interactions. The evidence for the 1.7 kcal/mol cutoff was based on mutations that had been deliberately designed to be unsuitable for Φ-value analysis because they are structurally disruptive. We confirm that reliable Φ values can be derived from the recommended mutations in suitable proteins with $0.6 < \Delta\Delta G_{D\text{-}N} < 1.7$ kcal/mol, and there are many reliable high Φ values. Transition states vary from being rather diffuse to being well formed with islands of near-complete secondary structure. We also confirm that the structures of transition-state ensembles can be perturbed by mutations with $\Delta\Delta G_{D\text{-}N} \gg 2$ kcal/mol and that protein-folding transition states do move on the energy surface on mutation.

barnase | protein A | nucleation–condensation | framework | Hammond

The Φ-value analysis is a particular set of protein-engineering methods that is used to map the structures of transition states and intermediates in protein-folding, catalysis, binding, and conformational transitions of proteins at the level of individual residues (1–5). Φ is the ratio of change of free energy of activation for folding, $\Delta\Delta G_{\ddagger\text{-D}}$, to the equilibrium free energy of folding, $\Delta\Delta G_{N\text{-}D}$,[†] and scores the extent of formation of structure on a scale of 0 to 1 at the level of individual residues. Φ is similar but not identical to the constants α or β of classical rate-equilibrium-free-energy relationships (REFERs) of covalent-bond chemistry. Linear free-energy relationships are the classical means of analyzing the structures of transition states. The structure of a reagent is subtly altered by small changes, and the consequent perturbations of the kinetics and equilibrium of the reaction are measured. Under certain circumstances, which often rely on the chemist's judgement in making sensible structural changes, there can be a linear relationship between $\Delta G^{\ddagger}$, the change in activation energy, and ΔG^0, the change in equilibrium free energy; i.e., $(\partial\Delta G^{\ddagger}/\partial\text{structure})/(\partial\Delta G^0/\partial\text{structure}) = \alpha$ in an REFER (6) (or $= \beta$ in the earlier Brønsted plots for catalysis). The α (or β) value is a measure of extent of covalent-bond making or breaking in the reaction, with $\alpha = 0$ implying no bond making (or breaking) and $\alpha = 1$ implying complete making (or breaking). Intermediate values of α imply partial bond making or breaking. Protein engineering allowed the equivalent of REFERs to be applied to changes in noncovalent interactions of protein side chains as true multipoint plots and as a collection of two-point Φ values (1, 2). In general, protein engineering gives two-point, mutationally specific plots, and the existence of multipoint plots is a bonus when all mutations respond with the same α or β (2)

There are important differences between Φ and α or β. Many chemical processes respond smoothly to changes in structure that are remote from the seat of reaction, but Φ can depend on the specific interactions that are mutated and the change in energetics of the denatured state, including its solvation energy, on mutation (3, 4, 7). These energies affect the interpretation of Φ and require that Φ-value analysis is best applied to certain types of mutation, interpreted within various constraints, and can be difficult for fractional values of Φ. In the purest form of Φ-value analysis, mutations are made that delete interactions that stabilize the native state of the protein without disrupting the structure of the protein, introducing new interactions or changing stereochemistry: "nondisruptive deletion mutations" (2), preferably of hydrophobic moieties, although more radical mutations of surface residues are permitted (4).

Φ-value analysis should be applied with the following caveats. (*i*) In general, fractional values of Φ are not linear with the extent of bond formation, and only the extreme values of 0 and 1 are fully interpretable *per se* as being completely denatured-like and completely native-like, respectively [the term "denatured" is used because denatured states can have residual structure, and thus changes are measured relative to the residual structure (8)]. (*ii*) However, Φ for nondisruptive deletion mutations of hydrophobic side chains is approximately linear with the extent of formation of the bonds between denatured and native conformations, and thus fractional values are a good indicator of the extent of noncovalent-bond formation. (*iii*) Deletion of large side chains can alter many interactions between different substructures and thus give an average value of Φ for all of those interactions [a fine structure analysis is required to dissect the different components (9)]. (*iv*) Deletion of large side chains may also move the transition state by Hammond and anti-Hammond effects (10–12). (*v*) Because of the uncertainties in interpretation of Φ, as many Φ values as possible should be made and the results divided into classes of "weak," "medium," and "strong" (as with nuclear Overhauser effects in NMR spectroscopy) for purposes of combining with simulation to obtain atomic-level resolution of transition states (13).

Φ values *per se* can give extensive atomic-level information on structures of transition states. For example, Φ-value analysis on chymotrypsin-inhibitor 2 (CI2) provided the experimental evidence for the nucleation–condensation mechanism of folding in which secondary and tertiary interactions form together in the transition state, which appears to form around an extended nucleus that has moderate Φ values, with Φ falling off with distance from the nucleus (5, 14, 15). In general, Φ values are the

Abbreviations: REFER, rate-equilibrium-free-energy relationship; CI2, chymotrypsin-inhibitor 2.

*To whom correspondence should be addressed. E-mail: arf25@cam.ac.uk.

[†]Φ can be measured for folding or unfolding. $\Delta\Delta G_{N\text{-}D}$, the free energy of folding, is equal to $-\Delta\Delta G_{D\text{-}N}$, the free energy of denaturation.

 www.pnas.org/cgi/doi/10.1073/pnas.0402684101

experimental data for benchmarking simulation and for reconstructing protein-folding transition states and pathways at atomic resolution by combining experiment and simulation, and Φ-value analysis is at its most powerful when in combination with simulation (16–21). The theoretical studies use a global analysis of the Φ values, not just a selected few.

A recent appraisal of the results of Φ-value analysis concluded that measurements should be restricted to those for $\Delta\Delta G_{\text{D-N}}$ values >1.7 kcal/mol (7 kJ/mol) (22). It also was concluded that most high values of Φ were artifacts of $\Delta\Delta G_{\text{D-N}}$ being <7 kJ/mol. On discarding the "artifactually" high Φ values, it was concluded that "diffuse" protein-folding transition states are uniformly diffuse, which is counter to much of theory, experiment, and simulation (16, 23–27). The authors' reasoning was based on the premise that Φ should be the same for all mutations at the same site, and thus deviations from this will show which Φ values are inaccurate (22). They tested the linearity hypothesis on a data set (28), which appeared to give the 1.7 kcal/mol cutoff point. However, in many cases, Φ is mutation-specific, because each mutation removes different interactions and has different effects on the denatured state (3, 4) and the data set used to calibrate Φ values had been deliberately designed by Davidson and coworkers (28) to be unsuitable for Φ-value analysis.

We formally demonstrate the flaws in the Φ-value reappraisal and confirm from experimental data that Φ values from non-disruptive mutations can be adequately reliable down to a $\Delta\Delta G_{\text{D-N}}$ value of ≈0.6 kcal/mol (2.5 kJ/mol) for suitable proteins. We show how the crucial probes for the formation of local secondary structure tend to have $\Delta\Delta G_{\text{D-N}}$ values of ≈0.6–2 kcal/mol (2.5–8 kJ/mol), and thus discarding their values discards the evidence for nucleation sites. We also refute the proposal (22, 29) that structures of transition-state ensembles are not affected by mutation. We use the REFER methods that supposedly indicated the 1.7 kcal/mol cutoff point to show that the cutoff is closer to 0.6 kcal/mol for suitable proteins.

Formal Analysis of Φ Values.

Relationship Between Φ and α. The premise of Sanchez and Kiefhaber (22) is that Φ is identical to the Leffler α and that all mutations at a particular position should give the same value of Φ. However, there is a crucial difference between the Leffler α and Φ, which may be shown formally. Let the free energy of the native state *N* be G_{N}, that of the denatured state be G_{D}, that of the transition state be G_{TS}, and mutant states be denoted by a prime (Fig. 1). The free energy of denaturation $\Delta G_{\text{D-N}} = G_{\text{D}} - G_{\text{N}}$; the free energy of folding $\Delta G_{\text{N-D}} = G_{\text{N}} - G_{\text{D}}$; the activation energy of unfolding $\Delta G_{\text{TS-N}} = G_{\text{TS}} - G_{\text{N}}$; and the activation energy of folding $\Delta G_{\text{TS-D}} = G_{\text{TS}} - G_{\text{D}}$. Abbreviating structure to "S":

$$\partial(G_{\text{TS}} - G_{\text{D}})/\partial S = \partial G_{\text{TS}}/\partial S - \partial G_{\text{D}}/\partial S \quad [1]$$

$$\partial(G_{\text{N}} - G_{\text{D}})/\partial S = \partial G_{\text{N}}/\partial S - \partial G_{\text{D}}/\partial S. \quad [2]$$

Thus,

$$\alpha = (\partial G_{\text{TS}}/\partial S - \partial G_{\text{D}}/\partial S)/(\partial G_{\text{N}}/\partial S - \partial G_{\text{D}}/\partial S). \quad [3]$$

The change in free energy of the denatured state with mutation, $\partial G_{\text{D}}/\partial S$, is an important component of α. It was the essential difference between the classical Leffler α or Brønsted β that led us to give the free-energy constant a new name, Φ (3).

The experimentally accessible quantities used in Φ-value analysis are equilibrium and kinetic data of denaturation measured on wild-type and mutant proteins separately, $\Delta G_{\text{D-N}}$, $\Delta G_{\text{D'-N'}}$, $\Delta G_{\text{TS-N}}$, $\Delta G_{\text{TS'-N'}}$, $\Delta G_{\text{TS-D}}$, and $\Delta G_{\text{TS'-D'}}$ (Fig. 1). The observed difference in free energy of denaturation of wild-type protein and that of a mutant (denoted by ′) $\Delta\Delta G_{\text{D-N}} = \Delta G_{\text{D'-N'}} - \Delta G_{\text{D-N}}$. The difference in free energy of activation of unfolding $\Delta\Delta G_{\text{TS-N}} = \Delta G_{\text{TS'-N'}} - \Delta G_{\text{TS-N}}$. The difference in free energy of activation of folding $\Delta\Delta G_{\text{TS-D}} = \Delta G_{\text{TS'-D'}} - \Delta G_{\text{TS-N}}$, which for two-state kinetics has the same transition state. The observed experimental quantities are related to the changes in the individual states on mutation for virtual thermodynamic cycles based on Fig. 1 (3, §): $\Delta\Delta G_{\text{D-N}} = \Delta G_{\text{D'-D}} - \Delta G_{\text{N'-N}}$; $\Delta\Delta G_{\text{TS-N}} = \Delta G_{\text{TS'-TS}} - \Delta G_{\text{N'-N}}$; and $\Delta\Delta G_{\text{TS-D}} = \Delta G_{\text{TS'-TS}} - \Delta G_{\text{D'-D}}$.

The experimentally determined Φ value for folding, Φ_{F},¶ is defined by

$$\Phi_{\text{F}} = \Delta\Delta G_{\text{TS-D}}/\Delta\Delta G_{\text{N-D}}. \quad [4]$$

In terms of the other changes in the cycle,

$$\Phi_{\text{F}} = (\Delta G_{\text{TS'-TS}} - \Delta G_{\text{D'-D}})/(\Delta G_{\text{N'-N}} - \Delta G_{\text{D'-D}}). \quad [5]$$

Eq. **5** is the two-point version of Eq. **3** for a finite change in structure.

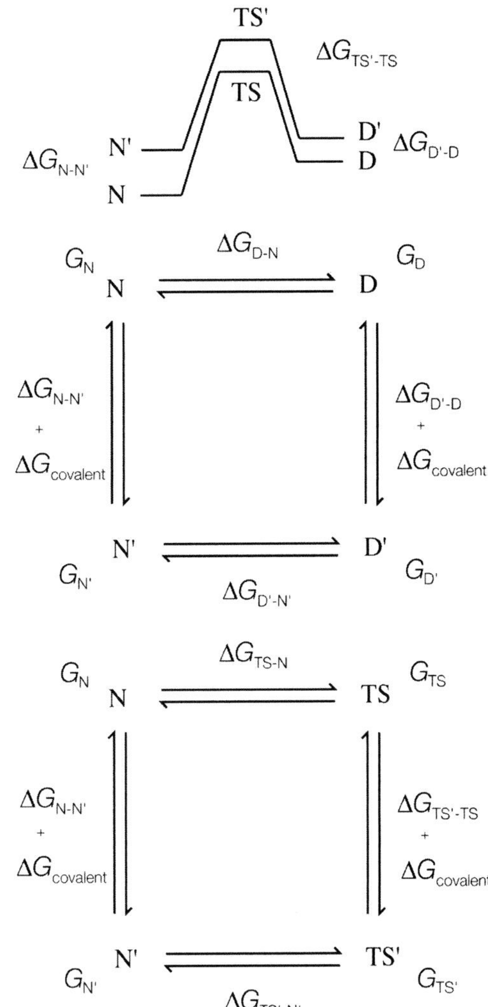

Fig. 1. Schematics of thermodynamic cycles and free-energy profiles.

§Although of routine use in physical chemistry, the application of thermodynamic cycles to protein folding was described as incorrect because they have hypothetical steps (30).

¶Folding can be more complicated than unfolding because of (unknown) residual structure in the denatured state and changes in rate-determining steps, and Φ-value analysis is applied more often to unfolding kinetics.

The change in free energy on mutating N to N′, $\Delta G_{N'-N}$, can be split up into notional components: the change in energy of the covalent bond that is mutated, ΔG_{cov}, the change in noncovalent interactions at the site of mutation, ΔG_{noncov} (without any reorganization of the structure of the protein on mutation), plus any additional changes because of the reorganization of the protein, ΔG_{reorg}, and the change in solvation energy, ΔG_{solv} (4, 7). ΔG_{cov} is the same for all states and drops out of the equations.

$$\Phi_F = [\Delta G_{noncov(TS'-TS)} + \Delta G_{reorg(TS'-TS)} + \Delta G_{solv(TS'-TS)} - \Delta G_{noncov(D'-D)} - \Delta G_{reorg(D'-D)} - \Delta G_{solv(D'-D)}]/[\Delta G_{noncov(N'-N)} + \Delta G_{reorg(N'-N)} + \Delta G_{solv(N'-N)} - \Delta G_{noncov(D'-D)} - \Delta G_{reorg(D'-D)} - \Delta G_{solv(D'-D)}] \quad [6]$$

The presence of the $\Delta G_{D'-D}$ term in Eq. **5** [= $\Delta G_{reorg(D'-D)}$ + $\Delta G_{solv(D'-D)}$ in Eq. **6**] can lead to Φ being uninterpretable, and the ΔG_{noncov} terms can be mutation-specific (2, 3, 7).

Φ is identical to the classical α or β under two extreme circumstances. (*i*) The target region is as unstructured in the transition state as in the denatured state. Under these circumstances, all the ΔG terms in TS′-TS are the same as those of D′-D, and Eqs. **5** and **6** reduce to $\Phi_F = 0$ (and for two-state folding, Φ for unfolding, Φ_U, = 1). (*ii*) The target region is as structured in the transition state as in the native state. Under these circumstances; all the ΔG terms in TS′-TS are the same as those of N′-N, and Eqs. **5** and **6** reduce to $\Phi_F = 1$ (and for two-state folding, $\Phi_U = 1$). The extreme values of 0 and 1 should be interpretable, therefore, for all mutations.

Fractional values of Φ are readily interpretable for the chemically sensible nondisruptive deletion mutations, especially of hydrophobic side chains, because the ΔG_{reorg} terms are minimized (2, 4) as are also the ΔG_{solv} for aliphatic to aliphatic mutations. For example, mutations of Ile → Ala and Val have values of ΔG_{solv} in water for fully exposed side chains of only −0.21 and −0.16 kcal/mol, respectively (31). Thus, when $\Delta\Delta G_{D-N}$ is significant and ΔG_{reorg} and ΔG_{solv} are low, $\Phi_U = [\Delta G_{noncov(TS'-TS)} - \Delta G_{noncov(N'-N)}]/[\Delta G_{noncov(D'-D)} - \Delta G_{noncov(N'-N)}]$, which is analogous to α. Additionally, the energetics for deletion of hydrophobic elements of side chains is dominated by van der Waals' interactions, which are approximately additive, as is found experimentally (32). Thus, for nondisruptive deletions of hydrophobic side chains, $\Phi_F \approx (n_{TS} - n_D)/(n_N - n_D)$, where n_{TS} is the number of van der Waals' interactions made by the target portion of the side chain in the transition state, n_N is in the native state, and n_D in the denatured state. For two-state kinetics, $\Phi_U \approx (n_{TS} - n_N)/(n_D - n_N)$. Importantly, Φ reports back on the interactions made in the native protein.

Fractional values can arise from either genuinely weakened interactions in a single transition-state ensemble or from a mixture of states in parallel pathways. The folding of the barley CI2, for example, has predominantly fractional values of Φ (14, 33). The conforming to an REFER shows that there are not parallel pathways (34).

Fractional values may also arise if a side chain makes interactions with multiple elements of substructures that have varying degrees of structure formation. The overall Φ value is then the weighted mean value of all the interactions. This possibility was noted for barnase and CI2, and a fine structure analysis was performed by making systematic mutations in the side chains concerned: e.g., Ile → Val → Ala → Gly, which gave the individual Φ values (9, 32, 35).

Accordingly, our preferred strategy for Φ-value analysis is to (*i*) mutate buried hydrophobic moieties by nondisruptive deletions (preferably Ile → Val → Ala → Gly; Leu → Ala → Gly; Thr → Ser; and Phe → Ala → Gly, avoiding Phe → Leu because of the change in stereochemistry); (*ii*) make a wider range of surface mutations, because larger side chains may be acceptable; (*iii*) use double-mutant cycles in which changes in solvation and reorganization energies tend to cancel out (4); and (*iv*) mutate Ala → Gly at solvent-exposed positions in secondary structural regions ("Ala → Gly scanning"), especially in α-helices, because they provide an exquisite probe of secondary structure (12). (*v*) Perform additional fine structure analyses by deleting different parts of larger side chains.

The magnitude of $\Delta\Delta G_{D-N}$ on mutation is a compromise: small values represent the smallest perturbation to the structure but have more attendant errors; larger values can be measured more accurately but often involve large changes that have more artifacts from reorganization energy changes on mutation and removing dispersed interactions. Our lower limit of acceptability for $\Delta\Delta G_{D-N}$ has generally been ≈0.6 kcal/mol (2.5 kJ/mol) for small, nondisruptive deletions (9, 14, 36).

Experimental Evidence for Lower Limit of $\Delta\Delta G_{D-N}$ for Φ Values. Comparing individual Φ values for mutations at a particular site with a multipoint Leffler plot would be a good means of detecting deviations for particular mutations (22), but for the problem of the ΔG_{solv} terms involved in $\Delta G_{D'-D}$ and ΔG_{reorg} in the native state, unless carefully chosen, the different mutants will have different values of ΔG_{solv} and could have very different values of ΔG_{reorg} in the native and transition-state structures. Additionally, there must also be a relatively linear function of ΔG_{noncov} with formation of structure over the range of structural transition.

An earlier attempt to calibrate two-point Φ plots against a multipoint Leffler plot (37) used mutations that are specifically not recommended, with only 5 of 47 being nondisruptive hydrophobic deletions: Ser → Asp, Glu; Ile → Thr, Tyr; Val → Ala, Lys; Leu → Ala, Thr, Ile; Ala → Cys, Pro, Ser, Gly, Thr, Leu, Asn; Gln → Ala, Leu, Ser, Thr, Lys; Ser → Ala, Gly, Trp, Cys, Thr, Ile, Tyr, Val, Asn, Gln, Lys; Val → Ala, Cys, Leu, Thr; Leu → Ser, Gln, Pro, Phe, Asn, Glu, Tyr; Leu → Ala, Ser; and Ser → Ala, Leu. Sanchez and Kiefhaber (22) showed that two-point Φ values deviated from a multipoint Leffler plot for $\Delta\Delta G_{D-N} <$ 1.7 kcal/mol by using data from a study by Davidson and coworkers (28). However, the title of that study was "Protein Folding Kinetics Beyond the Φ-Value: Using Multiple Amino Acid Substitutions to Investigate the Structure of the SH3 Domain Folding Transition State," and the rationale was described by the authors in the abstract as: "In contrast to most other folding kinetic studies which have focused primarily on nondisruptive substitutions with Ala or Gly, here we have examined the effects of substitutions with diverse amino acid residues." They mutated Glu → Asp, Gln, His, Lys, Ala, Ser, Val, Pro, Gly, Arg, Ile, Leu, and Ser → Lys, Arg, Leu, Ala, His, Val, Ile, Asn, Asp, Gly, Phe, Tyr, which are virtually all disruptive mutations of a mainly buried hydrogen bond. These mutations are by intent and definition unsuitable for calibrating Φ-value analysis (28).

Leffler Plots and Ala → Gly Scanning for Exposed Surface Regions. The Davidson and coworkers (28) mutations are not suitable, because the side chains are buried, which complicates both the specific interactions involved for each chain as well as the changes in solvation on mutation. In contrast, surface-exposed residues, especially of α-helices, provide an opportunity for testing REFERS of more than two points. If the solvent-exposed end of the residue in the N state does not make any specific interactions with the rest of the protein, then it should make similar interactions with solvent in the TS and D states. Thus, the energetics of solvation of polar moieties of common surface residues such as Arg, Lys, Glu, Asp, Gln, and Asn, for

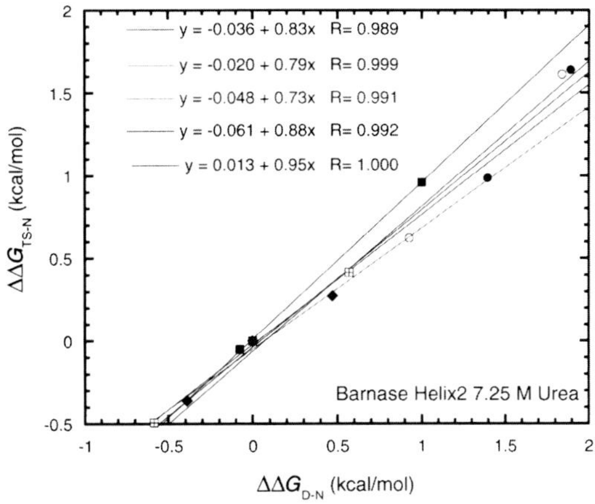

Fig. 2. Three-point Leffler plots for the unfolding of barnase. Mutations are as follows: Thr-26 → Ala → Gly, Lys-27 → Ala → Gly, Ser-28 → Ala → Gly, Glu-29 → Ala → Gly, and Gln-31 → Ala → Gly in helix 2.

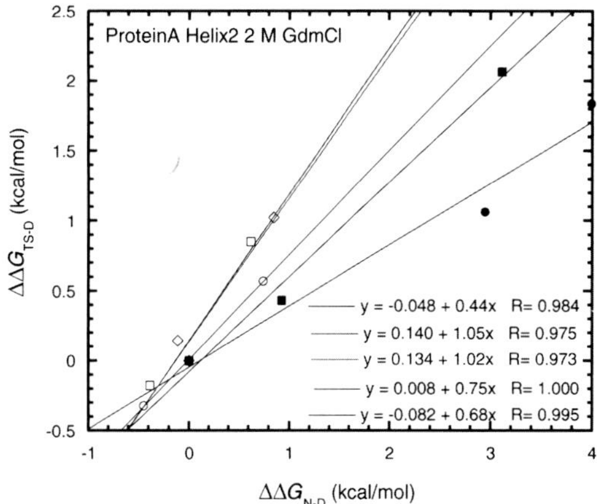

Fig. 4. Three-point Leffler plots for the following mutations of helix 2 of the B-domain of protein A: Gln-27 → Ala → Gly, Arg-28 → Ala → Gly, Asn-29 → Ala → Gly, Gln-33 → Ala → Gly, and Ser-34 → Ala → Gly.

example, will cancel out in the equations defining Φ. We have accumulated data for several helices in which surface-exposed residues are mutated to Ala and Gly. Helix 2 of barnase provides a good test for three-point Leffler plots, because the two-point Φ values show that the helix becomes highly unfolded in the transition state for unfolding (9) and thus the whole of the helix could constitute a multipoint REFER. Indeed, the energetics of mutation of Thr-26 → Ala → Gly, Lys-27 → Ala → Gly, Ser-28 → Ala → Gly, Glu-29 → Ala → Gly, and Gln-31 → Ala → Gly (Fig. 2) fit a good linear plot, with each three-point plot for an individual position having a correlation coefficient (R) between 0.99 and 1.0 and the slopes varying between 0.73 and 0.95 (mean = 0.84 ± 0.04 standard error). Individual values of Φ for all mutants (relative to wild type; Fig. 3) give a mean value of Φ of 0.86 ± 0.04 in a spread of 0.6 to 1.1. Ala → Gly scanning at each position gives a mean of 0.95 ± 0.08. The Φ values were derived from values of $\Delta\Delta G_{D\text{-}N}$ that have a standard error of ± 0.06 kcal/mol and values of $\Delta\Delta G_{TS\text{-}N}$ at 7.25 M urea that have a standard error of ±0.01–0.03 kcal/mol (12). Even with a $\Delta\Delta G_{D\text{-}N}$ value of 0.6 kcal/mol, the expected error in Φ should be only ≈10%. The linear plots in Fig. 2 and the values of Φ (Fig. 3) are nearly all obtained from values of $\Delta\Delta G_{D\text{-}N}$ below the 1.7 kcal/mol (7 kJ/mol) cutoff proposed by Sanchez and Kiefhaber (22).

Another suitable protein for testing the accuracy of values of Φ derived from low $\Delta\Delta G_{D\text{-}N}$ values is the B-domain of protein A from *Staphylococcus aureus*, a three-helix bundle protein. The data are slightly less accurate ($\Delta\Delta G_{D\text{-}N}$ ± 0.1 and $\Delta\Delta G_{TS\text{-}N}$ ± 0.08 kcal/mol), so that Φ for $\Delta\Delta G_{D\text{-}N}$ = 0.6 kcal/mol should have an error of ±20%, dropping to ±10% at $\Delta\Delta G_{D\text{-}N}$ = 1.2 kcal/mol (36). Individual Φ values show that the first and third helices are in the process of being formed in the transition state, whereas the second is nearly fully formed. There are good three-point REFERs (Fig. 4) for the mutations in helix 2 of Gln-27 → Ala → Gly, Arg-28 → Ala → Gly, Asn-29 → Ala → Gly, Gln-33 → Ala → Gly, and Ser-34 → Ala → Gly. The only low value of the slope is for Gln-27 → Ala → Gly, which results from specific interactions. Gln-27 makes a hydrogen bond with Asn-24, and there are large changes of $\Delta\Delta G_{D\text{-}N}$ (3 and 4 kcal/mol) on its mutation. REFERs for the energetics of Ala → Gly scanning (Fig. 5) show that helix 2 is ≈80% formed in the transition state, whereas the other two are not. The data for Ala → Gly scanning are clearly acceptable down to a $\Delta\Delta G_{D\text{-}N}$ value of ≈0.6 kcal/mol.

Schmid and coworkers (38), in a careful study of the folding of CspB protein, independently used a $\Delta\Delta G_{D\text{-}N}$ value of ≈0.6 kcal/mol as the lower limit for Φ-value analysis and also divided their results into weak, medium, and strong values. Radford and coworkers (25) used a cutoff of 0.7 kcal/mol for analyzing the

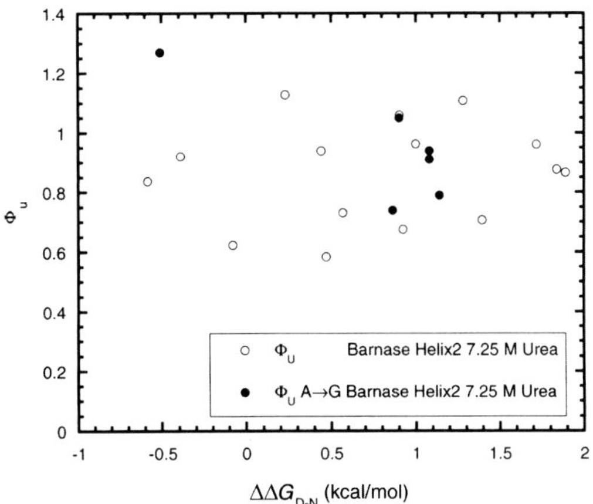

Fig. 3. Plot of Φ_U versus $\Delta\Delta G_{D\text{-}N}$ for mutations in helices 1 and 3 of barnase.

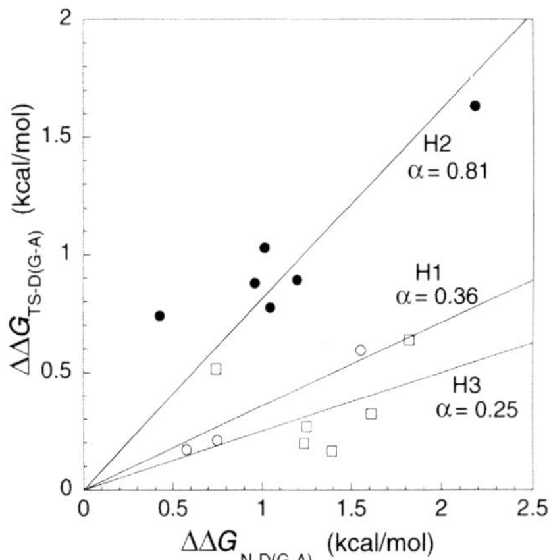

Fig. 5. Leffler plots of Ala → Gly scanning mutations in the three helices of the B-domain of protein A.

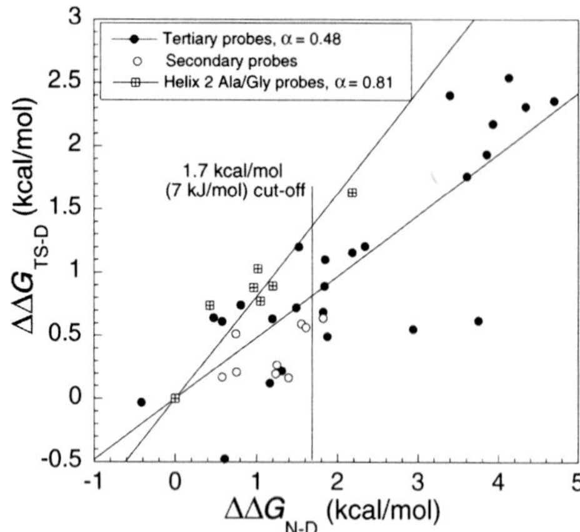

Fig. 6. Leffler plot for all mutations in the B-domain of protein A. Filled circles are used for the tertiary probes, and open circles are used for secondary structural probes.

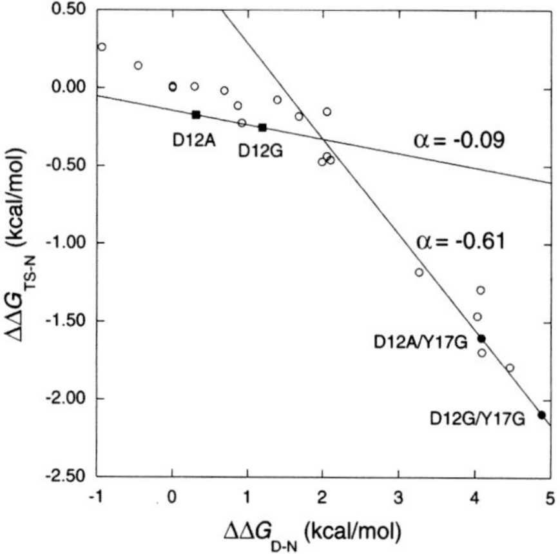

Fig. 7. Leffler plots for single, double, and triple mutations in helix 1 of barnase plus Ala → Gly scanning at position 12 in the mutant Tyr-17 → Gly. The mutants are as follows: Asp-8 → Ala, Asp-12 → Gly, Asp-12 → Ala, Tyr-13 → Ala, Tyr-13 → Ala/Thr-16 → Ser, Tyr-13 → Ala/Tyr-17 → Ala, Tyr-13 → Ala/Thr-16 → Ser/Tyr-17 → Ala, Gln-15 → Ile, Thr-16 → Ala, Thr-16 → Gly, Thr-16 → Ser, Thr-16 → Ser/Tyr-17 → Ala, Thr-16 → Arg, Tyr-17 → Ala, His-18 → Lys, His-18 → Gln, Tyr-17 → Ala, Tyr-17 → Gly, His-18 → Ala, His-18 → Gly, Asp-12 → Ala/Tyr-17 → Gly, and Asp-12 → Gly/Tyr-17 → Gly.

folding of the immunity proteins Im7 and Im9, with satisfactory agreement. [Sanchez and Kiefhaber (22) recalculated Φ values from ref. 25, but the data were incomplete; several values from larger destabilizing mutations, some of which result in Φ values >0.3, are omitted, and other values are shown that were not calculated in ref. 25 because the $\Delta\Delta G_{D\text{-}N}$ was below the cut off of 3 kJ/mol (0.7 kcal/mol) used (S. E. Radford, personal communication)].

Values of $\Delta\Delta G_{D\text{-}N}$ for Secondary and Tertiary Structure Probes. The fine structure probes that test specific interactions tend to be those that delete a small interaction; e.g., that of a single methylene group, or a single hydrogen bond, which both typically have $\Delta\Delta G_{D\text{-}N}$ values in the range of 1.5 ± 0.5 kcal/mol (39–41). Large changes of $\Delta\Delta G_{D\text{-}N}$ tend to be associated with the deletion of large side chains, especially in the hydrophobic core, or the disruption of buried salt bridges (32), as shown in Fig. 6 for the B-domain of protein A. By discarding the Φ values derived from a $\Delta\Delta G_{D\text{-}N}$ value of <1.7 kcal/mol, Sanchez and Kiefhaber (22) discarded most of the secondary structure probes and thus constructed REFER plots of mainly tertiary interactions, especially in the hydrophobic core. The formation of the core is always part of the rate-determining process and has fractional Φ values (3, 27, 36, 42, 43). Thus, by concentrating on such data, they concluded incorrectly that transition states are uniformly diffuse.

Movement of Transition-State Structure on Mutation. Sanchez and Kiefhaber (22, 29) claim that the structures of transition states do not change on mutation. They suggest that the observed movements of transition states along a reaction coordinate (10–12) are not a consequence of a change in transition-state structure via a Hammond effect but instead result from (partial) changes in rate-determining steps between formation and breakdown of intermediates or have complications from effects of mutation on the denatured state (44). It is very difficult to distinguish between true Hammond behavior and changes in the rate-determining step. However, there are well documented examples of anti-Hammond behavior (movement perpendicular to the reaction coordinate) that cannot be accounted for by changes in the rate-determining step along a reaction coordinate (12). A Leffler plot of successive mutations in helix 1 of barnase (Fig. 7) has a slope for unfolding of −0.09 for mutations with $\Delta\Delta G_{D\text{-}N} < 2$ kcal/mol, showing that it is ≈90% folded in the transition state, but for $\Delta\Delta G_{D\text{-}N} > 3$ kcal/mol, the slope gets steeper at −0.61, indicating it is only 40% folded and follows anti-Hammond behavior (12). The slopes are measured by Ala → Gly scanning at position 12 in wild-type protein and the same position in the mutant Tyr-17 → Gly, which is destabilized by 4.1 kcal/mol (12). Whereas the helix becomes less folded in the transition state on destabilization, the overall transition state follows Hammond and becomes more folded, with its relative surface exposure decreasing from 55% to 37% over a change of 5 kcal/mol in $\Delta\Delta G_{D\text{-}N}$ (Fig. 8). The data are for unfolding and are independent of mutations on the denatured state. The anti-Hammond behavior can be explained by either a gradual movement of the transition state or by a switch between parallel pathways (12). Simulation favors the gradual movement (45).

Sanchez and Kiefhaber propose that the larger the value of

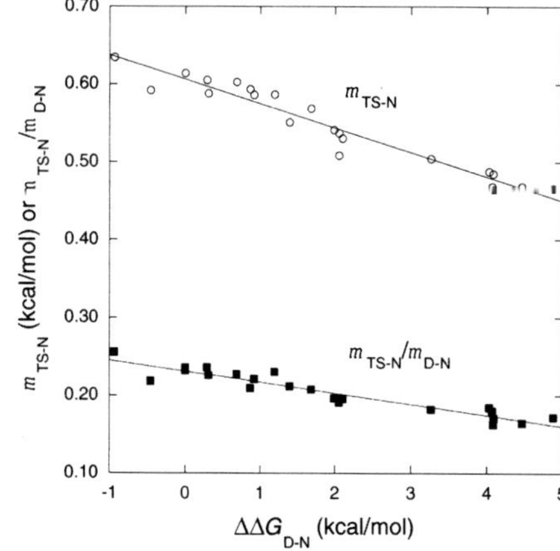

Fig. 8. Plots of data relating surface exposure on denaturation kinetics of barnase. $m_{TS\text{-}N}$ is the slope of the plot of $\Delta G_{TS\text{-}N}$ versus [urea], and $m_{D\text{-}N}$ is the slope of the plot of $\Delta G_{D\text{-}N}$ versus [urea]. $m_{TS\text{-}N}$ is the value at 7.25 M urea, measured for unfolding data acquired between 6 and 8.5 M urea, and it is accurate to ±2%. The ratio of $m_{TS\text{-}N}/m_{D\text{-}N}$ is a measure of the relative solvent exposure of the transition state to the denatured state. The value of $m_{TS\text{-}N}$ is a function of just the difference in solvent-accessible surface area of TS and N and does not depend on the properties of the denatured state.

$\Delta\Delta G_{D\text{-}N}$, the better for Φ-value analysis (22). However, Fig. 7 shows clearly that large changes of $\Delta\Delta G_{D\text{-}N}$ can lead to radical changes in structure in transition states. In general, the fine structural information requires specific probes, with energies of 0.6–2 kcal/mol. The more energy-disruptive probes can perturb the transition-state structure, which complicates a simple Φ-value analysis but can give information about the energy surface around the transition state.

Nature of Folding Transition States. Sanchez and Kiefhaber (22) use the classification of transition states being either diffuse, whereby most of the Φ values are fractional and polarized where there are regions that are fully formed or fully denatured. They suggest that there are never regions of high Φ value in the diffuse states, and thus all transition states are similar to that found for CI2. However, the transition state for the folding of the B-domain of protein A is not polarized, and there are regions, especially involving helix 2, that have Φ values approaching 1 (36). The Φ analyses of the Engrailed homeodomain family, although not extensive, show transition states that are basically structured all over and with regions of Φ values of 1 (27, 43). Transition states have a spectrum of structures, varying from the diffuse of pure nucleation–condensation to the more compact that approach the classical framework mechanism of folding, in which the repeating secondary structure is nearly fully formed and the core is in the process of consolidation (23–25, 38, 44–50).

In any case, it is not possible to divine from the transition-state structure *per se* whether there are nucleation sites, because the structure does not reveal *per se* the starting point of folding or the route by which the structure is formed (51). The experimental evidence for nucleation in CI2 folding, for example, came from ancillary studies that examined the denatured state of the protein and its fragments (52), and the evidence for framework for Engrailed homeodomain came from analyzing the structures of ground states as well as simulation (43). Additionally, high Φ values need not be associated with a nucleus, and low Φ values can be found in nuclei (19).

Conclusions

There are strong analogies between the determination of solution structures of proteins by NMR combined with simulated annealing and the determination of structures of transition states by Φ values and simulation. Just as there are nuclear Overhauser effects in NMR spectra of spurious intensity, there are undoubtedly some misleading Φ values, especially when $\Delta\Delta G_{D\text{-}N}$ is small. However, provided mutations are made within the prescribed rules and that a sufficient number are analyzed, then reliable results will be obtained down to changes in $\Delta\Delta G_{D\text{-}N}$ of ≈0.6 kcal/mol under optimal conditions. Higher values of $\Delta\Delta G_{D\text{-}N}$ do give statistically more precise data, but much larger values of $\Delta\Delta G_{D\text{-}N}$ may give less precise information, because they arise from dispersed interactions and may have higher contributions from perturbations of structure. Just as special methods are continually introduced to refine NMR methods, so ancillary methods such as Ala → Gly scanning are required for refining by Φ-value analysis.

1. Fersht, A. R., Leatherbarrow, R. J. & Wells, T. N. C. (1986) *Nature* **322,** 284–286.
2. Fersht, A. R., Leatherbarrow, R. & Wells, T. N. C. (1987) *Biochemistry* **26,** 6030–6038.
3. Matouschek, A., Kellis, J. T., Jr., Serrano, L. & Fersht, A. R. (1989) *Nature* **340,** 122–126.
4. Fersht, A. R., Matouschek, A. & Serrano, L. (1992) *J. Mol. Biol.* **224,** 771–782.
5. Fersht, A. R. (1995) *Proc. Natl. Acad. Sci. USA* **92,** 10869–10873.
6. Leffler, J. E. (1953) *Science* **117,** 340–341.
7. Fersht, A. R. (1988) *Biochemistry* **27,** 1577–1580.
8. Matouschek, A., Kellis, J. T., Jr., Serrano, L., Bycroft, M. & Fersht, A. R. (1990) *Nature* **346,** 440–445.
9. Serrano, L., Matouschek, A. & Fersht, A. R. (1992) *J. Mol. Biol.* **224,** 805–818.
10. Matouschek, A. & Fersht, A. R. (1993) *Proc. Natl. Acad. Sci. USA* **90,** 7814–7818.
11. Matouschek, A., Otzen, D. E., Itzhaki, L. S., Jackson, S. E. & Fersht, A. R. (1995) *Biochemistry* **34,** 13656–13662.
12. Matthews, J. M. & Fersht, A. R. (1995) *Biochemistry* **34,** 6805–6814.
13. Fersht, A. R. (1995) *Curr. Opin. Struct. Biol.* **5,** 79–84.
14. Itzhaki, L. S., Otzen, D. E. & Fersht, A. R. (1995) *J. Mol. Biol.* **254,** 260–288.
15. Fersht, A. R. (1997) *Curr. Opin. Struct. Biol.* **7,** 3–9.
16. Fersht, A. R. & Daggett, V. (2002) *Cell* **108,** 573–582.
17. Paci, E., Vendruscolo, M., Dobson, C. M. & Karplus, M. (2002) *J. Mol. Biol.* **324,** 151–163.
18. Klimov, D. K. & Thirumalai, D. (2002) *J. Mol. Biol.* **317,** 721–737.
19. Hubner, I. A., Shimada, J. & Shakhnovich, E. I. (2004) *J. Mol. Biol.* **336,** 745–761.
20. Weikl, T. R., Palassini, M. & Dill, K. A. (2004) *Protein Sci.* **13,** 822–829.
21. Onuchic, J. N. & Wolynes, P. G. (2004) *Curr. Opin. Struct. Biol.* **14,** 70–75.
22. Sanchez, I. E. & Kiefhaber, T. (2003) *J. Mol. Biol.* **334,** 1077–1085.
23. Daggett, V. & Fersht, A. (2003) *Nat. Rev. Mol. Cell Biol.* **4,** 497–502.
24. Daggett, V. & Fersht, A. R. (2003) *Trends Biochem. Sci.* **28,** 18–25.
25. Friel, C. T., Capaldi, A. P. & Radford, S. E. (2003) *J. Mol. Biol.* **326,** 293–305.
26. Li, L. & Shakhnovich, E. I. (2001) *Proc. Natl. Acad. Sci. USA* **98,** 13014–13018.
27. Gianni, S., Guydosh, N. R., Khan, F., Caldas, T. D., Mayor, U., White, G. W. N., DeMarco, M. L., Daggett, V. & Fersht, A. R. (2003) *Proc. Natl. Acad. Sci. USA* **100,** 13286–13291.
28. Northey, J. G. B., Maxwell, K. L. & Davidson, A. R. (2002) *J. Mol. Biol.* **320,** 389–402.
29. Sanchez, I. E. & Kiefhaber, T. (2003) *J. Mol. Biol.* **327,** 867–884.
30. Buchner, J. & Kiefhaber, T. (1990) *Nature* **343,** 601–602.
31. Wolfenden, R., Anderson, L., Cullis, P. M. & Southgate, C. C. B. (1981) *Biochemistry* **20,** 849–855.
32. Serrano, L., Kellis, J. T., Cann, P., Matouschek, A. & Fersht, A. R. (1992) *J. Mol. Biol.* **224,** 783–804.
33. Otzen, D. E., Itzhaki, L. S., Elmasry, N. F., Jackson, S. E. & Fersht, A. R. (1994) *Proc. Natl. Acad. Sci. USA* **91,** 10422–10425.
34. Fersht, A. R., Itzhaki, L. S., Elmasry, N., Matthews, J. M. & Otzen, D. E. (1994) *Proc. Natl. Acad. Sci. USA* **91,** 10426–10429.
35. Matouschek, A., Serrano, L. & Fersht, A. R. (1992) *J. Mol. Biol.* **224,** 819–835.
36. Sato, S., Religa, T. E. & Fersht, A. R. (2004) *Proc. Natl. Acad. Sci. USA* **101,** 6952–6956.
37. Cymes, G. D., Grosman, C. & Auerbach, A. (2002) *Biochemistry* **41,** 5548–5555.
38. Garcia-Mira, M. M., Bohringer, D. & Schmid, F. X. (2004) *J. Mol. Biol.*, in press.
39. Fersht, A. R., Shi, J. P., Knill-Jones, J., Lowe, D. M., Wilkinson, A. J., Blow, D. M., Brick, P., Carter, P., Waye, M. M. Y. & Winter, G. (1985) *Nature* **314,** 235–238.
40. Kellis, J. T. J., Nyberg, K. & Fersht, A. R. (1989) *Biochemistry* **28,** 4914–4922.
41. Serrano, L., Sancho, J., Hirshberg, M. & Fersht, A. R. (1992) *J. Mol. Biol.* **227,** 544–559.
42. Jackson, S. E., Elmasry, N. & Fersht, A. R. (1993) *Biochemistry* **32,** 11270–11278.
43. Mayor, U., Guydosh, N. R., Johnson, C. M., Grossmann, J. G., Sato, S., Jas, G. S., Freund, S. M. V., Alonso, D. O. V., Daggett, V. & Fersht, A. R. (2003) *Nature* **421,** 863–867.
44. Sanchez, I. E. & Kiefhaber, T. (2003) *J. Mol. Biol.* **325,** 367–376.
45. Daggett, V., Li, A. J. & Fersht, A. R. (1998) *J. Am. Chem. Soc.* **120,** 12740–12754.
46. Kragelund, B. B., Osmark, P., Neergaard, T. B., Schiodt, J., Kristiansen, K., Knudsen, J. & Poulsen, F. M. (1999) *Nat. Struct. Biol.* **6,** 594–601.
47. Chiti, F., Taddei, N., White, P. M., Bucciantini, M., Magherini, F., Stefani, M. & Dobson, C. M. (1999) *Nat. Struct. Biol.* **6,** 1005–1009.
48. Riddle, D. S., Grantcharova, V. P., Santiago, J. V., Alm, E., Ruczinski, I. & Baker, D. (1999) *Nat. Struct. Biol.* **6,** 1016–1024.
49. Martinez, J. C. & Serrano, L. (1999) *Nat. Struct. Biol.* **6,** 1010–1016.
50. Lindberg, M., Tangrot, J. & Oliveberg, M. (2002) *Nat. Struct. Biol.* **9,** 818–822.
51. Fersht, A. R. (2000) *Proc. Natl. Acad. Sci. USA* **97,** 1525–1529.
52. Itzhaki, L. S., Neira, J. L., Ruiz-Sanz, J., Gay, G. D. & Fersht, A. R. (1995) *J. Mol. Biol.* **254,** 289–304.

Chapter 13

Chymotrypsin Inhibitor 2: Two-State Folding and the Nucleation-Condensation Mechanism

'Seek simplicity, and distrust it.'

Alfred North Whitehead

Discovery of Two-State Folding

I have always been lucky with my choice of proteins, but it is more correct to say that they have more often than not chosen me. At the 1985 Keystone meeting, I shared a room with Peter Shewry who told me about chymotrypsin inhibitor 2 (CI2), and gave me its clone. This chance encounter was crucial for our development of protein folding. Colin Longstaff and my student Alison Campbell analysed the interactions of the inhibitor with enzymes (Alison has subsequently had a career in technology transfer, beginning in the MRC head office, where she was in charge of Greg and me). My student Sophie Jackson subsequently used CI2 for protein folding studies. She found that it had perfect chevron folding plots with long straight limbs, and the ratio of the rate constants and all of the associated kinetic parameters were consistent with two-state folding between native and denatured states without the accumulation of an intermediate. Until then, it had been thought that all proteins folded via intermediates. We had discovered two-state folding, but were careful enough to write that there could be high-energy intermediates that did not accumulate and were kinetically silent. The finding of two-state folding was a seminal contribution[120].

13.1 1991: Mike Lubienski and Sophie Jackson.

13.2 2003: Jane Clarke, Sophie Jackson and Gideon Schreiber.

10428 Reprinted from Biochemistry, 1991, 30.

Folding of Chymotrypsin Inhibitor 2. 1. Evidence for a Two-State Transition

Sophie E. Jackson[‡] and Alan R. Fersht*

MRC Unit for Protein Function and Design, Cambridge IRC for Protein Engineering, University Chemical Laboratory, Lensfield Road, Cambridge CB2 1EW, U.K.

Received April 25, 1991; Revised Manuscript Received July 18, 1991

ABSTRACT: The reversible folding and unfolding of barley chymotrypsin inhibitor 2 (CI2) appears to be a rare example in which both equilibria and kinetics are described by a two-state model. Equilibrium denaturation by guanidinium chloride and heat is completely reversible, and the data can be fitted to a simple two-state model involving only native and denatured forms. The free energy of folding in the absence of denaturant, ΔG_{H_2O}, at pH 6.3, is calculated to be 7.03 ± 0.16 and 7.18 ± 0.43 kcal mol^{-1} for guanidinium chloride and thermal denaturation, respectively. Scanning microcalorimetry shows that the ratio of the van't Hoff enthalpy of denaturation to the calorimetric enthalpy of denaturation does not deviate from unity, the value observed for a two-state transition, over the pH range 2.2–3.5. The heat capacity change for denaturation is found to be 0.789 kcal mol^{-1} K^{-1}. The rate of unfolding of CI2 is first order and increases exponentially with increasing guanidinium chloride concentration. Refolding, however, is complex and involves at least three well-resolved phases. The three phases result from heterogeneity of the unfolded form due to proline isomerization. The fast phase, 77% of the amplitude, corresponds to the refolding of the fraction of the protein that has all its prolines in a native trans conformation. The rate of this major phase decreases exponentially with increasing guanidinium chloride concentration. The unfolding and refolding kinetics can also be fitted to a two-state model. Importantly, ΔG_{H_2O} and m, the constant of proportionality of the free energy of folding with respect to guanidinium chloride concentration, calculated from the kinetic experiments, 7.24 ± 0.22 kcal mol^{-1} and 1.86 ± 0.05 kcal mol^{-1} M^{-1}, respectively, agree, within experimental error, with the values measured from the equilibrium experiments. This is perhaps the strongest evidence that the unfolding of CI2 follows a simple two-state transition.

Small monomeric proteins lacking disulfides bonds and cis proline peptide bonds provide the simplest starting point for investigating the rules that govern protein folding. In proteins that contain disulfide bridges or proline peptide bonds in a cis conformation, the rate-determining step of the protein-folding process is frequently that of disulfide bond rearrangement (Creighton, 1988) or proline isomerization (Kiefhaber et al., 1990a,b,c), respectively. The folding of larger multidomain proteins is still more complex (Jaenicke, 1987), and frequently there is more than one transition between native and denatured states as domains fold independently. This complicates both analysis and interpretation of results. Many small monomeric proteins have simple, single, unfolding transitions between native and denatured states. However, there is now increasing evidence for complex behavior within such simple systems. Studies on staphylococcal nuclease, for example, have shown that there is residual structure in the denatured state (Shortle, 1989), and scanning microcalorimetry studies on T4 lysozyme have shown that there is deviation from a two-state model at some values of pH (Kitamura & Sturtevant, 1989). Both events complicate the analysis of denaturation curves. Here, we present the characterization of the denaturation and renaturation of chymotrypsin inhibitor 2 and show it is a good model system for both equilibrium and kinetic studies.

Chymotrypsin inhibitor 2, found in the albumin fraction of seeds from the *Hiproly* strain of barley, is a member of the potato inhibitor I family of serine protease inhibitors (Jonassen, 1980). It is a small monomeric protein of 83 residues, with M_r = 9250 (Svendsen et al., 1980). It is a single-domain protein containing both α-helix and β-sheet structural elements. The tertiary structure of CI2 consists of a four-stranded mixed parallel and antiparallel β-sheet against which an α-helix packs to form the hydrophobic core. Between parallel strands 2 and 3, on the opposite side of the β-sheet to the α-helix, is a wide loop in extended conformation that contains the reactive site bond. Unlike many other serine protease inhibitors, CI2 contains no disulfide bonds. Four of the five prolines are well defined in the crystal structure and are in the trans conformation. The fifth proline is in a region not defined in either the crystal structure or the NMR solution structure. CI2 has one tryptophan residue that is buried in the hydrophobic core and that is responsible for a large change in the fluorescence spectrum of the protein on denaturation.

The crystal structure of CI2 has been solved to high resolution, 2.0 Å (McPhalen & James, 1987), which is a prerequisite for the rational design of site-directed mutants. The solution structure has been determined by NMR spectroscopy (Kjaer et al., 1987; Kjaer & Poulsen, 1987; Clore et al., 1987a,b). The gene encoding for the protein has been cloned and can be expressed at high levels in *Escherichia coli*. (Longstaff et al., 1990). This will allow future studies to probe the nature of protein stability and folding by using protein engineering techniques.

We now present a study on the equilibrium unfolding transition of CI2. The nature of the transition between unfolded and folded protein is characterized by a number of different techniques in order to detect stable folding intermediates. Guanidinium chloride and heat are used to denature the protein, and fluorescence is used as a probe of the state of the protein. In addition, differential scanning microcalorimetry experiments are performed. The kinetics of unfolding and refolding can be used to detect intermediates which are not stable enough to be detected under equilibrium conditions. The kinetics of unfolding and refolding are investigated using [GdnHCl] jump and pH-jump experiments.

* Author to whom correspondence should be addressed.
[‡]S.E.J. was supported by a SERC studentship.

Experimental Procedures

Materials

Chemicals. The buffer used in denaturation experiments was 2-(*N*-morpholino)ethanesulfonic acid (MES) purchased from Sigma. Guanidinium chloride was sequanal grade purchased from Pierce Chemicals. Water used in equilibrium and kinetic experiments was purified to 15 MΩ resistance by an Elgastat system. Subtilisin BPN′ used in the activity measurements was expressed and isolated from *E. coli* harboring the plasmid pPT30 (Thomas et al., 1985). Yeast extract and tryptone were purchased from Lab M, Bury, U.K. Ammonium sulfate was enzyme grade from BDH. All other chemicals were purchased from Sigma.

Recombinant CI2. The wild-type CI2 gene has been cloned into the high level expression vector pINIIIompA3, in which the structural gene is fused to the N-terminal signal sequence of *ompA* and is under control of the *llp* promoter and *lac* operator sequences (Longstaff et al., 1990). The *E. coli* strain TG2 containing pINIIIompA3 was grown in 2×TY medium (Maniatis et al., 1982) containing ampicillin (50 μg mL^{-1}) and IPTG (0.5 mM) for induction of expression. Cultures were grown overnight at 37 °C with vigorous shaking to an absorbance of 6–8 at 600 nm. Cells were harvested and concentrated by centrifugation (6500*g*), and the cell paste was resuspended in 10 mM Tris-HCl, pH 8.0, and 1 mM EDTA, then sonicated three times for 0.5 min on maximum power at 4 °C with a Heats Systems ultrasonic sonicator. Cell debris was removed by centrifugation (25000*g*) before the pH of the supernatant was lowered to 4.4 by addition of acetic acid. The precipitate was removed by centrifugation (25000*g*). Ammonium sulfate was added to 40% (w/v) to precipitate unwanted proteins, which were removed by centrifugation (25000*g*). CI2 was precipitated from this supernatant by the addition of ammonium sulfate to 65% (w/v) and collected by centrifugation (25000*g*). Protein pellets were resuspended in 10 mM sodium acetate, pH 4.4, and then separated on a Sephadex G-75 superfine column. Fractions were collected, and those containing inhibitory activity were pooled before cation-exchange chromatography on a Pharmacia FPLC Mono S column. CI2 was bound to the column in 10 mM acetate, pH 4.4, and eluted around 200 mM salt with a linear NaCl gradient. Purified CI2 was dialyzed extensively against water to remove salt, flash frozen, and stored at −70 °C. The purified protein was homogeneous as judged by $NaDodSO_4$–polyacrylamide gel electrophoresis.

Methods

Spectroscopy. The intrinsic fluorescence of CI2 increases on denaturation, which allows unfolding and refolding to be monitored by fluorescence spectroscopy. The fluorescence yield of unfolded and folded forms of the protein is constant with respect to guanidinium chloride concentration (up to 6.5 M). This gives flat baselines, and the fraction of unfolded protein at a given guanidinium chloride concentration is simply the ratio of the observed fluorescence change to the maximal fluorescence change. Fluorescence spectroscopy was used in guanidinium chloride and thermal denaturation studies as well as in the kinetic studies. The maximal change in fluorescence upon denaturation is obtained with an excitation wavelength of 280 nm and an emission wavelength of 356 nm. A Perkin-Elmer LS5B Luminescence spectrometer was used for the equilibrium studies, with slit widths of 2 mm.

Chemical Denaturation Experiments. Guanidinium chloride solutions were prepared gravimetrically in volumetric flasks. The guanidinium chloride solutions were divided into 800-μL aliquots with an SMI (American Hospital Supply) positive displacement pipetter with repetitive pipetting attachment and stored at −20 °C. For each data point in the denaturation experiment, 100 μL of CI2 stock solution in 450 mM MES was diluted into 800 μL of the appropriate denaturant concentration, with a SMI positive displacement pipetter. The final CI2 concentration was approximately 2.5 μM. The protein/denaturant solutions were preequilibrated at 25 °C for approximately 1 h. Spectroscopic measurements were carried out in a thermostated cuvette holder at 25 °C, the temperature being monitored throughout the experiment by a thermocouple immersed in a neighboring cuvette in the cell holder. The unfolding was found to be completely reversible. The activity of a sample of CI2 denatured in 6.5 M GdnHCl was found to be completely restored after dilution 100-fold into 10 mM MES, pH 6.3. Activity was measured by the inhibition by native CI2 of the serine protease subtilisin BPN′ (Longstaff et al., 1990).

Thermal Denaturation Experiments. Thermally induced unfolding was also monitored by fluorescence spectroscopy using 2.5 μM CI2 in 50 mM MES, pH 6.3. The pK_a of this buffer is temperature dependent, and the pH varies from 6.3 to 5.6 over the temperature range used. It was found that the stability of CI2 does not change significantly over this pH range. Jacketed cuvettes were heated by a circulating water bath. The temperature was monitored by a thermocouple immersed in the cuvette under observation above the light beam. Reversibility was shown by the regain in initial fluorescence after cooling.

Calorimetry. Measurements were carried out in 10 mM glycine-HCl solution at pH 2.2, 2.5, 2.8, 3.2, and 3.5. Under these conditions, the protein is soluble and no aggregation is observed on heating. Solutions were carefully dialyzed, before measurements, against solvent with the solvent being replaced several times to ensure complete equilibration. Protein concentrations in solution were determined spectrophotometrically with correction for light scattering (Winder & Gent, 1971). The concentration of protein varied from 0.5 to 1.5 mg mL^{-1}. The extinction coefficient of CI2 was determined in a separate experiment using the method of Gill and von Hippel (1989). The extinction coefficient of native CI2 is based on a comparison of the spectra for native and denatured (6 M GdnHCl) protein measured at identical protein concentrations. The concentration of denatured protein can be calculated from the known amino acid composition and extinction coefficients of tyrosine, tryptophan, and cysteine obtained from model compound studies (Edelhoch, 1967). The extinction coefficient for the native protein can then be calculated by using this concentration. The experiment was performed in triplicate, and an extinction coefficient of 7418 M^{-1} cm^{-1} at 282 nm was calculated.

Calorimetric measurements were performed with a DASM-4A capillary scanning microcalorimeter equipped with gold cells of 1.0-mL volume (Bureau of Biological Instruments of the Academy of Sciences of the USSR). Measurements were performed under an excess pressure of 2 bar. The heating rate was 1 K min^{-1}. A baseline scan with dialysis buffer filled cells was performed prior to measurements. The calorimeter was calibrated by applying an electrical current at 20 μW for 5 min. Repeated measurements with the same protein sample demonstrated reversibility of ≥90%. The areas under the transition curves were determined with a planimeter.

Kinetics. Reactions were followed with a Perkin-Elmer MPF-44B fluorescence spectrophotometer equipped with a rapid mixing head. The mixing device contained a T-jet mixing chamber followed by a 30-ms delay loop ensuring complete mixing of the solutions before observation. A Hellma flow through compact cell, 10 mm × 3 mm × 3 mm, was used. The solutions were driven through the mixing chamber manually from two Hamilton syringes resulting in a mixing ratio of 1:10. Data were acquired with a Tandon Target microcomputer, a DT2801 data translation board, and the Bio-kine data acquisition software package by Bio-logic and analyzed with the program ENZFITTER (Elsevier Biosoft, Cambridge) by R. J. Leatherbarrow.

Unfolding was initiated by diluting the aqueous protein solution (~45 μM) into guanidinium chloride solutions of different concentrations with final concentrations between 4.0 and 7.35 M. Unfolding data were fitted to a single-exponential rate equation.

Refolding was initiated, after preequilibration of the protein in 6.5 M GdnHCl, by diluting it into low concentrations of GdnHCl. Final guanidinium chloride concentrations were between 0.59 and 4.0 M. Points between 0 and 0.59 M GdnHCl were obtained by a pH-jump experiment. The protein was initially acid denatured by lowering the pH of the solution to 1.7 by addition of 5 M HCl. At this pH, it was found that the protein is completely denatured as determined by fluorescence spectroscopy. The subsequent addition of GdnHCl makes no further difference to the fluorescence spectrum. The protein was refolded by rapid mixing (1:1) with a strongly buffered solution at high pH. The final pH of the solution was pH 6.3. An Applied Photophysics stopped-flow spectrophotometer model SF 17MV was used to monitor the fast reaction. Refolding is a triphasic process with a very fast major phase and two slow minor phases. The split time base facility of the stopped-flow was used to monitor the three phases simultaneously. Typically split time bases of 0–500 ms and 1–500 s were used to monitor the fast and slow phases, respectively. The slow phases were fitted to a double exponential with floating end point by using the SF 17MV software package by P. J. King, Applied Photophysics. The fast phase was then fitted to a double exponential with floating end point with the second rate fixed to the value obtained for the faster of the two slow phases.

A Radiometer PHM64 research pH meter was used for pH measurements. The pH-jump experiment was performed in the absence and in the presence of low concentrations of denaturant. All kinetic experiments were performed at 25 °C.

Data Analysis

Guanidinium Chloride Denaturation. This analysis is for a two-state model of denaturation where only the native and the denatured states are populated. The equilibrium constant for unfolding, K_U, and free energy of unfolding, ΔG_U, in the presence of a denaturant may be calculated from

$$K_U = (F_N - F)/(F - F_U) = \exp(-\Delta G_U/RT) \quad (1)$$

where F is the observed fluorescence and F_N and F_U are the values of the fluorescence of the native and denatured forms of the protein, respectively, and R is the gas constant, 8.314 J mol^{-1} K^{-1}. The values of F_N and F_U are independent of [GdnHCl] up to 6.5 M so eq 1 can be applied directly. It has been found experimentally that the free energy of unfolding of proteins in the presence of guanidinium chloride is linearly related to the concentration of denaturant (Tanford, 1968; Pace, 1986):

$$\Delta G_U = \Delta G_{H_2O} - m[\text{denaturant}] \quad (2)$$

where ΔG_U is the free energy of unfolding at a particular denaturant concentration, ΔG_{H_2O} is the free energy of unfolding in water, and m is a constant that is proportional to the increase in degree of exposure of the protein on denaturation. The entire data set from the fluorescence-monitored guanidinium denaturation was fitted with the nonlinear regression analysis program ENZFITTER by using eq 3, which is derived from eqs 1 and 2.

$$F = F_N - (F_N - F_U)\frac{\exp(m[\text{GdnHCl}] - \Delta G_{H_2O})/RT}{1 + \exp(m[\text{GdnHCl}] - \Delta G_{H_2O})/RT} \quad (3)$$

Thermal Denaturation. K_U as a function of temperature was calculated in a similar way to the guanidinium chloride mediated denaturation, assuming a two-state transition. Outside the transition region, the fluorescence of the native and denatured forms varies linearly with temperature, and so the values of F_N and F_U are estimated at each temperature by extrapolation [see Pace (1986) for a review of this method]. The extrapolation lowers the accuracy of the data compared with that of the guanidinium chloride mediated denaturation.

Calculation of ΔG from Thermal Denaturation. The enthalpy, $\Delta H_U(T)$, and the entropy, $\Delta S_U(T)$, of unfolding at temperature T are given by classical thermodynamics by

$$\Delta H_U(T) = -R[\partial \ln K_U/\partial(1/T)] = RT^2(\partial \ln K_U/\partial T) \quad (4)$$

$$\Delta S_U(T_m) = \Delta H_U(T_m)/T_m \quad (5)$$

where T_m is the midpoint of thermal denaturation. van't Hoff plots ($\ln K_U$ vs $1/T$) of thermal denaturation are approximately linear through the T_m region, allowing an estimation of the enthalpy and entropy of unfolding at T_m. Plots over a wider range of temperatures are expected to be curved because $\Delta H_U(T)$ and $\Delta S_U(T)$ vary with temperature according to (Privalov, 1979)

$$\Delta C_p = [\partial \Delta H_U(T)/\partial T]_p = T[\partial \Delta S_U(T)/\partial T]_p \quad (6)$$

where ΔC_p is the change in heat capacity at constant pressure that accompanies the unfolding of a protein. Thus the value of $\Delta G_U(T)$ at a given temperature cannot be simply calculated from $\Delta H_U(T)$ and $\Delta S_U(T)$ without allowing for their variation because of ΔC_p. There is evidence that ΔC_p is independent of temperature between 20 and 80 °C (Privalov, 1979). Consequently, $\Delta H_U(T)$, the enthalpy of unfolding at any temperature T, $\Delta S_U(T)$, the entropy of unfolding at T, and $\Delta G_U(T)$, the free energy of unfolding at T, can be calculated at temperatures other than T_m from (Baldwin, 1986)

$$\Delta H_U(T) = \Delta H_U(T_m) + \Delta C_p(T - T_m) \quad (7)$$

$$\Delta S_U(T) = \Delta S_U(T_m) + \Delta C_p \ln (T/T_m) \quad (8)$$

$$\Delta G_U(T) = \Delta H_U(T) - T\Delta S_U(T) \quad (9)$$

RESULTS

Guanidinium Chloride Denaturation. Figure 1A shows the change of fluorescence of CI2 upon titration with guanidinium chloride. There is a single sharp transition between the initial and final states. Data from the transition region (3.2–4.6 M) were transformed to give a linear plot of ΔG_U against [GdnHCl] shown in Figure 1B. The entire data set can be fitted directly to eq 3, from which the value of ΔG_{H_2O} and m can be calculated. For wild-type CI2, $\Delta G_{H_2O} = 7.03 \pm 0.16$ kcal mol^{-1} and $m = 1.79 \pm 0.04$ kcal mol^{-1} M^{-1}. The midpoint of unfolding ($K_U = 1$) is at 3.92 ± 0.18 M GdnHCl. Figure 1A,B shows the best fits to a two-state model of protein folding.

In order to ascertain that the transition monitored fluorimetrically is due to the global denaturation of the protein, and not caused by some local unfolding event in the proximity of the tryptophan, a series of 1-D ^{1}H NMR experiments were

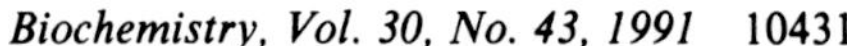

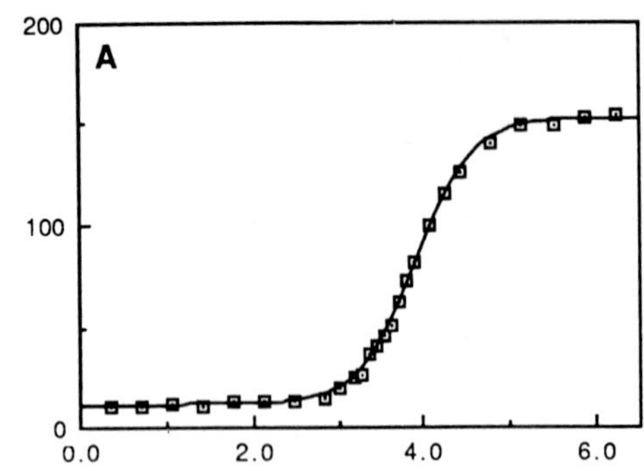

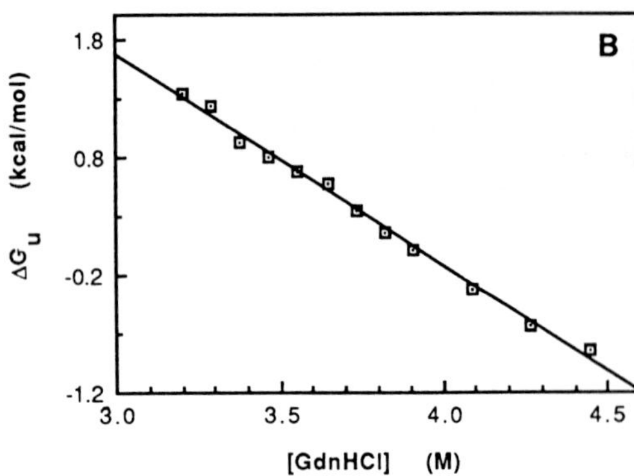

FIGURE 1: GdnHCl-induced denaturation of CI2 at 25 °C, in 50 mM MES, pH 6.3. (A) Intrinsic fluorescence monitored at 356 nm, with an excitation wavelength of 280 nm, as a function of [GdnHCl]. The solid curve shows the best fit of the data to eq 3. (B) Transformation of the fluorescence data to give a linear plot of change in free energy versus [GdnHCl] around the midpoint of the unfolding transition. The solid line indicates the best fit of the data to eq 3. Linear extrapolation of this plot to 0 M GdnHCl results in the free energy change of denaturation in the absence of denaturant.

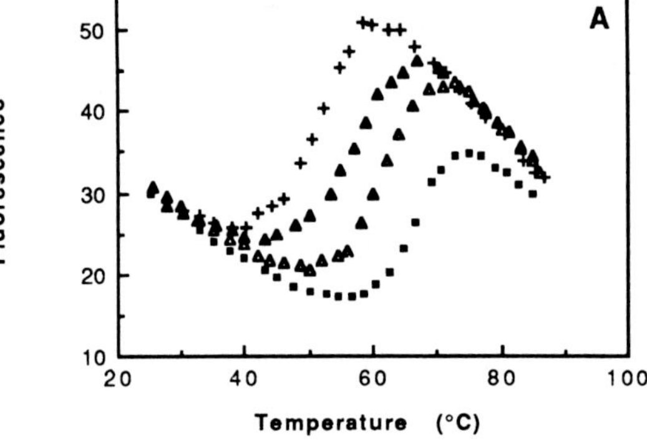

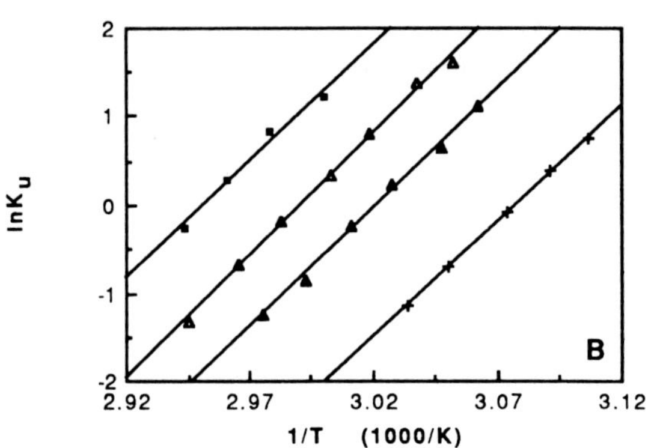

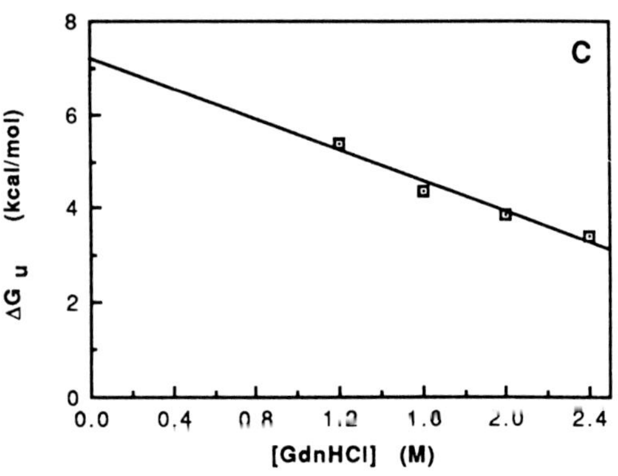

FIGURE 2: Thermal denaturation of CI2 in the presence of low concentrations of GdnHCl, in 50 mM MES, pH 6.3. (A) Intrinsic fluorescence of CI2 as a function of temperature, with (■) 1.2 M GdnHCl, (△) 1.6 M GdnHCl, (▲) 2.0 M GdnHCl, and (+) 2.4 M GdnHCl. (B) van't Hoff plots of the thermal transition curves around the midpoint of the transition, with (■) 1.2 M GdnHCl, (△) 1.6 M GdnHCl, (▲) 2.0 M GdnHCl, and (+) 2.4 M GdnHCl. The solid line indicates the best fit of the data to eq 4. (C) Plot of the free energy of denaturation at 25 °C, calculated from the van't Hoff plots, and using eqs 7–9, versus [GdnHCl]. Linear extrapolation to 0 M GdnHCl results in the free energy of denaturation in the absence of denaturant.

conducted at various GdnHCl concentrations, at 25 °C, using a Bruker AM500. From such a series, it was possible to follow peaks that are characteristic of the native structure of the protein as a function of guanidinium chloride concentration. All the peaks that were followed were observed to disappear/appear concomitantly in a single transition with a midpoint of unfolding at approximately 3.8 M GdnHCl (data not shown). This is consistent with the transition observed by fluorescence spectroscopy and with the transition being due to a global denaturation.

Thermal Denaturation. The midpoint for thermal denaturation in 50 mM MES, pH 6.3, in the absence of denaturant, was found to be over 80 °C, which is too high to allow an accurate direct measurement of ΔG_{H_2O} at 25 °C. The T_m was lowered by addition of GdnHCl to final concentrations of 1.2, 1.6, 2.0, and 2.4 M. Figure 2A shows the thermal denaturation curves at the four GdnHCl concentrations. The thermal denaturation curves show a single sharp transition between native and denatured forms typical of a two-state process. Figure 2B shows the corresponding van't Hoff plots from which ΔH_{T_m}, ΔS_{T_m}, and T_m can be calculated according to eqs 4 and 5. These values, in turn, can be used to calculate ΔG_U at 25 °C at the four different GdnHCl concentrations according to eqs 7, 8, and 9. In the calculations, a value of $\Delta C_p = 0.789$ kcal deg^{-1} mol^{-1} was used, based on the result from the calorimetric studies. Figure 2C shows the corresponding linear plot of the free energy of unfolding against GdnHCl concentration, from which the value of ΔG_{H_2O} at 25 °C can be calculated by extrapolation back to 0 M (see eq 2). This gives a value of 7.18 ± 0.43 kcal mol^{-1} for ΔG_{H_2O} and 1.62 ± 0.23 kcal mol^{-1} M^{-1} for m. These values are consistent, within experimental error, with the values obtained by guanidinium chloride denaturation.

Calorimetry. From the transition curves T_m, the transition temperature, $\Delta H_{cal}(T_m)$, the molar calorimetric enthalpy of denaturation at T_m, and $\Delta H_{vH}(T_m)$, the molar van't Hoff

10432 Biochemistry, Vol. 30, No. 43, 1991 Jackson and Fersht

Table I: Scanning Microcalorimetry Experiments

pH	T_m (K^{-1})[a]	$\Delta H_{cal}(T_m)$ (kcal mol^{-1})[a]	$\Delta H_{vH}(T_m)$ (kcal mol^{-1})[b]	$\Delta H_{vH}(T_m)/\Delta H_{cal}(T_m)$[c]	$\Delta S_{cal}(T_m)$ (kcal mol⁻ K^{-1})[d]	ΔG_{cal}(25 °C) (kcal mol^{-1})[e]
2.2	314.6	43.5	42.3	0.97	0.138	1.944
2.5	320.2	49.2	45.5	0.93	0.154	2.787
2.8	328.2	52.8	51.1	0.97	0.161	3.728
3.2	337.0	61.0	60.8	1.00	0.181	5.201
3.5	344.0	67.5	68.8	1.02	0.196	6.475

[a] T_m, the transition temperature, and $\Delta H_{cal}(T_m)$, the molar calorimetric enthalpy of denaturation at T_m, are measured directly from the transition curves (change in specific heat with temperature). [b] $\Delta H_{vH}(T_m)$, the molar van't Hoff enthalpy of denaturation at T_m, is calculated by using eq 10. [c] $\Delta H_{vH}(T_m)/\Delta H_{cal}(T_m)$ = 1.0 only for two-state transitions. [d] $\Delta S_{cal}(T_m)$ is calculated from eq 5. [e] ΔC_p can be calculated by using eq 6; ΔG_{cal} (25 °C) can then be calculated by using this value and eqs 7–9.

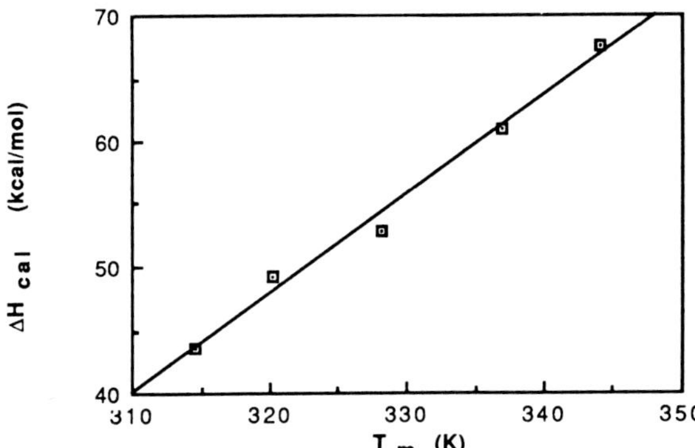

FIGURE 3: Scanning microcalorimetry of CI2. Plot of the calorimetric transition enthalpy $\Delta H_{cal}(T_m)$ versus the transition temperature T_m.

enthalpy of denaturation at T_m can be calculated as a function of pH. Calorimetric enthalpies for unfolding, $\Delta H_{cal}(T_m)$, are calculated numerically from the area under the transition curve. Assuming a two-state transition, the van't Hoff enthalpies at T_m, $\Delta H_{vH}(T_m)$, can be calculated from the calorimetric transition curves by using (Privalov, 1986; Shrake & Ross, 1990; Kitamura & Sturtevant, 1989)

$$\Delta H_{vH}(T_m) = 4RT_m^{\ 2}\frac{C_p(T_m)}{\Delta H_{cal}(T_m)} \qquad (10)$$

where $C_p(T_m)$ is the molar excess heat capacity at T_m. Transition temperatures, T_m, are taken as the peak of the calorimetric transition curve. Thermodynamic parameters determined from the data are summarized in Table I. The ratio of the van't Hoff enthalpy of denaturation to the calorimetric enthalpy of denaturation is shown, and the ratio is unity over the pH range studied. This shows that the two-state assumption for the unfolding transition applies under these conditions.

The variation of the calorimetric transition enthalpies with temperature is shown in Figure 3. As expected over this temperature range, ΔC_p is invariant and $\Delta H_U(T_m)$ is linearly dependent on T_m. By use of eq 6, the calorimetric value for the change in heat capacity on denaturation, ΔC_p, can be calculated from the gradient of such a plot. For CI2 the value of ΔC_p is 0.789 kcal mol^{-1} K^{-1}.

This corresponds to a specific heat capacity change of 0.0834 cal g^{-1} K^{-1}. This is slightly less than the average value found for globular proteins of about 0.1 kcal g^{-1} K^{-1} (Privalov, 1979) and may result from the long unstructured tail of some 20 amino acids at the N-terminus of CI2, which should not contribute to a change in heat capacity on denaturation.

The unfolded state of a protein has a much larger heat capacity than the native state (Privalov et al., 1989). This is mainly due to the increased hydration of hydrophobic residues that are exposed in the unfolded state but buried in the native state. The change in heat capacity on denaturation therefore reflects the change in the degree of exposure of hydrophobic residues (Privalov & Gill, 1989). The large positive value of ΔC_p for CI2 suggests that the unfolding reaction starts from a compact native structure, in which many of the hydrophobic residues are buried, and proceeds to an unfolded state in which such hydrophobic residues are highly solvated.

Both $\Delta H_U(T_m)$ and $\Delta S_U(T_m)$, measured from scanning microcalorimetry experiments (see Table I) are large and positive as has been observed for the high-temperature melting of a number of proteins (Privalov, 1979). At 25 °C, ΔH_U is 30.5 kcal mol^{-1} and ΔS_U is 102.3 cal mol^{-1} K^{-1}, which is comparable to the values found of 35 kcal mol^{-1} and 116 cal mol^{-1} K^{-1} for T4 lysozyme at 28 °C (Chen & Schellman, 1989). Unfortunately, it was not possible to measure the thermal denaturation of CI2 by scanning microcalorimetry at pH 6.3 because of its high T_m. This means it is not possible to compare the value for ΔG_{H_2O} calculated from calorimetric data to those values obtained by other methods.

Unfolding Kinetics. The rate constant for unfolding, k_u, is generally found to increase with increasing final [GdnHCl] according to

$$\ln k_u = \ln k_u^{H_2O} + m_{k_u}[\text{GdnHCl}] \qquad (11)$$

where $k_u^{H_2O}$ is the rate constant for unfolding in water and m_{k_u} is a constant. The unfolding of CI2 is monophasic, and the data can be fitted to a single-exponential rate equation. The observed first-order rate constant is found to increase exponentially with increasing [GdnHCl] over the range 4.5–7.35 M, in accordance with eq 11 (see Figure 4).

Refolding Kinetics. The refolding of CI2, by either [GdnHCl] jump or by pH-jump experiments, is a triphasic process with the three phases well-resolved at 25 °C and at low concentrations of denaturant. The multiphasic nature of the refolding reaction results from a heterogeneous population in the denatured state due to proline isomerization. The fast phase, 77% of the amplitude measured by stopped-flow spectroscopy, corresponds to the folding of the fraction of unfolded protein that has all its prolines in the trans conformation in solution. The following discussion applies only to the fast phase; characterization of the two slower phases and evidence that these correspond to rate-limiting proline isomerization reactions are discussed in the following paper (Jackson & Fersht, 1991).

If the process is a reversible transition between just two states, native, N, and denatured, U,

$$\mathrm{U} \underset{k_u}{\overset{k_f}{\rightleftharpoons}} \mathrm{N}$$

eqs 2 and 11 imply that the rate constant for folding, k_f, must follow the rate law

$$\ln k_f = \ln k_f^{H_2O} - m_{k_f}[\text{GdnHCl}] \qquad (12)$$

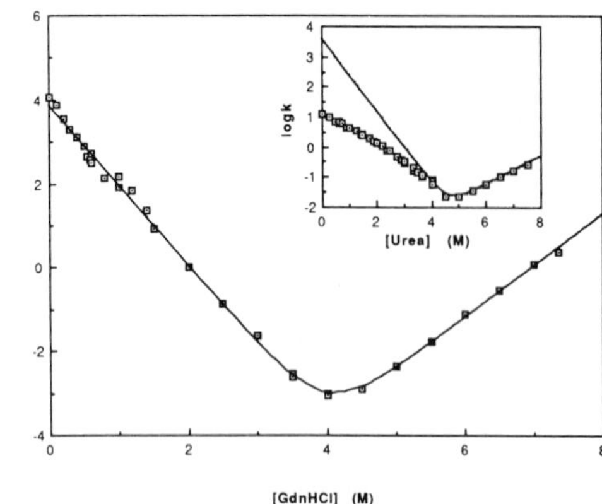

FIGURE 4: [GdnHCl] dependence of natural logarithm of the rate constants for denaturation and renaturation of CI2. Rate constants measured at 25 °C, in 50 mM MES pH 6.3. Points between 0 and 0.59 M GdnHCl were obtained by pH-jump experiments. Points between 0.59 and 7.35 M GdnHCl were obtained from [GdnHCl] jump experiments. Each point is obtained from the average of at least three separate experiments. The solid curve is the best fit of the data to a two-state model using eq 13. The insert shows the [urea] dependence of the rate constants for denaturation and renaturation for wild-type barnase. The solid curve is that calculated for a two-state system on the basis of the unfolding and equilibrium data (Matouschek et al., 1990).

where $m = m_{k_f} + m_{k_u}$ and $K_U = k_f/k_u$. These relationships are perhaps the most important criteria that show that a system follows the two-state model.

The rate constant for the fast phase, the rate of refolding, decreases exponentially with [GdnHCl] over the range 0–3.5 M (Figure 4). The complete kinetics of folding and unfolding can be fitted to eq 13, which is derived from eqs 11 and 12,

$$\ln k = \ln [k_f^{H_2O} \exp(-m_{k_f}[\text{GdnHCl}]) + k_u^{H_2O} \exp(m_{k_u}[\text{GdnHCl}])] \quad (13)$$

and based on a two-state transition, where k is the rate of unfolding or refolding at a particular GdnHCl concentration. The kinetic data for unfolding and refolding of CI2 fit to this two-state model. The calculated fit of the data to eq 13 is shown by the solid curve in Figure 4. The values of $k_f^{H_2O}$, $k_u^{H_2O}$, m_{k_f}, and m_{k_u}, calculated from the kinetic data according to eq 13, are 47.8 s^{-1}, 1.81 × 10^{-4} s^{-1}, −1.90 M^{-1}, and 1.24 M^{-1}, respectively. From these results, the equilibrium values of ΔG_{H_2O} and m can be calculated, taking into account the equilibria due to proline isomerization

$$U_c \underset{23\%}{\overset{K_{iso}}{\rightleftharpoons}} U_t \underset{k_u^{H_2O}}{\overset{k_f^{H_2O}}{\rightleftharpoons}} N$$
$$77\%$$

where U_t is the denatured protein with all its prolines in a trans conformation, U_c is the denatured protein with a proline in a cis conformation, and N is the native state, which has all its prolines in a trans conformation. Defining the equilibrium constant for isomerization, K_{iso}, as $[U_c]/[U_t]$, and the apparent equilibrium constant for unfolding calculated from the kinetic experiments, K, as $k_f^{H_2O}/k_u^{H_2O}$, it follows that the equilibrium constant for unfolding, K_U, is $K/(1 + K_{iso})$. From the relative amplitudes of the slow and fast phases, 23% and 77%, respectively, K_{iso} = 0.299 and K_U = 2.03 × 10^5 M. This gives a value for ΔG_{H_2O} of 7.24 ± 0.22 kcal mol^{-1}. Similarly m can be calculated from the kinetic data by using the relationship $m = m_{k_f} + m_{k_u}$. This gives an m value of 1.86 ± 0.05 kcal mol^{-1} M^{-1} and a midpoint of unfolding of 3.89 M GdnHCl. Within experimental error, these values agree with the values of 7.03 ± 0.16 kcal mol^{-1} and 1.79 ± 0.04 kcal mol^{-1} M^{-1} for ΔG_{H_2O} and m and a midpoint of unfolding at 3.92 ± 0.18 M GdnHCl, calculated from equilibrium experiments.

In addition to showing that the kinetics of unfolding and refolding of CI2 can be fitted a two-state transition, we have shown that $m = m_{k_f} + m_{k_u}$, and $K_U = k_f/k_u$. This is very strong evidence that CI2 behaves kinetically, as well as thermodynamically, as a two-state system of folding.

DISCUSSION

Criteria for a Two-State Transition. For the equilibrium denaturation of a protein to be described as two state, (i) the unfolding data must fit to a single transition curve as described by eqs 1–3, (ii) this transition must be independent of the probe used to determine the state of the protein, i.e., fluorescence, absorbance, circular dichroism, and NMR, etc, and (iii) the ratio of the van't Hoff enthalpy of denaturation to the calorimetric enthalpy of denaturation must be unity.

For the kinetics of unfolding and refolding to be described by a two-state model, (i) both unfolding and refolding must be monophasic processes (allowing for the possible heterogeneity of the unfolded form due to proline isomerization), (ii) if the logarithm of the equilibrium constant for unfolding and that for the rate constant for unfolding is linearly dependent on the concentration of denaturant, then the logarithm of the rate constant for refolding should also be linearly dependent on denaturant concentration, i.e., the data should fit to a two-state transition as described by eq 13, and (iii), most importantly, the values for ΔG_{H_2O} and m calculated from kinetic experiments must be the same as the values calculated from the equilibrium experiments, i.e., $m = m_{k_f} + m_{k_u}$, and $K_U = k_f/k_u$. This is perhaps the strongest evidence for a two-state transition.

Kinetic experiments are more sensitive than equilibrium experiments for observing intermediate states. Intermediate, which may not be stable under the conditions used in equilibrium experiments, may be metastable in the native conditions used in refolding experiments and, therefore, be detectable by kinetics. As a consequence, although the equilibrium properties of many proteins show two-state behavior, the kinetic properties of much fewer proteins also show two-state behavior. CI2 is a rare example of a protein in which both equilibria and kinetics are described by a two-state transition. The evidence for this, from both equilibrium and kinetic experiments, is discussed below

Evidence from Equilibrium Experiments. In all cases, the equilibrium denaturation of CI2 appears to be a two-state process where only the native and the denatured states are significantly populated. Both the GdnHCl-induced denaturation and the thermal denaturation curves show single sharp transitions between native and denatured forms that can be fitted to a two-state model. In addition, scanning microcalorimetry, which does not assume a mechanistical model for the transition, also shows that thermal denaturation is a two-state process. The ratio of the van't Hoff enthalpy of denaturation, ΔH_{vH}, to the calorimetric enthalpy of denturation, ΔH_{cal}, measured from such experiments is unity over the pH range 2.2–3.5. The calorimetric enthalpy of denaturation is the same as the van't Hoff enthalpy of denaturation only for two-state systems (Privalov, 1979, 1986; Shrake & Ross, 1990; Kitamura & Sturtevant, 1989). The average ratio of $\Delta H_{vH}/\Delta H_{cal}$ for CI2 is 0.98 ± 0.03, which is within the range observed for the two-state denaturation of several globular proteins of 1.00 ± 0.05 (Privalov, 1979).

In addition, the transition monitored by fluorescence spectroscopy is the same as that measured by NMR spectroscopy. All the NMR peaks that were followed as a function of [GdnHCl] were observed to disappear/appear concomitantly in a single transition which followed that observed by fluorescence spectroscopy. This also shows that the transition monitored by fluorescence spectroscopy corresponds to a global denaturation of the protein and not a local unfolding event.

Evidence from Kinetic Experiments. Unfolding is a first-order reaction whose rate increases exponentially with increasing GdnHCl concentration. Refolding, however, is more complex and involves at least three well-resolved phases. The three phases result from heterogeneity in the unfolded form due to proline isomerizations [see following paper (Jackson & Fersht, 1991)]. The fast phase corresponds to the refolding of the fraction of denatured protein that has all its prolines in a native trans conformation. It represents, therefore, a rate-limiting step involving folding of the protein and not proline isomerization. This phase is also first order, and the rate decreases exponentially with increasing GdnHCl concentration. This results in linear plots of both ln k_f and ln k_u against [GdnHCl]. The data for unfolding and refolding can be fitted to a scheme, described by eq 13, in which only two states are populated. This is a marked difference from the results found for the folding of other small monomeric proteins, such as barnase, where there is significant nonlinearity in the plot of ln k_f against urea concentration (see the insert of Figure 4) (Matouschek et al., 1990) and also parvalbumin (Kuwajima et al., 1988) and α-lactalbumin and lysozyme (Ikeguchi et al., 1986).

It has not yet been established where on the folding pathway the change in fluorescence that is used to monitor the folding state of the protein occurs. If the change occurs before the rate-limiting step, it is possible that the only way of detecting deviation from a two-state model may be calculating the equilibrium constant from the kinetics and comparing it to the equilibrium data, rather than by curvature in the plots of rate constants. Because of the inaccuracies inherent in kinetic measurements, this approach leaves intermediates with stabilities of less than 10% of the folded form undetected.

For barnase, in addition to the curvature in the plot of ln k_f against denaturant concentration, there is significant deviation between the values of $k_f^{H_2O}$ based on the two-state model and calculated from the equilibrium and unfolding data and that measured experimentally. The measured value is three orders of magnitude lower than the predicted value. Both the curvature in the ln k_f plot and the deviation between the measured and calculated value for $k_f^{H_2O}$ are the result of the accumulation of an intermediate on the folding pathway at low concentrations of denaturant where the intermediate is stable. In contrast, for CI2 not only is there no curvature in the plot of ln k_f against [GdnHCl] but the values for ΔG_{H_2O} and m calculated from the kinetic experiments agree with those obtained from equilibrium experiments. This shows that only two states, the native and denatured, are significantly populated on either the unfolding or refolding pathways. There is no evidence from the kinetics of refolding of CI2 (Figure 4) for the existence of an intermediate state on the folding pathway.

It is now generally believed that proteins must fold through intermediate states (Ptitsyn, 1987), and there is increasing evidence for the existence of transient intermediates on the folding pathway [Röder et al. (1988), Udgaonkar and Baldwin (1988), Bycroft et al. (1990), Matouschek et al. (1990), and see Kim and Baldwin (1990) for a review on folding intermediates]. Since intermediates states are not detected in either the equilibrium experiments or the kinetic experiments for CI2, this suggests that they are high in energy and only transiently populated.

How Unfolded Is the Denatured State? There is some evidence suggesting that the nature of the denatured state depends upon the denaturation conditions (Creighton, 1990). It is thought that the GdnHCl denatured state is the most extended and has the least order of all denatured states, including the thermally denatured state (Tanford, 1968). For CI2, the values of ΔG_{H_2O} calculated from both GdnHCl denaturation experiments and thermal denaturation experiments are 7.03 ± 0.16 and 7.18 ± 0.43 kcal mol^{-1}, respectively, at 25 °C. This shows that the denatured states formed by GdnHCl and thermal denaturation are, if not structurally, thermodynamically equivalent.

Values for ΔG_{H_2O} and m, measured by GdnHCl-induced denaturation, are 7.03 ± 0.16 kcal mol^{-1} and 1.79 ± 0.04 kcal mol^{-1} M^{-1}, respectively. Whereas the value of ΔG_{H_2O} lies within the range of values that have been measured for small globular proteins (Crighton, 1990), the m value seems small in comparison to many others. Barnase, for example, has an m value from GdnHCl denaturation studies of 4.51 kcal mol^{-1} M^{-1} (Kellis et al., 1989), a mutant of T4 lysozyme has an m value of 4.5 kcal mol^{-1} M^{-1} (Chen & Schellman, 1989), RNase A and RNase T$_1$ have values of 3.0 and 2.6, respectively, at pH 7 (Pace et al., 1990), and α-chymotrypsin and β-lactoglobulin have values of 4.1 and 3.9, respectively (Greene & Pace, 1974). The m value is proportional to $\Delta\alpha$, the average fractional change in degree of exposure of residues on denaturation. CI2 contains a large reactive site loop, with many residues accessible to solvent and, in addition, an unstructured N-terminus of some 20 amino acids that are completely exposed to solvent even in the native state. Therefore, the average change in degree of exposure on unfolding will be less than for other, more globular, proteins

A value of $\Delta\alpha$ can be calculated both from the crystal structure and from the experimental m value. $\Delta\alpha$ is calculated from the experimental m value by using the amino acid composition of the protein and the known free energies of transfer of amino acids from water into various concentrations of denaturant, δg_{tr}, (Pace, 1986). The total free energy of transfer, ΔG, can be calculated as a function of [GdnHCl] by summation of the free energy of transfer over all residues (see eq 14). The m value can then be related directly to $\Delta\alpha$ using

$$\Delta G = \Delta G_{H_2O} + \Delta\alpha_i \sum n_i \delta g_{tr,i} \quad (14)$$

eqs 2 and 14. The value for $\Delta\alpha$ for CI2 determined in this way is 0.38, compared with a value for barnase of 0.53 calculated by using the same method (Pace et al., 1990).

$\Delta\alpha$ can also be calculated from the crystal structure by using solvent accessible surface area calculations. The solvent accessible surface area of all the residues in native CI2 was calculated relative to an extended chain conformation that represents the unfolded form (Miller et al., 1987). This gives a value for $\Delta\alpha$ of 0.40. Within the accuracy of the calculations, the two values for $\Delta\alpha$ agree. The low experimental m value, therefore, seems to be a consequence of the less globular nature of CI2 relative to other small proteins and not a result of significant residual structure in the unfolded form.

Conclusions. In conclusion, CI2 is representative of the simplest possible system for studying protein folding. It is a rare example of a classical two-state system where only the native and denatured states are significantly populated, either under equilibrium conditions or, on the unfolding or refolding pathways, under nonequilibrium conditions. In addition, there is no evidence for significant residual structure in the denatured state. CI2 is, therefore, an ideal system for a simple analysis of both the equilibrium and kinetics of protein folding.

ACKNOWLEDGMENTS

We gratefully acknowledge Dr. Yuri Griko and Andreas Matouschek for performing the calorimetric experiments in the laboratory of Professor P. L. Privalov. We are also grateful to Dr. Tim Hubbard and Dr. Cyrus Chothia from the IRC, Cambridge, for calculations on the solvent accessible surface area of CI2.

REFERENCES

Baldwin, R. L. (1986) *Proc. Natl. Acad. Sci. U.S.A. 83*, 8069–8072.

Bycroft, M., Matouschek, A., Kellis, J. T., Serrano, L., & Fersht, A. R. (1990) *Nature 346*, 488–490.

Chen, B., & Schellman, J. A. (1989) *Biochemistry 28*, 685–691.

Clore, G. M., Gronenborn, A. M., Kjaer, M., & Poulsen, F. M. (1987a) *Protein Eng. 1*, 305–311.

Clore, G. M., Gronenborn, A. M., James, M. N. G., Kjaer, M., McPhalen, C. A., & Poulsen, F. M. (1987b) *Protein Eng. 1*, 313–318.

Creighton, T. E. (1978) *Prog. Biophys. Mol. Biol. 33*, 231–297.

Creighton, T. E. (1988) *Proc. Natl. Acad. Sci. U.S.A 85*, 5082–5086.

Creighton, T. E. (1990) *Biochem. J. 270*, 1–16.

Edelhoch, H. (1967) *Biochemistry 6*, 1948–1954.

Edelhoch, H., & Osbourne, J. (1976) *Adv. Protein Chem. 30*, 183–245.

Gill, S. C., & von Hippel, P. H. (1989) *Anal. Biochem. 182*, 319–326.

Greene, R. F., Jr., & Pace, C. N. (1974) *J. Biol. Chem. 249*, 5388–5393.

Ikeguchi, M., Kuwajima, K., Mitani, M., & Sugai, S. (1986) *Biochemistry 25*, 6965–6972.

Jackson, S. E., & Fersht, A. R. (1991) *Biochemistry* (following paper in this issue).

Jaenicke, R. (1987) *Prog. Biophys. Mol. Biol. 49*, 117–237.

Jonassen, I. (1980) *Carlsberg Res. Commun. 45*, 47–48.

Kellis, J. T., Nyberg, K., & Fersht, A. R. (1989) *Biochemistry 28*, 4914–4922.

Kiefhaber, T., Quaas, R., Hahn, U., & Schmid, F. X. (1990a) *Biochemistry 29*, 3053–3061.

Kiefhaber, T., Quaas, R., Hahn, U., & Schmid, F. X. (1990b) *Biochemistry 29*, 3061–3070.

Kiefhaber, T., Grunert, H. P., Hahn, U., & Schmid, F. X. (1990c) *Biochemistry 29*, 6475–6480.

Kim, P. S., & Baldwin, R. L. (1990) *Annu. Rev. Biochem. 59*, 631–660.

Kitamura, S., & Sturtevant, J. M. (1989) *Biochemistry 28*, 3788–3792.

Kjaer, M., & Poulsen, F. M. (1987) *Carlsberg Res. Commun. 52*, 355–362.

Kjaer, M., Ludvigsen, S., Sorensen, O. W., Denys, L. A., Kindtler, J., & Poulsen, F. M. (1987) *Carlsberg Res. Commun. 52*, 327–354.

Kuwajima, K., Sakuraoka, A., Fueki, S., Yoneyama, M., & Sugai, S. (1988) *Biochemistry 27*, 7419–7428.

Longstaff, C., Campbell, A., & Fersht, A. R. (1990) *Biochemistry 29*, 7339–7347.

McPhalen, C. A., & James, M. N. G. (1987) *Biochemistry 26*, 261–269.

Maniatis, T., Fritsch, E. F., & Sambrook, J. (1982) in *Molecular Cloning: A Laboratory Manual*, Cold Spring Harbor Laboratory, Cold Spring Harbor, NY.

Matouschek, A., Kellis, J. T., Serrano, L., Bycroft, M., & Fersht, A. R. (1990) *Nature 346*, 440–445.

Miller, S., Janin, J., Lesk, A. M., & Chothia, C. (1987) *J. Mol. Biol. 196*, 641–656.

Pace, C. N. (1986) *Methods Enzymol. 131*, 266–279.

Pace, C. N., Laurents, D. V., & Thomson, J. A. (1990) *Biochemistry 29*, 2564–2572.

Privalov, P. L. (1979) *Adv. Protein Chem. 33*, 167–241.

Privalov, P. L. (1986) *Methods Enzymol. 131*, 4–51.

Privalov, P. L., & Gill, S. J. (1989) *Adv. Protein Chem. 39*, 191–234.

Ptitsyn, O. B. (1987) *J. Protein Chem. 6*, 272–293.

Röder, H., Elöve, G. A., & Englander, S. W. (1988) *Nature 335*, 700–704.

Shortle, D. (1989) *J. Biol. Chem. 264*, 5215–5318.

Shrake, A., & Ross, P. D. (1990) *J. Biol. Chem. 265*, 5055–5059.

Svendsen, I., Jonassen, I., Hejgaard, J., & Boisen, S. (1980) *Carlsberg Res. Commun. 45*, 389–395.

Tanford, C. (1968) *Adv. Protein Chem. 23*, 121–282.

Thomas, P. G., Russell, A. J., & Fersht, A. R. (1985) *Nature 318*, 375–376.

Udgaonkar, J. B., & Baldwin, R. L. (1988) *Nature 335*, 694–699.

Winder, A. F., & Gent, W. L. G. (1971) *Biopolymers 10*, 1243–1251.

The Nucleation-Condensation Mechanism for Folding

We then applied an extensive Φ-value analysis to the folding, which we could do in the directions of both folding and unfolding, and discovered an entirely novel mechanism of folding — the nucleation-condensation mechanism. It was then thought that proteins folded either by a framework or hierarchical mechanism in which layers of secondary structure were first formed or by a molten globule mechanism whereby there was a hydrophobic-driven collapse to form native tertiary interactions. Nucleation mechanisms had gone out of favour because they are not consistent with the formation of stable intermediates whereas the other mechanisms are consistent. Our mechanism, however, is not the classical nucleation-growth but a novel mechanism whereby the protein collapses around a nucleus, which is in the process of being formed. The transition state for protein folding is extended and encompasses most of the molecule, being a distorted native state-like structure. The first paper in 1994 outlined the transition state structure as the result of a global collapse of the denatured state[121].

Daniel Otzen and Laura Itzhaki, who were the joint first authors of the papers, then did an exhaustive and exquisite Φ-value analysis[122]. We sent off the paper to *JMB* as the 'nucleation-collapse' mechanism, but a referee thought it might be confused with hydophobic collapse and so I coined the term 'nucleation-condensation'. The paper also contains explicit descriptions of how nucleation-condensation could change to framework as the stability of elements of structure became independent of that of the rest of the protein.

13.3 2003: Laura Itzhaki and Daniel Otzen.

13.4 2003: Jacqui Matthews, Laura Itzhaki and Vic Arcus.

13.5 2003: Andreas Ladurner.

J. Mol. Biol. (1995) **254**, 260–288

JMB

The Structure of the Transition State for Folding of Chymotrypsin Inhibitor 2 Analysed by Protein Engineering Methods: Evidence for a Nucleation-condensation Mechanism for Protein Folding

Laura S. Itzhaki, Daniel E. Otzen and Alan R. Fersht*

[1]*MRC Unit for Protein Function and Design Cambridge Centre for Protein Engineering, University Chemical Laboratory Lensfield Road, Cambridge CB2 1EW, U.K.*

**Corresponding author*

The 64-residue protein chymotrypsin inhibitor 2 (CI2) is a single module of structure. It folds and unfolds as a single co-operative unit by simple two-state kinetics *via* a single rate determining transition state. This transition state has been characterized at the level of individual residues by analysis of the rates and equilibria of folding of some 100 mutants strategically distributed at 45 sites throughout the protein. Only one residue, a helical residue (Ala16) buried in the hydrophobic core, has its full native interaction energy in the transition state. The only region of structure which is well developed in the transition state is the α-helix (residues 12 to 24). But, the interactions within it are weakened, especially at the C-terminal region. The rest of the protein has varying degrees of weakly formed structure. Thus, secondary and tertiary interactions appear to form concurrently. These data, reinforced by studies on the structures of peptide fragments, fit a "nucleation-condensation" model in which the overall structure condenses around an element of structure, the nucleus, that itself consolidates during the condensation. The high energy transition state is composed of the whole of the molecule making a variety of weak interactions, the nucleus being those residues that make the strongest interactions. The nucleus here is part of the α-helix and some distant residues in the sequence with which it makes contacts. The remainder of the protein has to be sufficiently ordered that it provides the necessary interactions to stabilize the nucleus. The nucleus is only weakly formed in the denatured state but develops in the transition state. The onrush of stability as the nucleus consolidates its local and long range interactions is so rapid that it is not yet fully formed in the transition state. The formation of the nucleus is thus coupled with the condensation. These results are consistent with a recent simulation of the folding of a computer model protein on a lattice which is found to proceed by a nucleation-growth mechanism. We suggest that the mechanism of folding of CI2 may be a common theme in protein folding whereby fundamental folding units of larger proteins, which are modelled by the folding of CI2, form by nucleation-condensation events and coalesce, perhaps in a hierarchical manner.

Keywords: protein folding; transition state; CI2; barnase; protein engineering

Introduction

Protein folding is the process by which a protein progresses from its denatured state to its specific biologically active conformation. It has been proposed that this process has to follow a specific pathway or set of pathways in order to fold in a finite time (Levinthal, 1968), and several different models have been suggested to describe the reaction. According to the hydrophobic collapse model, folding is initiated by burial of hydrophobic

Abbreviations used: CI2, chymotrypsin inhibitor 2; GdmCl, guanidinium chloride; D, denaturant.

side-chains (Dill *et al.*, 1993). The driving force for folding is visualised as the squeezing out of water from a rapidly formed hydrophobic core within which secondary and tertiary structure is subsequently formed. According to the jigsaw puzzle model (Harrison & Durbin, 1985), however, there is no preferential starting point for folding, and each folding attempt may follow a different path. Recent computer simulations based on lattice models support this idea (Sali *et al.*, 1994). Other models envisage pre-formed secondary structural elements diffusing together in collision, and stabilizing each other by local docking rearrangements; the rate-limiting step has been proposed to be variously diffusion (the diffusion-collision-adhesion model; Karplus & Weaver, 1976, 1994) or docking (the framework model; Ptitsyn, 1973, 1991; Kim & Baldwin, 1990) of these secondary elements. These latter models can be distinguished from an early nucleation model (Wetlaufer, 1973), in which the earliest structures are proposed to be formed by a slow random search followed by rapid growth and coalescence into the native folded structure.

Ultimately, the folding pathway of a particular protein is defined when all conformational states of the pathway, including the denatured state, possible intermediates, the major transition state and the native state, as well as their rates of interconversion, are known in detail at the level of individual residues (Fersht, 1993). Owing to the small difference in stability between the folded and unfolded form of a protein and the crudity of currently available potential energy functions, it is difficult to use computer calculations to predict the stability of a specific protein conformation, let alone the feasibility of different pathways, with confidence. It is necessary, therefore, to gather a large experimental database on the stabilizing features of protein structure, and to monitor these interactions, in structural and energetic terms, during folding. This laboratory employs a procedure ("the protein engineering method") for studying the structure of transition state and unstable intermediates at the level of individual residues by making kinetic measurements on the folding and unfolding of mutant proteins, and relating the changes in rate constants to the changes in equilibrium constants on mutation (Matouschek *et al.*, 1989, 1992; Fersht *et al.*, 1992; Serrano *et al.*, 1992a; Fersht, 1993, 1995a). The principle of the method is that the change in stability of the protein on mutation ($\Delta\Delta G_{F\text{-}U}$) is measured, as is also the change in stability of the transition state for folding ($\Delta\Delta G_{\ddagger\text{-}U}$), and the two values are compared. If the region of the protein at the site of mutation is as folded in the transition state as in the final folded structure, then $\Delta\Delta G_{\ddagger\text{-}U} = \Delta\Delta G_{F\text{-}U}$. If that region is as unfolded as in the denatured protein, then $\Delta\Delta G_{\ddagger\text{-}U} = 0$. The ratio of the two energies, $\Delta\Delta G_{\ddagger\text{-}U}/\Delta\Delta G_{F\text{-}U}$, is defined as Φ_F, which varies from 0 for the example of being completely denatured in the transition state to 1.0 for being completely folded. The large scale application of this procedure to a protein is a major undertaking, which can be simplified by working with small monomeric proteins, especially those that have minimal residual structure in the unfolded state. The method has been employed in depth to the analysis of barnase, a small ribonuclease, which folds *via* a metastable folding intermediate (reviewed by Fersht, 1993).

Here, we present a detailed description of the transition state for folding of the 64-residue monomeric protein chymotrypsin inhibitor 2 (CI2). The folding of CI2 is simplified by the lack of disulphide bridges and *cis*-peptidyl-prolyl bonds in the native state. Its native structure is known in the crystal state (McPhalen & James, 1987; Harpaz *et al.*, 1994) and in solution (Ludvigsen *et al.*, 1991). The protein consists of a single domain (or module) that does not have readily discernible substructures that make interactions primarily within themselves. Previous studies have established that the folding and unfolding of wild-type CI2 and a range of mutants conform to a single two-state model under both equilibrium and non-equilibrium conditions (Jackson & Fersht, 1991a,b; Jackson *et al.*, 1993a,b). In particular, the ratio of rate constants for unfolding and refolding give the correct equilibrium constant for unfolding (Jackson & Fersht, 1991a; Jackson *et al.*, 1993b). This does not mean that there are no folding intermediates in the pathway but simply that any such intermediates are not significantly populated at equilibrium or during the approach to equilibrium. There is thus a single rate-determining transition state for both unfolding and refolding (for a discussion of transition states in protein folding see Fersht, 1995a). A thermodynamic characterization of the transition state for folding/unfolding of CI2 indicates that 75% of the total decrease in heat capacity, ΔC_P, between denatured and native states occurs on going from the denatured to the transition state, and this indicates a significant burial of hydrophobic side-chains in the transition state (Jackson & Fersht, 1991b). This is consistent with the large positive enthalpy of activation of folding at lower temperatures. The occurrence of simple two-state kinetics has the important consequence that the single rate-determining transition state can be studied in the direction of both refolding and unfolding.

A preliminary description of this transition state, derived from the protein engineering method, showed a structure that is like an expanded form of the folded structure in which most interactions in the protein have been greatly weakened (Otzen *et al.*, 1994). The major element of structure that is best, but not completely, formed energetically is the single α-helix. This contrasts with the structure for the major transition state for the folding of barnase, which has many elements of structure either fully formed or fully denatured (Fersht, 1993). It was proposed that the folding of CI2 is representative of the folding of a single module protein and is, perhaps, a model for the folding of a domain (module) in a larger protein. The small size of CI2 makes it an attractive target for computer simulations, and the first such work on the transition state for folding/unfolding gives results in agreement with experiment (Li & Daggett, 1994, and unpublished data). Since it is likely that the folding of CI2 will be used to benchmark and refine continuing theoretical studies on protein folding, it is necessary to define the structure of the transition state of CI2 with precision and present it in detail. It is important to obtain as much data as possible for this process, especially as any one single mutation does have the possibility of intrusion from artifacts. We now report an extensive set of experimental data, including double mutant cycles and the recently introduced Ala → Gly scanning procedure (Matthews & Fersht, 1995), to delineate in detail the transition state for the folding of the barley chymotrypsin inhibitor, analysed from both folding and unfolding kinetics. The structure of the transition state points towards a mechanism for the folding, namely nucleation-condensation.

Results

CI2 has been shown to follow the two-state model of protein folding, under both equilibrium and non-equilibrium conditions (Jackson & Fersht, 1991a; Jackson *et al.*, 1993b), i.e. it has only one kinetically significant transition state. Data analysis has been described in detail previously (Jackson *et al.*, 1993a,b), and is briefly summarised here.

GdmCl equilibrium denaturation

It is usually assumed that there is a linear relationship between the free energy of unfolding in the presence of GdmCl (abbreviated here to D) and the concentration of denaturant (Tanford, 1968; Pace, 1986):

$$\Delta G^{D}_{U\text{-}F} = \Delta G^{H_2O}_{U\text{-}F} - m_{U\text{-}F}[\text{denaturant}] \quad (1)$$

where $\Delta G^{D}_{U\text{-}F}$ is the free energy of unfolding at a particular denaturant concentration, D, $\Delta G^{H_2O}_{U\text{-}F}$ is the free energy of unfolding in water, and $m_{U\text{-}F}$ is a constant that is proportional to the increase in the degree of exposure of the protein on denaturation. From equation (1), it is apparent that at $[D]_{50\%}$, the concentration of GdmCl at which 50% of the protein is denatured, $\Delta G^{H_2O}_{U\text{-}F} = m_{U\text{-}F}[D]_{50\%}$; thus:

$$\Delta G^{D}_{U\text{-}F} = m_{U\text{-}F}([D]_{50\%} - [D]) \quad (2)$$

The denaturation curves were fitted to an equation, derived from (2) above (Clarke & Fersht, 1993), which yields the values for $[D]_{50\%}$ and $m_{U\text{-}F}$ and their standard errors (Table 1). All errors are calculated from the best fit of the data and are not standard errors from repetitive runs, unless otherwise stated. Values of $\Delta G^{H_2O}_{U\text{-}F}$ are also given in Table 1.

Repetitive measurements of $m_{U\text{-}F}$ for an individual mutant have a variability of ± 5 to 10%, whereas $[D]_{50\%}$ is very reproducible at ± 0.02 M because $[D]_{50\%}$ is insensitive to small errors in baselines. It is clear that the $m_{U\text{-}F}$-values for wild-type and mutants are the same within experimental error, apart from a few outliers, which are statistically expected in a large data base. Therefore, the value of $\Delta\Delta G^{[D]50\%}_{U\text{-}F}$, the difference in the free energy of unfolding between wild-type and mutant proteins at a mean value of the $[D]_{50\%}$ for the two proteins, can be calculated from the equation:

$$\Delta\Delta G^{[D]50\%}_{U\text{-}F} = \langle m_{U\text{-}F} \rangle \Delta[D]_{50\%} \quad (3)$$

where $\langle m_{U\text{-}F} \rangle$ is the average value of m, obtained from measurements on all the mutant proteins of 1.90($\pm$0.03) kcal/mol^{-1} M^{-1} ($n = 124$, including unpublished mutants). The use of a mean value of $m_{U\text{-}F}$ allows calculation of the change in the free energy of unfolding on mutation with a low standard error. Equation (3) is very robust since it gives accurate values of $\Delta\Delta G^{[D]50\%}_{U\text{-}F}$ even if the linearity assumption of equation (1) breaks down but there is a small curvature in the plots that is the same for each mutant (Matouschek *et al.*, 1994).

A value of $\Delta\Delta G_{U\text{-}F}$ can also be calculated using the individual m values:

$$\Delta\Delta G^{H_2O}_{U\text{-}F} = \Delta G^{H_2O}_{U\text{-}F} - \Delta G'^{H_2O}_{U\text{-}F} \quad (4)$$

where $\Delta G^{H_2O}_{U\text{-}F}$ and $\Delta G'^{H_2O}_{U\text{-}F}$ are the free energies of unfolding in water for wild-type and mutant, respectively, or at any denaturant concentration using the more general equation:

$$\Delta\Delta G^{[D]}_{U\text{-}F} = m_{U\text{-}F}([D]_{50\%} - [D]) - m'_{U\text{-}F}([D]'_{50\%} - [D]) \quad (5)$$

where $m'_{U\text{-}F}$ and $[D]'_{50\%}$ are the m-value and midpoint of denaturation, respectively, for the mutant protein, and $m_{U\text{-}F}$ and $[D]_{50\%}$ are the values for wild-type (Clarke & Fersht, 1993). The relative merits of these three methods for calculating $\Delta\Delta G_{U\text{-}F}$ and the errors of the resulting values have been discussed in detail previously (Jackson *et al.*, 1993a). $\Delta\Delta G^{[D]50\%}_{U\text{-}F}$, however, has the smallest errors and, for mutations in the hydrophobic core of CI2, shows the best agreement with $\Delta\Delta G_{U\text{-}F}$ measured directly in water by calorimetry (Jackson *et al.*, 1993a). Values of $\Delta\Delta G^{[D]50\%}_{U\text{-}F}$ are given in Table 1. Values of $\Delta G^{H_2O}_{U\text{-}F}$ have not been given because the long extrapolation results in very large errors associated with this parameter.

Unfolding kinetics

Plots of the natural logarithm of the rate constants of unfolding against the final GdmCl concentration are linear, within experimental error over the experimentally accessible range of concentrations, conforming to the equation:

$$\ln k_u = \ln k_u^{H_2O} + m_{ku}[D] \quad (6)$$

where k_u is the rate constant of unfolding at a given GdmCl concentration, $k_u^{H_2O}$ is the rate constant of unfolding in water, m_{ku} is the slope, and [D] is the

Table 1. changes in the free energies of unfolding upon mutation determined by reversible guanidinium chloride denaturation experiments

Mutant	$m_{U\text{-}F}$ (kcal mol^{-1} M^{-1})	$[D]_{50\%}$ (M)	$\Delta G_{U\text{-}F}^{H_2O}$ [a] (kcal mol^{-1})	$\Delta\Delta G_{U\text{-}F}^{[D]_{50\%}}$ [b] (kcal mol^{-1})
Wild-type	1.90 ± 0.03	4.00 ± 0.01	7.60 ± 0.12	0
KA2	1.77 ± 0.05	3.72 ± 0.01	6.56 ± 0.18	0.55 ± 0.04
KA2/EA7	1.87 ± 0.07	3.43 ± 0.02	6.43 ± 0.24	1.10 ± 0.04
KM2	1.88 ± 0.05	3.66 ± 0.01	6.87 ± 0.19	0.67 ± 0.03
TA3[e]	1.83 ± 0.09	3.56 ± 0.02	6.51 ± 0.31	0.85 ± 0.05
TV3[e]	1.67 ± 0.08	3.83 ± 0.03	6.39 ± 0.32	0.32 ± 0.07
TG3[e]	1.95 ± 0.12	3.40 ± 0.03	6.65 ± 0.41	1.16 ± 0.06
PA6	3.46 ± 0.75	3.19 ± 0.05	11.0 ± 2.40	1.57 ± 0.10
PA6/AG16	2.39 ± 0.14	2.64 ± 0.02	6.30 ± 0.37	2.65 ± 0.06
EA7	1.75 ± 0.06	3.76 ± 0.02	6.57 ± 0.24	0.47 ± 0.05
EQ7	1.91 ± 0.12	3.68 ± 0.03	7.03 ± 0.44	0.62 ± 0.07
LA8[c]	2.15 ± 0.13	2.62 ± 0.02	5.63 ± 0.34	2.68 ± 0.06
KA11	1.52 ± 0.08	4.22 ± 0.02	6.40 ± 0.34	−0.42 ± 0.05
SG12[d]	2.10 ± 0.07	3.59 ± 0.02	7.54 ± 0.25	0.80 ± 0.05
SA12[d]	1.86 ± 0.08	3.54 ± 0.02	6.58 ± 0.29	0.89 ± 0.05
EQ14[d]	1.87 ± 0.10	3.85 ± 0.03	7.20 ± 0.39	0.29 ± 0.06
ED14[d]	1.83 ± 0.11	3.73 ± 0.04	6.83 ± 0.42	0.52 ± 0.08
EN14[d]	2.01 ± 0.10	3.64 ± 0.02	7.32 ± 0.37	0.70 ± 0.05
EQ15[d]	1.84 ± 0.09	3.76 ± 0.03	6.92 ± 0.34	0.47 ± 0.06
ED15[d]	2.20 ± 0.16	3.62 ± 0.03	7.96 ± 0.58	0.74 ± 0.06
EN15[d]	2.18 ± 0.11	3.45 ± 0.02	7.52 ± 0.38	1.07 ± 0.05
EA14/EA15[d]	2.11 ± 0.18	3.61 ± 0.03	7.62 ± 0.65	0.76 ± 0.06
SG12/EA14/EA15[d]	2.15 ± 0.16	3.16 ± 0.03	6.79 ± 0.51	1.63 ± 0.07
SA12/EA14/EA15[d]	2.18 ± 0.14	3.14 ± 0.02	6.85 ± 0.44	1.67 ± 0.05
AG16	1.80 ± 0.07	3.44 ± 0.02	6.19 ± 0.25	1.09 ± 0.05
KA17	1.73 ± 0.05	3.75 ± 0.01	6.47 ± 0.17	0.49 ± 0.03
KG17	1.99 ± 0.10	2.80 ± 0.02	5.59 ± 0.27	2.32 ± 0.06
KA18	1.61 ± 0.13	4.11 ± 0.07	6.61 ± 0.53	−0.21 ± 0.13
KG18	1.73 ± 0.13	3.49 ± 0.04	6.02 ± 0.48	0.99 ± 0.08
VA19[d]	2.01 ± 0.15	3.75 ± 0.03	7.54 ± 0.57	0.49 ± 0.06
IV20[d]	1.99 ± 0.12	3.33 ± 0.02	6.63 ± 0.40	1.30 ± 0.05
LA21	1.93 ± 0.15	3.32 ± 0.03	6.39 ± 0.49	1.33 ± 0.07
LG21	1.68 ± 0.07	3.29 ± 0.02	5.51 ± 0.23	1.38 ± 0.05
QA22	1.77 ± 0.15	3.99 ± 0.05	7.06 ± 0.60	0.02 ± 0.10
QG22	1.81 ± 0.12	3.69 ± 0.03	6.68 ± 0.43	0.60 ± 0.06
DA23	1.87 ± 0.05	3.51 ± 0.01	6.55 ± 0.16	0.96 ± 0.03
KA24	1.57 ± 0.08	3.67 ± 0.03	5.75 ± 0.28	0.65 ± 0.06
KG24	1.75 ± 0.17	2.36 ± 0.05	4.11 ± 0.40	3.19 ± 0.11
PA25	2.07 ± 0.10	3.09 ± 0.02	6.38 ± 0.32	1.76 ± 0.05
EA26	1.78 ± 0.06	3.83 ± 0.01	6.82 ± 0.24	0.32 ± 0.03
EM28/ML40	1.78 ± 0.05	4.16 ± 0.02	7.42 ± 0.21	−0.32 ± 0.05
IV29[c]	1.99 ± 0.13	3.43 ± 0.02	6.83 ± 0.45	1.11 ± 0.05
IA29[c]	2.07 ± 0.19	1.99 ± 0.03	4.12 ± 0.38	3.90 ± 0.09
IA29/IV57[c]	1.83 ± 0.10	1.90 ± 0.02	3.48 ± 0.19	4.08 ± 0.08
IV30[e]	1.77 ± 0.05	4.04 ± 0.04	7.17 ± 0.19	−0.08 ± 0.08
IA30[c]	2.22 ± 0.09	2.91 ± 0.02	6.45 ± 0.27	2.12 ± 0.06
IG30[e]	2.13 ± 0.08	2.19 ± 0.01	4.65 ± 0.17	3.52 ± 0.07
IT30[e]	1.97 ± 0.07	3.31 ± 0.02	6.53 ± 0.24	1.34 ± 0.04
LA32	2.11 ± 0.09	2.78 ± 0.02	5.86 ± 0.27	2.37 ± 0.05
LI32	1.70 ± 0.08	3.87 ± 0.03	6.59 ± 0.33	0.26 ± 0.07
LV32	1.76 ± 0.10	3.74 ± 0.03	6.58 ± 0.36	0.50 ± 0.06
LV32/FL50	2.37 ± 0.13	2.75 ± 0.02	6.52 ± 0.37	2.42 ± 0.06
LV32/FA50	1.94 ± 0.10	2.24 ± 0.02	4.34 ± 0.22	3.41 ± 0.07
LA32/FL50	1.94 ± 0.09	2.24 ± 0.02	4.35 ± 0.19	3.42 ± 0.07
LA32/FA50	1.59 ± 0.14	1.53 ± 0.08	2.44 ± 0.25	4.79 ± 0.19
LA32/VA38	1.96 ± 0.06	2.37 ± 0.01	4.65 ± 0.14	3.16 ± 0.06
LV32/VA38	1.86 ± 0.06	3.05 ± 0.01	5.67 ± 0.20	1.85 ± 0.05
LA32/VA38/FL50	1.93 ± 0.06	2.21 ± 0.01	4.26 ± 0.13	3.48 ± 0.07
LV32/VA38/FL50	1.83 ± 0.05	2.60 ± 0.01	4.76 ± 0.13	2.72 ± 0.06
PA33	1.79 ± 0.07	3.91 ± 0.02	7.02 ± 0.27	0.17 ± 0.05
VT34[e]	1.65 ± 0.08	3.47 ± 0.02	5.74 ± 0.26	1.03 ± 0.05
VA34[e]	1.70 ± 0.15	3.67 ± 0.05	6.24 ± 0.57	0.64 ± 0.11
VG34[e]	2.00 ± 0.08	2.75 ± 0.01	5.50 ± 0.22	2.43 ± 0.05
TS36	1.66 ± 0.03	3.99 ± 0.01	6.62 ± 0.13	0.02 ± 0.02
TV36	1.67 ± 0.12	3.61 ± 0.04	6.03 ± 0.43	0.76 ± 0.08
TA36	1.50 ± 0.05	4.12 ± 0.01	6.18 ± 0.21	−0.23 ± 0.03
IA37	1.59 ± 0.06	3.98 ± 0.03	6.32 ± 0.25	0.03 ± 0.06
IA37Δ38	1.90 ± 0.05	2.52 ± 0.01	4.78 ± 0.13	2.88 ± 0.06
VA38	2.14 ± 0.12	3.24 ± 0.02	6.93 ± 0.38	1.47 ± 0.05

continued

Table 1. *continued*

Mutant	$m_{U\text{-}F}$ (kcal mol^{-1} M^{-1})	$[D]_{50\%}$ (M)	$\Delta G^{H_2O}_{U\text{-}F}$ [a] (kcal mol^{-1})	$\Delta\Delta G^{[D]50\%}_{U\text{-}F}$ [b] (kcal mol^{-1})
VA38/FL50	2.02 ± 0.08	2.67 ± 0.01	5.40 ± 0.20	2.58 ± 0.06
VA38/FA60	2.19 ± 0.12	1.75 ± 0.02	3.82 ± 0.21	4.37 ± 0.09
TA39[f]	1.89 ± 0.14	3.63 ± 0.04	6.86 ± 0.51	0.72 ± 0.08
TD39[f]	1.87 ± 0.08	4.01 ± 0.03	7.50 ± 0.33	−0.02 ± 0.06
TA39/EA41[f]	1.94 ± 0.10	3.54 ± 0.02	6.87 ± 0.36	0.89 ± 0.05
TD39/EA41[f]	1.89 ± 0.09	3.86 ± 0.02	7.30 ± 0.35	0.27 ± 0.04
EA41	1.78 ± 0.16	3.64 ± 0.05	6.48 ± 0.59	0.70 ± 0.10
RA43	1.78 ± 0.10	3.70 ± 0.03	6.58 ± 0.38	0.58 ± 0.07
RA43/DA45	1.53 ± 0.10	3.37 ± 0.04	5.14 ± 0.36	1.22 ± 0.08
DA45	1.83 ± 0.08	3.59 ± 0.02	6.58 ± 0.28	0.80 ± 0.05
VA47[c]	1.76 ± 0.22	1.46 ± 0.10	2.57 ± 0.37	4.93 ± 0.21
LA49[c]	1.98 ± 0.12	2.02 ± 0.03	4.00 ± 0.25	3.84 ± 0.09
FL50	1.95 ± 0.11	2.91 ± 0.02	5.68 ± 0.33	2.11 ± 0.06
FV50	2.24 ± 0.18	2.77 ± 0.03	6.19 ± 0.51	2.39 ± 0.07
FA50	2.17 ± 0.13	2.02 ± 0.03	4.39 ± 0.26	3.84 ± 0.08
VA51[c]	2.15 ± 0.19	2.98 ± 0.03	6.41 ± 0.57	1.98 ± 0.07
DA52	1.99 ± 0.21	2.24 ± 0.04	4.46 ± 0.47	3.41 ± 0.10
DN53	1.60 ± 0.03	4.00 ± 0.01	6.39 ± 0.12	−0.004 ± 0.03
ND56	1.80 ± 0.07	3.38 ± 0.02	6.09 ± 0.24	1.21 ± 0.05
NA56	1.79 ± 0.04	3.57 ± 0.01	6.38 ± 0.14	0.83 ± 0.03
IV57[c]	1.82 ± 0.12	4.10 ± 0.05	7.46 ± 0.50	−0.19 ± 0.10
IA57[c]	1.93 ± 0.21	1.79 ± 0.05	3.45 ± 0.39	4.29 ± 0.12
AG58[e]	2.19 ± 0.22	3.03 ± 0.04	6.65 ± 0.67	1.88 ± 0.08
VT60[e]	1.63 ± 0.14	3.80 ± 0.06	6.20 ± 0.53	0.38 ± 0.11
VA60[e]	1.92 ± 0.11	3.22 ± 0.03	6.17 ± 0.37	1.51 ± 0.06
VG60[e]	2.61 ± 0.11	2.33 ± 0.01	6.08 ± 0.27	3.24 ± 0.06
PA61	1.80 ± 0.14	2.28 ± 0.04	4.10 ± 0.32	3.34 ± 0.09
VT63[e]	1.79 ± 0.13	3.41 ± 0.03	6.08 ± 0.45	1.15 ± 0.07
VA63[e]	1.95 ± 0.14	3.25 ± 0.03	6.33 ± 0.45	1.45 ± 0.07
VG63[e]	2.10 ± 0.06	2.20 ± 0.01	4.61 ± 0.14	3.50 ± 0.07

Measured at 25°C, 50 mM Mes (pH 6.25).

[a] $\Delta G^{H_2O}_{U\text{-}F} = m_{U\text{-}F}[D]_{50\%}$.

[b] $\Delta\Delta G^{[D]50\%}_{U\text{-}F} = \langle m_{U\text{-}F}\rangle([D]_{50\%} - [D]'_{50\%})$, where $\langle m_{U\text{-}F}\rangle$ is the mean value of $m_{U\text{-}F}$, from measurements on all the mutant proteins and repetitive runs on wild-type, of 1.94(±0.033) kcal mol^{-2}, and $[D]'_{50\%}$ and $[D]_{50\%}$ are the concentrations of GdmCl at which 50% of mutant protein and wild-type protein, respectively, are denatured. Note that $\Delta\Delta G^{[D]50\%}_{U\text{-}F}$ is defined as $\Delta G_{U\text{-}F}$ (wild type) − $\Delta G_{U\text{-}F}$ (mutant).

[c] From Jackson *et al.* (1993a).

[d] From elMasry & Fersht (1994).

[e] From Otzen & Fersht (1995).

[f] From Jackson & Fersht (1994).

GdmCl concentration. If $\ln k_u$ is plotted against [D], then the plots can be extrapolated to obtain $\ln k_u^{4M}$. $\ln k_u^{H_2O}$, $\ln k_u^{4M}$ and m_{ku} for all mutants are given in Table 2. The standard errors are significantly lower for $\ln k_u^{4M}$ compared with $\ln k_u^{H_2O}$ because of the shorter extrapolation.

The stability of the transition state of the mutant protein relative to that of the wild-type protein is calculated from the unfolding kinetics from the equation:

$$\Delta\Delta G_{\ddagger\text{-}F} = -RT \ln(k_u/k_u') \quad (7)$$

where $\Delta\Delta G_{\ddagger\text{-}F}$ is the difference in energy of the transition state of unfolding relative to the folded state between wild-type and mutant, and k_u and k_u' are the rate constants of unfolding for the wild-type and mutant, respectively. The values of $\Delta\Delta G_{\ddagger\text{-}F}$ are given in Table 2.

Refolding kinetics

Jackson & Fersht (1991b) have shown that there is a series of slow refolding phases that arise from *cis-trans* isomerisation of peptidyl-prolyl bonds. The analysis of energetics and pathway in this study concerns only the major fast refolding phase of the all-*trans* peptidyl-proline from of the unfolded state to the all-*trans* folded state that accounts for approximately 70% of the total amplitude.

Plots of the natural logarithm of the rate constants of folding against the final GdmCl concentration are linear, conforming to the equation:

$$\ln k_f = \ln k_f^{H_2O} + m_{kf}[D] \quad (8)$$

where k_f is the rate constant of folding at a given GdmCl concentration, $k_f^{H_2O}$ is the rate constant of folding in water, m_{kf} is the slope, and [D] is the final GdmCl concentration. $\ln k_f^{H_2O}$ and m_{kf} for all mutants are given in Table 3.

The stability of the transition state of the mutant protein relative to that of the wild-type protein is calculated from the folding kinetics, in a similar manner to the analysis of the unfolding data, using the equation:

$$\Delta\Delta G_{\ddagger\text{-}U} = -RT \ln(k_f/k_f') \quad (9)$$

where $\Delta\Delta G_{\ddagger\text{-}U}$ is the difference in energy of the transition state of unfolding relative to the folded

Table 2. Rates and thermodynamic data from unfolding kinetics of CI2 mutants

Mutant	$\ln k_u^{H_2O}$	$\ln k_u^{4M}$	m_{ku} (M^{-1})	$\Delta\Delta G_{\ddagger\text{-F}}^{H_2O\,a}$ (kcal mol^{-1})	$\Delta\Delta G_{\ddagger\text{-F}}^{4M\,a}$ (kcal mol^{-1})
Wild type	−9.04 ± 0.07	−3.90 ± 0.02	1.31 ± 0.01	0	0
KA2	−7.83 ± 0.08	−2.72 ± 0.03	1.28 ± 0.01	−0.72 ± 0.06	−0.67 ± 0.02
KA2/EA7	−7.36 ± 0.22	−2.36 ± 0.07	1.25 ± 0.04	−1.00 ± 0.14	−0.91 ± 0.04
KA2/DA23	−6.74 ± 0.13	−1.57 ± 0.04	1.29 ± 0.02	−1.36 ± 0.09	−1.38 ± 0.02
KM2	−8.25 ± 0.13	−2.93 ± 0.04	1.33 ± 0.02	−0.47 ± 0.09	−0.57 ± 0.03
TA3	−7.98 ± 0.10	−2.73 ± 0.03	1.31 ± 0.02	−0.63 ± 0.07	−0.69 ± 0.02
TV3	−8.58 ± 0.06	−3.41 ± 0.02	1.29 ± 0.01	−0.27 ± 0.05	−0.29 ± 0.01
TG3	−7.68 ± 0.16	−2.51 ± 0.05	1.29 ± 0.03	−0.81 ± 0.10	−0.83 ± 0.03
PA6	−7.62 ± 0.49	−1.24 ± 0.11	1.60 ± 0.10	−0.84 ± 0.29	−1.58 ± 0.07
PA6/AG16	−6.65 ± 0.06	−1.28 ± 0.02	1.34 ± 0.01	−1.42 ± 0.05	−1.55 ± 0.01
EA7	−8.59 ± 0.13	−3.39 ± 0.05	1.30 ± 0.02	−0.27 ± 0.09	−0.30 ± 0.03
LA8[b]	−4.42 ± 0.14	−0.12 ± 0.03	1.05 ± 0.03	−2.73 ± 0.09	−2.24 ± 0.02
KA11	−9.39 ± 0.05	−4.32 ± 0.02	1.27 ± 0.01	0.21 ± 0.05	0.24 ± 0.01
SG12	−8.42 ± 0.22	−3.24 ± 0.08	1.29 ± 0.04	−0.36 ± 0.14	−0.39 ± 0.05
SA12	−8.53 ± 0.22	−3.37 ± 0.08	1.29 ± 0.04	−0.30 ± 0.14	−0.32 ± 0.05
EQ14	−8.97 ± 0.12	−3.70 ± 0.04	1.32 ± 0.02	−0.04 ± 0.08	−0.12 ± 0.03
ED14	−8.76 ± 0.29	−3.43 ± 0.09	1.33 ± 0.05	−0.16 ± 0.18	−0.28 ± 0.05
EN14	−9.01 ± 0.13	−3.51 ± 0.05	1.38 ± 0.02	−0.02 ± 0.09	−0.23 ± 0.03
EQ15	−8.71 ± 0.16	−3.53 ± 0.05	1.30 ± 0.03	−0.20 ± 0.10	−0.22 ± 0.03
ED15	−8.28 ± 0.08	−2.82 ± 0.03	1.37 ± 0.01	−0.45 ± 0.06	−0.64 ± 0.02
EN15	−7.95 ± 0.10	−2.89 ± 0.06	1.26 ± 0.02	−0.65 ± 0.07	−0.60 ± 0.04
EA14/EA15	−8.17 ± 0.23	−2.95 ± 0.08	1.31 ± 0.04	−0.52 ± 0.14	−0.56 ± 0.05
SG12/EA14/EA15	−6.66 ± 0.28	−1.95 ± 0.09	1.18 ± 0.05	−1.41 ± 0.17	−1.16 ± 0.05
SA12/EA14/EA15	−7.03 ± 0.10	−2.18 ± 0.03	1.21 ± 0.02	−1.19 ± 0.07	−1.02 ± 0.02
AG16	−9.47 ± 0.14	−4.24 ± 0.05	1.31 ± 0.02	0.26 ± 0.09	0.20 ± 0.03
KA17	−7.83 ± 0.12	−2.87 ± 0.04	1.24 ± 0.02	−0.72 ± 0.08	−0.61 ± 0.03
KG17	−6.13 ± 0.11	−1.63 ± 0.03	1.13 ± 0.02	−1.72 ± 0.08	−1.35 ± 0.02
KA18	−8.89 ± 0.13	−3.64 ± 0.05	1.31 ± 0.02	−0.09 ± 0.09	−0.15 ± 0.03
KG18	−8.35 ± 0.11	−3.13 ± 0.04	1.30 ± 0.02	−0.41 ± 0.08	−0.46 ± 0.03
VA19[b]	−7.80 ± 0.13	−2.45 ± 0.05	1.35 ± 0.02	−0.73 ± 0.09	−0.86 ± 0.03
IV20[b]	−8.02 ± 0.11	−2.94 ± 0.03	1.27 ± 0.02	−0.60 ± 0.08	−0.57 ± 0.02
LA21	−7.44 ± 0.09	−2.77 ± 0.02	1.17 ± 0.01	−0.95 ± 0.07	−0.67 ± 0.01
LG21	−7.19 ± 0.04	−2.61 ± 0.01	1.14 ± 0.01	−1.10 ± 0.05	−0.76 ± 0.01
QA22	−8.98 ± 0.06	−3.72 ± 0.02	1.32 ± 0.01	−0.03 ± 0.06	−0.11 ± 0.02
QG22	−8.27 ± 0.09	−3.21 ± 0.03	1.27 ± 0.01	−0.46 ± 0.07	−0.41 ± 0.02
DA23	−7.87 ± 0.21	−2.36 ± 0.06	1.38 ± 0.04	−0.69 ± 0.13	−0.91 ± 0.04
KA24	−7.90 ± 0.06	−2.72 ± 0.02	1.30 ± 0.01	−0.68 ± 0.06	−0.70 ± 0.02
KG24	−4.48 ± 0.10	1.00 ± 0.01	1.37 ± 0.03	−2.70 ± 0.07	−2.90 ± 0.01
PA25	−6.43 ± 0.19	−1.78 ± 0.06	1.16 ± 0.03	−1.55 ± 0.12	−1.25 ± 0.04
EA26	−8.61 ± 0.09	−3.49 ± 0.03	1.28 ± 0.01	−0.25 ± 0.07	−0.25 ± 0.02
IV29[b]	−8.00 ± 0.09	−2.73 ± 0.03	1.32 ± 0.01	−0.62 ± 0.07	−0.69 ± 0.02
IA29[b]	−4.30 ± 0.18	0.10 ± 0.03	1.10 ± 0.04	−2.80 ± 0.11	−2.37 ± 0.02
IA29/IV57[b]	−4.33 ± 0.15	0.04 ± 0.04	1.09 ± 0.03	−2.79 ± 0.10	−2.33 ± 0.03
IV30	−8.99 ± 0.13	−4.13 ± 0.05	1.22 ± 0.02	−0.03 ± 0.09	0.13 ± 0.03
IA30	−7.50 ± 0.08	−2.52 ± 0.03	1.25 ± 0.01	−0.91 ± 0.06	−0.82 ± 0.02
IG30	5.29 ± 0.13	−0.19 ± 0.04	1.28 ± 0.03	−2.22 ± 0.09	−2.20 ± 0.02
IT30	−8.19 ± 0.09	−3.25 ± 0.03	1.24 ± 0.02	−0.50 ± 0.07	−0.39 ± 0.02
LA32	−6.34 ± 0.23	−1.34 ± 0.06	1.25 ± 0.04	−1.60 ± 0.14	−1.51 ± 0.04
LI32	−8.54 ± 0.03	−3.32 ± 0.01	1.31 ± 0.01	−0.29 ± 0.05	−0.35 ± 0.01
LV32	−8.22 ± 0.04	−2.96 ± 0.01	1.31 ± 0.01	−0.48 ± 0.05	−0.56 ± 0.01
LV32/FL50	−6.17 ± 0.04	−1.19 ± 0.01	1.25 ± 0.01	−1.70 ± 0.05	−1.61 ± 0.01
LV32/FA50	−5.08 ± 0.11	−0.52 ± 0.03	1.14 ± 0.02	−2.34 ± 0.08	−2.00 ± 0.02
LA32/FL50	−5.07 ± 0.07	−0.56 ± 0.02	1.13 ± 0.01	−2.35 ± 0.06	−1.98 ± 0.01
LA32/FA50	−3.65 ± 0.10	0.55 ± 0.03	1.05 ± 0.02	−3.19 ± 0.07	−2.64 ± 0.02
LA32/VA38	−5.34 ± 0.10	−0.52 ± 0.02	1.20 ± 0.02	−2.19 ± 0.07	−2.00 ± 0.02
LV32/VA38	−6.44 ± 0.09	−1.51 ± 0.03	1.23 ± 0.02	−1.54 ± 0.07	−1.42 ± 0.02
LA32/VA38/FL50	−5.06 ± 0.06	−0.68 ± 0.02	1.10 ± 0.01	−2.35 ± 0.06	−1.91 ± 0.01
LV32/VA38/FL50	−5.81 ± 0.04	−1.05 ± 0.01	1.19 ± 0.01	−1.91 ± 0.05	−1.69 ± 0.01
VT34	−7.78 ± 0.03	−3.07 ± 0.01	1.18 ± 0.01	−0.75 ± 0.04	−0.50 ± 0.01
VA34	−8.21 ± 0.03	−3.35 ± 0.07	1.22 ± 0.03	−0.49 ± 0.05	−0.33 ± 0.04
VG34	−6.05 ± 0.04	−1.67 ± 0.02	1.09 ± 0.01	−1.77 ± 0.05	−1.32 ± 0.01
TV36	−7.83 ± 0.08	−3.11 ± 0.03	1.18 ± 0.01	−0.71 ± 0.06	−0.47 ± 0.02
IA37Δ38	−5.25 ± 0.05	−0.18 ± 0.01	1.27 ± 0.01	−2.24 ± 0.05	−2.20 ± 0.01
VA38	−7.22 ± 0.03	−2.43 ± 0.01	1.20 ± 0.01	−1.08 ± 0.04	−0.87 ± 0.01
VA38/FL50	−6.31 ± 0.05	−1.59 ± 0.02	1.18 ± 0.01	−1.62 ± 0.05	−1.37 ± 0.01
VA38/FL50	−4.12 ± 0.06	0.32 ± 0.02	1.11 ± 0.01	−2.91 ± 0.05	−2.50 ± 0.01
TA39[c]	−8.39 ± 0.08	−3.13 ± 0.03	1.32 ± 0.01	−0.38 ± 0.06	0.46 ± 0.02
TD39[c]	−9.30 ± 0.17	−3.97 ± 0.07	1.35 ± 0.03	0.15 ± 0.11	0.04 ± 0.04

continued

Table 2. *continued*

Mutant	$\ln k_u^{H_2O}$	$\ln k_u^{4M}$	m_{ku} (M^{-1})	$\Delta\Delta G_{\ddagger\text{-F}}^{H_2O\,a}$ (kcal mol^{-1})	$\Delta\Delta G_{\ddagger\text{-F}}^{4M\,a}$ (kcal mol^{-1})
TA39/EA41[c]	−7.99 ± 0.05	−2.82 ± 0.03	1.30 ± 0.01	−0.62 ± 0.05	0.64 ± 0.02
TD39/EA41[c]	−8.95 ± 0.09	−3.65 ± 0.03	1.34 ± 0.02	−0.05 ± 0.07	0.15 ± 0.02
EA41[c]	−8.10 ± 0.08	−3.01 ± 0.03	1.28 ± 0.01	−0.56 ± 0.06	0.53 ± 0.02
RA43	−8.25 ± 0.11	−3.05 ± 0.04	1.30 ± 0.02	−0.47 ± 0.08	−0.50 ± 0.03
RA43/DA45	−6.37 ± 0.23	−1.69 ± 0.08	1.17 ± 0.04	−1.58 ± 0.14	−1.31 ± 0.05
DA45	−8.08 ± 0.08	−3.01 ± 0.03	1.27 ± 0.01	−0.57 ± 0.06	−0.53 ± 0.02
VA47[b]	−3.08 ± 0.08	2.54 ± 0.03	1.41 ± 0.02	−3.53 ± 0.06	−3.81 ± 0.02
LA49[b]	−6.05 ± 0.15	−1.46 ± 0.05	1.15 ± 0.03	−1.77 ± 0.10	−1.45 ± 0.03
FL50	−7.13 ± 0.04	−2.03 ± 0.01	1.28 ± 0.01	−1.13 ± 0.05	−1.11 ± 0.01
FV50	−6.26 ± 0.04	−1.50 ± 0.01	1.19 ± 0.01	−1.64 ± 0.05	−1.42 ± 0.01
FA50	−5.12 ± 0.04	−0.33 ± 0.01	1.20 ± 0.01	−2.32 ± 0.05	−2.12 ± 0.01
VA51[b]	−7.20 ± 0.05	−2.21 ± 0.03	1.25 ± 0.01	−1.09 ± 0.05	−1.00 ± 0.02
DA52	−4.14 ± 0.07	−0.31 ± 0.01	0.96 ± 0.02	−2.90 ± 0.06	−2.13 ± 0.01
DN52	−8.93 ± 0.11	−3.89 ± 0.04	1.26 ± 0.02	−0.07 ± 0.08	−0.01 ± 0.02
ND56	−7.92 ± 0.07	−3.17 ± 0.02	1.19 ± 0.01	−0.66 ± 0.06	−0.43 ± 0.02
NA56	−8.12 ± 0.08	−3.00 ± 0.03	1.28 ± 0.01	−0.55 ± 0.06	−0.53 ± 0.02
IV57[b]	−9.31 ± 0.15	−3.56 ± 0.05	1.28 ± 0.03	0.16 ± 0.10	−0.20 ± 0.03
IA57[b]	−2.67 ± 0.16	1.94 ± 0.03	1.17 ± 0.04	−3.77 ± 0.10	−3.46 ± 0.02
AG58	−6.46 ± 0.06	−1.78 ± 0.02	1.17 ± 0.01	−1.52 ± 0.05	−1.26 ± 0.01
VT60	−8.42 ± 0.04	−3.44 ± 0.01	1.24 ± 0.01	−0.37 ± 0.05	−0.27 ± 0.01
VA60	−6.93 ± 0.08	−1.67 ± 0.03	1.32 ± 0.01	−1.25 ± 0.06	−1.32 ± 0.02
VG60	−3.93 ± 0.17	1.45 ± 0.02	1.34 ± 0.04	−3.02 ± 0.11	−3.17 ± 0.02
VM60/ML39	−7.43 ± 0.04	−2.10 ± 0.01	1.33 ± 0.01	−0.95 ± 0.05	−1.07 ± 0.01
PA61	−4.03 ± 0.36	1.71 ± 0.04	1.43 ± 0.10	−2.97 ± 0.22	−3.32 ± 0.03
VT62	−7.35 ± 0.07	−2.28 ± 0.02	1.27 ± 0.01	−1.00 ± 0.06	−0.96 ± 0.02
VA82	−7.19 ± 0.06	−1.77 ± 0.02	1.36 ± 0.01	−1.09 ± 0.05	−1.26 ± 0.01
VG82	−3.98 ± 0.06	1.59 ± 0.01	1.39 ± 0.01	−3.00 ± 0.05	−3.25 ± 0.01

[a] $\Delta\Delta G_{\ddagger\text{-F}} = -RT\ln(k_u/k_u')$, where k_u and k_u' are the rate constants of unfolding for wild-type and mutant, respectively (in s^{-1}).
[b] From Jackson *et al.* (1993b).
[c] From Jackson & Fersht (1994).

state between wild-type and mutant, and k_f and k_f' are the rate constants of folding for the wild-type and mutant, respectively. The values of $\Delta\Delta G_{\ddagger\text{-U}}$ are given in Table 3.

Analysis of two-state behaviour

The complete kinetics of folding and unfolding can be fitted to the equation (10), which is derived from equations (6) and (8), and based on a two-state transition:

$$\ln k = \ln(k_f^{H_2O}\exp(-m_{U\text{-}\ddagger}[D]) + k_u^{H_2O}\exp(m_{\ddagger\text{-F}}[D])) \quad (10)$$

For wild-type and the mutants, the kinetic data for unfolding and refolding fit well to this model (see Jackson *et al.*, 1993b for representative curves). The values $k_f^{H_2O}$, $k_u^{H_2O}$, $m_{U\text{-}\ddagger}$ and $m_{\ddagger\text{-F}}$ obtained from this equation can be used to calculate equilibrium parameters, and can then be compared with those obtained directly from equilibrium measurements (Table 1). $\Delta G_{U\text{-F}}^{H_2O}$ can be calculated from the kinetic data using the ratios of the unfolding and refolding rate constants (Jackson *et al.*, 1993b). $[D]_{50\%}$ can be calculated from kinetic experiments, using the equation:

$$[D]_{50\%} = (\ln(k_f^{H_2O}/k_u^{H_2O}))/(m_{ku} - m_{kf}) \quad (11)$$

We have determined m_{kf} between 0 and 0.6 M GdmCl, and also from 0.5 M GdmCl through the transition region. There are, on average, slightly higher values of m_{kf} for the range 0 to 0.6 M than calculated from the linear regions of the plots of $\ln k_f$ *versus* [GdmCl] for the higher concentrations, indicative of slight curvature in the plots of $\ln k_f$ *versus* [GdmCl]. Slight curvature in plots of $\Delta\Delta G_{U\text{-F}}$ *versus* [GdmCl] for protein denaturation has been noted by Santoro & Bolen (1992) at low [GdmCl] (<1.5 M), attributable to changes in ionic strength, but with excellent linearity at higher concentrations. In accord with this, we find that the values of m_{kf} at the higher concentrations fit the theoretical equations better. For example, a plot of $[D]_{50\%}$ calculated from equation (11) *versus* $[D]_{50\%}$ measured from equilibrium unfolding (Figure 1) has intercept −0.06(±0.04) M and slope 1.05(±0.01), using m_{kf} determined for [GdmCl] > 0.5 M, whereas just using data for m_{kf} in the range 0 to 0.6 M gives values of 0.27(±0.05) M and 1.01(±0.01), respectively. The values of $m_{U\text{-F}}$ for equilibrium denaturation (equation (1)) are related to those from kinetics by: $m_{U\text{-F}} = RT(m_{ku} - m_{kf})$. Substituting values of $m_{ku} - m_{kf}$ determined above 0.5 M GdmCl gives a mean value of 1.86(±0.01) kcal mol^{-1} M^{-1}, compared with the measured value of 1.90(±0.03). Using m_{kf} in the range 0 to 0.6 M GdmCl gives: $m_{U\text{-F(calc)}}$ = 2.05(±0.02).

There is a virtually perfect fit to a two-state mechanism for [GdmCl] > 0.5 M, and an acceptable fit over the whole range. The slight positive

Table 3. Rates and thermodynamic data from refolding kinetics of CI2 mutants

Mutant	ln $k_f^{H_2O}$	m_{ku}[a] (M^{-1})	$\Delta\Delta G_{\ddagger-U}^{H_2O}$[b] (kcal mol^{-1})
Wild-type	4.03 ± 0.04	−1.82 ± 0.12	0
KA2	4.21 ± 0.02	−2.24 ± 0.05	−0.11 ± 0.03
KM2	4.01 ± 0.01	−1.90 ± 0.05	0.02 ± 0.03
TA3	3.84 ± 0.01	−1.86 ± 0.03	0.11 ± 0.03
TV3	3.75 ± 0.02	−1.69 ± 0.06	0.17 ± 0.03
TG3	3.94 ± 0.02	−2.16 ± 0.06	0.06 ± 0.03
PA6	3.86 ± 0.01	−2.06 ± 0.04	0.10 ± 0.03
PA6/AG16	1.95 ± 0.04	−1.63 ± 0.11	1.23 ± 0.03
EA7	3.72 ± 0.01	−1.95 ± 0.02	0.19 ± 0.03
LA8[b]	3.36 ± 0.04	−2.02 ± 0.10	0.40 ± 0.03
KA11	3.59 ± 0.01	−1.62 ± 0.02	0.26 ± 0.03
SG12	3.65 ± 0.04	−2.33 ± 0.12	0.23 ± 0.04
SA12	3.39 ± 0.05	−2.05 ± 0.14	0.38 ± 0.04
EQ14	3.43 ± 0.03	−2.06 ± 0.08	0.36 ± 0.03
ED14	3.86 ± 0.04	−2.42 ± 0.11	0.10 ± 0.04
EN14	3.14 ± 0.02	−2.28 ± 0.05	0.53 ± 0.03
EQ15	3.62 ± 0.09	−2.23 ± 0.24	0.25 ± 0.06
ED15	3.76 ± 0.03	−2.17 ± 0.09	0.16 ± 0.03
EN15	3.07 ± 0.07	−2.13 ± 0.19	0.57 ± 0.05
EA14/EA15	3.10 ± 0.04	−1.69 ± 0.11	0.55 ± 0.04
SG12/EA14/EA15	2.73 ± 0.02	−1.84 ± 0.07	0.77 ± 0.03
SA12/EA14/EA15	2.90 ± 0.05	−2.14 ± 0.14	0.67 ± 0.04
AG16	2.09 ± 0.02	−2.46 ± 0.06	1.15 ± 0.03
KA17	3.81 ± 0.04	−2.20 ± 0.11	0.14 ± 0.04
KG17	2.56 ± 0.03	−2.04 ± 0.08	0.87 ± 0.03
KA18	3.90 ± 0.01	−1.77 ± 0.03	0.08 ± 0.03
KG18	2.88 ± 0.01	−1.86 ± 0.27	0.68 ± 0.03
VA19[c]	4.25 ± 0.02	−1.88 ± 0.04	−0.13 ± 0.03
IV20[c]	3.16 ± 0.03	−2.25 ± 0.07	0.52 ± 0.03
LA21	3.48 ± 0.02	−2.20 ± 0.06	0.33 ± 0.03
LG21	3.23 ± 0.03	−2.32 ± 0.08	0.48 ± 0.03
QA22	4.24 ± 0.02	−2.15 ± 0.05	−0.12 ± 0.03
QG22	3.92 ± 0.02	−2.27 ± 0.05	0.07 ± 0.03
DA23	4.43 ± 0.02	−2.16 ± 0.06	−0.23 ± 0.03
KA24	4.42 ± 0.05	2.31 ± 0.13	−0.23 ± 0.04
KG24	3.52 ± 0.05	−2.47 ± 0.31	0.31 ± 0.04
PA25	3.44 ± 0.03	−2.50 ± 0.08	0.35 ± 0.03
EA26	3.81 ± 0.04	−2.20 ± 0.11	0.14 ± 0.04
IV29[c]	3.72 ± 0.03	−2.32 ± 0.07	0.19 ± 0.03
IA29[c]	2.37 ± 0.05	−2.73 ± 0.13	0.98 ± 0.04
IA29/IV57[c]	2.03 ± 0.05	−2.85 ± 0.13	1.18 ± 0.04
IV30	3.92 ± 0.02	−2.23 ± 0.06	0.07 ± 0.03
IA30	2.94 ± 0.05	−2.90 ± 0.14	0.65 ± 0.04
IC30	2.48 ± 0.06	−2.88 + 0.16	0.92 ± 0.04
IT30	3.25 ± 0.05	−2.42 ± 0.18	0.47 ± 0.04
LA32	3.29 ± 0.11	−3.01 ± 0.37	0.44 ± 0.07
LI32	4.17 ± 0.02	−2.14 ± 0.06	−0.08 ± 0.03
LV32	4.07 ± 0.02	−2.15 ± 0.04	−0.02 ± 0.03
LV32/FL50	3.18 ± 0.04	−2.36 ± 0.10	0.51 ± 0.03
LV32/FA50	2.48 ± 0.03	−2.55 ± 0.10	0.92 ± 0.03
LA32/FL50	2.52 ± 0.04	−2.40 ± 0.12	0.90 ± 0.04
LA32/FA50	1.84 ± 0.11	−2.84 ± 0.28	1.30 ± 0.07
LA32/VA38	3.13 ± 0.10	−2.82 ± 0.28	0.54 ± 0.06
LV32/VA38	3.82 ± 0.04	−2.43 ± 0.10	0.13 ± 0.03
LA32/VA38/FL50	2.47 ± 0.07	−2.40 ± 0.19	0.92 ± 0.05
LV32/VA38/FL50	3.01 ± 0.07	−2.25 ± 0.18	0.60 ± 0.05
VT34	3.63 ± 0.02	−2.21 ± 0.06	0.24 ± 0.03
VA34	4.05 ± 0.03	−2.48 ± 0.10	−0.01 ± 0.03
VG34	3.36 ± 0.04	−2.48 ± 0.10	0.40 ± 0.03
TV36	3.79 ± 0.02	−2.29 ± 0.05	0.14 ± 0.03
IA37Δ38	3.69 ± 0.04	−2.65 ± 0.11	0.20 ± 0.03
VA38	3.73 ± 0.03	−2.40 ± 0.09	0.18 ± 0.03
VA38/FL50	2.90 ± 0.03	−2.58 ± 0.07	0.67 ± 0.03
VA38/FA50	2.02 ± 0.05	−2.76 ± 0.13	1.19 ± 0.04
TA39[d]	3.97 ± 0.03	−2.15 ± 0.08	0.04 ± 0.03
RA43	3.95 ± 0.03	−2.12 ± 0.08	0.05 ± 0.03
RA43/DA45	3.90 ± 0.05	−2.55 ± 0.14	0.08 ± 0.04
DA45	3.51 ± 0.02	−2.28 ± 0.06	0.31 ± 0.03
VA47[c]	2.31 ± 0.05	−2.56 ± 0.13	1.02 ± 0.04

continued

Table 3. *Continued*

Mutant	$\ln k_f^{H_2O}$	m_{ku}[a] (M^{-1})	$\Delta\Delta G_{\ddagger\text{-U}}^{H_2O}$[b] (kcal mol^{-1})
LA49[c]	0.61 ± 0.05	−2.81 ± 0.17	2.03 ± 0.04
FL50	3.03 ± 0.04	−2.70 ± 0.11	0.59 ± 0.04
FV50	3.03 ± 0.03	−2.31 ± 0.08	0.60 ± 0.03
FA50	2.08 ± 0.11	−2.73 ± 0.28	1.16 ± 0.07
VA51[c]	3.20 ± 0.04	−2.54 ± 0.11	0.49 ± 0.03
DA52	3.34 ± 0.38	−2.75 ± 0.11	0.41 ± 0.03
ND56	3.65 ± 0.04	−2.56 ± 0.10	0.23 ± 0.03
NA56	3.91 ± 0.02	−2.39 ± 0.06	0.07 ± 0.03
IV57[c]	3.85 ± 0.01	−2.01 ± 0.04	0.11 ± 0.03
IA57[c]	3.43 ± 0.02	−2.45 ± 0.07	0.36 ± 0.03
AG58	3.68 ± 0.02	−1.88 ± 0.06	0.21 ± 0.03
VT60	3.71 ± 0.03	−1.89 ± 0.09	0.19 ± 0.03
VA60	4.11 ± 0.04	−2.31 ± 0.11	−0.04 ± 0.04
VG60	3.82 ± 0.03	−2.10 ± 0.07	0.13 ± 0.03
VM60/ML39	3.94 ± 0.05	−2.07 ± 0.13	0.05 ± 0.04
VT63	3.89 ± 0.01	−1.99 ± 0.01	0.09 ± 0.03
VA63	3.96 ± 0.02	−2.14 ± 0.05	0.04 ± 0.03
VG63	3.86 ± 0.04	−2.25 ± 0.11	0.10 ± 0.03

[a] Measured in the range 0 to 0.6 M GdmCl.
[b] $\Delta\Delta G_{\ddagger\text{-U}}^{H_2O} = -RT \ln(k_F/k_F')$, where k_f and k_f' are the rate constants of folding in water for wild-type and mutant, respectively, measured from pH-jump experiments (in s^{-1}).
[c] From Jackson *et al*. (1993b).
[d] From Jackson & Fersht (1994).

deviation in m_{kf} at lower [GdmCl] is not due to the two-state mechanism becoming three state, since the presence of an intermediate leads to a lowering of m_{kf}, as illustrated for barnase by Matouschek *et al*. (1990). Instead, the deviation is likely to result from a genuine non-linearity of response of free energy to [GdmCl], as demonstrated by Santoro & Bolen (1992). The value of m_{kf} is not used in any of the calculations used for analysing the transition state in this study; the only experimental quantities used are: $k_f^{H_2O}$, which is measured directly in water; k_u^{4M}, which is obtained by a short extrapolation; and $\Delta\Delta G_{U\text{-}F}^{[D]50\%}$, which has been checked by differential scanning calorimetry. In other studies, where we wish to see how m_{kf} changes on mutation, we prefer the data set determined for 0 to 0.6 M GdmCl, since it measured the more precisely, and systematic deviations tend to cancel out during comparisons.

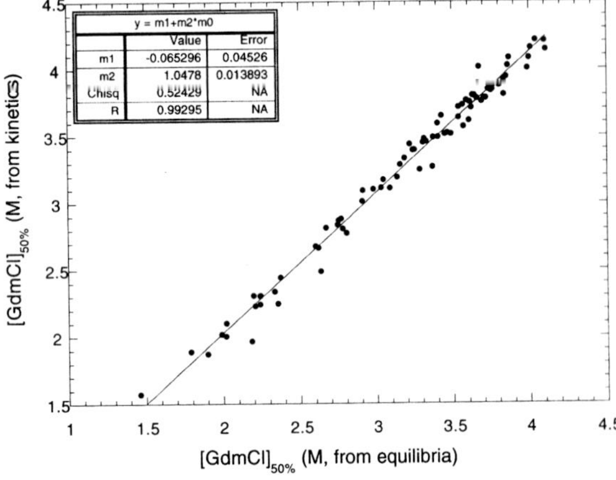

Figure 1. Plot of the concentration of GdmCl required for 50% denaturation of CI2 calculated from the ratio of kinetic rate constants (equation (11)) against that measured by equilibrium denaturation, showing the excellence of the fit to two-state kinetics.

The protein engineering method

The procedure we use for studying folding pathways by protein engineering has been discussed extensively elsewhere (Fersht *et al*., 1992; Fersht, 1993, 1995a). Briefly, a suitable side-chain interaction is truncated by mutagenesis and the resulting mutant characterized by equilibrium denaturation and kinetic unfolding and refolding experiments to determine the extent of this interaction at different stages on the folding pathway. The interaction is determined by the following ratio:

$$\Phi_F = \Delta\Delta G_{\ddagger\text{-U}} / \Delta\Delta G_{F\text{-U}} \quad (12)$$

where $\Delta\Delta G_{\ddagger\text{-U}}$ is the change in free energy of folding of the transition state on mutation, and $\Delta\Delta G_{F\text{-U}}$ is the change in equilibrium free energy of folding on mutation. A Φ_F of 1 shows that the transition state is disrupted by mutation by the same energy as is the fully folded protein and so indicates complete formation of native structure in the transition state, while $\Phi_F = 0$ shows that the transition state is as insensitive to mutation as is the fully denatured state and so is indicative of complete lack of native structure in the transition state. Fractional Φ-values can be difficult to interpret for several reasons. They can arise if a protein folds by parallel pathways, in which parts of the protein are native-like in the transition-state of one pathway (i.e. $\Phi_F = 1$) but unfolded in the transition-state of another pathway (i.e. $\Phi_F = 0$). However, CI2 was shown recently to fold along a single pathway involving one or at the most a closely related set of transition states (Fersht *et al.*, 1994).

Analysis of fractional Φ-values can also be complicated by factors such as access of water to the site of mutation. This influences the observed Φ-value by introducing solvation energy terms during the folding reaction. In the case where water does not enter the site of mutation, there may be a more linear relationship between the Φ-value and the extent of formation of non-polar interactions (Fersht *et al.*, 1992). Double mutant cycles allow us to focus on the specific interaction between two residues in which the effect of surrounding residues and solvent interactions tend to cancel out (Horovitz & Fersht, 1990; Fersht *et al.*, 1992). Interactions of individual moieties within a side-chain can be probed during folding by making a series of mutations at a single site (Fersht *et al.*, 1992; Serrano *et al.*, 1992a). For example, the series of mutations Val → Ala → Gly allows us to attribute structure formation to each of the C^{β}, $C^{\gamma1}$ and $C^{\gamma2}$ methyl(ene) groups.

Choice of mutation

Our approach for designing mutations is as follows. The pseudo-wild-type crystal structure of CI2 (Harpaz *et al.*, 1994) is examined for side-chain atoms that probe a particular structural interaction. The ideal mutation, which has been termed "non-disruptive" (Fersht *et al.*, 1992), deletes only a small part of the side-chain, removing defined interactions without introducing new ones, altering the stereochemistry or perturbing the structure. Thus it acts as a probe of the folding pathway. Our favourite mutations are: Ile → Val → Ala →Gly; Thr → Ser; Ser → Ala; Tyr → Phe. When in doubt, we mutate larger side-chains to Ala.

Since Φ-value analysis monitors formation of interactions of side-chains, it is not a direct measure of the extent of secondary structure formation, but Φ-values can probe the consequences of formation of secondary structure. We have made, for example, a series of mutations at sites in the α-helix and in the β-sheet where the side-chain makes interactions almost exclusively within the secondary structural element, and thus the Φ-values which we obtain at these sites are an indirect probe of the formation of secondary structure interactions. In addition, we have made mutations at side-chains whose interactions are predominantly tertiary, for example within the core and the minicore. The results should, therefore, provide us with a very comprehensive picture of the transition state in the folding pathway of CI2.

Structure of CI2

The secondary structure of CI2, defined from an NMR analysis of the solution structure (Ludvigsen *et al.*, 1991) and in agreement with the crystal structure of the mutant EA14EA15 (which we call the pseudo-wild-type, Harpaz *et al.*, 1994) is as follows (Figure 2): residues 3 to 5, β-strand 1; 5 to 8, type III reverse turn; 8 to 11, type II reverse turn; residues 10 to 11, β-strand 2; residues 12 to 24, α-helix; residues 25 to 28, type I reverse turn; 28 to 34 β-strand 3; 35 to 44 reactive site loop (extended structure); residues 45 to 52, β-strand 4; residues 52 to 54, turn; residues 55 to 58, β-strand 5 and residues 60 to 64, β-strand 6. The numbering is based on the truncated version of CI2 lacking the first 19 residues found in the original CI2. These residues are devoid of fixed structure and their removal does not change the folding properties of CI2.

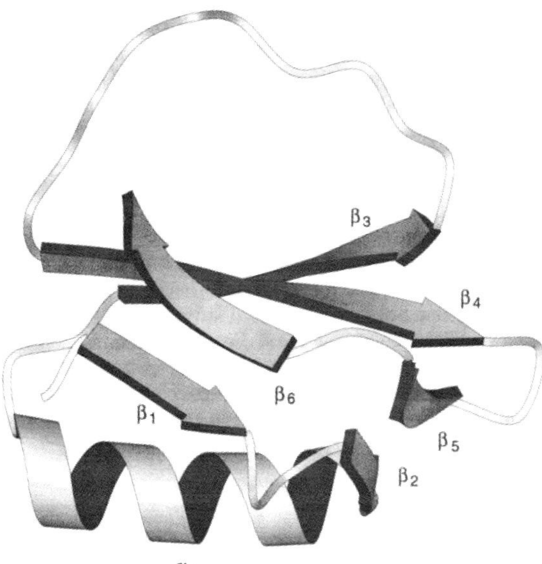

Figure 2. Schematic representation of the structure of CI2. The diagram was produced using the program MolScript (Kraulis, 1991).

The β-sheet packs against the α-helix to form the hydrophobic core. The reactive-site loop projects out from the other side of the β-sheet, in an extended conformation. In contrast to the hydrophobic character of the interface of the β-sheet with the α-helix, the other side of the sheet forms an extensive network of hydrogen bonds and electrostatic interactions with the reactive-site loop.

α-Helix

The α-helix runs from 13 to 23 according to Φ and Ψ angles (Li & Daggett, 1994), but we use the definition of caps by Richardson & Richardson (1988) to designate Ser12 as the N-cap of the helix and Lys24 as its C-cap. The α-helix consists of three turns and is somewhat irregular in its geometry. From the crystal structure (McPhalen & James, 1987) and from the recent higher-resolution crystal structure of the pseudo-wild-type (Harpaz *et al.*, 1994), it appears that some of the main-chain NH–CO hydrogen bonds are longer than is usual for an α-helix, while others are not formed at all.

These are NH18–CO14, NH19–CO15, NH22–CO18 and NH23–CO19. The N-cap of the helix at position 12 is occupied by a Ser residue. The OH group of the side-chain of Ser12 makes a hydrogen bond with the main-chain NH group of residue 15 (N-cap + 3) and with a water molecule. This end of the helix is very solvent-exposed, with many strong hydrogen bonds between main-chain CO and NH groups and water molecules. On the buried side of the α-helix, there are four hydrophobic residues (Val13, Ala16, Val19, Ile20) which pack against the β-sheet to form the hydrophobic core. On the solvent-exposed face are residues Lys17, Lys18, Leu21 and Gln22. The C-cap at position 24 is occupied by a Lys residue, followed by an α-helix stop at Pro25. This end of the helix is much less solvated than the other, with the C-cap almost completely buried.

β-Sheet

The six-stranded β-sheet of CI2 is classified as a pseudo-β-sheet because the components of the main pair of strands (strands 3 and 4) are joined by a β-bridge (the two β-sheet hydrogen bonds Ala58–Phe50 and Gly64–Arg46) to the other strands. The two parallel strands 3 and 4 (28 to 34 and 45 to 51) form the main body of the β-sheet, displaying a regular β-strand hydrogen bond pattern. Shorter segments of anti-parallel structures are formed between 55-57 in strand 5 and 11-13 in strand 2, 56-58 in strand 5 and 50-52 in strand 4, 62-64 in strand 6 and 46-48 in strand 4 as well as between 61-63 in strand 6 and 3-5 in strand 1. The β-sheet shows deviation from canonical structure in that an array of ordered water molecules are observed in the crystal structure which mediate hydrogen bonding between strand 4 and strand 6. These bridging water molecules are not visible in the solution structure determined by NMR (Ludvigsen *et al.*, 1991), where the backbone of Arg48 forms a hydrogen bond with that of Gly64 directly, rather than by bridging water molecules seen in the crystal structure. A recent molecular dynamics study (Nanzer *et al.*, 1994) demonstrates the dynamic nature of the β sheet structure of CI2. The hydrogen bond between the N-H of residue 48 and the C = O of residue 64 is present only some 20% of the time. The hydrogen bonding between two strands comprising residues 28 to 30 (strand 3) and 45 to 47 (strand 4) as well as residues 3 to 5 (strand 1) and 61 to 63 (strand 6) was also shown to be mobile: part of the time they form a four-stranded β-sheet in the core, and part of the time the strands move apart to form hydrogen bonds with nearby water molecules (Nanzer *et al.*, 1994).

Hydrophobic cores

The hydrophobic core is formed by packing of one face of the α-helix onto the β-sheet, and consists of 12 hydrophobic residues, namely Trp5, Leu8, Ala16, Val19, Ile20, Ala27, Ile29, Val47, Leu49, Val51, Ile57 and Pro61. The core buries almost 1900 Å^2 of surface, and exposes only 133 Å^2 to solvent. In addition, the side-chains of residues Leu32, Val38 and Phe50 form a hydrophobic pocket near one end of the reactive loop, on the other side of the β-sheet relative to the hydrophobic core. This cluster is more solvent-accessible than the main core, burying 425 Å^2 and exposing to solvent 133 Å^2 of hydrophobic surface area. We call this cluster the minicore. These three residues are highly conserved among different members of the inhibitor family to which CI2 belongs.

Turns and ends

Hydrogen bonds are prominent in the turns and in the terminal residues of CI2. While the free amino group of Met1, an artifact of the truncated construct which we are studying, does not link to any group, the N^{ζ} atom of Lys2 is bonded with $O^{\epsilon 2}$ of Glu7, and this interaction (or the intra-residual hydrogen bond between $O^{\epsilon 1}$ and N of Glu7) may contribute to the stability of the turn after β-strand 1. The pK_a of Glu7 is lowered from 4.3 to 3.1 (B. Davis & A.R.F., unpublished). In the reactive site loop, the backbone of Val34 and Gly35 forms a small bulge out of the loop, and this appears to be stabilized by a hydrogen bond between $O^{\gamma 1}$ of Thr36 and the carbonyl oxygen of Pro33. At the other end of the loop, Arg43 and Asp45 form two hydrogen bonds between their respective side-chains. The charged carboxyl group of the C-terminal Gly64 forms a strong hydrogen bond with N^{H2} of Arg46. In addition, N^{ϵ} of Arg46 also hydrogen bonds with the O of Gly64. Finally, there is a very extensive hydrogen-bonding network in the β-hairpin loop joining β-strands 4 and 5, where the side-chain $O^{\delta 1}$ of Asp52 forms simultaneous hydrogen bonds with the amides of Leu54 and Asn56 (as well as Asp55 in the wild-type crystal structure). This network decreases the pK_a of the side-chain from 3.9 to 2.7 (B. Davis & A.R.F., unpublished).

Structure of the transition state

Relationship of Φ_F and Φ_U

The two-state nature of the folding kinetics of CI2 means that the strength of interactions in the transition state can be studied by kinetics directly from both directions of folding (Φ_F from refolding kinetics in water) and unfolding (Φ_U by extrapolating the unfolding kinetics to zero molar denaturant (water), where $\Phi_U = \Delta\Delta G_{\ddagger\text{-}F}/\Delta\Delta G_{U\text{-}F}$), and thus provides us with independent means of verifying results and a means of checking some of the assumptions made in the analysis. For a two-state reaction, $\Phi_F + \Phi_U = 1$, so that $\Phi_F = 1 - \Phi_U$.

The Φ-values obtained from these two approaches, Φ_F and Φ_U, respectively, are listed in Table 4. $\Phi_F^{H_2O}$ is measured accurately with no extrapolation, and so is the most reliable quantity. $1 - \Phi_U^{H_2O}$ is inaccurate due to a long extrapolation

Table 4. Values of Φ_F and Φ_U calculated from refolding and unfolding kinetics, respectively

Mutant	$\Phi_F^{H_2O}$ [a]	$1-\Phi_U^{H_2O}$ [b] ($=\Phi_F^{H_2O}$)	$1-\Phi_U^{4M}$ [c] ($=\Phi_F^{4M}$)
Core			
LA8[d]	0.15 ± 0.01	−0.02 ± 0.04	0.24 ± 0.05
AG16	1.06 ± 0.05	1.24 ± 0.08	1.20 ± 0.03
VA19[d]	−0.26 ± 0.07	−0.51 ± 0.26	−0.71 ± 0.14
IV20[d]	0.40 ± 0.03	0.54 ± 0.06	0.56 ± 0.04
IV29[d]	0.17 ± 0.03	0.44 ± 0.06	0.39 ± 0.04
IA29[d]	0.25 ± 0.01	0.28 ± 0.03	0.43 ± 0.05
IA29/IV57[d]	0.29 ± 0.01	0.32 ± 0.03	0.39 ± 0.03
VA47[d]	0.21 ± 0.01	0.28 ± 0.03	0.14 ± 0.11
LA49[d]	0.53 ± 0.02	0.54 ± 0.03	0.63 ± 0.02
VA51[d]	0.25 ± 0.02	0.45 ± 0.03	0.54 ± 0.04
IV57[e]	—	—	—
IA57[d]	0.08 ± 0.01	0.12 ± 0.03	0.18 ± 0.09
PA61	0.02 ± 0.01	0.11 ± 0.07	−0.07 ± 0.08
Minicore			
LA32	0.19 ± 0.03	0.33 ± 0.06	0.41 ± 0.01
LI32	−0.31 ± 0.14	−0.16 ± 0.36	−0.55 ± 0.09
LV32	−0.04 ± 0.05	0.04 ± 0.15	−0.22 ± 0.07
LV32/FL50	0.21 ± 0.01	0.30 ± 0.03	0.46 ± 0.03
LV32/FA50	0.27 ± 0.01	0.31 ± 0.03	0.41 ± 0.03
LA32/FL50	0.26 ± 0.01	0.31 ± 0.02	0.42 ± 0.03
LA32/FA50	0.27 ± 0.02	0.33 ± 0.03	0.33 ± 0.07
LA32/VA38	0.17 ± 0.02	0.31 ± 0.02	0.37 ± 0.02
LV32/VA38	0.07 ± 0.02	0.17 ± 0.04	0.20 ± 0.03
LA32/VA38/FL50	0.26 ± 0.01	0.32 ± 0.02	0.45 ± 0.02
LV32/VA38/FL50	0.22 ± 0.02	0.30 ± 0.02	0.34 ± 0.02
VA38	0.12 ± 0.02	0.27 ± 0.04	0.46 ± 0.03
VA38/FL50	0.26 ± 0.01	0.37 ± 0.02	0.49 ± 0.02
VA38/FL50	0.26 ± 0.01	0.33 ± 0.02	0.49 ± 0.03
FL50	0.28 ± 0.02	0.47 ± 0.03	0.48 ± 0.03
FV50	0.25 ± 0.01	0.31 ± 0.03	0.48 ± 0.04
FA50	0.30 ± 0.02	0.40 ± 0.02	0.51 ± 0.03
Helix			
SG12	0.29 ± 0.05	0.54 ± 0.18	0.54 ± 0.06
SA12	0.43 ± 0.05	0.66 ± 0.16	0.63 ± 0.06
EQ14	1.23 ± 0.27	0.86 ± 0.29	0.57 ± 0.09
ED14	0.20 ± 0.07	0.69 ± 0.34	0.43 ± 0.11
EN14	0.75 ± 0.06	0.97 ± 0.13	0.68 ± 0.05
EQ15	0.53 ± 0.14	0.58 ± 0.22	0.50 ± 0.07
ED15	0.22 ± 0.05	0.39 ± 0.10	0.23 ± 0.06
EN15	0.53 ± 0.05	0.39 ± 0.07	0.50 ± 0.04
EA14/EA15	0.73 ± 0.08	0.32 ± 0.20	0.32 ± 0.08
SG12/EA14/EA15	0.47 ± 0.03	0.14 ± 0.11	0.36 ± 0.06
SA12/EA14/EA15	0.40 ± 0.02	0.29 ± 0.05	0.46 ± 0.04
KA17	0.28 ± 0.08	−0.48 ± 0.19	−0.42 ± 0.07
KG17	0.38 ± 0.02	0.26 ± 0.01	0.44 ± 0.03
KA18[e]	—	—	—
KG18	0.70 ± 0.06	0.58 ± 0.09	0.48 ± 0.05
LA21	0.25 ± 0.03	0.28 ± 0.06	0.49 ± 0.04
LG21	0.35 ± 0.03	0.21 ± 0.05	0.36 ± 0.03
QA22[e]		—	—
QG22	0.12 ± 0.05	0.25 ± 0.13	0.27 ± 0.06
DA23	−0.25 ± 0.03	0.28 ± 0.14	0.01 ± 0.05
KA24	−0.35 ± 0.07	−0.04 ± 0.13	−0.34 ± 0.07
KG24	0.10 ± 0.01	0.16 ± 0.04	−0.01 ± 0.10
β-Sheet			
TA3	0.13 ± 0.03	0.26 ± 0.10	0.13 ± 0.05
TV3	0.52 ± 0.14	0.17 ± 0.24	−0.05 ± 0.06
TG3	0.05 ± 0.03	0.30 ± 0.10	0.29 ± 0.05
IV30[e]	—	—	—
IA30	0.31 ± 0.02	0.57 ± 0.03	0.66 ± 0.02
IG30	0.26 ± 0.01	0.37 ± 0.03	0.43 ± 0.02
IT30	0.35 ± 0.03	0.63 ± 0.03	0.72 ± 0.02
VT34	0.23 ± 0.03	0.27 ± 0.06	0.44 ± 0.03
VA34	−0.01 ± 0.05	0.23 ± 0.15	0.42 ± 0.09
VG34	0.16 ± 0.01	0.27 ± 0.03	0.48 ± 0.02
AG58	0.11 ± 0.02	0.19 ± 0.04	0.41 ± 0.06
VT60	0.51 ± 0.17	0.03 ± 0.31	0.14 ± 0.08
VA60	−0.03 ± 0.02	0.18 ± 0.05	0.12 ± 0.05

continued

Table 4. *continued*

Mutant	$\Phi_F^{H_2O\,a}$	$1-\Phi_U^{H_2O\,b}$ $(=\Phi_F^{H_2O})$	$1-\Phi_U^{4M\,c}$ $(=\Phi_F^{4M})$
VG60	0.04 ± 0.01	0.06 ± 0.04	0.27 ± 0.03
VM60/ML39	0.05 ± 0.03	0.16 ± 0.05	0.08 ± 0.04
VT63	0.08 ± 0.02	0.13 ± 0.07	0.09 ± 0.07
VA63	0.03 ± 0.02	0.25 ± 0.05	0.13 ± 0.06
VG63	0.03 ± 0.01	0.14 ± 0.02	0.14 ± 0.03
Reactive site loop			
TV36	0.19 ± 0.04	0.06 ± 0.13	0.29 ± 0.06
IA37/Δ38	0.09 ± 0.01	0.22 ± 0.02	0.22 ± 0.02
TA39	0.19 ± 0.07	0.46 ± 0.11	0.33 ± 0.01
EA41	0.32 ± 0.09	0.20 ± 0.15	0.15 ± 0.16
TA39/EA41	0.18 ± 0.06	0.30 ± 0.07	0.27 ± 0.06
TD39/EA41	0.82 ± 0.24	0.80 ± 0.25	0.38 ± 0.17
YG42	0.07 ± 0.01	0.16 ± 0.04	0.08 ± 0.03
RA43	0.09 ± 0.06	0.19 ± 0.17	0.05 ± 0.07
RA43/DA45	0.06 ± 0.03	−0.29 ± 0.14	−0.36 ± 0.11
DA64	0.44 ± 0.03	0.29 ± 0.09	0.30 ± 0.04
Turns			
KA2	−0.19 ± 0.05	−0.30 ± 0.14	−0.39 ± 0.05
KA2/EA7	0.17 ± 0.02	0.10 ± 0.13	0.14 ± 0.05
KM2	0.03 ± 0.04	0.29 ± 0.14	0.11 ± 0.05
PA6	0.07 ± 0.02	0.47 ± 0.19	0.44 ± 0.12
EA7	0.40 ± 0.07	0.44 ± 0.19	0.30 ± 0.07
KA11	−0.49 ± 0.07	0.51 ± 0.13	0.26 ± 0.06
PA25	0.20 ± 0.02	0.12 ± 0.12	0.33 ± 0.04
EA26	0.42 ± 0.12	0.22 ± 0.22	0.18 ± 0.07
DA52	0.12 ± 0.01	0.15 ± 0.03	0.39 ± 0.06
ND56	0.19 ± 0.03	0.45 ± 0.05	0.61 ± 0.02
NA56	0.09 ± 0.04	0.34 ± 0.08	0.30 ± 0.03

Φ_F and Φ_U are related by the expression $\Phi_F + \Phi_U = 1$, so that $\Phi_F = 1 - \Phi_U$. $\Phi_F^{H_2O}$ from the folding experiments is the most accurate, since data measured in H_2O are directly compared. Φ_U^{4M} is less accurate, since unfolding data are extrapolated to 4 M GdmCl. $\Phi_U^{H_2O}$ is the least accurate, since there is a long extrapolation of unfolding data at high [GdmCl] to 0 M. The presence of 4 M GdmCl is expected to perturb the structure of the transition state.

[a] $\Phi_F^{H_2O} = \Delta\Delta G_{\ddagger\text{-}U}^{H_2O} / \Delta\Delta G_{F\text{-}U}^{[D]50\%}$ (where $\Delta\Delta G_{F\text{-}U}^{[D]50\%} = -\Delta\Delta G_{U\text{-}F}^{[D]50\%}$).

[b] $\Phi_U^{H_2O} = \Delta\Delta G_{\ddagger\text{-}F}^{H_2O} / \Delta\Delta G_{U\text{-}F}^{[D]50\%}$.

[c] $\Phi_U^{4M} = \Delta\Delta G_F^{4M} / \Delta\Delta G_{U\text{-}F}^{4M}$.

[d] From Jackson & Fersht (1993b).

[e] Φ-Values were not calculated for these mutants because $\Delta\Delta G_{U\text{-}F}^{[D]50\%} \approx 0$.

from the concentrations of GdmCl over which the rates of unfolding were measured. $1 - \Phi_U^{4M}$ is accurate, being determined only from a small extrapolation from higher values of [GdmCl], but is measured in 4 M denaturant. It is expected that there will be some perturbation of the structure of the transition state in 4 M GdmCl because of the presence of denaturant that stabilizes the denatured state of proteins. In particular, there should be a general effect according to the Hammond postulate (Hammond, 1955), which states that as a state on a reaction pathway becomes destabilized, the transition state for its formation should move closer to it in structure. This postulate holds for protein folding (Matouschek & Fersht, 1993) and so it is likely that the structure of the transition state for protein folding moves closer to that of the folded protein with increasing concentration of denaturant, as has been found for barnase (Matthews & Fersht, 1995). The values of $\Phi_F^{H_2O}$ and $1 - \Phi_U^{H_2O}$ should be identical for CI2 folding according to the two-state model, and there is, indeed, good agreement within experimental error for the majority of mutations, with the best agreement observed for mutations in the hydrophobic core and minicore. In general, the values of $1 - \Phi_U^{4M}$ are somewhat higher than $\Phi_F^{H_2O}$, as expected. The differences are not sufficiently high to change radically the description of energies. Poorer agreement may result where there are smaller equilibrium destabilization energies in certain regions of the protein on mutation and the consequent large errors in the calculation of Φ-values, rather than deviation from the two-state behaviour.

The major assumptions of the protein engineering approach are that (1) the mutation does not alter the folding pathway, (2) the mutation does not significantly change the structure of the folded protein, (3) the mutation does not perturb the structure of the unfolded state, and (4) the target groups do not make new interactions with new partners during the course of the reaction (Fersht *et al.*, 1992). Evidence that these assumptions are valid for CI2 is provided by the observation of good agreement between Φ-values when different mutations are constructed at the same site. Thus, the overall formation of structure in the transition state parallels the formation of interactions between individual side-chains. We now describe the structures of individual regions of CI2 in the transition state, as elucidated from the protein engineering approach.

Hydrophobic core

Some of these results have been described previously (Jackson *et al.*, 1993b). The results for the major hydrophobic core of CI2 can be split into three categories: those mutations that result in a Φ_F of close to zero, those mutations that result in fractional Φ_F, and one mutation that results in a Φ_F of 1. The residues with low Φ_F (between 0 and 0.2) are located on the edge of the core, in turns or at the beginning or end of a β-strand. These include the following mutations: Leu → Ala8, Val → Ala47, Ile → Ala57, Pro → Ala61, in β-strand 6. The residues with fractional values of Φ_F (between 0.3 and 0.55) are all located in the centre of the core, either in the α-helix or in the centre of a β-strand. These include the following mutations: Ile → Val20, Ile → Val29, Ile → Ala29, Leu → Ala49, Val → Ala51, Ile → Ala29/Ile → Val57. Val → Ala19 has a negative Φ_F-value, suggesting that the side-chain of Val19 makes more contacts, some of them non-native, in the transition state than in the native state. There must, therefore, be some structural rearrangement on going from the transition state to the native state. Ala16, in the second turn of the α-helix, is completely buried and probes interactions with the first turn of the α-helix and β-strand 4. The mutation Ala → Gly16 has a Φ_F of 1, a result which has not been found for any other residue of CI2, including the residues with which Ala16 interacts in the native state (Ile57, Leu8 and Leu49). However, while the mutation Ala → Gly16 very specifically probes the interactions involving just the methyl group (C^{β}), the mutations that have been made at the residues with which it interacts (Ile → Ala57, Leu → Ala8 and Leu → Ala49) delete several atoms and, thus, may obscure very specific atom-to-atom interactions during folding.

Minicore

All values of Φ_F for mutations in the minicore are fractional, and most lie between 0.2 and 0.4. Fractional Φ-values are difficult to interpret because they may arise from a variety of effects. However, the rate constants for the 16 minicore mutants and wild-type protein fit beautifully to a Brønsted plot of $\ln k_f$ (or $\ln k_u$) *versus* $\Delta G_{F\text{-}U}/RT$ (Fersht *et al.*, 1994) with a Brønsted β-value ($=RT\partial \ln k_f/\partial\Delta G_{F\text{-}U}$) of 0.3. This uniform response of an element of structure to a large number of mutations is best explained by the structure having 30% of energy of interaction in the transition state (Fersht *et al.*, 1994). This interpretation is supported by analysis using double mutant cycles (Carter *et al.*, 1984; Horovitz & Fersht, 1990, 1992; Fersht *et al.*, 1992). The coupling energy, $\Delta\Delta G_{int}$, between two residues can be calculated from a double-mutant cycle. (The coupling energy between two residues is the energy change on mutating two residues simultaneously minus the sum of the change in energies on mutating them separately, and is the quantitative measure of the co-operativity of the two residues concerned in stabilising a structure). The coupling energy between two residues in the transition state may also be calculated from kinetics and manipulated to give a Φ-value, Φ_F^{int}, (Table 7, Horovitz *et al.*, 1991; Fersht *et al.*, 1992; Horovitz & Fersht, 1992). Values of Φ_F^{int} also cluster around 0.3 showing that the co-operativity of formation over this region is about 30% of the energetics in the fully folded state.

The families of mutants constructed at single sites allow us to perform a fine-structure analysis, in which Φ_F for specific parts of a single side-chain in

Table 5. Coupling energies for double mutants in CI2

Triple mutant	$\Delta\Delta G_{int}^{U\text{-}F}$ ([D]50%) (kcal mol^{-1})	$\Delta\Delta G_{int}^{\ddagger\text{-}U}$ (4 M GdmCl) (kcal mol^{-1})	$\Delta\Delta G_{int}^{\ddagger\text{-}U}$ (H_2O) (kcal mol^{-1})	$\Phi_{int}^{F(H_2O)a}$	$\Phi_{int}^{F(4M)a}$
KA2/EA7	−0.08 ± 0.07	0.16 ± 0.05	−0.02 ± 0.05	b	b
KA2/DA23	0.35 ± 0.08	0.23 ± 0.05	−0.06 ± 0.05	−0.17 ± 0.15	0.35 ± 0.11
RA43/DA45	0.15 ± 0.11	−0.28 ± 0.06	0.28 ± 0.04	b	b
LV32/FL50	0.19 ± 0.10	0.24 ± 0.02	0.23 ± 0.07	b	b
LV32/FA50	0.93 ± 0.13	0.26 ± 0.03	0.30 ± 0.07	0.32 ± 0.08	0.28 ± 0.04
LA32/FL50	1.06 ± 0.11	0.32 ± 0.04	0.40 ± 0.07	0.38 ± 0.07	0.40 ± 0.05
LA32/FA50	1.42 ± 0.21	0.43 ± 0.04	0.55 ± 0.07	0.39 ± 0.07	0.30 ± 0.05
VA38/FL50	1.00 ± 0.10	0.39 ± 0.02	0.22 ± 0.07	0.22 ± 0.07	0.39 ± 0.04
VA38/FA50	0.94 ± 0.13	0.45 ± 0.02	0.32 ± 0.05	0.34 ± 0.08	0.48 ± 0.07
LA32/VA38	0.68 ± 0.10	0.30 ± 0.04	0.31 ± 0.06	0.45 ± 0.11	0.44 ± 0.08
LV32/VA38	0.13 ± 0.09	0.12 ± 0.02	0.19 ± 0.06	b	b

Derived from double mutant cycles. The cycle indicated by KA2/EA7, for example, consists of wild-type, the single mutants Lys → Ala2 and Glu → Ala7 and the double mutant Lys → Ala2/Glu → Ala7. The coupling energy, $\Delta\Delta G_{int}^{U\text{-}F}$, is calculated from: $\Delta\Delta G_{int}^{U\text{-}F} = \Delta\Delta G^{U\text{-}F}_{E-XY\to E-Y} + \Delta\Delta G^{U\text{-}F}_{E-XY\to E-X} - \Delta\Delta G^{U\text{-}F}_{E-XY\to E}$, where the subscript E − XY → E − Y indicates the mutant Lys → Ala21, E − XY → E − X indicates the mutant Glu → Ala26, etc. (Fersht *et al.*, 1992). $\Delta\Delta G_{int}^{\ddagger\text{-}F}$ (H_2O) and $\Delta\Delta G_{int}^{\ddagger\text{-}F}$ (4 M GdmCl) are calculated from the values of $\Delta\Delta G^{\ddagger\text{-}U}$ in a similar manner (Fersht *et al.*, 1992). $\Delta\Delta G_{int}^{\ddagger\text{-}U}$ at 4 M GdnHCl is calculated from: $\Delta\Delta G_{int}^{\ddagger\text{-}U} = \Delta\Delta G_{int}^{U\text{-}F} - \Delta\Delta_{int}^{\ddagger\text{-}F}$.

[a] $\Phi_{int}^{F} = \Delta\Delta G_{int}^{\ddagger\text{-}U}/\Delta\Delta G_{int}^{U\text{-}F}$.

[b] Φ-Values are not calculated for these mutants, because $\Delta\Delta G_{U\text{-}F}^{[D]50\%} \approx 0$.

Table 6. Fine-structure analysis of mutants in CI2

Mutation	$\Delta\Delta G_{U-F}^{H_2O}$ (kcal mol^{-1})	$\Delta\Delta G_{\ddagger-F}^{4M}$ (kcal mol^{-1})	$\Delta\Delta G_{\ddagger-U}^{H_2O}$ (kcal mol^{-1})	$1 - \Phi_U^{4M}$	$\Phi_F^{H_2O}$
VG3	0.83 ± 0.09	0.53 ± 0.04	−0.11 ± 0.04	0.36 ± 0.08	−0.14 ± 0.05
VA3	0.52 ± 0.09	0.40 ± 0.02	−0.06 ± 0.04	0.23 ± 0.13	−0.11 ± 0.08
VA29	2.80 ± 0.10	1.68 ± 0.03	0.80 ± 0.05	0.40 ± 0.02	0.29 ± 0.02
VA30	2.20 ± 0.10	0.68 ± 0.04	0.58 ± 0.05	0.69 ± 0.02	0.27 ± 0.03
AG30	1.40 ± 0.09	1.38 ± 0.03	0.27 ± 0.06	0.02 ± 0.07	0.19 ± 0.04
VT30	1.42 ± 0.09	0.26 ± 0.04	0.40 ± 0.05	0.82 ± 0.03	0.28 ± 0.04
IV32	0.25 ± 0.09	0.21 ± 0.02	0.06 ± 0.04	0.16 ± 0.32	0.23 ± 0.18
VA32	1.87 ± 0.08	0.96 ± 0.04	0.46 ± 0.08	0.49 ± 0.03	0.25 ± 0.04
VA3/FL50	1.00 ± 0.09	0.37 ± 0.02	0.39 ± 0.05	0.62 ± 0.04	0.39 ± 0.06
VA32/VA38	1.31 ± 0.08	0.58 ± 0.02	0.41 ± 0.07	0.56 ± 0.03	0.31 ± 0.06
VA32/VA38/FL50	0.76 ± 0.09	0.22 ± 0.02	0.32 ± 0.07	0.71 ± 0.04	0.42 ± 0.10
TA34	−0.39 ± 0.12	−0.17 ± 0.04	−0.25 ± 0.04	0.57 ± 0.17	0.63 ± 0.23
LV50	0.28 ± 0.09	0.31 ± 0.02	0.00 ± 0.05	−0.12 ± 0.38	0.01 ± 0.17
VA50	1.45 ± 0.11	0.69 ± 0.02	0.56 ± 0.07	0.52 ± 0.04	0.39 ± 0.06
LA50	1.73 ± 0.10	1.01 ± 0.02	0.57 ± 0.08	0.42 ± 0.04	0.33 ± 0.05
LA50/VA38	1.79 ± 0.11	1.13 ± 0.02	0.41 ± 0.07	0.37 ± 0.04	0.23 ± 0.04
LA50/LA32	1.37 ± 0.20	0.66 ± 0.02	0.40 ± 0.08	0.52 ± 0.07	0.29 ± 0.07
LA50/LV32	0.99 ± 0.09	0.40 ± 0.02	0.42 ± 0.05	0.60 ± 0.04	0.42 ± 0.06
TA60	1.13 ± 0.13	1.05 ± 0.02	−0.24 ± 0.05	0.08 ± 0.10	−0.21 ± 0.05
TA63	0.30 ± 0.10	0.31 ± 0.02	−0.05 ± 0.04	−0.02 ± 0.34	−0.15 ± 0.14

Comparison of different mutations at the same positions in order to analyse the contributions of individual parts of each side-chain.

the transition state can be isolated (Fersht *et al.*, 1992; Serrano *et al.*, 1992a) (Table 5). The Φ_F for composite mutation Val → Ala32 is higher than that for the composite mutation Ile → Val32, arguing that the C^δ methyl group is less structured (i.e. interacting with other atoms to a lesser degree) in the transition state than the two C^γ methyl groups. We cannot correlate this with the structure of the native protein, since the crystal structure has a Leu and not an Ile at this position. For the mutation Val → Ala32, the interaction of $C^{\gamma 1}$ and $C^{\gamma 2}$ with other residues in the transition state, as measured by $\Phi_F H_2O$, is unchanged, within error, by the presence of other mutations such as at positions 38 and 50. In the case of the mutation Leu → Ala50, Φ-values vary between 0.23 and 0.42, depending on the presence of mutations at positions 32 and 38. Thus, these mutations subtly affect the structure formation at position 50 in the transition state.

The Φ-values for mutations in the minicore are not significantly larger or smaller than those at other sites in the protein. The minicore is not, therefore, a nucleation centre for folding *via* hydrophobic clustering, nor do tertiary interactions appear to form later than secondary interactions; tertiary interactions form in parallel with secondary structure interactions. The minicore interactions in the transition state are consistent with the overall structure of the transition state of folding of CI2 as an expanded loose version of the native state.

α-Helix

Folding of the α-helix of CI2 was probed with mutations at 12 of the 13 sites. Three of the α-helix residues, Ala16, Val19 and Ile20, are involved in the packing of the α-helix against the β-sheet to form the major hydrophobic core, and are described in detail under the hydrophobic core. The side-chains of the other sites mutated in the α-helix are all solvent-exposed, and make interactions almost exclusively with other helix residues only. They are all charged groups, with the exception of Leu21. Mutations were made to Ala and to Gly, at Ser12 (the N cap), Lys17, Lys18, Leu21, Gln22 and the C-cap, Lys24. All these mutations result in fractional Φ_F-values. At positions 14 and 15, the Glu side-chain has been mutated to Gln and to Asp, both of which are conservative mutations. There is, however, considerable inconsistency in the values of Φ_F for the different mutations at each of these two positions, which may indicate that these mutations cause some structural reorganisation in the folded protein, or that the folding/unfolding pathway is altered. However, it is more likely that the Φ_F-values obtained for the mutations to Gln and to Asp are not very reliable because of the relatively small values of $\Delta\Delta G_{U-F}$ ($\leqslant$0.5 kcal/mol). Mutation to Asn at these positions had a larger effect on the equilibrium free energy of unfolding, and the Φ_F-values of 0.72 (for Glu → Asn14) and 0.52 (for Glu → Asn15) are, therefore, a more reliable measure of the degree of structure formation at these sites.

A generally benign probe for solvent-exposed positions in α-helices is the comparison of Ala and Gly ("Ala → Gly scanning", Matthews & Fersht, 1995). The difference in stability of Ala *versus* Gly appears to depend solely on the solvent-accessible surface area (Serrano *et al.*, 1992b,c). Application to the helix here gives data in good agreement with the other probes (Table 6). The structure in the transition state becomes progressively weaker towards the C-cap, which itself is almost completely unstructured in the transition state. We see a gradation, though somewhat irregular, of Φ_F along the helix. It is clear that the N-cap box (residues 12, 14 and 15) is one of the most ordered regions of the protein in the transition state for folding, with Φ_F-values greater than 0.5.

β-Sheet

The structure of the β-sheet in the transition state of refolding for CI2 has been probed at six different sites, namely Thr3 (β-strand 1), Ile30 (β-strand 3), Val34 (β-strand 3), Ala58 (β-strand 5), Val60 (β-strand 6) and Val63 (β-strand 6). These residues were chosen because their side-chains interact primarily with other β-sheet residues in the native state. Since the mutation Ile → Val30 has only a very small effect on the equilibrium free energy of unfolding (it stabilises the protein by 0.08 kcal/mol), it was not subjected to Φ-value analysis.

Mutations in β-strands 1, 5 and 6 (residues 3, 58, 60 and 63) show very low Φ_F-values, indicating that these strands are not structured in the transition state. Indeed, these strands comprise the only part of CI2 where there is a very clear absence of native structure in the transition state. These residues constitute the N and C termini of the protein, and in the unfolded state they are likely to be very far apart. It is not surprising, therefore, that the coming together of the two separate ends of the protein constitutes a very late stage of the folding process. Further, the two C-terminal strands of the β-sheet of CI2 lack the completely regular hydrogen-bonding network of the canonical β-sheet. Water molecules form bridging hydrogen bonds between strands 4, 5 and 6, for example between the carbonyl oxygen atoms of Gln59 and Val60 and the amide nitrogen of Phe50 (McPhalen & James, 1987). In contrast, strand 4 is hydrogen-bonded in a regular fashion to strand 3, to form a relatively long stretch of parallel β-sheet.

The mutation Thr → Val3 displays a Φ_F-value of 0.52, which is considerably higher than those of the other two mutations made at position 3. However, when we combine the data we have at this site to construct the composite mutations Val → Ala3 and Ala → Gly3, we obtain low Φ_F-values (0.07 and −0.19) similar to the mutations Thr → Ala3 and Thr → Gly3. The anomalous behaviour of the Thr → Val mutation may be a result of a hydrophilic/hydrophobic switch involved which introduces a complication in the interpretation of Φ_F-values, since the possibility arises that the mutated site will be solvated in a manner different from that of the wild-type residue, leading to a different free energy of solvation. There is no general solution to this problem. It is apparent that more consistent results in terms of Φ_F-values are achieved when we focus on mutations where hydrophobic groups are removed, e.g. Val → Ala3 and Thr → Ala3. This pattern is repeated at residues 34 and 60, where the mutations Val → Thr (and, in the case of residue 34, Thr → Ala) have Φ_F-values which are significantly higher than those observed for mutations involving hydrophobic deletions. The $C^{\gamma 1}$ and $C^{\gamma 2}$ atoms of these residues are not completely buried and the solvation energies are again likely to contribute to the deviation in Φ_F-values. Val63, on the other hand, is completely buried and Φ_F for the mutation Val → Thr63 is in good agreement with those for the other mutations at this site. On the other hand, Ile30 has well-exposed $C^{\gamma 1}$ and $C^{\gamma 2}$ atoms, yet the Val → Thr30 mutation does not deviate from the other mutations at residue 30 in terms of Φ_F-values. Obviously it is not straightforward to generalize the impact of Val↔Thr mutations on Φ_F-value interpretation.

The central residues of β-strands 3 and 4 interact with the α-helix to form the major hydrophobic core of CI2. While the side-chain of Ile30 in β-strand 3 points away from the core and up towards the loop, it remains within 4.5 Å of the side-chain atoms of several core residues like Ile29 and Val47. Indeed, Φ_F-values for mutations at this position are higher than those for mutations in strands 1, 5 and 6. Judging from the fine-structure analysis of the secondary mutant Ala → Gly30 (Table 6), the C^{β} atom is less structured than the rest of the side-chain atoms, with a Φ_F-value of 0.15.

Val34 at the very edge of β-strand 4 appears to be structured to a similar degree to Ile30 in the transition state: Val → Ala34 displays a Φ_F-value of 0, but higher values, comparable to those at position 30, are obtained from unfolding studies and from a fine-structure analysis of the mutant Ala → Gly34 (Table 6).

Reactive site loop

The reactive site loop consists of ten residues between Gly35 and Ile44. This loop is very solvent-exposed, and mutations do not destabilize the protein to any great extent, except for that of Val38 whose side-chain takes part in tertiary interactions with Leu32 and Phe50 to form the hydrophobic minicore. Access of water to the site of mutation in conjunction with low values of $\Delta\Delta G^{H_2O}_{U\text{-}F}$ make Φ_F-value analysis difficult. Thr36 has been subjected to extensive mutagenesis, but only Thr → Val36 is sufficiently destabilized to yield a reliable value of Φ_F, of 0.19 (mutations Thr → Ser36 and Thr → Ala36 have insignificant effect on both $\Delta\Delta G^{H_2O}_{U\text{-}F}$ and $\Delta\Delta G^{H_2O}_{\ddagger\text{-}U}$). However, the mutation involves a hydrophilic/hydrophobic switch, and the interpretation of this figure is subject to the same limitations as discussed in the case of the β-sheet mutants. The mutation removes the hydrogen bond between $O^{\gamma 1}$ of Thr36 and the backbone O of Pro33, an interaction which does not appear to be very strong (our unpublished results). Thus, the hydrogen bond is hardly formed at all in the transition state of folding.

We have also performed a double-mutant cycle on the two residues Arg43 and Arg45. According to the crystal structure of pseudo-wild-type CI2, the two side-chains are aligned so that N^{H2} and N^{ϵ} of Arg43 can form hydrogen bonds with $O^{\delta 2}$ and $O^{\delta 1}$, respectively, of Asp45. In addition, there is extensive packing interaction between the two residues. Yet the interaction energy between the two residues is very low (0.15 kcal/mol), and prevents us from calculating values of Φ_F^{int} with any precision. The values of Φ_F for the individual mutants are low for Arg → Ala43 and Arg → Ala43/Asp → Ala45 (less than 0.1), but considerably higher for Asp → Ala43 alone (0.39). Presumably, this reflects interactions with other residues, namely Ile44, Glu26 or Arg46. The mutations Tyr → Gly42 and Tyr → Ala42 also have low value of Φ_F (0.07).

Turns

Lys2 is not part of a turn, but its side-chain N^{ζ} forms a hydrogen bond with $O^{\epsilon 2}$ of Glu7 (part of the type III turn between β-strand 1 and the α-helix). The interaction with Glu7 is very weak (0.07 kcal/mol) (Table 7), and so we cannot measure Φ_F^{int}. Thus, the free energy of refolding and unfolding for the double mutant Lys → Ala2/Glu → Ala7 is simply the sum of two independent contributions from the mutations Lys → Ala2 and Glu → Ala7.

Lys2 is also close to the side-chain of Asp23, and this interaction is stronger, namely 0.58 kcal/mol, though no hydrogen bond is implicated in the pseudo-wild-type crystal structure. Asp23 appears to be involved in a non-native interaction during the transition state for the folding of CI2, since the mutant Asp → Ala23 folds considerably faster than wild-type, although the mutation destabilizes the protein overall by 0.96 kcal/mol. This non-native interaction probably involves Lys2, since the mutation Lys → Ala2 also refolds faster than wild-type, and has a value of Φ_F of –0.19. Probing the interaction energy between Lys2 and Asp23 also gives a negative value of Φ_F^{int} (–0.17). The negative Φ_F-values are reflected in both the folding and the unfolding data (except for Asp → Ala23). Interestingly, the mutant Lys → Met2, which retains the long non-polar side-chain but removes the polar hydrogen-bonding N^{ζ} atom, has a value of Φ_F of 0, indicating that any hydrogen bond involving Lys2 is completely absent in the transition state of refolding. Asp23 and Lys2 may be involved in a non-native interaction during refolding, which does not involve the hydrogen bond found in the native state.

Glu7 by itself is fairly structured in the transition state, with a Φ_F-value of 0.40, significantly larger than those for mutation of the surrounding residues Pro6 and Leu8 (Pro → Ala6 and Leu → Ala8 have Φ_F-values of 0.07 and 0.15, respectively). This may reflect early interactions with Trp5 and Glu4.

The residues in the β-hairpin loop between strands 4 and 5, as probed by mutations of Asp52 and Asn56, are almost completely unstructured in the transition state for refolding. Such low values of Φ_F were also observed in β-strands 5 and 6. Thus the C-terminal 13 residues from Asp52 to Gly64 only attain native structure after the transition state.

Interestingly, the mutant Asp → Asn52, whose equilibrium free energy of unfolding is identical to that of wild-type, refolds with a significantly lower rate constant than wild-type, giving $\Delta\Delta G_{\ddagger\text{-U}}^{H_2O}$ = 0.24(±0.02) kcal/mol. At 4 M GdmCl, the unfolding rate constant is identical to wild-type (Table 2); however, since the value of m_{ku} for Asn52 is slightly lower than the value for wild-type, Asn52 unfolds more slowly at higher values of GdmCl. Mutants with stability identical to wild-type may thus be affected differently during the folding pathway.

Similarly, Lys → Ala11 stabilizes CI2 by 0.42 kcal/mol, yet increases $\Delta G_{\ddagger\text{-U}}^{H_2O}$ by 0.21 kcal/mol. This is the opposite effect to that seen for Lys2 and Asp23, where mutations destabilize the protein but increase folding. All other mutations that have little effect on $\Delta\Delta G_{\text{U-F}}^{H_2O}$ (Gln → Ala22, Thr → Ala36, Thr → Ser36, Ile → Ala37, Thr → Asp39, Met → Leu40, Met → Leu40/Lys → Met53) have positive fractional Φ-values (Tables 1 to 3).

Table 7. Ala → Gly scanning for the α-helix and β-sheet of CI2

Mutation	$\Delta\Delta G_{\text{U-F}}^{H_2O}$[a] (kcal mol^{-1})	$\Delta\Delta G_{\ddagger\text{-F}}^{4M}$[b] (kcal mol^{-1})	$\Delta\Delta G_{\ddagger\text{-U}}^{H_2O}$[c] (kcal mol^{-1})	$1 - \Phi_U^{4M}$	$\Phi_F^{H_2O}$
α-Helix					
AG17	1.84 ± 0.06	0.73 ± 0.03	0.74 ± 0.05	0.60 ± 0.02	0.40 ± 0.03
AG18	1.20 ± 0.15	0.30 ± 0.04	0.60 ± 0.04	0.75 ± 0.05	0.50 ± 0.07
AG22	0.58 ± 0.12	0.31 ± 0.03	0.19 ± 0.04	0.46 ± 0.12	0.33 ± 0.10
AG24	2.54 ± 0.12	2.20 ± 0.02	0.54 ± 0.06	0.14 ± 0.04	0.21 ± 0.02
β-Sheet					
AG30	1.40 ± 0.09	1.38 ± 0.03	0.27 ± 0.06	0.02 ± 0.07	0.19 ± 0.04
AG34	1.80 ± 0.12	0.99 ± 0.04	0.41 ± 0.05	0.45 ± 0.04	0.23 ± 0.03
AG60	1.72 ± 0.08	1.85 ± 0.02	0.17 ± 0.05	−0.07 ± 0.05	0.10 ± 0.03
AG63	1.99 ± 0.09	1.98 ± 0.02	0.06 ± 0.05	0.01 ± 0.05	0.03 ± 0.02

The data are for Gly *versus* Ala at each position, where the rate and equilibrium data for the Ala mutant are used in the previous equations as "wild-type".

[a] Change in the free energy of unfolding.

[b] Change in the free energy of the transition state of unfolding in 4 M GdmCl.

[c] Change in the free energy of the transition state of folding in water.

Discussion

Necessity and reliability of the protein engineering procedure

There are no stable intermediates that can be detected on the folding pathway of CI2 and so the only structure to be analysed is the transition state†. The only way of analysing transition states experimentally is by kinetics; and the only procedure for extracting structural information at the level of individual residues in the transition state is the use of kinetic measurements on mutants. There is thus currently only one procedure available for the experimental study of structural events on the folding pathway of CI2: the protein engineering method. The structure of the transition state for the folding and unfolding of CI2 has been so mapped by comparing the kinetics and equilibria of folding of a large number of mutants. There is satisfying agreement between the values of Φ_F measured directly from folding kinetics in water and Φ_F calculated from $1 - \Phi_U$ measured from unfolding kinetics in GdmCl solutions and those values extrapolated to water. This shows that the structure of the transition state is the same when measured in the directions of unfolding and refolding. Multiple mutations at the same site, mutations in contiguous regions, and the use of more advanced procedures such as double mutant cycles and the comparison of Ala and Gly (Ala → Gly scanning) at the same positions as the single mutations also give highly consistent data. We are confident, therefore, that the results of the Φ-value analysis are not artifactual but are satisfactorily probing the energetic changes and, by inference, the structural properties of the transition state. A further facet of Φ-value analysis is that it provides measurements of changes in energy that can be used to compare experiment with theory. The combined data are summarised in Table 8. It is worth emphasizing that the values of Φ_F are measured directly from reforming rate constants in water and do not require extrapolation from measurements in the presence of denaturants.

Gross nature of the transition state of folding and unfolding

Although values of Φ_F of 0 and 1 correspond to completely denatured and completely folded structures, respectively, there is not, in general, a linear relationship between a fractional value of Φ and the extent of formation of non-covalent bonds. There should, however, be an approximately linear relationship for the special case of the mutation of larger hydrophobic side-chains to smaller ones (Matouschek *et al.*, 1989, 1990; Fersht *et al.*, 1992). Many of the mutations in this study are of that type so Φ in those cases is a good indication of the extent of structure formation. Fractional Φ-values may result from genuinely weakened interactions or the reaction proceeding by parallel pathways, some of which have the element of structure tested by the Φ-value fully formed and others fully unfolded (or a mixture of the different processes), as discussed by Fersht (1995). A test has been devised to distinguish between the two mechanisms which, when applied to the folding of CI2, indicated genuine weakened interactions (Fersht *et al.*, 1994).

The structure of the transition state is, to a first approximation, a relatively uniformly expanded form of the folded structure. This may be illustrated by a classical linear free energy plot: a plot of $\Delta\Delta G_{\ddagger\text{-F}}$ *versus* $\Delta\Delta G_{\text{U-F}}$ (Figure 3) is approximately linear, with slope 0.7. This is equivalent to a mean value of Φ_U of 0.7 or $\Phi_F = 0.3$. This means that, on average, 70% of the free energy of the interactions is lost on reaching the transition state for unfolding from the folded state. This relatively uniform behaviour may be contrasted with that of the transition state for the unfolding of barnase: the plot of $\Delta\Delta G_{\ddagger\text{-F}}$ *versus* $\Delta\Delta G_{\text{U-F}}$ (Figure 3) is scattered but has a set of points clustering around the line of slope 1, i.e. regions that are completely unfolded ($\Phi_U = 1$); another set around the line of slope 0, i.e. regions that are full-folded ($\Phi_U = 0$); and others values in between. The mean value of Φ_U is similar, however, at 0.69.

The 70% loss of energy on going from the folded state of CI2 to the transition state may be compared with a increase of 40% in the average degree of exposure (calculated from $m_{\ddagger\text{-F}}/m_{\text{U-F}}$), and an increase of 25% in ΔC_P (Jackson & Fersht, 1991b), which is usually taken as a measure of the change in the surface exposure of hydrophobic residues.

Local variations in formation of structure

The α-helix shows a gradation of structure along its length with the N-cap >50% formed in terms of energy, and the C-cap almost completely unstructured. Thus, formation of the α-helix is on-going throughout the folding reaction and starts from the N-cap. The structure of the α-helix in the transition state may be a looser form of that in the folded protein, with the tertiary interactions that are likely to stabilize this secondary structure in the folded protein by packing against the β-sheet to form the core similarly weakened (see section on the hydrophobic core). A plot (not shown) of $\Delta\Delta G_{\ddagger\text{-F}}$ *versus* $\Delta\Delta G_{\text{U-F}}$ for the helical residues is quite linear, however, with slope 0.65($\pm$0.08), correlation coefficient 0.87, indicating that the helix, when judged as a whole, is in the process of being formed during the transition state.

† The transition state for protein folding has characteristics similar to those in simple chemistry. Probing the energy surfaces of protein folding transition states by structure-reactivity relationships of physical-organic chemistry reveals behaviour consistent with their being at a saddle point (Matouschek & Fersht, 1993; Matthews & Fersht, 1995; Matouschek *et al.*, 1995).

Table 8. Summary of the sites of mutation in CI2 and their Φ-values

Residue	Description	Solvent-accessible area (Å^2)			Buried area (Å^2)			Φ_F[a] (for mutation)
		Total	Side-chain	Non-polar side-chain	Total	Side-chain	Non-polar side-chain	
Whole protein		4183	3352	2398	6884	4730	3823	
A. *Core*[b]	The core consists of 12 residues with hydrophobic side-chains; Trp5, Leu8, Ala16, Val19, Ile20, Ala27, Ile29, Val47, Leu49, Val51, Ile57 and Pro61	133	81	81	1888	1420	1393	
Leu8[b]	On edge of hydrophobic core, in a type III reverse turn between Trp5 and Leu8, and also a type II reverse turn between Leu8 and Lys 11. C^{γ}, $C^{\delta1}$, and $C^{\delta2}$ atoms make contact with side-chains of Trp5, Ala16, Val19, Ile20, Pro61 and Lys11. Leu → Ala deletes interactions with α-helix and β-strand 6	6	6	6	174	131	131	0.2 (L → A)
Ala16	α-Helix residue, in centre of hydrophobic core. Completely buried. C^{β} makes contact with side-chains of Leu8, Leu49, and Ile57 and with backbone of Val13 and Glu15. Ala → Gly deletes interactions with N-terminal part of α-helix and β-strand 4	0	0	0	113	67	67	1.1 (A → G)
Val19[b]	α-Helix residue, on edge of hydrophobic core. Side-chain partially exposed. Interacts with Trp5, Leu8, Ile19 and Asp23. Val → Ala deletes interactions between β-strand 1 and the α-helix	62	59	59	98	58	58	−0.3 (V → A)
Ile20[b]	α-Helix residue, in centre of hydrophobic core. Completely buried. $C^{\delta1}$ methyl group packs against Trp5, Leu8, Ala16, Val47, Leu49 and Pro61. Ile → Val deletes interactions between α-helix and β-strands 1, 4 and 6	0	0	0	182	140	140	0.4 (I → V)
Ile29[b]	Ile29 is second residue of β-strand 3, at the edge of the core. Ile → Val deletes contacts of $C^{\delta1}$ methyl group with side-chains of Lys17, Ile20 and Leu21, removing interactions between β-strand 2 and α-helix. Ile → Ala deletes contacts between $C^{\gamma1}$ and $C^{\gamma2}$ methyl groups and side-chains of Ala27, Val31, Val47, and Leu49, removing interactions from β-strand 3 with α-helix and β-strand 4	27	5	5	155	135	135	0.2 (I → V) 0.3 (I → A)
Val47[b]	On edge of hydrophobic core but completely buried. First residue in β-strand 4. Val → Ala deletes contacts with Trp5, Ile20, Lys17, Ala27, Leu49, Pro61 and Val63, removing contacts with β-strand 1, middle of α-helix and β-strand 6	0	0	0	160	117	117	0.2 (V → A)
Leu49[b]	In centre of hydrophobic core, central residue of β-strand 4. Completely buried. Leu → Ala deletes interactions with residues Val13, Ala16, Ile20, Ile29, Val31, Val47, Ile57 and Pro61, removing interactions with α-helix, and β-strands 3, 4, 5 and 6	0	0	0	180	137	137	0.5 (L → A)
Val51[b]	Last residue in β-strand 4, at edge of core. Val → Ala deletes interactions with Val13, Val31, Leu47, Asp55 and Ile57, removing interactions with α-helix and β-strands 3, 4 and 5	13	11	11	147	106	106	0.3 (V → A)
Ile57[b]	Second residue in β-strand 5, at edge of core. $C^{\delta1}$ makes contacts with Val13, Ala16, Leu47 and Val51, the $C^{\gamma1}$ and $C^{\gamma2}$ methyl groups make contacts with Leu8, Val9, Ala16, Leu49 and Pro61. Ile → Val deletes interactions between β-strand 5 and α-helix and β-strand 4, Ile → Ala removes interactions with β-strand 1	3	0	0	179	140	140	0.1 (I → A)
Pro61	Second residue of β-strand 6 (O H-bonds to N^{H} of Trp5). Side-chain points into centre of core. Side-chain atoms contact Trp5, Leu8, Ile20, Val47, Arg48, Leu49, Ile57, Glu59 and Arg62, all in the core except Arg48, Glu59 and Arg62. Pro → Ala mutation removes contact with β-strands 4 and 5	2	0	0	141	105	105	0.0 (P → A)

continued overleaf

Table 8. *continued*

Residue	Description	Solvent-accessible area (Å²)			Buried area (Å²)			Φ_F[a] (for mutation)
		Total	Side-chain	Non-polar side-chain	Total	Side-chain	Non-polar side-chain	
B. *α-Helix*	The α-helix consists of 13 residues between Ser12 and Lys24.	809	700	503	1307	843	711	
Ser12	N-cap of the α-helix. Solvent-accessible surface mainly C^{β} (36 Å2). Interacts mainly with residues 32 to 34. Ser → Ala deletes interactions at the N-cap with Lys11, Val13, Glu14 and Glu15. Ser → Gly removes further interactions of the C^{β} methylene with these residues	48	44	36	74	3	8	0.4 (S → A) 0.3 (S → G)
Glu14	(N-cap + 2) position of the α-helix. Side-chain points away from helix into solvent. Solvent-accessible surface area mainly C^{γ} (31 Å²), C^{β} (27 Å²) and $O^{\varepsilon 1}$ (23 Å²). Glu → Ala removes interactions with Ser12 and Val13 (for triple mutant SG12/EA14/EA15)	98	94	66	85	45	−5	1.2 (E → Q) 0.2 (E → D)
Glu15	(N-cap + 3) position of the α-helix. Side-chain directs towards N-cap of α-helix, Solvent-accessible surface mainly $O^{\varepsilon 1}$ (31 Å²), C^{γ} (20 Å²) and $O^{\varepsilon 2}$ (17 Å²). Glu → Ala removes interactions with Lys11, Ser12 and Glu14 (for triple mutant SG12/EA14/EA15)	89	83	35	94	55	26	0.5 (E → Q) 0.5 (E → D)
Lys17	(N-cap + 5) position of the α-helix. Side-chain points into solvent. Solvent-accessible surface entirely side-chain atoms, mainly N^{ζ} (42 Å²) and C^{δ} (20 Å²). Lys → Gly removes interactions with Val13, Glu14, Glu15, Ala16, Lys18, Ile20, all in the α-helix, and with Ile29 in β-strand 3	72	72	31	139	95	88	0.3 (K → A) 0.4 (K → G) 0.4 (A → G)
Lys18	Central position of the α-helix. Side-chain points into solvent. Highly solvent-accessible side-chain, mainly C^{ε} (51 Å²), C^{δ} (34 Å²), N^{ζ} (37 Å²) and (14 Å²). Lys → Gly removes interactions within the α-helix at Glu14, Glu15, Lys17 and Val19	148	140	104	63	27	15	−0.4 (K → A) 0.7 (K → G) 0.5 (A → G)
Leu21	(C-cap − 3) position of the α-helix. Solvent-accessible surface mainly side-chain, viz $C^{\delta 1}$ (49 Å²), C^{β} (11 Å²) and $C^{\delta 2}$ (10 Å²). Leu → Ala removes interactions within the helix at Lys17, Ile20, Lys24 and Pro25, with Ala27, and with Ile29 in β-strand 3. Leu → Gly removes interactions of the C^{β} methylene within the α-helix at Lys17, Lys18, Ile20 and Gln22	68	63	63	112	74	74	0.3 (L → A) 0.4 (L → G)
Gln22	(C-cap − 2) position of the α-helix. Side-chain points away from helix into solvent. Solvent-accessible surface mainly $N^{\varepsilon 2}$ (57 Å²), C^{γ} (25 Å²), O (20 Å²) and $O^{\varepsilon 1}$ (19 Å²). Gln → Gly removes interactions with Lys18, Val19, Leu21 and Asp23, all in the α-helix	149	124	38	39	20	15	0.1 (Q → G)
Asp23	(C-cap − 1) position of the α-helix. Side-chain points towards Trp5. Side-chain within 4.5 Å of Trp5, Lys2, Val19 and Gln22. $O^{\delta 2}$ within hydrogen-bonding distance of $N^{\varepsilon 1}$ of Trp5 (2.80 Å), but no direct hydrogen bonds to Lys2 (closest distance: 4.41 Å between Lys2 N^{ζ} and Asp23 $O^{\delta 2}$). 79 Å² solvent-accessible surface area, mainly $O^{\delta 2}$ (35 Å2), $O^{\delta 1}$ (5 Å²) and O (28 Å²). Asp → Ala removes interactions with Lys2 (N-terminal part of CI2), Trp5 and Val19 (both in core) and α-helix	79	47	40	79	66	4	−0.3 (D → A) 0.3 (A → G)
Lys24	The C-cap of the α-helix. Side-chain points towards end of reactive site loop (residues 62 to 64). Does not form salt-bridge with neighbouring Glu26. Solvent-accessible surface virtually entirely side-chain, mainly N^{ζ} (18 Å²) and C^{γ} (13 Å²). N^{ζ} H-bonds to $O^{\varepsilon 1}$ of Glu26. This H-bond is removed on mutation to Gly. Lys → Gly also removes interactions with Trp5 in β-strand 1, with Ile20, Asp23 in the helix, with Pro25, Glu26, Ala27 in a reverse turn following the C-cap of the helix, Ile44, Asp45, Arg46, Val47 in β-strand 4, and Val63	41	39	21	170	128	98	−0.4 (K → A) 0.1 (K → G) 0.2 (A → G)

C. β-*Sheet*	1280	1105	658	3740	2621	2255		
Thr3	First residue of β-strand 1 (CO H-bonds to NH of Val63 on β-strand 4). Side-chain oriented towards Val63. $C^{\gamma2}$ and C^{β} contact Val63 (β-strand 6), while $O^{\gamma1}$ contacts Lys2 and Glu4 (β-strand 1). Thr → Val replaces $O^{\gamma1}$ with $C^{\gamma2}$, and Thr → Ala removes part of contact to β-strand 6	98	76	52	48	26	22	0.1 (T → A) 0.5 (T → V) 0.1 (T → G)
Ile30	Third residue of β-strand 3 (NH H-bonds to Val47, CO H-bonds to Leu49, both on β-strand 4). Side-chain points away from core. Solvent-accessible surface mainly $C^{\delta1}$ (45 Å^2) and $C^{\gamma2}$ (17 Å^2). C^{β} contacts Ile48, Val47 and Arg48, $C^{\gamma1}$ contacts Gly28, Ile29 and Arg46, $C^{\gamma2}$ contacts Val31, Leu32 and Arg48 while $C^{\delta1}$ contacts Ile29. All these residues are in β-strands 3 and 4. Ile → Val, Ile → Ala and Ile → Gly remove contacts to β-strand 3 and β-strand 4, while Ile → Thr changes interaction from non-polar to polar in this region	54	50	50	128	90	90	0.3 (I → A) 0.3 (I → G) 0.4 (I → T) 0.2 (A → G)
Val34	Last residue of β-strand 2 (NH H-bonds to Val51 in β-strand 4). Just before reactive site loop (residues 35 to 44). Points towards side-chain of Ala58. Solvent-accessible surface area mainly O (17 Å^2), $C^{\gamma1}$ (29 Å^2) and $C^{\gamma2}$ (13 Å^2). C^{β} contacts Pro33, Gly35, Val51 and Glu59, $C^{\gamma1}$ contacts Gly35 and Glu59 while $C^{\gamma2}$ contacts Pro33, Phe50, Val51, Asp52, Ala58 and Glu59. All these residues except Gly35 (active loop) and Asp52 (β-turn) are in β-strands 3 to 5. Val → Ala and Val → Gly remove these contacts, while Val → Thr changes a non-polar interaction to a polar interaction in these regions	59	42	42	101	75	75	0 (V → A) 0.2 (V → G) 0.2 (V → T) 0.2 (A → G)
Ala58	Last residue of β-strand 5 (NH H-bonds to Phe50). C^{β} points towards side-chain of Val34, and contacts Val34, Phe50, Val51, Asp52, Asn56, Ile57 and Glu59 in β-strands 3 to 5. Ala → Gly removes these contacts	37	12	12	76	55	55	0.1 (A → G)
Val60	First residue of β-strand 6. Side-chain points towards Pro6. Solvent-accessible surface mainly $C^{\gamma2}$ (36 Å^2) C^{β} contacts Gly64, Arg62, Glu59, Pro6, and Val9, $C^{\gamma1}$ contacts Glu4, Trp5, Pro6 and Pro61 while $C^{\gamma2}$ contacts Pro6, Val9 and Glu59. All these residues except Pro6 are in β-strands 1,2 and 6. Val → Ala and Val → Gly remove these contacts and Val → Thr changes a non-polar interaction to a polar one in this region	60	51	51	100	66	66	0 (V → A) 0 (V → G) 0.1 (A → G)
Val63	Penultimate residue of β-strand 6 (NH H-bonds to Thr3). Side-chain entirely buried. Side-chain points towards Thr22. C^{β} contacts Thr3, Ile45, Arg46, Arg62 and Gly64, while $C^{\gamma1}$ contacts Lys24, Tyr42, Arg43 and Gly64 and $C^{\gamma2}$ contacts Lys2, Trp5, Lys24, Ile44, Arg46, Val47 and Arg62. These contacts are in β-sheets 1,3 and 6, the reactive loop and two turn regions. Val → Ala and Val → Gly remove contacts with all these regions. Val → Thr changes a non-polar interaction to a polar one	7	0	0	154	117	117	0.1 (V → T) 0 (V → A) 0 (V → G) 0 (A → G)
D. *Minicore*		133	106	106	425	323	323	
Leu32	Antepenultimate residue of β-strand 3 (NH H-bonds to CO of Leu49, CO H-bonds to Val51). Side-chain points away from β-sheet towards reactive site loop. Only side-chain is solvent-accessible, mainly $C^{\delta2}$ 35 Å^2). Side-chain atoms mainly pack to the two other minicore side-chains (Val38 and Phe50), but also contact Ile30, Val31, Pro33 (β-strand 3), Thr36 (loop) and Leu49 (β-strand 4 and core). Leu → Ile, Leu → Val and Leu → Ala modify and reduce contact to these regions	43	42	42	137	95	95	0.2 (L → A) −0.3 (L → I) −0.1 (L → V)
Val38	Two residues before scissile bond in reactive loop. Side-chain points from loop down to β-sheet. Side-chain solvent-accessible surface area mainly $C^{\gamma2}$ (28 Å^2). Side-chain contacts to minicore as well as Thr36, Ile37, Thr39 (loop) and Arg48 (β-strand 4) are removed by the mutation Val → Ala	58	33	33	102	84	84	0.1 (V → A)

continued overleaf

Table 8. *continued*

Residue	Description	Solvent-accessible area (Å²)			Buried area (Å²)			Φ_F[a]
		Total	Side-chain	Non-polar side-chain	Total	Side-chain	Non-polar side-chain	(for mutation)
Phe50	Penultimate residue in β-strand 4 (CO H-bonds to NH of Ala58). Side-chain points up from β-sheet towards the reactive site loop. Only its side-chain is solvent-accessible, mainly $C^{\epsilon 1}$ (17 Å²), $C^{\delta 1}$ (6 Å²) and C^{ζ} (5 Å²). Side-chain contacts to minicore and also to Pro33, Val34 (β-strand 3), Thr36 (loop), Arg48 (β-strand 4), Leu49, Val51 and Glu59 (core). Phe → Leu, Phe → Val and Phe → Ala (which successively remove the ring atoms) change these packing interactions	32	31	31	186	144	144	0.3 (F → L, F → V, F → A)
E. *Turns & loop*								
Lys2	In coil at N terminus just before β-strand 1. All side-chain atoms interact with Trp5. N^{ζ} just above the pyrrole moiety of the indole atom. All except C^{β} interact with Glu4. In addition, C^{δ}, C^{ϵ} and N^{ζ} interact with Asp23 (a hydrogen bond to $O^{\delta 2}$ of Asp42 may be mediated by a bridging water molecule). C^{ϵ} and N^{ζ} interact with Glu7. N^{ζ} is within hydrogen-bonding distance of $O^{\epsilon 2}$ of Glu7 (2.62 Å). C^{β} and C^{γ} interact with Thr3. Lys → Met maintains interactions with all the residues (β-strand 1, α-helix and turn before helix), but removes the H-bonds to Glu7 and Asp23 and the polar interaction with Trp5, while Lys → Ala removes all the interactions	168	80	44	43	87	75	−0.2 (K → A) 0 (K → M)
Pro6	In type III turn between β-strand 1 and α-helix. Side-chain points into solvent and contacts with Glu4 and Trp5 (β-strand 1), Glu7 (turn before helix) and Val60 (β-strand 6). Pro → Ala removes contact with β-strands 1 and 6 and the turn	75	65	65	68	41	41	0.1 (P → A)
Glu7	In type III turn betwen β-strand 1 and α-helix. Side-chain points back along the backbone towards the N terminus and contacts Lys2, Glu4, Trp5, Pro6 and Leu8. $O^{\epsilon 2}$ hydrogen-bonds to N^{ζ} of Lys2. Glu → Ala removes interactions with β-strand 1, turn before helix and the hydrogen bond to Lys2. Glu → Gln prevents deprotonation at low pH, but should otherwise not perturb interactions	94	66	56	89	72	5	0.4 (E → A)
Glu26	In type I reverse turn between α-helix and β-strand 3. Side-chain sticks straight into solvent and has contacts to Lys24 (helix), Pro25, Ala27 (same turn) and Asp45 (β-strand 4). $O^{\epsilon 1}$ H-bonds to N^{ζ} of Lys24. Glu → Ala removes these interactions, including the H-bond to Lys24	157	132	56	26	6	5	0.4 (E → A)
Thr36	At N-terminal part of reactive loop. Side-chain points from loop down towards the β-sheet. C^{β} interacts with Leu32, Pro33, G135, Ile37 and Phe50, $O^{\gamma 1}$ H-bonds to O of Pro33. This hydrogen bond appears to stabilize the segment of the reactive site loop composed of residues 34 and 35, which partly projects out. $O^{\gamma 1}$ also interacts with Leu32 and Gly35, while $C^{\gamma 2}$ interacts with Leu51, Ile56 and Val57. Thr → Val, Thr → Ser and Thr → Ala remove the H-bond to Pro52, and change interaction with loop and minicore	54	36	31	92	66	43	0.2 (T → V)
Ile37	In beginning of reactive loop. Highly exposed, mainly side-chain. Side-chain atoms contact backbone of Gly33, Thr36 and Val38 (all in loop). Ile → Ala removes contact with reactive loop only	163	141	141	19	−1	−1	

Thr39	Residue before scissile residue. Side-chain points down towards the β-sheet. Side-chain solvent-accessible surface mainly $C^{\gamma 2}$ (67 Å^2). Side-chain atoms contact side-chain of Glu41 and backbone of Met40 and Val38. Hydrogen-bond between $O^{\gamma 1}$ of Thr39 and $O^{\gamma 1}$ of Glu51. Thr → Ala and Thr → Asp remove this H-bond and a few contacts with reactive loop	94	82	78	52	20	−4	
Met40	Scissile residue. Hyperexposed. Side-chain atoms contact backbone of Thr39 and Glu41. Met → Ala removes contact with reactive loop only	199	162	162	5	−2	−2	
Glu41	Residue after scissile residue. Side-chain points down towards β-sheet. $O^{\epsilon 2}$ H-bonds to N^{H1} of Arg46. Other side-chain atoms contact Thr39 ($O^{\gamma 1}$ H-bonds to $O^{\gamma 1}$ of Thr39), Met40, Tyr42 and Arg43. Glu → Ala removes H-bond to Arg46 and contacts with reactive loop, including the H-bond	80	62	50	103	76	11	
Arg43	At C-terminal part of reactive loop, where loop makes turn into β-strand 3. End of side-chain solvent exposed, mainly N^{H1} (51 Å^2), Cδ (26 Å^2) and N^{H2} (23 Å^2). C^{γ} interacts with Glu41, Tyr42 and Arg46 while C^{δ}, N^{ϵ}, C^{ζ}, N^{H1} and N^{H2} all interact with Asp45 only. Only interactions with reactive loop (mainly Asp45, but also Glu41, Tyr42 and Arg46). N^{ϵ} H-bonds to $O^{\delta 1}$ of Asp45, possibly stabilizing turn between loop and β-strand 3. Arg → Ala removes these interactions	89	87	40	152	109	49	0.1 (R → A)
Asp45	Last residue in the loop. Solvent-accessible surface mainly C^{β} (26 Å^2). $O^{\delta 1}$ H-bonds to N^{ϵ} of Arg43. Other side-chain atoms interact with side-chain of Arg43 and backbone of Ile44, Glu26 and Arg46. Asp → Ala removes these interactions	49	44	33	102	63	15	
Asp52	In turn between β-strand 4 and β-strand 5. Solvent-accessible surface area mainly $O^{\delta 2}$ (25 Å^2) and C^{β} (19 Å^2). Side-chain only contacts Lys53, Leu54 (same turn), Asp55 and Asn56 (β-strand 5). $O^{\delta 2}$ interacts closely with the NH groups of Lys53, Leu54, Asp55 and Asn26, stabilizing the β-hairpin turn. In addition, C^{γ} and $O^{\delta 1}$ interact with Asp55 and Asn56. Asp → Ala removes these interactions, including the H-bonding network, while Asn → Ala only affects protonation at lower pH	54	45	21	97	61	27	0.1 (D → A)
Asn56	First residue of β-strand 5. Side-chain points away into solvent. Solvent-accessible surface area mainly $N^{\delta 2}$ (27 Å^2) and $O^{\delta 1}$ (11 Å^2). C^{γ} interacts with Ile57, Asp55, Lys11, Asp52, Leu54 and Gly10, $O^{\delta 1}$ interacts with NH on Asn56, Asp55, Lys11, Ser12, Leu54 and Gly10 while $N^{\delta 2}$ interacts with Asp52, Leu54 and Gly10. Side-chain contacts Gly10, Lys11 (turn), Ser12 (helix N-cap), Leu54, Asp55 (turn) and Ile57 (β-strand 5). Asn → Ala removes these interactions, while Asn → Asp introduces a charge	41	41	4	117	72	40	0.2 (N → D) 0.1 (N → A)

Area of extended surface (i.e. the completely unfolded protein) calculated according to values from Miller *et al.* (1987). In this study, accessible surface area of each amino acid residue was calculated for residue X in in a Gly-X-Gly tripeptide, with the main-chain in an extended conformation (Miller *et al.*, 1987).

[a] Values of Φ_F are measured from refolding rate constants that are directly measured in water.

[b] From Jackson *et al.* (1993b).

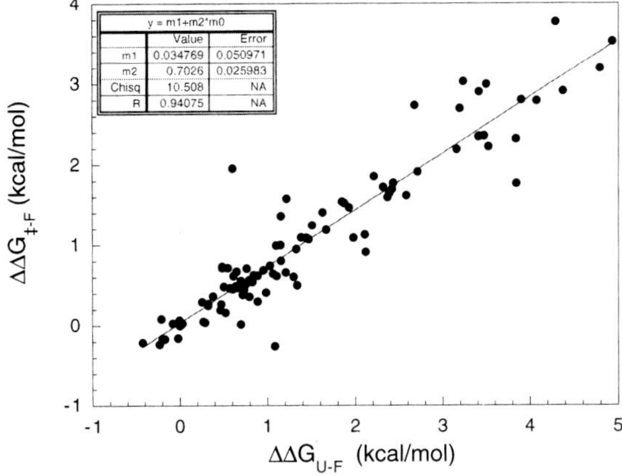

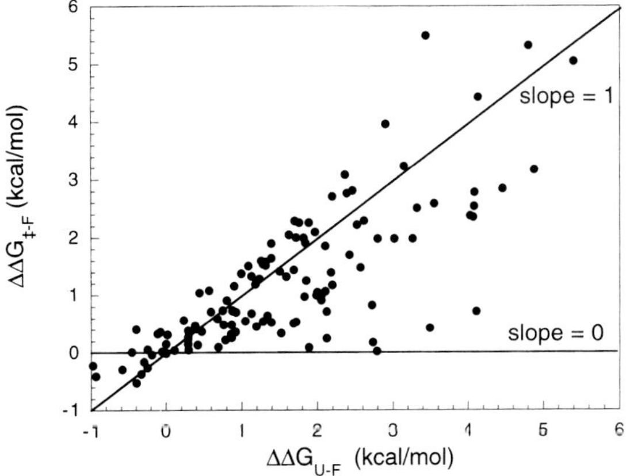

Figure 3. Plots of $\Delta\Delta G_{\ddagger\text{-F}}$ *versus* $\Delta\Delta G_{\text{U-F}}$ for the unfolding in water of (top) CI2 and (bottom) barnase. The linear regression line for the data for CI2 is given. Lines of slope 0 and 1 are indicated for barnase.

The degree of formation of structure is very low in the β-sheet. Φ-Value analysis allows us to monitor formation of interactions of side-chains, and is, therefore, not a direct measure of the extent of secondary structure formation, which rather depends on backbone-backbone hydrogen-bonding. However, secondary structure formation is a prerequisite for the side-chains to be able to establish contacts within the β-sheet. The side-chains of the residues probed in the β-sheet, namely Thr3, Ile30, Val34, Ala58, Val60 and Val63, all interact primarily with other β-sheet side-chains. Their structure in the transition state thus reports on the overall β-sheet structure.

In β-strand 3, there is a similar extent of formation of structure in the transition state for positions 34 and 30. We observe similar values at position 29, at the edge of the core. In strand 4, Φ_F-values at position 43 (end of reactive site loop), 47 (edge of core), 50 (minicore) and 51 (edge of core) are close in magnitude to each other and to those in strand 3. It is evident that the folding process shows no clear differentiation between core residues and non-core residues in the β-strands 3 and 4. Rather, the data give excellent support to the model that the main body of CI2 folds as a single co-operative unit, with secondary and tertiary interactions forming concomitantly with few exceptions, namely at residues 49 and 16. Strands 1, 5 and 6 appear to be completely unfolded in the transition state. Despite the formation of the β-sheet requiring that residues far apart in sequence (at opposite ends of the molecule) be brought in close proximity, their alignment is not part of the rate-limiting step of the folding process, and once again demonstrates that formation of intact secondary structure is not a prerequisite for further rearrangements in folding.

The three-residue hydrophobic patch in the reactive site loop, which we call the minicore, has been thoroughly examined by double mutant cycles. All the kinetic data converge to suggest that the minicore is ca 20 to 40% structured in the transition state relative to the folded state. Further, it is formed along a single folding pathway, rather than parallel pathways (Fersht *et al.*, 1994). This conclusion is reached both by mutation of individual residues, double mutant cycle analysis and fine-structure analysis.

Relationship to structures of peptide fragments of CI2

In studies elsewhere, the structures of peptides corresponding to the progressive elongation of CI2 from its N terminus have been analysed (Prat Gay *et al.*, 1995; unpublished work, this laboratory). The native secondary structural elements develop fully only in parallel with the formation of tertiary interactions (Prat Gay *et al.*, 1995). Fragment (1 to 50) is not compact and requires elongation to (1 to 53) before compact structure can be detected. The helix does not form significantly until the protein is nearly intact, and even the fragment 1 to 60, which lacks only the last four residues, has fluctuating structure and some non-native interactions (Prat Gay *et al.*, 1995). Small fragments which include the residues corresponding to the α-helix of the intact protein have been shown to have negligible structure in water, and in the helix-promoting solvent trifluoroethanol, only nascent helix is formed (Itzhaki *et al.*, 1995). Titration of the formation of nascent helix with trifluoroethanol and extrapolation to 0 M trifluoroethanol does suggest, however, that the nascent helix is present to the order of ~3% in water. Further, the transition state for the association of two complementary fragments of CI2 (fragment 1 to 40 and fragment 41 to 64), analysed using the protein engineering method (Prat Gay *et al.*, 1994; Ruiz-Sanz *et al.*, 1995), is remarkably similar in structure to the transition state for folding of the intact protein. Therefore, the co-operative folded structure is assembled only as the two fragments associate. The overwhelming conclusion that can be drawn from the study of fragments of CI2 is that isolated sequences of this protein cannot form significant fractions of native-like structure independently and, therefore, that tertiary interactions in this small protein are both crucial for the native fold and to the transition state for its formation. (See Creighton (1995) for an analysis of these phenomena in terms of co-operativity of interactions.)

Relationship to theoretical models: nucleation-condensation

The structure of the transition state for the folding of intact protein and the structures of fragments support a folding pathway for CI2 in which there is no early, independent, formation of native-like secondary structure. Rather, folding of CI2 is a very concerted reaction; secondary and tertiary structure are formed concomitantly and are consolidated throughout the folding reaction. These observations do not fit to a simple framework model in which extensively formed elements of structure are formed early and then coalesce, or even to any model that requires a pre-formed initiation site.

Based on the observation that the co-operativity of multiple interactions is required to stabilize both the native structure and the transition state for its formation, we and Creighton (1995) envisage the folding of CI2 to proceed qualitatively, as follows, in general terms. There is rapid random searching of conformations in the unfolded state of the protein under conditions that favour folding until sufficient tertiary interactions are made that stabilize certain elements of secondary structure. When sufficient interactions are made, the transition state is reached and there follows the rapid formation of the final structure.

Nucleation-condensation

Our data fit an even more specific model, which we term nucleation-condensation. This involves a nucleus that consists primarily of adjacent residues. The nucleus does not form stable structure without assistance from interactions made with residues that are distant in sequence. Formation of the small nucleus cannot be solely rate determining since a significant fraction of the overall structure must be in the approximately correct conformation to provide the long-range contacts to stabilize the nucleus. In CI2, for example, the α-helix has a very weak tendency to be formed in a nascent manner, driven by local interactions. The helix remains embryonic until sufficient long range interactions are built up that it becomes stable. The rest of the protein then condenses around the helix and the other native interactions that are developed during the stabilization of the helix. The onset of stability as the multiple interactions are made is so rapid that the helix is still in the process of being consolidated as the transition state is reached. Thus, formation of the nucleus is coupled with more general formation of structure and so nucleation is just part of the rate determining step. The coupling of nucleation and condensation leads to a relatively compact transition state. The whole of the molecule is thus involved in forming the transition state, the nucleus is simply the best formed part of the structure in the transition state.

This nucleation-condensation mechanism has certain characteristics. First, is that the nucleation site does not need to be extensively preformed in the denatured state. A theoretical analysis of the optimization of the rates of protein folding suggests that the less it is formed in the denatured state, the better (Fersht, 1995b). It may or may not be detectable in isolated peptides, but it is only essential that it is extensively formed in the transition state. Further, it need not be formed completely in the transition state but, because of the onrush of co-operative interactions that stabilize it, it may be in the process of being formed in the transition state. Second, although much of the structure is from local, contiguous residues, there are important stabilizing contributions from long range interactions, i.e. from contacts with residues that are far removed in sequence.

Abkevich *et al.* (1994) have performed Monte Carlo simulations of the folding of a 36-mer "protein" on a lattice. They observed in their simulations the formation of a specific nucleus of interactions whose formation constituted the "transition state" of the reaction and precipitated subsequent rapid folding, and the "protein" was found to fold *via* a nucleation growth mechanism†. Further, without prior knowledge of our experimental data, E. Shakhnovich and co-workers (E. Shakhnovich, personal communication) have

† A referee has pointed out that the nucleation-growth mechanism of Abkevich *et al.* (1994) is for the formation of a molten globule and not native protein. However, those authors now believe their proposal is more general and have given us permission to quote: "In the original paper (Abkevich *et al.*, 1994) it was suggested that the observed nucleation mechanism applies to formation of the "native-like" Molten Globule with the native backbone conformation already formed but with rotating side-chains. This assertion was based on the hypothesis that the rate-limiting step in folding is associated with tight-packing of side-chains while backbone conformation is already close to native. The existing experimental data do not support this hypothesis. Rather they suggest that the rate-limiting step is the formation of the backbone fold, probably with simultaneous packing of side-chains. In this case, the nucleation mechanism of the formation of backbone fold found in lattice models applies to the whole process of folding. This is especially clear in the case when folding is kinetically two-state, as in the case of CI2" (E. Shakhnovich, personal communication).

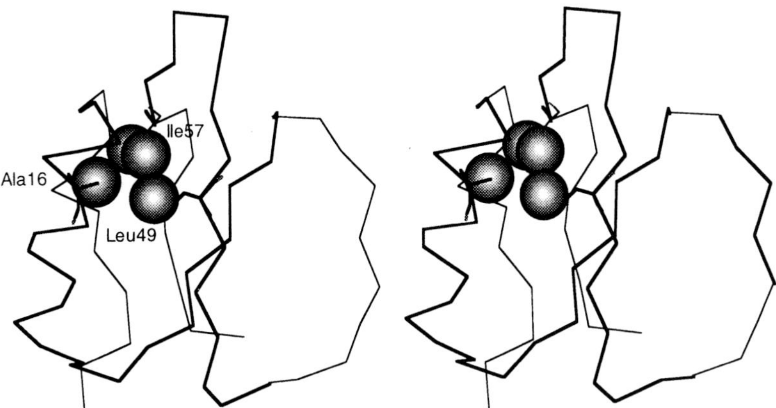

Figure 4. Stabilization of the formation of the N-terminal region of the α-helix in the transition state by the interaction of the side-chain of Ala16 with residues that will form the hydrophobic core. The filled spheres (C^{β} of Ala16; $C^{\delta 1}$ and $C^{\gamma 1}$ of Ile57; and $C^{\delta 1}$ of Ile57) are drawn with the full van der Waals' radii. The N-terminal region of the helix and the ancillary interactions constitute the proposed nucleation site.

predicted that Ala16 is a key residue in nucleation (Figure 4). The overall agreement between simulation and experiment is so encouraging that experiment and theory could well combine to give a satisfactory computer model for folding.

It is now common practice to search for initiation sites in the folding of proteins by searching for structure in isolated peptides by experiment or computation (Hirst & Brooks, 1995). The results of the experiments on CI2 and the lattice simulations suggest that this may be an inconclusive process since the nucleation sites do not necessarily emerge until the transition state is reached. Further, even if significant structure is found, it does not mean that it is a nucleation site or that it is formed at the earliest stages of folding.

Concerted *versus* stepwise folding mechanisms in general

Barnase is larger than CI2, comprising separate modules (Yanagawa *et al.*, 1993), and folds *via* a distinct folding intermediate in which the regions containing the major α-helix and β-sheet are reasonably well formed and connected by a loose and weak hydrophobic core (Fersht, 1993). In the absence of each other, i.e. in peptide fragments, the structures are formed only weakly under folding conditions (Kippen *et al.*, 1994a) and have only very weak structure under denaturing conditions (Kippen *et al.*, 1994b). Further, destabilizing mutations in the helix cause it to be less folded in the intermediate and the subsequent transition state for folding, and the extent of formation of the overall

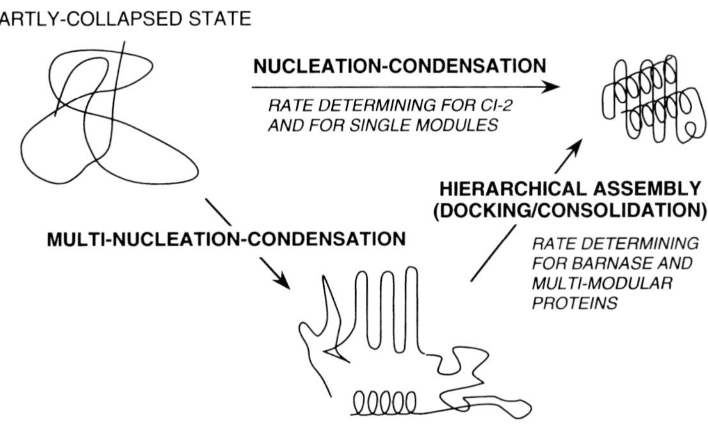

Figure 5. A unified scheme for the initiation of folding in CI2 and barnase, extending the mechanism presented by Otzen *et al.* (1994) using the additional data and analysis in this study. The "partly-collapsed" state is the "unfolded" state under conditions that favour folding. CI2, which folds as a single co-operative unit, folds in one step, which is the nucleation-condensation process. The folding of barnase is consistent with the formation of a folding intermediate by nucleation steps and then the structure consolidating in the rate determining step. Although the folding of barnase is drawn as stepwise, the nucleation-condensation and docking-consolidation processes do tend to merge under certain conditions (M. Oliveberg, P. Dalby and A.R.F., unpublished).

structure is coupled to the stability of the helix (Matthews & Fersht, 1995). This shows that the folding intermediate of barnase is also formed in a concerted manner. Thus barnase fits also a nucleation-condensation folding mechanism (Figure 5). We suggest that the folding of CI2 is a model for the folding of modules in larger proteins.

Whether or not folding is a concerted or stepwise process depends on the stability of individual substructures of the protein when they are considered in isolation or loosely complexed with each other. For example, since the α-helix and β-sheet of CI2 are unstable in the absence of the rest of the protein, they are formed in a concerted manner. If they were stable, then the protein could form by a framework or collision/diffusion model. The same is true for individual modules within a protein. If a module is stable in the absence of the other modules, then it could form independently. The more the individual structures are stable, the more stepwise and hierarchical the folding process. We thus speculate that small proteins tend to fold by nucleation-condensation mechanisms but larger proteins will have a tendency to form in their later stages by a hierarchical assembly of smaller units, but the folding of the small units will tend to be nucleation-condensation processes (Figure 5).

Materials and Methods

Materials

All chemicals are as described in previous papers (Jackson & Fersht, 1991a; Jackson *et al.*, 1993a). Mutagenesis, expression and purification of the wild-type and mutant proteins are as described previously (Jackson *et al.*, 1993a). The buffer used in the equilibrium and kinetic folding experiments was 2-(*N*-morpholino)ethanesulfonic acid (Mes) purchased from Sigma. A 1 M stock solution of Mes (pH 6.25) contained 415 mM of the free acid and 585 mM of the sodium salt. 16 novel mutant proteins are presented in this paper: Lys → Met2, Lys → Ala2, Glu → Ala7, LysA → Ala2/Glu → Ala7, Lys → Ala2/Asp → Ala23, Ala → Gly16, Asp → Ala23, Thr → Ala36, Tyr → Ala42, Tyr → Gly42, Arg → Ala43, Asp → Ala45, Arg → Ala43/Asp → Ala45, Asp → Ala52, Asp → Asn52, Pro → Ala61.

GdmCl equilibrium denaturation experiments

The intrinsic fluorescence of CI2 increases on denaturation, allowing unfolding and refolding to be monitored by fluorescence spectroscopy. The maximal change in fluorescence upon denaturation is obtained with an excitation wavelength of 280 nm and an emission wavelength of 356 nm. All experiments were performed at 25°C.

A Perkin Elmer LS5B luminescence spectrometer was used for the GdmCl equilibrium denaturation experiments, with a bandpass of 10 nm. The GdmCl solutions were prepared using a Hamilton Microlab M robot by aliquoting appropriate volumes of a solution of 7.5 M GdmCl containing 50 mM Mes (pH 6.25) and a solution of 50 mM Mes (pH 6.25). For each data point in the denaturation experiment, 100 μl of CI2 stock solution in 50 mM Mes (pH 6.25) were diluted into 800 μl of the appropriate denaturant concentration, using a SMI positive displacement pipetter. Final concentrations were 2.5 μM CI2 and 50 mM Mes (pH 6.25); final concentrations of GdmCl were 0 M to 5.5 M, in 0.1 M or 0.2 M increments. The protein-denaturant solutions were equilibrated at 25°C for approximately one hour before measuring their fluorescence. Many experiments were repeated with GdmCl solutions that had been prepared in the conventional manner by weighing the appropriate amount for each individual concentration into a volumetric flask. Identical results were obtained.

Kinetic unfolding experiments

Reactions were followed with a Perkin-Elmer MPF-44B fluorescence spectrophotometer equipped with a rapid mixing head. The mixing device contained a T-jet mixing chamber followed by a 30 ms delay loop ensuring complete mixing of the solutions before observation. A Hellma flow-through cell (10 mm × 3 mm × 3 mm) was used. The solutions were driven through the mixing chamber manually from two Hamilton syringes resulting in a mixing ratio of 1:10. The observation cell and reservoir syringes were thermostatted separately using two water baths and the temperature of each was monitored using an Edale Instrument Thermometer and maintained to ±0.1 deg.C. Data were acquired with a Tandon Target Microcomputer, a DT2801 Data Translation board and the Bio-kine data acquisition software package (Bio-logic), and analysed using the programme Kaleidagraph (Abelbeck Software). Unfolding was initiated by rapidly diluting one volume of the protein solution (approximately 20 μM) in 50 mM Mes (pH 6.25) with ten volumes of concentrated GdmCl solution (containing 50 mM Mes, pH 6.25). This resulted in final GdmCl concentrations of between 3 and 7 M. The lowest denaturant concentration was chosen to result in at least 98% unfolded protein. An excitation wavelength of 280 nm and an emission wavelength of 356 nm were used. Excitation and emission slit bandwidths of 10 nm were used.

Kinetic refolding experiments

Refolding reactions above 0.5 M GdmCl were monitored using the same apparatus set-up as for unfolding experiments. Refolding was initiated by rapid 1:11 dilution of unfolded protein (approximately 20 μM) in 6.5 M GdmCl and 50 mM Mes (pH 6.25) into different concentrations of GdmCl solution in 50 mM Mes (pH 6.2). For lower final concentrations of GdmCl, an Applied Photophysics Stopped-Flow Spectrophotometer Model SF 17MV was used. The temperature of the cell and reservoir syringes was maintained by thermostatting with a Grant LTD6 water bath. Temperatures were maintained to ±0.1 deg.C based on an internal temperature probe in the stopped flow apparatus which had previously been calibrated against an Edale Instrument Thermometer. Protein was denatured for these studies by addition of HCl to a concentration of 20 mM (pH 1.7). Refolding was initiated by rapid mixing of one volume of the protein with one volume of 100 mM Mes (pH 6.65) (consisting of 21.5 mM of the free acid and 78.5 mM of the sodium salt). The final pH of the solution was pH 6.25, 50 mM Mes, and the final protein concentration was approximately 10 μM. Refolding was performed in the absence and in the presence of low concentrations (up to 0.6 M, with 0.1 M increments) of GdmCl. An excitation wavelength of 280 nm was used, and slit bandwith of 2 nm. A glass cut-off filter was used to allow emission above 315 nm to be monitored. Refolding rates obtained from the two different sets of refolding experiments were identical within error when unfolded protein was refolded into GdmCl concentrations that could be monitored by both the Perkin-Elmer MPF-44B fluorescence spectrophotometer and the Applied Photophysics Stopped-Flow Spectrophotometer.

Acknowledgements

We thank Eugene Shakhnovich for prior communication of results about Ala16 and for clarifying comments about nucleation-growth mechanisms.

References

Abkevich, V. I., Gutin, A. M. & Shakhnovich, E. I. (1994). Specific nucleus as the transition state for protein folding: Evidence from the lattice model. *Biochemistry*, **32**, 10026–10036.

Carter, P. J., Winter, G., Wilkinson, A. J. & Fersht, A. R. (1984). The use of double mutants to detect structural changes in the active site of the tyrosyl-tRNA synthetase (*Bacillus stearothermophilus*). *Cell*, **38**, 835–840.

Clarke, J. & Fersht, A. R. (1993). Engineered disulfide bonds as probes of the folding pathway of barnase – increasing the stability of proteins against the rate of denaturation. *Biochemistry*, **32**, 4322–4329.

Creighton, T. E. (1995). Protein-folding – an unfolding story. *Curr Biol.* **5**, 353–356.

Dill, K. A., Fiebig, K. M. & Chan, H. S. (1993). Cooperativity in protein-folding kinetics. *Proc. Natl Acad. Sci. USA*, **90**, 1942–1946.

elMasry, N. F. & Fersht, A. R. (1994). Mutational analysis of the N-capping box of chymotrypsin inhibitor-2. *Protein Eng.* **4**, 777–782.

Fersht, A. R. (1993). Protein folding and stability – the pathway of folding of barnase. *FEBS Letters*, **325**, 5–16.

Fersht, A. R. (1995a). Characterizing transition states in protein-folding – an essential step in the puzzle. *Curr. Opin. Struct. Biol.* **5**, 79–84.

Fersht, A. R. (1995b). Optimization of rates of protein folding: The nucleation-condensation mechanism and its implications. *Proc. Natl Acad. Sci. USA*, in the press.

Fersht, A. R., Matouschek, A. & Serrano, L. (1992). The folding of an enzyme. 1. Theory of protein engineering analysis of stability and pathway of protein folding. *J. Mol. Biol.* **224**, 771–782.

Fersht, A. R., Itzhaki, L. S., Matthews, J. M. & Otzen, D. E. (1994). Single versus parallel pathways of protein folding and fractional formation of structure in the transition state. *Proc. Natl Acad. Sci. USA*, **91**, 10426–10429.

Hammond, G. S. (1955). A correlation of reaction rates. *J. Am. Chem. Soc.* **77**, 334–338.

Harpaz, Y., elMasry, N. F., Fersht, A. R. & Henrick, K. (1994). Direct observation of better hydration at the N-terminus of an α-helix with glycine rather than alanine as the N-cap residue. *Proc. Natl Acad. Sci. USA*, **91**, 311–315.

Harrison, S. C. & Durbin, R. (1985). Is there a single pathway for the folding of a polypeptide chain? *Proc. Natl Acad. Sci. USA*, **82**, 4028–4030.

Hirst, J. D. & Brooks, C. L. I. (1995). Molecular-dynamics simulations of isolated helices of myoglobin. *Biochemistry*, **34**, 7614–7621.

Horovitz, A. & Fersht, A. R. (1990). Strategy for analysing the cooperativity of intramolecular interactions in peptides and proteins. *J. Mol. Biol.* **214**, 613–617.

Horovitz, A. & Fersht, A. R. (1992). Co-operative interactions during protein folding. *J. Mol. Biol.* **224**, 733–740.

Horovitz, A., Serrano, L. & Fersht, A. R. (1991). COSMIC analysis of the major α-helix of barnase during folding. *J. Mol. Biol.* **219**, 5–9.

Itzhaki, L. S., Neira, J.-L., Prat Gay, G. de, Ruiz-Sanz, J. & Fersht, A. R. (1995). Search for nucleation sites in smaller fragments of chymotrypsin inhibitor 2. *J. Mol. Biol.* **254**, 289–304.

Jackson, S. E. & Fersht, A. R. (1991a). Folding of chymotrypsin inhibitor-2. 1. Evidence for a two-state transition. *Biochemistry*, **30**, 10428–10435.

Jackson, S. E. & Fersht, A. R. (1991b). Folding of chymotrypsin inhibitor-2. 2. Influence of proline isomerization on the folding kinetics and thermodynamic characterization of the transition state of folding. *Biochemistry*, **30**, 10436–10443.

Jackson, S. E. & Fersht, A. R. (1994). Contribution of residues in the reactive-site loop of chymotrypsin inhibitor-2 to protein stability and activity. *Biochemistry*, **33**, 13880–13887.

Jackson, S. E., Moracci, M., elMasry, N., Johnson, C. & Fersht, A. R. (1993a). The effect of cavity creating mutations in the hydrophobic core of chymotrypsin inhibitor 2. *Biochemistry*, **32**, 11262–11269.

Jackson, S. E., elMasry, N. & Fersht, A. R. (1993b). Structure of the hydrophobic core in the transition state for folding of chymotrypsin inhibitor 2: a critical test of the protein engineering method of analysis. *Biochemistry*, **32**, 11270–11278.

Karplus, M. & Weaver, D. C. (1976). Protein folding dynamics. *Nature*, **260**, 404–406.

Karplus, M. & Weaver, D. L. (1994). Protein folding dynamics – the diffusion-collision model and experimental data. *Protein Sci.* **3**, 650–668.

Kim, P. S. & Baldwin, R. L. (1990). Intermediates in the folding reactions of small proteins. *Annu. Rev. Biochem.* **59**, 631–660.

Kippen, A., Sancho, J. & Fersht, A. R. (1994a). Structural studies on peptides corresponding to mutants of the major alpha-helix of barnase. *Biochemistry*, **33**, 3778–3786.

Kippen, A., Arcus, V. L. & Fersht, A. R. (1994b). Folding of barnase in parts. *Biochemistry*, **33**, 10013–10021.

Kraulis, P. J. (1991). Molscript, a program to produce both detailed and schematic plots of protein structures. *J. Appl. Crysallog.* **24**, 946–950.

Levinthal, C. (1968). Are there pathways for protein folding? *J. Chim. Phys.* **85**, 44–45.

Li, A. & Daggett, V. (1994). Characterization of the transition-state of protein unfolding by use of molecular-dynamics – chymotrypsin inhibitor-2. *Proc. Natl Acad. Sci. USA*, **91**, 10430–10434.

Ludvigsen, S., Shen, H., Kjaer, M., Madsen, J. C. & Poulsen, F. M. (1991). Refinement of the three-dimensional solution structure of barley serine proteinase inhibitor 2 and comparison with the structures in crystals. *J. Mol. Biol.* **222**, 621–635.

Matouschek, A. & Fersht, A. R. (1993). Application of

physical organic chemistry to engineered mutants of proteins – hammond postulate behavior in the transition state of protein folding. *Proc. Natl Acad. Sci. USA*, **90**, 7814–7818.

Matouschek, A., Kellis, J. T., Jr, Serrano, L. & Fersht, A. R. (1989). Mapping the transition state and pathway of protein folding by protein engineering. *Nature*, **342**, 122–126.

Matouschek, A., Kellis, J. T., Jr, Serrano, L., Bycroft, M. & Fersht, A. R. (1990). Transient folding intermediates characterized by protein engineering. *Nature*, **346**, 440–445.

Matouschek, A., Serrano, L. & Fersht, A. R. (1992). The folding of an enzyme. 4. Structure of an intermediate in the refolding of barnase analysed by a protein engineering procedure. *J. Mol. Biol.* **224**, 819–835.

Matouschek, A., Matthews, J. M., Johnson, C. M., & Fersht, A. R. (1994). Extrapolation to water of kinetic and equilibrium data for the unfolding of barnase in urea solutions. *Protein Eng.* **7**, 1089–1095.

Matouschek, A., Otzen, D. E., Itzhaki, L. S., Jackson, S. E., & Fersht, A. R. (1995). Movement of the position of the transition state in protein folding. *Biochemistry*, **34**, in the press.

Matthews, J. M. & Fersht, A. R. (1995). Exploring the energy surface of protein-folding by structure-reactivity relationships and engineered proteins – observation of Hammond behaviour for the gross structure of the transition-state and anti-Hammond behaviour for structural elements for unfolding of barnase. *Biochemistry*, **34**, 6805–6814.

McPhalen, C. A. & James, M. N. G. (1987). Crystal and molecular structure of the serine proteinase inhibitor CI-2 from barley seeds. *Biochemistry*, **26**, 261–269.

Miller, S., Janin, J., Lesk, A. M. & Chothia, C. (1987). Interior and surface of monomeric proteins. *J. Mol. Biol.* **196**, 641–656.

Nanzer, A. P., Poulsen, F. M., van Gunsteren, W. F. & Torda, A. E. (1994). A reassessment of the structure of chymotrypsin inhibitor-2 (CI-2) using time-averaged NMR restraints. *Biochemistry*, **33**, 14503–14511.

Otzen, D. E., Itzhaki, L. S., elMasry, N. F., Jackson, S. E. & Fersht, A. R. (1994). Structure of the transition state for the folding/unfolding of the barley chymotrypsin inhibitor 2 and its implications for mechanisms of protein folding. *Proc. Natl Acad. Sci. USA*, **91**, 10422–10425.

Otzen, D. E. & Fersht, A. R. (1995). Side-chain determinants of β-sheet stability. *Biochemistry*, **34**, 5718–5724.

Pace, C. N. (1986). Determination and analysis of urea and guanidinium denaturation curves. *Methods Enzymol.* **131**, 266–279.

Prat Gay, G. de, Ruiz-Sanz, J., Davis, B. J. & Fersht, A. R. (1994). The structure of the transition state for the association of 2 fragments of the barley chymotrypsin inhibitor-2 to generate native-like protein and implications for mechanisms of protein-folding. *Proc. Natl Acad. Sci. USA*, **91**, 10943–10946.

Prat Gay, G. de, Ruiz-Sanz, J., Neira, J. L., Itzhaki, L. S. & Fersht, A. R. (1995). Folding of a nascent polypeptide chain *in-vitro* – cooperative formation of structure in a protein module. *Proc. Natl Acad. Sci. USA*, **92**, 2683–3686.

Ptitsyn, O. B. (1973). Stage mechanism of the self-organization of protein molecules. *Dokl. Acad. Nauk.* **210**, 1213–1215.

Ptitsyn, O. B. (1991). How does protein synthesis give rise to the 3D-Structure. *FEBS Letters*, **285**, 176–181.

Richardson, J. S. & Richardson, D. C. (1988). Amino acid preferences for specific locations at the ends of α-helices. *Science*, **240**, 1642–1648.

Ruiz-Sanz, J., Prat Gay, G. de, Otzen, D. E. & Fersht, A. R. (1995). Protein-fragments as models for events in protein-folding pathways – protein engineering analysis of the association of two complementary fragments of the barley chymotrypsin inhibitor-2 (CI-2). *Biochemistry*, **34**, 1695–1700.

Sali, A., Shakhnovich, E. & Karplus, M. (1994). Kinetics of protein folding. A lattice model study of the requirements for folding to the native state. *J. Mol. Biol.* **235**, 1614–1636.

Santoro, M. M. & Bolen, D. W. (1992). A test of the linear extrapolation of unfolding free energy changes over an extended denaturant concentration range. *Biochemistry*, **31**, 4901–4907.

Serrano, L., Matouschek, A. & Fersht, A. R. (1992a). The folding of an enzyme. 3. Structure of the transition state for unfolding of barnase analysed by a protein engineering procedure. *J. Mol. Biol.* **224**, 805–818.

Serrano, L., Neira, J. L., Sancho, J. & Fersht, A. R. (1992b). Effect of alanine versus glycine in α-helices on protein stability. *Nature*, **356**, 453–455.

Serrano, L., Sancho, J., Hirshberg, M. & Fersht, A. R. (1992c). α-helix stability in proteins. 1. Empirical correlations concerning substitution of side-chains at the N and C-caps and the replacement of alanine by slycine or serine at solvent-exposed surfaces. *J. Mol. Biol.* **227**, 544–559.

Tanford, C. (1968). Protein denaturation. *Advan. Protein Chem.* **23**, 121–282.

Wetlaufer, D. B. (1973). Nucleation, rapid folding, and globular intrachain regions in proteins. *Proc. Natl Acad. Sci. USA*, **70**, 697–701.

Yanagawa, H., Yoshida, K., Torigoe, C., Park, J. S., Sato, K., Shirai, T. & Go, M. (1993). Protein anatomy – functional roles of barnase module. *J. Biol. Chem.* **268**, 5861–5865.

Edited by J. Karn

(Received 16 August 1995; accepted 15 September 1995)

I quickly published a follow-up paper in *PNAS*[123] explaining the energetic importance of the nucleation-condensation. But, what had really become clear from those first applications of Φ-analysis to two-state folding is that the transition state for folding of most domains is an extended structure that looks like a distorted native state structure. Once this feature is realized, many other factors considered important in folding mechanisms fall simply into place. I summarized some of these in another deceptively simple paper[124]. The little derivation of an equation relating folding rates to contact order gave me much pleasure!

Transition-state structure as a unifying basis in protein-folding mechanisms: Contact order, chain topology, stability, and the extended nucleus mechanism

Alan R. Fersht*

Cambridge University Chemical Laboratory and Cambridge Centre for Protein Engineering, Lensfield Road, Cambridge CB2 1EW, United Kingdom

Contributed by Alan Fersht, December 9, 1999

I attempt to reconcile apparently conflicting factors and mechanisms that have been proposed to determine the rate constant for two-state folding of small proteins, on the basis of general features of the structures of transition states. Φ-Value analysis implies a transition state for folding that resembles an expanded and distorted native structure, which is built around an extended nucleus. The nucleus is composed predominantly of elements of partly or well-formed native secondary structure that are stabilized by local and long-range tertiary interactions. These long-range interactions give rise to connecting loops, frequently containing the native loops that are poorly structured. I derive an equation that relates differences in the contact order of a protein to changes in the length of linking loops, which, in turn, is directly related to the unfavorable free energy of the loops in the transition state. Kinetic data on loop extension mutants of CI2 and α-spectrin SH3 domain fit the equation qualitatively. The rate of folding depends primarily on the interactions that directly stabilize the nucleus, especially those in native-like secondary structure and those resulting from the entropy loss from the connecting loops, which vary with contact order. This partitioning of energy accounts for the success of some algorithms that predict folding rates, because they use these principles either explicitly or implicitly. The extended nucleus model thus unifies the observations of rate depending on both stability and topology.

nucleation-condensation | diffusion-collision | SH3 | CI2 | loops

To understand pathways of protein folding, experimentalists and theoreticians have, over the past decade, focused their efforts on analyzing small proteins. Many of these fold very rapidly with simple two-state kinetics. The structures of the rate-determining transition states have been analyzed in increasing numbers at atomic resolution by protein engineering and Φ-values and by various types of computer simulation. A recent development has been to correlate rate constants of folding (k) of the two-state proteins with their topology by using the gross parameter of the contact order (CO) defined by:

$$CO = \frac{1}{LN}\sum^{N} \Delta Z_{i,j}, \qquad [1]$$

where N is the total number of contacts in the protein, $\Delta Z_{i,j}$ is the number of residues separating contacts i and j, and L is the number of residues in the protein. In a protein with low contact order, residues interact, on average, with others that are close in sequence. A high contact order implies that there is a large number of long-range interactions (1). That is, residues interact frequently with partners that are far apart in sequence. There is a statistically significant correlation between ln$k$ and CO, whereby the rate constant of folding decreases with increasing contact order (Fig. 1). This correlation points to topology being an important factor in the rate of folding. The questions are why and what does it tell us?

The structure of the rate-determining transition state in protein folding can be derived by Φ-value analysis (2, 3). This procedure uses protein engineering to make suitable mutants of the protein, and changes in the free energy of activation ($\Delta\Delta G^{\ddagger}$) and equilibrium ($\Delta\Delta G$) on mutation are measured. Φ is defined by $\Delta\Delta G^{\ddagger}/\Delta\Delta G$. A value of Φ for folding of 0 means that the interaction measured is as poorly formed in the transition state as it is in the denatured state. A value of one means that it is as well formed in the transition state as in the native structure. Exactly the same approach had been used previously (4, 5) to understand changes in enzyme-substrate reactions during binding and catalysis, and the analogous equation was used to define the equivalent of Φ (5).

Very recently, Φ-value analysis has been applied to three proteins to support the contact order theory and the role of topology (6–8). But, in apparent contradiction, it has been found that three members of a family of the same topology fold with rate constants that correlate with stability and not contact order (9). There is strong evidence that many proteins fold by a nucleation mechanism, whereas arguments have been made in favor of hierarchical (framework) mechanisms in which preformed elements of secondary structure associate (10, 11). I wish now to present arguments that there are no real conflicts among these proposals, and that each of these mechanisms is accommodated in existing schemes that invoke general features of transition states for folding determined by Φ-value analysis.

Nature of Transition State for Protein Folding. Φ-Value analysis of CI2 (12, 13) shows that:

(*i*) The protein folds around an extended nucleus that is composed of a contiguous region of structure (for CI2, an α-helix) and long-range native interactions with groups distant in sequence.

(*ii*) The transition state for folding is a distorted form of the native structure, which appears to be more distorted and weakened the further away from the nucleus. There is a gradation of Φ-values, the ones in the nucleus tending to be 0.5–0.7 and the more distal ones, from 0.1 to 0.3.

(*iii*) It was reasoned that the mechanism did not involve the association of preformed elements of secondary structure, but that the secondary and tertiary interactions are formed in parallel because the Φ-values in the nucleus were significantly less than 1. A mechanism was proposed, nucleation-condensation (or nucleation-collapse), that involves the simultaneous

*To whom reprint requests should be addressed. E-mail: arf10@cam.ac.uk.

The publication costs of this article were defrayed in part by page charge payment. This article must therefore be hereby marked "*advertisement*" in accordance with 18 U.S.C. §1734 solely to indicate this fact.

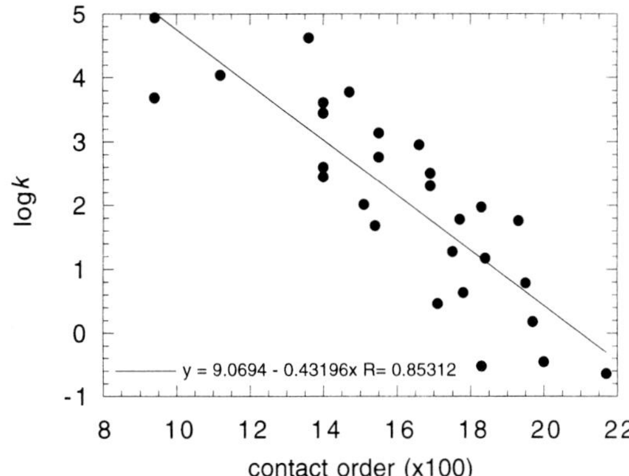

Fig. 1. Plot of log*k* vs. 100 × *CO* for two-state folding proteins listed in Table 18.1 of ref. 3 and unpublished data from this laboratory.

collapse or condensation of the tertiary structure around the extended nucleus as it is formed.

(*iv*) Mutations, especially in the folding nucleus, affect the folding rate, and there is a relation between folding rate and stability [a Brønsted plot (14)].

The rate-determining transition state in the multistep pathway for the folding of barnase is more polar (15). Many of the Φ-values are very close to 0 or 1, and so the rate-determining step is the docking of preformed structural domains. Mutations in the regions of barnase that are unstructured in the transition state have folding rates that are insensitive to mutation. The transition state structures of many small proteins may be classified into "CI2" (a gradation of Φ-values) or "barnase" (polarized, with a significant number of Φ-values close to 0 or 1) and are listed in ref. 3, Table 19.2. Nucleation-condensation appears to be a widespread mechanism. Indeed, lattice simulations independently showed that a specific nucleus is an optimal mechanism for folding model proteins (16).

Implications of a Native-Like Transition State. The transition state resembles native-like structural elements with the connecting loops tending to be poorly structured (some structured loops are formed in the rate-determining transition state for barnase). A consequence of the extended nucleus is that the overall topology of the transition state must resemble that of the native structure. The correlation between folding rate constant, which depends on transition state structure, and contact order of the native state for a large number of proteins (Fig. 1) implies that the topology of the transition state resembles the topology of the native chain in general.

Relationship of Contact Order to Loop Length and Configurational Entropy. It has been speculated that the dependence of rate constant on contact order may be a consequence of the relative importance of short-range and long-range interactions, or it might somehow relate to the length of the connecting loops in proteins (17). To resolve this, I derive a simple equation relating contact order and loop length for a specific case and then apply this to kinetic data. The analysis is done for just that specific case, but it illustrates the general principles.

Suppose two segments of a protein, A and B, are connected by a loop of length n_l, and that there are $N_{A/B}$ interactions across the A/B interface. Suppose we insert l residues in the loop that makes no interaction with other residues (Fig. 2). The total number of interactions in the protein does not change, but the value of Z for the interaction across the A/B interfaces increases

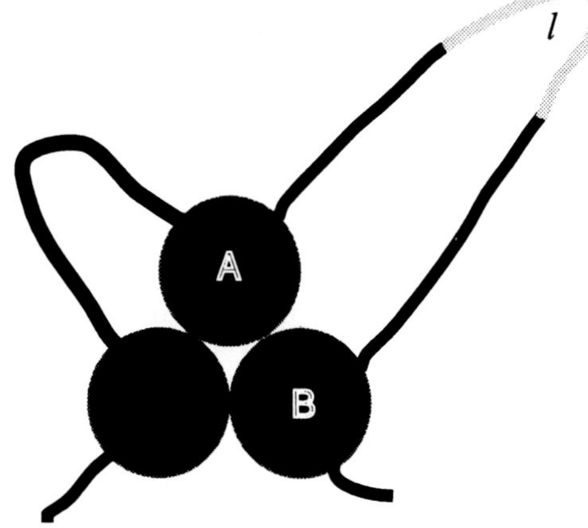

Fig. 2. Cartoon of the extended (specific) nucleus mechanism of the nucleation-condensation mechanism. This is for the extreme case of the connecting loops being unstructured. The filled circles represent native-like elements of secondary structure that interact mainly by native tertiary interactions. The shaded part of the loop illustrates an insertion of length *l*.

by $l \cdot N_{A/B}$, because each of the $N_{A/B}$ interactions is displaced by l residues in sequence. Thus:

$$CO = \frac{1}{(L+l)N}\left(\sum^{N} \Delta Z_{i,j} + l \cdot N_{A/B}\right). \quad [2]$$

Thus, contact order increases with increasing loop size.

The loss of configurational entropy of closing an unstructured loop of $n_l + l$ residues relative to one of n_l residues is calculated from standard polymer theory to be:

$$\Delta\Delta S = -\frac{3}{2}R \ln\left(1+\frac{l}{n_l}\right). \quad [3]$$

Eqs. **2** and **3** show that changes in contact order and configurational entropy are directly related. The relationship becomes simpler when $l \ll n_l$. Then

$$\Delta CO \approx \frac{l \cdot N_{A/B}}{LN} \quad [4]$$

and

$$\Delta\Delta S \approx -\frac{3}{2}R\frac{l}{n_l} \quad [5]$$

To a first approximation:

$$\Delta\Delta S \approx -\frac{3RLN}{2N_{A/B}n_l}\Delta CO \quad [6]$$

or

$$\Delta CO \approx -\frac{2N_{A/B}n_l}{3LN} \times \frac{\Delta\Delta S}{R}. \quad [7]$$

There are several assumptions in Eq. **3** that cause considerable uncertainty in the exact relationship between $\Delta\Delta S$ and CO. As discussed (18), the exact value of n_l is not known, and interactions within the loop and with neighboring residues will alter its energy. Further, the numerical term 3/2 may be an underestimate, and values up to 2.4 may be appropriate, depending on the nature of the side chains. Conversely, if the loop is partly structured in the transition state, then 3/2 will be an overestimate.

Effect of Unstructured Closed Loop on Folding Kinetics. A simple way of analyzing reaction kinetics is the application of transition state theory or analogous theories that assume that a transition state or activated complex is in (virtual) rapid equilibrium with the ground state. The rate constant is then given by an equation that is the product of the virtual equilibrium constant and the frequency (ν) and transmission coefficient (κ) of passing over the energy barrier. Suppose that the transition state has a free energy that is higher than the ground state by $\Delta\Delta G^{\ddagger}$, then the rate constant is given by:

$$k = \kappa\nu \exp(-\Delta\Delta G^{\ddagger}/RT). \quad [8]$$

Energies that affect the equilibrium are directly manifested in the activation energy, $\Delta\Delta G^{\ddagger}$. The product $\kappa\nu$ cancels out when comparing rates of reaction under identical conditions, as in Φ-value analysis. We can calculate the effect of loop entropy on rate by assuming that folding follows Eq. **8**. Suppose that the loop of n_l residues is unstructured in the transition state but is closed at either end by tertiary interactions made as two elements of structure associate in the transition state. Increasing the loop by l residues increases the loss of configurational entropy by $\Delta\Delta S$ according to Eq. **3**. The increase in free energy of activation because of loop size is given by $T\Delta\Delta S$, and the change in ln$k$ by $-\Delta\Delta S/R$. Thus, according to Eq. **6**,

$$\Delta \ln k \approx -\frac{3LN}{2N_{A/B}n_l}\Delta CO. \quad [9]$$

This equation predicts an approximately linear relationship between changes in k and changes in contact order resulting from changes in loop size. The equation is derived for a specific case for changes within a single protein, but it does illustrate some general points. Importantly, the slope is a function of the number of interactions between the elements connected by the loop and the nature and degree of structure formation within the loop in the transition state, which will alter the factor of 3/2. The slope is thus unlikely to be a constant.

Diffusion Control of Protein Folding? The effects of inserting residues into loops of CI2 and the α-spectrin SH3 domain have been systematically analyzed and have provided an excellent system for benchmarking the above equations (18, 19). Before applying Eq. **9** to the data, the question has to be addressed of whether the folding reaction is diffusion controlled. The usual meaning of "diffusion control" in chemical kinetics is that a reaction is limited by the rate of diffusion together of molecules. The characteristics of such a diffusion-controlled reaction are that it has a low activation energy, the rate decreases with increasing viscosity ($k \alpha 1/\eta$), and the reaction rate is not affected by the activation energy of the chemical steps. It was proposed from the effects of viscogenic agents that the folding of CspB is diffusion controlled, that is, compaction of the polypeptide chain is rate determining because the rate constant for folding is inversely proportional to viscosity (20). Plaxco and Baker showed that folding of the 62-residue IgG binding domain of protein L is not diffusion controlled (21), as did Bhattacharya and Sosnick the α-helical GCN4-p2′ (22). The kinetics was analyzed by Kramers' theory (23), a more fundamental version of Eq. **8** that does not invoke a fixed value of ν. Kramers' theory invokes an inverse dependence of $\kappa\nu$ on viscosity under "high-viscosity" conditions because of frictional effects on passage over the transition state barrier. At "high viscosity," the system moves many times back and forth over the top of the barrier in a diffusion-like way before it can escape to give products, and the rate constant becomes inversely proportional to viscosity. Under these conditions, the rate constant varies with both viscosity and the change in chemical activation energy. Plaxco and Baker (21) pointed out that the interpretation of the observed inverse dependence of folding rate on viscosity is ambiguous and is consistent both with diffusion-limited chain collapse and frictional effects on the transition state, because both predict that rates vary as $1/\eta$. But numerical simulation demonstrated that the effects of viscosity were at the transition state and not at the early diffusive steps (21). The assertions of Baker and Plaxco are directly supported from observations on CI2 that the folding of wild-type and loop insertion mutants of CI2 is on a much longer time scale (>10 ms) than chain compaction events, which are on a time scale of 100 μs or less (24). Indeed, one mutant of CI2 (RF48) folds some 40 times faster than wild type, with a half life of 0.3 ms, showing directly that diffusional events per se are not rate limiting for wild type. Similarly, some mutants of the SH3 domain fold faster than wild type (120 s^{-1} vs. 3 s^{-1}; Luis Serrano, personal communication) so that its folding is not diffusion limited. Further, a compact transition state is not consistent with an early transition state that is expected from collapse being rate determining (21, 24).

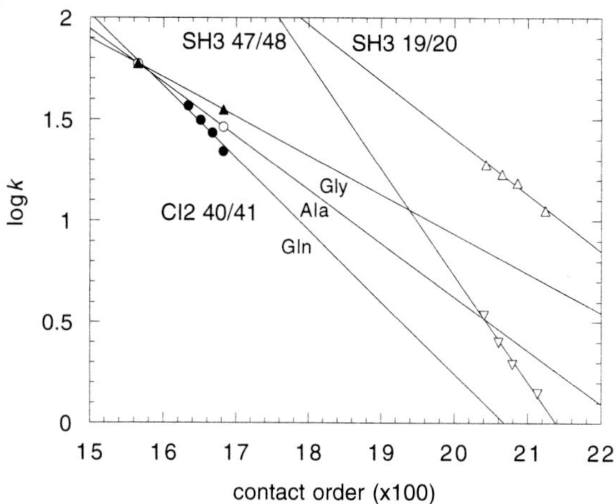

Fig. 3. Plots of log$k$ vs. 100 × CO for loop insertion mutants of CI2 and the α-spectrin SH3 domain.

Experimental Tests of the Contact Order Relationships. Plotted in Fig. 3 are the logarithms ($\log_{10}$) of the rate constants vs. 100 × CO for mutants of CI2 containing 7, 9, 11, and 13 residues that are mainly Gln, and 13 residues that are predominantly Ala or Gly inserted between residues 40 and 41 of its loop. The slopes are −0.36, −0.26, and −0.19, respectively, compared with a value of −0.12 calculated from Eq. **9** (converted to $\log_{10}$), assuming a value of n_l of 15, which is the length of the loop in the native protein. Also plotted are data for 2, 4, 6, 10 residues inserted between residues 19 and 20 or 47 and 48 of α-spectrin SH3. The slopes are −0.28 and −0.53, respectively, compared with values of −0.07 and −0.47 calculated from Eq. **8** and the native loop lengths. These values span that of −0.43 found for the proteins plotted in Fig. 1. Given the uncertainties in the theory behind Eq. **3** and the assumptions about the nature of the loops in the transition state, the agreement is probably as good as can be expected. Nevertheless, the observations that logarithms of folding rate constants are linear with contact order for these specific systems, and that the slopes are of the right order of magnitude and vary qualitatively as predicted, are consistent with and lend strong support for the importance of contact order in protein folding.

How Precise Is the Dependence of ln*k* on Contact Order? The rate constant for folding depends on all the energy differences between the transition and ground states. The energies first analyzed are those caused by direct interactions of residues, especially those in the folding nucleus. These can dominate the rate equation. For example, the rate constants for the folding of CI2 span three orders of magnitude: wild type folds at 25°C at 56 s^{-1}, the double mutant AG16/IA57 in the folding nucleus at 2.4 s^{-1}, and RF48 at 2300 s^{-1}. Consequently, the contact order equation cannot by itself accurately predict rate constants. But the correlation between log*k* and contact order is truly remarkable and points to general principles about the nature of folding that must be included in simulations. Further, the correlation also implies that the free energy of forming the nucleus in the transition state has the component from direct interaction constant within a few kcal/mol so that the entropy terms from contact order appear above the "noise" level from differences in specific interactions.

Pathway to the Extended Nucleus: Nucleation-Condensation vs. Diffusion-Collision. The increasing accumulation of Φ-value data and the correlation of the contact order plot with native state topology are strong evidence for the mechanism involving an extended nucleus in the transition state being quite general. There is persuasive evidence both for the nucleation-condensation mechanism for CI2 and allied proteins with simultaneous formation of secondary and tertiary interactions and some evidence for a diffusion-collision model for the folding of the small α-helical fragment of λ repressor by preformed elements of secondary structure associating (25, 26). The strict diffusion-collision model predicts that local interactions everywhere in helices and strands define the folding rate. Nucleation-condensation predicts that some native tertiary interactions are crucial as well as the native interactions' secondary structure in the nucleus. In practice, the transition states for both processes involve extended structures with a mixture of tertiary and secondary interactions, the secondary structural elements in the diffusion-collision mechanism forming the tertiary interactions as the secondary structures coalesce. Nucleation-condensation and diffusion-collision mechanisms are basically extremes of the same process, with the elements of secondary structure being inherently more stable and better formed in the diffusion model than in nucleation-condensation (27) (Fig. 4).

The transition states for the stepwise and nucleation mechanisms are thus qualitatively similar, which leads to a dilemma in interpreting kinetic data and simple models of folding. According to the transition state equation (Eq. **8**), the rate folding constant depends on just the energy difference between the transition state and the ground state, provided any preequilibria are rapid (Fig. 3). Thus, kinetic data relating changes in structure and kinetics respond in a qualitatively similar manner to changes in structure. But they can be distinguished between by quantitative data: the diffusion-collision mechanism is predicted to have Φ-values of 1 for the relevant secondary structure, which is found for a model system (28), whereas nucleation-condensation has mainly fractional values that can tend to 1.0 for especially stable elements.

Baldwin and Rose (10, 11) have argued that all proteins fold in a hierarchical model, with the successive docking of elements of native structure. But the predictive success of their hierarchical model does not have implications for the kinetic mechanism, because any mechanism that invokes rapid preequilibria and an extended nucleus containing native-like secondary structure will fit the observed kinetics.

Simulations of Folding. The mechanistic features of an extended folding nucleus that is stabilized directly by native-like secondary-structure interactions and destabilized by loop entropy are explicitly or implicitly included in the simple algorithms for folding of Muñoz and Eaton (29) and Baker and coworkers (6). Muñoz and Eaton successfully calculated the folding rate constants of 22 proteins using an elementary statistical mechanical model and the known distribution of interactions in their three-dimensional structures. They assumed residues come into contact only after all of the intervening chain is in the native conformation, and that native structure grows from localized regions that then fuse to form the complete native molecule. The relative success of their calculations suggested that folding rate constants are largely determined by the distribution and strength of contacts in the native structure, that is, topology is important.

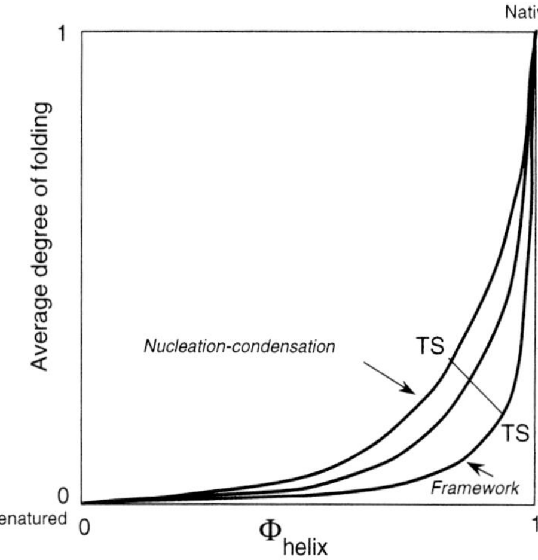

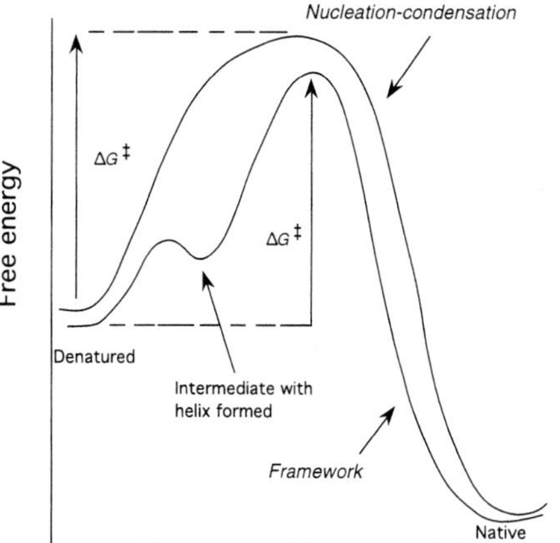

Fig. 4. (*Lower*) Simplified energy diagrams for true two-state folding via nucleation-condensation and apparent two-state kinetics for a framework mechanism that involves the formation of, say, an α-helix, at a higher energy than the denatured state. If both mechanisms involve an extended network of long-range native-like tertiary interactions around the helix, then the free energy of activation, $\Delta G^{\ddagger}$, responds to changes in structure in a similar manner for both mechanisms, because $\Delta G^{\ddagger}$ depends just on the difference in energy between similar transition states and the denatured state. (*Upper*) Two-dimensional representation of the merging of the nucleation-condensation and framework mechanisms. In the framework mechanism, the Φ-values for the formation of the helix are close to 1, because it is relatively stable and can form to an appreciable extent in the absence of tertiary interactions. As the helix becomes less stable, it requires more tertiary interactions to become stable in the transition state, and so the formation of helix is coupled with that of tertiary structure. The Φ-values for formation of the helix can then be appreciably less than 1.

The Baker model invokes similar features and is successful in predicting structures as well as rates (6, 30). Very recently, Debe and Goddard (31) have calculated accurately the rates of folding of 21 of the two-state folding proteins, on the basis of the nucleation-condensation mechanism. The extended nucleus mechanism of the nucleation-condensation mechanism is clearly a very robust basis for calculating folding rate constants.

Although there is no single mechanism for protein folding, the extended transition state provides a unifying feature in the two-state folding of small domains. An extended nucleus is necessary for the folding of these domains, because a large number of interactions have to be made for an energetically downhill passage after the transition state.

I thank David Baker, Hans Frauenfelder, Kevin Plaxco, Luis Serrano, and Eugene Shakhnovich for helpful discussion, Kevin Plaxco for the computer program for calculating contact orders, and Eugene Shakhnovich for the suggestion of Fig. 2.

1. Plaxco, K. W., Simons, K. T. & Baker, D. (1998) *J. Mol. Biol.* **277,** 985–994.
2. Matouschek, A., Kellis, J. T., Jr., Serrano, L. & Fersht, A. R. (1989) *Nature (London)* **340,** 122–126.
3. Fersht, A. (1999) *Structure and Mechanism in Protein Science: A Guide to Enzyme Catalysis and Protein Folding* (Freeman, New York).
4. Fersht, A. R., Leatherbarrow, R. J. & Wells, T. N. C. (1986) *Nature (London)* **322,** 284–286.
5. Fersht, A. R., Leatherbarrow, R. & Wells, T. N. C. (1987) *Biochemistry* **26,** 6030–6038.
6. Riddle, D. S., Grantcharova, V. P., Santiago, J. V., Alm, E., Ruczinski, I. & Baker, D. (1999) *Nat. Struct. Biol.* **6,** 1016–1024.
7. Martinez, J. C. & Serrano, L. (1999) *Nat. Struct. Biol.* **6,** 1010–1016.
8. Chiti, F., Taddei, N., White, P. M., Bucciantini, M., Magherini, F., Stefani, M. & Dobson, C. M. (1999) *Nat. Struct. Biol.* **6,** 1005–1009.
9. Clarke, J., Cota, E., Fowler, S. B. & Hamill, S. J. (1999) *Structure (London)* **7,** 1145–1153.
10. Baldwin, R. L. & Rose, G. D. (1999) *Trends Biochem. Sci.* **24,** 26–33.
11. Baldwin, R. L. & Rose, G. D. (1999) *Trends Biochem. Sci.* **24,** 77–83.
12. Otzen, D. E., Itzhaki, L. S., elMasry, N. F., Jackson, S. E. & Fersht, A. R. (1994) *Proc. Natl. Acad. Sci. USA* **91,** 10422–10425.
13. Itzhaki, L. S., Otzen, D. E. & Fersht, A. R. (1995) *J. Mol. Biol.* **254,** 260–288.
14. Fersht, A. R., Itzhaki, L. S., elMasry, N. F., Matthews, J. M. & Otzen, D. E. (1994) *Proc. Natl. Acad. Sci. USA* **91,** 10426–10429.
15. Fersht, A. R., (1993) *FEBS Lett.* **325,** 5–16.
16. Abkevich, V. I., Gutin, A. M. & Shakhnovich, E. I. (1994) *Biochemistry* **33,** 10026–10036.
17. Goldenberg, D. P. (1999) *Nat. Struct. Biol.* **6,** 987–990.
18. Ladurner, A. G. & Fersht, A. R. (1997) *J. Mol. Biol.* **273,** 330–337.
19. Viguera, A. R. & Serrano, L. (1997) *Nat. Struct. Biol.* **4,** 939–946.
20. Jacob, M., Schindler, T., Balbach, J. & Schmid, F. X. (1997) *Proc. Natl. Acad. Sci. USA* **94,** 5622–5627.
21. Plaxco, K. W. & Baker, D. (1998) *Proc. Natl. Acad. Sci. USA* **95,** 13591–13596.
22. Bhattacharyya, R. P. & Sosnick, T. R. (1999) *Biochemistry* **38,** 2601–2609.
23. Kramers, H. A. (1940) *Physica* **7,** 284–304.
24. Ladurner, A. G. & Fersht, A. R. (1999) *Nat. Struct. Biol.* **6,** 28–31.
25. Burton, R. E., Huang, G. S., Daugherty, M. A., Fullbright, P. W. & Oas, T. G. (1996) *J. Mol. Biol.* **263,** 311–322.
26. Burton, R. E., Huang, G. S., Daugherty, M. A., Calderone, T. L. & Oas, T. G. (1997) *Nat. Struct. Biol.* **4,** 305–310.
27. Fersht, A. R. (1997) *Curr. Opin. Struct. Biol.* **7,** 3–9.
28. Kippen, A. D. & Fersht, A. R. (1995) *Biochemistry* **34,** 1464–1468.
29. Munoz, V. & Eaton, W. A. (1999) *Proc. Natl. Acad. Sci. USA* **96,** 11311–11316.
30. Simons, K. T., Bonneau, R., Ruczinski, I. & Baker, D. (1999) *Proteins Struct. Funct. Genet.* Suppl. 3, 171–176.
31. Debe, D. A. & Goddard, W. A. (1999) *J. Mol. Biol.* **294,** 619–625.

Chapter 14

Plasticity of the Transition State in Folding: Structure–Reactivity Relationships

'Πάντα ῥεῖ (panta rhei — everything flows).'

attributed to Heraclitus

We took the structure–activity relationships to the next stage by showing that we could detect movements of the transition state along the reaction co-ordinate as found for simple organic chemical reactions according to the Hammond postulate[125,126]. Jaqui Matthews also found perpendicular movement — the 'Anti-Hammond' effect[127], for which Valerie Daggett later simulated the effects and the movements[128]. In effect, we could map out changes in the energy landscape around the transition state by using changes in Φ-values on rounds of mutation and also changes in another parameter, β_T, the β-Tanford value. In these papers, I paid homage to Charles Tanford, whose work I greatly admired, by lending his name to a term, β_T, which relates the relative change in solvent accessible surface area on going from the denatured to transition states. That free energy relationship term is based on Tanford's seminal work on the effects of denaturants on protein folding.

Mikael Oliveberg refused to work on Φ-values with me and followed his own path, producing some clever papers[129–139] but eventually saw the light[140–142], and subsequently has taken the analysis of transition state plasticity to new levels. His paper on negative enthalpy values in protein folding[131] is a classic. There are claims by Thomas Kiefhaber that some of the apparent movements of transition states are, in fact, movements of ground states. But, for the ones we described, the movements are genuine transition state effects[119].

14.1 2003: Gun Stenberg, Mikael Oliveberg and Pia Harryson.

14.2 2003: Gonzalo se Prat Gay, Stefan Freund and Ralph Golbik.

Proc. Natl. Acad. Sci. USA
Vol. 90, pp. 7814–7818, August 1993
Biochemistry

Application of physical organic chemistry to engineered mutants of proteins: Hammond postulate behavior in the transition state of protein folding

(protein engineering/linear free-energy relationships/barnase)

ANDREAS MATOUSCHEK*† AND ALAN R. FERSHT*†

*Cambridge Centre for Protein Engineering, University Chemical Laboratory, Lensfield Road, Cambridge, CB2 1EW, United Kingdom; and †Medical Research Council Centre, Hills Road, Cambridge, CB2 2QH, United Kingdom

Contributed by Alan R. Fersht, May 26, 1993

ABSTRACT Transition states in protein folding may be analyzed by linear free-energy relationships (LFERs) analogous to the Brønsted equation for changes in reactivity with changes in structure. There is an additional source of LFERs in protein folding: the perturbation of the equilibrium and rate constants by denaturants. These LFERs give a measure of the position of the transition state along the reaction coordinate. The transition state for folding/unfolding of barnase has been analyzed by both types of LFERs: changing the structure by protein engineering and perturbation by denaturants. The combination has allowed the direct monitoring of Hammond postulate behavior of the transition state on the reaction pathway. Movement of the transition state has been found and analyzed to give further details of the order of events in protein folding.

Structure–activity studies are fundamental to the methods of physical organic chemistry. Some of the basic concepts have been found to be applicable to the study of proteins whose structures have been subtly altered by site-directed mutagenesis, both for the analysis of catalysis (1–6) and for mapping the pathway of protein unfolding and folding at the level of individual atomic interactions (7–11). In physical organic chemistry, structure–activity relationships are often analyzed by linear free-energy relationships (LFERs) such as the Brønsted equation for changes in reactivity with changes in structure. That is, a rate constant k responds to changes in an associated equilibrium constant K according to $\partial \log k/\partial \log K = \beta$. An analogous equation may be applied to the analysis of transition states and intermediates in protein folding where the structure of the protein is altered by site-directed mutagenesis. There is an additional way of constructing LFERs to study the rates of protein unfolding and folding: the rates and equilibria of unfolding may be perturbed by the addition of denaturants. Both the equilibrium and activation free energies of unfolding have generally been found to be linearly related to the concentration of denaturant (urea or guanidinium chloride) (12, 13):

$$\Delta G_{\text{U-F}} = \Delta G_{\text{U-F}}^{\text{H}_2\text{O}} - m_{\text{U-F}}[\text{denaturant}] \quad [1]$$

$$\Delta G_{\ddagger\text{-F}} = \Delta G_{\ddagger\text{-F}}^{\text{H}_2\text{O}} - m_{\ddagger\text{-F}}[\text{denaturant}], \quad [2]$$

where $\Delta G_{\text{U-F}}$ is the difference in free energy between folded and unfolded protein at a given denaturant concentration, $\Delta G_{\ddagger\text{-F}}$ is the difference in free energy between the transition state and the folded protein at a given denaturant concentration, $\Delta G_{\text{U-F}}^{\text{H}_2\text{O}}$ and $\Delta G_{\ddagger\text{-F}}^{\text{H}_2\text{O}}$ are the values in water, and $m_{\text{U-F}}$ and $m_{\ddagger\text{-F}}$ are constants for a particular protein. The quantities m are very informative *per se*. Proteins are denatured by solvents such as urea and guanidinium chloride solutions because all parts of proteins, but especially hydrophobic side chains, are more soluble in denaturant solutions than in water. $m_{\ddagger\text{-F}}$ and $m_{\text{U-F}}$ are proportional to the change in exposure of amino acids as the structure of the folded protein changes to that of the transition or unfolded state. The average fractional increase in exposure to solvent of the protein in the transition state relative to that on complete unfolding is given by $m_{\ddagger\text{-F}}/m_{\text{U-F}}$ (12). The ratio $m_{\ddagger\text{-F}}/m_{\text{U-F}}$ is thus an index of the position of the transition state on the reaction coordinate. Eqs. **1** and **2** may be combined to give $\partial(\Delta G_{\ddagger\text{-F}})/\partial(\Delta G_{\text{U-F}}) = m_{\ddagger\text{-F}}/m_{\text{U-F}}$, so that the quantity $m_{\ddagger\text{-F}}/m_{\text{U-F}}$ corresponds also to a Brønsted β value.

The position of the transition state on the reaction coordinate can thus be determined by using the slopes of unfolding kinetics and equilibrium denaturation. The sensitivity of the rate and equilibrium constants to the concentration of denaturants provides additional probes of the position of the transition state on the reaction pathway. We now analyze the transition state for the unfolding of the small ribonuclease barnase, which has been subject to considerable earlier studies (7, 10).

MATERIALS AND METHODS

All the mutations discussed here have been described before (9) or will be discussed elsewhere. The proteins were expressed and purified and their free energies of unfolding were determined by urea denaturation as published (9). The rate constants of protein unfolding were measured as described (7, 10).

RESULTS

The rate constants of unfolding of barnase and its mutants were generally measured at six equally spaced urea concentrations from 6.0 to 8.5 M. Fig. 1 shows plots of log k_{u} against [urea] for wild-type barnase and all the Ile → Val mutations in the main hydrophobic core. All plots seem perfectly linear. However, just as classical LFERs can deviate from linearity when plotted over a very wide range of changes in activity, the relationship between log k_{u} and [urea] for barnase deviates from linearity when the rate constants are analyzed over a very wide range of urea concentrations. The data are fully described by

$$\log k_{\text{u}} = \log k_{\text{u}}^{\text{H}_2\text{O}} + m_{k_{\text{u}}}[\text{urea}] - 0.014[\text{urea}]^2, \quad [3]$$

so that $m_{\ddagger\text{-F}} = 2.303RT(m_{k_{\text{u}}} - 0.014[\text{urea}])$. We therefore choose a fixed concentration of urea for the determination of

The publication costs of this article were defrayed in part by page charge payment. This article must therefore be hereby marked "*advertisement*" in accordance with 18 U.S.C. §1734 solely to indicate this fact.

Abbreviation: LFER, linear free-energy relationship.

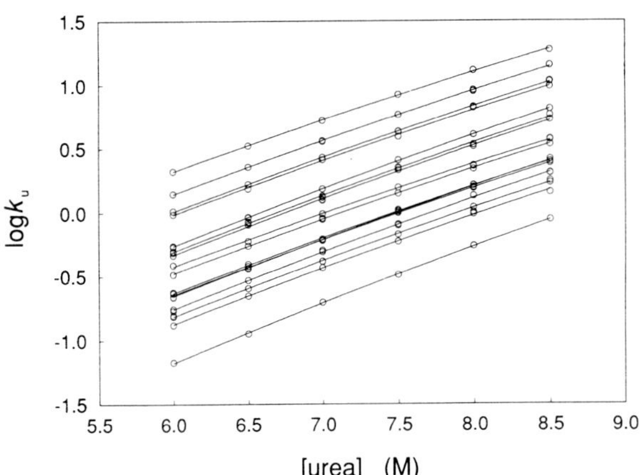

FIG. 1. Plots of log k_u versus [urea] for wild-type barnase and all Ile → Val mutants in the main hydrophobic core.

$m_{\ddagger\text{-F}}$. Here, the value is 7.25 M, which is the middle of the concentration range generally used. The slopes $m_{\ddagger\text{-F}}$ are found to vary significantly with mutation.

The change in free energy of activation, $\Delta\Delta G_{\ddagger\text{-F}}$, on mutation is defined by

$$\Delta\Delta G_{\ddagger\text{-F}} = -RT\ln(k'_u/k_u), \quad [4]$$

where k'_u is the first-order rate constant for the unfolding of a mutant and k_u is that for wild-type barnase.

The values of $m_{\ddagger\text{-F}}$, $m_{\text{U-F}}$, $m_{\ddagger\text{-F}}/m_{\text{U-F}}$, and $\Delta\Delta G_{\ddagger\text{-F}}$ for 7.25 M urea are listed in Table 1. The values in water should not be used for the following analysis because errors in the experimental determination of $m_{\ddagger\text{-F}}$ and $\Delta\Delta G_{\ddagger\text{-F}}$ will be statistically correlated: errors leading to lower values of $m_{\ddagger\text{-F}}$ will automatically yield lower extrapolated values of $\Delta\Delta G_{\ddagger\text{-F}}$ and vice versa. At 7.25 M urea, the errors in the two parameters will not be significantly correlated.

Fig. 2*a* shows a plot of $m_{\ddagger\text{-F}}/m_{\text{U-F}}$ against $\Delta\Delta G_{\ddagger\text{-F}}$ for all mutations in the main hydrophobic core of barnase, with the values of $\Delta\Delta G_{\ddagger\text{-F}}$ and $m_{\ddagger\text{-F}}$ at 7.25 M. The plots show that $m_{\ddagger\text{-F}}/m_{\text{U-F}}$ is correlated with $\Delta\Delta G_{\ddagger\text{-F}}$ (correlation coefficient $r = 0.70$, number of points $n = 20$) with the intercept on the y axis at 0.306 ± 0.005 and a slope of 0.019 ± 0.005 mol·kcal^{-1} for a linear fit. This analysis does not depend on the fitting of the unfolding rate constants to the quadratic Eq. **3** but is also observed when a linear approximation is used. The same type of dependence between $m_{\ddagger\text{-F}}/m_{\text{U-F}}$ and $\Delta\Delta G_{\ddagger\text{-F}}$ is found to hold for all but one (Thr → Ala 6) of the mutations affecting the major α-helix (Fig. 2*b*). The correlation coefficient is $r = 0.93$ ($n = 10$, y-axis intercept $= 0.297 \pm 0.004$, and slope $= 0.037 \pm 0.005$ mol·kcal^{-1}). However, this behavior is not general for all mutants. An equivalent plot for mutations outside the hydrophobic cores and not affecting α-helices shows little correlation (Fig. 2*c*; $r = 0.20$, $n = 25$).

We used for the above calculations the same value of $m_{\text{U-F}}$ for each mutant, 1.973 ± 0.007 kcal·mol^{-1}·M^{-1}, which is the mean value from all the unfolding curves. There is no material difference in using the observed individual values of $m_{\text{U-F}}$ for each point. The correlation coefficients in this case for the main core, the major helix, and the mutations affecting neither cores nor helices are 0.64, 0.92, and 0.40, respectively. We prefer to use the mean value of $m_{\text{U-F}}$ because the individual measurements have large experimental errors and are the least accurate values in the study (9).

DISCUSSION

There are important differences between structure–activity relationships as applied to small molecules and as part of the protein engineering procedure. (*i*) In simple physical organic chemistry, one looks systematically at the effects of substituents on different parts of the reactant. The division and attribution of the effects are usually clear. In protein folding and enzyme catalysis, the changes are distributed widely over the macromolecule and have to be grouped by experiment and analysis. (*ii*) The reactions in classical physical organic chemistry generally involve covalent bond changes that are subject to quantum theory. In particular, electron transfer reactions in biological systems can be analyzed by simple equations, such as the Marcus relationship (14). The changes in proteins that we have examined involve noncovalent interactions which may be analyzed by classical statistical mechanics. (*iii*) Reaction coordinate diagrams for protein folding must contain a large number of intermediates, both sequential and in parallel, that differ only slightly in energy.

There is a linear dependence of $m_{\ddagger\text{-F}}/m_{\text{U-F}}$ on $\Delta\Delta G_{\ddagger\text{-F}}$ for specific sets of mutations (Fig. 2 *a* and *b*). As the value of $\Delta\Delta G_{\ddagger\text{-F}}$ increases (i.e., the energy difference between the transition state and the ground state increases), the value of $m_{\ddagger\text{-F}}/m_{\text{U-F}}$ also increases; i.e., the position of the transition state on the reaction coordinate moves away from the ground state. This is the experimental observation of the Hammond postulate (15). Hammond formulated the following postulate in order to describe the experimentally observed correlations between changes in equilibrium and rate constants upon modification of the reagent: "If two states, as for example, a transition state and an unstable intermediate, occur consecutively during a reaction process and have nearly the same energy content, their interconversion will involve only a small reorganization of the molecular structure" (15). The closer the two states are in energy, the more their structures resemble each other. Mutations in the main hydrophobic core and in the major α-helix show clear correlations of the energy difference between transition state and folded state with the position on the reaction coordinate of the two states relative to each other.

The mutations of barnase can be separated into three groups, those in the main hydrophobic core, those in the major α-helix, and those affecting neither hydrophobic cores nor helices. Hammond-type behavior is exhibited clearly by mutations in the major α-helix (Fig. 2*b*) and strongly implied for those in the major core (Fig. 2*a*). The mutations outside cores or α-helices show considerably less evidence for Hammond-type behavior (Fig. 2*c*). The weak correlation between $m_{\ddagger\text{-F}}/m_{\text{U-F}}$ and $\Delta\Delta G_{\ddagger\text{-F}}$ could be due to a set of mainly uncorrelated events but with a few mutations having some real correlation.

The transition state of unfolding is also the final rate-limiting transition state of barnase folding (7, 11). We have inferred from our previous data that the breaking of the hydrophobic core is part of the rate-determining step for folding and unfolding, possibly the major factor. Here we find that mutations that weaken the core result in an apparent movement of the transition state and folded state towards each other. This supports our view that events in the hydrophobic cores are important in the rate-limiting transition state.

The major α-helix is not stable by itself in solution. Experiments on fragments of barnase have shown that although its amino acid sequence has a tendency to form an α-helical configuration in solution, this tendency is very small [<5% in water at 4°C (16)]. The helix becomes formed in the protein through its interaction with the rest of the protein. The hydrophobic face of the major α-helix in barnase is an important part of the main hydrophobic core. The data for mutations in the main hydrophobic core and in the major α-helix in Fig. 2 *a* and *b* can be superimposed, and the linear fits of both data sets coincide within experimental error. In

Table 1. Values of $\Delta\Delta G_{\ddagger-F}$, m_{U-F}, $m_{\ddagger-F}$, and $m_{\ddagger-F}/m_{U-F}$ at 7.25 M urea of wild-type barnase and its mutants

Protein	$\Delta\Delta G_{\ddagger-F}$, kcal·mol^{-1}	m_{U-F}, kcal·mol^{-1}·M^{-1}	$m_{\ddagger-F}$, kcal·mol^{-1}·M^{-1}	$m_{\ddagger-F}/m_{U-F}$
Wild type		1.92 ± 0.03	0.614 ± 0.000	0.311 ± 0.001
IV 4	−0.45	1.88 ± 0.19	0.577 ± 0.007	0.292 ± 0.004
IA 4	−0.95	2.08 ± 0.21	0.536 ± 0.004	0.272 ± 0.002
NA 5	−1.51	2.00 ± 0.20	0.548 ± 0.004	0.278 ± 0.002
TA 6	−1.53	1.97 ± 0.20	0.570 ± 0.010	0.289 ± 0.005
TG 6*	−0.68	2.08 ± 0.21	0.544 ± 0.005	0.276 ± 0.003
TD 6*	0.05	1.93 ± 0.19	0.604 ± 0.008	0.306 ± 0.004
DA 8	−0.11	2.08 ± 0.21	0.593 ± 0.008	0.301 ± 0.004
VA 10†	−1.67	1.89 ± 0.19	0.499 ± 0.005	0.253 ± 0.003
VT 10†	−1.13	2.07 ± 0.21	0.536 ± 0.008	0.272 ± 0.004
YA 13*	−1.18	2.03 ± 0.20	0.505 ± 0.005	0.256 ± 0.003
YA 13, YA 17*	−1.69	2.00 ± 0.20	0.465 ± 0.005	0.236 ± 0.003
QI 15	0.26	1.82 ± 0.18	0.634 ± 0.005	0.321 ± 0.003
TS 16*	−0.26	2.03 ± 0.20	0.575 ± 0.004	0.292 ± 0.002
TS 16, YA 17*	−0.46	1.99 ± 0.20	0.530 ± 0.007	0.269 ± 0.004
TR 16*	0.14	1.99 ± 0.20	0.592 ± 0.005	0.300 ± 0.003
YA 17*	−0.47	2.02 ± 0.20	0.541 ± 0.007	0.274 ± 0.004
HQ 18*	−0.07	1.91 ± 0.19	0.551 ± 0.000	0.279 ± 0.001
KR 19	0.07	1.81 ± 0.18	0.610 ± 0.011	0.309 ± 0.006
NA 23	−1.78	1.94 ± 0.19	0.485 ± 0.004	0.246 ± 0.002
YF 24	0.25	1.95 ± 0.20	0.535 ± 0.005	0.271 ± 0.003
IV 25	−1.07	2.00 ± 0.20	0.554 ± 0.005	0.281 ± 0.003
IA 25	−2.85	1.85 ± 0.19	0.251 ± 0.014	0.127 ± 0.007
TA 26	−1.64	2.00 ± 0.20	0.529 ± 0.008	0.268 ± 0.004
TG 26	−0.98	2.03 ± 0.20	0.488 ± 0.010	0.247 ± 0.005
KG 27	−0.41	1.94 ± 0.19	0.528 ± 0.003	0.267 ± 0.002
EG 29	−1.65	1.84 ± 0.18	0.566 ± 0.011	0.287 ± 0.006
QA 31	0.09	1.98 ± 0.20	0.580 ± 0.005	0.294 ± 0.003
QS 31	−0.26	2.00 ± 0.20	0.571 ± 0.007	0.290 ± 0.004
LQ 33	−1.42	1.90 ± 0.19	0.597 ± 0.008	0.303 ± 0.004
VA 36	−1.20	1.83 ± 0.18	0.560 ± 0.010	0.284 ± 0.005
VT 36	−1.27	1.95 ± 0.20	0.605 ± 0.008	0.307 ± 0.004
ND 41	−2.86	1.89 ± 0.19	0.622 ± 0.014	0.315 ± 0.007
VA 45	−1.82	2.08 ± 0.21	0.550 ± 0.005	0.279 ± 0.003
VT 45	−2.62	1.99 ± 0.20	0.595 ± 0.015	0.301 ± 0.008
IV 51	−1.74	2.11 ± 0.21	0.543 ± 0.007	0.275 ± 0.004
DN 54	−2.62	2.11 ± 0.21	0.551 ± 0.010	0.279 ± 0.005
DA 54	−3.32	1.95 ± 0.20	0.526 ± 0.052	0.267 ± 0.026
IV 55	−0.03	1.85 ± 0.19	0.581 ± 0.003	0.294 ± 0.002
IT 55	−0.18	1.84 ± 0.18	0.537 ± 0.004	0.272 ± 0.002
IA 55	−0.23	1.84 ± 0.18	0.551 ± 0.004	0.279 ± 0.002
NA 58	−0.21	1.97 ± 0.20	0.610 ± 0.010	0.309 ± 0.005
KR 62	−0.05	1.95 ± 0.20	0.601 ± 0.008	0.305 ± 0.004
IV 76†	0.67	1.95 ± 0.03	0.581 ± 0.004	0.294 ± 0.002
IV 76, 88†	0.67	1.86 ± 0.07	0.565 ± 0.005	0.286 ± 0.003
IV 76, 96†	−0.87	2.03 ± 0.03	0.570 ± 0.010	0.289 ± 0.005
IV 76, 109†	−1.21	1.93 ± 0.05	0.586 ± 0.001	0.297 ± 0.001
IV 76, 88, 96†	−0.94	1.90 ± 0.04	0.540 ± 0.005	0.274 ± 0.003
IV 76, 88, 109†	−1.50	2.02 ± 0.06	0.550 ± 0.007	0.279 ± 0.004
IV 76, 96, 109†	−1.72	1.91 ± 0.09	0.545 ± 0.004	0.276 ± 0.002
IV 76, 88, 96, 109†	−1.94	2.10 ± 0.09	0.525 ± 0.004	0.266 ± 0.002
IA 76†	−1.03	1.99 ± 0.20	0.584 ± 0.008	0.296 ± 0.004
NA 77	−1.58	1.98 ± 0.20	0.551 ± 0.003	0.279 ± 0.002
YF 78	−1.31	1.98 ± 0.20	0.577 ± 0.018	0.292 ± 0.009
TV 79	0.48	1.88 ± 0.19	0.540 ± 0.005	0.274 ± 0.003
NA 84	−1.87	2.04 ± 0.20	0.584 ± 0.005	0.296 ± 0.003
IV 88†	−0.36	2.03 ± 0.09	0.575 ± 0.008	0.292 ± 0.004
IV 88. 96†	−0.69	1.98 ± 0.06	0.565 ± 0.005	0.286 ± 0.003
IV 88, 109†	−1.11	1.96 ± 0.05	0.575 ± 0.007	0.292 ± 0.004
IV 88, 96, 109†	−1.55	2.07 ± 0.04	0.555 ± 0.003	0.281 ± 0.002
LV 89	−0.20	1.89 ± 0.19	0.588 ± 0.004	0.298 ± 0.002
LT 89	−0.17	1.78 ± 0.18	0.634 ± 0.003	0.321 ± 0.002
SA 91	0.11	1.79 ± 0.18	0.586 ± 0.012	0.297 ± 0.006
SA 92	−0.16	1.78 ± 0.18	0.611 ± 0.010	0.310 ± 0.005
IV 96†	−0.44	1.97 ± 0.03	0.574 ± 0.004	0.291 ± 0.002

Table 1. (*continued*)

Protein	$\Delta\Delta G_{\ddagger\text{-F}}$, kcal·mol^{-1}	$m_{\text{U-F}}$, kcal·mol^{-1}·M^{-1}	$m_{\ddagger\text{-F}}$, kcal·mol^{-1}·M^{-1}	$m_{\ddagger\text{-F}}/m_{\text{U-F}}$
IV 96, 109†	−1.15	1.98 ± 0.04	0.574 ± 0.007	0.291 ± 0.004
TV 99	−2.04	1.84 ± 0.18	0.580 ± 0.005	0.294 ± 0.003
YF 103	0.15	1.89 ± 0.19	0.571 ± 0.010	0.290 ± 0.005
TV 105	−1.16	1.95 ± 0.20	0.612 ± 0.003	0.310 ± 0.002
KR 108	0.25	1.71 ± 0.17	0.610 ± 0.008	0.309 ± 0.004
IV 109†	−0.54	2.14 ± 0.09	0.586 ± 0.004	0.297 ± 0.002
IA 109†	−0.95	2.15 ± 0.22	0.607 ± 0.008	0.308 ± 0.004

Mutations are indicated with one-letter amino acid symbols; e.g., IV 4 indicates Ile → Val at position 4. All measurements were made at 25°C in 50 mM Mes buffer (pH 6.3). Folding and unfolding were monitored by fluorescence spectroscopy (excitation wavelength, 290 nm; emission wavelength, 315 nm). $\Delta\Delta G_{\ddagger\text{-F}}$ is the difference in free energy of activation of unfolding between mutant and wild-type barnase at 7.25 M urea. $m_{\text{U-F}}$ is the slope of plots of the free energy of equilibrium unfolding against urea concentration (Eq. **1**). $m_{\ddagger\text{-F}}$ is calculated by using Eq. **3** at 7.25 M urea. $m_{\ddagger\text{-F}}/m_{\text{U-F}}$ is calculated by using the average of $m_{\text{U-F}}$ for all mutants (1.973 ± 0.007 kcal·mol^{-1}·M^{-1}). The errors quoted are standard errors.

*Mutations affecting mainly the major α-helix of barnase.

†Mutations affecting mainly the major hydrophobic core.

mutants that contain very destabilized helices the helix structure is loosened in the transition state. Thus, in the unfolding of mutants with a destabilized helix, the loosening of the helix becomes an earlier event, and the formation of the helix a later event, in the folding process.

The three-dimensional energy surface representing the folding pathway can be expected to be highly crinkled with many local minima. Superimposed on this pattern are the minima that represent the unfolded, intermediate, and folded states of the protein. The very broad minimum representing the unfolded state allows the protein to take up many different conformations. The minimum becomes narrower along the folding pathway as the structure of the protein becomes increasingly well defined. Depending on which aspects of this model are emphasized, the observed Hammond behavior can be explained in two ways. (*i*) Mutation causes the highest point on the profile to move smoothly along the pathway. The pathway itself does not change. The protein still passes through the same ensemble of structures, but the energies of the structures along the reaction coordinate change. (*ii*) Alternatively, the energy profile can be seen as a manifestation of many converging parallel pathways. The pathways are rather similar but differ in, among other things, the position of the rate-limiting transition state. The mutations change the energies of the pathways and thus their relative occupancies. Seen in this way, the folding pathway changes with mutation. Different parts of the folding process fit the different models without contradictions. For example, the folding pathway is

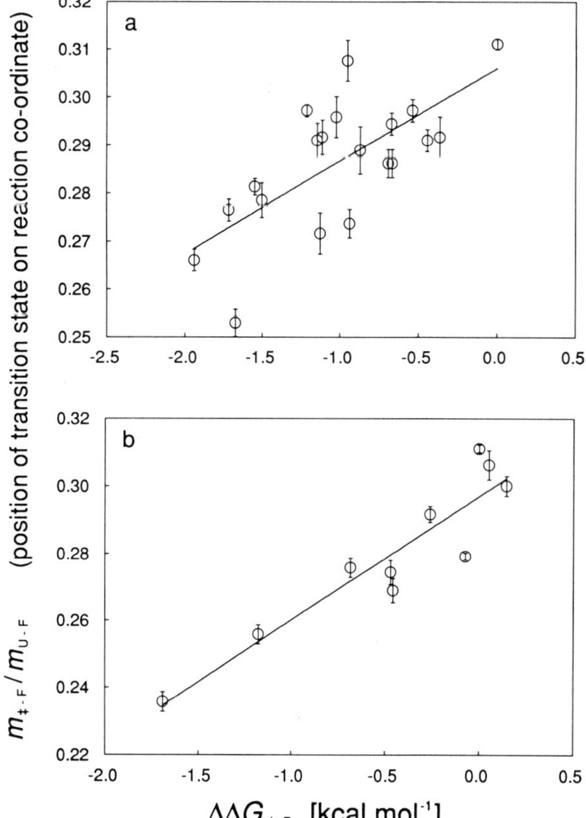

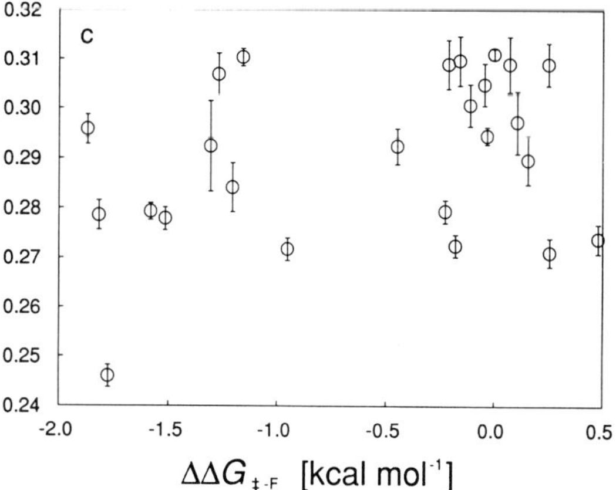

FIG. 2. Plots of $m_{\ddagger\text{-F}}/m_{\text{U-F}}$ against $\Delta\Delta G_{\ddagger\text{-F}}$ at 7.25 M urea for mutants of barnase with mutations in the main hydrophobic core (*a*), mutations affecting the major α-helix (*b*), or mutations affecting neither α-helices nor cores (*c*). The mutations in *b* are those that affect mainly interactions within the helix.

sequential for the formation of major structural elements; e.g., the β-sheet is formed before many of the loops. On the other hand, within an element of structure that is formed late on the pathway, the folding proceeds by many different parallel pathways until that structure is formed.

In summary, we have observed Hammond behavior in the folding of the small ribonuclease barnase. Mutations that reduce the energy difference between the folded state and the transition state lead to the two states being closer to each other in structure. The results on mutations of barnase show that the transition state of protein folding and unfolding is strongly influenced by events in the main hydrophobic core of the protein and in its major helix. It appears that the position of the transition state on the reaction coordinate is sensitive to mutation.

1. Fersht, A. R., Leatherbarrow, R. J. & Wells, T. N. C. (1986) *Nature (London)* **322**, 284–286.
2. Wells, T. N. C. & Fersht, A. R. (1986) *Biochemistry* **25**, 1881–1886.
3. Fersht, A. R., Leatherbarrow, R. & Wells, T. N. C. (1987) *Biochemistry* **26**, 6030–6038.
4. Wells, T. N. C. & Fersht, A. R. (1989) *Biochemistry* **28**, 9201–9209.
5. Toney, M. D. & Kirsch, J. F. (1991) *Biochemistry* **30**, 7456–7461.
6. Avis, J. & Fersht, A. R. (1993) *Biochemistry* **32**, 5321–5326.
7. Matouschek, A., Kellis, J. T., Jr., Serrano, L. & Fersht, A. R. (1989) *Nature (London)* **342**, 122–126.
8. Matouschek, A., Kellis, J. T., Jr., Serrano, L., Bycroft, M. & Fersht, A. R. (1990) *Nature (London)* **346**, 440–445.
9. Serrano, L., Kellis, J. T., Jr., Cann, P., Matouschek, A. & Fersht, A. R. (1992) *J. Mol. Biol.* **224**, 783–804.
10. Serrano, L., Matouschek, A. & Fersht, A. R. (1992) *J. Mol. Biol.* **224**, 805–818.
11. Serrano, L., Matouschek, A. & Fersht, A. R. (1992) *J. Mol. Biol.* **224**, 847–859.
12. Tanford, C. (1970) *Adv. Prot. Chem.* **24**, 1–95.
13. Creighton, T. E. (1988) *Proc. Natl. Acad. Sci. USA* **85**, 5082–5086.
14. Marcus, R. A. & Sutin, N. (1985) *Biochim. Biophys. Acta* **811**, 265–322.
15. Hammond, G. S. (1955) *J. Am. Chem. Soc.* **77**, 334–338.
16. Sancho, J., Neira, J. L. & Fersht, A. R. (1992) *J. Mol. Biol.* **224**, 749–758.

Chapter 15

Combining Experiment and Simulation

'An error does not become truth by reason of multiplied propagation, nor does truth become error because nobody sees it.'

Mahatma Gandhi

Scientific meetings are frequently more important for meeting people than listening to lectures, to whit the encounter with Peter Shewry, resulting in the CI2 work. By good fortune, in 1993, I met a young theoretician, Valerie Daggett, who had pioneered the atomistic simulation of protein unfolding and was interested in studying CI2. Thus began a very fruitful collaboration in which Valerie did the simulation without knowing our Φ data and then we compared results. Now, this might sound like the way simulation should always be done, but it was very rare to find someone willing to chance their arm because of the difficulties in simulation, which needed to be checked against experiment. The first papers on CI2 were very important because they showed the power of using experiment to benchmark simulation, which could then be used with confidence to describe the structure of the whole transition state at true atomic resolution, and then describe the whole of the pathway. The first of our more than 25 joint papers[91,128,143–164] was published in 1996 on CI2[143]. Valerie invented a method of determining the position of a transition state in a molecular dynamics simulation of protein (un)folding, which could then be tested against our Φ-values. There was always satisfactory agreement and so we could then use the whole of the unfolding simulation in reverse to describe the productive folding pathways of barnase, CI2 and several other proteins.

The simulations were criticised because they were done at 498 K to speed up unfolding *in silico* so that it would occur in the few nanoseconds (ns) then possible for simulation. Although the CI2 simulations were fine, and have indeed stood the test of time and have been extensively validated by Valerie, I decided to cut the Gordian knot of the dilemma of simulations being limited to less than a microsecond (μs), and experiments being milliseconds (ms). Bengt Nolting[165,166] had done the first studies on sub-ms folding in our lab using our ancient Temperature-jump (T-jump) apparatus. We found a suitable protein that would fold in μs, the engrailed homeodomain, EnHD. EnHD and its family members proved to be more 'wonder proteins' for folding and opened another very rich vein to mine. We were able to describe with confidence the pathway of folding of the protein from ns to μs[147,151,156].

I received a knighthood from Her Majesty Queen Elizabeth at the beginning of 2003 for my work on protein engineering. It was completely unexpected as I had no idea that I was being considered. Most knighthoods are awarded for serving on committees or preparing reports and rarely go to scientists who eschew such efforts. When I saw an envelope addressed to me in November 2002 from 10 Downing Street, I realised that Mr. Tony Blair was informing me of an award. I thought, 'oh dear, I have been awarded an OBE or maybe even a CBE,' and I near fell over when I saw 'knighthood' shouting at me in the letter.

Sending a paper to *Nature* seemed a good way to celebrate. Ugo Mayor, my graduate student, showed experimentally that EnHD folded by a folding intermediate and Nick Guydosh, a master's student, mapped the structure of the transition state by Φ-values. Valerie was able to fill in the rest of the details. Unlike CI2, this protein folds by a framework mechanism. I chose this paper to illustrate our collaboration[156]. The earlier work is reviewed in Refs. 151, 153 and 154.

15.1 2003: The Queen has a very gentle touch.

15.2 2003: Ugo Mayor, Stefano Gianni and Valerie Daggett.

letters to nature

The complete folding pathway of a protein from nanoseconds to microseconds

Ugo Mayor*, Nicholas R. Guydosh*, Christopher M. Johnson*, J. Günter Grossmann†, Satoshi Sato*, Gouri S. Jas‡§, Stefan M. V. Freund*, Darwin O. V. Alonso||, Valerie Daggett|| & Alan R. Fersht*

* *MRC Centre for Protein Engineering, Hills Road, Cambridge CB2 2QH, UK*
† *CLRC Daresbury Laboratory, Daresbury, Warrington WA4 4AD, UK*
‡ *Laboratory of Chemical Physics, NIDDK, NIH, Bethesda, Maryland 20892, USA*
|| *Department of Medicinal Chemistry, University of Washington, Seattle, Washington 98195-7610, USA*

Combining experimental and simulation data to describe all of the structures and the pathways involved in folding a protein is problematical. Transition states can be mapped experimentally by ϕ values[1,2], but the denatured state[3] is very difficult to analyse under conditions that favour folding. Also computer simulation at atomic resolution is currently limited to about a microsecond or less. Ultrafast-folding proteins fold and unfold on timescales accessible by both approaches[4,5], so here we study the folding pathway of the three-helix bundle protein Engrailed homeodomain[6]. Experimentally, the protein collapses in a microsecond to give an intermediate with much native α-helical secondary structure, which is the major component of the denatured state under conditions that favour folding. A mutant protein shows this state to be compact and contain dynamic, native-like helices with unstructured side chains. In the transition state between this and the native state, the structure of the helices is nearly fully formed and their docking is in progress, approximating to a classical diffusion–collision model. Molecular dynamics simulations give rate constants and structural details highly consistent with experiment, thereby completing the description of folding at atomic resolution.

Engrailed homeodomain (En-HD) from *Drosophila melanogaster* is a 61-residue α-helical protein[6]. Measurement of folding kinetics by temperature jump through the thermal denaturation transition region shows En-HD to be the fastest unfolding and one of the fastest folding proteins so far recorded directly[4], and its unfolding and folding are compatible with the present time regime of molecular dynamics simulation. Our previous simulations of the unfolding of En-HD predict the correct order of magnitude of the unfolding rate constant and also significant residual native α-helical structuring in the denatured state ensemble[4]. We have now been able to generate a denatured state, by means of mutation, that is stable under physiological conditions, to analyse it in depth and to show by NMR that the wild-type protein populates an equivalent denatured state. We have measured rate constants for folding and unfolding from some hundreds of nanoseconds to microseconds to show that the protein folds by an intermediate that is, in fact, the denatured state under physiological conditions, and we have simulated the unfolding of the protein by molecular dynamics to reconstruct the structures of the transition and denatured states. We now describe each state along the folding pathway, first by experiment and then by simulation.

We first generated a denatured state that was stable under physiological conditions. The native state for most proteins is favoured by only 5–15 kcal mol^{-1}. This balance has been exploited to analyse native and denatured states in equilibrium under

§ Present address: Biosciences Center, Kansas University, Lawrence, Kansas 66047, USA.

letters to nature

folding conditions by NMR[7]. In principle, however, the equilibrium position for any protein can be shifted by mutation so that the denatured state largely predominates under physiological conditions[8]. We lowered the stability of En-HD by mutating a leucine residue to alanine at position 16 (L16A), which deleted stabilizing tertiary interactions without introducing a side chain that was likely to make new specific interactions. Mutation of leucine to alanine tends to lower the stability of proteins by up to 4–5 kcal mol^{-1} (ref. 9). Wild-type En-HD had a free energy of folding of only 2.5 kcal mol^{-1}. A change in free energy of 4–5 kcal mol^{-1} therefore should, and did, radically tip the balance so that the protein was predominantly denatured under physiological conditions.

There were no detectable native tertiary interactions in L16A under physiological conditions; see for example, the aromatic region of the ^{1}H-NMR spectra (Fig. 1). The backbone was very dynamic, but the protein had most of its native α-helical structure, as monitored by circular dichroism (CD) and Hα chemical shifts (Fig. 1). The denatured state revealed by L16A had the characteristics of a folding intermediate with much of the correct native secondary structure. Its radius of gyration increased by 33% compared with the native state (17.0 and 12.8 Å, respectively), as monitored by solution X-ray scattering experiments. The CD and NMR spectra of L16A changed non-cooperatively with increasingly denaturing conditions towards those of a random coil (data not shown).

We then probed the denatured state of wild-type En-HD directly under physiological conditions using native state ^{1}H/^{2}H-exchange experiments[10]. The free energy of what is effectively the complete unfolding of regions of the protein is measured by the rate of exchange of buried backbone amide groups with solvent. The free energy of denaturation of En-HD in ^2H_2O measured by conventional procedures for thermal melting was 2.8 ± 0.2 kcal mol^{-1} at 5 °C. However, the free energy of opening of most of the amide groups in the helices was significantly higher; on average the helices unfolded with a free energy of 3.4 ± 0.1 kcal mol^{-1}, with helix I being the most protected (Fig. 2). There was little protection against exchange for residues 50–55, residues that were highly flexible in L16A (Fig. 1). Thus, the completely open denatured state of En-HD was ~1 kcal mol^{-1} less stable than the dominant denatured state, which had considerable native α-helical structure and closely resembled the denatured state of the L16A mutant, namely a folding intermediate. Folding intermediates are often the predominant denatured state of a protein under physiological conditions[11].

Folding intermediates generated in thermal denaturation simulations, and then quenched to 25 °C, of both the wild-type and L16A proteins had extensive secondary structure and few tertiary contacts (Fig. 3). The overall helical content was 57% for the wild-type intermediate ensemble and 64% for the crystal structure. Breakdown of the helix content along the sequence was consistent with the hydrogen protection results: helix I was highly populated with a helix content of 76%, helix II was less stable at 38%, and helix III had a helix content of 51%. Interestingly, however, the intermediate contained two helices much of the time, as the loop between helices I and II adopted helical structure and served to unite the two segments (Fig. 3). Non-native helical structure in the loop between these helices is supported by the NMR chemical shift deviations for L16A (Fig. 1). The radius of gyration of this intermediate state was 30% greater than the native state (14.4 Å for the wild type, 14.2 Å for L16A and 11 Å for the native state), in agreement with experiments on L16A. Further unfolding at high temperature produced a very open unfolded state with little helical structure (21% on average) (Fig. 3). This highly unfolded form had a radius of gyration of 20.0 and 18.8 Å by simulation and experiment (as measured on the acid-denatured state), respectively. Similarly, experiments on En-HD at high temperature or in high concentrations of urea failed to detect regular secondary structure (Fig. 1).

We also detected a folding intermediate from ultra-fast kinetic

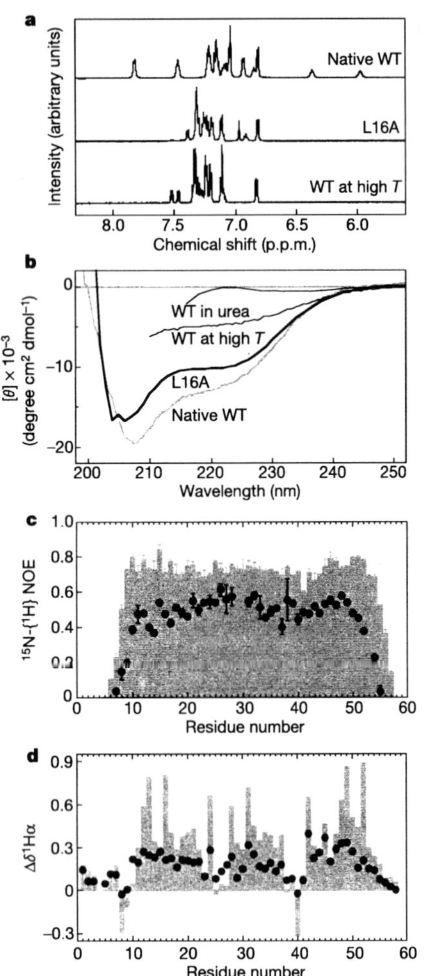

Figure 1 Characterization of wild-type En-HD (WT) and the L16A mutant. **a**, Aromatic region of the ^{1}H-NMR spectra of WT and L16A. Dispersion of the aromatic chemical shifts was lost in L16A (25 °C). **b**, Far-ultraviolet CD spectra of WT (25 and 95 °C) and L16A (25 °C). **c**, ^{15}N-{^{1}H} heteronuclear NOE data (I/I_0) of WT (grey bars) and L16A (black circles) at 25 °C. The low NOE values for L16A indicate its higher flexibility. **d**, Deviations in chemical shift from reported random coil values (Δδ ^{1}Hα) in p.p.m. of WT (grey bars) and L16A (black circles) at 25 °C.

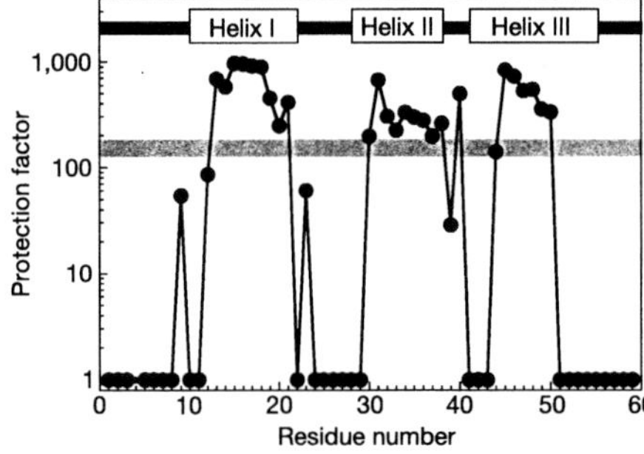

Figure 2 Protection factors for the amide groups of wild-type En-HD, calculated from the ^{1}H/^{2}H-exchange experiments at 5 °C. A value of 1 is given to those residues for which protection was not observed. The expected protection factor from a two-state fit of the denaturation curves is the thick grey line.

measurements. We showed previously that we could measure the rate constants for folding and unfolding by changing the temperature of a solution of En-HD rapidly (temperature jump) and monitoring the fluorescence of the single tryptophan residue[4]. We have now increased the resolution of our experiments with the use of laser *T*-jump[12] and improvements to the classical electrical discharge apparatus[4]. In so doing, we detected two-step folding kinetics (Fig. 4). At 25 °C, there was a fast phase of $t_{1/2} \approx 1.5\,\mu s$, in addition to the previously reported phase of 15 μs (ref. 4). We were able to assign these phases by two means. First, *T*-jump kinetics on L16A unfolding and refolding exhibited just the fast relaxation time (Fig. 4), indicating that it corresponded to the transition between its equilibrium denatured state (folding intermediate) and a more unfolded denatured state. Indeed, with increasing temperature the tryptophan fluorescence in both the mutant and the native proteins approached that of an exposed tryptophan residue both in equilibrium and *T*-jump experiments. Second, we measured the rate of formation of the wild-type native structure by the change in line shape[13] of the leucine-16 side-chain resonances between 43 and 56 °C. The rate constants corresponded to those of the slower phase in the *T*-jump kinetics (Fig. 4).

If the fast phase corresponds to the full unfolding of the protein detected by the solvent exchange experiments, then combining the values of the measured equilibrium constants for the overall unfolding and the exchange experiments with the standard formulae for relaxation times gives a minimal kinetic scheme at 25 °C of

$$\mathrm{U} \underset{0.7\times10^5\,\mathrm{s}^{-1}}{\overset{4\times10^5\,\mathrm{s}^{-1}}{\rightleftharpoons}} \mathrm{I} \underset{0.2\times10^4\,\mathrm{s}^{-1}}{\overset{4\times10^4\,\mathrm{s}^{-1}}{\rightleftharpoons}} \mathrm{N}$$

where U is the unfolded denatured state, I is the folding intermediate with considerable α-helical secondary structure (or D, the denatured state under physiological conditions) and N is the native state. The results of the various kinetic experiments are compiled in Fig. 4. The unfolding experiments were conducted at up to 65 °C and extrapolated to 100 °C. The times to reach the transition state in the unfolding simulations at 75 and 100 °C fit well with experiment: at 75 °C, the simulated time was approximately 60 ns and the extrapolated $t_{1/2}$ was 330 ns; at 100 °C the times were 2 and 5 ns, respectively (Fig. 4).

We analysed the structure of the transition state between N and I by means of ϕ values (a measure of the degree of native structure in the transition state; 0 implies denatured, 1 implies fully native)

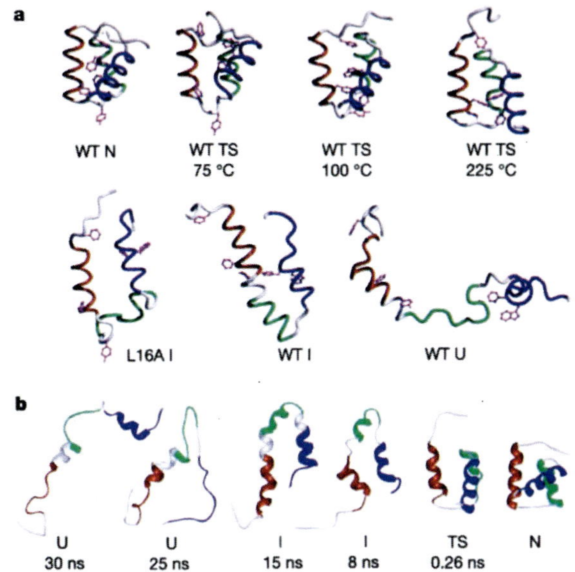

Figure 3 Representative structures from the molecular dynamics simulations. **a**, Snapshots for the transition-state (TS) ensembles identified from wild-type (WT) simulations at different temperatures (61 ns at 75 °C, 1.985 ns at 100 °C and 0.26 ns at 225 °C), the intermediate (I) state (10-ns snapshots from the wild-type and L16A 225 °C trajectories quenched to 25 °C) and the unfolded (U) state (represented by the 40-ns WT 225 °C structure). The native helical segments are coloured as follows: red, residues 10–22; green, residues 28–38; blue, residues 42–55. Aromatic side chains are shown in magenta. **b**, Structures from the wild-type 225 °C denaturation simulation shown in reverse, to illustrate a probable folding pathway of the protein to reach the native (N) state.

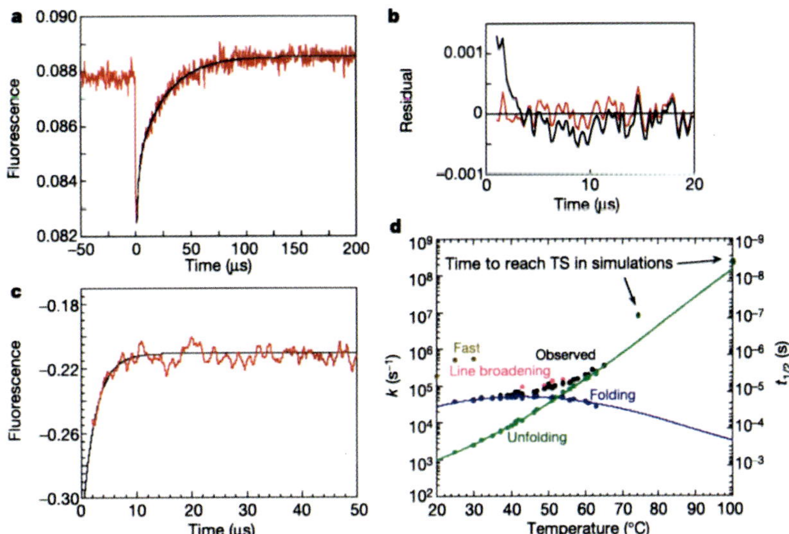

Figure 4 Kinetics of folding and unfolding. **a**, Laser-heating *T*-jump of En-HD to 25 °C; data after 800 ns fitted to a double exponential. Details of output and errors are available from the authors. **b**, Residuals from **a** (red line), randomly distributed around zero, unlike those of a single-exponential fit (black line). **c**, The L16A mutant of En-HD, *T*-jumped to 25 °C (Joule heating), showed only the fast phase. En-HD that had been *T*-jumped with the Joule-heating apparatus gave very similar rate constants to those from the laser apparatus. Details of output and errors are available from the authors. **d**, Observed relaxation rate constants (*k*) versus *T*. Solid curves were derived from combining *k* with equilibrium data from calorimetry. The filled red circles are results from NMR line-broadening analysis. TS, transition state.

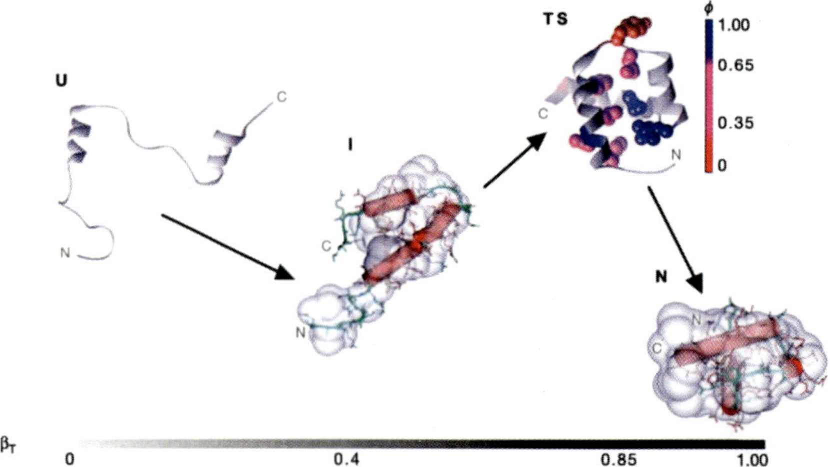

Figure 5 The complete folding pathway of En-HD by experiment and simulation. U is the quenched 40-ns snapshot from the simulation at 225 °C; I is the denatured state under 'physiological conditions'. The 10-ns structure from the high-temperature molecular dynamics simulation is enclosed in the grey molecular envelope obtained by X-ray scattering. TS is the transition state between I and N, the native state. ϕ values are colour-coded, with residues used as probes of secondary structure coloured directly on the backbone. The simulated structure is from the transition state ensemble for the wild-type at 100 °C (1.985 ns). Structures are positioned according to their relative solvent accessibilities (β_T), as obtained from the dependence of the rate constants on urea concentration. Molecular dynamics simulation gave consistent results.

calculated from the unfolding kinetics[14] for 16 mutations. The transition-state ensembles were also identified from the unfolding simulations at different temperatures, ranging from 75 to 225 °C, and they were very similar (Fig. 3). The experimentally and theoretically derived ϕ values are in good agreement: for example, the correlation coefficient between the values calculated for the 100 °C transition-state ensembles and experiment was 0.86. The combined picture is that the main transition state for folding and unfolding was very native-like, as reflected by both the ϕ values and the molecular-dynamics-generated structures (Fig. 5), as expected for a transition from a structured intermediate. Helix I, helix III and the turn between helices II and III were largely formed (Fig. 3) (for example, ϕ values: A14G in H1 = 0.79; G39A in the turn, 0.92; A43G in H3, 1.05)[14]. The loop between helices I and II was less structured (for example, ϕ for A25G was 0.17). The hydrophobic core had fractional ϕ values indicative of partial structure (typically more than 0.4). This structural pattern is just what is expected for a framework[15] or diffusion–collision mechanism[16], in which native-like secondary structure forms and then docks in the rate-determining step (Fig. 3). The use of coarse-grained models has shown that En-HD can fold by diffusion–collision[17]. The results here show that a diffusion–collision mechanism is approximated at atomic resolution, both by experiment and simulation by comparing the intermediate and transition states.

The helical contents of isolated synthetic peptides corresponding to helices I, II and III were roughly 30%, 15% and 25%, respectively, as determined both by trifluoroethanol titration experiments[18] and directly from the CD ellipticity at 222 and 208 nm (ref. 19). These abnormally high values, which are probably increased by tertiary contacts in the intact protein, as indicated by our simulations, precipitate a tendency towards the diffusion–collision model. We have argued elsewhere[20] that there is a continuum of mechanisms between the classical stepwise diffusion–collision mechanism and nucleation–condensation, in which there is concurrent consolidation of secondary and tertiary structure. It is probably correct to describe En-HD as being close to the diffusion–collision limit, because the pure mechanism of totally independent formation of secondary and tertiary interactions is most unlikely.

En-HD is an informative protein for studying unfolding and folding. Fast events can be observed directly at the upper limits predicted for folding[21,22] that enter the time range currently accessible by molecular dynamics simulation. The agreement between molecular dynamics simulation and experiment is very encouraging and on the same timescale. Combining molecular dynamics and experiment has now allowed us to characterize all of the necessary structures along the folding pathway. One of the longest-running controversies in mechanistic studies is whether intermediates are on the pathway or are unwanted side products. Here, simulation shows that the intermediate is on the pathway and that the protein does not fold by an intermediate that must unfold to be productive. Simulation could well be the answer in general for solving such problems in mechanism. □

Methods

Materials

^{15}N-labelled and ^{13}C–^{15}N-labelled wild-type En-HD were prepared[4] to >99% homogeneity as determined by both mass spectrometry and SDS chromatography, as was the L16A mutant. CD spectra of 30-μM samples in 50 mM sodium acetate, pH 5.7, 100 mM NaCl were recorded in an AVIV 202SF instrument. NMR experiments on 0.5–2-mM samples were performed with Bruker DRX 500 and DRX 600 spectrometers equipped with inverse triple-resonance probes and single-axis gradients.

Peptides of sequence AcNH-SSEQLARLKREFNENR-NH_2, AcNH-YLTERRRQQLSSELGL-NH_2 and AcNH-NEAQIKIWFQNKRAKI-NH_2, corresponding to helices I, II and III of En-HD, were synthesized chemically by G. Bloomberg (University of Bristol). Peptide purity and concentration were determined by mass spectroscopy and amino acid analysis, respectively.

NMR Methods

NMR resonances of both wild-type and L16A mutant En-HD were assigned by using standard triple-resonance techniques[23]. ^1H-NMR spectra in ^2H_2O, 50 mM sodium acetate-d_3 and 100 mM NaCl were recorded from 1 to 90 °C, with weak presaturation of the residual solvent peak. The spectra were internally referenced to 3-(trimethylsilyl)propionate at 0 p.p.m. Series of ^1H,^{15}N heteronuclear single-quantum-coherence spectra for solvent exchange experiments into ^2H_2O were acquired with 1024 × 90 complex points, four scans per increment and a dead time of 4 min. Solvent exchange data were analysed with standard equations for EX2 conditions[24]. The apparent equilibrium constant for the formation of the open state was calculated as $K_{op} = k_{ex}/k_{int}$, where k_{ex} is the observed exchange rate and k_{int} is the intrinsic exchange rate calculated from model peptide studies[25], with the SPHERE online software[26]. Apparent free energies of opening were calculated from this equilibrium constant. Enhancements of the ^{15}N-{^1H} nuclear Overhauser effect (NOE) were determined from spectra obtained with and without presaturation of amide proton resonances. Presaturation was achieved by applying 120° decoupling pulses in 5-ms intervals with a field strength of 25 kHz for 3 s.

Reference spectra were obtained by replacing the presaturation sequence with a simple delay of the same length. For both experiments, the overall recycle delay was set to 5 s. Two sets of interleaved NOE and control experiments were added to obtain the intensities used in the calculation.

The ^{1}H-NMR lineshapes of the L16 δ protons from 43 to 56 °C were fitted to a two-site exchange model[13,27]. Relative populations of the native and the denatured protein were determined independently from differential scanning calorimetry measurements; other parameters were obtained from the spectra. The independent variables were a normalization factor, the exchange rate and the frequency for L16 in the denatured state.

Solution X-ray scattering

Solution X-ray scattering data for wild-type En-HD and the L16A mutant were collected at station 2.1 of the Daresbury Synchrotron Radiation Source, UK, and processed with procedures described previously[28,29]. Several shape restorations for both protein states have been undertaken to evaluate the similarity and reliability of the low-resolution models. Indeed, the calculations resulted in similar shapes that were consistent with those presented here as semitransparent surfaces.

Kinetics methods

A DIA-RT T-Jump (DiaLog) electrical-discharge temperature-jump instrument[12] and a laser *T*-jump instrument[12] were used. Protein concentrations were in the range 15–300 μM, buffered with 50 mM HEPES buffer, pH 8.0, 100 mM NaCl. The rate constants were independent of concentration.

Computer simulations

Previous molecular dynamics simulations of the wild-type protein were at 100 and 225 °C for 76 and 42 ns, respectively[4]. New simulations presented here were of the wild type (15 ns at 25 °C, 100 ns at 75 °C, 24 ns at 100 °C, and 21 ns at 225 °C) and the L16A mutant (42 ns at 25 °C, 10 ns at 100 °C, and two independent simulations of 10 and 22 ns at 225 °C). There were further simulations in which structures from these high-temperature simulations were quenched to 25 °C and the water density was adjusted correspondingly (from 0.829 g ml^{-1} at 225 °C to 0.997 g ml^{-1}). We performed the following quenches: the 10-ns snapshot from the wild-type simulation at 225 °C was simulated at 25 °C for 40 ns to represent the folding intermediate; the 40-ns snapshot from the wild-type simulation at 225 °C was run for 10 ns to represent the unfolded state; and the 10-ns snapshot from the L16A simulation at 225 °C was simulated for 10 ns to model the equilibrium intermediate state. Many of these simulations were performed as controls and to ensure the reproducibility of the results. The various En-HD simulations amounted to 0.422 μs.

Received 18 October 2002; accepted 13 January 2003; doi:10.1038/nature01428.

1. Fersht, A. R., Leatherbarrow, R. J. & Wells, T. N. Structure-activity relationships in engineered proteins: analysis of use of binding energy by linear free energy relationships. *Biochemistry* **26,** 6030–6038 (1987).
2. Fersht, A. R., Matouschek, A. & Serrano, L. The folding of an enzyme. I. Theory of protein engineering analysis of stability and pathway of protein folding. *J. Mol. Biol.* **224,** 771–782 (1992).
3. Shortle, D. The denatured state (the other half of the folding equation) and its role in protein stability. *FASEB J.* **10,** 27–34 (1996).
4. Mayor, U., Johnson, C. M., Daggett, V. & Fersht, A. R. Protein folding and unfolding in microseconds to nanoseconds by experiment and simulation. *Proc. Natl Acad. Sci. USA* **97,** 13518–13522 (2000).
5. Ferguson, N. *et al.* Using flexible loop mimetics to extend phi-value analysis to secondary structure interactions. *Proc. Natl Acad. Sci. USA* **98,** 13008–13013 (2001).
6. Clarke, N. D., Kissinger, C. R., Desjarlais, J., Gilliland, G. L. & Pabo, C. O. Structural studies of the engrailed homeodomain. *Prot. Sci.* **3,** 1779–1787 (1994).
7. Blanco, F. J., Serrano, L. & Forman-Kay, J. D. High populations of non-native structures in the denatured state are compatible with the formation of the native folded state. *J. Mol. Biol.* **284,** 1153–1164 (1998).
8. Spector, S., Rosconi, M. & Raleigh, D. P. Conformational analysis of peptide fragments derived from the peripheral subunit-binding domain from the pyruvate dehydrogenase multienzyme complex of *Bacillus stearothermophilus*: evidence for nonrandom structure in the unfolded state. *Biopolymers* **49,** 29–40 (1999).
9. Serrano, L. & Fersht, A. R. Principles of protein stability derived from protein engineering methods. *Curr. Opin. Struct. Biol.* **3,** 75–83 (1993).
10. Englander, S. W. & Krishna, M. M. G. Hydrogen exchange. *Nature Struct. Biol.* **8,** 741–742 (2001).
11. Oliveberg, M. & Fersht, A. R. Thermodynamics of transient conformations in the folding pathway of barnase: reorganization of the folding intermediate at low pH. *Biochemistry* **35,** 2738–2749 (1996).
12. Thompson, P. A., Eaton, W. A. & Hofrichter, J. Laser temperature jump study of the helix ⇔ coil kinetics of an alanine peptide interpreted with a 'kinetic zipper' model. *Biochemistry* **36,** 9200–9210 (1997).
13. Huang, G. S. & Oas, T. G. Submillisecond folding of monomeric lambda repressor. *Proc. Natl Acad. Sci. USA* **92,** 6878–6882 (1995).
14. Matouschek, A., Kellis, J. T. Jr, Serrano, L. & Fersht, A. R. Mapping the transition state and pathway of protein folding by protein engineering. *Nature* **340,** 122–126 (1989).
15. Baldwin, R. L. & Rose, G. D. Is protein folding hierarchic? II. Folding intermediates and transition states. *Trends Biochem. Sci.* **24,** 77–83 (1999).
16. Karplus, M. & Weaver, D. L. Protein-folding dynamics—the diffusion–collision model and experimental-data. *Prot. Sci.* **3,** 650–668 (1994).
17. Islam, S. A., Karplus, M. & Weaver, D. L. Application of the diffusion–collision model to the folding of three-helix bundle proteins. *J. Mol. Biol.* **318,** 199–215 (2002).
18. Kippen, A. D., Arcus, V. L. & Fersht, A. R. Structural studies on peptides corresponding to mutants of the major alpha-helix of barnase. *Biochemistry* **33,** 10013–10021 (1994).
19. Greenfield, N. & Fasman, G. D. Computed circular dichroism spectra for the evaluation of protein conformation. *Biochemistry* **8,** 4108–4116 (1969).
20. Daggett, V. & Fersht, A. R. Is there a unifying mechanism of protein folding? *Trends Biochem. Sci.* **28,** 19–26 (2003).
21. Hagen, S. J., Hofrichter, J., Szabo, A. & Eaton, W. A. Diffusion-limited contact formation in unfolded cytochrome c: estimating the maximum rate of protein folding. *Proc. Natl Acad. Sci. USA* **93,** 11615–11617 (1996).
22. Bieri, O. *et al.* The speed limit for protein folding measured by triplet-triplet energy transfer. *Proc. Natl Acad. Sci. USA* **96,** 9597–9601 (1999).
23. Cavanagh, J., Fairbrother, W. J., Palmer, A. G. & Skelton, N. J. *Protein NMR Spectroscopy: Principles and Practice* (Academic, San Diego, 1996).
24. Itzhaki, L. S., Neira, J. L. & Fersht, A. R. Hydrogen exchange in chymotrypsin inhibitor 2 probed by denaturants and temperature. *J. Mol. Biol.* **270,** 89–98 (1997).
25. Bai, Y., Milne, J. S., Mayne, L. & Englander, S. W. Primary structure effects on peptide group hydrogen exchange. *Proteins* **17,** 75–86 (1993).
26. Zhang, Y.-Z. *Structural Biology and Molecular Biophysics.* PhD thesis, Univ. Pennsylvania (1995).
27. Sandstrom, J. *Dynamic NMR Spectroscopy* (Academic, New York, 1982).
28. Grossmann, J. G., Hall, J. F., Kanbi, L. D. & Hasnain, S. S. The N-terminal extension of rusticyanin is not responsible for its acid stability. *Biochemistry* **41,** 3613–3619 (2002).
29. Witty, M. *et al.* Structure of the periplasmic domain of *Pseudomonas aeruginosa* TolA: evidence for an evolutionary relationship with the TonB transporter protein. *EMBO J.* **21,** 4207–4218 (2002).

Acknowledgements This work was supported with an 'Ikertzaileen prestakuntza' grant (U.M.) from the Government of the Basque Country and with a Winston Churchill Scholarship (N.R.G). The computational studies were supported by the National Institutes of Health (to V.D.).

Competing interests statement The authors declare that they have no competing financial interests.

Correspondence and requests for materials should be addressed to A.R.F. (e-mail: arf25@cam.ac.uk).

Ugo subsequently was able to stabilise the folding intermediate of EnHD by protein engineering, and Tomasz Religa, my first Polish graduate student, solved its structure by NMR. It is surprisingly difficult to prove that folding intermediates are on-, rather than off-pathway and also to solve their structures[167]. Tomasz went on to prove definitively that the intermediate was on-pathway by laser T-jump methods. We had built an instrument in-house. For several years we had given up rapid mixing methods and used electrical discharge T-jump machines. Although I had designed the earlier mechanical machines, the more modern ones were designed by Chris Johnson, who worked with me for more than 15 years, aided by the LMB electronics and machine shops. Building our own apparatus has always been important.

15.3 2003: Tomasz Religa.

15.4 2003: Chris Johnson.

Vol 437|13 October 2005|doi:10.1038/nature04054 nature

LETTERS

Solution structure of a protein denatured state and folding intermediate

T. L. Religa[1], J. S. Markson[1], U. Mayor[1]†, S. M. V. Freund[1] & A. R. Fersht[1]

The most controversial area in protein folding concerns its earliest stages. Questions such as whether there are genuine folding intermediates, and whether the events at the earliest stages are just rearrangements of the denatured state[1] or progress from populated transition states[2], remain unresolved. The problem is that there is a lack of experimental high-resolution structural information about early folding intermediates and denatured states under conditions that favour folding because competent states spontaneously fold rapidly. Here we have solved directly the solution structure of a true denatured state by nuclear magnetic resonance under conditions that would normally favour folding, and directly studied its equilibrium and kinetic behaviour. We engineered a mutant of *Drosophila melanogaster* Engrailed homeodomain that folds and unfolds reversibly just by changing ionic strength. At high ionic strength, the mutant L16A is an ultra-fast folding native protein, just like the wild-type protein; however, at physiological ionic strength it is denatured. The denatured state is a well-ordered folding intermediate, poised to fold by docking helices and breaking some non-native interactions. It unfolds relatively progressively with increasingly denaturing conditions, and so superficially resembles a denatured state with properties that vary with conditions. Such ill-defined unfolding is a common feature of early folding intermediate states and accounts for why there are so many controversies about intermediates versus compact denatured states in protein folding.

Drosophila melanogaster Engrailed homeodomain (En-HD) is a 61-residue, three-helix bundle protein with helices spanning residues 10–22 (H1), 28–37 (H2) and 42–56 (H3)[3]. It folds on the microsecond timescale via a folding intermediate, and its pathway is well studied by experiment and simulation[4]. We have engineered En-HD to be denatured under conditions that favour folding by introducing the mutation L16A (ref. 5). L16 is a highly conserved, fully buried residue located in the middle of H1. The L16A mutation removes a number of local interactions and numerous long-range interactions with residues in H2, the turn between H2 and H3, but mainly with H3. Φ-analysis and simulation show that the core becomes disrupted in this region in the transition state for unfolding, but the helical secondary structure of H1 remains intact, as does that of H2 and H3 and the turn connecting them[4]. The L16A mutation is likely to destabilize the native structure substantially but not folding intermediates. All evidence so far shows that L16A is not only a genuine model for the denatured state of En-HD under folding conditions but is also a denatured state in its own right. The mutation acts to tip the energetic balance between native and denatured states so that L16A is denatured at physiological ionic strength[5]. Addition of salt stabilizes the protein and screens charge–charge repulsions so that it refolds[5].

It is important first to define the term 'denatured state' explicitly. We adopt the convention that the maximally unfolded state of a protein, in which the backbone NH groups have little protection against $^1H/^2H$-exchange, is U. The denatured state, D, is the lowest energy non-native state under a defined set of conditions[6]. Under physiological conditions that favour folding, the lowest energy denatured state, D_{phys}, can be a folding intermediate[7]. Native state $^1H/^2H$-exchange kinetics has shown, for example, that the denatured state of En-HD has protected helices H1, H2 and H3, and the U state is at 1 kcal mol^{-1} higher energy[4].

Chemical shifts[8], nuclear Overhauser effects (NOEs)[9,10], paramagnetic relaxation[11,12] and residual dipolar couplings (RDCs)[13] are used to infer the structural properties of the denatured state ensemble and calculate its three-dimensional structure[14]. We solved the structure of the denatured En-HD mutant L16A at physiological ionic strength (~150 mM) using NOEs and classical nuclear magnetic resonance (NMR) procedures[15] (Protein Data Bank code 1ZTR; Fig. 1a). Over 800 intra-residue NOEs were identified. Over 90 of them were long-range NOEs, with most of them occurring between H2 and H3 (Supplementary Table 1). Experiments on deuterated samples identified only one significant long range contact between I45 HN and S35 HN. We found no NOEs between H1 and H2 or H3. The detected sequential NOEs together with Hα, Cα and Cβ chemical shifts (see Fig. 1b, c) implied that the protein displays helical properties at least between residues 10–25, 28–38 and 42–53. These regions differ from the wild type, which is not helical around residues 22–28 and has a longer H3. The structure of the region spanning H2 and H3 and the connecting loop is as well defined as many NMR structures of native proteins (backbone root mean square deviation (r.m.s.d.) of 0.87 Å) and similar to that of the wild-type protein. But, although H1 was found to be helical, the angle between it and the rest of the protein is only approximate because of the lack of long-range NOEs between them. It may well be that the angle between H1 and the rest of the protein does indeed fluctuate.

H1 is longer in the L16A mutant than in the wild-type protein by at least one turn. Numerous hydrophobic residues become solvent exposed and form contacts not observed in the wild type. Residues F8 and L13 of the L16A mutant, which interact with H3, turn residues and L16 in the wild-type protein, have NOEs between each other which are unambiguously different from those in the wild-type protein. F20, which interacts mainly with the carboxy terminus of H3 in the wild type, bends back towards H1 in the L16A mutant, showing numerous non-wild-type NOEs with residues close in sequence. Y25, which is solvent-exposed in the wild type, where it forms long-range interactions with R53, has numerous non-wild-type NOEs to residues in the wild-type loop region. T27 $H\gamma_2$, fully exposed in the wild type, has an unambiguous NOE with the F49 aromatic ring in the L16A mutant. Significant portions of W48 and F49 are buried by interactions with H1 in the wild type. In the L16A mutant, both residues bend back to the face of H3 and become partially buried in the H2-loop-H3 hydrophobic core.

[1]MRC Centre for Protein Engineering and Cambridge University Chemical Laboratories, MRC Centre, Hills Road, Cambridge CB2 2QH, UK. †Present address: Wellcome Trust Cancer Research UK Gurdon Institute, Tennis Court Road, Cambridge CB2 1QN, UK.

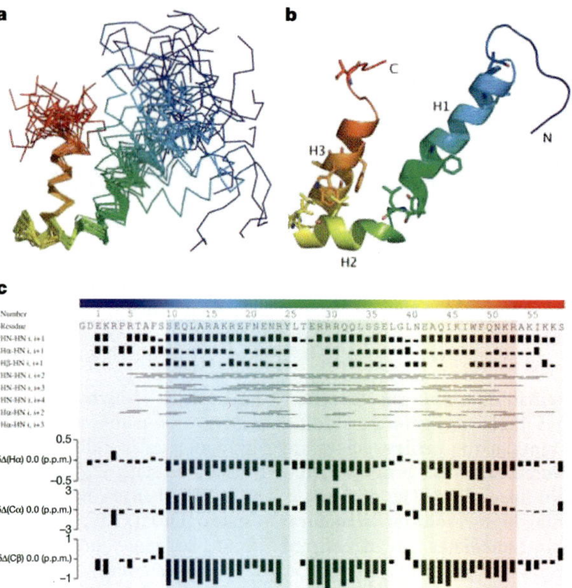

Figure 1 | Structure of En-HD L16A at low ionic strength and summary of sequential NOEs and secondary chemical shift values (Protein Data Bank code 1ZTR). **a**, Bundle of the 25 lowest-energy conformers with backbone atoms N, Cα and C′ of residues 28–53 superimposed for a minimal r.m.s.d. The different conformations of residues 1–28 are a consequence of the lack of a significant number of long-range NOEs between them and residues 28–53. **b**, Lowest energy structure with location of hydrophobic side chains. **c**, Sequential NOEs and Hα, Cα and Cβ chemical shift deviations. Negative, positive and negative deviations from random-coil values for Hα, Cα and Cβ chemical shifts, respectively, indicate helical structure.

We examined the structure of the backbone by HN-N RDC measurements using acrylamide and Pf1 phage as alignment media (Supplementary Information). The RDCs were fully consistent with the NOEs. The structures encompassing helices 2 and 3 superposed well on those derived from NOEs, and the orientation of H1 in the RDC structures was well within the envelope of the NOE structures. We analysed ^{15}N relaxation data for the L16A mutant with the reduced spectral density approach at 25 °C (refs 16–18) to measure the degree of mobility along its chain. The results show that it is a well-ordered protein (Fig. 2; see also Supplementary Information).

The structure of L16A is fully consistent with previous experimental data, including its circular dichroism, fluorescence spectra and dimensions and shape from small angle X-ray scattering, and also with the native-state hydrogen/deuterium (H/D)-exchange pattern[4,5] and multiple molecular dynamics simulations of the unfolding of wild-type En-HD[4]. The structure of L16A is also perfectly consistent with the Φ-analysis of the structure of the transition state for unfolding of En-HD[4], with the helix-turn-helix motif around H2 and H3 being fully formed in the transition state ($\Phi \approx 1$), as is the secondary structure in H1, with the core between H1 and the rest of the protein coming apart.

Having found that the L16A mutant is a good model for the denatured state of En-HD and is a well-structured intermediate that is poised for folding, we investigated the L16A mutant as a genuine denatured state and folding intermediate in its own right in equilibrium with its native state. L16A went from having little detectable native tertiary structure at low ionic strength (145 mM) (ref. 5) to being ~98% folded at 3.5 M NaCl, with a free energy of denaturation of 2.3 kcal mol^{-1} (Fig. 3a, c). Denaturation of L16A in 0.1 M NaCl had an apparently non-cooperative transition for melting of the helical structure[5], monitored by circular dichroism. However, in 4 M NaCl, L16A melted with increasing temperature according to a classical cooperative two-state transition that paralleled the thermal denaturation of the wild-type protein (Fig. 3c). The heteronuclear single quantum correlation (HSQC) NMR spectrum of the L16A mutant in 4 M NaCl had a chemical shift dispersion almost as high as that of wild type (data not shown). A plot of the circular dichroism signal at 222 nm or of tryptophan fluorescence against the logarithm of the ionic strength showed a sharp cooperative transition for formation of native structure, with a mid-point of about 1 M (Fig. 3d). Comparison of the stability of L16A and wild type at high ionic strength implies that the denatured state of L16A is favoured by about 2 kcal mol^{-1} at the ionic strength of 145 mM used in NMR studies. Wild-type En-HD folds according to apparent two-step kinetics, with an unassigned ultra-fast phase at the 1-μs region and a very fast phase at the 10-μs region that has been assigned by NMR line broadening to correspond to the formation of native structure[4]. At low ionic strength, L16A exhibits the 1-μs phase only[4]. We saw (Fig. 4) both slow and fast phases for L16A on addition of 500 mM NaCl, and mainly the slower phase at 2 M NaCl. The data agree well with those of wild-type En-HD at low salt, with the unfolding limb of the plot being observable at a lower temperature.

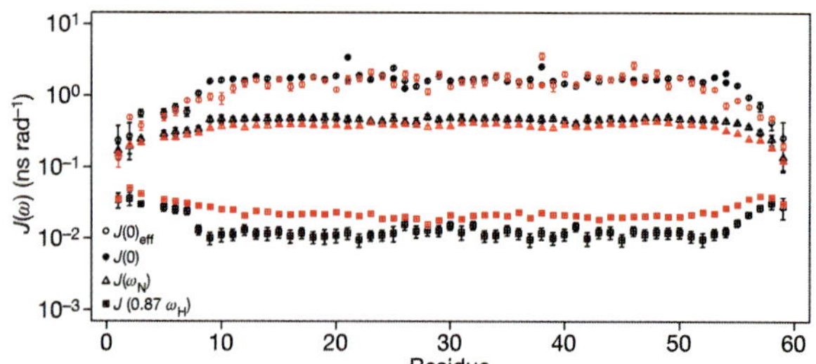

Figure 2 | Backbone dynamics of wild-type and En-HD L16A protein. Reduced spectral densities for En-HD L16A (red) and wild type (black) at 25 °C. $J(0)$ relaxation rates compensated for chemical exchange using η/R_2 (ref. 23) are shown with filled circles.

NATURE|Vol 437|13 October 2005 **LETTERS**

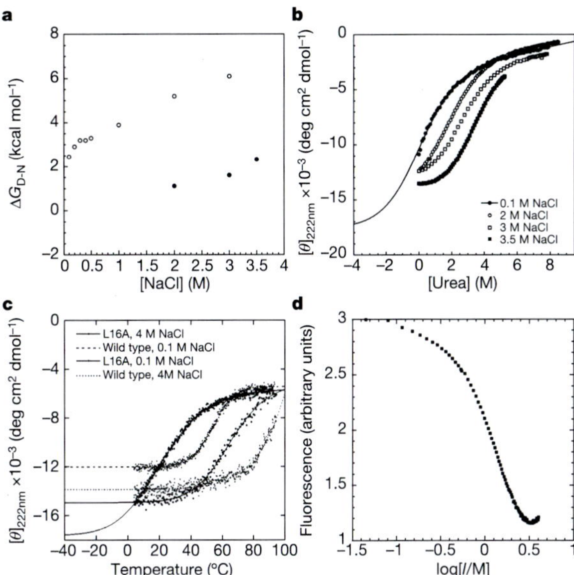

Figure 3 | Effects of salt on the properties of wild-type En-HD and L16A, and the difficulties of distinguishing between cooperative and non-cooperative transitions. a, Free energies of denaturation, $\Delta G_{\text{D-N}}$, for wild-type and L16A at 25 °C, as a function of NaCl concentration. Open circles, wild-type $\Delta G_{\text{D-N}}$ calculated from urea denaturation experiments; filled circles, L16A $\Delta G_{\text{D-N}}$ calculated from urea denaturation. **b,** Urea denaturation of L16A monitored by circular dichroism in the presence of 0.1, 2, 3, or 3.5 M NaCl, at 25 °C. Data were fitted to the standard two-state denaturation equation. At 0.1 M NaCl, we would need experimental data at negative concentrations of urea to fit the curve with any confidence—we used the limiting value of the circular dichroism signal from **c** and assumed a zero baseline slope in the absence of the necessary data to give a value of m of 0.5 with $[\text{urea}]_{50\%}$ of 0 M (compare with $m = 0.71$ for wild type or the L16A mutant at high salt). **c,** Thermal denaturation of wild-type En-HD and L16A, monitored by circular dichroism at various salt concentrations. We can fit the data at 0.1 M to a cooperative transition by assuming that the baseline at low temperature has zero slope, to give a T_m of 21 °C and $\Delta H_{\text{D-N}}$ of 12 kcal mol^{-1}. We would need to measure down to −20 to −40 °C for a confident fit. **d,** Titration of En-HD L16A with NaCl, at 25 °C, with fluorescence amplitude plotted against log[I] (ref. 24), where I is the ionic strength.

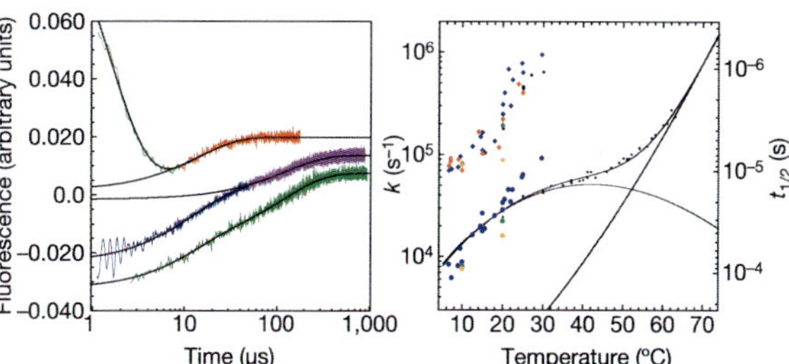

Figure 4 | Relaxation kinetics of wild-type En-HD and the L16A mutant at various concentrations of NaCl as a function of temperature at 50 mM NaAc, pH 5.7. The left panel shows time courses for the relaxation kinetics of L16A. The red curve is for 100 mM NaCl; the green at the beginning is the response time of the T-jump machine, which is long at this ionic strength. The theoretical curves are for a single exponential through the folding relaxation and a double to include the response time. Blue and purple are fast and slow phases at 500 mM NaCl. Dark green is the trace with 1 M NaCl. The fits in the latter two are double exponentials for the actual two phases. The right panel shows relaxation rate constants versus temperature. Diamonds, fast phase; circles, slow phase; red, En-HD L16A in 100 mM NaCl; blue, En-HD L16A in 500 mM NaCl; dark green, En-HD L16A in 1 M NaCl; orange, En-HD L16A in 2 M NaCl; black, En-HD wild type in 50 mM NaAc pH 5.7. The thin black lines are the deconvoluted folding and unfolding rate constants for the slow folding phase of the wild-type protein.

At 25 °C and 500 mM salt, where it is ~10% folded, the L16A mutant unfolded at 60,000 s^{-1} and folded at 6,000 s^{-1}. In 2 M salt, refolding increased to ~20,000 s^{-1} and unfolding decreased to ~3,000 s^{-1}. L16A is thus a fast folding protein with rate constants comparable to those of wild type. The L16A mutant at low ionic strength is not just a model for the denatured state of En-HD but is a true denatured state in equilibrium with its native state in a conventional folding reaction for which the equilibrium position changes with ionic strength.

We can now use the properties of L16A to shed light on controversies about intermediates and denatured states in general because the denatured state of L16A is a well-defined and chemically reasonable high-resolution structure. One conventional test for a folded structure when it cannot be directly characterized is the observation of a cooperative folding transition. Larger proteins that have been traditionally studied usually display sharp folding transitions. But, with very small proteins the transitions become very broad. For example, a protein that has an enthalpy of denaturation ($\Delta H_{\text{D-N}}$) of only 12 kcal mol^{-1} would require a temperature range of about the melting temperature (T_m) ±60–70 °C to go from 98–99% native to 98–99% denatured; such a low and spread-out value of $\Delta H_{\text{D-N}}$ would be difficult to detect by calorimetry. Similarly, an m (that is, the rate of change of free energy of denaturation with concentration of urea, $\partial \Delta G_{\text{D-N}}/\partial[\text{urea}]$) of 0.5 would require a denaturant concentration range of 9–11 M to go from 98–99% native to 98–99% denatured. In practice, a weakly stable small protein can be studied over only such a limited range of temperatures and denaturant such that it is not possible to determine with confidence whether a transition is cooperative and to determine T_m values with precision because the baselines of curves are inaccessible. The folding transition of the L16A mutant under physiological conditions is a typical ill-defined example, and we have equivocated on whether it is a discrete state because it was difficult to argue whether the transition was cooperative or whether we were just fitting the middle regions of a non-cooperative transition with a mixture of states[5]. Now that we have characterized L16A as being folded, we have fitted its denaturation at physiological ionic strength to cooperative denaturation transitions to give a T_m of ~33–36 °C and $\Delta H_{\text{D-N}}$ of ~13 kcal mol^{-1} (Fig. 3). In particular, the thermal titration of more than 30 residues monitored by NMR fits the middle of sigmoidal transition curves (data not shown; $T_m = 36.3 \pm 4.3$ °C and $\Delta H_{\text{D-N}}$ of 13.4 ± 3.4 (±s.d.)), the caveat being that there is a small spread in T_m that may be accounted for by the uncertainties in not knowing the endpoints or slopes of baselines of the transitions. It is very rare to find a folding intermediate as well ordered and stable as that of the L16A mutant. Most other examples of intermediates cannot be isolated for direct study and will have even lower energy transitions that will require impossibly large temperature ranges to distinguish between cooperative and non-cooperative transitions. It is, accordingly, nearly impossible in most cases to distinguish experimentally between discrete intermediates and denatured states of varying compaction.

METHODS

NMR methods. NMR experiments were performed in 50 mM sodium acetate-d_3 buffer, pH 5.7, 100 mM NaCl and at temperatures of 25 and 5 °C on Bruker DRX500 and Avance 800 spectrometers equipped with 5-mm triple-resonance probes and single-axis gradients. Protein concentrations were 600 μM for nuclear Overhauser effect spectroscopy (NOESY) experiments and 100 μM for the RDC measurements using protein prepared as described in the Supplementary Information. There were no changes in chemical shifts and NOESY spectra within this range. Resonance assignments and NOESY spectra were obtained at 5 °C. At this temperature, the amide chemical shift dispersion for En-HD L16A was 1.4 p.p.m. with most of the residues located within a 0.8-p.p.m. window. Backbone assignments were obtained with standard triple-resonance experiments[15], and two-dimensional constant-time-[^{13}C,^{1}H]-HSQC and three-dimensional

[^{13}C,^{15}N,^{1}H]-(H)CC(CO)NH-TOCSY (TOCSY, total correlated spectroscopy) spectra enabled a 95% complete side-chain assignment, including stereo-specific assignments of the methyl groups. Structural NOEs were obtained from two-dimensional [^{1}H,^{1}H]-NOESY as well as three-dimensional [^{15}N,^{1}H,^{1}H]-NOESY and [^{13}C,^{1}H,^{1}H]-NOESY spectra. The mixing time in all experiments was 150 ms to minimize spin diffusion. A 100% deuterated ^{15}N-labelled sample was used to record three-dimensional [^{15}N,^{15}N,^{1}H]-HSQC-NOESY-HSQC and two-dimensional [^{1}H,^{1}H]-NOESY experiments with a mixing time of 600 ms[10,19]. These experiments, optimized for long-range (>5 Å) contacts, confirmed weak NOEs obtained with 150 ms mixing time but detected no additional long-range interactions between H1 and H2/H3. A 600-ms three-dimensional [^{15}N,^{1}H,^{1}H]-NOESY and two-dimensional [^{1}H,^{1}H]-NOESY were used to confirm amide to aromatic side-chain interactions in a 100% deuterated, Phe, Tyr back-protonated sample. Cross-peaks from the 150-ms NOESY spectra were divided into three bins with cutoffs at 2.7, 3.6 and 5.5 Å based on their volumes. Amide–amide NOEs from the 600-ms three-dimensional [^{15}N,^{15}N,^{1}H]-HSQC-NOESY-HSQC and amide–aromatic NOEs from the NOESY spectra were divided into four bins with cutoffs at 2.7, 3.6, 4.7 and 7.2 Å. On the basis of these restraints, 50 structures were calculated using the CNS[20] package and the 25 lowest energy structures are presented here.

We checked whether assuming a single entity for the L16A mutant affected the structure determination. We applied ensemble averaging as implemented in CNS[21,22] to account for possible ensemble solution to our NOEs. We obtained the same structure but with lower precision, decreasing to 2.1 Å for the H2/H3 fragment. The NMR data are consistent with a single entity, but, as is true for all structures, cannot exclude an ensemble of rapidly interconverting structures.

^{15}N backbone amide relaxation data for En-HD L16A and wild type were acquired and analysed as described previously[5].

Thermodynamic and kinetic characterization. The standard buffer was 100 mM NaCl, 50 mM NaAc pH 5.7 ($I = 145$ mM), with I increased by addition of NaCl. Folding equilibria and kinetics of L16A and wild-type protein were studied as described[4,5]. Urea denaturations and salt titrations were performed at 25 °C using an Aviv 202-SF instrument, with fluorescence recorded through a 320-nm long-pass filter and a protein concentration of ~8 μM. Thermal denaturations were performed on a Jasco J-720 spectrometer with a heating rate of 60–80 °C per hour and a protein concentration generally ~10–20 μM. Kinetic experiments were performed on a DIA-RT T-jump (DiaLog) or a TJ-64 T-jump (Hi-Tech) electrical-discharge temperature-jump instrument. Protein concentrations were 15–300 μM in acetate or HEPES[4].

Received 11 March; accepted 19 July 2005.

1. Krantz, B. A., Mayne, L., Rumbley, J., Englander, S. W. & Sosnick, T. R. Fast and slow intermediate accumulation and the initial barrier mechanism in protein folding. *J. Mol. Biol.* **324**, 359–371 (2002).
2. Yang, W. Y. & Gruebele, M. Folding at the speed limit. *Nature* **423**, 193–197 (2003).
3. Clarke, N. D., Kissinger, C. R., Desjarlais, J., Gilliland, G. L. & Pabo, C. O. Structural studies of the engrailed homeodomain. *Protein Sci.* **3**, 1779–1787 (1994).
4. Mayor, U. *et al.* The complete folding pathway of a protein from nanoseconds to microseconds. *Nature* **421**, 863–867 (2003).
5. Mayor, U., Grossmann, J. G., Foster, N. W., Freund, S. M. & Fersht, A. R. The denatured state of Engrailed Homeodomain under denaturing and native conditions. *J. Mol. Biol.* **333**, 977–991 (2003).
6. Dill, K. A. & Shortle, D. Denatured states of proteins. *Annu. Rev. Biochem.* **60**, 795–825 (1991).
7. Oliveberg, M. & Fersht, A. R. Thermodynamics of transient conformations in the folding pathway of barnase: Reorganization of the folding intermediate at low pH. *Biochemistry* **35**, 2738–2749 (1996).
8. Korzhnev, D. M. *et al.* Low-populated folding intermediates of Fyn SH3 characterized by relaxation dispersion NMR. *Nature* **430**, 586–590 (2004).
9. Neri, D., Billeter, M., Wider, G. & Wuthrich, K. NMR determination of residual structure in a urea-denatured protein, the 434-repressor. *Science* **257**, 1559–1563 (1992).
10. Mok, Y. K., Kay, C. M., Kay, L. E. & Forman-Kay, J. NOE data demonstrating a compact unfolded state for an SH3 domain under non-denaturing conditions. *J. Mol. Biol.* **289**, 619–638 (1999).
11. Gillespie, J. R. & Shortle, D. Characterization of long-range structure in the denatured state of staphylococcal nuclease. I. Paramagnetic relaxation enhancement by nitroxide spin labels. *J. Mol. Biol.* **268**, 158–159 (1997).
12. Gillespie, J. R. & Shortle, D. Characterization of long-range structure in the denatured state of staphylococcal nuclease. II. Distance restraints from paramagnetic relaxation and calculation of an ensemble of structures. *J. Mol. Biol.* **268**, 170–184 (1997).
13. Mohana-Borges, R., Goto, N. K., Kroon, G. J., Dyson, H. J. & Wright, P. E. Structural characterization of unfolded states of apomyoglobin using residual dipolar couplings. *J. Mol. Biol.* **340**, 1131–1142 (2004).
14. Dyson, H. J. & Wright, P. E. Nuclear magnetic resonance methods for elucidation of structure and dynamics in disordered states. *Methods Enzymol.* **339**, 258–270 (2001).
15. Wuthrich, K. *NMR of Proteins and Nucleic Acids* (John Wiley & Sons, New York, 1986).
16. Farrow, N. A., Zhang, O., Szabo, A., Torchia, D. A. & Kay, L. E. Spectral density function mapping using ^{15}N relaxation data exclusively. *J. Biomol. NMR* **6**, 153–162 (1995).
17. Farrow, N. A., Zhang, O., Forman-Kay, J. D. & Kay, L. E. Comparison of the backbone dynamics of a folded and an unfolded SH3 domain existing in equilibrium in aqueous buffer. *Biochemistry* **34**, 868–878 (1995).
18. Ishima, R. & Nagayama, K. Protein backbone dynamics revealed by quasi spectral density function analysis of amide N-15 nuclei. *Biochemistry* **34**, 3162–3171 (1995).
19. Koharudin, L. M. I., Bonvin, A., Kaptein, R. & Boelens, R. Use of very long-distance NOEs in a fully deuterated protein: an approach for rapid protein fold determination. *J. Mag. Res.* **163**, 228–235 (2003).
20. Brunger, A. T. *et al.* Crystallography & NMR system: A new software suite for molecular structure determination. *Acta Crystallogr. D* **54**, 905–921 (1998).
21. Bonvin, A. M. J. J. & Brunger, A. T. Conformational variability of solution nucelar magnetic resonance structures. *J. Mol. Biol.* **250**, 80–93 (1995).
22. Bonvin, A. M. J. J. & Brunger, A. T. Do NOE distances contain enough information to assess the relative populations of multi-conformer structures? *J. Biomol. NMR* **7**, 72–76 (1996).
23. Fushman, D., Tjandra, N. & Cowburn, D. Direct measurement of N-15 chemical shift anisotropy in solution. *J. Am. Chem. Soc.* **120**, 10947–10952 (1998).
24. Rios, M. A. D. & Plaxco, K. W. Apparent Debye-Huckel electrostatic effects in the folding of a simple, single domain protein. *Biochemistry* **44**, 1243–1250 (2005).

Supplementary Information is linked to the online version of the paper at www.nature.com/nature.

Acknowledgements T.L.R. was supported by a Trinity College Internal Graduate Studentship, and J.S.M. by a Gates Cambridge Scholarship. We would like to thank T. J. Rutherford and L. Kowalik for help with NMR, and J. M. Perez Canadillas for help with protein expression.

Author Contributions T.L.R. was responsible for the major experimental work and planning of this study, with further contributions from J.S.M., U.M. and S.M.V.F. A.R.F. wrote the paper with T.L.R. and performed data analysis.

Author Information Reprints and permissions information is available at npg.nature.com/reprintsandpermissions. The authors declare no competing financial interests. Correspondence and requests for materials should be addressed to A.R.F. (arf25@cam.ac.uk).

Unifying Protein Folding Mechanisms

By studying a series of members of the homeodomain superfamily of proteins, we were able to observe a slide from framework mechanism to nucleation-condensation with change of stability of substructures, as hypothesized in the original 1995 paper[154,155]. We speculated, on good evidence, that there is a continuum of mechanisms. There is pure nucleation-condensation at one extreme and a pure stepwise folding by association of preformed elements of secondary structure (the framework mechanism) at the other. The most common mechanism is closer to nucleation-condensation[155]. Stefano Gianni returned to Rome, where he continues to collaborate with Per Jemth, who is now at Uppsala, his fellow post-doctoral in my lab.

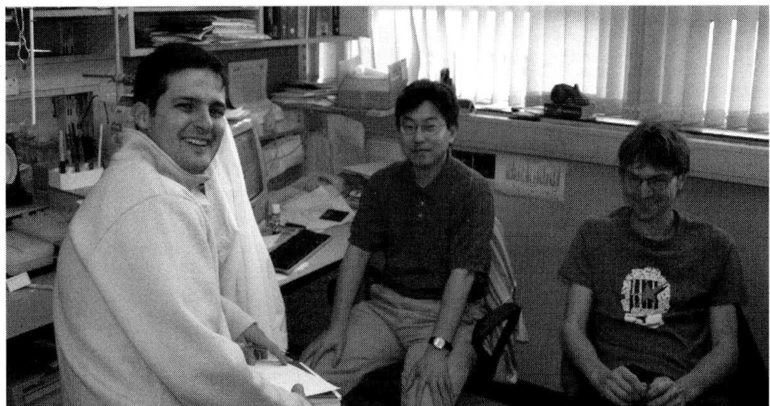

15.5 2003: Stefano Gianni, Satoshi Sato and Per Jemth.

15.6 2003: Zoryana Oliynyk and Ugo Mayor.

Unifying features in protein-folding mechanisms

Stefano Gianni*, Nicholas R. Guydosh*, Faaizah Khan*, Teresa D. Caldas*, Ugo Mayor*, George W. N. White†, Mari L. DeMarco†, Valerie Daggett†‡, and Alan R. Fersht*‡

*Medical Research Council Centre for Protein Engineering, Hills Road, Cambridge CB2 2QH, United Kingdom; and †Department of Medicinal Chemistry, University of Washington, Seattle, WA 98195-7610

Contributed by Alan R. Fersht, September 9, 2003

We compare the folding of representative members of a protein superfamily by experiment and simulation to investigate common features in folding mechanisms. The homeodomain superfamily of three-helical, single-domain proteins exhibits a spectrum of folding processes that spans the complete transition from concurrent secondary and tertiary structure formation (nucleation–condensation mechanism) to sequential secondary and tertiary formation (framework mechanism). The unifying factor in their mechanisms is that the transition state for (un)folding is expanded and very native-like, with the proportion and degree of formation of secondary and tertiary interactions varying. There is a transition, or slide, from the framework to nucleation–condensation mechanism with decreasing stability of the secondary structure. Thus, framework and nucleation–condensation are different manifestations of an underlying common mechanism.

two-state | three-state | framework | nucleation | homeodomain

A Holy Grail of protein folding is to find a single mechanism. Given the diversity of protein structure and the evolutionary pressure on function and not on folding rates, a unique mechanism for folding would seem unlikely. If there are simplifying features, then small, single-domain proteins may be the most likely to exhibit them. But such proteins seem to fold by two distinct mechanisms. The λ_{6-85} repressor fragment (1) and the engrailed homeodomain (En-HD; ref. 2) seem to fold by a classical diffusion–collision mechanism (3–5) whereby secondary structural elements form independently and then dock to form the tertiary structure. Chymotrypsin inhibitor 2, on the other hand, folds by nucleation–condensation, which is characterized by concerted consolidation of secondary and tertiary interactions as the whole domain collapses around an extended nucleus (6). It has been argued on general grounds that nucleation–condensation and diffusion–collision are different manifestations of a common mechanism in which secondary structure and tertiary structure form in parallel (7, 8). Nucleation–condensation reflects the situation when secondary structure is inherently unstable in the absence of tertiary interactions whereas diffusion–collision becomes more probable with increasing stability of secondary structure.

Studies of the folding of point mutants of a prototype protein are essential for discovering atomic level details of folding mechanisms and kinetics. Single-point mutants may even cause gross changes in the kinetics of folding, such as the transition from three-state to two-state folding (9). But, to extrapolate a general understanding of folding mechanisms, studies on members of the same fold family (different homologues sharing the same overall topology but with different primary structures) can be useful in finding correlations between amino acid sequences and three-dimensional structures (10–16). Although there can be different folding routes through different transition states for some proteins (17), it seems that mechanisms of folding are often conserved in a protein family although there may be variations in the population of intermediates (11, 13, 16). Here, we extend our earlier studies of the En-HD and characterize and compare the kinetics and folding pathways of three members of the homeodomain-like protein family that share the same overall topology: human TRF1 Myb domain (hTRF1), human RAP1 Myb domain (hRAP1), and c-Myb-transforming protein (c-Myb).

The three-dimensional structures of those sequence-specific, DNA-binding proteins consist of three-helix bundle structures with a characteristic helix–turn–helix motif (helices II and III, Fig. 1). Their backbone architecture is highly conserved. Minor differences are: (*i*) the loop that connects helix I and helix II in hRAP1, which is slightly longer and less ordered than in the other domains (11 residues vs. 4 residues for En-HD); and (*ii*) the fine structural organization of the turn in the helix–turn–helix motif, which is generally longer (up to 8 residues for hTRF1) than the prototypic 4-residue length in En-HD. Further, the structural superposition of c-Myb, En-HD, and hTRF1 shows fine differences in the architecture of their three helical bundles that stem mainly from a different angle between helices I and III (Fig. 1*a*). Importantly, the program AGADIR predicts that the first and second α-helices of En-HD should have much higher propensity for formation than in c-Myb, hTRF1, and hRAP1 (Fig. 1*b*). Similarly, the algorithm of Dodd and Eagan (18) predicted En-HD to have a high tendency for formation of its helix–turn–helix motif, compared with a very low probability for the other members of the homeodomain-like family studied here (data not shown).

Materials and Methods

Proteins. The synthetic genes coding for hRAP1, hTRF1, and c-Myb were synthesized, and their gene products were expressed and purified to homogeneity by standard procedures (2).

Equilibrium and Kinetic Measurements. All experiments were carried out in the presence of 100 mM NaCl and 50 mM sodium acetate buffer (pH 5.7) at 25 ± 0.1°C. Equilibrium denaturation with urea was followed by fluorescence using an appropriate cut-off filter (>320 nm for c-Myb, En-HD, and hRAP1 and >360 nm for hTRF1, excitation at 280 nm) and by circular dichroism at 222 nm. Stopped-flow experiments were carried out on an Applied Photophysics (Letherhead, U.K.) SX-18MV instrument. Folding and unfolding were initiated by an 11-fold dilution of the denatured or the native protein in the appropriate urea solution. Temperature-jump measurements were made by using a DIA-RT capacitor-discharge T-jump apparatus (Dia-Log, Düsseldorf, Germany) and 10- to 70-μM protein concentration; 25-kV discharges were used to effect 2°C jumps, by using a 20-nF capacitor, to a final temperature of 25 ± 0.1°C, with complete heating within 5 μs under optimal conditions. Continuous-flow measurements were carried out as described (19). For hRAP1 and c-Myb, which each contain a Pro residue, there was a slow urea-independent phase with rate constants of $\approx 1 \times 10^{-2}$·sec^{-1} for hRAP1 and 7×10^{-3}·sec^{-1} for c-Myb of 5% of the

Abbreviations: En-HD, engrailed homeodomain; c-Myb, c-Myb-transforming protein; hTRF1, human TRF1 Myb domain; β_T, β Tanford value; hRAP1, human RAP1 Myb domain; MD, molecular dynamics.

‡To whom correspondence should be addressed. E-mail: daggett@u.washington.edu or arf25@cam.ac.uk.

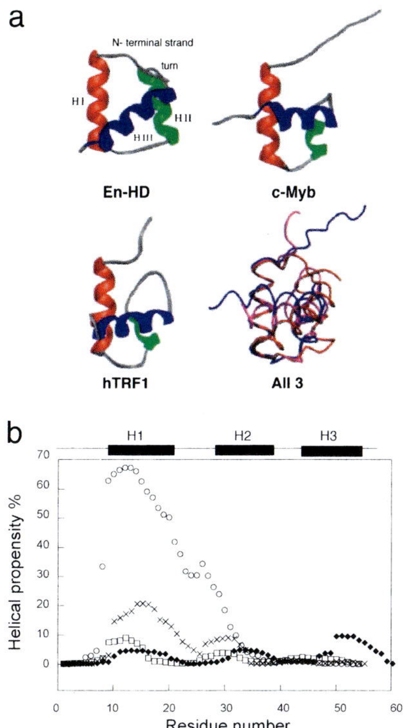

Fig. 1. (*a*) Individual and superposed structures of hTRF1, c-Myb, and En-HD, with alignment of each structure about helix I. (*b*) *Ab initio* secondary structure prediction for En-HD (○), c-Myb (×), hRAP1 (♦), and hTRF1 (□). The three helices of each protein (En-HD, residues 10–22, 28–38, and 43–55; c-Myb, residues 149–162, 166–172, and 178–189; and hTRF1, residues 8–19, 26–32, and 41–51) are marked above the graph. The helical propensities were calculated by using the AGADIR program (32) (www.embl-heidelberg.de/cgi/agadir-wrapper.pl).

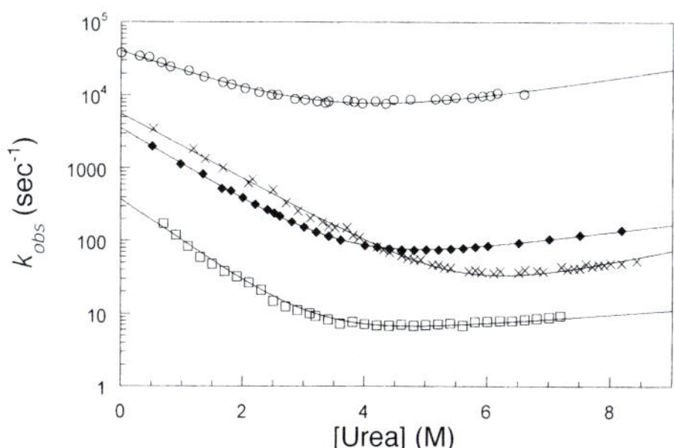

Fig. 2. Chevron plots of wild-type En-HD (○), c-Myb (×), hRAP1 (♦), and hTRF1 (□) measured in 50 mM sodium acetate buffer and 100 mM NaCl at 25°C. The lines are the best fit for a kinetic two-state model (27).

total folding amplitudes, which presumably arose from a cis-trans isomerization in the denatured state. We did not analyze this phase further.

Molecular Dynamics (MD) Simulations. We extended one of the previously described simulations of the unfolding of En-HD at 498 K (2, 20) from 40 to 65 ns for better sampling of the denatured state. All calculations were performed by using the program ENCAD (21–23). Seven independent simulations of c-Myb were performed at 498 K: one proceeded for 60 ns and six were 2 ns in duration. The initial structure for the 60-ns simulation of c-Myb and four of the 2-ns simulations was the average NMR structure [PDB code 1IDY (24)]. Structure 1 of the NMR ensemble was used for the other 2-ns simulations. The c-Myb domain of this study contains an additional Met at the N terminus, which was built onto the NMR structure. His residues (H159 and H184) were protonated, and Glu and Asp were negatively charged, reflecting a pH of ≈4–7.

The starting structure for the 60-ns c-Myb simulation was minimized for 1,000 steps. After minimization, water molecules were added to solvate the protein in a rectangular box extending at least 8 Å in all directions, resulting in 3,461 water molecules. The water density was set to the experimental value [0.829 g/ml at 225°C (25)] by adjusting the volume of the box. The solvent water was then subjected to the conjugate gradient minimization of 1,000 cycles followed by 1,000 steps of molecular dynamics. The water was then minimized again for 1,000 cycles. Finally, the protein was minimized for 1,000 steps, followed by 1,000 steps of minimization of the entire protein–water system. After these preparatory steps, the system was heated to 225°C, and the simulation was allowed to evolve over time by using a 2-fs integration time step. Atoms were allowed to move according to Newton's equations of motion, and the velocities of the atoms were adjusted intermittently until the system reached the desired temperature, after which the NVE ensemble (microcanonical ensemble in which the number of particles, volume, and energy are constant) was used. In all calculations, an 8-Å, force-shifted, nonbonded interaction cut-off was used. The nonbonded list was updated every two cycles. The short simulations for c-Myb were performed similarly except that the number of steps in the preparatory dynamics and minimization routines was varied between 990 and 1,100 to produce independent trajectories. The random number see for MD was also varied. The number of water molecules ranged from 3,461 to 3,486. The same protocols were used in two simulations of S16A hTRF1 at 498 K, beginning with the NMR averaged structure (26). These simulations were each 60 ns in duration. The number of water molecules was 2,666.

Results and Discussion

Transition from Two- to Three-State Kinetics. There is a progression from three-state to two state kinetics across the family. Apart from the proline phase, the folding of hTRF1, hRAP1, and c-Myb fitted a single exponential process under all conditions, unlike the multiphasic kinetics of En-HD (2). The folding/unfolding rate constants for hTRF1, hRAP1, and c-Myb fitted the canonical chevron plots with urea concentration (Fig. 2), characteristic of a two-state process (27). Apart from En-HD, all of the values of $\Delta G^0_{\text{D-N}}$ and $m_{\text{D-N}}$ calculated from kinetics were in good agreement with those calculated from equilibrium experiments (data not shown), consistent with a two-state model (27). Further, the kinetic $m_{\text{D-N}}$ value of 0.59 (±0.03) kcal·mol^{-1}·M^{-1} for En-HD is significantly lower than that measured at equilibrium, 0.80 (±0.05) kcal·mol^{-1}·M^{-1}, consistent with a compact folding intermediate accumulating on the pathway (2).

The refolding rate constant of each protein (Table 1) had a similar strong dependence on denaturant concentration whereas the unfolding rate changed only slightly (see *m* values in Table 1). The compactness of the transition state, relative to the native state, can be estimated by the β Tanford (β_T) value [$\beta_T = m_F/(m_U + m_F)$], which is one measure of the position of the transition state on the reaction coordinates (28). The β_T values of 0.90 for hTRF1, 0.82 for hRAP1, and 0.79 (±0.02) for c-Myb (Table 1) indicated that they had a compact native-like transition state.

There was no correlation between stability and folding rate, in contrast to the folding of Src homology 3 (29) and immunoglobin-like domains (12). The variation in helical and turn propensities for En-HD and its homologs may reconcile the different kinetic mechanisms for folding (7, 8, 30, 31). Calculations of the

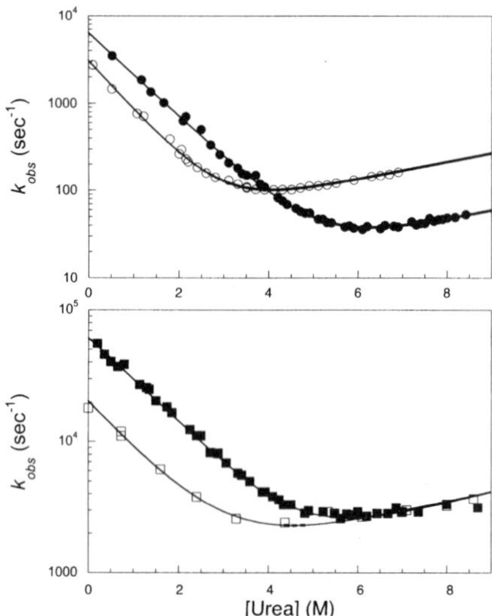

Fig. 3. (*Upper*) Chevron plot of wild-type c-Myb (●) compared with a representative secondary probe in the turn region, namely G175A mutant (○). Fits follow a simple two-state model. (*Lower*) Chevron plots of the pseudo-wild type of the En-HD (■) compared with a representative secondary probe mutant in the turn, namely G39A (□). Fits are the result of a three-state model. Quantitative comparison of these homologous positions in the two proteins suggests that the turn of En-HD is fully structured in the folding transition state whereas that of c-Myb is mainly unstructured.

helical stabilities by AGADIR (ref. 32 and Fig. 1) suggest that the regions spanning helices I and II of En-HD have a high and continuous helical propensity whereas all of the other proteins have a much lower helical propensity, with c-Myb being second at 20% for helix I. Thus, the strong inherent propensity to form native secondary structure in the absence of tertiary interactions seemed to dominate the folding kinetics of En-HD. Conversely, the proteins of inherently weaker secondary structure folded in a two-state manner.

Φ-Value Analysis of En-HD and c-Myb. We investigated by Φ-value analysis (33) the folding transition states of c-Myb, the most stable protein in this study, and a stabilized pseudo-wild-type mutant of En-HD, K52A, which is more tolerant to mutations (Table 2). For a two-state system, $\Phi_F = 1 - \Phi_U$, where F = folding and U = unfolding. Generally, the equation $\Phi_F = 0$ implies that the residue in question in the transition state is structurally similar to the denatured state; and the equation $\Phi_F = 1$ implies full extent of native interactions in the transition state (34). Owing to the complexities in analyzing the multistate kinetics of folding of En-HD, we concentrated only on its rate-determining transition state for unfolding, that is, its final transition state for folding.

We made two classes of mutation, either perturbing primarily secondary structure or primarily tertiary structure (Table 2). Identical Φ values were calculated from the folding and unfolding rate constants for c-Myb, confirming that the folding and unfolding kinetics of all of the mutants of c-Myb obey the classical behavior for a two-state system. Similarly, the observed $m_{D\text{-}N}$ values were all similar to the value obtained for wild-type c-Myb. Moreover, the values of β_T were all very similar (Table 1).

The Φ values for En-HD and c-Myb were quite similar (Table 2), but there was one significant difference in the pattern of Φ_F between the two proteins (Table 2): the region around the turn in helix II–turn–helix III in En-HD had Φ_F values that were close to 1 whereas the values for c-Myb were close to zero (Table 2 and Fig. 3). The Φ_F values parallel the prediction of a high probability for the helix–turn–helix in En-HD but not c-Myb. The folding transition state of c-Myb had an extended, loosely packed hydrophobic nucleus located at the interface between the three helices. The folding nucleus of En-HD is similar but seems to be more structured, by virtue of a cluster of native-like interactions throughout the turn.

It is difficult to distinguish between nucleation–condensation and diffusion—collision solely on the basis of a limited number of Φ_F values. Both mechanisms are consistent with an extended nucleus in the transition state, but diffusion–collision should have a significant number of Φ_F values that probe secondary stucture close to 1 (35). There is considerable ancillary evidence for diffusion–collision for En-HD: the presence of a folding intermediate with much native secondary structure and the sequence of events observed in the MD simulations (described below), as well as several Φ_F values in the secondary structure close to 1 (ref. 2 and Table 2). Conversely, chymotrypsin inhibitor 2 folds from an unstructured denatured state, with a very clear pattern of fractional Φ_F values from a very large number of mutants for which a double logarithmic plot of the unfolding rate constant vs. equilibrium constant follows a linear Brønsted plot of slope 0.64 (6, 36). The Brønsted plot for c-Myb is very similar, with a slope of 0.69 (Fig. 4), which represents the average degree of formation of interaction energies within the whole protein. The fractional slope and reasonable fit to a single line for nearly all mutants implies that secondary structure and tertiary structure are being formed in parallel, according to the nucleation–condensation mechanism (37). In contrast, nearly all of the positions probing secondary structure in the Brønsted plot for En-HD fall on a line with a slope of 0, as one would expect for helices that are fully formed before docking in the transition state (Fig. 4). In contrast, the majority of the positions probing tertiary structure are similar to c-Myb, as expected for docking of side chains in the rate-determining step. Further evidence comes from MD simulation.

MD Simulations of En-HD, c-Myb, and hTRF1. We performed MD simulations of the thermal denaturation of En-HD, c-Myb, and hTRF1 to obtain atomic resolution information for unfolding pathways and associated transition, intermediate, and denatured states. Multiple independent simulations of each protein were performed at different temperatures. We have already shown that the unfolding times in simulations at 75 and 100°C are in good agreement with experiment (2). Also, we have shown that the overall unfolding process for this (2, 20) and other proteins

Table 1. Kinetic folding parameters for the homeodomain-like proteins

	k_F, s^{-1}	k_U, s^{-1}	m_F, kcal·mol^{-1}·M^{-1}	m_U, kcal·mol^{-1}·M^{-1}	$m_{D\text{-}N}$, kcal·mol^{-1}·M^{-1}	$\Delta G_{D\text{-}N}$, kcal·mol^{-1}	β_T
hTRF1	370 ± 20	3.20 ± 0.30	0.86 ± 0.02	0.09 ± 0.01	0.95 ± 0.02	2.82 ± 0.05	0.90 ± 0.02
c-Myb	6,200 ± 200	5.30 ± 0.60	0.65 ± 0.01	0.17 ± 0.01	0.82 ± 0.01	4.17 ± 0.07	0.79 ± 0.02
hRAP1	3,600 ± 100	18.0 ± 2.0	0.68 ± 0.01	0.15 ± 0.01	0.83 ± 0.01	3.12 ± 0.06	0.82 ± 0.02
En-HD	39,900 ± 700	2,100 ± 500	0.49 ± 0.02	0.10 ± 0.02	0.59 ± 0.03	1.7 ± 0.2	0.83 ± 0.06

Calculated according to a two-state model (27); SE are given. k_F and k_U are the rate constants for folding and unfolding, respectively. m is the slope of (log k)/RT vs. urea concentration.

Table 2. Φ and S values for the transition states of En-HD and c-Myb*

Mutant	Location	Interaction probed	Φ_F[†]	S value[‡]
En-HD				
F8A	N-term.	3°: packing throughout core	0.42	0.35
L13A	HI	3°: core packing around the N-term. strand	0.51	0.63
A14G	HI	2°	0.79	0.85
L16V	HI	3°: packing throughout the core	0.39	0.56
F20A	HI	3°: packing throughout the core	0.36	0.20
Y25G	Loop	2° and 3°	0.28	0.08
A25G	Loop	2° and 3°	0.17	0.08
L26A	Loop	2° and 3° packing around loop	0.46	0.31
L38A	HII	3°: core packing around turn	0.48	0.62
L38V	HII	3°: core packing around turn	0.83	0.62
G39A	Turn	2° and 3°	0.92	0.79
L40A	Turn	2° and 3°: core packing around turn	0.95	0.58
A43G	HIII	2°	1.05	0.73
I45V	HIII	2° and 3°: core packing around turn	0.69	0.62
A54G	HIII	2°	0.62	0.59
c-Myb				
W147F	N-term.	3°: core packing and H-bond between N-term. strand and HIII	0.11	0.21
E151Q	HI	3°: region around turn	0.16	0.81
D152N	HI	3°: interactions between HI and HIII	0.18	0.87
I154V	HI	2° and 3°: interaction between HI and HII/turn	0.84	0.82
LI55A	HI	3°: packing of core	0.34	0.63
A158G	HI	2°	0.66	0.67
G163A	Loop	2° and 3°	0.46	0.47
A167G	HII	2°	0.92	0.66
I169V	HII	3°: core packing between HI HII in loop	0.34	0.25
A170G	HII	2° and 3°: interaction between HII and HIII/turn	0.48	0.32
L173V	Turn	3°: interactions between HI and HIII	0.34	0.39
P174G	Turn	2° and 3°: packing around turn	0.08	0.15
G175A	Turn	2° and 3°: packing around turn	0.32	0.58
R176K	Turn	3°: region between turn and the N-term./HI	0.31	0.46
A180G	HIII	2° and 3°: between HIII and N-term. strand	0.55	0.75
I181A	HIII	3°: packing around turn.	0.64	0.52
W185F	HIII	3°: core packing and H-bond between HI and HIII	−0.04	0.44
M189A	C-term.	3°: packing between HI and HIII	0.01	0.27

*S value is the simulated equivalent of Φ. HI, helix I; HII, helix II; HIII, helix III; 2°, secondary probe; 3°, tertiary probe; term., terminal.

[†]Because En-HD folds from its open denatured state, U, via a compact highly helical intermediate, I, Φ_F values were calculated for the transition state that leads to the formation of the fully native state, using the relationship $\Phi_F = 1 - \Phi_U$. c-Myb folds by simple kinetics, with a single transition state. Φ_F was found to be insensitive to the concentration of urea, and each value has a SE of ± 0.04–0.06.

[‡]Average S values were calculated for the trajectories and time periods described in the legend to Fig. 5. For the residues in turns and loops, the $S_{3°}$ values are given because the main chain became distorted but side chain interactions were frequently maintained. In such cases, the product of $S_{2°}$ and $S_{3°}$ failed to capture this retention of structure. In particular, this was the case with residues 8, 26, and 40 of En-HD and residues 163, 173, 175, and 176 of c-Myb.

(38) is essentially independent of temperature and that increasing the temperature merely accelerates the process. So, we focus on very high temperature simulations (498 K, or 225°C, and elevated pressure, ≈26 atm, to maintain water in the liquid state) because of their improved sampling of the denatured state although some results for lower experimental temperatures are also presented to show the generality and applicability of the findings.

Putative transition state ensembles were identified for the various simulations by using a conformational clustering procedure (39, 40). Representative structures from these ensembles for the independent trajectories (two each at 373 and 498 K for En-HD; seven at 498 K for c-Myb; and two at 498 K for hTRF1) are presented in Fig. 5*a*. Although the structures are native-like, the average solvent-accessible surface area increased by ≈17% for En-HD and c-Myb, or $\beta_T = 0.83$, and ≈20% for hTRF1, or $\beta_T = 0.80$. This degree of exposure is consistent with the β_T value determined experimentally: 0.83, 0.79, and 0.90 for En-HD, c-Myb, and hTRF1, respectively.

We calculated S values, the semiquantitative MD equivalents of Φ values (41). The S and Φ values are in reasonable agreement (Table 2), with correlation coefficients of 0.79 and 0.74 for En-HD and c-Myb (excluding the E151Q and D152N mutants), respectively. Φ_F is a good measure for fractional bond formation when determined for mutations of larger to smaller hydrophobic side chains but not necessarily so for mutation of charged or polar residues (33, 34). We found the best agreement for the hydrophobic mutations (the correlation coefficient given for c-Myb decreases by 0.28 if the E151 and D152 mutations are considered). The transition state structures were effectively insensitive to changes in temperature and sequence, considering the global fold of the protein (Fig. 5*a*). On closer inspection, however, it can be seen that the transition state for En-HD is more structured than that of c-Myb. In particular, helix II is

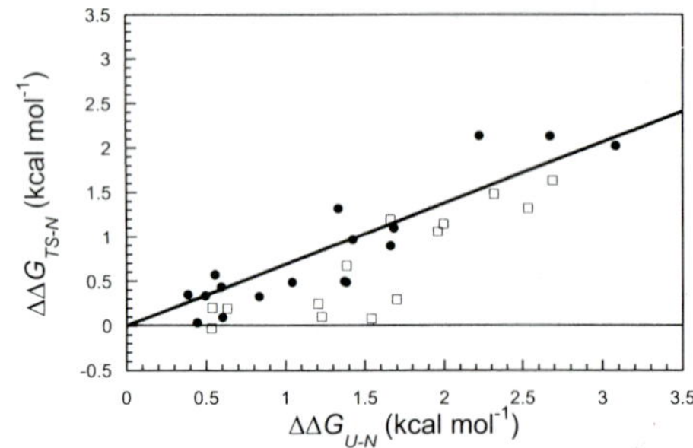

Fig. 4. Brønsted plot for c-Myb (●) and En-HD (□). The line is the best linear fit to the Brønsted plot for the c-Myb mutants ($r = 0.90$). We noted that, for En-HD, the mutants along the line with slope 0 are involved in the stabilization of H1 (A14G) and the HII–turn–HIII motif, indicating that these secondary elements are fully formed before the docking of the transition state (see text).

highly variable in c-Myb, it experiences more fraying at the ends of helices I and III, and the loop between helices II and III is distorted. Thus, the transition state of En-HD contains a high degree of secondary structure and consolidation of the packing interactions and the final expulsion of water occurs in the transition state and during the progression to the native state. In contrast, both secondary and tertiary interactions are in the process of being formed in the transition states of c-Myb and hTRF1.

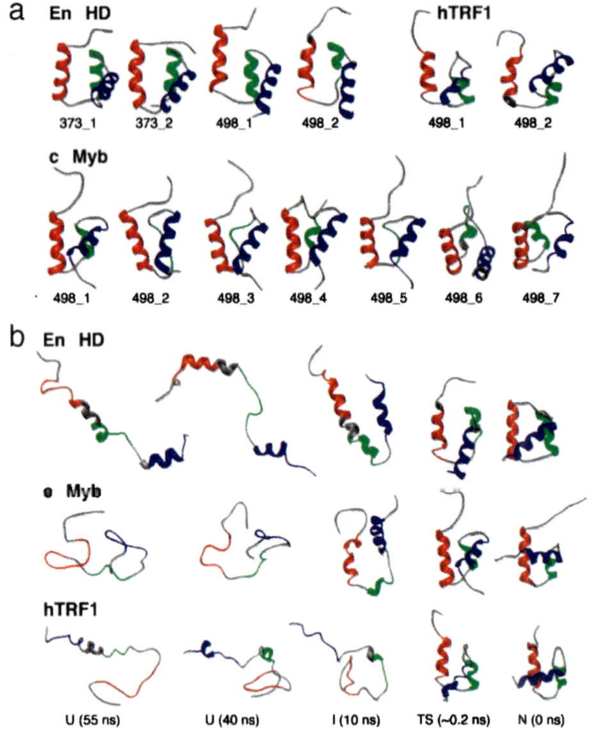

Fig. 5. (*a*) Representative transition-state (TS) structures for independent unfolding simulations of En-HD, c-Myb, and hTRF1 at different temperatures. The TS ensembles correspond to the following time intervals: En-HD, 373_1, 1.715–1.72 ns; 373_2, 1.98–1.985 ns; 498_1, 0.165–0.17 ns; and 498_2, 0.255–0.26 ns; c-Myb, 498_1, 0.1–0.105 ns; 498_2, 0.275–0.28 ns; 498_3, 0.315–0.32 ns; 498_4, 0.225–0.23 ns; 498_5, 0.50–0.505 ns; 498_6, 0.55–0.555 ns; and 498_7, 0.145–0.150 ns; and hTRF1, 498_1, 0.14–0.145 ns; and 498_2, 0.125–0.13 ns. The structure corresponding to the final time point for each TS ensemble is presented. (*b*) Structures from 498-K simulations of En-HD (498_2), c-Myb (498_1), and hTRF1 (498_1).

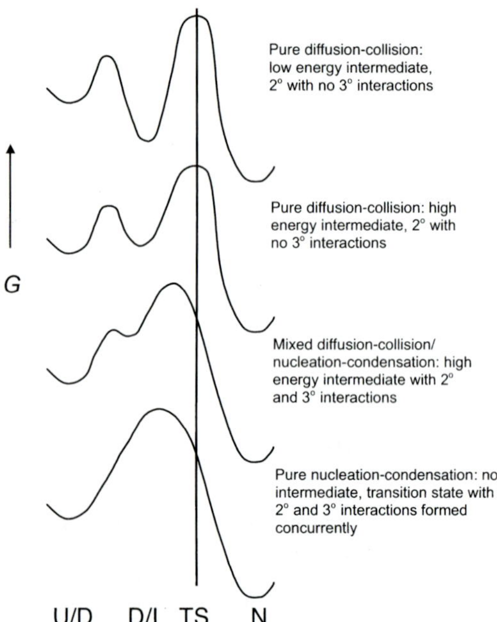

Fig. 6. Simplified energy diagrams for folding of small single-domain proteins. Pure nucleation–condensation implies that secondary and tertiary structures are formed simultaneously in the absence of intermediates, as observed for hTRF1. As the propensity for forming secondary structure increases, the mechanism slides from the nucleation–condensation to the pure diffusion-collision model, as observed for En-HD.

Importantly, we also detected an intermediate in the unfolding pathway of c-Myb that was not apparent in the experiments. c-Myb and En-HD both populate intermediates with considerable helical structure in the MD simulations (Fig. 5*b*). This finding is consistent with experiment on En-HD, which folds via a highly helical intermediate, whereas c-Myb folds in an apparent two-state manner. However, the population of the c-Myb intermediate increases in single-site mutants that specifically increase the helical and turn propensities such that the intermediate becomes experimentally visible (unpublished data). The unfolded states of these two proteins are quite different. The unfolded state of En-HD contains residual, dynamic helical structure whereas that of c-Myb seems to be a compact random coil (Fig. 5*b*). Finally, we did not detect intermediates during the unfolding of hTRF1, and its denatured state is quite disordered, like that of c-Myb (Fig. 5*b*). These findings are consistent with the lower helical propensities of hTRF1 compared with c-Myb and En-HD.

Unifying Features in Folding Mechanisms. The transition state of En-HD, as reflected in both the Φ value analysis and the simulated transition state ensembles, and the actual simulated unfolding pathway show what is expected for a diffusion–collision mechanism (4), with nearly fully formed elements of secondary structure coalescing via their hydrophobic side chains in the transition state (2, 7). The transition state structures of c-Myb and hTRF1 are similar except that the helices are not fully formed, particularly helix II, and instead are in the process of being consolidated as would be expected for the nucleation–condensation mechanism. But, it must be noted that movements from diffusion–collision to nucleation–condensation are not detected simply by the helical content of the folding transition states but by the careful analysis of whether the secondary and tertiary structures are formed simultaneously or not (Figs. 5*b* and 6). This point can be addressed only by careful characterization of denatured and intermediate states, as discussed by Mayor *et al.* (2) for En-HD, such as investigation of the formation of secondary structure elements in the denatured state (5), Φ values, and Brønsted plot analysis that specifically monitors the differences in free energy between the ground states and the transition state between them, as well as direct MD simulation.

The nucleation–condensation model postulates the existence of a folding nucleus whose formation stabilizes the transition state. However, formation of a small folding nucleus is probably not solely rate determining because a significant fraction of the overall structure must be in approximately the correct conformation for the discontinuous network of residues in the nucleus to come together. An important corollary of this observation is that formation of the nucleus (nucleation) is coupled with a more general formation of structure (condensation), giving rise to a roughly linear Brønsted profile as observed for c-Myb.

The Src homology 3 domain (42) and WW domains (43, 44) seem to fold via an alternative folding mechanism that implies the presence of a structurally polarized transition state, perhaps caused by loop or hairpin nucleation events that seem to initiate the folding of these all β-sheet proteins (45). Polarized transition states represent a hybrid between nucleation–condensation and diffusion–collision. The structure of the transition state, which resembles an expanded version of the native state, with fractional and low Φ values throughout the hydrophobic core, is consolidated by a discrete and localized cluster of preformed secondary elements, with high Φ values, giving rise to a structurally polarized transition state.

hTRF1 and En-HD represent two extreme variations of the folding process for this particular topology. As the propensity for forming secondary structure increases, the mechanism slides from nucleation–condensation to the diffusion–collision/framework model (Fig. 6). Based on our results, En-HD seems to fold via the framework model whereas c-Myb seems to fold via a mixed framework/nucleation–condensation model with a high energy intermediate, and hTRF1 seems to fold via a pure nucleation–condensation process. The common feature in the un(folding) of these proteins is a transition state that is very native-like, with a mixture of tertiary and secondary interactions. But the balance of tertiary and secondary interactions and the route of reaching the transition state depend crucially on the inherent propensities for secondary and tertiary structure.

We thank Dr. Christopher M. Johnson for skillful technical assistance. S.G. is supported by a fellowship from the Istituto Pasteur-Fondazione Cenci Bolognetti (Rome), N.R.G. is supported by a Winston Churchill Scholarship, and U.M. is supported by an "Ikertzaileen prestakuntza" grant from the Government of the Basque Country. V.D. is grateful for support of the computational work provided by Grant GM 50789 from the National Institutes of Health (NIH). This work is supported by an NIH Pharmacological Sciences Training Grant 5 TG32 GM07750 (to M.L.D.).

1. Myers, J. K. & Oas, T. G. (1999) *Biochemistry* **38,** 6761–6768.
2. Mayor, U., Guydosh N. R., Johnson C. M., Grossmann J. G., Sato S., Jas G. S., Freund S. M., Alonso D. O., Daggett V. & Fersht, A. R. (2003) *Nature* **421,** 863–867.
3. Islam, S. A., Karplus, M. & Weaver, D. L. (2002) *J. Mol. Biol.* **318,** 199–215.
4. Karplus, M. & Weaver, D. L. (1994) *Protein Sci.* **3,** 650–668.
5. Kim, P. & Baldwin, R. L. (1990) *Annu. Rev. Biochem.* **59,** 631–660.
6. Itzhaki, L. S., Otzen, D. E. & Fersht, A. R. (1995) *J. Mol. Biol.* **254,** 260–288.
7. Daggett, V. & Fersht, A. R. (2003) *Trends Biochem. Sci.* **28,** 18–25.
8. Fersht, A. R. (1999) *Structure and Mechanism in Protein Science* (Freeman, New York).
9. Matouschek, A., Kellis, J. T., Jr., Serrano, L., Bycroft, M. & Fersht, A. R. (1990) *Nature* **346,** 440–445.
10. Chiti, F., Taddei, N., White, P. M., Bucciantini, M., Magherini, F., Stefani, M. & Dobson, C. M. (1999) *Nat. Struct. Biol.* **6,** 1005–1009.
11. Ferguson, N., Capaldi, A. P., James, R., Kleanthous, C. & Radford, S. E. (1999) *J. Mol. Biol.* **286,** 1597–1608.
12. Fowler, S. B. & Clarke, J. (2001) *Structure.* **9,** 355–366.
13. Gianni, S., Travaglini-Allocatelli, C., Cutruzzola, F., Bigotti, M. G. & Brunori, M. (2001) *J. Mol. Biol.* **309,** 1177–1187.
14. Martinez, J. C. & Serrano, L. (1999) *Nat. Struct. Biol.* **6,** 1010–1016.
15. Riddle, D. S., Grantcharova, V. P., Santiago, J. V., Alm, E., Ruczinski, I. & Baker, D. (1999) *Nat. Struct. Biol.* **6,** 1016–1024.
16. Staniforth, R. A., Giannini, S., Bigotti, M. G., Cutruzzolà, F., Travaglini-Allocatelli, C. & Brunori, M. (2000) *J. Mol. Biol.* **297,** 1231–1244.
17. Ternstrom, T., Mayor, U., Akke, M. & Oliveberg, M. (1999) *Proc. Natl. Acad. Sci. USA* **96,** 14854–14859.
18. Dodd, I. B. & Egan, J. B. (1990) *Nucleic Acids Res.* **18,** 5019–5026.
19. Ferguson, N., Johnson, C. M., Macias, M., Oschkinat, H. & Fersht, A. R. (2001) *Proc. Natl. Acad. Sci. USA* **98,** 13002–13007.
20. Mayor, U., Johnson, C. M., Daggett, V. & Fersht, A. R. (2000) *Proc. Natl. Acad. Sci. USA* **97,** 13518–13522.
21. Levitt, M. (1990) ENCAD, Computer Program for Energy Calculations and Dynamics (Molecular Applications Group, Palo Alto, CA).
22. Levitt, M., Hirshberg, M., Sharon, R. & Daggett, V. (1995) *Comput. Phys. Commun.* **91,** 215–221.
23. Levitt, M., Hirshberg, M., Sharon, R., Laidig, K. E. & Daggett, V. (1997) *J. Phys. Chem. B* **101,** 5051–5061.
24. Furukawa, K., Oda, M. & Nakamura, H. (1996) *Proc. Natl. Acad. Sci. USA* **93,** 13583–13588.
25. Kell, G. S. (1967) *J. Chem. Eng.* **12,** 66–68.
26. Nishikawa, T., Nagadoi, A., Yoshimura, S., Aimoto, S. & Nishimura, Y. (1998) *Structure (London)* **6,** 1057–1065.
27. Jackson, S. E. & Fersht, A. R. (1991) *Biochemistry* **30,** 10428–10435.
28. Tanford, C. (1968) *Adv. Protein Chem.* **23,** 121–282.
29. Plaxco, K. W., Simons, K. T., Ruczinski, I. & Baker, D. (2000) *Biochemistry* **39,** 11177–11183.
30. Muñoz, V. & Serrano, L. (1996) *Folding Des.* **1,** 71–77.
31. López-Hernández, E., Cronet, P., Serrano, L. & Muñoz, V., (1997) *J. Mol. Biol.* **266,** 610–620.
32. Muñoz, V. & Serrano, L. (1997) *Biopolymers* **41,** 495–509.
33. Fersht, A. R., Matouschek, A. & Serrano, L. (1992) *J. Mol. Biol.* **224,** 771–782.
34. Matouschek, A., Kellis, J. T., Jr., Serrano, L. & Fersht, A. R. (1989) *Nature* **340,** 122–126.
35. Fersht, A. R. (2000) *Proc. Natl. Acad. Sci. USA* **97,** 1525–1529.
36. Otzen, D. E., Itzhaki, L. S., ElMasry, N. F., Jackson, S. E. & Fersht, A. R. (1994) *Proc. Natl. Acad. Sci. USA* **91,** 10422–10425.
37. Fersht, A. R., Matouschek, A., Serrano, L. (1997) *Curr. Opin. Struct. Biol.* **7,** 3–9.
38. Day, R., Bennion, B. J., Ham, S. & Daggett, V. (2002) *J. Mol. Biol.* **322,** 189–203.
39. Li, A. & Daggett, V. (1994) *Proc. Natl. Acad. Sci. USA* **91,** 10430–10434.
40. Li, A. & Daggett, V. (1996) *J. Mol. Biol.* **257,** 412–429.
41. Daggett, V., Li, A., Itzhaki, L. S., Otzen, D. E. & Fersht, A. R. (1996) *J. Mol. Biol.* **257,** 430–440.
42. Grantcharova, V. P., Riddle, D. S., Santiago, J. V. & Baker, D. (1998) *Nat. Struct. Biol.* **5,** 714–720.
43. Jager, M., Nguyen, H., Crane, J. C., Kelly, J. W. & Gruebele, M. (2001) *J. Mol. Biol.* **311,** 373–393.
44. Ferguson, N., Pires, J. R., Toepert, F., Johnson, C. M., Pan, Y. P., Volkmer-Engert, R., Schneider-Mergener, J., Daggett, V., Oschkinat, H. & Fersht, A. R. (2001) *Proc. Natl. Acad. Sci. USA* **98,** 13008–13013.
45. Ferguson, N. & Fersht, A. R. (2003) *Curr. Opin. Struct. Biol.* **13,** 75–81.

Downhill Folding Controversy

I became drawn, unwillingly, into a controversy over the existence of 'downhill folding'. The proteins studied in the early years all fold slowly and on a time scale that allows identification of the separate native and denatured states by, for example, NMR spectroscopy. The proteins that we have worked on since the EnHD all fold on very rapid time scales as we wished to do experiments that were amenable to the time scale of atomistic simulation. These proteins fold and unfold so fast that NMR spectroscopy just sees the average structure of native and denatured states when they are both present during denaturation. Victor Muñoz proposed that one of the proteins that Neil Ferguson was working on in my lab folded 'downhill', that is there is a continuous slide between native and denatured states, with only one structural ensemble present at any one time. Neil's very carefully executed data were entirely consistent with classical barrier limited folding. The underlying problem is that it is nigh impossible to distinguish unambiguously between classical barrier limited folding of very fast folding proteins and downhill folding because they are kinetically equivalent in most cases[168]. Fang Huang constructed a confocal fluorescence microscope for very sensitive single-molecule fluorescence energy transfer experiments. He decisively resolved the dispute, to my satisfaction, by observing separate states, on the 100 μs time scale[169]. All our other data are consistent with a classical energy barrier limited folding of the peripheral subunit binding domains[170–176]. Hannes Neuweiler implemented single molecule fluorescence correlation spectroscopy in my lab, based on methods he had developed earlier, and he used it to measure the segmental chain diffusion in the denatured state of the domain, which occurs on the μs time scale, ten times faster than the overall folding[177].

When I entered the field of folding, there was little direct experimental measurement of the contributions of different interactions to stability, there was no description at atomic level of protein folding and no detailed description of transition states or on-pathway non-covalent intermediates, and proteins were thought to fold either by framework or molten-globule collapse. My group was the first to quantify experimentally individual contributions to energetics, provided the first detailed descriptions of transition states, characterised on-pathway intermediates, discovered two-state folding, solved, in collaboration with Valerie Daggett's simulation, many complete folding pathways and hence proposed a new and unifying mechanism of nucleation-condensation.

Many of the members of the early folding group have become famous in their own right and are now well-established professors, or the equivalent: students such as Andreas Matouschek, Sophie Jackson, Liz Meiering, Jane Clarke, Vic Arcus, Paul Dalby, Sarah Perrett, Jaqui Matthews, Daniel Otzen and Andreas Ladurner; post-doctorals such as Laura Itzhaki, Amnon Horovitz, Gideon Schreiber and Gonzalo de Prat Gay; and many others who have forged successful careers in science. There is a very strong presence in Spain with Luis Serrano, Javi Sancho, Jose Luis Neira and Jesus Sanz from the early days, as well as Ugo Mayor, and those who worked on other projects who include Fernando Corrales, Jose Manuel Cañadillas, Charo Fernandez-Fernandez and Bego Sot in recent years.

One good outcome of the folding studies is that Chris Dobson and I became close friends. Chris was our major rival. Nevertheless, we used to have joint group meetings, twice yearly with buses from Oxford to Cambridge and vice versa. I proposed Chris for a John Humphrey Plummer Professorship in Cambridge, which he was awarded, and we have adjoining offices and shared secretarial offices. Some people were surprised that I would want my main 'rival' so close. The answer is clear: you want to be in the best department with best possible colleagues. You want to see your colleagues do well and prosper, and you all do better. There was easily room for both of us, and we transformed the chemistry department. My students Jane Clarke and Sophie Jackson, have been appointed faculty members there, as well Paul Barker, a former CPE member. Michele Vendrusculo and Carol Robinson from Chris's lab are also faculty members.

15.7 2008: Alan and Victor Muñoz at the Colosseum in Rome.

15.8 2003: Chris Dobson and Marilyn Fersht.

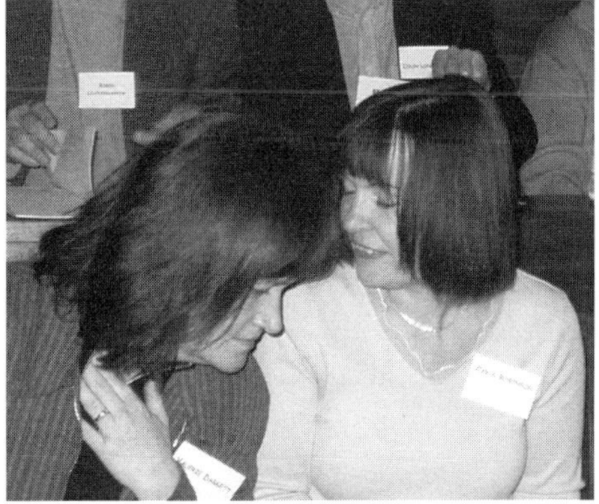

15.9 2003: Valerie Daggett and Carole Robinson.

Chapter 16

The Molecular Chaperone GroE

'A new scientific truth does not triumph by convincing its opponents and making them see the light, but rather its opponents eventually die, and a new generation grows up that is familiar with it.'

Max Planck

We did good pioneering work on the molecular chaperone GroE, but it is one area where our efforts do not seem to have been appreciated, and I always felt an interloper in the field. But, we discovered or demonstrated many of the key features of its mechanism. The GroEL/GroES complex is one of a variety of proteins that assist in the folding of proteins, during or after biosynthesis, and also prevent irreversible denaturation and aggregation after heat shock. GroEL is a 14-mer, comprised of two 7-membered rings that form a cavity. The ends can be capped by GroES, a 7-mer, each subunit of which binds to the apical domain of a GroEL subunit. I was persuaded to work on GroE by my research assistant, Tammy Gray, who wanted to do a Ph.D. on it under the auspices of the Open University. Her initial data got me quickly interested as it showed sigmoid plots of rate against substrate concentration, the hallmark of co-operative binding. We, thus, discovered in our first paper in 1991 that GroE is an allosteric protein[178]. The paper, which was perfectly sound and had an extremely important novel result, was rejected by *Biochemistry* and published in *FEBS Letters*, where it quickly racked up more than 100 citations. The cavity in GroE was thought to be too small to accommodate unfolded proteins and so denatured proteins were assumed to fold in solution after transient interaction with GroE. But we demonstrated that barnase folds while bound to GroE[179,180].

16.1 1991: Tammy Gray.

J. Mol. Biol. (1993) **232**, 1197–1207

Refolding of Barnase in the Presence of GroE

Tamara E. Gray† and Alan R. Fersht

MRC Unit for Protein Function and Design
Cambridge IRC for Protein Engineering, Department of Chemistry
University of Cambridge, Lensfield Road, Cambridge CB2 1EW, U.K.

(Received 29 January 1993; accepted 20 April 1993)

The refolding of barnase in the presence of GroEL has been monitored on the millisecond to seconds time scale using stopped-flow kinetics. GroEL binds rapidly and tightly to the denatured enzyme with a second-order rate constant of greater than $1{\cdot}3\times10^{8}\ s^{-1}\ M^{-1}$ and slows down greatly the rate of barnase refolding. However, addition of ever increasing concentrations of GroEL does not prevent barnase refolding completely, as would be expected from mass action if folding of barnase could proceed only in free solution. At saturating concentrations of GroEL, barnase refolds with a half-life of 30 s, compared with 50 ms for refolding of free enzyme. The rate-determining step in the refolding of free barnase is the reaction of a "late" folding intermediate. A mutant of barnase that folds more slowly (Ser→Ala91), refolds at a correspondingly lower rate when bound to GroEL, suggesting that formation of the fully folded state may be rate limiting for folding on GroEL. For the slow-folding Ser→Ala91 mutant, the rate-determining refolding step has a half-life of 180 ms. In sequential mixing experiments, a delay was introduced to allow the Ser→Ala91 mutant to refold for 30 ms before being mixed with GroEL. This reduces by 50% the amount of mutant barnase initially bound by GroEL. As only 11% of this mutant barnase is fully refolded from the late intermediate in 30 ms, there is preferential binding of an earlier refolding state to GroEL. We show by single mixing experiments that binding, not hydrolysis, of ATP reduces the lag in regain of barnase activity seen with GroEL alone. In the presence of high concentrations of ATP and GroEL the rate constant for refolding of barnase approaches that found in their absence, probably because ATP reduces the affinity of GroEL for refolding barnase, such that bound barnase is released and refolds unhindered. The addition of exceedingly small quantities of GroES in the presence of excess GroEL and a moderate amount of ATP also has a marked effect on the barnase refolding rate constant, suggesting that GroES may have higher affinity for the barnase : GroEL complex than for GroEL.

Keywords: GroEL; GroES; protein folding; chaperonin; barnase

1. Introduction

The molecular chaperone GroEL has been demonstrated to interact with a wide variety of proteins in non-native conformations (Viitanen *et al.*, 1992), showing little if any discrimination amongst substrates. It has been shown for several systems that GroEL can effectively block "off-pathway" folding reactions, thus preventing folding polypeptides from aggregating (Buchner *et al.*, 1991; Holl-Neugebauer *et al.*, 1991). Recent studies *in vitro* (Langer *et al.*, 1992*a*) also indicate that GroEL may be the last in a series of chaperone molecules to interact with newly folding polypeptide chains, with the first being DnaK (the *Escherichia coli* Hsp70 homologue), which coats the polypeptide chain as it emerges from the ribosome. According to this sequential model for chaperone action (Manning-Krieg *et al.*, 1991; Langer *et al.*, 1992*a*), all polypeptides may interact with DnaK as they are translated in *E. coli* and thus enter a chaperone-mediated folding pathway *in vivo*.

Detailed studies *in vitro* on the interaction of GroEL with polypeptide substrates have tended to employ polypeptide substrates that are prone to aggregation and tend not to fold spontaneously, except at low concentrations and low temperatures, for example rhodanese (Mendoza *et al.*, 1991*a*). On the other hand, detailed protein folding studies *in vitro*, in the absence of other factors, have often focussed on small monomeric polypeptides that readily refold *in vitro*. Here, we analyse the effect of GroEL on the refolding *in vitro* of barnase, a small

† Present address: Department of Chemical Immunology, Weizmann Institute of Science, Rehovot, 76100, Israel.

monomeric ribonuclease from *Bacillus amyloliquefaciens*. Barnase was chosen as a model substrate because its refolding and unfolding kinetics have been extensively studied in the past using a combination of NMR and protein engineering techniques (Matouschek *et al.*, 1992*a*,*b*). These studies have shown that barnase has a kinetically significant late-folding intermediate. There is much detailed structural information on this intermediate that precedes the formation of the transition state for folding, in which many of the tertiary interactions that are only partly formed in the intermediate state become fully formed (Matouschek *et al.*, 1992*a*,*b*; Serrano *et al.*, 1992). In particular, these studies have shown that interactions involving the five-stranded anti-parallel β-sheet in barnase are formed very early on the refolding pathway (Matouschek *et al.*, 1992*b*). One barnase mutant used in this study, where such an interaction has been disrupted, is at position 91 where alanine is substituted instead of a serine residue. Serine 91 is thought to be important in the formation of the tight β-turn between strands 3 and 4 of the β-sheet. The mutant at this position refolds more slowly than wild-type barnase. Formation of the late intermediate of the Ser→Ala91 mutant is retarded such that an extra phase in refolding is just detectable by fluorescence and the rate of folding to the native state from the late intermediate is reduced approximately threefold compared to wild-type (Matouschek *et al.*, 1992*a*).

It has been suggested that one of the factors likely to be important in determining whether a chaperone binds to a particular protein *in vivo* is the rate at which that protein folds to attain its native conformation (Hardy & Randall, 1991). This kind of kinetic partitioning between protein folding and binding to a molecular chaperone, as described by Hardy & Randall (1991), is analysed here using barnase, which folds rapidly with a half-life of approximately 50 ms. To simplify the data interpretation, recovery of barnase activity is used to monitor refolding instead of the intrinsic barnase fluorescence, which to date has generally been used in barnase folding studies (Matouschek *et al.*, 1992*a*). The data are analysed with respect to the major refolding phase of barnase, which represents the conversion of a late-folding intermediate or intermediates, in which all the proline residues are in the native *trans* conformation, into the fully active state. The minor slower-refolding phases, due to a small fraction of the denatured barnase having *cis* peptidyl–prolyl bonds, are not considered here. The effects of other factors that are important in regulating GroEL action are also investigated.

2. Experimental Procedures

(a) *Materials*

The buffer used in the folding experiments was 2-[*N*-morpholino]ethanesulphonic acid (Mes) from Sigma. GpUp (guanylyl(3′-5′)uridine 3′-monophosphate) was from Sigma and the concentration was determined spectrophotometrically using an extinction coefficient of 20·6 mM^{-1} cm^{-1} (Stanley & Bock, 1965).

(b) *Expression and purification of proteins*

Barnase was expressed and purified as described previously (Serrano *et al.*, 1990). The concentration of barnase was determined spectrophotometrically at 280 nm using $E_{0\cdot1\%}=2\cdot2$ (Loewenthal *et al.*, 1991). GroEL and GroES were expressed from plasmid pOF39 (Fayet *et al.*, 1986), which was a generous gift from C. Georgopoulos, and purified according to methods described previously (Gray & Fersht, 1991). The concentration of GroEL was determined by quantitative amino acid analysis and agreed well with values obtained with the BioRad protein assay kit. The GroES concentration was determined with the BioRad protein assay kit. All molar GroEL and GroES concentrations refer to the oligomeric forms (i.e. 14mer and 7mer, respectively).

(c) *Stopped-flow refolding (single mixing experiments)*

Barnase (0·12 μM, unless indicated otherwise) was denatured in the absence of buffer by lowering the pH to 1·5 using HCl, followed by equilibration for 5 min at 25°C. The final concentration of HCl in the denatured protein solution was 32 mM. Refolding was initiated by rapid mixing of equal volumes of the denatured protein with 100 mM Mes buffer at pH 7·22, containing various concentrations of GroEL, 400 μM GpUp, 4 mM KCl and 4 mM $MgCl_2$, which gave a final pH of 6·3. All refolding experiments were performed in an Applied Photophysics SF.17MV stopped-flow apparatus at 25°C. The flow cell was completely cleared of the previous solution, by flushing with at least 5 cell volumes of fresh solutions (mixed) between the experiments. Rate constants obtained for barnase refolding from the acid or urea-denatured state have been found to be the same (Matouschek *et al.*, 1990, 1992*a*). Here, denaturation by acid was chosen as the experiments were easier to perform and it circumvents problems that can arise when mixing solutions of different viscocity.

After "pH-jump" in the stopped-flow apparatus to initiate barnase refolding, regain of barnase activity was followed by monitoring the change in absorbance at 275 nm on transesterification of GpUp (Day *et al.*, 1992). This wavelength produces the maximal change in absorbance, with the lowest background. The K_m for GpUp binding to barnase is around 20 μM (Day *et al.*, 1992) and the final substrate concentration in the refolding assays was $10\times K_m$ (200 μM). A 0·2 cm pathlength cell was used in combination with a 240 to 400 nm bandpass filter to reduce background absorbance and stray light. Where indicated, ATP, ADP and AMP-PNP were added either to the denatured barnase solution before mixing, or to the solution containing refolding buffer. We checked that there is no detectable pH change in the presence of the highest ATP concentration, and found that the kinetics of refolding were the same regardless of whether ATP was added to the acid solution or the refolding buffer solution. The concentration of any additional nucleotide was limited to about 500 μM, since at higher concentrations the background absorbance at 275 nm becomes unacceptably high. GroES was added to the solution containing refolding buffer just before mixing. It was established that there was no residual barnase left in the flow cell after exchange of solutions, by mixing GpUp with buffer (pH 6·3 final). No barnase activity was detectable, only a slow decrease in absorbance due to photolysis was observed under these conditions (see Results, section (b)).

Some experiments involved a higher barnase concentration (6 μM). This was required in order to obtain a measurable "burst" in the regain of barnase activity in the absence of any additional factors.

The refolding of barnase may also be followed by monitoring the intrinsic fluorescence of barnase (Matouschek *et al.*, 1992*a*). To establish that GpUp does not affect barnase refolding, fluorescence traces were measured as described by Matouschek *et al.* (1992*a*), with excitation at 290 nm, 10 nm bandpass and emission collected at >305 nm using a glass cut-off filter. The barnase concentration was 1 μM and the GpUp concentration was 80 μM (final). In the case where native barnase (in water) was mixed with GroEL, the "refolding" buffer contained 100 mM Mes (38·7 mM acid form and 61·3 mM salt form) which gives a pH of 6·3 on dilution to 50 mM at 25°C.

(d) *Stopped-flow refolding (sequential mixing experiments)*

Experiments involving sequential mixing were performed in an Applied Photophysics DX.17MV apparatus (Fig. 1), (Fersht & Jakes, 1975). Barnase refolding was initiated by premixing 0·24 μM acid-denatured Ser→Ala91 mutant barnase with refolding buffer (pH 7·22) described in section (c) above, minus GpUp substrate, with or without GroEL. After various ageing times, this premixed refolding solution was further mixed 1:1 with 50 mM Mes buffer (19·35 mM acid form and 30·65 mM salt form), pH 6·3, 2 mM $MgCl_2$, 2 mM KCl, 400 μM GpUp and various concentrations of GroEL. The absorbance change at 275 nm was monitored only after the second mix with buffer containing GpUp substrate. The final barnase concentration was 0·06 μM and GpUp concentration was 200 μM. As ageing times differ slightly between runs, the data obtained from these experiments were not averaged. The ageing times quoted here are the delay times between the triggering of the premix and the later triggering of the second mix (Fig. 1).

(e) *Data Fitting*

In the analysis of data obtained for barnase refolding, the folding of barnase is approximated by a single exponential function, and it is assumed that GpUp only interacts with fully folded barnase, eqn (1):

$$B_u \xrightarrow{k} B_f \xrightarrow[k_T]{\text{GpUp}} \text{Gp}+\text{Up}, \qquad (1)$$

where B_u represents unfolded barnase, B_f represents fully folded active barnase, k is the overall refolding rate constant and k_T is k_{cat} for the transesterification reaction.

Absorbance curves, obtained from single mixing experiments, were the average of at least 5 independent traces (each trace comprising 400 data points). There is some photolysis of GpUp at the high light intensities. This was corrected for by adding the photolysis rate from control experiments (see Results, section (b)) to each averaged curve (or to the raw data in the sequential mixing experiments). The initial lag and linear portion of the data set was then fitted using the computer program Kaleidagraph (version 2.1 Synergy Software PCS Inc.) to eqn (2), which describes lag kinetics:

$$[\text{Product}] = c - \frac{k_T}{k}([B_T - B_O]) + k_T[B_T]t + \frac{k_T}{k}([B_T - B_O])\exp(-kt), \qquad (2)$$

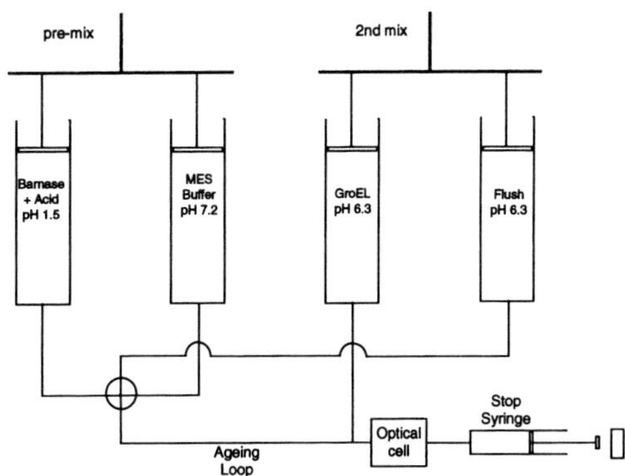

Figure 1. Schematic diagram of the Applied Photophysics stopped-flow apparatus (based on the pulsed-quenched flow apparatus of Fersht & Jakes (1975) used in the sequential mixing experiments). For details see Experimental Procedures, sections (c) and (d).

where c is the offset, k_T is the catalytic rate constant (k_{cat}) for the transesterification reaction, B_T is the total barnase concentration, B_O is the free (active) barnase concentration at $t=0$, k is the overall refolding rate constant, and t is the time in seconds. Eqn (2) was arranged in Kaleidagraph such that the data set was fitted to the minimum number of variables. The offset was changed iteratively until the best fit was obtained. All the curves were fitted with 3 variables corresponding to $k_T[B_T]$, $k_T[B_T - B_O]$ and k. In cases where it was found that $k_T[B_T] = k_T[B_T - B_O]$, these data would also fit well to a simpler form of equation (2), shown below:

$$[\text{Product}] = c - \frac{k_T}{k}([B_T]) + k_T[B_T]t + \frac{k_T}{k}([B_T])\exp(-kt), \qquad (3)$$

where the term $[B_O]$ from eqn (2) was neglected and there are only two variables. In cases where there was no observable lag phase, and the refolding rate was too fast to determine with the activity assay, the transesterification rate (k_T $[B_T]$) was determined by fitting the intial linear portion of the curve to a straight line.

3. Results

(a) *Effect of GpUp on barnase refolding*

The major refolding phase of wild-type barnase is characterized by a large change in fluorescence that, under the conditions described here, has a rate constant of 12·7 s^{-1} (Matouschek *et al.*, 1992*a*). In the presence of GpUp, a similar rate constant of 12 s^{-1}, determined from fitting the data points shown in Figure 2(a) to a single exponential process, could still be observed by fluorescence. The inset in Figure 2(a) shows that the deviation of calculated from experimental data appears to be random. The change in signal, however, is in the opposite direction and of different amplitude from that normally obtained for barnase refolded in the absence of GpUp. Binding of GpUp to native barnase was found to quench the intrinsic barnase fluorescence (data not shown) and this change is greater than the increase in fluorescence observed when unfolded barnase refolds, resulting in an overall decrease in fluorescence. These results are most easily inter-

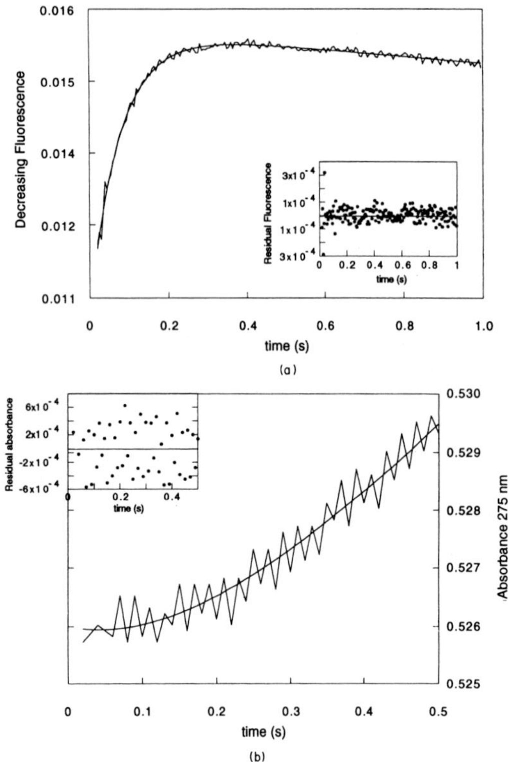

Figure 2. (a) Refolding of wild-type barnase (1 μM) in the presence of GpUp (80 μM) monitored by fluorescence. Barnase in 32 mM HCl (pH 1·5) is mixed with GpUp in Mes buffer to give a final pH of 6·3. The decrease in fluorescence at 315 nm is measured on excitation at 290 nm. The data are fitted to a curve describing a single exponential process plus a linear drift. The inset shows the residual difference in the experimental data from the theoretical curve. (b) Refolding of SA91 mutant barnase (6 μM) in the presence of 200 μM GpUp. Recovery of activity is monitored by following the change in absorbance at 275 nm as the barnase substrate GpUp undergoes transesterification. The fit to equation (2) is shown by the continuous line and the residual difference between the experimental data and the theoretical curve is shown in the inset.

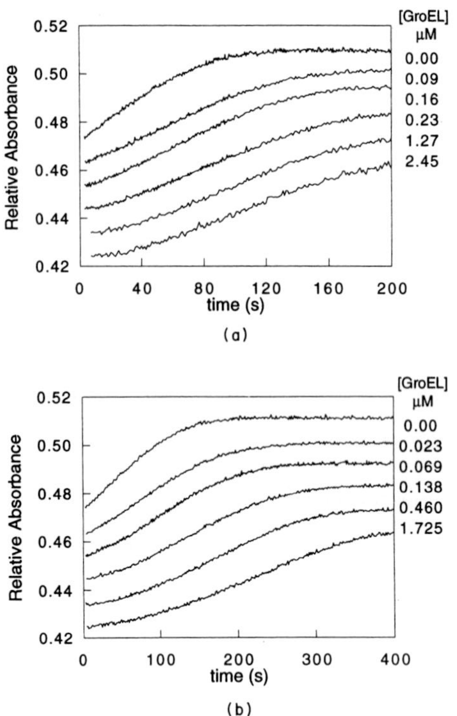

Figure 3. Regain of (a) wild-type and (b) SA91 mutant barnase activity at different GroEL concentrations, monitored by the absorbance change at 275 nm, on transesterification of GpUp (200 μM, final). Barnase (0·12 μM) in 32 mM HCl (pH 1·5) is mixed with GpUp in Mes buffer to give a final pH of 6·3. Each curve is the average of at least 5 traces and has been corrected for photolysis. The curves have been displaced from each other on the y axis for clarity. The numbers to the right of the figures represent the final GroEL concentration (μM).

preted as very rapid binding of GpUp to native barnase, such that in the pH-jump experiments the refolding rate is limiting in terms of the change in fluorescence observed. In Figure 2(a), there is also a slow linear change in fluorescence, which follows the rapid exponential change. This may reflect hydrolysis of the GpUp substrate.

As a further demonstration that barnase refolding is unaffected by the presence of GpUp, 6 μM Ser→Ala91 mutant barnase was refolded and the regain of activity monitored as described in Experimental Procedures, section (c). The high barnase concentration ensured that lag kinetics, reflecting the burst in recovery of activity upon refolding, could be detected when the reaction was monitored on a short time-scale (Fig. 2(b)). The barnase mutant, Ser→Ala91, refolds with an overall refolding rate constant of $3{\cdot}6\,s^{-1}$. This value was determined from the fit of the data shown in Figure 2(b) to equation (2). It agrees well with the refolding rate constant of $3{\cdot}8\,s^{-1}$ determined by monitoring the intrinsic fluorescence of mutant barnase, Ser→Ala91 (Matouschek *et al.*, 1992*a*).

(b) *Effect of GroEL on recovery of barnase activity*

Results of rapid mixing experiments involving wild-type barnase and Ser→Ala91 mutant barnase with various concentrations of GroEL are shown in Figure 3(a) and (b), respectively. A lag phase in recovery of activity is found in the presence of GroEL, which becomes more pronounced with increasing GroEL concentrations. This indicates that GroEL interacts with the refolding polypeptide chain. The lag in recovery of activity is significantly longer for the slower folding Ser→Ala91 mutant compared with wild-type barnase at the same GroEL concentration. It can be seen from Figure 3(a) and (b) that, after the lag period, all the activity curves are nearly parallel with the control in the absence of GroEL, indicating a high regain of activity.

Some photolysis of GpUp occurs under the experi-

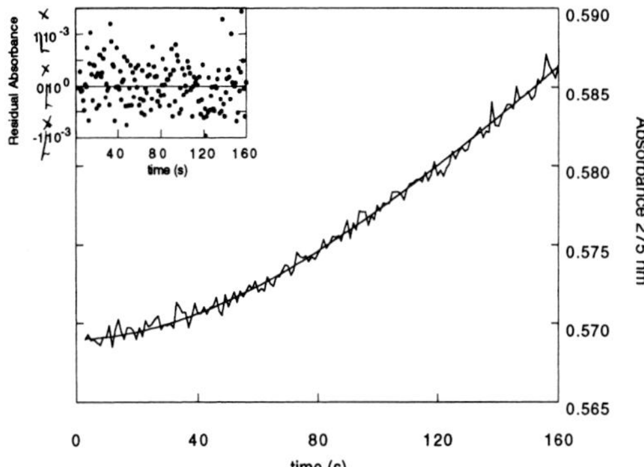

Figure 4. A typical reactivation curve for SA91 mutant barnase (0·06 μM), in the presence of GroEL (0·81 μM), fitted to eqn (2), (continuous line). The residual difference between the experimental data and the theoretical curve is shown in the inset.

mental conditions, as indicated by a linear decrease in observed absorbance after all the substrate has undergone transesterification. This photolysis rate is approximately linear over the time course of 500 seconds, with the decrease in absorbance units occurring with a rate constant of $2{\cdot}55\times10^{-5}\ s^{-1}$. The photolysis rate was found to be independent of GroEL concentration and corresponds well to the initially linear decrease in absorbance found with only GpUp substrate in the flow cell. The linear rate attributed to photolysis was added to all the data obtained from activity measurements before fitting.

The curves for the reactivation of barnase obtained in the presence of GroEL were fitted to equation (2), which describes lag kinetics. An example of a typical fit to equation (2) for Ser→Ala91 mutant barnase is shown in Figure 4 as the continuous line. The inset in Figure 4 shows that deviation of the experimental data from the theoretical curve is random. The inclusion of GroEL upon initiation of refolding of either mutant or wild-type barnase leads to a marked reduction in the overall refolding rate constant, k, by several hundredfold (Tables 1 and 2). Approximately equimolar concentrations of GroEL and barnase appear to be sufficient to bring about this marked effect on the refolding rate constant, k, for both mutant and wild-type, indicating that binding of GroEL to barnase must be very tight. Further increase in GroEL concentration to as much as 50 times that of barnase does not reduce k below $0{\cdot}03\ s^{-1}$ for wild-type and $0{\cdot}01\ s^{-1}$ for Ser→Ala91. We observe approximately the same relative reduction in the overall refolding rate constant for wild-type and the slower-folding mutant barnase in the presence of similar GroEL concentrations.

Greater than 65% of the activity for wild-type and 50% of Ser→Ala91 mutant barnase is recovered with excess GroEL (Tables 1 and 2). There are three proline residues in barnase all being in *trans* conformation in the native state. Previously it has been shown (Matouschek *et al.*, 1990) that around 25% of the fluorescence signal observed for barnase refolding in the absence of other factors may be attributed to slow events arising from the isomerization of *cis* prolines in a subpopulation of the denatured protein with half-lives of 10 to 15 seconds and around one second. Isomerization from *cis* to *trans* of peptidyl–prolyl peptide bonds may be the reason for the recovery of less than 100% activity upon refolding in the presence of GroEL.

The initial GpUp concentration in the refolding experiments is 200 μM ($\sim10\times K_m$), so that refolded barnase becomes completely saturated with substrate, and a true steady state is reached before the substrate becomes depleted. It is crucial for the data analysis that the ratio of [GpUp]/[barnase] is

Table 1

Refolding of wild-type barnase at various GroEL concentrations

GroEL† (μM)	Barnase (μM)	Refolding rate constant k (s^{-1})	% Regain of activity	$k_T\ [B_T]$ ($\mu M\ s^{-1}$)	$k_T\ [B_T-B_O]$ ($\mu M\ s^{-1}$)	% Barnase initial bound
0	3·50	12·7‡				0
0	1·00	12·73§				0
0	0·06	‖	100			0
0·09	0·06	0·082±0·016	69	1·55±0·03	1·15±0·01	74±10
0·12	0·06	0·069±0·014	77	1·73±0·04	1·27±0·01	71±8
0·16	0·06	0·076±0·009	73	1·63±0·02	1·44±0·01	88±7
0·35	0·06	0·059±0·007	63	1·41±0·02	1·65±0·01	119±6
0·69	0·06	0·035±0·004	59	1·35±0·04	1·34±0·04	100±4
1·27	0·06	0·028±0·004	66	1·48±0·06	1·44±0·04	97±5
1·38	0·06	0·028±0·002	65	1·45±0·06	1·50±0·04	103±5
1·96	0·06	0·035±0·004	60	1·38±0·04	1·73±0·08	125±5
2·42	0·06	0·030±0·005	62	1·39±0·07	1·57±0·06	110±6
2·76	0·06	0·029±0·004	63	1·41±0·06	1·56±0·04	111±5

Conditions as in Experimental Procedures, section (c). Data analysed according to eqn (2). Standard errors are given.

† Concentration of Oligomer.

‡ Data from (Matouschek *et al.*, 1992*a*), where k is the rate constant determined for the major refolding phase in the absence of GpUp. The barnase concentration was 3·5 μM.

§ k is the rate constant for refolding determined from the data in Fig. 2.

‖ The refolding rate was too fast to determine with the activity assay at this low concentration of barnase.

Table 2

Refolding of SA91 barnase at various GroEL concentrations

GroEL† (μM)	Barnase (μM)	Refolding rate constant k (s^{-1})	% Regain of activity	k_T $[B_T]$ (μM s^{-1})	k_T $[B_T - B_O]$ (μM s^{-1})	Initial % barnase bound
0	3·50	3·8‡				0
0	0·06	§	100			0
0·02	0·054	0·072±0·013	72	1·10±0·02	0·90±0·10	82±11
0·07	0·06	0·022±0·005	63	1·05±0·01	0·65±0·03	62±4
0·14	0·054	0·018±0·002	60	0·90±0·03	0·75±0·02	83±4
0·35	0·06	0·015±0·002	63	1·05±0·04	0·90±0·02	86±4
0·46	0·054	0·017±0·001	53	0·80±0·01	0·80±0·02	100±2
0·81	0·06	0·018±0·001	53	0·80±0·01	0·90±0·02	113±3
1·15	0·06	0·016±0·001	45	0·75±0·01	0·85±0·02	113±3
1·73	0·054	0·013±0·001	46	0·70±0·01	0·70±0·01	100±2
2·07	0·06	0·010±0·001	53	0·80±0·03	0·90±0·02	113±4
2·65	0·06	0·016±0·001	36	0·60±0·01	0·75±0·03	125±4
2·99	0·06	0·014±0·002	33	0·55±0·02	0·65±0·02	118±5

Conditions as in Experimental Procedures, section (c). Data analysed according to eqn (2). Standard errors are given.

† Concentration of Oligomer.

‡ Data from (Matouschek *et al.*, 1992*a*), where k is the rate constant determined for the major refolding phase in the absence of GpUp.

§ The refolding rate was too fast to determine with the activity assay at this low concentration of barnase.

high enough for sufficient GpUp to remain after ten half-lives of refolding, so that the reaction rate is still linear with time. If [GpUp]/[barnase] is too low, then too much GpUp is hydrolysed to give a true steady-state region in the lag kinetics. This gives artefactual results that overestimate the refolding rate constant and underestimate the transesterification rates. Less than 50% recovery of activity is found only in cases where the lag phase is sufficiently long such that substrate depletion becomes significant and is more evident with the data for the slower-folding Ser→Ala91 barnase mutant at the highest GroEL concentrations where the data were fitted over as much as 350 seconds. In absolute terms, the activity of the Ser→Ala91 mutant barnase in the absence of other factors appears to be 75% of wild-type barnase under the experimental conditions used here.

From the fits to equation (2) an estimate of the amount of barnase bound to GroEL at $t = 0$ (i.e. within 1 ms, the deadtime of the stopped-flow machine) was determined by dividing $k_T[B_T - B_O]$ by $k_I[B_I]$. A substantial amount of barnase is bound at the lowest GroEL concentrations tested (Tables 1 and 2). When the GroEL concentration (0·023 μM) is half that of Ser→Ala91 mutant barnase (0·054 μM), substantially more than 50% of the barnase appears to be bound to GroEL (Table 2), indicating that each GroEL oligomer can probably bind at least two barnase molecules.

(c) *Kinetic Analysis*

Initial attempts were made to analyse the data according to a scheme in which only free barnase could fold to yield active enzyme, scheme (I):

$$\begin{array}{l} B_u + \text{GroEL} \underset{k_{off}}{\overset{k_{on}}{\rightleftharpoons}} B_u.\text{GroEL}, \\ \downarrow k \\ B_f \end{array} \qquad \text{(I)}$$

where B_u and B_f represent unfolded and folded barnase, respectively. The data in Tables 1 and 2, however, do not fit to an equation derived from scheme (I), where it is assumed that the equilibrium between unfolded barnase and the barnase : GroEL complex is achieved very quickly. According to scheme (I) (and the law of mass action), at very high GroEL concentrations the observed refolding rate constant should tend to zero. For both wild-type and mutant barnase, the observed refolding rate constant, k, levels off at some value above zero (Tables 1 and 2).

An alternative scheme, (II), that fully accounts for our observations, involves refolding of barnase on GroEL:

$$\begin{array}{l} B_u + \text{GroEL} \underset{k_{off}}{\overset{k_{on}}{\rightleftharpoons}} B_u.\text{GroEL} \overset{k_b}{\rightarrow} B_f + \text{GroEL}, \\ \downarrow k_f \\ B_f \end{array} \qquad \text{(II)}$$

where k_f and k_b refer to the refolding rate constants of free and GroEL bound barnase. k_{on} is the second-order association rate constant for binding of denatured barnase to GroEL and k_{off} is the first-order dissociation rate constant for the complex. In both schemes (I) and (II), it is assumed that native, active, barnase is not bound by GroEL. We find that the activity of native barnase is unaffected after incubation with GroEL overnight, at 25°C (data not shown).

The refolding rate constants for barnase bound to GroEL(k_b), estimated in the presence of excess GroEL for both wild-type and mutant barnase (Tables 1 and 2) are 0·03 s^{-1} and 0·01 s^{-1}, respectively. The relative ratio of k_b (mutant) : k_b (wild-type) is similar to the ratio of k_f (mutant) : k_f (wild-type), where k_f is the overall refolding rate constant in the absence of chaperone. This is as would be expected if binding to GroEL is before the stage when the particular mutation has an influence on

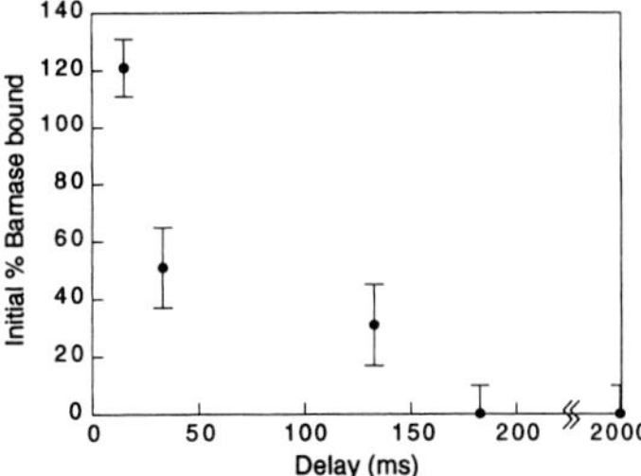

Figure 5. Refolding of barnase SA91 after delayed addition of GroEL. The delay time represents the time allowed for SA91 mutant barnase to refold before the addition of GroEL. The final concentrations of barnase and GroEL are 0·06 μM and 1 μM, respectively. The initial % barnase bound to GroEL was calculated after fitting the data to eqn (2), as described in Experimental Procedures, section (e). For data obtained after delays of 183 and 1985 ms there was no detectable lag phase, and so the data were fitted to a straight line.

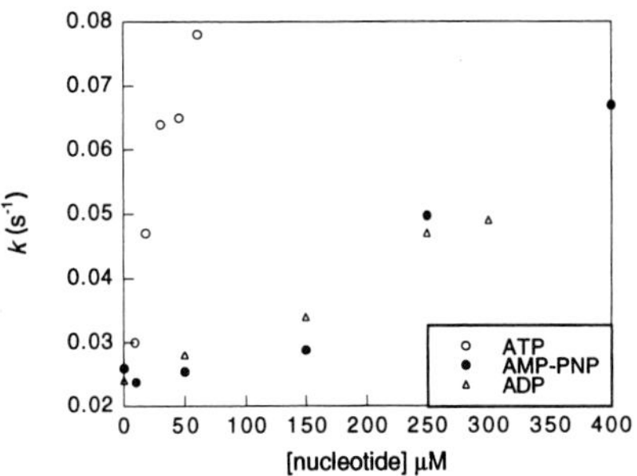

Figure 6. Effect of ATP, AMP-PNP and ADP on the wild-type barnase refolding rate constant in the presence of GroEL (1 μM). Barnase (0·054 μM) in 32 mM HCl (pH 1·5) and different concentrations of ATP, AMP-PNP or ADP is mixed with GpUp (400 μM) and GroEL in Mes buffer to give a final pH of 6·3, as described in detail in Experimental Procedures, section (c). Values for k were determined from fits to eqn (2), as described in Experimental Procedures, section (e).

folding and assumes that folding of GroEL bound barnase proceeds along the same pathway as does free barnase. Until more mutants have been analysed, it is impossible to speculate on whether this is likely to be the case.

A value for the second-order rate constant (k_{on}), for binding of denatured barnase to GroEL may be estimated assuming pseudo first-order kinetics. For binding of barnase to GroEL to compete with the refolding rate constant of $12\ s^{-1}$ for wild-type barnase, $k_{on} \times [\text{GroEL}] > 12\ s^{-1}$. From the data in Table 1, where the GroEL concentration is 0·09 μM (i.e. greater than [barnase]), with 70% of the barnase bound, k_{on} must be greater than $1{\cdot}3 \times 10^8\ s^{-1}\ M^{-1}$, indicating that binding of barnase by GroEL approaches the limit of diffusion control.

(d) *The effect of increasing delay times before GroEL is mixed with refolding barnase*

The slow-folding barnase mutant (Ser→Ala91) was used in these experiments, as it has been noted previously (Matouschek *et al.*, 1992*a*) that there may be a fast process proceeding with a half-time of around 20 ms detectable by fluorescence that precedes the major slow-refolding phase, which has a half-time of around 180 ms. The results obtained by delaying the addition of GroEL to the refolding barnase are shown in Figure 5. The shortest measurable delay times are about 10 ms, which is the time taken for the solution to reach the optical cell. The introduction of a delay time of around 30 ms before GroEL addition, has a marked effect on the kinetics of Ser→Ala91 mutant barnase refolding (Fig. 5). The amount of barnase initially bound to GroEL is reduced from approximately 100% to 50% merely by lengthening the delay time from 15 ms to 30 ms. This suggests that GroEL binds most effectively to some form of barnase that is present before the formation of the well-characterized late intermediate. With longer delay times, the lag phase is reduced so much that the data do not fit to equation (2). We cannot exclude, however, that there is some binding of GroEL to the late intermediate, but it is likely to be weaker.

(e) *Effect of nucleotides on the kinetics of refolding*

The effects of various concentrations of ATP, ADP and AMP-PNP on wild-type barnase refolding in the presence of 1 μM GroEL were investigated by mixing acid-denatured barnase containing various nucleotide concentrations with GroEL in refolding buffer. The results obtained after fitting the data to equation (2) are shown in Figure 6. The characteristic lag phase is reduced, and the observed barnase refolding rate increases, with increasing ATP concentration, until the kinetics show no lag phase, as if in the absence GroEL. Similar effects were obtained with ADP and AMP-PNP but only at markedly higher concentrations, they appear, therefore, to bind to GroEL with significantly less affinity. As the non-hydrolysable ATP analogue, AMP-PNP, and ADP have a similar effect to ATP in this assay, albeit at higher concentrations, it is likely that nucleotide binding rather than ATP hydrolysis is responsible for the effects observed here. These experiments suggest that binding of ATP to GroEL prevents the binding of barnase. Similar results were obtained when ATP was added to GroEL in refolding buffer before mixing with acid-denatured barnase, so that GroEL would be bound with nucleotide before mixing with barnase (data not shown). Increases in the barnase refolding rate constant in the presence of ATP (and GroEL) could, therefore, be due to a reduction in the concentration of unligated GroEL, which has significantly higher affinity for refolding barnase than the nucleotide-bound form of GroEL. However, the initial percentage of wild-type barnase bound to GroEL in the presence of nucleotides was estimated from fits to equation (2) (data not shown) and in all cases was

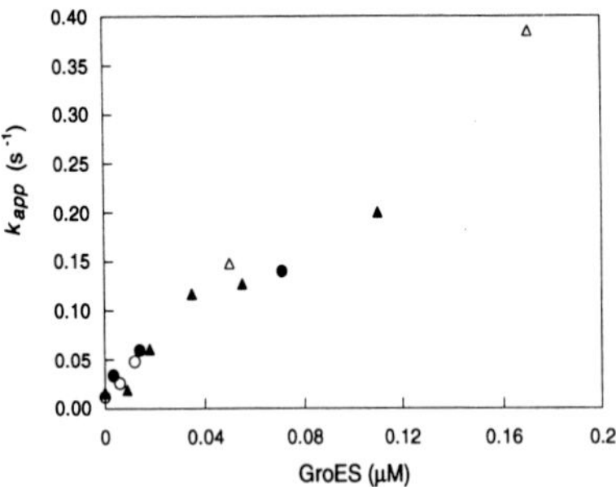

Figure 7. Effect of GroES on mutant barnase, SA91, refolding in the presence of GroEL and ATP. The refolding rate constants are referred to as apparent, since the barnase concentration in some cases is sufficiently high that there may be some depletion of GpUp substrate. The rate constants are determined from fitting data obtained from single mixing experiments (described in Experimental Procedures, sections (c) and (e)). The final reagent concentrations are: 1 μM GroEL, 50 μM ATP and 0·06 μM barnase, k_{app} is determined from the fit to eqn (2) (open circles); 0·5 μM GroEL, 50 μM ATP and 0·06 μM barnase, k_{app} is determined from the fit to eqn (2) (filled triangles); 0·5 μM GroEL, 50 μM ATP and 0·24 μM barnase, k_{app} is determined from the fit to eqn (2) (filled circles); 3 μM GroEL, 500 μM ATP and 0·75 μM barnase, k_{app} is determined from the fit to eqn (3) (open triangles).

found to be close to 100%. This suggests that there is release of barnase bound to GroEL as a consequence of ATP binding to the barnase : GroEL complex, but attempts to demonstrate this directly have been unsuccessful.

(f) *The effect of the co-chaperonin GroES*

Sub-stoichiometric amounts of GroES (compared with GroEL) were included in single mixing experiments. This results in a marked reduction in the lag phase for recovery of barnase activity compared with that found in the presence of GroEL and ATP alone. After fitting to equation (2), the overall refolding rate constant is found to increase with GroES concentration in the presence of GroEL and ATP (Fig. 7). GroES binds to the barnase : GroEL complex, which facilitates barnase release. The low GroES concentrations involved suggest that it may recycle during the course of the experiment. In the absence of nucleotides, GroES has a negligible effect on barnase refolding in the presence of GroEL (data not shown).

4. Discussion

(a) *The method*

GroEL has been shown to interact with a wide variety of substrate proteins, but detailed studies have focussed mainly on substrates that are prone to aggregation and tend not to fold spontaneously *in vitro*, except at very low concentrations and low temperature, for example rhodanese (Mendoza *et al.*, 1991*a*). Here, coupling stopped-flow mixing with a continuous activity assay has enabled us to study the effect of GroEL on a fast-folding ribonuclease, barnase. Stopped-flow mixing ensured that mixing was reproducible as well as rapid, and data could be acquired on a millisecond time scale. The activity assay has the advantage that it is an unambiguous monitor of the formation of folded active barnase. The presence of the substrate GpUp used in the activity assays does not affect the kinetics of barnase refolding (Fig. 2(a) and (b)).

(b) *Barnase folds on GroEL*

There is a clear lag period before barnase refolds in the presence of GroEL and recovers most of its activity (Fig. 3(a) and (b)). The length of the lag period depends on the GroEL concentration and on the intrinsic rate of barnase folding. There are longer lag periods observed with the slow folding of mutant barnase (Ser→Ala91). The lag in recovery of barnase activity implies that GroEL interacts with refolding barnase. The data, when fitted to equation (2), which describes lag kinetics, yield valuable information on the refolding rate of barnase in the presence of GroEL. Most importantly, in the presence of increasing GroEL concentrations, the observed rate constant for barnase refolding levels off at a significant non-zero value (Tables 1 and 2). This means that there must be at least partial refolding of barnase while still bound to GroEL. If the unfolded state of barnase that is initially bound to GroEL has to dissociate from GroEL to refold, then by mass action, at high concentrations of GroEL, the concentration of that unfolded state would become vanishingly small and so the rate constant for refolding would tend to zero. Therefore, a form of barnase that cannot be bound to GroEL is released from the GroEL : barnase complex, either a late-folding intermediate or the fully refolded barnase. The results of folding the mutant Ser→Ala91 on GroEL suggest that it is the formation of the fully folded state that is rate limiting on GroEL. The rate-determining step for refolding of this mutant and wild-type barnase in solution is for the final intermediate converting to the folded state. The rate constants for refolding of both on GroEL are reduced by similar factors (Tables 1 and 2).

Results presented here differ from those of Zahn & Pluckthun (1992), Viitanen *et al.* (1990), Martin *et al.* (1991) and Mendoza *et al.* (1991*b*), who found that only a small amount of protein substrate is able to fold in the presence of GroEL alone. Instead, we find for barnase at least 50% recovery of activity within three minutes of initiating refolding in the presence of excess GroEL. Recently, Schmidt & Buchner (1992) have also noted incomplete inhibition of the folding of an immunoglobulin F_{ab} fragment by GroEL alone. We have shown that barnase can fold slowly on GroEL and other factors are not required for its release. The mechanism of GroEL action appears to vary with the substrate. Our observations may be reconciled with results for other substrates that fold less readily than barnase, in that there is folding to some intermediate state on GroEL, as has been shown for rhodanese (Martin *et al.*, 1991), lactate dehydrogenase (Badcoe *et al.*, 1991) and pre-β-lactamase (Zahn & Pluckthun, 1992). In these cases the energy of the transition state for folding (on GroEL) may be so high that, in the absence of other factors such as ATP and GroES, the substrates remain tightly bound to GroEL.

Results from refolding DHFR (dihydrofolate reductase) in the presence of GroEL (Viitanen *et al.*, 1991) indicate that a DHFR:GroEL complex is exceedingly stable. However, the presence of DHFR substrates can alter the equilibrium between bound and free DHFR such that over a period of hours significant activity is recovered (Viitanen *et al.*, 1991). Although GpUp (the barnase substrate used in our activity assays) could destabilize a barnase:GroEL complex, this does not affect the refolding rate constants that we have measured as they are independent of GpUp concentration.

(c) *Folding of slow mutant* versus *folding of wild-type barnase*

It is interesting that even when folding of barnase is slowed down by introducing an amino acid substitution, the interaction with GroEL still appears to be a (relatively) transient one. As the refolding rates on GroEL for mutant and wild-type barnase are reduced by the same relative amount, we implied above that GroEL interacts with barnase early on the folding pathway (that is, before the mutation affects refolding). This was confirmed with the results obtained by delayed addition of GroEL to refolding Ser→Ala91 mutant barnase (Fig. 5). Barnase was allowed to refold for 30 milliseconds before mixing with GroEL. Since the refolding rate constant (from the late-folding intermediate to the fully folded state) is $3{\cdot}8\,s^{-1}$ under these conditions (Matouschek *et al.*, 1992*a*), only 11% of the late intermediate is converted during this period. Yet, there is a drop in the percentage of barnase that binds to GroEL by 50%. GroEL binds preferentially, therefore, to a form of barnase that precedes the late intermediate. The recent results of Schmidt & Buchner (1992), showing that GroEL can bind to an all-β-protein, indicate that GroEL does not just recognize specific elements of secondary structure but, more probably, interactive residues exposed early in folding. From extensive studies on barnase folding using both NMR and protein engineering techniques (Matouschek *et al.*, 1992*a*,*b*), the earliest events are partial α-helix formation and β-strand formation. The tight β-turn directly after residue 91 is thought to form at least partially at this time (Matouschek *et al.*, 1992*a*,*b*). Our findings that barnase binds to GroEL at a very early stage (possibly even before formation of the tight β-turn directly after residue 91), is consistent with the suggestions of Schmidt & Buchner (1992) and similar to those of Badcoe *et al.* (1991), who found that GroEL binds most tightly to completely denatured lactate dehydrogenase.

(d) *Nucleotides prevent folding on GroEL*

The inclusion of ATP in the refolding experiments (with GroEL) increases the rate of barnase refolding (Fig. 6). ATP binding may compete simply to prevent barnase binding to GroEL. This is consistent with the findings that ATP is bound (Bochkareva *et al.*, 1992) and hydrolysed by GroEL co-operatively (Gray & Fersht, 1991), which must necessarily involve different conformational states of GroEL. However, since all the barnase is initially bound by GroEL even in the presence of nucleotides, ATP probably also binds to the barnase:GroEL complex, inducing a conformational change that facilitates the release and subsequent fast folding of barnase.

We suspected that ATP hydrolysis was not required for the inhibitory effect on the ability of GroEL to bind barnase and this was confirmed using the non-hydrolysable ATP analogue, adenylyl imidodiphosphate (AMP-PNP; Fig. 6). However, significantly higher AMP-PNP concentrations are required to increase the observed barnase refolding rate constant appreciably. This suggests that AMP-PNP binds to GroEL more weakly than ATP, with possibly the orientation of the gamma phosphate being crucial in binding. ADP has a similar effect to that of AMP-PNP; at least ten times more ADP than ATP is needed to increase the observed barnase refolding rate constant by the same amount under our conditions. Similar results with other substrate proteins have been found (Badcoe *et al.*, 1991; Viitanen *et al.*, 1990; Baneyx & Gatenby, 1992; Fisher, 1992; Schmidt & Buchner, 1992). It is possible that GroEL, used in the experiments in the absence of additional factors, may have some residual ADP bound, but this does not affect the interpretation of our results, as ADP at low concentrations has virtually no effect on the affinity of GroEL for barnase (Fig. 6).

(e) *Effects of GroES*

GroES increases the effect of ATP, in abolishing lag kinetics (Fig. 7) by binding to the barnase:GroEL complex. In combination with ATP, GroES promotes the release and subsequent folding of barnase. It seems likely that a more effective release of barnase from GroEL in the presence of its effectors involves a GroEL conformational change. In the experiments shown in Figure 7 some proportion of GroES must bind to unligated GroEL in the presence of ATP, reducing the amount of GroES that is free to bind to the barnase:GroEL complex. GroES may bind with higher affinity to the barnase:GroEL complex, thus accounting for the relatively large effects observed at low GroES concentrations (Fig. 7).

(f) *Conclusions*

Here we have shown that the small ribonuclease barnase can fold with or without the involvement of a chaperonin. The rate of association of GroEL with unfolded barnase is close to being diffusion controlled, which suggests that, under certain conditions, even faster folding proteins than barnase may be unlikely to escape interaction with GroEL. It seems likely that, *in vivo*, there is the potential for GroEL to interact with all folding polypeptide chains, but this is tightly regulated by the local concentrations of ATP and GroES. This is entirely consistent with the sequential model proposed for chaperone action (Manning-Krieg *et al.*, 1991; Langer *et al.*, 1992*a*), where the polypeptide chain is "handed on" from one type of chaperone molecule to another before the fully folded protein appears.

Slowing down folding, to give GroEL more time to bind to the substrate, does not ensure that folding will be prevented by the formation of a tighter complex: we find correspondingly slower folding on GroEL of a slower-folding barnase mutant. It appears that kinetic partitioning in terms of barnase folding *versus* binding to GroEL is an oversimplification of the situation *in vitro*: first, barnase is bound so rapidly by GroEL at equimolar concentrations, that virtually no barnase escapes interaction; and, second, binding to the chaperonin does not prevent folding completely. Kinetic partitioning is, therefore, probably of less importance than the stability of intermediates on the folding pathway in determining whether the folding of a fast-folding polypeptide is arrested or just retarded by GroEL in the absence of other factors.

Athough there is some evidence for partial refolding of polypeptides whilst bound to GroEL (Martin *et al.*, 1991; Badcoe *et al.*, 1991), there is only one other report of substrate protein folding, with reasonable yield to the native state, in the presence of GroEL alone (Schmidt & Buchner, 1992). Here, we demonstrate using kinetics that there is significant folding of barnase on GroEL. The idea that a polypeptide can fold whilst tightly bound to GroEL is more readily understood when one considers the unusual quaternary structure of GroEL. The "double doughnut" structure (Hendrix, 1979; Saibil *et al.*, 1991) with large cavities at either end of the molecule may provide the unique microenvironment that can tightly bind barnase and yet allow slow folding to proceed. It is tempting to speculate that barnase is bound in the cavities at either end of the GroEL structure. This would be consistent both with the findings that more than one barnase molecule can bind to each GroEL oligomer and with a stoichiometry of 2 for the binding of rhodanese to GroEL (Bochkareva *et al.*, 1992). It is conceivable that weak binding of barnase at many sites within the GroEL cavity is not sufficient to prevent the highly co-operative folding of barnase, but the folded polypeptide remains mostly trapped within the cavity until GroEL effectors trigger the conformational change that promotes efficient substrate release. ATP and GroES can promote the release of barnase from GroEL, such that the barnase refolding rate approaches that found in the absence of other factors. These effects are consistent with major conformational changes in GroEL in the presence of these effectors, which have previously been demonstrated, directly by electron microscopy (Saibil *et al.*, 1991; Langer *et al.*, 1992*b*) and indirectly by kinetics (Gray & Fersht, 1991).

Superficially, our results show some similarity to those found by Wiech *et al.* (1992), where Hsp90 is shown to increase the reactivation of two model proteins in the absence of nucleoside triphosphates, and it is suggested that release of the substrate protein may be driven by folding of the non-native protein. Our results, however, highlight why it is so difficult to generalize about the mechanism of chaperone action, as it depends so much on the stability and folding kinetics of the particular substrate.

We are grateful to Dr A. Matouschek for the gift of purified wild-type and mutant barnase used in the experiments described here, and also to Drs A. Day, J. Eder and A. Horovitz for critically reading the manuscript.

References

Badcoe, I. G., Smith, C. J., Wood, S., Halsall, D. J., Holbrook, J. J., Lund, P. & Clarke, A. R. (1991). Binding of a chaperonin to the folding intermediates of lactate dehydrogenase. *Biochemistry*, **30**, 9195–9200.

Baneyx, F. & Gatenby, A. A. (1992). A mutation in GroEL interferes with protein folding by reducing the rate of discharge of sequestered polypeptides. *J. Biol. Chem.* **267**, 11637–11644.

Bochkareva, E. S., Lissin, N. M., Flynn, G. C., Rothman, J. E. & Girshovich, A. S. (1992). Positive cooperativity in the functioning of molecular chaperone GroEL. *J. Biol. Chem.* **267**, 6796–6800.

Buchner, J., Schmidt, M., Fuchs, M., Jaenicke, R., Rudolph, R., Schmid, F. X. & Kiefhaber, T. (1991). GroE facilitates the refolding of citrate synthase by suppressing aggregation. *Biochemistry*, **30**, 1586–1591.

Day, A. G., Parsonage, D., Ebel, S., Brown, T. & Fersht, A. R. (1992). Barnase has subsites that give rise to large rate enhancements. *Biochemistry*, **31**, 6390–6395.

Fayet, O., Louarn, J. M. & Georgopoulos, C. (1986). Supression of the *Escherichia coli dnaA46* mutation by amplification of the *groEL* and *groES* genes. *Mol. Gen. Genet.* **202**, 435–445.

Fersht, A. R. & Jakes, R. (1975). Demonstration of the two reaction pathways for the aminoacylation of tRNA. Application of the pulsed quenched flow technique. *Biochemistry*, **14**, 3350–3356.

Fisher, M. T. (1992). Promotion of the in vitro renaturation of dodecameric glutamine synthetase from *Escherichia coli* in the presence of GroEL (chaperonin-60) and ATP. *Biochemistry*, **31**, 3955–3963.

Gray, T. E. & Fersht, A. R. (1991). Cooperativity in ATP hydrolysis by GroEL is increased by GroES. *FEBS Letters*, **292**, 254–258.

Hardy, S. J. S. & Randall,L. L. (1991). A kinetic partitioning model of selective binding of nonnative proteins by the bacterial chaperone SecB. *Science*, **251**, 439–443.

Hendrix, R. W. (1979). Purification and properties of GroE, a host protein involved in bacteriophage assembly. *J. Mol. Biol.* **129**, 375–392.

Holl-Neugebauer B., Rudolph, R., Schmidt, M. & Buchner, J. (1991). Reconstitution of a heat-shock effect *in vitro*. Influence of GroE on the thermal aggregation of α-glucosidase from yeast. *Biochemistry*, **30**, 11609–11614.

Langer, T., Lu, C., Echols, H., Flanagan, J., Hayer, M. K. & Hartl, F. U. (1992*a*). Successive action of DnaK, DnaJ and GroEL along the pathway of chaperone-mediated protein folding. *Nature (London)*, **356**, 683–689.

Langer, T., Pfeifer, G., Martin, J., Baumeister, W. & Hartl, F. U. (1992*b*). Chaperonin-mediated protein folding: GroES binds to one end of the GroEL cylinder, which accomodates the protein substrate within its central cavity. *EMBO J.* **11**, 4757–4765.

Loewenthal, R., Sancho, J. & Fersht, A. R. (1991). Fluorescence spectrum of barnase: contributions of three tryptophan residues and a histidine-related pH-dependence. *Biochemistry*, **30**, 6775–6779.

Manning-Krieg, U. C., Scherer, P. E. & Schatz, G. (1991). Sequential action of mitochondrial chaperones in protein import into the matrix. *EMBO J.* **10**, 3273–3280.

Martin, J., Langer, T., Boteva, R., Schramel, A., Horwich, A. L. & Hartl, F. U. (1991). Chaperonin-mediated protein folding at the surface of groEL through a 'molten globule'-like intermediate. *Nature (London)*, **352**, 36–42.

Matouschek, A., Kellis, J. T., Jr, Serrano, L. & Fersht, A. R. (1990). Transient folding intermediates characterised by protein engineering. *Nature (London)*, **346**, 440–445.

Matouschek, A., Serrano, L. & Fersht, A. R. (1992*a*). The folding of an enzyme. IV. Structure of an intermediate in the refolding of barnase analysed by a protein engineering procedure. *J. Mol. Biol.* **224**, 819–835.

Matouschek, A., Serrano, L., Meiering, E. M., Bycroft, M. & Fersht, A. R. (1992*b*). The folding of an enzyme V. $H/^2H$ exchange-nuclear magnetic resonance studies on the folding pathway of barnase: complementarity to and agreement with protein engineering studies. *J. Mol. Biol.* **224**, 837–845.

Mendoza, J. A., Rogers, E., Lorimer, G. H. & Horowitz, P. M. (1991*a*). Unassisted refolding of urea unfolded rhodanese. *J. Biol. Chem.* **266**, 13587–13591.

Mendoza, J. A., Lorimer, G. H. & Horowitz, P. M. (1991*b*). Intermediates in the chaperonin-assisted refolding of rhodanese are trapped at low temperature and show a small stoichiometry. *J. Biol. Chem.* **266**, 16973–16976.

Saibil, H., Dong, Z., Wood, S. & Auf Der Mauer, A. (1991). Binding of chaperonins. *Nature (London)*, **353**, 25–26.

Schmidt, M. & Buchner, J. (1992). Interaction of GroE with an All-β-protein *J. Biol. Chem.* **267**, 16829–16833.

Serrano, L., Horovitz, A., Avron, B., Bycroft, M. & Fersht. A. R. (1990). Estimating the contribution of engineered surface electrostatic interactions to protein stability using double mutant cycles. *Biochemistry*, **29**, 9343–9352.

Serrano, L., Matouschek, A. & Fersht, A. R. (1992). The folding of an enzyme III. Structure of the transition state for unfolding of barnase analysed by a protein engineering procedure. *J. Mol. Biol.* **224**, 805–818.

Stanley, W., Jr & Bock, R. (1965). The spectrophotometric constants of di- and trinucleotides in pancreatic ribonuclease digests of ribonucleic acid. *Anal. Biochem.* **13**, 43–65.

Viitanen, P. V., Lubben, T. H., Reed, J., Goloubinoff, P., O'Keefe, D. P. & Lorimer, G. H. (1990). Chaperonin-facilitated refolding of ribulosebisphosphate carboxylase and ATP hydrolysis by chaperonin 60 (groEL) are K^+ dependent. *Biochemistry*, **29**, 5665–5671.

Viitanen, P. V., Donaldson, G. K., Lorimer, G. H., Lubben, T. H. & Gatenby, A. A. (1991). Complex interactions between the chaperonin 60 molecular chaperone and dihydrofolate reductase. *Biochemistry*, **30**, 9716–9723.

Viitanen, P. V., Gatenby, A. A. & Lorimer, G. H. (1992). Purified chaperonin 60 (groEL) interacts with the nonnative states of a multitude of *Escherichia coli* proteins. *Protein Sci.* **1**, 363–369.

Wiech, H., Buchner, J., Zimmermann, R. & Jakob, U. (1992). Hsp90 chaperones protein refolding *in vitro* *Nature (London)*, **358**, 169–170.

Zahn, R. & Pluckthun, A. (1992). GroE prevents accumulation of early folding intermediates of pre β lacta mase without changing the folding pathway. *Biochemistry*, **31**, 3249–3255.

Edited by R. Huber

Ashley Buckle solved the first X-ray structural model for peptide substrates bound to GroEL[181]. Somewhat later, Paul Sigler, a very good friend, published an essentially identical structure in *Cell*, which was considered so novel by the journal to be chosen as the cover illustration! Paul immediately sent me a reprint, dedicated in admiration of my work and apologised that *Cell* had embarrassed him. Paul was one of the most honest and big-hearted scientists, who died too young, and one of my fans. We also noted that the ring of subsites around the neck of GroEL would spring apart during the allosteric transition of the protein, weakening the binding and possibly forcing the unfolding of a peptide bound across several sub-sites.

16.2 1999: Ashley Buckle, Qinghua Wang and Jean Chatelier.

Proc. Natl. Acad. Sci. USA
Vol. 94, pp. 3571–3575, April 1997
Biochemistry

A structural model for GroEL–polypeptide recognition

(protein folding/hsp60/cpn60/mini-chaperone/crystallography)

ASHLEY M. BUCKLE, RALPH ZAHN*, AND ALAN R. FERSHT

Cambridge University Chemical Laboratory and Cambridge Centre for Protein Engineering, Medical Research Council Centre, Hills Road, Cambridge CB2 2QH, United Kingdom

Contributed by Alan R. Fersht, January 31, 1997

ABSTRACT A monomeric peptide fragment of GroEL, consisting of residues 191–376, is a mini-chaperone with a functional chaperoning activity. We have solved the crystal structure at 1.7 Å resolution of GroEL(191–376) with a 17-residue N-terminal tag. The N-terminal tag of one molecule binds in the active site of a neighboring molecule in the crystal. This appears to mimic the binding of a peptide substrate molecule. Seven substrate residues are bound in a relatively extended conformation. Interactions between the substrate and the active site are predominantly hydrophobic, but there are also four hydrogen bonds between the main chain of the substrate and side chains of the active site. Although the preferred conformation of a bound substrate is essentially extended, the flexibility of the active site may allow it to accommodate the binding of exposed hydrophobic surfaces in general, such as molten globule-type structures. GroEL can therefore help unfold proteins by binding to a hydrophobic region and exert a binding pressure toward the fully unfolded state, thus acting as an "unfoldase." The structure of the mini-chaperone is very similar to that of residues 191–376 in intact GroEL, so we can build it into GroEL and reconstruct how a peptide can bind to the tetradecamer. A ring of connected binding sites is noted that can explain many aspects of substrate binding and activity.

The molecular chaperone GroEL assists the folding of many newly synthesized proteins in *Escherichia coli* and facilitates the refolding *in vitro* of several proteins that would otherwise misfold or aggregate. GroEL is a cylinder made from two heptameric rings stacked back to back, creating a central cavity ≈45 Å wide (1, 2). Each 57-kDa subunit consists of three domains: the ATP-binding equatorial domain (residues 6–133 and 409–523) connects the two rings; the apical domain (residues 191–376) forms the flexible opening of cylinder and contains the putative polypeptide-binding and GroES-binding sites, which line the inner wall of the cavity; and the intermediate domain (residues 134–190 and 377–408) makes intersubunit contacts within a ring and transmits ATP- and GroES-mediated allosteric effects. The nature of binding of nonnative proteins by GroEL is, so far, unresolved (3–8). But, residues in the region of 199–264 were postulated from a site-directed mutagenesis study to be involved in the binding of polypeptides (9).

The action of GroEL *in vivo* is multifaceted (10–16). Studies on smaller, isolated domains of the molecule can elucidate the functions of individual components in a detail that may not be achievable when using the intact molecule. For example, isolated monomeric polypeptide-binding fragments of GroEL (residues 191–345 and 191–376) exhibit high chaperone activity *in vitro*, effecting the refolding of rhodanese and cyclophilin, and the unfolding of barnase (17). These "mini-chaperones" are active without the allosteric properties or the central cavity, which has been proposed to act as a "folding cage" *in vivo* (18, 19). The crystal structure of GroEL(191–345) at 2.5 Å resolution (17) is very similar to its region in the intact chaperone.

We have investigated the structural details of the polypeptide-binding site of GroEL by solving the crystal structure at 1.7 Å resolution of the mini-chaperone corresponding to residues 191–376 fused to a 17-residue N-terminal tag (which we number as residues −17 to −1). The tail of one molecule is seen to bind in the active site of a neighbor so that we construct a model for the chaperone–substrate complex.

MATERIALS AND METHODS

Protein Expression and Purification. The mini-chaperone (GroEL191–376) was produced by subcloning the apical domain of GroEL (residues 191–376) by polymerase chain reaction into the polylinker site of a pRSET A vector (Invitrogen), coding for an N-terminal histidine tag, which contained an engineered thrombin cleavage site (17). The histidine tag was composed of 17 amino acids (−17 MRGSHHHHHH-GLVPRGS −1). Expression in *E. coli* TG2 cells and purification of the mini-chaperone was as described (17).

Crystallization. Crystals of the mini-chaperone were obtained from hanging drops initially containing protein at 23 $mg \cdot ml^{-1}$/0.5 M NaCl/50 mM Tris·HCl, pH 8.5/25% glycerol, equilibrated against reservoirs consisting of 1.0 M NaCl/100 mM Tris·HCl, pH 8.5/25% glycerol. Crystals grew in space group $P2_12_12_1$, with cell dimensions a = 47.72 Å, b = 63.81 Å, and c = 75.10 Å.

Structure Determination and Refinement. X-ray data were collected from a crystal flash-frozen in liquid N_2 at 100 K, using a 15 cm MAR Research image plate detector at Deutsches Elektronen Synchrotron (Hamburg; station X31, λ = 1.07 Å). Data processing, data reduction, electron density syntheses, and structural analyses were carried out using CCP4 software (20). The structure was solved by molecular replacement using the program AMORE (20) and a search model consisting of residues 191–345 of the refined structure of a recently solved mini-chaperone (17). The asymmetric unit contains one protein monomer. Model rebuilding was performed with the program O (21), and the structure was refined using X-PLOR (22), followed by REFMAC (20).

RESULTS

Three-Dimensional Structure of the Mini-Chaperone. The refined model contained 292 water molecules and was com-

0027-8424/97/943571-5$2.00/0
PNAS is available online at **http://www.pnas.org**.

Data deposition: The atomic coordinates and structure factors have been deposited in the Protein Data Bank, Chemistry Department, Brookhaven National Laboratory, Upton, NY 11973 (reference 1KID).

*Present address: Institut für Molekularbiologie, Biophysik, Eidgenössische Technische Hochschule Honggerberg (HPM G5), CH-8093 Zürich, Switzerland.

Table 1. Summary of crystallographic data

Data collection statistics	
Resolution, Å	12.8–1.7
Measured reflections	260,633
Unique reflections	25,290
Completeness of data, %*	97.6 (85.6)
R_{merge}, %*†	4.7 (21.7)
$\langle I/\sigma I\rangle$*	11.8 (3.2)
Multiplicity*	4.9 (4.4)
Refinement statistics	
Resolution, Å	12.8–1.7
R factor/free R factor, %, $F > 0$‡	18.0/22.4
rmsd bond length, Å	0.016
rmsd bond angle, deg	2.7

rmsd, root-mean-square deviation.

*Values given in parentheses are for the highest resolution shell.

†Agreement between intensities of repeated measurements of the same reflections and can be defined as: $\Sigma(I_{\mathrm{h,i}} - \langle I_{\mathrm{h}}\rangle)/\Sigma I_{\mathrm{h,i}}$, where $I_{\mathrm{h,i}}$ are individual values and $\langle I_{\mathrm{h}}\rangle$ is the mean value of the intensity of reflection h.

‡The free R factor was calculated with the 10% of data omitted from the refinement.

plete. Crystallographic data are summarized in Table 1. The quality of the electron density was excellent throughout (Fig. 1). Overall, the structure was almost identical to the corresponding region of the intact protein (1, 2), and could be described as a β-sandwich scaffold flanked by helical and loop regions (Fig. 2 *Top*). A least-squares fit of the structures of mini-chaperones GroEL191–376 and GroEL191–345 (17), using backbone atoms of β-sheet residues, gave a root-mean-square deviation (rmsd) of 0.3 Å (for the β-sheet residues). The largest differences were found in helices H8 and H9 (differences in positions of Cα atoms in the range 0.5–1.7 Å) (Fig. 3). A least-squares fit of mini-chaperone GroEL191–376 and residues 191–376 in a subunit (chain A) of intact GroEL (Protein Data Bank code 1OEL), using backbone atoms of β-sheet residues, gave an rmsd of 0.7 Å (for the β-sheet residues). Again, the largest differences were found in helices H8 and H9 (Cα differences in the range 0.8–2.0 Å) (Fig. 3).

Binding of the N-terminal Tag to the Active Site. Residues −7 to −1 (sequence GLVPRGS) of the N-terminal tag could be built into the electron density map, and adopt a relatively extended conformation (Fig. 1 *Lower*; Fig. 2). Electron density for the remaining 10 residues of the tag (−17 to −8) was not observed. The tag projected away from the protein surface, and residues −7 to −1 associated with a neighboring molecule in the crystal lattice, binding to residues that have been implicated in the interaction with polypeptide substrates and with GroES (Fig. 2; ref. 9). These residues are nonpolar and are located within a shallow cleft between α-helices H8 and H9, and in an adjacent surface that is formed by the packing between helix H9 and a neighboring loop (residues 199–204) (Fig. 2; ref. 9). We could identify in the crystal structure the binding of residues −7 to −1 of the tag. The side chain of L-6 fit snugly inside a hydrophobic pocket within the interhelical cleft, the remaining residues running diagonally along the cleft and then across the ridge formed by helix H9 (Fig. 2). Most interactions were nonpolar (Table 2), although there were four hydrogen bonds between the main chain of the tag and three side chains (E257, N265, and T261) of helix H9. Model building and an earlier mutagenesis study (9) imply that the substrate-binding site extends further (to include Y199, Y203,

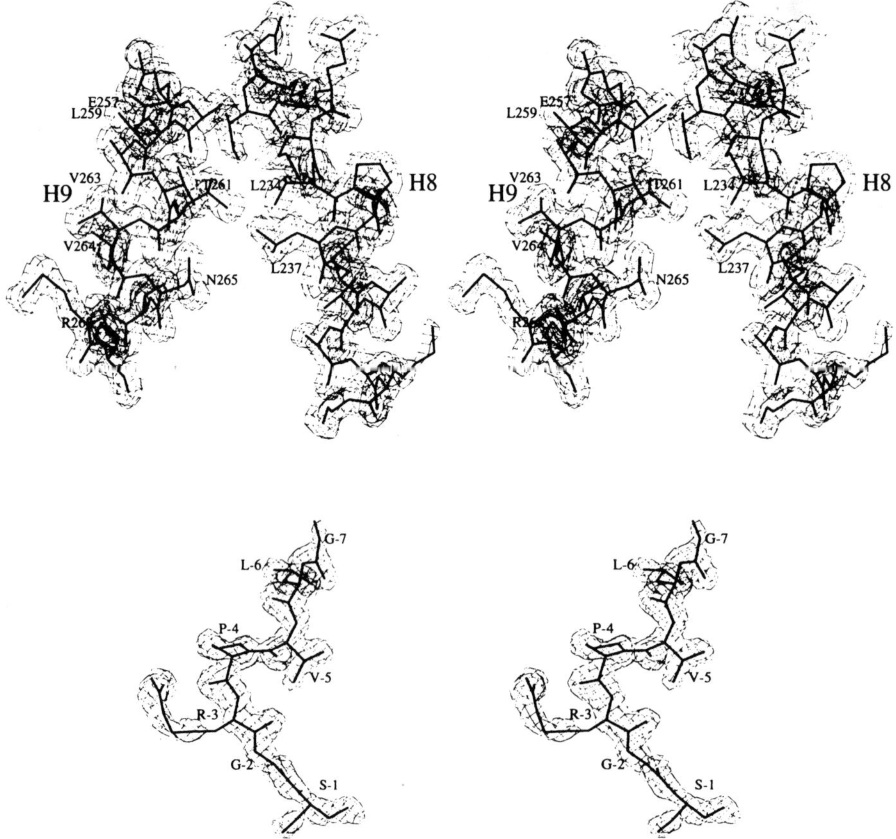

FIG. 1. $2F_o$-F_c electron density map calculated with SIGMAA coefficients (23), contoured at 1σ (σ is the root-mean-square deviation from the mean electron density in the unit cell). Superimposed is the refined atomic model. (*Upper*) Helices H8 and H9. (*Lower*) The N-terminal tag. Drawn with the BOBSCRIPT (extensions to the program MOLSCRIPT; ref. 24).

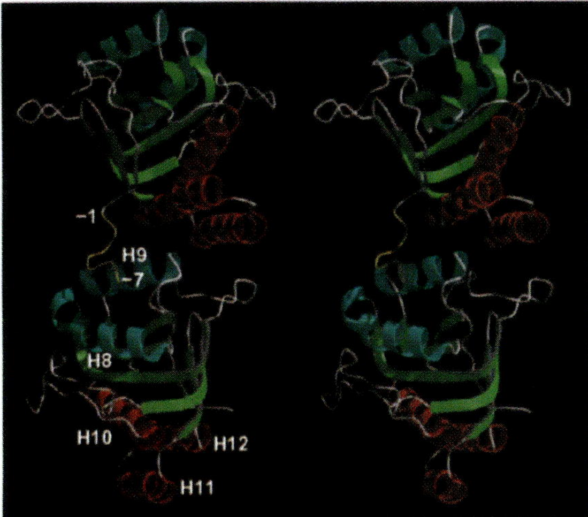

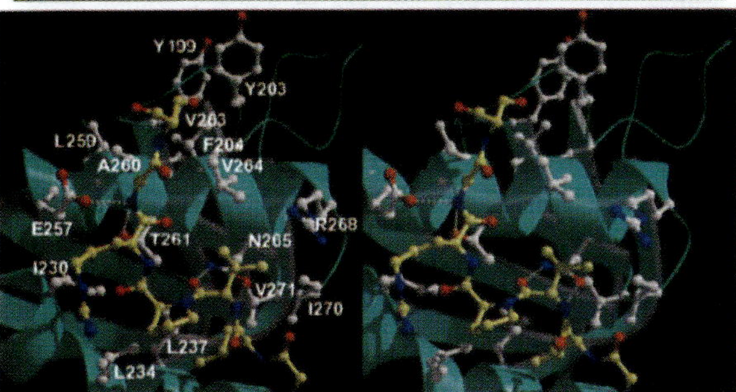

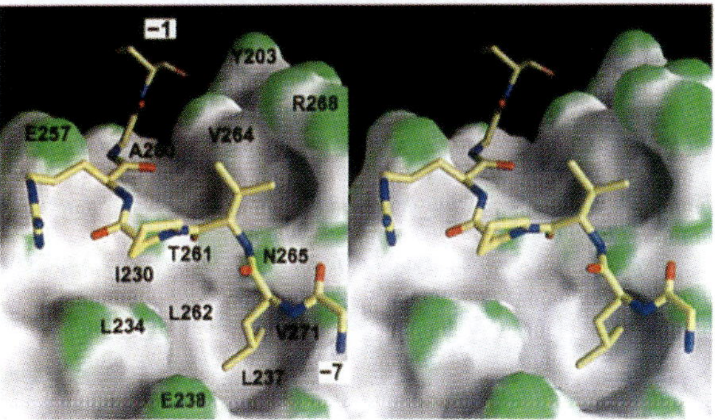

FIG. 2. (*Top*) Stereo cartoon representation of the structure of the mini-chaperone (GroEL191–376), showing the interaction between the N-terminal tag and a neighboring molecule in the crystal lattice (related by a crystallographic two-fold screw operation along the *c* axis, positioned approximately vertical and in the plane of the paper). The N-terminal tag (residues −1 to −7) is colored yellow. (*Middle*) Close-up of peptide-binding site interactions, in stereo. The peptide is represented by yellow bonds; neighboring residues are represented by white bonds. Hydrogen bonds are represented by broken white lines. Drawn with BOBSCRIPT (extensions to the program MOLSCRIPT; ref. 24) and RASTER3D (25). (*Lower*) As in *Middle* but showing the molecular surface of the mini-chaperone. The surface is colored according to surface curvature to highlight concave surface pockets. Convex, concave, and flat surfaces are colored green, grey, and white, respectively. Residues underlying the surface are labeled. Drawn with GRASP (26). All three figures show the model in approximately the same orientation.

F204, and V263). Approximately 10–11 residues may be accommodated in the active site in an extended conformation.

Comparison with Site-Directed Mutagenesis Results. Fenton *et al.* (9) postulated that residues Y199, Y203, F204, L234, L237, L259, V263, and V264 were involved in binding polypeptide substrates (Table 2). Residues Y199, Y203, and F204 would be sterically inaccessible to the N-terminal tag, so we did

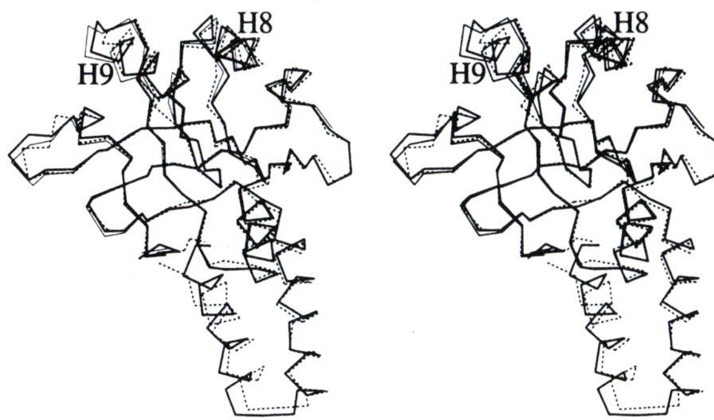

FIG. 3. Stereo representation of the overlay of Cα atoms of the apical domain in intact GroEL (residues 191–376; broken bonds), mini-chaperone GroEL191–376 (thick bonds), and mini-chaperone GroEL191–345 (thin bonds) (17). Structures were fitted using backbone atoms from β-sheet residues. Drawn with MOLSCRIPT (24).

not observe binding close to them. Residues L234, L237, L259, V263, and V264 were in the binding site. Residues I230 and E238 were tested by mutagenesis (9) and found not to affect binding, but we found that they did interact with the tag. Possibly, the binding assay was not sensitive enough to detect the changes on mutation. In addition, we have identified a series of residues that were not mutated but are in the binding site (Table 2) and extend it to residue V271.

DISCUSSION

The Nature of GroEL–Polypeptide Binding. The most important observation from this study is that the polypeptide tag of one molecule binds in the region of a neighboring molecule that has been identified (9) as being involved in binding polypeptide substrates. The N-terminal tag does not interact with the rest of its own chain. Further, the majority of interactions involved in the packing of crystal molecules are

Table 2. Interactions between mini-chaperone and the N-terminal tag

Mini-chaperone residue	Closest distance of interaction, Å
Residues postulated to interact from sdm (9)	
Y199	17.3
Y203	7.5
F204	10.9
L234	4.3
L237	3.8
L259	6.4
V263	5.4
V264	3.5
Residues not found to interact from sdm (9)	
I230	3.3
E238	3.5
Residues not mutated (9)	
A241	4.3
E257*	2.7
A260	3.6
T261*	3.1
N265*	2.7
R268	3.8
I270	3.6
V271	3.7

The table comprises all the residues in the apical domain of GroEL (residues 191–376) that have been found by site-directed mutagenesis (sdm) to appear to be involved in the binding of peptide substrates (9) and all those detected in the crystal structure.
*Hydrogen-bonded interaction.

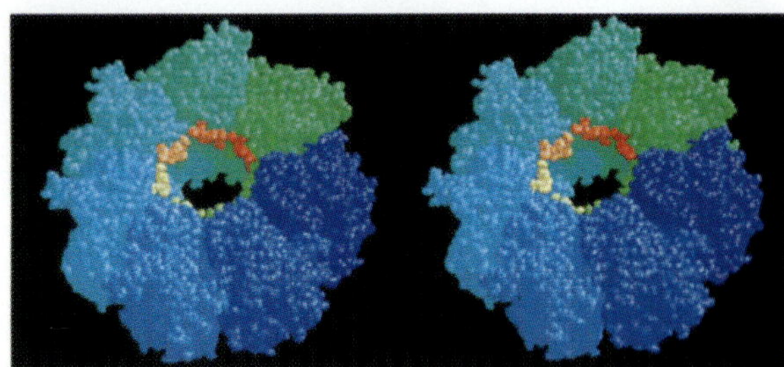

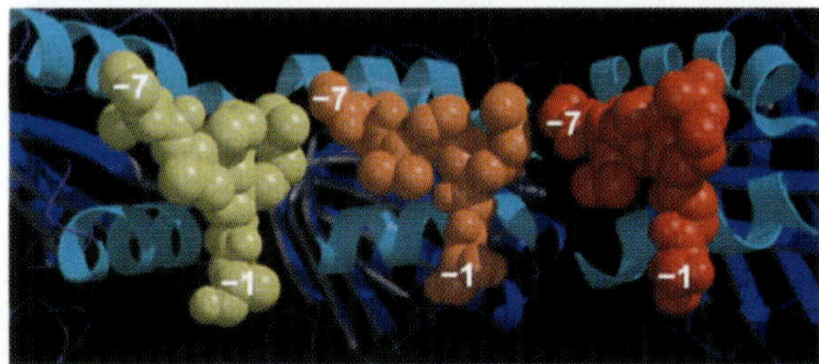

FIG. 4. (*Upper*) Stereoview of one heptameric ring of the GroEL tetradecamer showing the position of the N-terminal tag bound to each apical domain, near the opening to the central cavity. This model is generated by the superposition of the mini-chaperone GroEL(191–376) with each corresponding apical domain (residues 191–376) in intact GroEL (the second ring of the GroEL cylinder, generated by a two-fold symmetry operation, is not shown, but stacks against the underside of the drawn ring). GroEL subunits are colored around the ring going from blue to green. Superimposed "bound peptides" are colored from green to red. Drawn with RASMOL (32). (*Lower*) Cross-section of the model shown in *Upper* looking directly at the inner wall of the cavity, and showing the apical domains (cartoon with helices H8 and H9 colored cyan) from three subunits with modeled peptide (shown as space-filling models colored yellow, orange, and red, respectively). Drawn with MOLSCRIPT (24) and RASTER3D (25). Each of the separate peptides (residues −1 to −7) could be linked together by small fragments of peptides so that a longer peptide could bind from one contiguous site to the next.

located elsewhere on the protein surface. Accordingly, it is most likely that the nature of the interaction between the tag and the putative polypeptide binding site is not restricted to a crystalline environment. Thus, the mode of binding should provide structural information that is relevant to GroEL–peptide interactions. The binding site is strongly hydrophobic, but polar residues are also available for hydrogen bonding with the main chain of a peptide substrate. The polar residues are arranged such that an extended conformation of the substrate is bound.

Flexibility of the Mini-Chaperone and Unfoldase Activity. The peptide binding site is the most flexible region in the structure of intact GroEL (refs. 1 and 2; Fig. 3). We propose that the binding site can accommodate a wide range of substrate side chains because of variation of the dimensions of the active site through movements of helices H8 and H9 and surrounding loops. Interestingly, there is a structural similarity between the mini-chaperone and the enzymes pyruvate phosphate dikinase and aconitase, both of which possess a well-ordered β-sandwich surrounded by more flexible helical and loop regions (A. G. Murzin, personal communication). The dynamic behavior of the helical and loop structures of these enzymes is crucial to their function (27, 28) by an induced fit mechanism (29).

Thus, although the preferred polypeptide conformation bound by GroEL is essentially extended, the flexible surface may also accommodate the binding of exposed hydrophobic surfaces in general, such as the face of an amphipathic helix (3, 4, 8) and molten globule-type structures (5–7, 30) as well as the more fully unfolded states, which are implied by hydrogen-exchange studies (11). GroEL can thus help unfold proteins by binding to a hydrophobic region and exert a binding pressure toward the fully unfolded state. By this means, GroEL can act as an unfoldase by a "strain" or "stress" mechanism (31).

Binding of a Polypeptide Substrate to the GroEL Tetradecamer. Building this binding site into the structure of intact GroEL reveals a ring of sites around the GroEL apex (Fig. 4). Each site is arranged at approximately 45° to the axis of the ring so that a long denatured protein could bind as a series of parallel segments, connected by loops that extend into the cavity. Rigid-body motions of the apical domains relative to each other are seen by electron microscopy (33, 34) in the ATP-mediated allosteric transitions that regulate the binding, folding, and release of polypeptide substrates (14, 16, 35). Such movements on the binding of ATP would cause the binding sites to move away from each other, weakening binding and resulting in release of polypeptide. It is also possible that direct competitive inhibition between the denatured peptide and the "mobile loop" (residues 16–34) in GroES could displace bound polypeptide from GroEL (36–38), especially as the conserved tripeptide I25-V26-L27 in this loop could bind to GroEL in the mode suggested from our model.

Subsites. The binding of one molecule of denatured barnase to an excess of GroEL fits to a model of very tight 1:1 binding, whereas four molecules of denatured barnase bind when it is present in excess over GroEL (39). This can be interpreted in light of the present model, which has a ring of seven identical sites: in a 1:1 complex, barnase binds to more than one subsite, and therefore has tight binding because of the synergistic effects; whereas in a 4:1 complex there are fewer subsites available for each barnase substrate molecule.

A postdoctoral Liebig fellowship to R.Z. is gratefully acknowledged.

1. Braig, K., Otwinowski, Z., Hegde, R., Boisvert, D. C., Joachimiak, A., Horwich, A. L. & Sigler, P. B. (1994) *Nature (London)* **371,** 578–86.
2. Braig, K., Adams, P. D. & Brünger, A. T. (1995) *Nat. Struct. Biol.* **2,** 1083–1094.
3. Landry, S. J. & Gierasch, L. M. (1991) *Biochemistry* **30,** 7359–7362.
4. Landry, S. J., Jordan, R., McMacken, R. & Gierasch, L. M. (1992) *Nature (London)* **355,** 455–457.
5. Hayer-Hartl, M. K., Ewbank, J. J., Creighton, T. E. & Hartl, F.-U. (1994) *EMBO J.* **13,** 3192–3202.
6. Martin, J., Langer, T., Boteva, R., Schramel, A., Horwich, A. L. & Hartl, F. U. (1991) *Nature (London)* **352,** 36–42.
7. Mendoza, J. A., Butler, M. C. & Horowitz, P. M. (1992) *J. Biol. Chem.* **267,** 24648–24654.
8. Schmidt, M. & Buchner, J. (1992) *J. Biol. Chem.* **267,** 16829–16833.
9. Fenton, W. A., Kashi, Y., Furtak, K. & Horwich, A. L. (1994) *Nature (London)* **371,** 614–619.
10. Ellis, R. J. & Hartl, F. U. (1996) *FASEB J.* **10,** 20–26.
11. Zahn, R., Perrett, S., Stenberg, G. & Fersht, A. R. (1996) *Science* **271,** 642–645.
12. Zahn, R., Perrett, S. & Fersht, A. R. (1996) *J. Mol. Biol.* **261,** 43–61.
13. Gray, T. E., Eder, J., Bycroft, M., Day, A. G. & Fersht, A. R. (1993) *EMBO J.* **12,** 4145–4150.
14. Weissman, J. S., Kashi, Y., Fenton, W. A. & Horwich, A. L. (1994) *Cell* **78,** 693–702.
15. Todd, M. J., Viitanen, P. V. & Lorimer, G. H. (1994) *Science* **265,** 659–666.
16. Corrales, F. J. & Fersht, A. R. (1996) *Proc. Natl. Acad. Sci. USA* **93,** 4509–4512.
17. Zahn, R., Buckle, A. M., Perrett, S., Johnson, C. M. J., Corrales, F. J., Golbik, R. & Fersht, A. R. (1996) *Proc. Natl. Acad. Sci. USA* **93,** 15024–15029.
18. Agard, D. A. (1993) *Science* **260,** 1903–1904.
19. Ellis, R. J. (1994) *Curr. Opin. Struct. Biol.* **4,** 117–122.
20. Anonymous (1994) *Acta Crystallogr.* **D50,** 760–763.

21. Jones, T. A., Zou, J.-Y., Cowan, S. W. & Kjeldgaard, M. (1991) *Acta. Crystallogr.* **A47,** 110–119.
22. Brünger, A. T. (1992) XPLOR Manual (Yale University, New Haven, CT), Version 3.0.
23. Read, R. (1986) *Acta Crystallogr.* **A42,** 140–149.
24. Kraulis, P. (1991) *J. Appl. Crystallogr.* **24,** 946–950.
25. Merrit, E. A. & Murphy, M. E. P. (1994) *Acta Crystallogr.* **D50,** 869–873.
26. Nicholls, A. (1992) GRASP, Graphical Representation and Analysis of Surface Properties (New York).
27. Herzberg, O., Chen, C. C., Kapadia, G., McGuire, M., Carroll, L. J., Noh, S. J. & Dunaway, M. D. (1996) *Proc. Natl. Acad. Sci. USA* **93,** 2652–2657.
28. Lauble, H. & Stout, C. D. (1995) *Proteins* **22,** 1–11.
29. Koshland, D. E., Jr., Nemethy, G. & Filmer, D. (1966) *Biochemistry* **5,** 365–385.
30. Robinson, C. V., Gross, M., Eyles, S. J., Ewbank, J. J., Mayhew, M., Hartl, F. U., Dobson, C. M. & Radford, S. E. (1994) *Nature (London)* **372,** 646–651.
31. Fersht, A. R. (1985) *Enzyme Structure and Mechanism* (W. H. Freeman and Company, New York).
32. Sayle, R. A. & Milnerwhite, E. J. (1995) *Trends Biochem. Sci.* **20,** 374–376
33. Langer, T., Pfeifer, G., Martin, J., Baumeister, W. & Hartl, F. U. (1992) *EMBO J.* **11,** 4757–4765.
34. Chen, S., Roseman, A. M., Hunter, A. S., Wood, S. P., Burston, S. G., Ranson, N. A., Clarke, A. R. & Saibil, H. R. (1994) *Nature (London)* **371,** 261–264.
35. Todd, M. J., Viitanen, P. V. & Lorimer, G. H. (1993) *Biochemistry* **32,** 8560–8567.
36. Zeilstra, R. J., Fayet, O. & Georgopoulos, C. (1991) *Annu. Rev. Microbiol.* **45,** 301–325.
37. Zeilstra, R. J., Fayet, O. & Georgopoulos, C. (1994) *J. Bacteriol.* **176,** 6558–6565.
38. Landry, S. J., Zeilstra, R. J., Fayet, O., Georgopoulos, C. & Gierasch, L. M. (1993) *Nature (London)* **364,** 255–258.
39. Corrales, F. J. & Fersht, A. R. (1995) *Proc. Natl. Acad. Sci. USA* **92,** 5326–5330.

We then found that the same site binds an α-helix[182]. Subsequently, we went on to show that the substrate binding site on GroEL recognises, in general, sequential and non-sequential linear structural motifs compatible with both extended β-strands and α-helices[183], with predominantly hydrophobic side chains extending from the substrate into the site, but can also accommodate polar residues[184]. A student from Peking University, Qinghua Wang, wrote me such a beautiful series of e-mails that I was persuaded to take her on as a Ph.D. student in competition from fully funded local rivals. She then completed her Ph.D. in record time and discovered from X-ray crystallographic studies a larger binding site on the apical domain that forms a continuous surface around the opening of the central cavity of GroEL that can accommodate a wide range of non-native protein conformations[184].

I believe that the binding modes we found are the key to the mechanism of action of GroEL: those motifs, especially the extended β-strands, are regions in proteins that are prone to aggregation. GroEL can act as a temporary parking spot for partly unfolded proteins by binding to those motifs. The apical domain alone, the 'minichaperone' has chaperone activity[185] and it can rescue the heat shock activity of GroE in *E. coli* so that the complex quaternary structure of GroEL, its central cavity, and the structural allosteric changes that take place on the binding of nucleotides and GroES are not essential for all of its functions *in vivo*[186]. Jean Chatellier worked out the activity of different constructs of GroE in *E. coli*[187] and we proposed an evolutionary route for the activity of GroE:

> The minichaperone core acts as a primitive chaperone by providing a binding surface for denatured states that prevents their self-aggregation. The assembly of seven minichaperones into a ring then enhances substrate binding by introducing avidity. The acquisition of binding sites for ATP then allows the modulation of substrate binding by introducing the allosteric mechanism that causes cycling between strong and weak binding sites. This is accompanied by the acquisition by the heptamer of the binding of GroES, which functions as a lid to the central cavity and competes for peptide binding sites. Finally, dimerization of the heptamer enhances its biological activity.[187]

The refolding activity of the minichaperone was disputed by two leading groups. Those claims were decisively refuted by our making a 'refolding column' using the minichaperone immobilised on beads[188,189] as well as our showing the minichaperone is active *in vivo*, taking over the heat shock role of GroEL in *E. coli*[186]. Our 'refolding chromatography' was subsequently used to purify human CD1[190,191] and solve their crystal structures[192,193].

The MRC started a spin-out company, *Avidis*, in Clermont Ferrand, with French investment and Jean Chatellier as CEO, based on my refolding technology and John Walker's expression systems. It was an interesting experience. The company still exists. I stopped work on GroE as it was a sideline and I had learned enough about the protein and had run out of ideas. In any case, Amnon Horovitz had fallen in love with Tammy, married her, and then took the allosteric mechanism of GroEL to new levels of sophistication by discovering 'nested' co-operativity, amongst other fine work. Qinghua met her husband Jianpeng Ma at a meeting, 'chaperoned' by GroEL as he had performed molecular dynamics simulations on the allosteric transition. Although our work is not as cited as it should be, I consider GroEL as a very successful project, having produced surrogate grandchildren.

Chapter 17

Protein–Protein Interactions

'A false balance is abomination to the LORD: but a just weight is his delight.'

Proverbs 11.1

I have had a long association with the Weizmann Institute of Science. In 1968, I had applied to do my post-doctoral work there, but the recipient failed to answer for several months, and Bill Jencks, in the meanwhile, offered me a position by return of post, most fortunately. Subsequently, I became the co-chairman of their Scientific and Advisory Committee, and, as mentioned, three of my Israeli post-doctorals have become full professors there, and a fourth their senior administrator. Gideon Schreiber applied Φ-value and double-mutant cycle analysis to protein–protein interactions, and produced some classic papers on the subject[194–200] in addition to a succession of other papers. Double-mutant cycles are a precision tool for analysing interactions by site-directed mutagenesis as they pinpoint both interacting partners and artefacts often cancel out.

17.1 1989: Amnon Horovitz.

17.2 2003: Andreas Matouschek, Paul Barker and Gideon Schreiber.

J. Mol. Biol. (1995) **248**, 478–486

JMB

Energetics of Protein–Protein Interactions: Analysis of the Barnase–Barstar Interface by Single Mutations and Double Mutant Cycles

Gideon Schreiber and Alan R. Fersht

Cambridge Centre for Protein Engineering, Medical Research Coucil Centre, Hills Road, Cambridge CB2 2QH U.K.

The interaction of barnase, an extracellular RNase of *Bacillus amyloliquefaciens,* with its intracellular inhibitor barstar is a suitable paradigm for protein–protein interactions, since the structures of both the free and the complexed proteins are available at high resolution. The contributions of residues from both proteins to the energetics of kinetics and thermodynamics of binding were measured by double mutant cycle analysis. Such cycles reveal whether the contributions from a pair of residues are additive, or the effects of mutations are coupled. The aim of the study was to determine which of the interactions are co-operative. Double mutant cycles were constructed between a subset of five barnase and seven barstar residues, which were shown by structural and mutagenesis studies to be important in stabilising the complex. The coupling energy between two residues was found to decrease with the distance between them. Generally, residues separated by less than 7 Å interact co-operatively. At greater separations, the effects of mutation are additive, and the energetics of the interactions are independent of each other. The highest coupling energies are found between pairs of charged residues (1.6 to 7 kcal mol^{-1}). Three of the six most important interactions detected by double mutant cycle analysis (with coupling energies of more than 3.0 kcal mol^{-1}) had not been noted previously from examination of the crystal structure. The effects of mutation on the kinetics of association are all additive, apart from charged residues located at distances of up to 10 Å apart, which are co-operative. This can be explained by the fact that the transition state for association occurs before most interactions are formed.

Keywords: protein–protein interaction; protein engineering; molecular recognition

Introduction

Understanding the interaction between two proteins depends on characterising the individual interactions at their interface. Such a study is most feasible when the structures of the individual components have been solved, both separately and in their complex, so that all components are known. The interaction of barstar (b*), the intracellular polypeptide inhibitor of barnase (bn), with barnase, an extracellular RNase of *Bacillus amyloliquefaciens*, is a suitable paradigm for protein–protein interactions (Hartley, 1989). The two proteins form a very tight complex, with a K_d of 10^{-14} M (Schreiber & Fersht, 1993*a*). The structure of barstar in solution has been solved by NMR (Lubienski *et al.*, 1994), and the structure of the barnase–barstar complex determined to high resolution using X-ray crystallography (Guillet *et al.*, 1993; Buckle *et al.*, 1994). The structure of barnase has been determined in the crystalline form and in solution (Mauguen *et al.*, 1982; Bycroft *et al.*, 1991). Mutagenesis studies have shown that the residues that contribute most to the tight interaction of the complex are, in barnase: Lys27bn, Arg59bn, Arg87bn, and His102bn (the changes of free energy of association on mutation, $\Delta\Delta G_{mut\text{-}wt}$, are some 5 to 6 kcal mol^{-1} on mutation of each residue), and in barstar: Asp35b*, Asp39b*, and Glu76b* (with Asp39b* alone contributing 7.7 kcal mol^{-1} to $\Delta\Delta G_{mut\text{-}wt}$) (Hartley, 1993; Schreiber & Fersht, 1993*a*; Schreiber *et al.*, 1994). (The suffix b* indicates that the residue is in barstar, and bn indicates that it is in barnase.)

The now conventional procedure for experimentally analysing interactions is by site-directed mutagenesis, usually using single mutations. If a mutated residue is directly or indirectly involved in

Abbreviations used: bn, barnase; b*, barstar.

binding, the mutation can lead to a change in the binding energy of the protein–protein complex. A change in free energy on the mutation of a residue X to A, $\Delta\Delta G_{X\rightarrow A}$, is not, in general, the intrinsic binding energy of the side-chain of X that is mutated, but simply a quantitative measure of the relative binding energy of X rather than A being present; that is the specificity of the interaction (Fersht *et al.*, 1985, 1987, 1992; Fersht, 1987, 1993). A more advanced procedure for mutagenesis is to mutate, both singly and doubly, pairs of residues (X and Y); that is the double mutant cycle method (Carter *et al.*, 1984; Horovitz, 1987; Fersht *et al.*, 1992). This gives a coupling energy, $\Delta\Delta G_{int}$, defined by:

$$\Delta\Delta G_{int} = \Delta\Delta G_{X\rightarrow A,Y\rightarrow B} - \Delta\Delta G_{X\rightarrow A} - \Delta\Delta G_{Y\rightarrow B} \quad (1)$$

where $\Delta\Delta G_{Y\rightarrow B}$ is the change in binding energy on mutation of Y to B, and $\Delta\Delta G_{X\rightarrow A,Y\rightarrow B}$ the change on the simultaneous mutation of X to A and Y to B. $\Delta\Delta G_{int}$ is a measure of the co-operativity of interaction of the two components that are mutated. If the effects of the mutations are independent (non-co-operative), the change in free energy for the double mutant is the sum of those for the two single mutations, but if the mutated residues are coupled, then the change in free energy for the double mutant differs from the sum of the two single mutants. In some circumstances, $\Delta\Delta G_{int}$ can be reduced to the interaction energy between two residues (Horovitz *et al.*, 1990; Serrano *et al.*, 1990; Loewenthal *et al.*, 1992).

In this paper, we analyse experimentally the energetics of binding and specificity in the association of barnase and barstar, using the double mutant cycle method. We examine how pairs of residues contribute to the binding, and how co-operative the interactions are over the binding surface.

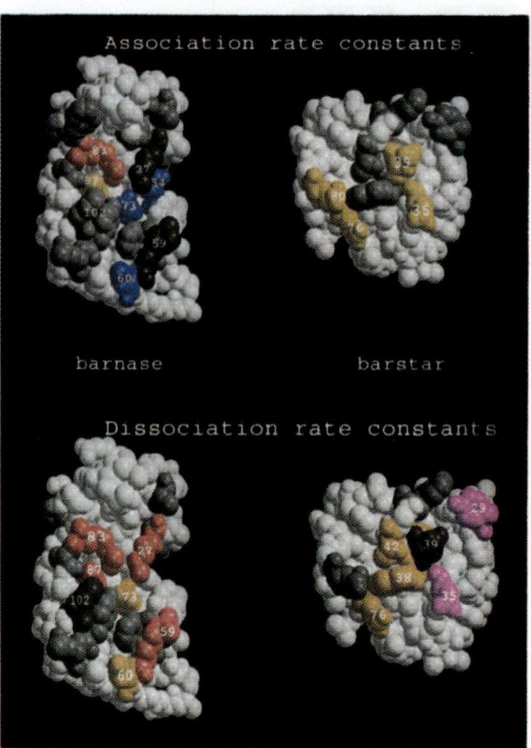

Figure 1. Mutational scanning of association and dissociation rate constants for the barnase–barstar complex. Single residues on the surface of barnase and barstar were mutated (mostly to Ala, and to Phe for Trp). The residues were coloured in respect of the measured change in the association or dissociation rate constant upon mutagenesis. For residues coloured grey the measured change was less than twofold. For association rate constant: Yellow, 2 to 3-fold decrease; red, 4 to 6-fold decrease; brown, 7 to 10-fold decrease, and blue 2 to 4-fold increase. For dissociation rate constant: Yellow, 1 to 2 orders of magnitude, decrease; red, 3 orders of magnitude, decrease; brown, 4 to 5 orders of magnitude, decrease. The data are from Table 1 and former research (Schreiber & Fersht, 1993*a*; Schreiber *et al.*, 1994). This Figure was drawn with the MolScript program (Kraulis, 1991).

Results

Residues located at the interface between barnase and barstar were subjected to screening by single mutations to determine their contribution to association and binding (Figure 1) (Schreiber & Fersht, 1993*a*; Schreiber *et al.*, 1994; and current research).

Residues contributing to the stability of the complex

Barstar inhibits the RNase activity of barnase by binding to its active site. The site of phosphodiester cleavage (the P_1 phosphate site) is occupied by Asp39b* in this barnase–barstar complex (Buckle & Fersht, 1994; Buckle *et al.*, 1994). Asp39b* forms hydrogen bonds with Arg83bn, Arg87bn and His102bn. The OD1 atom of Asp39b* is 4.5 Å away from the NZ group of Lys27bn (Figure 2). A single mutation of any one of these residues in barnase or barstar results in a decrease in the binding energy of the complex by 5.4 to 7.7 kcal mol^{-1} (Table 1). Lys27bn also forms a hydrogen bond with Thr42b*, and the mutation Thr42b* → Ala reduces the binding energy by 1.8 kcal mol^{-1}. His102bn acts as a general acid/base group during RNA cleavage by barnase. In the barnase–barstar complex, the side-chain of His102bn interacts with three barstar residues: the NE2 atom of His102bn forms a hydrogen bond with Asp39*, ND1 forms a hydrogen bond with the backbone amide of Gly31b*, and the His side-chain interacts face-to-edge with the aromatic ring of Tyr29b* (Figure 2). The single mutations His102bn → Ala and Tyr29b* → Ala result in a loss of 6.1 kcal mol^{-1} and 3.4 kcal mol^{-1} binding energy, respectively. Interestingly, the mutation Tyr29b* → Phe does not destabilise the complex, despite the two hydrogen bonds of the OH group of Tyr29b* with the carbonyl groups of Asn83bn andAsn84bn: a possible explanation could be that the hydrogen bond angles are not favourable (Buckle *et al.*, 1994). At the edge

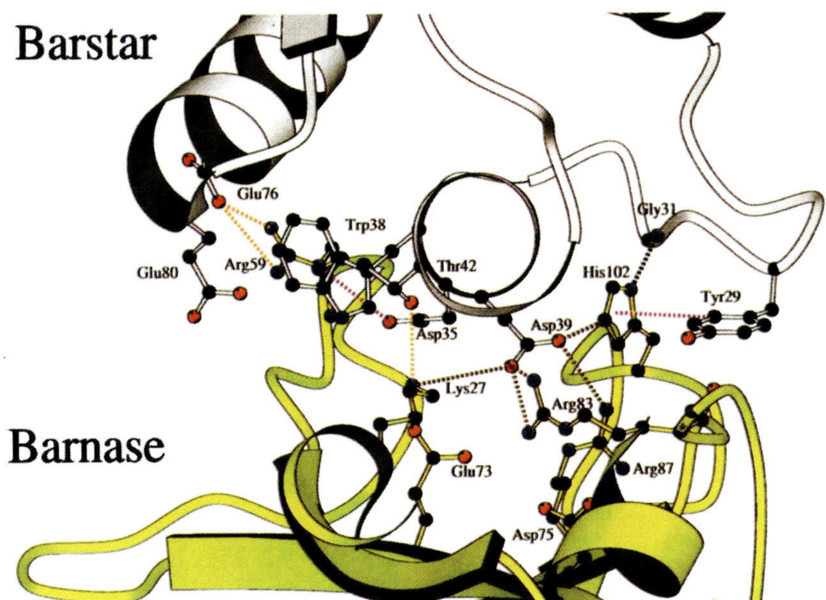

Figure 2. Cross-section through the barnase–barstar interface, showing some important protein–protein interactions and the residues mutated in this study. Major interactions are shown by broken lines with colours according to the apparent strength of the interaction: Brown, 4.5 to 6.0 kcal mol^{-1}; purple, 3.0 to 4.5 kcal mol^{-1}; orange; 1.3 to 3.0 kcal mol^{-1}; black indicates a potentially important hydrogen bond which could not be measured. This Figure was drawn with the MolScript program (Kraulis, 1991).

of the interface, Arg59bn forms hydrogen bonds with Asp35b* and Glu76b*, and packs closely against the aromatic ring of Trp38b*. Mutating these residues to Ala decreases the binding energy by 5.2, 4.5, 1.6 and 1.4 kcal mol^{-1}, respectively.

pH dependence of the dissociation rate constant

The pK_a value of His102bn is lowered from pH 6.3 in the free enzyme (Sali *et al*., 1988) to pH $\leqslant$ 5 in the barnase–barstar complex, as measured by the pH dependence of the dissociation rate constants of the complexes of wild-type and His102bn → Ala with barstar (Schreiber & Fersht, 1993*a*). The residue His102bn has a very low pK_a in the complex because its ND1 forms a hydrogen bond with the backbone amide of Gly31b*, and a second proton cannot be accepted at this position without propagating a structural change (Buckle *et al*., 1994). Therefore, the pH dependence of the dissociation rate constant can serve as a probe for structural changes in the vicinity of His102bn caused by mutations. Within the range pH 8 to 5, the dissociation rate constant of the wild-type complex increases about 20-fold with no levelling off at pH 5 (Figure 3). Mutating Tyr29b* → Ala and Arg87bn → Ala does not cause a large change in the pH profile (Figure 3), indicating that these two mutations do not significantly alter the structure in the vicinity of His102bn. On mutation of Asp39b* → Ala, the pK_a of the complex is shifted to about pH 7.3. The increase in the pK_a of His102bn in the complex from the value of $\leqslant$5 measured for wild-type is expected for this mutation, since His102bn is no longer constrained by Asp39b*.

Double mutant cycle analysis of the barnase–barstar interface

Double mutant cycles were constructed for all of the residues described above (five from barnase and seven from barstar) in order to measure the degree of co-operativity between their interactions within the complex.

Interactions between side-chains in barnase and barstar can be grouped according to their coupling energies (Figure 4). The highest coupling energies are found for Asp39b* interacting with Lys27bn, Arg83bn, Arg87bn and His102bn (5 to 7 kcal mol^{-1}, Table 1). The pairs Asp35b* ↔ Arg59bn and Tyr29b* ↔ His102bn have coupling energies of 3 to 4 kcal mol^{-1}. Coupling energies of 1 to 2 kcal mol^{-1} were found for Thr42b* interacting with Lys27bn and Arg83bn, and for Glu76b* interacting with Arg59bn. Coupling energies below 1 kcal mol^{-1}, but above the experimental error of about $\pm$0.35 kcal mol^{-1} (calculated for two standard deviations from the mean), were found for many of the other interactions examined.

Double mutant cycle analysis of barnase–barstar complex formation

Double mutant cycles can be used to measure the co-operativity of the kinetics of association between barnase and barstar, since the association rate constant reflects the transition state energy for complex formation. The transition state energy can be calculated from the association rate constant applying transition state theory (Fersht, 1985). The change in transition state energy on mutation ($\Delta\Delta G^{\ddagger}_{mut\text{-}wt}$) is calculated from

$$\Delta\Delta G^{\ddagger}_{mut\text{-}wt} = -RT\ \ln(k_{1\,(wt)}/k_{1\,(mut)}) \quad (2)$$

Table 1

Kinetics and equilibrium of association of barstar and barnase mutants

Barnase	Barstar	k_1 $\times10^{-8}$ (s^{-1} M^{-1})	k_{-1} $\times10^{3}$ (s^{-1})	K_d (pM)	ΔG (kcal mol^{-1})	$\Delta\Delta G$[a] (kcal mol^{-1})	$\Delta\Delta G_{int}$[b] (kcal mol^{-1})	$\Delta\Delta G^{\ddagger}_{ass}$[c] (kcal mol^{-1})	Distance (Å)[d]
wt	wt	3.7	0.0037	0.01	19.0				
K27A	wt	0.51	4.5	88	13.6	5.4			
R59A	wt	0.34	2.4	70	13.8	5.2			
R83Q	wt	1.1	10	94	13.6	5.4			
R87A	wt	1.4	17	120	13.5	5.5			
H102A	wt	4.0	129	320	12.9	6.1			
wt	Y29F	3.0	0.0024	0.008	19.1	−0.1			
wt	Y29A	2.9	1.0	3.5	15.6	3.4			
wt	D35A	1.9	3.8	20	14.5	4.5			
wt	W38F	4.3	0.07	0.16	17.4	1.6			
wt	D39A	1.9	900	4060	11.3	7.7			
wt	T42A	3.2	0.072	0.23	17.2	1.8			
wt	W44F	3.1	0.0034	0.011	19.0	0.0			
wt	E76A	2.0	0.021	0.1	17.65	1.4			
wt	E80A	2.1	0.0052	0.025	18.5	0.5			
K27A	Y29A	0.46	970	21,000	10.4	8.6	0.2 (0.24)	0.1 (0.23)	11.2 (NZ ↔ OH)
K27A	D35A	0.37	3600	1×10^5	9.5	9.5	0.4 (0.22)	0.2 (0.17)	9.6 (NZ ↔ OD1)
K27A	W38F	0.38	21	56	12.6	6.4	0.6 (0.33)	−0.3 (0.29)	3.8 (NZ ↔ NE1)
K27A	D39A	0.58	680	11,700	10.8	8.2	4.8 (0.21)	0.5 (0.17)	4.5 (NZ ↔ OD1)
K27A	T42A	0.43	6.8	157	13.3	5.7	1.5 (0.28)	0.0 (0.21)	3.0 (NZ ↔ OG1)
K27A	E76A	0.16	13	810	12.3	7.7	0.1 (0.25)	−0.3 (0.18)	10.1 (NZ ↔ OE1)
K27A	E80A	0.32	3.5	110	13.5	5.5	0.4 (0.29)	0.1 (0.23)	7.1 (NZ ↔ OE1)
R59A	Y29A	0.28	250	9000	10.9	8.1	0.6 (0.28)	0.0 (0.25)	15.5 (NE ↔ OH)
R59A	D35A	0.31	14	440	12.7	6.3	3.4 (0.2)	0.4 (0.15)	4.4 (NE ↔ OD1) 2.9 (N ↔ OD1)
R59A	W38F	0.33	13	400	12.8	6.2	0.6 (0.32)	−0.1 (0.29)	3.4 (NH2 ↔ ○)[e]
R59A	T42A	0.21	23	1000	12.2	6.8	0.2 (0.29)	−0.2 (0.18)	8.0 (NH2 ↔ OG1)
R59A	E76A	0.36	1.6	44	14.1	4.9	1.7 (0.24)	0.4 (0.21)	3.0 (NH1 ↔ OE1) 3.0 (NH2 ↔ OE1)
R59A	E80A	0.31	2.0	65	13.9	5.1	0.6 (0.31)	0.3 (0.25)	5.8 (NH2 ↔ OE2)
R83Q	Y29A	0.93	1100	12,000	10.7	8.3	0.6 (0.30)	0.1 (0.24)	6.0 (NH1 ↔ OH)
R83Q	D35A	0.63	7100	110,000	9.4	9.6	0.3 (0.30)	0.1 (0.19)	11.6 (NH2 ↔ OD1)
R83Q	W38F	1.0	91	900	12.3	6.7	0.3 (0.37)	−0.1 (0.32)	8.6 (NH1 ↔ NE1)
R83Q	D39A	1.0	53	525	12.6	6.4	6.7 (0.25)	0.3 (0.14)	2.5 (NH2 ↔ OD1) 2.9 (NH1 ↔ OD1)
R83Q	T42A	1.2	35	300	12.9	6.1	1.2 (0.28)	0.1 (0.22)	5.8 (NH1 ↔ OG1)
R83Q	E76A	0.4	34	845	12.3	6.7	0.1 (0.21)	−0.2 (0.19)	14.9 (NH1 ↔ OE1)
R83Q	E80A	0.63	11	168	13.3	5.7	0.2 (0.33)	0.0 (0.24)	12.1 (NH2 ↔ OE1)
R87A	Y29A	1.6	1300	8000	11.0	8.0	1.0 (0.25)	0.2 (0.21)	4.3 (NH2 ↔ OH)
R87A	W38F	1.9	275	1450	12.0	7.0	0.2 (0.37)	0.2 (0.2)	9.7 (NE ↔ NE)
R87A	D39A	1.6	300	1850	11.9	7.1	6.1 (0.26)	0.1 (0.18)	2.9 (NH2 ↔ OD1)
R87A	T42A	1.1	310	1400	12.0	7.0	0.4 (0.29)	0.5 (0.24)	8.6 (NH2 ↔ OG1)
R87A	E76A	0.69	74	1080	12.2	6.8	0.1 (0.26)	0.0 (0.22)	16.6 (NH2 ↔ OE1)
R87A	E80A	0.76	24	310	12.9	6.1	0.0 (0.31)	−0.1 (0.25)	14.5 (NH ↔ OE1)
H102A	Y29A	3.6	150	420	12.7	6.3	3.3 (0.30)	0.0 (0.27)	3.4 (○ ↔ CE1)[e]
H102A	Y29F	3.9	45	117	13.5	5.5	0.5 (0.36)	0.1 (0.27)	
H102A	W38F	3.2	1280	4000	11.4	8.6	0.2 (0.39)	−0.2 (0.35)	9.3 (NE2 ↔ CG)
H102A	D39A	4.4	17,000	39,000	10.1	8.9	4.9 (0.35)	−0.1 (0.26)	2.8 (NE2 ↔ OD2)
H102A	T42A	2.7	2400	9200	10.9	8.1	−0.1 (0.31)	−0.2 (0.26)	8.7 (NE2 ↔ OG1)
H102A	E76A	1.7	590	3500	11.5	7.5	0.0 (0.29)	−0.1 (0.24)	15.3 (NE2 ↔ OE1)
H102A	E80A	2.0	180	880	12.3	6.7	−0.1 (0.33)	0.1 (0.27)	15.4 (NE2 ↔ OE1)

All rate constants were measured in 50 mM Tris-HCl buffer (pH 8) at 25°C. The numbers in parentheses are the calculated errors (2σ) for $\Delta\Delta G_{int}$ and $\Delta\Delta G^{\ddagger}_{ass}$ for the individual data.

[a] The difference in free energy of binding wild-type (wt) and mutant proteins $\Delta\Delta G = \Delta G_{mut} - \Delta G_{wt}$.

[b] $\Delta\Delta G_{int}$ is the measure of co-operativity between 2 residues in kcal mol^{-1} calculated from a double mutant cycle (equation (1)).

[c] $\Delta\Delta G^{\ddagger}_{ass}$ is the coupling energy between 2 residues during association of the complex. It is calculated in a similar way to $\Delta\Delta G_{int}$, but from k_1 according to the transition state theory (see the text).

[d] Distances between side-chains in the complex (between the indicated atoms) as measured in the crystal structure between chains A and D of the asymmetric unit (Buckle *et al.*, 1994).

[e] Distances were measured from the centre of the aromatic or histidine rings, to simulate an interaction of the π electron cloud.

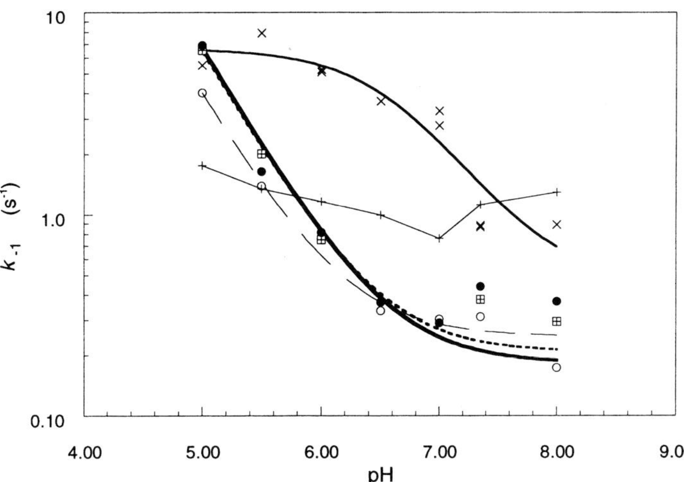

Figure 3. Dissociation rate constants measured between pH 5 and 8 for wild-type and various mutants. Rate constants were measured in 50 mM buffer (acetate (pH 5.0), Mes (pH 5.5 and 6.0), Mops (pH 7.0), Tris base (pH 7.35 and 8.0). The dissociation rate constants of the different protein pairs are all scaled to fit on one axis. (●) Dissociation rate constant for wild-type barnase and barstar, ×100,000; (+) His102bn → Ala barnase with barstar, ×10; (⊞) wt barnase with Tyr29b* → Ala barstar, ×300; (×) wild-type barnase with Asp39b* → Ala barstar; (○) Arg87bn → Ala barnase with wild-type barstar, ×10. The standard error of the mean for association and dissociation rate constants is about ±12%.

Single mutations result in a change of the association rate constant for mutations of charged residues only. A mutation of any one of these residues in barstar decreases the association rate constant twofold, irrespective of the residue mutated (Schreiber *et al.*, 1994). Single mutations of positively charged to neutral residues located in the barstar-binding surface of barnase result in a two- to tenfold decrease in the association rate constant, dependent on the mutation (Figure 1; and Schreiber & Fersht (1993*a*)). Coupling energies between pairs of residues during the association of barnase and barstar are shown in Figure 5. No coupling energies were observed for pairs of residues in which at least one of them was uncharged, and for pairs of distant charged residues. Pairs of ionic residues located in close proximity in the complex showed co-operativity during the association process, with coupling energies of around 0.4 kcal mol^{-1}.

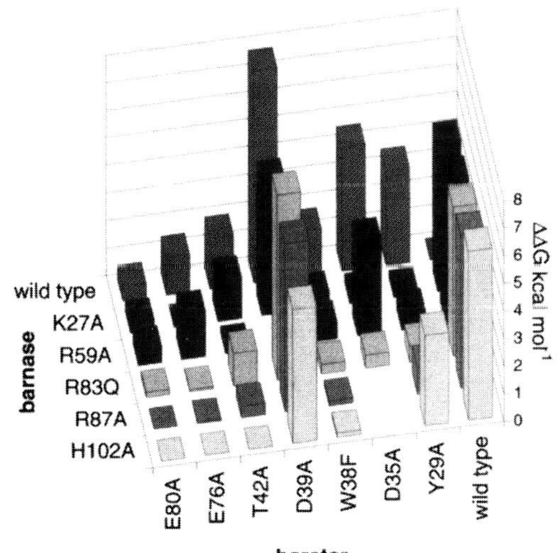

Figure 4. Co-operativity between residue pairs in the interface of the barnase–barstar complex. The rows marked as wild-type represent $\Delta\Delta G_{mut\text{-}wt}$ of the various single mutations. All the other fields in the matrix show the calculated interaction energy between the marked residue pairs as calculated by double mutant cycles. Experimental errors for $\Delta\Delta G_{mut\text{-}wt}$ are about ± 0.2 kcal mol^{-1} (2σ), and for $\Delta\Delta G_{int}$ are shown in Table 1.

Discussion

The complex formed by barnase and barstar is one of the tightest protein complexes reported in the literature. Most of the barstar residues that contribute to binding are located on one continuous polypeptide chain between residues 29 and 42, which consists of a loop and its adjacent helix. Only a single interaction (Arg59bn with Glu76b*) is made with helix$_4$ of barstar (residues 68 to 81). The localisation of most important residues for binding on one continuous polypeptide chain has been observed for protease inhibitors; however, in antibody–antigen complexes, the interaction surfaces of both components are formed by several regions of peptides (Janin & Chothia, 1990).

In this work, we measured coupling energies between residues in both proteins by double mutant cycle analysis on a subset of barnase and barstar residues that were identified by mutagenesis as contributing to binding. Mutations that were likely to lead to significant structural changes in the proteins were not further analysed. Five barnase residues and seven barstar residues were considered to be suitable targets for double mutant cycles analyses. Double mutant cycles were constructed between most residues in the subset to enable us to quantify the energetics of the interaction without being biased by structural information about the interface. We then investigated the extent to which coupling energies could be derived from an examination of the structure of the complex. At the centre of the binding surface, Asp39b* binds to Arg83bn, Arg87bn and His102bn, with Asp39b* mimicking the P_1 phosphate

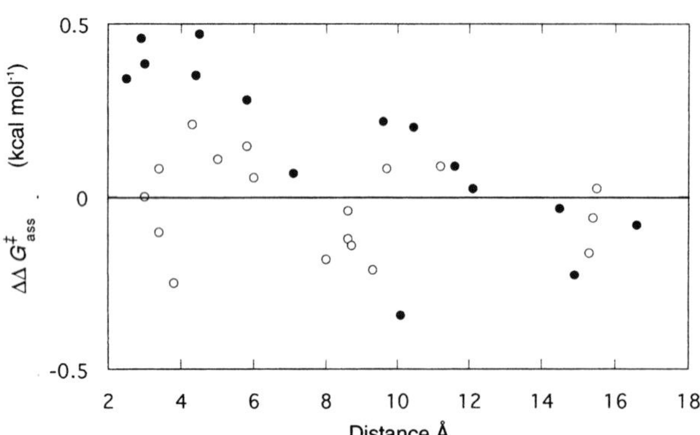

Figure 5. Double mutant cycle analysis for kinetics of association *versus* the distance between the pair of residues in the complex. Activation energies for the rate constant of association for different mutants and the co-operativity between them ($\Delta\Delta G^{\ddagger}_{ass}$) were calculated as explained in the text. A $\Delta\Delta G^{\ddagger}_{ass}$ of zero indicates that the pair of residues analysed do not interact during the association reaction. A positive $\Delta\Delta G^{\ddagger}_{ass}$ value means that the 2 residues interact favourably during the association process, and a negative value that the 2 residues oppose each other. Filled circles indicate the $\Delta\Delta G^{\ddagger}_{ass}$ between 2 charged mutants, and open circles indicate all other mutant pairs analysed. Calculated errors (2σ) are shown in Table 1.

site of an RNA substrate, and being within close hydrogen-bonding distance from those residues. The coupling energies between Asp39b* with Arg83bn and Arg87bn are over 6 kcal mol^{-1}, making them among the strongest interactions yet reported. At the other side of the interaction surface, the salt-bridge between Glu76b* and Arg59bn shows a coupling energy of only 1.6 kcal mol^{-1}, the distance between the residues in the pairs in both cases being about 3 Å. Three of the six most important interactions detected by double mutant cycle analysis (with coupling energies of more than 3.0 kcal mol^{-1}) were not noted previously as being potentially important from the analysis of the crystal structure (Guillet *et al.*, 1993; Buckle *et al.*, 1994). The coupling energies for Lys27bn with Asp39b* and for Arg59bn with Asp35b* are 4.8 and 3.4 kcal mol^{-1}, respectively, despite the distances of 4.5 Å and 4.4 Å between the two residues from each pair. Interactions of this magnitude were not expected at this distance. His102bn and Tyr103bn are the only barnase residues that move significantly upon forming the complex (Buckle & Fersht, 1994; Buckle *et al.*, 1994). His102bn makes face-to-edge interactions with both Tyr29b* and Tyr103bn, with both Tyr residues being perpendicular to the face of His102bn, locking it into position. The strong interaction between His102bn and Tyr29b* (coupling energy of 3.3 kcal mol^{-1}) cannot be simply explained as a base-to-aromatic interaction, because for such an interaction we would expect the edge of the histidine to be perpendicular to the face of a tyrosine, and hence the histidine would serve as a hydrogen donor, which it is not the case here (Burley & Petsko, 1986; Loewenthal *et al.*, 1992; Perutz, 1993).

Figure 6 shows the coupling energies between barnase and barstar residues as a function of the distance between them. Overall, the coupling energy between pairs of residues decreases with distance, with significant values of free energy for all pairs of residues separated by up to 7 Å. Ionic interactions are known to extend to long distances (Russell *et al.*, 1987; Jackson & Fersht, 1993), but Figure 6 shows that coupling energies of between 0.35 and 1.2 kcal mol^{-1} are also found for interactions of uncharged residues separated by more then 4 Å. The pairs of residues in this group are of a varied nature: ionic interactions (Lys27bn ↔ Glu80b* and Arg59bn ↔ Glu80b*), interactions between ionic and aromatic groups (Lys27bn ↔ Trp38b*, Arg83bn ↔ Tyr29b* and Arg87bn ↔ Tyr29b*) and interactions between ionic and polar side-chains as for Arg83bn ↔ Thr42b* and Arg87bn ↔ Thr42b*. None of these interactions are simple water-mediated hydrogen bonds. It is

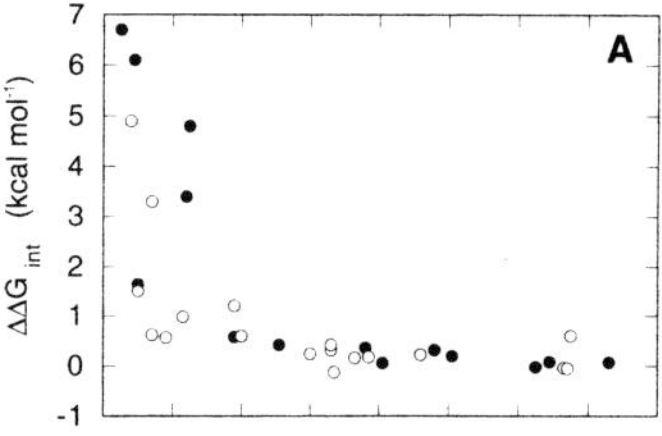

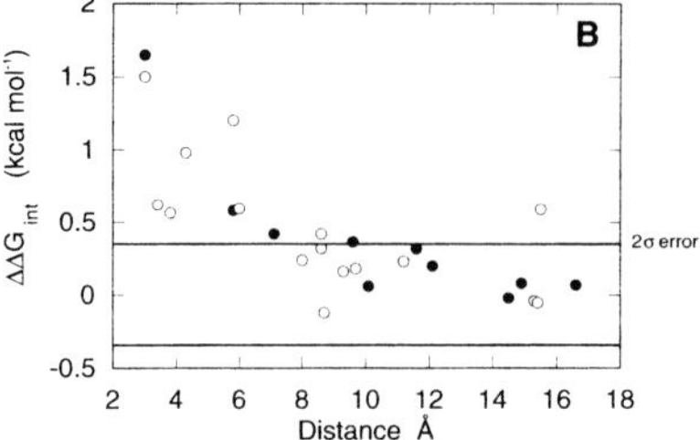

Figure 6. Calculated coupling energies between different residue pairs in the interface of barnase and barstar *versus* the distance between them in Å. Filled circles indicate $\Delta\Delta G_{int}$ between 2 charged residues, and open circles indicate that at least 1 of the 2 residues is not charged. B, As A, but with a different scale for $\Delta\Delta G_{int}$. The data are taken from Table 1. Errors for $\Delta\Delta G_{int}$ are shown in Table 1.

difficult, at this stage, to analyse the nature of those weak long-range interactions. Hydrogen bonds are partly of an electrostatic nature, so they might show longer-range contributions for binding. Some of the measured coupling energies may also result from secondary interactions, through a third side-chain. It could well be that the coupling energies result indirectly from local rearrangements of the solvation shells on mutating, rather than from direct interactions. No significant coupling energies are measured between pairs of residues separated by >8 Å, apart from the coupling energy of Arg59bn with Tyr29b* of 0.64 kcal mol^{-1}. The important conclusions from this are: (1) that the range of amino acid interaction distances in a protein is more than 4 Å, even for interactions in which one or more of the components is not a charged residue; and (2) that the energetics of an interaction cannot be simply assessed by the distance between the two partners, even for the simple case of an electrostatic interaction, which is independent of the orientation of the interacting residues. The comparison between potentially important bonds derived from structural information with the measured coupling energies of those bonds emphasises that the nature of polar interactions is very dependent on their local environment. This was suggested also by the analysis of mutational data from the complex of human growth hormone binding with its receptor (Cunningham & Wells, 1993). The change in free energy of binding did not correlate with the change in buried surface area, number of van der Waals' contacts or side-chain accessibility changes upon mutagenesis. This is in contrast with the good fit found between the number of methyl side-chain groups found within 6 Å of the deleted methyl group in the core of barnase and their contribution to the free energy of unfolding (Serrano *et al*., 1992).

Two central questions in analysing coupling energy data are: to what extent do they reflect the real interaction energies, and to what extent do they reflect the structural rearrangement of the protein and solvent? Additivity of mutational effects in proteins was observed for distant mutations at rigid molecular interfaces, such as in some enzyme–substrate interactions and some protein–protein interactions (Carter *et al*., 1984; Horovitz & Rigbi, 1985; Wells, 1990). This general rule seems to apply also to our measurements. The changes in the free energies of binding of all the pairwise mutations for residues separated by more than 8 Å are equal to the sums of the two single mutants, and no significant co-operativity was detected between those pairs of residues (Figure 6). These results suggest that no major rearrangement of the interface occurs as a result of mutagenesis. The question becomes more complicated for residues located in close proximity to each other, since they show co-operativity in double mutant cycles. The pH dependence of the dissociation rate constant of the complex results from the steric inability of His102bn to accept a second proton. This pH dependency may be used as a probe for structural rearrangements in the vicinity of His102 upon introducing mutation. Mutations of Arg87bn → Ala and Tyr29b* → Ala do not alter the pH dependence of dissociation of the complex, which suggests that the structure of His102bn in the complex in the presence of those mutations is not altered. As expected, mutating Asp39b* → Ala results in a substantial shift in the pK_a of the complex.

The rate of a chemical reaction is determined by the difference in free energy between the transition state of the reaction and its ground state. That energy can be calculated by transition state theory (Fersht, 1985). However, the calculation of the change in energy on mutation is simpler, since the absolute value of the free energy is not important, but only the change in it according to equation (2). The reference states for this calculation are the free proteins in solution, which have a coupling energy $\Delta\Delta G_{int}$ of zero. Double mutant cycles can be applied to $\Delta\Delta G^{\ddagger}$ as for any other thermodynamic measurement (Fersht *et al*., 1992; Horovitz & Fersht, 1992). Where two residues interact during the transition for complex formation, their contribution to the free energy of the transition state should be non-additive. The coupling energy will be reflected in the non-additivity of the association rate constants.

The association rate constant for the formation of the wild-type complex is 3.7×10^8 M^{-1} s^{-1}, which is too fast to be explained solely by diffusion without electrostatic effects. This, and the presence of multiple charges (mostly positive on barnase and negative on barstar), suggest that steering effects are important during the association. Co-operativity is found only for pairs of charged residues located close to each other in the complex. The lack of co-operativity during association for all unchanged residues in the interface means simply that the transition state for association occurs before the co-operative effects come into play. The existence of an intermediate on the association pathway was suggested from the analysis of the pH dependence of the association and dissociation rate constants (Schreiber & Fersht, 1993*a*). The lack of dependence of the association rate constant on pH suggests that the interaction of barstar with His102bn is formed after the transition state. The current results suggest that all of the non-electrostatic interactions are formed after the transition state of binding barnase to barstar. This can be explained in physical terms by the kinetics of protein–protein association using a computer simulation of Brownian dynamics (Northrup & Erickson, 1992). The association of two proteins can be divided into two steps. The first step is the translation of the two molecules until they are within colliding distance. During this step it is expected that the contribution of a charged residue to the association rate constant will be proportional to its electrostatic field potential. The second step is a rotational one, which results in the final docking of the two proteins. It was suggested that pairs of proteins surrounded by water undergo multiple collisions, with rotational reorientation during each encounter (Northrup & Erickson, 1992). At this stage, oppositely charged residues that are located in close proximity in the complex can assist in the guiding and docking of the two proteins. A mutation of one of those charged residues will result in the neutralisation of the second charged residue of the pair in the final docking steps, and so mutations of closely interacting residues are not additive during the association of the two molecules. In other words, those residues show co-operative behaviour during the association of the complex.

The aim of this work was to construct a map of the interaction energies at the surface of contact between two proteins, barnase and barstar. The comparison between the energetic and structural binding surfaces emphasises that structural details by themselves are not sufficient to evaluate the energetics of binding. Coupling energies exist for residues up to 7 Å apart. The most important interactions are between Asp39b* and the four barnase side-chains that surround it. Interestingly, Lys27bn is 4.5 Å distant from Asp39*, but still interacts strongly with it. On the other hand, the salt-bridge between Glu76b* and Arg59bn is relatively weak, showing the importance of the local environment to binding. The general architecture of the interaction seems to be rigid, even for mutations that destabilise the interaction significantly. This can be explained by the fit between the proteins being lock-and-key, rather than an induced fit, with very similar structures of the free and the complexed proteins. The interactions between residues that we tried to quantify in this study add up to one of the tightest interactions between proteins found so far.

Materials and Methods

Protein expression and purification

Site-directed mutagenesis of barstar was performed by the method of Sayers *et al.* (1988). The oligonucleotides that were used for mutagenesis of barstar residues were: Tyr29 → Ala, TTC ACC GTA GGC TTC CGG AAG; Tyr29 → Phe, TTC ACC GTA GAA TTC CGG AAG; Trp38 → Phe, TCA GAC AAT CGA ATA AAG CGT; Thr42 → Ala, CCA TCC GGC CAG ACA ATC; and for Trp44 → Phe, GGT ACT CCA CAA ATC CGG TCA GA: For mutagenesis of barnase Phe82 → Ala, CTG AAT TTC TGG CGC CTG ATG T. Mutant plasmids were identified by direct sequencing. All the other mutations used in this work have been described (Mossakowska *et al.*, 1989; Meiering *et al.*, 1992; Schreiber *et al.*, 1994). The expression and purification of barstar and barnase have been described (Serrano *et al.*, 1990; Schreiber & Fersht, 1993*b*). Barstar has two Cys residues (residues 40 and 82). Barstar was unfolded in urea in the presence of 10 mM DTT for 30 minutes, refolded and extensively dialysed before use to eliminate the presence of oxidised or dimeric proteins. Protein concentrations were determined by measuring the absorbance at 280 nm using an extinction coefficient of 27,400 $M^{-1} cm^{-1}$ for barnase wild-type and mutants, 23,000 $M^{-1} cm^{-1}$ for barstar wild-type, 20,600 $M^{-1} cm^{-1}$ for Tyr29b* → Ala, 16,000 $M^{-1} cm^{-1}$ for Trp38b* → Phe, and 16,170 $M^{-1} cm^{-1}$ for Trp44b* → Phe, estimated by the method of Gill & von Hippel (1989).

Kinetic and equilibrium measurements

The association and dissociation rate constants for the barnase–barstar complexes (wild-type and mutants) were measured in 50 mM Tris-HCl (pH 8) buffer at 25°C, as described previously (Schreiber & Fersht, 1993*a*; Schreiber *et al.*, 1994). The association of barnase and barstar was measured under second-order conditions, the concentrations of the proteins being equal (Fersht, 1985; Schreiber & Fersht, 1993*a*; Schreiber *et al.*, 1994). The formation of the complex was followed by investigation of the change in fluorescence upon association, using a stopped-flow fluorimeter with a 315 nm cut-off filter for measuring emission. The association of reduced wild-type barstar with barnase (as well as some of the mutants) could be measured only upon excitation of 230 nm, not 280 nm. The biophysical basis of this phenomenon will be discussed elsewhere (G. Schreiber, R. Golbik & A. R. Fersht, unpublished results). Similar rate constants of association were measured for mutants that show changes in fluorescence intensity on excitation at both 230 nm and 280 nm. Dissociation rate constants ("off rates") were measured under first-order conditions by first allowing the barnase–barstar complex to form, then "chasing" off the mutant barstar with five to ten times the concentration of wild-type barstar. A [^{3}H]barnase–barstar or barnase–[^{3}H]-barstar complex was used for complexes with dissociation rate constants of $5 \times 10^{-3} s^{-1}$ or more. Dissociation rate constants faster than $5 \times 10^{-3} s^{-1}$ were measured with a stopped-flow fluorimeter. This method takes advantage of the differences in fluorescence behaviour between wild-type and mutant complexes, relative to the free form, as described previously (Schreiber & Fersht, 1993*a*). Dissociation rate constants were calculated by fitting the obtained data set to an exponential equation.

Very small dissociation constants (K_d), such as that for the complex of wild-type barstar and barnase, cannot be measured directly, but may be derived from kinetic measurements of the dissociation rate constant (k_{-1}) and association rate constant (k_1), so that the dissociation constant is calculated as $K_d = k_{-1}/k_1$. The free energy of an interaction is calculated as $\Delta G = -RT \ln K_d$, and the interaction energy between two residues is calculated by the double mutant cycle method, as in equation (1).

The standard error of the mean (σ) was calculated from two to five repetitions of each experiments for all of the data for mutant proteins, and for more measurements on wild-type proteins. The standard error of the mean for association and dissociation rate constants is about 12%. Accordingly, 2σ, the 95% confidence level for ΔG is ± 0.2 kcal mol^{-1}, and for $\Delta\Delta G_{mut-wt}$ the 2σ is ± 0.2 kcal mol^{-1} (since the error for the wild-type data is substantially smaller). For $\Delta\Delta G_{int}$ and $\Delta\Delta G^{\ddagger}_{ass}$, the individual errors (2σ) are shown in Table 1.

References

Buckle, A. M. & Fersht, A. R. (1994). Subsite binding in an RNase: structure of a barnase–tetranucleotide complex at 1.76 Å resolution. *Biochemistry*, **33**, 1644–1653.

Buckle, M., Schreiber, G. & Fersht, A. R. (1994). Protein–protein recognition: Crystal structural analysis of a barnase–barstar complex at 2.0 Å resolution. *Biochemistry*, **33**, 8878–8889.

Burley, S. K. & Petsko, G. A. (1986). Amino-aromatic interactions in proteins. *FEBS Letters*, **203**, 139–143.

Bycroft, M., Ludvingen, S., Fersht, A. R. & Poulsen, F. M. (1991). Determination of the three-dimensional solution structure of barnase using nuclear magnetic resonance spectroscopy. *Biochemistry*, **30**, 8697–8701.

Carter, P. J., Winter, G., Wilkinson, A. J & Fersht, A. R. (1984). The use of double mutants to detect structural changes in the active site of the tyrosyl-tRNA synthetase. (*Bacillus stearothermophilus*). *Cell*, **38**, 835–840.

Cunningham, B. C. & Wells, J. A. (1993). Comparison of a structural and a functional epitope. *J. Mol. Biol.* **234**, 554–563.

Fersht, A. R. (1985. *Enzyme Structure and Mechanism*, W. H. Freeman and Company, New York.

Fersht, A. R. (1987). Dissection of the structure and activity of the tyrosyl-tRNA synthetase by site-directed mutagenesis. *Biochemistry*, **26**, 8031–8037.

Fersht, A. R. (1993). The sixth Datta lecture. Protein folding and stability: The pathway of folding of barnase. *FEBS Letters*, **325**, 5–16.

Fersht, A. R., Wilkinson, A. J., Carter, P. & Winter, G. (1985). Fine structure–activity analysis of mutations of position 51 of tyrosyl-tRNA synthetase. *Biochemistry*, **23**, 5858–5861.

Fersht, A. R., Leatherbarrow, R. & Wells, T. N. C. (1987). Structure–activity relationships in engineering proteins: Analysis of use of binding energy by linear free energy relationships. *Biochemistry*, **26**, 6030–6038.

Fersht, A. R., Matouschek, A. & Serrano, L. (1992). The folding of an enzyme I: Theory of protein engineering analysis of stability and pathway of protein folding. *J. Mol. Biol.* **224**, 771–782.

Gill, S. C. & von Hippel, P. H. (1989). Calculation of protein extinction coefficients from amino acid sequence data. *Anal. Biochem.* **182**, 319–326.

Guillet, V., Lapthorn, A., Hartley, R. W. & Maugen, Y. (1993). Recognition between a bacterial ribonuclease, barnase, and its natural inhibitor, barstar. *Structure*, **1**, 165–177.

Hartley, R. W. (1989). Barnase and barstar: Two small proteins to fold and fit together. *Trends Biochem. Sci.* **14**, 450–454.

Hartley, R. W. (1993). Directed mutagenesis and barnase–barstar recognition. *Biochemistry*, **32**, 5978–5984.

Horovitz, A. (1987). Non-additivity in protein–protein interactions. *J. Mol. Biol.* **196**, 733–735.

Horovitz, A. & Fersht, A. R. (1992). Co-operative interactions during protein folding. *J. Mol. Biol.* **224**, 733–740.

Horovitz, A. & Rigbi, M. (1985). Protein–protein interactions: Additivity of the free energies of association of amino acid residues. *J. Theor. Biol.* **116**, 149–159.

Horovitz, A., Serrano, L., Avron, B., Bycroft, M. & Fersht, A. R. (1990). Strength and cooperativity of contributions of surface salt bridges to protein stability. *J. Mol. Biol.* **216**, 1031–1044.

Jackson, S. E. & Fersht, A. R. (1993). Contribution of long-range electrostatic interactions to the stabilization of the catalytic transition state of the serine protease subtilisin BPN'. *Biochemistry*, **32**, 13909–13916.

Janin, J. & Chohia, C. (1990). The structure of protein–protein recognition sites. *J. Biol. Chem.* **265**, 16027–16030.

Kraulis, P. (1991). MolScript, a program to produce both detailed and schematic plots of protein structures. *J. Appl. Crystallogr.* **24**, 946–950.

Loewenthal, R., Sancho, J. & Fersht, A. R. (1992). Histidine-aromatic interactions in barnase: Elevation of histidine pK_a and contribution to protein stability. *J. Mol. Biol.* **224**, 759–770.

Lubienski, M. J., Bycroft, M., Freund, S. M. V. & Fersht, A. R. (1994). ^{13}C Assignments and three-dimensional solution structure of barstar using nuclear magnetic resonance spectroscopy. *Biochemistry*, **33**, 8866–8877.

Mauguen, Y., Hartley, R. W., Dodson, E. J., Dodson, G. G., Bricogne, G., Chothia, C. & Jack, A. (1982). Molecular structure of a new family of ribonucleases. *Nature (London)*, **29**, 162–164.

Meiering, E. M., Serrano, L. & Fersht, A. R. (1992). Effect of active site residues in barnase on activity and stability. *J. Mol. Biol.* **225**, 585–589.

Mossakowska, D. E., Nyberg, K. & Fersht, A. R. (1989). Kinetic characterization of the recombinant ribonuclease from *Bacillus amyloliquefaciens* (barnase) and investigation of key residues in catalysis by site-directed mutagenesis. *Biochemistry*, **28**, 3843–50.

Northrup, S. H. & Erickson, H. P. (1992). Kinetics of protein–protein association explained by Brownian dynamic computer simulation. *Proc. Nat. Acad. Sci., U.S.A.* **89**, 3338–3342.

Perutz, M. F. (1993). The role of aromatic rings as hydrogen bond acceptors in molecular recognition. *Phil. Trans. Roy. Soc.* **345**, 105–112.

Russell, A. J., Thomas, P. G. & Fersht, A. R. (1987). Electrostatic effects on modification of charged groups in the active site cleft of subtilisin by protein engineering. *J. Mol. Biol.* **193**, 803–813.

Sali, D., Bycroft, M. & Fersht, A. R. (1988). Stabilization of protein structure by interaction of an α-helix dipole with a charged side-chain. *Nature (London)*, **335**, 496–500.

Sayers, J. R., Schmidt, W. & Eckstein, F. (1988). 5'–3' Exonucleases in phosphorothioate-based oligonucleotide directed mutagenesis. *Nucl. Acids Res.* **16**, 791–802.

Schreiber, G. & Fersht, A. R. (1993*a*). The interaction of barnase with its polypeptide inhibitor barstar studied by protein engineering. *Biochemistry*, **32**, 5145–5150.

Schreiber, G. & Fersht, A. R. (1993*b*). The refolding of *cis*- and *trans*-peptidylprolyl isomers of barstar. *Biochemistry*, **32**,

Schreiber, G., Buckle, A. M. & Fersht, A. R. (1994). Stability versus function: Two competing forces in the evolution of barstar. *Structure*, **2**, 945–951.

Serrano, L., Horovitz, A., Avron, B., Bycroft, M. & Fersht, A. R. (1990). Estimating the contribution of engineered surface electrostatic interactions to protein stability using double mutant cycles. *Biochemistry*, **29**, 9343–9352.

Serrano, L., Kellis, J. T., Jr Cann, P., Matouschek, A. & Fersht, A. R. (1992). The folding of an enzyme II. Substructure of barnase and the contribution of different interactions to protein stability. *J. Mol. Biol.* **224**, 783–804.

Wells, J. A. (1990). Additivity of mutational effects in proteins. *Biochemistry*, **29**, 8509–8517.

Edited by R. Huber

(Received 2 December 1994; accepted 1 February 1995)

Chapter 18

Tumour Suppressor p53: Protein Stability and Cancer

'He who saves a single life saves the entire world.'

The Talmud

The Tumour Suppressor p53

Having spent 12 years putting mutations into proteins to understand them, I decided it was time to analyse the effects of mutations in proteins that lead to disease. The obvious target was cancer, which is a disease of mutation. One of my lifelong friends from Imperial College days is Sir David Lane, the discoverer of the tumour suppressor p53, which is the first line of defence in the cell against cancer. As p53 is inactivated by mutation in some 50% of human cancers, I chose it to be the subject of the whole panoply of technology and ideas from our protein folding studies. Once again, it was a most fortunate choice. The protein itself has unveiled itself to be one of the most fascinating multifunctional proteins, which has a large number of crucial roles in processes ranging from fertility to senescence. We now know that it is a hub protein at the centre of a large number of cycles in the cell. It functions mainly as a transcription factor, leading to the transcription of genes of a multitude of proteins, including feedback loops. Although discovered as a tumour suppressor, it has a general role of disposing of unwanted or rogue cells as an organism develops. There is a battle between senescence, as p53 can destroy stem cells, and cancer as p53 destroys cancer cells. Importantly, the protein inducted me into the p53 community of scientists, which is exceptionally friendly and supportive as well, introducing me to very smart biologists.

As usual, my studies were initiated by a very simple series of experiments. p53 is a multidomain protein, consisting of four chains held together by a small tetramerization domain. Each chain has an unstable core (DNA binding) domain, which is the site of most oncogenic mutations, and about 40% disordered structure. The structure of the core domain of p53 had been solved as part of a series of brilliant studies by Nikola Pavletich. This domain recognises the DNA response elements that target p53 to different genes. The reason why some of the many inactivating mutations function was clear: they are residues in contact with DNA, so mutation weakens the binding. But the cause of inactivation by the remainder of the mutations was somewhat mysterious as many were far removed from the binding site of the domain. I set two new graduate students,

Alex Bullock and Julia Henckel, and two post-doctorals Penka Nikolova and Brian Dedecker to analyse those 'structural mutants', initially by thermodynamic measurements. By applying the methods we had used for studying protein folding in model systems, we found that some 30% of the oncogenic mutants are simply unstable at body temperature but are structurally intact at lower temperatures[201].

18.1 1997: *Top*: Alex Bullock and Julia Henckel. *Bottom*: Penka Nikolova and Brian DeDecker.

18.2 2003: Assaf Friedler.

18.3 2003: Trevor Rutherford and Dima Veprintsev.

Proc. Natl. Acad. Sci. USA
Vol. 94, pp. 14338–14342, December 1997
Biochemistry

Thermodynamic stability of wild-type and mutant p53 core domain

(tumor suppressor/denaturation/folding/zinc/protein)

ALEX N. BULLOCK*, JULIA HENCKEL*, BRIAN S. DEDECKER*, CHRISTOPHER M. JOHNSON*, PENKA V. NIKOLOVA*, MARK R. PROCTOR*, DAVID P. LANE†, AND ALAN R. FERSHT*‡

*Cambridge University Chemical Laboratory and Cambridge Centre for Protein Engineering, Medical Research Council Centre, Hills Road, Cambridge CB2 2QH, United Kingdom; and †Cancer Research Campaign Laboratories, Department of Biochemistry, University of Dundee, Dundee DD1 4HN, United Kingdom

Contributed by Alan R. Fersht, October 20, 1997

ABSTRACT **Some 50% of human cancers are associated with mutations in the core domain of the tumor suppressor p53. Many mutations are thought just to destabilize the protein. To assess this and the possibility of rescue, we have set up a system to analyze the stability of the core domain and its mutants. The use of differential scanning calorimetry or spectroscopy to measure its melting temperature leads to irreversible denaturation and aggregation and so is useful as only a qualitative guide to stability. There are excellent two-state denaturation curves on the addition of urea that may be analyzed quantitatively. One Zn^{2+} ion remains tightly bound in the holo-form of p53 throughout the denaturation curve. The stability of wild type is 6.0 kcal (1 kcal = 4.18 kJ)/mol at 25°C and 9.8 kcal/mol at 10°C. The oncogenic mutants R175H, C242S, R248Q, R249S, and R273H are destabilized by 3.0, 2.9, 1.9, 1.9, and 0.4 kcal/mol, respectively. Under certain denaturing conditions, the wild-type domain forms an aggregate that is relatively highly fluorescent at 340 nm on excitation at 280 nm. The destabilized mutants give this fluorescence under milder denaturation conditions.**

The tumor suppressor protein p53 is a sequence-specific transcription factor that functions to maintain the integrity of the genome (1). On its induction in response to DNA damage, p53 promotes cell cycle arrest in G_1 phase (2) and apoptosis if DNA repair is not possible (3). Negative regulation occurs by the synthesis and subsequent binding of the oncoprotein Mdm2 to the transactivation domain of p53. This targets it for degradation and ensures that the cellular stability of p53 is low (4, 5). About 50% of human cancers and 95% of lung cancers are associated with mutations in p53. The majority of these map to its core domain, which is responsible for binding DNA (6). The crystal structure of the core domain bound to DNA has been determined (7). A number of the tumorigenic mutants affect residues that contact the DNA, but many are not directly involved in binding and appear to affect the thermodynamic stability of the protein (8, 9).

p53 is a possible target for cancer therapy, including drugs that can stabilize it or using superstable p53 variants that would be suitable for gene therapy applications. There is a lack of quantitative information on the stability of p53 on which to base experiments measuring its change in stability on mutation. Data tend to be restricted so far to measurements of the temperature dependence of transactivation or PAb 1620 binding (8, 9), a monoclonal antibody specific for the native state of wild-type p53 (10). These suggest that p53 is relatively unstable. We find in this study that the core domain denatures irreversibly with temperature, and so the T_m measured by differential scanning calorimetry or spectroscopy cannot be used quantitatively for analyzing structure–activity relationships of p53. We have turned instead to studying the stability of the isolated core domain by using urea-mediated denaturation, which is of proven use for systematic measurements of the effects of mutation on stability (11–14). There are potential problems because of Zn^{2+} ions being bound to p53 and the resultant complications of various holo- and apo-forms of the native and denatured state. We have found procedures for measuring the equilibria between the holo-denatured and native states, which may be used to measure the free energy of unfolding in water, $\Delta G^{H_2O}_{D\text{-}N}$.

MATERIALS AND METHODS

Gene Subcloning and Mutagenesis. The portion of the human p53 core domain encoding residues 94–312 was amplified by PCR from the plasmid pT7hp53 (15) by using the oligonucleotides 5′-GGG AAT TCC ATA TGT CAT CTT CTG TCC CTT CCC AGA AAA CCT ACC AG-3′ and 5′-GGG AAT TCA GGT GTT GTT GGG CAG TGC TCG CTT AGT GCT CCC-3′. This was digested with *Nde*I and *Eco*RI restriction enzymes and subcloned into the polylinker region of pRSET(A) (Invitrogen). The ligated plasmid was transformed into *Escherichia coli* strain DH5α. The subsequently recovered plasmid, pRSET(A)-corehp53, was sequenced by using T7 promoter and T7 terminator primers (5′-TAA TAC GAC TCA CTA TAG GG-3′ and 5′-CGT AGT TAT TGC TCA GCG G-3′, respectively). Mutagenesis was performed with inverse PCR or the QuikChange site-directed mutagenesis kit (Stratagene).

Protein Expression and Purification. Protein was expressed in *E. coli* strain BL21(DE3), which was grown at 37°C to an OD_{600} = 1.2 before overnight induction at 25°C with 1 mM isopropyl β-D-thiogalactoside. Cells were harvested by centrifugation and sonicated in 50 mM Tris, pH 7.2/5 mM DTT/1 mM phenylmethylsulfonyl fluoride. Soluble lysate was loaded onto a SP-Sepharose cation exchange column (Pharmacia) and eluted with a NaCl gradient (0–1 M). Further purification was achieved by affinity chromatography with a heparin-Sepharose column (Pharmacia) in 50 mM Tris, pH 7.2/5 mM DTT with a NaCl gradient (0–1 M) for elution. Protein concentration was measured spectrophotometrically by using an extinction coefficient of ε_{280} = 17,130 $cm^{-1} \cdot M^{-1}$ calculated by the method of Gill and von Hippel (16). The accuracy of this method was confirmed by amino acid analysis. Electrospray mass spectrometry showed that the initiator methionine residue is cleaved after translation.

The publication costs of this article were defrayed in part by page charge payment. This article must therefore be hereby marked "*advertisement*" in accordance with 18 U.S.C. §1734 solely to indicate this fact.

Abbreviation: DSC, differential scanning calorimetry.

‡To whom reprint requests should be addressed at: University Chemical Laboratory, University of Cambridge, Lensfield Road, Cambridge CB2 1EW, U.K. e-mail: arf10@cam.ac.uk.

Differential Scanning Calorimetry (DSC). DSC experiments were performed by using a Microcal VPDSC (Microcal, Amherst, MA) with a cell volume of 0.5 ml. Temperatures from 4 to 42°C or 95°C were scanned at a rate of 60°C/h. Dialyzed samples (7–36 μM protein) were degassed and flushed with argon for 10–15 min on ice. The dialysis buffer was used for baseline scans. A pressure of 25 psi (1 psi = 6.89 kPa) was applied to the cell, and the system was allowed to equilibrate at 4°C for 20 min before scanning. The data were analyzed with ORIGIN software (Microcal).

DNA binding experiments with DSC were carried out in 50 mM Tris, pH 7.2/1 mM DTT. The apparent T_m of p53 core domain, the double-stranded oligonucleotide consensus sequence (22-mer) (7), and the complex of double-stranded DNA–protein in equimolar amounts (7 μM) were determined with and without 0.3 M NaCl in solution. Consensus DNA was synthesized as single-stranded oligonucleotides (5′-ATA ATT GGG CAA GTC TAG GAA A-3′ and 5′-TTT CCT AGA CTT GCC CAA TTA T-3′), which were then annealed for DNA binding studies.

Equilibrium Denaturation. Denaturation was monitored by fluorescence using an Aminco-Bowman Series 2 luminescence spectrofluorimeter with excitation at 280 nm (bandpass, 4 nm), and this emission scanned from 300 to 370 nm (bandpass, 4 nm). Stock urea solutions were prepared gravimetrically. One-hundred microliters of protein (≈18 μM) in the relevant buffer was added to 800 μl of buffer containing the appropriate concentration of urea. Each sample was incubated at 10°C for at least 5 h before fluorescence was measured with four matched 1-ml cuvettes in thermostatted cuvette holders at the same temperature. The temperature was monitored by a thermocouple in the cell.

The data were fitted to Eq. **1**, which assumes a two-state model in which the fluorescence of the folded and unfolded states is dependent on denaturant concentration (11):

$$F = \frac{(\alpha_N + \beta_N[D]) + (\alpha_D + \beta_D[D])\exp\{m([D] - [D]_{50\%})/RT\}}{1 + \exp\{m([D] - [D]_{50\%}/RT\}} \qquad [1]$$

where F is the fluorescence at the given [denaturant], α_N and α_D are the intercepts, and β_N and β_D are the slopes of the baselines at low (N) and high (D) denaturant concentrations, respectively, [D] is the concentration of denaturant, $[D]_{50\%}$ is the concentration of denaturant at which half of the protein is denatured, m is the slope of the transition, R is the gas constant, and T is the temperature in K. The data were fitted to this equation by nonlinear least squares analysis by using the general curve fit option of the KALEIDAGRAPH program (Abelbeck Software, Reading, PA), which gives the calculated values for individual experimental measurements of m and $[D]_{50\%}$ together with their standard errors.

The free energy of denaturation of proteins in the presence of denaturant $\Delta G^{D}_{D\text{-}N}$ is, to a first approximation, linearly related to the concentration of denaturant (17) (Eq. **2**):

$$\Delta G^{D}_{D\text{-}N} = \Delta G^{H_2O}_{D\text{-}N} - m[D]. \qquad [2]$$

The difference in free energy of denaturation between wild-type and mutant proteins may be calculated from:

$$\Delta\Delta G^{[D]50\%}_{D\text{-}N} = \langle m\rangle\Delta[D]_{50\%} \qquad [3]$$

where $\Delta[D]_{50\%}$ is the difference between the value of $[D]_{50\%}$ for wild-type and mutant and $<m>$ is the average value of $\langle m\rangle$ (12).

RESULTS

DSC. Thermal denaturation of p53 core domain was found to be irreversible by using a range of pH values, buffers, and other reagents and so can be used only as a qualitative guide to stability. Irreversible denaturation occurred during a single DSC scan, and the formation of an aggregate was seen in attempts to measure the T_m by circular dichroism. The DSC exotherms show a strong pH dependence with apparent values of T_m of 42°C in 20 mM sodium phosphate, pH 7.4/150 mM NaCl/1 mM DTT and 36°C in 40 mM Mes, pH 6.0/1 mM DTT. DSC was used qualitatively to examine the DNA binding of p53 core domain (Fig. 1). The addition of a 22-mer double-stranded DNA p53 consensus sequence was found to raise the apparent T_m in 50 mM Tris buffer from 42 to 49°C. This result shows all the protein is able to bind DNA, which has a significant stabilizing effect. This effect was reduced in the presence of 300 mM NaCl.

Fluorescence Spectra. On excitation at 280 nm, the single tryptophan residue in denatured p53 fluoresces strongly at 350 nm, whereas it fluoresces very weakly in native p53. The difference in fluorescence is at a maximum at 356 nm. The fluorescence spectra during urea denaturation in phosphate or Tris buffer show a clear isofluorescent point (around 316–321 nm, dependent on spectrofluorimeter and mutant used), suggesting that only two species are present during denaturation (Fig. 2), apart from when aggregation occurs (see below). The isofluorescent point can be used to normalize fluorescence data for small differences in protein concentration or cuvette size. The weaker fluorescence of tyrosine residues at 305 nm decreases on denaturation.

Zinc Binding During Equilibrium Denaturation. Biochemical studies on the binding of DNA show that DTT is required to maintain p53 in the reduced state (18), so DTT has been used throughout the folding experiments described here. The use of DTT with zinc-binding proteins can create problems, as DTT forms a stable Zn–DTT complex of low solubility (19). But equilibrium measurements on mutants studied here at 10°C and wild type at 25°C appear independent of the concentration of DTT in the range of 40 μM to 40 mM. This suggests that there is no loss of Zn^{2+} ion during denaturation, with the two large loop elements (loops 2 and

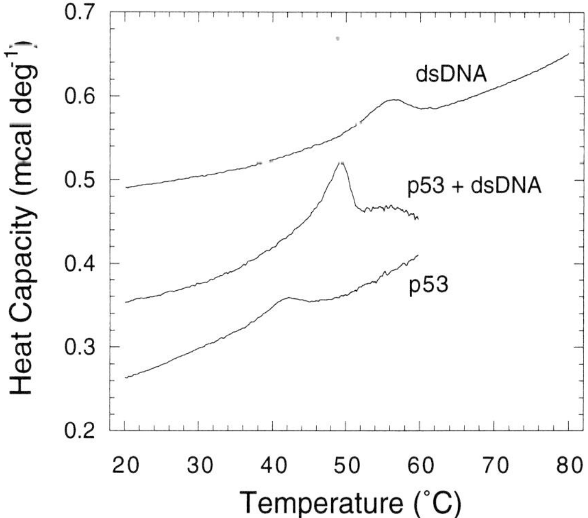

FIG. 1. DNA binding analysis of p53 core domain from DSC. Thermograms were determined in 50 mM Tris, pH 7.2/1 mM DTT of (top trace) double-stranded DNA (dsDNA) oligonucleotide (7 μM) (T_m = 56°C), (middle trace) p53 core domain + dsDNA oligonucleotide equimolar (T_m = 49°C), (bottom trace) p53 core domain (7 μM) (T_m = 42°C) alone. Data are shown without correction for the buffer baseline and are offset for clarity.

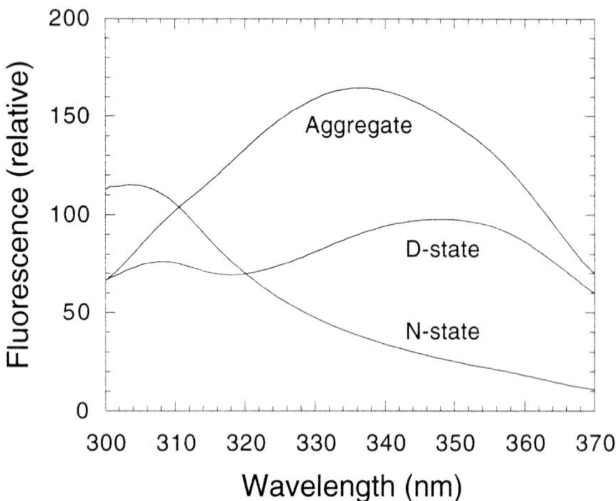

FIG. 2. Fluorescence of p53 core domain in 50 mM sodium phosphate, pH 7.2/5 mM DTT at 25°C on excitation at 280 nm. The native state (N) has a tyrosine emission maximum at 305 nm, which shifts to 310 nm on denaturation. There is a clear tryptophan emission maximum for the denatured state (D) at 350 nm. The tryptophan fluorescence of N is very weak. D and N have the same fluorescence at 321 nm. Aggregated states have a fluorescence maximum at 340 nm; shown is the emission of p53 core domain after being heated to 60°C and cooled to 25°C.

3) binding Zn^{2+} in the denatured state with a higher affinity than DTT. Direct measurement of Zn^{2+} by using a spectrophotometric assay with the metallochromic indicator 4-(2-pyridylazo)resorcinol (20) and mercurial-promoted Zn^{2+} release shows that 1 mol is bound per mol of native core domain and that this remains bound in 5 M urea. The chemical equilibrium is just between the two species at the top of Fig. 3 with no significant accumulation of the other possible equilibrium species.

Aggregation of p53 Core Domain. During urea-induced denaturation at 37°C, or even 25°C in non-phosphate buffers, there is a transient species that has an emission maximum at 340 nm with an intensity greater than that from the denatured state at 356 nm (Fig. 2). Gel filtration revealed this species to be an aggregate. The 340 nm fluorescence is also observed during the thermal unfolding of the protein or under moderately denaturing conditions (1–2 M urea) in the presence of 1 mM EDTA, which chelates Zn^{2+}. A visible aggregate is formed at increased protein concentrations (≈25 μM); the 340 nm fluorescence is removed on centrifugation. The same fluorescence is observed when mutants of the core domain are subjected to milder denaturation conditions.

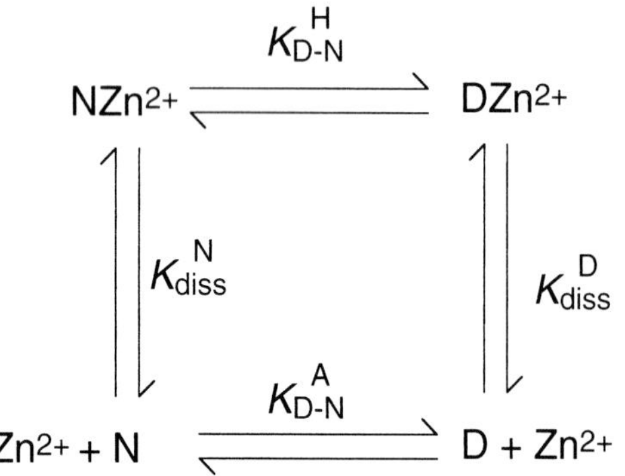

FIG. 3. Scheme for the equilibria between the native and denatured states of p53 core domain with and without bound zinc. The equilibrium that has been measured is shown by $K^{H}_{D\text{-}N}$.

Urea Denaturation. The denaturation of wild-type and mutant p53 core domains was measured from the changes in fluorescence on the addition of urea. A complete emission spectrum was recorded at each concentration of urea so denaturation curves could be analyzed from changes in tryptophan and tyrosine fluorescence, any aggregation could be detected at 340 nm, and the isofluorescent point checked. There was excellent agreement between the changes at 305 nm for tyrosines and 356 nm for the lone tryptophan (which is 25% solvent exposed), showing a cooperative transition. The denaturation curves are entirely reversible at 10°C, identical results being obtained on the renaturation of a denatured sample. Therefore, equilibrium measurements can be made from denaturation curves, which fit to a two-state transition (Fig. 4). Denaturation of wild-type protein is also fully reversible at 25°C, but a standard temperature of 10°C was chosen because some mutants do not denature reversibly at higher temperatures. R175H appears to be at the stability limit for our folding studies because of its degree of degradation on purification. Gel filtration was used as a third column step to purify folded R175H from smaller fragments, although resolution was not complete. The raw data at 356 nm without normalization are shown in Fig. 4 and give a good measurement of $[urea]_{50\%}$. Results of denaturation are presented in Table 1.

DISCUSSION

We have found conditions to measure the reversible denaturation of p53 core domain and many of its mutants with urea denaturation. Importantly, a single Zn^{2+} ion is bound throughout the reaction, because its loss would generate denaturation curves that are very difficult to analyze because of changes in molecularity. Aggregation was observed under certain conditions, monitored from fluorescence at 340 nm. We suggest the 340 nm fluorescence is the signature of the aggregated state of p53 core domain. At 10°C, aggregation is avoided and unfolding is entirely reversible with denaturation curves starting from folded or unfolded protein overlaying to give the same *m* and $[urea]_{50\%}$ values (Eq. **1**), thus allowing the free energy of unfolding $\Delta G_{D\text{-}N}$, to be calculated. Most proteins are stable by some 5–15 kcal·mol^{-1} at 25°C (21). Here, we show that the isolated core domain is of moderate thermodynamic stability ($\Delta G^{H_2O}_{D\text{-}N}$ = 6.0 kcal·mol^{-1}, 25°C). The domain will be even less stable at 37°C so only small changes in stability will lead to loss of function.

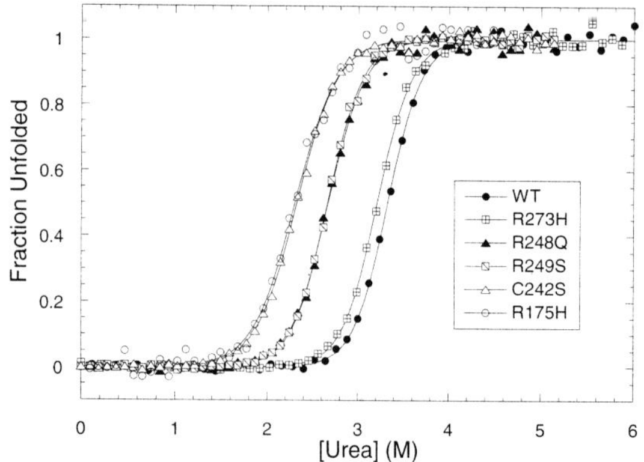

FIG. 4. Urea denaturation of wild-type (WT) and mutant p53 core domain monitored by normalized fluorescence emission at 356 nm (data without normalization are shown for R175H; excellent data are obtained in general without normalization). Data are plotted as fraction unfolded and fitted to Eq. **1**. Protein was buffered in 50 mM sodium phosphate, pH 7.2/5 mM DTT. An excitation wavelength of 280 nm was used.

Table 1. Equilibrium denaturation of p53 core domain

p53 variant*	m, kcal·mol^{-1}·M^{-1}	[Urea]$_{50\%}$, M	$\Delta G_{D\text{-}N}^{H_2O}$†, kcal·mol^{-1}	$\Delta\Delta G_{D\text{-}N}^{3M\ urea}$†, kcal·mol^{-1}
Wt, 25°C (1)	−2.24 ± 0.008	2.66 ± 0.01	5.96 ± 0.20	
Wt, 10°C (2)	−2.97 ± 0.18	3.33 ± 0.01	9.76 ± 0.24	
R175H, 10°C (1)	−2.59 ± 0.19	2.30 ± 0.02	6.75 ± 0.17	3.01 ± 0.08
C242S, 10°C (2)	−2.68 ± 0.01	2.33 ± 0.01	6.82 ± 0.16	2.94 ± 0.06
R248Q, 10°C (1)	−2.91 ± 0.15	2.67 ± 0.01	7.82 ± 0.19	1.94 ± 0.05
R249S, 10°C (3)	−3.09 ± 0.09	2.66 ± 0.01	7.81 ± 0.19	1.95 ± 0.04
R273H, 10°C (1)	−3.11 ± 0.10	3.21 ± 0.01	9.41 ± 0.23	0.35 ± 0.04

*The number of determinations is shown in parentheses. Data are from normalized fluorescence emission at 356 nm (except for R175H). The standard errors of the data are shown. [Urea]$_{50\%}$ values can be determined with a high degree of reproducibility but m values are hard to determine accurately, particularly where the [urea]$_{50\%}$ value is low (12). Protein was buffered in 50 mM sodium phosphate, pH 7.2/5 mM DTT. Wt, wild type.

†The mean m value of 10 urea denaturation curves at 10°C is −2.93 (±0.07) kcal·mol^{-1}·M^{-1}, which was used to calculate $\Delta G_{D\text{-}N}^{H_2O}$ and $\Delta\Delta G_{D\text{-}N}^{3M\ urea}$ from Eq. **2** and Eq. **3**, respectively. The values of $\Delta\Delta G_{D\text{-}N}^{3M\ urea}$ are calculated for 3 M urea because there is only a short extrapolation of the data and so the standard error is small. There are higher standard errors at 0 M urea. Values of $\Delta\Delta G_{D\text{-}N}$ tend to change only slightly with temperature.

Mutants of p53 Core Domain Are Destabilized. Mutants of p53 core domain, corresponding to known "hot spots" associated with tumors, were constructed and their stabilities were determined (Fig. 5). Some of the residues mutated interact directly with DNA whereas others are part of the protein scaffold ("structural" mutants).

The stabilities fall into three classes. The DNA-binding mutant R273H is of similar stability to wild type, whereas C242S and R175H, which surround the Zn^{2+}-binding site, are destabilized by 2.9 and 3.0 kcal·mol^{-1}, respectively. R248Q and R249S are of intermediate stability. These results are qualitatively consistent with studies on antibody binding (8, 9). R175H reacts strongly with an antibody specific for the denatured p53 and does not react with an antibody specific for the folded state of the domain; and, conversely, R273H has the opposite specificity.

The distinction between structural (R249S) and DNA contact mutations (R248Q) is shown here to be less straightforward than previously thought. R248Q was presumed to have no role in stability yet fits both categories and demonstrates the general trend that the degree of destabilization is more severe the closer residues lie to the Zn^{2+}-binding site. A spectrophotometric assay (20) shows that C242S retains a bound Zn^{2+} ion throughout denaturation despite the loss of one cysteine ligand. The replacement of arginine with histidine at residue 175 creates the possibility of competition between Cys-176 and His-175 for Zn^{2+} ligation, which could create a His_2-Cys_2 motif. The results show the importance of residues in loops 2 and 3 and in particular those at the Zn^{2+}-binding site. These residues have been shown to form a network of salt bridges, hydrogen bonds, and van der Waals contacts, which help present DNA binding residues for their major and minor groove binding sites.

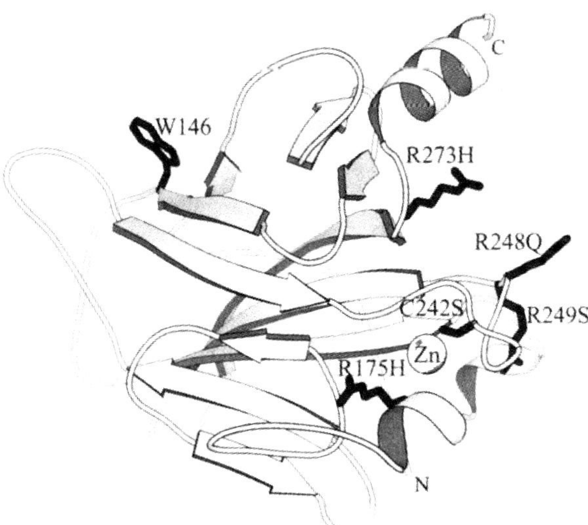

FIG. 5. MOLSCRIPT (22) picture of p53 core domain. Residues mutated are labeled and shown as stick models as is Trp-146, which was used as a fluorescent probe for denaturation.

Implications for Drug Therapy. Although the mutants R175H, C242S, R248Q, R249S, and R273H are less stable (Table 1), they do retain cooperative two-state unfolding transitions (Fig. 4), demonstrating defined stable structures at 10°C. The degree of destabilization of mutants reported here is consistent with protein engineering studies of other proteins (12). The changes in stability are sufficiently small to allow possible therapeutic use of small molecules to rescue p53 function by stabilizing it. Any protein-stabilizing drug for effective cancer treatment must, however, also create the correct geometry at the DNA binding surface as many mutants do not show wild-type specificity or binding levels even when folded at low temperature (9).

We thank Dr. Mark Bycroft for helpful advice. This work was supported by the Cancer Research Campaign of the United Kingdom. J.H. holds a Kekulé scholarship, Fonds der Chemischen Industrie, Germany. B.S.D. is a recipient of a Burroughs Wellcome Fund Hitchings–Elion Fellowship, and P.V.N. is a postdoctoral fellow of the Medical Research Council of Canada.

1. Lane, D. P. (1992) *Nature (London)* **358,** 15–16.
2. Kastan, M. B., Onyekwere, O., Sidransky, D., Vogelstein, B. & Craig, R. W. (1991) *Cancer Res.* **51,** 6304–6311.
3. Polyak, K., Xia, Y., Zweier, J. L., Kinzler, K. W. & Vogelstein, B. (1997) *Nature (London)* **389,** 300–305.
4. Ciechanover, A., Shkedy, D., Oren, M. & Bercovich, B. (1994) *J. Biol. Chem.* **269,** 9582–9589.
5. Haupt, Y., Maya, R., Kazaz, A. & Oren, M. (1997) *Nature (London)* **387,** 296–299.
6. Pavletich, N. P., Chambers, K. A. & Pabo, C. O. (1993) *Genes Dev.* **7,** 2556–2564.
7. Cho, Y., Gorina, S., Jeffrey, P. D. & Pavletich, N. P. (1994) *Science* **265,** 346–355.
8. Rolley, N., Butcher, S. & Milner, J. (1995) *Oncogene* **11,** 763–770.
9. Friedlander, P., Legros, Y., Soussi, T. & Prives, C. (1996) *J. Biol. Chem.* **271,** 25468–25478.
10. Milner, J., Cook, A. C. & Sheldon, M. (1987) *Oncogene* **1,** 453–455.
11. Clarke, J. & Fersht, A. R. (1993) *Biochemistry* **32,** 4322–4329.

12. Serrano, L., Kellis, J. T., Jr., Cann, P., Matouschek, A. & Fersht, A. R. (1992) *J. Mol. Biol.* **224,** 783–804.
13. Shortle, D., Stites, W. E. & Meeker, A. K. (1990) *Biochemistry* **29,** 8033–8041.
14. Santoro, M. M. & Bolen, D. W. (1992) *Biochemistry* **31,** 4901–4907.
15. Midgley, C. A., Fisher, C. J., Bartek, J., Vojtesek, B., Lane, D. & Barnes, D. (1992) *J. Cell Sci.* **101,** 183–189.
16. Gill, S. C. & von Hippel, P. H. (1989) *Anal. Biochem.* **182,** 319–326.
17. Pace, C. N. (1986) *Methods Enzymol.* **131,** 266–279.
18. Rainwater, R., Parks, D., Anderson, M. E., Tegtmeyer, P. & Mann, K. (1995) *Mol. Cell. Biol.* **15,** 3892–3903.
19. Cornell, N. W. & Crivaro, K. E. (1972) *Anal. Biochem.* **47,** 203–208.
20. Hunt, J. B., Neece, S. H. & Ginsburg, A. (1985) *Anal. Biochem.* **146,** 150–157.
21. Fersht, A. R. & Serrano, L. (1993) *Curr. Opin. Struct. Biol.* **3,** 75–83.
22. Kraulis, P. J. (1991) *J. Appl. Crystallogr.* **24,** 946–950.

The finding that about 30% of oncogenic mutants of p53 are simply temperature sensitive implies that there should be drugs that can bind to and stabilise p53 in many cancers[202,203]. These compounds would constitute a novel class of anti-cancer drugs. So, we set out to prove this in principle. An Israeli post-doctoral who had been trained in peptide chemistry, Assaf Friedler, was set the task of finding a peptide that would bind to and stabilise p53[204]. In collaboration with Swedish colleagues, Galina Selivanova and Klas Wiman, we found that our peptide had weak activity in a cancer cell line[205]. These exciting results set up a whole new career opportunity, which I grabbed with both hands: designing novel anti-cancer drugs based on thermodynamic, biophysical and structural studies.

18.4 2008: 65th Birthday Meeting in Shanghai. *Standing*: Eviatar Natan, Sidhar Rajagopalan, Alessandro Vezzoli, Grace Yu, Maria García-Alai, Fang Huang, Sarah Burge, Kian-Hoe Khoo, Marina Vaysburd, Frank Boeckler and Andreas Joerger. *Sitting*: Marilyn and Alan.

A peptide that binds and stabilizes p53 core domain: Chaperone strategy for rescue of oncogenic mutants

Assaf Friedler, Lars O. Hansson, Dmitry B. Veprintsev, Stefan M. V. Freund, Thomas M. Rippin, Penka V. Nikolova, Mark R. Proctor, Stefan Rüdiger, and Alan R. Fersht*

Cambridge University Chemical Laboratory and Cambridge Centre for Protein Engineering, Medical Research Council Centre, Hills Road, Cambridge CB2 2QH, United Kingdom

Contributed by Alan R. Fersht, November 27, 2001

Conformationally compromised oncogenic mutants of the tumor suppressor protein p53 can, in principle, be rescued by small molecules that bind the native, but not the denatured state. We describe a strategy for the rational search for such molecules. A nine-residue peptide, CDB3, which was derived from a p53 binding protein, binds to p53 core domain and stabilizes it *in vitro*. NMR studies showed that CDB3 bound to p53 at the edge of the DNA binding site, partly overlapping it. The fluorescein-labeled peptide, FL-CDB3, binds wild-type p53 core domain with a dissociation constant of 0.5 μM, and raises the apparent melting temperatures of wild-type and a representative oncogenic mutant, R249S core domain. gadd45 DNA competes with CDB3 and displaces it from its binding site. But this competition does not preclude CDB3 from being a lead compound. CDB3 may act as a "chaperone" that maintains existing or newly synthesized destabilized p53 mutants in a native conformation and then allows transfer to specific DNA, which binds more tightly. Indeed, CDB3 restored specific DNA binding activity to a highly destabilized mutant I195T to close to that of wild-type level.

More than 50% of human cancers have missense mutations in the gene coding for the tumor suppressor p53 that result in its inactivation (1). Nearly all such mutations are in the DNA-binding core domain (p53C) (1). The six most frequent cancer-associated mutations are the "hot-spots" R175H, G245S, R248Q, R249S, R273H, and R282W. These mutations can be divided into two categories: (*i*) DNA-contact mutations (R248 and R273) that result in loss of DNA-binding residues, and (*ii*) "structural mutations" that result in structural changes in p53C (2). An assessment of the mutation database (3, 4), based on thermodynamic stability and DNA binding properties of the mutants, classifies three broad phenotypes: (*i*) DNA-contact mutations that have little effect on folding or stability (e.g., R273H); (*ii*) mutations that cause a local distortion, mainly in proximity to the DNA-binding site (e.g., R249S), which are usually destabilized by <2 kcal/mol; and (*iii*) mutations that cause global unfolding (e.g., mutations in the β sandwich) that are destabilized by >3 kcal/mol.

Owing to the large frequency of p53 mutants in cancers, a promising strategy in cancer therapy is rescue of p53 mutants and restoration of the tumor suppression activity (reviewed in refs. 4 and 5). Different classes of mutants require different rescue strategies. DNA contact mutants need the introduction of functional groups that will establish new contacts with the DNA, compensating for the missing contacts. Globally unfolded mutants could be rescued by stabilizing agents that will lead to refolding of the mutant. In principle, any molecule that will bind the native, but not the denatured state, will cause the equilibrium to be shifted toward the native state. A small molecule, CP-31398, has been found by random screening to bind and stabilize the core domain and has been reported to be an effective anticancer agent (6).

Here, we test the feasibility of stabilizing p53C by small peptides that bind its native state and which may act as leads for drugs. Such peptides may be found rationally by examining known complexes of p53 with its binding proteins, and so avoid random screening. This strategy has been neglected because of the likelihood of the inhibition of the natural activity of p53 by the small molecule. Nevertheless, we have found a small peptide that did stabilize p53 by binding at the edge of its DNA-binding site, and could restore sequence-specific DNA-binding activity to the highly destabilized mutant I195T to near wild-type levels.

Materials and Methods

Peptide Synthesis. The peptides were synthesized using a 432A Synergy peptide synthesizer (Applied Biosystems), using standard Fmoc chemistry, and were purified using reverse-phase HPLC [Waters 600 equipped with a 996 PDA detector and a preparative reverse-phase C8 column (Vydac, Hesperia, CA)]. The gradient was 100%A to 100%B in 35 min (A = 0.1% TFA in water, B = 95% acetonitrile/5% water/0.1%TFA). The purified peptides were characterized by matrix-assisted laser desorption ionization–time-of-flight (MALDI-TOF) MS and had the expected M_r. The fluorescein-labeled core domain binding peptide FL-CDB3 was purchased from Graham Bloomberg (University of Bristol, U.K.).

Protein Expression and Purification. Human p53C wild type and mutants (residues 94–312) and human tetrameric p53 (residues 94–360) were cloned, expressed, and purified as described (7). ^{15}N-labeled human p53C was produced as described (8).

NMR Spectroscopy. Samples for NMR experiments contained ^{15}N-labeled p53C at a concentration of 225 μM and the corresponding CDB peptide at a final concentration of 2–2.5 mM in 150 mM KCl, 5 mM DTT, and 5% D_2O in 25 mM sodium phosphate buffer, pH 7.2. All ^{1}H–^{15}N heteronuclear sequential quantum correlation (HSQC) spectra were acquired as described (8).

Binding Studies by Surface Plasmon Resonance. A BIACORE 2000 equipped with a sensor chip SA (BIAcore, Uppsala, Sweden) was used to screen the peptides for p53 core domain binding, and to quantify the binding of p53 core domain to peptide CDB3. Biotinylated CDB peptides were immobilized and the binding of p53 was studied. Immobilization and binding measurements were performed at 10°C (unless otherwise stated) with 50 mM Hepes (pH 7.2)/5 mM DTT as running buffer. The streptavidin surface of the chip was activated as described by the manufacturer. Biotinylated peptides (1.5–4.0 mM peptide in buffer containing 0.13 M NaCl) were immobilized at 5 μl/min until the level of saturation. Flow cell 1 was used as a background for the change in bulk refractive index.

To screen for binding to immobilized peptides, we measured

Abbreviations: CDB, core domain binding; FL, fluorescein-labeled; p53C, p53 core domain.

*To whom reprint requests should be addressed. E-mail: arf25@cam.ac.uk.

The publication costs of this article were defrayed in part by page charge payment. This article must therefore be hereby marked "*advertisement*" in accordance with 18 U.S.C. §1734 solely to indicate this fact.

the association of p53 (0.36–18 μM) for 15 min at 10 μl/min. Bound protein was dissociated by a regeneration cycle of 1–3 min with 1 M NaCl.

The binding affinity of p53 for immobilized CDB3 was estimated from the half saturation concentration of binding isotherm with varying concentrations of p53 core domain (0.019–0.19 μM). The binding association was measured for 5 min at 30 μl/min and 20°C. A two-state equation was fitted to the relative responses versus the logarithm of the p53 concentrations (KALEIDAGRAPH, Abelbeck Software, Reading, PA).

The binding affinity of soluble, unlabeled CDB3 was determined by competition experiments (9). The samples contained 0.20 μM p53 with various concentrations of CDB3 (0.030–120 μM). Association data were collected after 1 h of incubation at 20°C (5 min, 30 μl/min). Control samples containing p53 only were run as references. The initial association rate (over the first 150 s) was estimated by fitting a linear equation to data (BIAEVALUATION 3.1, BIAcore AB, Uppsala, Sweden). The data were analyzed according to a 1:1 binding model (9), using KALEIDAGRAPH. Control experiment showed that association rate of binding is proportional to the concentration of p53 (0.19–1.9 μM).

Fluorescence Anisotropy Measurements. All anisotropy measurements were performed with fluorescein-labeled CDB3 (FL-CDB3, sequence FL-REDEDEIEW-NH_2) at 10°C, using a Perkin–Elmer LS-50b luminescence spectrofluorimeter equipped with a Hamilton microlab M dispenser controlled by laboratory software. The peptide ($\approx$5 μM, 900 μl) was dissolved in 50 mM Hepes buffer (pH 7.2)/5 mM DTT. Fluorescence anisotropy was measured on excitation at 480 nm (bandwidth 8 nm) and emission at 525 nm (bandwidth 2.5 nm).

To determine the dissociation constant for FL-CDB3 complexed with various p53C constructs FL-CDB3 (900 μl, $\approx$5 μM) was placed in the cuvette and the appropriate p53C construct (240 μl, $\approx$50 μM) was placed in the dispenser. Additions of 3 μl of protein were titrated into the peptide solution about every 1 min, the solution was stirred for 30 s, and the anisotropy was measured. Dissociation constants for the FL-CDB3-p53C complex were calculated by fitting the anisotropy and fluorescence titration curves (corrected for dilution) to a simple 1:1 equilibrium model (unpublished data).

DNA-binding experiments were performed similarly with 140 mM NaCl added to the buffer. Fluorescein-labeled gadd45 30-mer DNA (15 nM) and fluorescein-labeled random DNA (7 nM) were used.

Anisotropy was measured in competition experiments to study (indirectly) how CDB3 derivatives or gadd45 30-mer DNA compete with the fluorescein-labeled CDB3 for the binding site of p53 core domain (wavelengths as above, slit widths of excitation10 nm, and emission 8 nm). A stock solution of unlabeled CDB3 was titrated into a cuvette containing 900 μl 2.0 μM p53 and 0.50 μM FL-CDB3. For DNA, stock solutions of 5 and 25 μM were used.

The concentrations of p53-FL-CDB3 complex and free FL-CDB3 before addition of competitor were calculated using the K_d of 0.53 μM. The change in concentration of free FL-CDB3 after addition of competitor peptide aliquot was derived from the change in the anisotropy. The concentration of CDB3-p53C complex was calculated from the total concentration of p53C. When unlabeled CDB3 was in large excess over p53C, the fraction of free peptide could be approximated to its total concentration and K_d was derived from the equilibrium equation.

Differential Scanning Calorimetry (DSC). DSC experiments were performed using a Microcal VP-DSC microcalorimeter (Microcal, Amherst, MA). Temperatures from 5–95°C were scanned at a rate of 60°/h, using a Hepes buffer (pH 7.2)/1 mM DTT, which also served for baseline measurements. Samples of wild-type and mutant p53C (6–15 μM) in the presence or absence of FL-CDB3 (15–80 μM) in the above buffer were prepared and then degassed for 15 min before each experiment. A pressure of 25 psi (1.56 atm) was applied to the cell. The data were analyzed using ORIGIN software (Microcal).

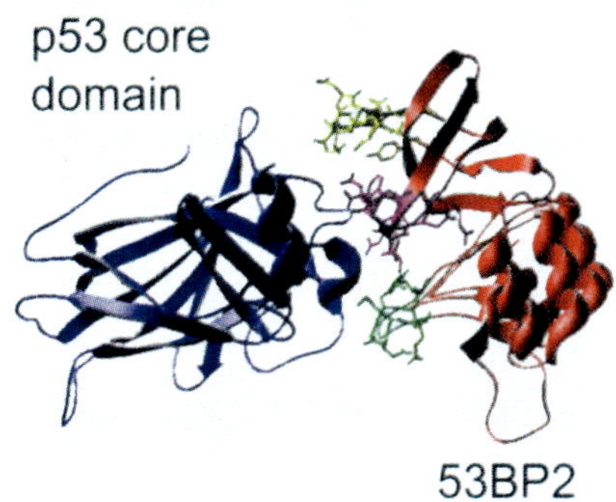

Fig. 1. Crystal structure of the p53C (blue)–53BP2 (red) complex (coordinates taken from ref. 10) with the three 53BP2-derived peptides synthesized for this study highlighted: CDB1 (residues 422–428), green; CDB2 (residues 469–477), yellow; CDB3 (residues 490–498), purple. The picture was generated using SWISSPDB VIEWER (23).

Results

Design of Potential p53C Binding Peptides. A prime source of peptides that could bind the native state of p53 is from p53-binding proteins. A rare example of a complex of a protein-p53 that has been solved at high resolution is the p53C-53BP2 complex (10). 53BP2 is a p53 binding protein (11) that enhances p53-mediated transactivation, and specifically stimulates the apoptotic function of p53 (12–14). 53BP2 binds p53C in its DNA binding site, with three loops making the contacts with p53 (ref. 10; Fig. 1). Three peptides corresponding to these three loops were synthesized and tested (CDB1–CDB3; see Table 1).

A second potential source for core domain binding peptides are sequences within p53 itself that bind the core domain and regulate its activity. Two such regions within p53 are the C-terminal domain (amino acids 363–393; ref. 15) and the proline-rich domain (amino acids 54–94; ref. 16). Several overlapping peptides corresponding to both regions were synthesized (CDB4 and CDB7–CDB10 in Table 1). Because Ser-378 within the C-terminal domain is known to undergo phosphorylation (17), phosphopeptides derived from this region were also synthesized (CDB5 and CDB6 in Table 1). It was suggested that the C-terminal and the proline-rich domains can bind the core domain only in presence of each other (18), and thus a fusion peptide between these domains was also designed (CDB11 in Table 1).

Screening of the CDB Peptides for Binding p53C. We used heteronuclear NMR spectroscopy for initial screening for binding of the peptides to p53C to monitor any changes in the backbone ^{1}H and ^{15}N resonances of ^{15}N-labeled p53C (8). Chemical shift changes were observed only with CDB2 and CDB3, implicating binding of only these peptides to p53C. We used surface plasmon resonance for initial estimates of the affinity for p53C. Peptides CDB1, -2, -3, -9, and -11 were resynthesized with a biotin label attached to their N terminus, and immobilized onto a streptavidin (SA) sensor chip. p53C had the tightest binding to CDB3, in good agreement with the NMR data. The concentration of

Table 1. Peptides tested for binding p53 core domain

Peptide	Derived from	Sequence
Peptides derived from 53BP2		
CDB1	53BP2 422–428	MTYSDMQ-NH_2
CDB2	53BP2 469–477	YEPQNDDEL-NH_2
CDB3	53BP2 490–498	REDEDEIEW-NH_2
Peptides derived from p53		
CDB4	p53 81–100	TPAAPAPAPSWPLSSSVPSQ-NH_2
CDB5	phospho-Ser-378 p53 369–383	LKSKKGQSTpSRHKKL-NH_2
CDB6	phospho-Ser-378 p53 361–383	GSRAHSSHLKSKKGQSTpSRHKKL-NH_2
CDB7	p53 369–383	LKSKKGQSTSRHKKL-NH_2
CDB8	p53 361–383	GSRAHSSHLKSKKGQSTSRHKKL-NH_2
CDB9	p53 81–94	TPAAPAPAPSWPLS-NH_2
CDB10	p53 76–94	APAAPTPAAPAPAPSWPLS-NH_2
CDB11	fusion p53 82–94 and 369–383 with a GG linker	PAAPAPAPSWPLSGGLKSKKGQSTSRHKKL-NH_2

p53C for 50% binding was estimated to be 200 nM. There was no significant binding to CDB1 or CDB9. CDB3 was chosen as a lead peptide for further experiments.

Characterization of CDB3-p53C Binding Site. Chemical shift differences between the spectra of the bound and unbound p53C were used to identify the site in p53C where CDB3 bound. Changes of backbone ^1H–^{15}N resonances for each residue between the bound and the unbound states were found mainly in loop 1, helix 2, and strand 8 (Fig. 2), which are located at one edge of the DNA-binding site. The residues that had a significant chemical shift (above five times the standard deviation: $\Delta\partial > 0.25$ ppm for ^{15}N and $\Delta\partial > 0.05$ ppm for ^1H, color-coded blue) were: F113, H115, G117, T118, V122, and T123 (from loop 1), H233 (from strand 8), G279 and R280 (both from helix 2), Y126, H178, H296, and E298 (from the rest of the protein). Residues that revealed moderate chemical shifts ($\Delta\partial$ differences between 2.5 times and 5 times the standard deviation: $0.125 < \Delta\partial < 0.25$ ppm for ^{15}N, and $0.025 < \Delta\partial < 0.05$ ppm for ^1H, color-coded purple) were: L114 and C124 (from loop 1), C229, T230, and T231 (all from strand 8), R282 and R283 (from helix 2), Y103, N131, T140, V143, C176, H179, I195, G199, R202, N239, I251, R267, C277, K291, and L299 (from the rest of the protein). CDB3, as a free peptide (color-coded red in Fig. 2), bound p53C in a different location from that of the original loop within 53BP2, which binds in the middle of the DNA-binding site (between loop 3 and the other side of helix 2; ref. 10).

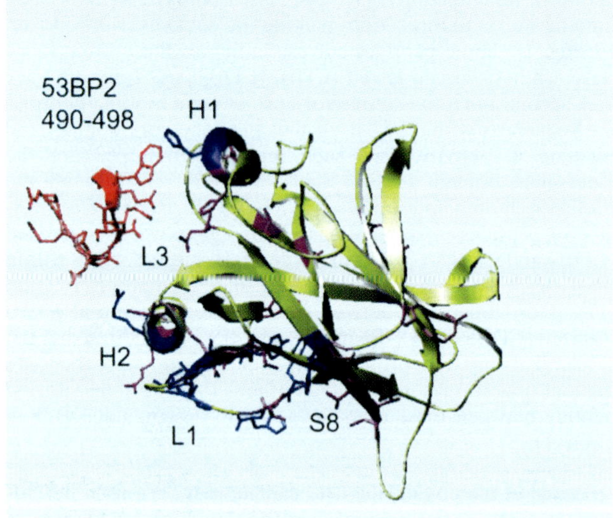

Fig. 2. Chemical shift changes ($\Delta\partial$) in p53C on binding to CDB3. Deviations above five times the standard deviation ($\Delta\partial > 0.25$ ppm for ^{15}N and $\Delta\partial > 0.05$ ppm for ^1H) were considered significant (color-coded blue). $\Delta\partial$ differences between 2.5 times and 5 times the standard deviation ($0.125 < \Delta\partial < 0.25$ ppm for ^{15}N, $0.025 < \Delta\partial < 0.05$ ppm for ^1H) were considered as minor (color-coded purple), and $\Delta\partial$ differences below 2.5 times the standard deviation ($\Delta\partial < 0.125$ ppm for ^{15}N and $\Delta\partial < 0.025$ ppm for ^1H) were considered insignificant (color-coded yellow). See text for residue number details. CDB3 in its original position in the 53BP2-p53 complex is shown in red (coordinates taken from ref. 10). The picture was generated using SWISSPDB VIEWER.

Fluorescence anisotropy titrations were used to determine accurately the dissociation constant for the p53C–CDB3 complex at 10°C. p53C (residues 94–312) was titrated into fluorescein-labeled CDB3 (FL-CDB3) and changes in anisotropy of the labeled peptide (Fig. 3) as well as the total fluorescence at 525 nm were monitored. The initial anisotropy value for the labeled peptide was 0.04, and the limiting value for the FL-CDB3-p53C complex was 0.20. The binding curve was fitted to a 1:1 simple equilibrium model, and K_d was found to be 0.53 ± 0.09 μM (Table 2). To confirm that FL-CDB3 binds tetrameric p53, and not only isolated core domain, K_d for the binding to the tetrameric p53 construct (residues 94–360) was determined in the same way and was found to be 0.77 ± 0.09 μM (data not shown).

The binding of FL-CDB3 to two p53C mutants was measured: to G245S, which is 95% folded at 37°C and is destabilized by 1.21 kcal/mol at 10°C; and to R249S, which is 85% folded at 37°C and is distorted and destabilized by 1.92 kcal/mol at 10°C (3, 7). At 10°C both mutants are expected to be in a native-like conformation. K_d values, from fluorescence anisotropy (see Fig. 3), were 0.57 ± 0.09 μM for G245S and 3.3 ± 0.5 μM for R249S.

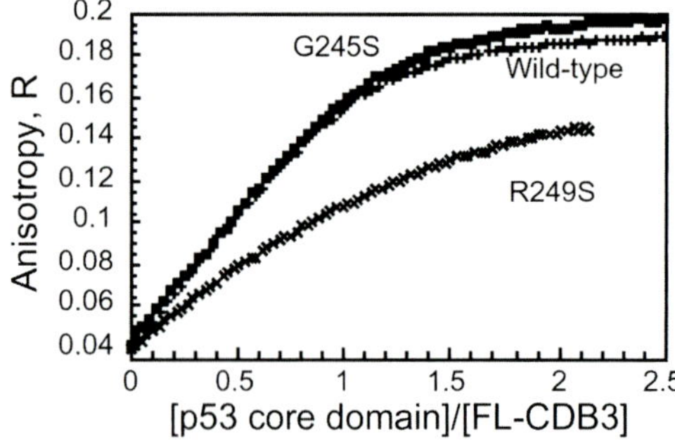

Fig. 3. CDB3 binding to p53C analyzed by fluorescence anisotropy. Wild-type and mutant p53C were titrated into a fluorescein-labeled CDB3 (4.6 μM; see *Materials and Methods* for details).

Table 2. K_d values for FL-CDB3 binding to wild-type and mutant p53

Protein	Conditions	K_d, μM
Wild-type core (94–312)		0.53 ± 0.09
Wild-type core + tet (94–363)		0.77 ± 0.09
Wild-type core	4 M urea	61 ± 10
Wild-type core	2 M Gdm Cl	>1,000
G245S (94–312)		0.57 ± 0.09
G245S (94–312)	4 M urea	39 ± 4
G245S (94–312)	2 M Gdm Cl	>1,000
R249S (94–312)		3.3 ± 0.5

K_d values were determined from the anisotropy and fluorescence at 525 nm following titration of p53 into fluorescein-labelled CDB3. The temperature was 10°C and the buffer was 50 mM Hepes (pH 7.2)/5 mM DTT.

Tighter Binding of Fluorescein-Labeled CDB3. To determine whether attaching the different labels (fluorescein and biotin) to CDB3 N terminus alters K_d, we measured the dissociation constants for the unlabeled peptide by competition BIAcore (Fig. 4*b*) and competition with FL-CDB3 analyzed by anisotropy (Fig. 4*a*). The unlabeled peptide had a K_d of 37 μM (Fig. 4*a*). Biotinylation of the N terminus improved the affinity (compared with the unlabeled peptide) 3-fold for solution measurements (K_d = 12 μM; see Fig. 4*a*) and even more for the immobilized sample – BIAcore assays (apparent K_d = 0.2 μM; data not shown). Perhaps the unlabeled peptide was bound more weakly because of the positive charge on its N-terminal α-amino group, which

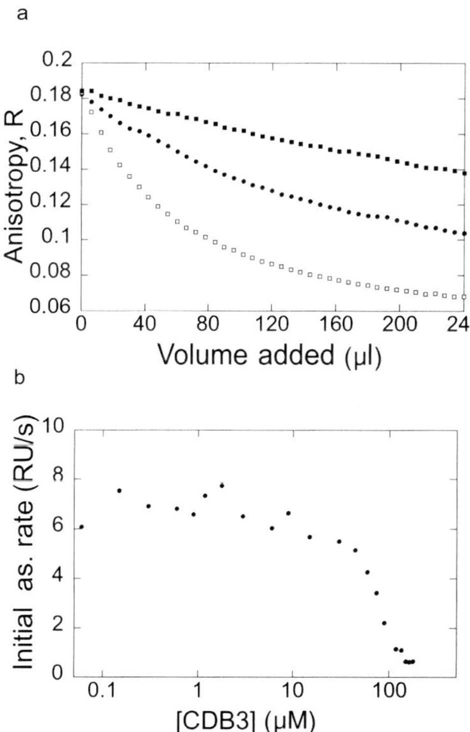

Fig. 4. Binding competition experiments. (*a*) Competition experiment where unlabeled or biotinylated CDB3 were titrated into 0.50 μM fluorescein-labeled CDB3 and 2.0 μM p53C wild-type (■ and □, 0.26 mM and 2.6 mM unlabeled CDB3, respectively, and ●, 0.24 mM biotinylated CDB3). (*b*) Titration of CDB3 binding to p53C by Competition BIAcore. The concentration of free p53C (reflected by association rate in binding to immobilized CDB3) was analyzed by BIAcore after incubation of 0.2 μM p53C and various concentrations of free CDB3.

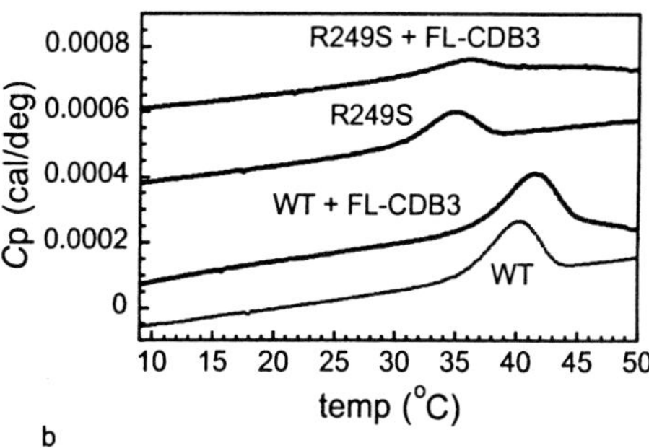

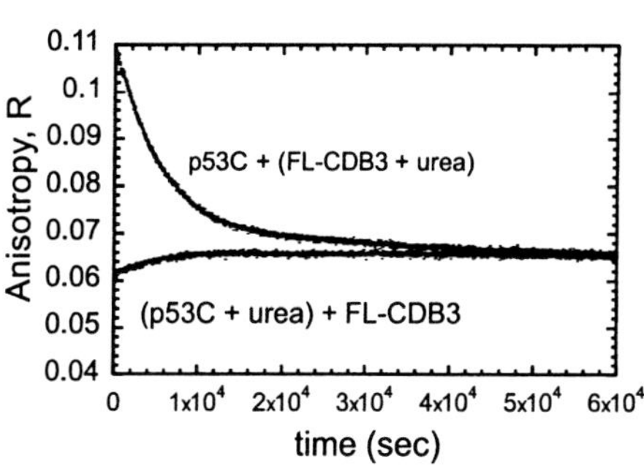

Fig. 5. Stabilization of p53C by FL-CDB3. (*a*) Differential scanning calorimetry (irreversible denaturation). The apparent T_m of wild-type and R249S core domain in the presence or absence of FL-CDB3 was determined as described in *Materials and Methods*. For the wild-type core domain, T_m = 40.1°C in the absence of the peptide and 41.6°C in its presence. For R249S, T_m = 34.9°C in the absence of the peptide and 35.9°C in its presence. Raw data are shown and are offset for clarity. (*b*) Time-dependent anisotropy studies of the binding of CDB3 to wild-type p53C in presence of 3 M urea (reversible denaturation). Wild-type p53C (5 μM) was preincubated overnight with 3 M urea, then mixed with FL-CDB3 (5 μM), and the anisotropy change over time was monitored. As a control, the same protein (5 μM) was mixed with 3 M urea and with FL-CDB3 (5 μM) without preincubation and anisotropy changes over time were monitored.

might cause an electrostatic repulsion form the positively charged protein surface. Attaching a label on the N terminus of the peptide eliminates this charge and might add hydrophobic interactions that improve binding.

FL-CDB3 Stabilized p53C and Raised Its Apparent T_m. Differential scanning calorimetry (DSC) was used to detect stabilization of p53C by FL-CDB3. The thermal denaturation of p53C is irreversible, and thus only an apparent melting temperature (T_m) can be determined (7), but increase in stability can be correlated with increase in the apparent T_m. The apparent T_m of wild-type p53C was 40.1°C, and increased by 1.5° in the presence of the peptide FL-CDB3 (Fig. 5*a*). The apparent T_m of the mutant R249S was raised from 34.9°C to 35.9°C by the peptide.

FL-CDB3 Induced Refolding of Equilibrium-Denatured p53C. The ability of FL-CDB3 to refold p53C under equilibrium denaturation conditions was monitored using fluorescence anisotropy. p53C was incubated overnight at 10°C in 3 M urea, under which conditions it is predominantly denatured. Then it was mixed with FL-CDB3 in 3 M urea, and the changes in anisotropy over time were monitored (Fig. 5*b*). The initial anisotropy value for the labeled peptide was 0.04. Upon mixing the peptide with p53C a rapid binding event took place, leading to the formation of a FL-CDB3-p53C complex. The anisotropy values for the complex following preincubation overnight with 3 M urea were 0.06–0.07, far below the limiting anisotropy value for the bound complex at

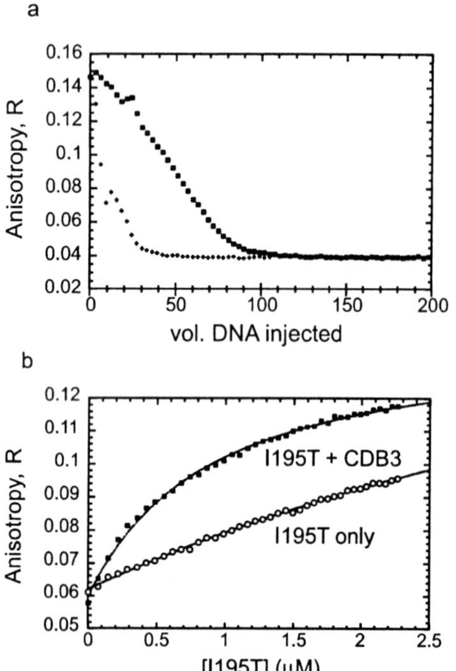

Fig. 6. Effect of CDB3 on DNA binding of p53C. (*a*) DNA competes with FL-CDB3 on p53C binding. 30-mer gadd45 DNA (+, 25 μM; □, 5 μM) was titrated into a mixture of p53C-FL-CDB3 as described in *Materials and Methods*. (*b*) CDB3 restores DNA binding to the I195T mutant. I195T (10 μM) was preincubated for 1 h in the presence (■) and the absence (○) of 100 μM CDB3 and titrated into 15 nM fluorescein-labeled 30-mer gadd45 DNA. Dissociation constants were calculated from a fit to a 1:1 binding model.

these concentrations, which was 0.17 (estimated from Fig. 3), because under these conditions most of the protein was denatured and did not bind the peptide. There was an increase in the anisotropy over time, as the peptide induced protein refolding by mass action (Fig. 5*b*). On mixing p53C and FL-CDB3 (5 μM each) with 3 M urea without preincubation overnight, unfolding took place reaching the same endpoint. Overall, FL-CDB3 induced refolding of p53C and in its presence the equilibrium shifted toward the native state. Binding of FL-CDB3 to p53C became progressively weaker with increasing [urea], consistent with both a lack of binding to the denatured state and weaker binding to the native state (data not shown). Thus, the stabilizing effects in 3 M urea were not as pronounced as they would be in water alone.

DNA Competed with FL-CDB3 for p53C Binding. From the NMR data (Fig. 2) it seems that CDB3 bound p53C at the edge of the DNA binding site, suggesting at least a partial overlap between the two binding sites. We measured the competition between the binding of FL-CDB3 and gadd45 DNA to p53C by using competition fluorescence anisotropy. DNA (5 and 25 μM) displaced the peptide completely from the binding site, indicating overlap between the DNA and peptide binding sites (Fig. 6*a*).

CDB3 Restored Sequence-Specific DNA Binding to the Highly Destabilized p53 Mutant I195T. We tested whether CDB3 can restore sequence-specific DNA binding activity to p53C mutants by observing its effect on the β-sandwich mutant I195T, which is highly destabilized by 4.1 kcal/mol (3) and has poor DNA-binding affinity. I195T (10 μM) was incubated for 1 h at 10°C in presence of CDB3 (100 μM) (or its absence) and titrated into fluorescein-labeled gadd45 DNA in presence of the same peptide concentration. In the absence of peptide, I195T bound gadd45 DNA with K_d = 6 μM (Fig. 6*b*). After incubation with CDB3, the binding improved 6-fold, and K_d was 1 μM, which is close to the value of 0.8 μM for the wild type. As expected, CDB3 did not affect DNA binding of the completely native wild-type p53C (data not shown).

To confirm that the restoration of DNA binding is sequence-specific, we repeated the experiments with the random double-stranded DNA sequence fluorescein-AATATGGTTTGAATA-AAGAGTAAAGATTTG. Binding of I195T to this sequence was very weak, and was not improved, but rather inhibited, by the peptide (data not shown).

Discussion

We report a 9-residue peptide, CDB3, and its biotin and fluorescein-labeled derivatives that bind and stabilize p53C. CDB3 is derived from residues 490–498 of 53BP2, which constitute one of its binding loops for p53. The most striking properties of CDB3 and its derivatives are their abilities to: stabilize wild-type and mutant p53C, as shown by raising their apparent melting temperatures; induce refolding of reversibly denatured p53C; and restore sequence-specific DNA binding to a highly destabilized p53C mutant. Thus, a small peptide can stabilize p53C and restore its sequence-specific DNA binding simply by binding its native state but not the denatured state and shifting the equilibrium toward the native form.

Defining a General Target Site Within p53C for Stabilizing Peptides. The search strategy used in this study was successful for discovering a peptide that binds p53C in a defined site. Being derived from a protein that binds p53C in its DNA-binding site, CDB3 itself binds also in the DNA-binding site, although in a slightly different location. The CDB3-binding site within p53C, as mapped by NMR chemical shift analysis, is situated at the edge of the DNA-binding site and consists of three structural elements (loop 1, helix 2, and the edge of strand 8), which are remote sequentially but close spatially. This site might serve as a specific target for core-domain-stabilizing molecules. Its advantage as such is its location in proximity to the DNA-binding site, enabling a local stabilizing effect in that site. Indeed, the NMR data shows that CDB3 binding generates a strong localized effect on the edge of the DNA-binding site within p53C, whereas DNA binding results in shifts that are spread throughout the whole protein structure (data not shown).

CDB3, as a free peptide, did not bind p53C exactly in the same location as the parent loop in the 53BP2 protein (10). The original 53BP2 loop binds the core domain between helix 2 and loop 3 (10), whereas CDB3 binds at the other side of helix 2, close to loop1. A possible explanation is that the CDB3-binding site might also be an alternative binding site for 53BP2, and the two binding sites might have a regulatory role. Alternatively, owing to its high negative charge, CDB3 as a free peptide might act partly as a "DNA-mimic" that binds the positively charged surface of the DNA-binding site.

Implications for Rescue of p53C Mutants. CDB3 was found to bind two p53C hot-spot mutants: G245S, which is weakly destabilized (3), and R249S, which is distorted in the DNA-binding region (3, 8). The affinity of FL-CDB3 to G245S was the same as for the wild type. Binding to R249S was weaker, but still in the low micromolar range.

CDB3-like compounds could be used in principle for the rescue of mutants that are weakly destabilized (e.g., G245S) and mutants that are at body temperature globally unfolded (e.g., I195T) and are unable to bind DNA (see above). Indeed, we demonstrated that CDB3 can restore sequence-specific DNA binding to the highly destabilized I195T mutant. Peptides such as CDB3 cannot be used to rescue DNA contact mutants. Other strategies, which involve introduction of residues or small molecules that contribute the missing interactions, should be used for rescue of these mutants.

The mode of action of CDB3 is different from that of the previously reported C-terminal peptides (19–22). CDB3 stabilizes p53 by binding its native but not its denatured state, whereas the C-terminal peptides specifically regulate the activity and the DNA binding of p53C. CDB3 and its labeled derivative FL-CDB3 are lead compounds, and they can be used as a basis for the future design of peptides and small molecules that have a larger stabilizing effect on p53C. We have applied the key biophysical measurements that successfully detected the binding and stabilization of p53C by CDB3 to the drug CP-31398 (6), but all of the results were negative (T.M.R, V. J. N. Bykov, S.M.V.F., G. Selivanova, K. G. Wiman, and A.R.F., unpublished data).

Chaperone Strategy. The inherent drawback of using a natural binding site for a drug is that it competes with the natural ligand. Thus, it might be thought that the competition between DNA and CDB3 peptide would preclude it from being of use as a lead. But this need not be so. Because the binding of DNA itself stabilizes p53C, and it binds very tightly, stabilization by a peptide such as CDB3 is needed only for mutants where DNA binding is impaired because mutant p53 is in denatured conformation. Once the protein has bound DNA, the peptide is not needed any more. The ability of CDB3 to induce refolding of p53C, together with the observation that DNA can displace it from p53, led us to propose a "chaperone" mechanism for rescuing a denatured mutant (Fig. 7). CDB3 binds an unfolded or distorted mutant that is unable to bind DNA, either immediately on biosynthesis or later for reversibly denatured mutants, and shifts the equilibrium toward the native state. Then DNA can bind the protein, displacing the peptide, which is free again to bind another protein molecule. We have demonstrated this mechanism for the highly destabilized mutant I195T. Binding of this mutant to the gadd45 DNA was improved 6-fold in presence of CDB3, to the level of wild-type p53C (Fig. 6*b*). Random DNA, on the other hand, remained unaffected. The competition between DNA and peptide should favor DNA even better *in vivo* because the peptide binds independently to each site in a tetramer of p53, but DNA binds far more tightly to the native tetramer because of cooperativity, thus allowing DNA to displace the drug more easily. CDB3 may well be a lead compound for designing drugs to stabilize p53.

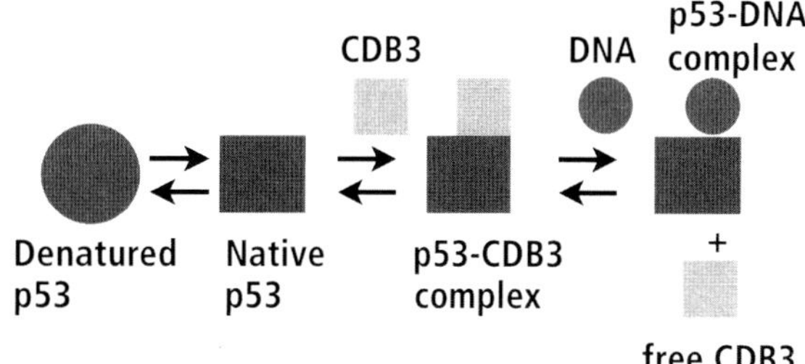

Fig. 7. "Chaperone" strategy for rescue of p53. A schematic model of the proposed mechanism of action for CDB3. See text for details.

We thank Dr Mark Bycroft for useful discussions. A.F. is supported by Long-Term Fellowship LT00056/2000-M from the Human Frontier Science Program. L.O.H. is supported by Wenner-Gren Foundations, Stockholm. T.M.R. is the recipient of Herchel Smith and Medical Research Council (UK) scholarships. This work was also partly supported by the Cancer Research Campaign (London). D.B.V. was supported by Postdoctoral Fellowship LT0318/1998-M from the Human Frontier Science Program.

1. Hainaut, P. & Hollstein, M. (2000) *Adv. Cancer Res.* **77,** 81–137.
2. Cho, Y., Gorina, S., Jeffrey, P. D. & Pavletich, N. P. (1994) *Science* **265,** 346–355.
3. Bullock, A. N., Henckel, J. & Fersht, A. R. (2000) *Oncogene* **19,** 1245–1256.
4. Bullock, A. N. & Fersht, A. R. (2001) *Nat. Cancer Rev.* **1,** 68–76.
5. Hupp, T. R., Lane, D. P. & Ball, K. L. (2000) *Biochem. J.* **352,** 1–17.
6. Foster, B. A., Coffey, H. A., Morin, M. J. & Rastinejad, F. (1999) *Science* **286,** 2507–2510.
7. Bullock, A. N., Henckel, J., DeDecker, B. S., Johnson, C. M., Nikolova, P. V., Proctor, M. R., Lane, D. P. & Fersht, A. R. (1997) *Proc. Natl. Acad. Sci. USA* **94,** 14338–14342.
8. Wong, K. B., DeDecker, B. S., Freund, S. M., Proctor, M. R., Bycroft, M. & Fersht, A. R. (1999) *Proc. Natl. Acad. Sci. USA* **96,** 8438–8442.
9. Nieba, L., Krebber, A. & Pluckthun, A. (1996) *Anal. Biochem.* **234,** 155–165.
10. Gorina, S. & Pavletich, N. P. (1996) *Science* **274,** 1001–1005.
11. Iwabuchi, K., Bartel, P. L., Li, B., Marraccino, R. & Fields, S. (1994) *Proc. Natl. Acad. Sci. USA* **91,** 6098–6102.
12. Samuels-Lev, Y., O'Connor, D. J., Bergamaschi, D., Trigiante, G., Hsieh, J. K., Zhong, S., Campargue, I., Naumovski, L., Crook, T. & Lu, X. (2001) *Mol. Cell.* **8,** 781–794.
13. Iwabuchi, K., Li, B., Massa, H. F., Trask, B. J., Date, T. & Fields, S. (1998) *J. Biol. Chem.* **273,** 26061–26068.
14. Lopez, C. D., Ao, Y., Rohde, L. H., Perez, T. D., O'Connor, D. J., Lu, X., Ford, J. M. & Naumovski, L. (2000) *Mol. Cell. Biol.* **20,** 8018–8025.
15. Bayle, J. H., Elenbaas, B. & Levine, A. J. (1995) *Proc. Natl. Acad. Sci. USA* **92,** 5729–5733.
16. Muller-Tiemann, B. F., Halazonetis, T. D. & Elting, J. J. (1998) *Proc. Natl. Acad. Sci. USA* **95,** 6079–6084.
17. Takenaka, I., Morin, F., Seizinger, B. R. & Kley, N. (1995) *J. Biol. Chem.* **270,** 5405–5411.
18. Kim, A. L., Raffo, A. J., Brandt-Rauf, P. W., Pincus, M. R., Monaco, R., Abarzua, P. & Fine, R. L. (1999) *J. Biol. Chem.* **274,** 34924–34931.
19. Abarzua, P., LoSardo, J. E., Gubler, M. L., Spathis, R., Lu, Y. A., Felix, A. & Neri, A. (1996) *Oncogene* **13,** 2477–2482.
20. Selivanova, G., Iotsova, V., Okan, I., Fritsche, M., Strom, M., Groner, B., Grafstrom, R. C. & Wiman, K. G. (1997) *Nat. Med.* **3,** 632–638.
21. Selivanova, G., Ryabchenko, L., Jansson, E., Iotsova, V. & Wiman, K. G. (1999) *Mol. Cell. Biol.* **19,** 3395–3402.
22. Hupp, T. R., Sparks, A. & Lane, D. P. (1995) *Cell* **83,** 237–245.
23. Guex, N. & Peitsch, M. C. (1997) *Electrophoresis* **18,** 2714–2723.

Becoming a Structural Biologist

To go further, we needed to know the structures of oncogenic mutants of p53. Unfortunately, p53 is so unstable that it is very difficult to handle and not suitable for extensive biophysical and structural biology studies. Using our old protein-engineering tricks[103], Penka Nikolova designed a fully biologically active superstable mutant[206], and Andreas Joerger solved the structures of many of the common oncogenic mutants[207–213]. Four decades after being recruited to work with structural biologists, I joined them!

18.5 2008: *Back row*: Cetin Baloglu, Wiktor Banachewicz, Jenifer Lum, Karoly von Glos, Fang Huang, Tuck Seng Wong, Jan van Dieck, Fiona Townsley, Caroline Blair, Kian Hoe Khoo, Dima Veprintsev, Andreas Joerger, Gianni Settanni, Jeremy Touati, Sarah Burge, Stephan Hammer, Eyal Arbely, Maria Garcia Alai. *Front row*: Tobias Brandt, Rainer Wilcken, Marina Vaysburd, Paula Murphy, Alan, Miriana Petrovich, Joel Kaar, Tony Andreeva, Agnes Jaulent, Hannes Neuweiler.

Structural basis for understanding oncogenic p53 mutations and designing rescue drugs

Andreas C. Joerger, Hwee Ching Ang, and Alan R. Fersht*

Cambridge University Chemical Laboratory and Cambridge Centre for Protein Engineering, Medical Research Council Centre, Hills Road, Cambridge CB2 2QH, United Kingdom

Contributed by Alan R. Fersht, August 22, 2006

The DNA-binding domain of the tumor suppressor p53 is inactivated by mutation in ≈50% of human cancers. We have solved high-resolution crystal structures of several oncogenic mutants to investigate the structural basis of inactivation and provide information for designing drugs that may rescue inactivated mutants. We found a variety of structural consequences upon mutation: (*i*) the removal of an essential contact with DNA, (*ii*) creation of large, water-accessible crevices or hydrophobic internal cavities with no other structural changes but with a large loss of thermodynamic stability, (*iii*) distortion of the DNA-binding surface, and (*iv*) alterations to surfaces not directly involved in DNA binding but involved in domain–domain interactions on binding as a tetramer. These findings explain differences in functional properties and associated phenotypes (e.g., temperature sensitivity). Some mutants have the potential of being rescued by a generic stabilizing drug. In addition, a mutation-induced crevice is a potential target site for a mutant-selective stabilizing drug.

cancer | crystal | structure | drug design | polymorphism

The tumor suppressor protein p53 is a 393-aa transcription factor that regulates the cell cycle and plays a key role in the prevention of cancer development. In response to oncogenic and other stresses, p53 induces the transcription of a number of target genes, resulting in cell-cycle arrest, senescence, or apoptosis (1, 2). In ≈50% of human cancers, p53 is inactivated as a result of missense mutation in the p53 gene (3, 4).

The multifunctionality of p53 is reflected in the complexity of its structure. Each chain in the p53 tetramer is composed of several domains. There are well defined DNA-binding and tetramerization domains and highly mobile, largely unstructured regions (5–9). Most p53 cancer mutations are located in the DNA-binding core domain of the protein (3). This domain has been structurally characterized in complex with its cognate DNA by x-ray crystallography (5, 10, 11) and in its free form in solution by NMR (12). It consists of a central β-sandwich that serves as a basic scaffold for the DNA-binding surface. The DNA-binding surface is composed of two large loops (L2 and L3) that are stabilized by a zinc ion and a loop–sheet–helix motif. Together, these structural elements form an extended surface that makes specific contacts with the various p53 response elements. The six amino acid residues that are most frequently mutated in human cancer are located in or close to the DNA-binding surface (compare release R10 of the TP53 mutation database at www-p53.iarc.fr) (3). These residues have been classified as "contact" (Arg-248 and Arg-273) or "structural" (Arg-175, Gly-245, Arg-249, and Arg-282) residues, depending on whether they directly contact DNA or play a role in maintaining the structural integrity of the DNA-binding surface (Fig. 1) (5).

Urea denaturation studies have shown that the contact mutation R273H has no effect on the thermodynamic stability of the core domain, whereas structural mutations substantially destabilize the protein to varying degrees, ranging from 1 kcal/mol for G245S and 2 kcal/mol for R249S up to >3 kcal/mol for R282W (13). The destabilization has severe implications for the folding state of these mutants in the cell. Because the wild-type core

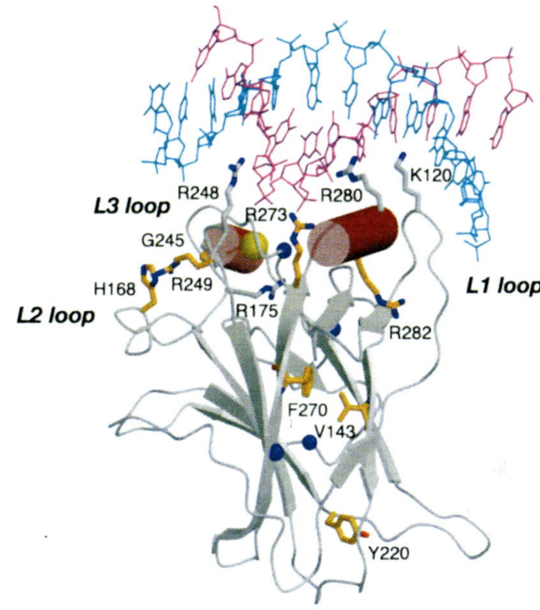

Fig. 1. Structure of the p53 core domain bound to consensus DNA (5). The two strands of bound consensus DNA are shown in blue and magenta. The bound zinc ion is displayed as a golden sphere. Cancer mutation sites that were structurally studied in this work and earlier work (16) are shown in orange. The blue spheres indicate the location of the mutation sites in the superstable quadruple mutant M133L/V203A/N239Y/N268D (*T*-p53C).

domain is only marginally stable and has a melting temperature only slightly above body temperature, highly destabilized mutants such as R282W are largely unfolded under physiological conditions and, hence, are no longer functional (14).

To understand the role of individual mutants in carcinogenesis and to assess the possibility of rescuing their function, it is important to know the effect of the mutation not only on the overall stability but also on the local structure. Qualitative NMR studies indicate that hotspot mutants evince characteristic local structural changes (15). We have recently elucidated the structural effects of the contact mutation R273H and the structural mutation R249S by x-ray crystallography (16). In addition, these structural studies revealed that the second-site suppressor mutation H168R rescues the function of R249S in a specific manner by mimicking the structural role of Arg-249 in the wild type (16).

Author contributions: A.C.J. and A.R.F. designed research; A.C.J. and H.C.A. performed research; A.C.J. and H.C.A. analyzed data; and A.C.J. and A.R.F. wrote the paper.

The authors declare no conflict of interest.

Abbreviation: PDB, Protein Data Bank.

Data deposition: The atomic coordinates and structure factors have been deposited in the Protein Data Bank, www.pdb.org (PDB ID codes 2J1W, 2J1X, 2J1Y, 2J1Z, 2J20, and 2J21).

*To whom correspondence should be addressed. E-mail: arf25@cam.ac.uk.

 www.pnas.org/cgi/doi/10.1073/pnas.0607286103

Here, we extend our crystallographic studies on the p53 core domain to obtain a more comprehensive picture of the effects of common cancer mutations on the structure and function of p53. Using a stabilized variant of the p53 core domain (*T*-p53C), we have elucidated the diverse structural effects of cancer hotspot mutations in the DNA-binding surface (G245S, R273C, and R282W) and at the periphery of the β-sandwich (Y220C), as well as the effects of highly destabilizing, cavity-creating mutations (V143A and F270L) in the hydrophobic core of the β-sandwich region of the protein (Fig. 1).

Results and Discussion

We focused our crystallographic studies on cancer-hotspot mutations in the DNA-binding surface (G245S, R273C, and R282W) and at the far end of the β-sandwich (Y220C), as well as on two potentially cavity-creating cancer mutations (V143A and F270L) in the hydrophobic core of the β-sandwich (Fig. 1). To do this, we introduced these mutations into a stabilized variant of p53 core domain (*T*-p53C). *T*-p53C contains four point mutations (M133L, V203A, N239Y, and N268D) that stabilize the core domain by 2.6 kcal/mol. It has wild-type-like DNA-binding properties, its full-length version is fully active in human cells (Sebastian Mayer and A.R.F., unpublished data), and the structure is essentially the same as that of the wild type, apart from the mutated side chains, which confer additional stability (17). The effects of the mutations are simply to raise the T_m of the protein and its mutants by 6°C, making them easier to handle (18).

The crystallization conditions for five of the six mutants were similar to those for *T*-p53C, and the obtained crystals were isomorphous. They belonged to space group $P2_12_12_1$ with two molecules in the asymmetric unit, and the corresponding structures were determined at high resolution ranging from 1.6 to 1.8 Å (Table 1, which is published as supporting information on the PNAS web site). In all of these five constructs containing an extended C terminus (residues 94–312), the C-terminal residues beyond Lys-291 were disordered, as found previously (5, 16, 17). For *T*-p53C-G245S, a shorter construct was used (residues 94–293), and we obtained crystals in a new crystal form (space group $P2_1$ with four molecules in the asymmetric unit), which allowed us to determine the structure at 1.69-Å resolution (Table 1). The structural studies were complemented by urea denaturation studies to determine the effects of mutation on the thermodynamic stability of the protein (Table 2, which is published as supporting information on the PNAS web site).

R273C and R273H Are Classic Contact Mutants. R273C and R273H are two of the five most frequent cancer-associated mutations. They affect Arg 273, which makes major contacts with the phosphate backbone of target DNA. Having previously solved the crystal structure of *T*-p53C-R273H (16), we have now determined the structure of *T*-p53C-R273C (at 1.8-Å resolution) to elucidate the structural changes in the cysteine variant. Overall, the structures of *T*-p53C, *T*-p53C-R273H, and *T*-p53C-R273C are virtually identical. A pairwise superposition of the C^{α} atoms from equivalent chains gives rmsds of <0.16 Å. The R273C mutation simply removed a DNA contact without perturbing the conformation of neighboring residues, such as Phe-134 and Asp-281 (Fig. 2). The preservation of the overall architecture of the DNA-binding surface in the structures of *T*-p53C-R273C and *T*-p53C-R273H explains why these mutants exhibit residual DNA-binding activity despite the loss of a crucial DNA contact (ref. 18 and unpublished data).

G245S Induces Small Conformational Changes in the L3 Loop. The G245S mutation is located in the L3 loop, which binds to the minor groove of DNA response elements via Arg-248 (Fig. 1). The crystal structure of *T*-p53C-G245S revealed distinct structural changes in the immediate environment of the mutation site. Observed conformational changes in some surface loops (e.g., in the L1 loop and the S7–S8 loop) can be attributed to differences in crystallization conditions and crystal packing for *T*-p53C and *T*-p53C-G245S and reflect the intrinsic conformational flexibility of these loop regions (12, 17, 19). In *T*-p53C-G245S, the hydroxyl group of Ser-245 points toward the zinc ligands Cys-238 and Cys-242 and displaces a structural water molecule that is observed in the wild-type and various mutant structures (Fig. 3*A*). The main-chain atoms of Ser-245 form the same hydrogen bonds with neighboring residues (Cys-242 and the side chain of Arg-249) as observed for the glycine in the wild type and *T*-p53C, although the backbone conformation of Ser-245 is slightly different. Both dihedral angles ϕ and ψ have changed by ≈20° and fall within a "generously allowed" region of the Ramachandran plot (ϕ = −141°, ψ = −102°). As such, Ser-245 adopts a moderately unfavorable main-chain conformation.

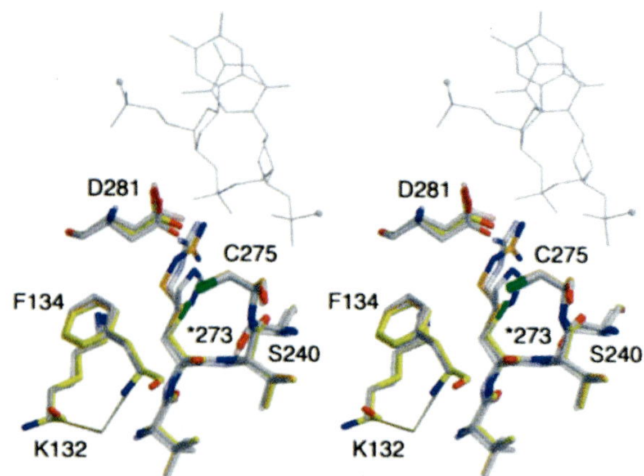

Fig. 2. Crystal structure of *T*-p53C-R273C. Stereoview of the mutation site in the structure of *T*-p53C-R273C [Protein Data Bank (PDB) ID code 2J20, yellow] superimposed on the structure of *T*-p53C (PDB ID code 1UOL, orange), *T*-p53C-R273H (PDB ID code 2BIM, gray), and DNA-bound wild type (PDB ID code 1TSR, light gray). One strand of bound DNA in the vicinity of Arg-273 is shown as a gray line, with small spheres indicating the continuation of the DNA backbone.

Most interestingly, the mutation appears to trigger a flip of the peptide bond between Met-243 and Gly-244, resulting in a displacement of the corresponding C^{α} atoms by 0.7 and 1.7 Å, respectively. This peptide flip relative to the conformation found in the structures of *T*-p53C and DNA-bound wild type was observed in all four molecules of the asymmetric unit. There is also a significant structural response from Pro-177, which follows the movement of Gly-244 to avoid steric clashes and is shifted to a similar extent. The backbone conformation of residues 246–250 of the L3 loop, which includes the DNA-contact residue Arg-248, is only marginally affected, and their C^{α} displacements are in the 0.5-Å range. Some of the residues with the most significant shifts in their C^{α} atoms (Met-243, Gly-244, and Pro-177), however, are key residues in the subunit interface of the core domain dimer bound to a DNA half-site (10, 20, 21). (Fig. 3*C*). Although the G245S mutation has no major direct effect on the DNA-contact residue Arg-248 in the DNA-free form of *T*-p53C-G245S, it presumably puts conformational strain on the subunit interface upon DNA binding. Such effects on the subunit interface would explain DNA-binding studies on full-length *T*-p53-G245S, which show that specific binding to the *gadd45* response element is ≈15-fold reduced relative to *T*-p53 at physiological ionic strength (18). In contrast, R249S, the other structural cancer hotspot mutation in the L3 loop, substantially

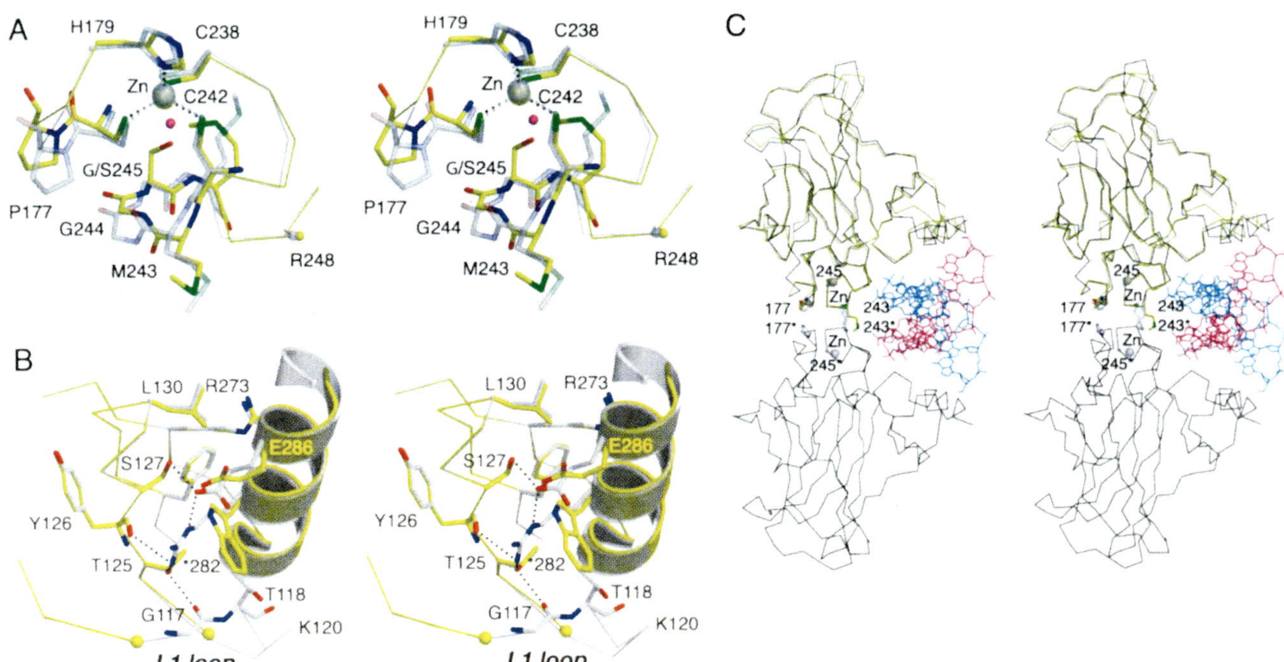

Fig. 3. Structural effects of the cancer mutations G245S and R282W. (*A*) Stereoview of the zinc-binding region of *T*-p53C-G245S (PDB ID code 2J1Y, molecule A, yellow) superimposed on the structure of *T*-p53C (PDB ID code 1UOL, molecule A, gray). The C^{α} atom of the DNA-contact residue Arg-248 is depicted as a small sphere in the color of the corresponding chain. A structural water molecule that is present in *T*-p53C but displaced by the side chain of Ser-245 in *T*-p53C-G245S is shown as a magenta sphere. (*B*) Stereoview of the loop–sheet–helix motif in the crystal structure of *T*-p53C-R282W (PDB ID code 2J21, molecule A, yellow) superimposed on *T*-p53C (PDB ID code 1UOL, molecule A, gray). Stabilizing interactions mediated via Arg-282 in *T*-p53C are shown as dotted lines. Residues 117–121 of *T*-p53C-R282W were disordered. The C^{α} atoms on both sides of the chain break are highlighted by small yellow spheres. (*C*) C^{α} trace of a wild-type core domain dimer bound to a DNA half-site (PDB ID code 2AC0, black) (10). The two strands of bound DNA are shown in blue and magenta. *T*-p53C-G245S, shown in yellow, is superimposed on one of the two monomers. Selected residues at the core–core domain interface are highlighted.

perturbs the L3 loop, induces high flexibility, and favors alternative conformations with much more severe consequences for the overall protein stability and DNA binding (16, 18). Hence, for these two structural hotspot mutations in the L3 loop, there is a direct correlation between the magnitude of loss of DNA binding and the extent of the observed structural perturbation.

Our structures also explain interactions of p53 mutants with other proteins, such as the apoptosis-stimulating protein ASPP2. Under conditions where the mutants are fully folded, several cancer mutants bind to the 53BP2 domain of the ASPP2 protein at wild-type levels, whereas no binding is detected for the mutants *T*-p53C-R249S and *T*-p53C-G245S, which have distinct structural changes in the region that forms the central part of the p53–53BP2 interface (22).

The R282W Mutation Affects the Packing of the Loop–Sheet–Helix Motif. Arg-282 plays a crucial role in maintaining the structural integrity of the loop–sheet–helix motif that binds to the major groove of DNA target sites (Fig. 1). Its side chain is involved in a network of interactions that pack helix H2 against the S2–S2′ β-hairpin and loop L1 (Fig. 3*B*). The 1.6-Å crystal structure of *T*-p53C-R282W revealed that the overall fold of the protein is not compromised by the R282W hotspot mutation. The C^{α} atoms of *T*-p53C-R282W and *T*-p53C can be superimposed with a rmsd of 0.26 Å. There are, however, substantial structural perturbations in the loop–sheet–helix motif (Fig. 3*B*), and several stabilizing interactions are lost, which accounts for the loss of thermodynamic stability of 3 kcal/mol. The region of the L1 loop next to Trp-282 is disordered because of steric hindrance, and no conclusive electron density was observed for several residues within this loop (residues 117–121, including the DNA contact Lys-120). As such, Trp-282 displaces what appears to be the weakest structural link in the loop–sheet–helix motif. The side chain of Trp-282 is solvent-exposed and much less well embedded within the protein structure than the arginine in *T*-p53C. The destabilization of the loop–sheet–helix motif is also reflected in the *B* factor profile, which shows an increased mobility for the main-chain atoms at both ends of the chain break in loop L1 and around the mutation site in the C-terminal helix, and this trend is propagated toward the chain end (Fig. 6, which is published as supporting information on the PNAS web site). The overall packing characteristics of the C-terminal helix containing the DNA-contact residues Cys-277 and Arg-280 against the core of the protein, however, are preserved. Hence, at the subphysiological temperatures at which the mutant R282W is largely folded, it is still able to bind *gadd45* DNA (14) despite the structural disorder in the L1 loop containing the DNA-contact residue Lys-120. This retention of binding activity is consistent with a recent mutational analysis of the L1 loop (23) and with the observation that this loop is a mutational "cold spot" that is only rarely mutated in human cancer (3), indicating that structural perturbations in the L1 loop are tolerated to a certain degree without abrogating p53 function.

Y220C Creates a Solvent-Accessible Cleft at the Periphery of the β-Sandwich. Y220C is the most common cancer mutation outside the DNA-binding surface (compare release R10 of the TP53 mutation database at www-p53.iarc.fr) (3). It is located at the far end of the β-sandwich, at the start of the loop connecting β-strands S7 and S8 (Fig. 1) and destabilizes the core domain by 4 kcal/mol (Table 2). The benzene moiety of Tyr-220 forms part of the hydrophobic core of the β-sandwich, whereas the hydroxyl group points toward the solvent. The 1.65-Å crystal structure of

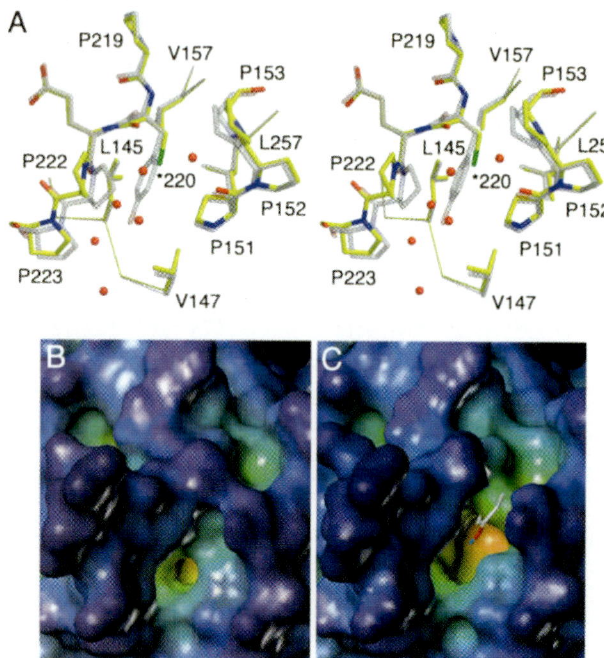

Fig. 4. Crystal structure of *T*-p53C-Y220C. (*A*) Stereoview of the mutation site at the periphery of the β-sandwich in *T*-p53C-Y220C (PDB ID code 2J1X, molecule A, yellow) superimposed on the structure of *T*-p53C (PDB ID code 1UOL, molecule A, gray). Several water molecules close to Cys-220 in *T*-p53C-Y220C that fill the cleft created by the mutation are shown as red spheres. (*B*) Molecular surface of *T*-p53C around Tyr-220. (*C*) Molecular surface of *T*-p53C-Y220C. The view is the same as in *B*. The position of the side chain of Tyr-220 in *T*-p53C is shown as a stick model.

T-p53C-Y220C showed that the Y220C mutation creates a solvent-accessible cleft that is filled with water molecules at defined positions but leaves the overall structure of the core domain intact (Fig. 4). The structural changes upon mutation link two rather shallow surface clefts, preexisting in the wild type, to form a long, extended crevice in *T*-p53C-Y220C, which has its deepest point at the mutation site (Fig. 4 *B* and *C*). Cys-220 occupies approximately the position of the equivalent atoms of Tyr-220 in the wild type. The positions of neighboring hydrophobic side chains located in the core of the β-sandwich have not shifted significantly. The mutation, however, results in a loss of hydrophobic interactions and a suboptimal packing of these hydrophobic core residues. The side chain of Leu-145, which was buried in the wild type, for instance, becomes partly solvent accessible in *T*-p53C-Y220C. The largest structural changes in the immediate environment of the mutation site are found in the S7–S8 loop for Pro-222. Throughout the structure, there is no C^{α} displacement >0.9 Å.

β-Sandwich Mutations and the Molecular Basis of Temperature Sensitivity. About one-third of the reported cancer mutations in the p53 core domain are located outside the structural elements that form the DNA-binding surface (loops L2 and L3 and the loop–sheet–helix motif). The V143A mutant is of particular interest because of the well documented temperature sensitivity of its binding to many response elements in both yeast and mammalian systems. At body temperature, the mutant is inactive and unfolded, whereas it retains transactivational activity at lower temperatures (24, 25). The structures of *T*-p53C-V143A and *T*-p53C-F270L provide the molecular basis for understanding the temperature-sensitive behavior of many β-sandwich mutants. Both mutations created internal

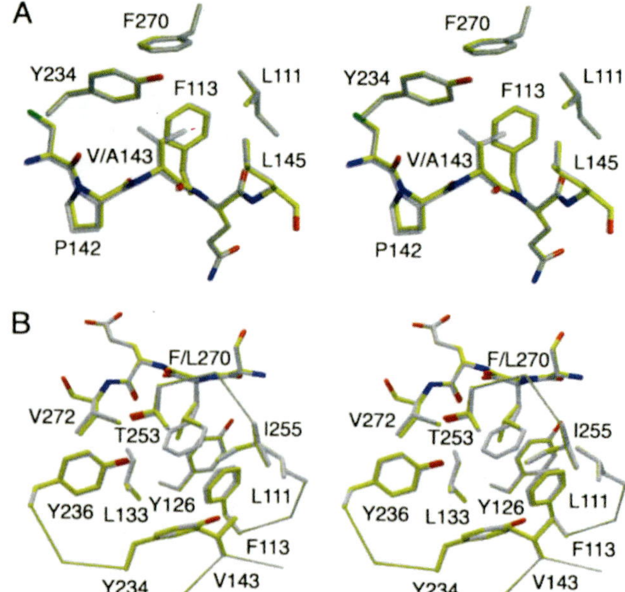

Fig. 5. Crystal structures of *T*-p53C-V143A and *T*-p53C-F270L. (*A*) Stereoview of the structure of *T*-p53C-V143A (PDB ID code 2J1W, yellow) superimposed on *T*-p53C (PDB ID code 1UOL, gray). All residues in the hydrophobic core of the β-sandwich within a 4.5-Å radius of the Val-143 side chain in *T*-p53C are shown. (*B*) Stereoview of the structure of *T*-p53C-F270L (PDB ID code 2J1Z, yellow) superimposed on *T*-p53C (PDB ID code 1UOL, gray). All residues within a 6-Å radius of the Phe-270 side chain in *T*-p53C are shown.

cavities in the hydrophobic core of the β-sandwich without collapse of the surrounding structure (Fig. 5). The cavity volume was 47 $Å^3$ (*T*-p53C-V143A) and 51 $Å^3$ (*T*-p53C-F270L), as defined by the volume that can be occupied by a probe mimicking the size of a water molecule (1.4-Å probe radius). Additional information on these cavities is provided in Table 3, which is published as supporting information on the PNAS web site. In agreement with the hydrophobic nature of the cavities, no ordered buried water molecules were detected in the crystal structure. Although the overall structure of the core domain was perfectly conserved, the creation of void volumes came at a high energetic cost of 3.7 and 4.1 kcal/mol. These observations are consistent with studies on "large-to-small" substitutions in the hydrophobic core of T4-lysozyme and barnase (26–28).

A recent study has identified a large number of temperature-sensitive p53 mutants (29). Most mutations were clustered in the β-sheet region of the protein, and the substitutions were mainly from large hydrophobic residues to smaller hydrophobic residues. Incidentally, V143A was not detected in this study, whereas mutations at residue 270 were (F270I and F270C). Interestingly, the Y220C mutation has also been reported to cause temperature-sensitive behavior (25). Again, this behavior is in agreement with our crystallographic data, which show that the mutation-induced structural changes are very localized, far away from the DNA-binding surface. A common structural feature of the β-sandwich mutants seemed to be that there were only minor structural disruptions upon mutation, although the effect on the thermodynamic stability of the protein was generally more severe than for the hotspot mutations in the DNA-binding surface. The much more compact and robust structural framework of the β-sandwich compared with the zinc-binding region and the loop–sheet–helix motif renders it generally much less susceptible to mutation-induced structural changes, in particular for large-to-small substitutions. The absence of structural changes in surface regions, especially in the DNA-binding surface, however, is key for functionality. Hence, temperature-sensitive behavior can be expected for all cancer mutations that destabilize the core domain without compromising the surface complementarity that is crucial to the function of p53, not only for binding to specific promoter sequences but also for interactions with regulatory proteins and for correct domain organization in tetrameric full-length p53 (9, 30–33).

Implications for Rescue Strategies. At first glance, the discussion of the local structural changes induced by highly destabilizing mutations appears to be rather academic when assessing the functional consequences under physiological conditions, because these mutant proteins will most likely be largely unfolded. Our structural observations, however, have profound implications for therapeutic strategies that aim to rescue the function of p53 with small-molecule drugs that stabilize p53 (reviewed in refs. 34 and 35). The mutations N239Y and N268D have been reported to restore transcriptional activity in a subset of cancer mutants, including G245S (36). Double-mutant cycles suggest that they act as global stability suppressors (37). As such, they mimic the effects of a hypothetical generic small-molecule drug. The structure of *T*-p53C-G245S, however, exhibits distinct structural perturbations despite the presence of these stabilizing suppressor mutations, concomitant with altered functional properties (e.g., impaired binding of 53BP2; see above). For mutants with local structural changes in the DNA-binding surface, simple stabilization by means of a generic small molecule may not be enough to fully restore activity. A prime example is the mutant R249S. Studies on second-site suppressor mutations have shown that full restoration of DNA-binding activity is found only in the presence of a mutation (H168R) that specifically reverses the structural changes induced by the oncogenic mutation (16, 36, 37).

On the basis of our structural studies, β-sandwich mutants, such as V143A and F270L, represent much more promising targets for rescue by generic small-molecule drugs, because, in this case, stabilizing the wild-type conformation of the protein by ligand binding may be sufficient to restore wild-type-like activity under physiological conditions. The Y220C mutant not only has the potential of being rescued by a generic wild-type-binding compound but also has the possibility of being a target for a mutant-selective drug that can bind in the crevice formed by the deletion (Fig. 4*C*). This crevice region is particularly attractive because it appears to be distant from the functional sites and interfaces of the protein.

What Makes Human Alleles Deleterious? It is estimated that ≈20% of common nonsynonymous single-nucleotide polymorphisms result in amino acid changes that damage the corresponding protein, resulting in ≈2,000 potentially deleterious alleles in the average human genotype (38). The structural and energetic response of the p53 core domain to the V143A and F270L mutations in the hydrophobic core of the β-sandwich suggests that this type of mutation results in deleterious phenotypes mainly because p53 is only marginally stable at body temperature. Equivalent mutations in a more stable structural scaffold may have only subtle effects on function and result in rather neutral phenotypes.

There is growing evidence that p53 has evolved to be highly dynamic and intrinsically unstable (12, 19, 39). As a negative side effect of this evolutionary process, the core domain of p53 has in a way become more susceptible to cancer-associated mutations in the β-sandwich. To our knowledge, there has been no comprehensive quantitative study on the stability of human proteins. It would be interesting to address the question of whether the low intrinsic thermodynamic stability of p53 is the exception or a common feature of highly regulated multifunctional proteins at pivotal cellular checkpoints, as also observed in the case of the tumor suppressor protein p16, for example (40). Such knowledge combined with more structural data on the effect of mutations would significantly improve predictions as to whether nonsynonymous single-nucleotide polymorphisms are benign or likely to cause functional disorders.

Conclusion

Our crystallographic studies paint an intriguing picture of the effects of p53 cancer mutations. They clearly show that the mutations result in a variety of characteristic structural changes, concomitant with distinct energetic responses, although the overall structural framework is largely conserved. Accordingly, general conclusions about "mutant p53" based on the results obtained for one particular mutant have to be treated with caution. Two aspects have to be considered when rationalizing functional properties of a particular cancer mutant, even though there is a direct causal connection between the two: (*i*) the effect of mutation on the overall thermodynamic stability of the protein and (*ii*) the nature of local conformational changes. The effects on thermodynamic stability will determine whether a mutant is folded at a particular temperature. Whether a mutant is functional in the folded state, however, depends on the extent and nature of the structural changes, in particular in the DNA-binding surface, which will eventually determine the selectivity for various response elements or the interactions with other proteins, such as apoptotic cofactors. Recent studies in yeast and mammalian cell lines have shown that different mutants exhibit distinct transactivation patterns that are directly connected with different phenotypes (41, 42). More importantly, a study involving breast cancer patients suggested that different p53 mutations are associated with different prognostic values (43). These observations highlight the need for a thorough understanding of the mutation–structure and structure–function relationship in p53 so that biological outcome and response to drug treatment can be predicted. The details of individual structures point toward the possibility of designing specific drugs to bind to them and stabilize the native conformation.

Materials and Methods

Mutagenesis and Protein Purification. Mutagenesis, gene expression, and protein purification were performed as described in ref. 16, with a modification for *T*-p53C-G245S, for which a shortened construct comprising residues 94–293 (instead of 94–312) was used. After the final purification step (gel filtration), the mutant proteins were concentrated to 6–7 mg/ml, flash-frozen, and stored in liquid nitrogen.

Equilibrium Denaturation. Urea denaturation studies were performed as described in ref. 16 by using the equations described in ref. 14 for data analysis.

Crystallization and Structure Determination. All crystals were grown at 17°C by using the sitting drop vapor-diffusion technique. Crystals of *T*-p53C-V143A, *T*-p53C-Y220C, *T*-p53C-F270L, and *T*-p53C-R282W were grown under the conditions described for *T*-p53C (17); crystals of *T*-p53C-R273C were grown under the conditions described for *T*-p53C-R273H (16). Crystals of *T*-p53C-G245S were grown by mixing 2 μl of protein solution (6 mg/ml) with 0.4 μl of water and 1.6 μl of the reservoir solution (19% polyethylene glycol 3350/0.19 M calcium acetate, pH 7.2). Crystals were flash-frozen in liquid nitrogen by using mother liquor with either 20% polyethylene glycol 200 or 20% glycerol as a cryoprotectant. X-ray data sets were collected at 100 K on beamlines 10.1 and 14.1 at the Synchrotron Radiation Source (Daresbury, U.K.). Data processing was performed by using Mosflm (44) and Scala (45). All crystals except *T*-p53C-G245S belonged to space group $P2_12_12_1$ with two molecules per asymmetric unit and were isomorphous to those obtained for *T*-p53C and *T*-p53C-R273H (16, 17). Structure solution and refinement were performed with CNS (46). After an initial round of rigid-body refinement using the structure of either *T*-p53C (PDB ID code 1UOL) or *T*-p53C-R273H (PDB ID code 2BIM) as a starting model, the structures were refined by iterative cycles of refinement with CNS and manual model-building with MAIN (47). Crystals of *T*-p53C-G245S belonged to space group $P2_1$ with four molecules per asymmetric unit. The structure was solved by molecular replacement with CNS using *T*-p53C as a search model. Subsequent refinement was performed as described above for the other mutants. The final models (1.6- to 1.8-Å resolution) had a crystallographic R factor of 18.3–19.8% (R_{free} = 20.6–22.3%) and excellent stereochemistry as verified with PROCHECK (48). Detailed data collection and refinement statistics are summarized in Table 1.

Structure Analysis. Unless otherwise stated, detailed descriptions of mutant structures are based on the comparison of molecule A of a particular mutant with molecule A of *T*-p53C. Numbering of secondary structure elements is as reported for the wild-type structure in complex with DNA (5) [e.g., loop L2 comprises residues 164–194, including a short helix (H1)]. Depending on the program used to assign secondary structure, the β-turn region at the beginning of the L2 loop is sometimes alternatively assigned as a short 3_{10}-helix. Volumes of internal cavities were calculated with VOIDOO (49) by using different probe sizes (1.4- and 1.2-Å radius). Calculations were performed for 10 random orientations of the molecule. The cavity refinement parameters were as described in ref. 28. Structural figures were prepared by using MOLSCRIPT (50), RASTER3D (51), and SYBYL 6.9 (Tripos, St. Louis, MO).

We thank Caroline Blair for protein purification, Fiona Sait for setting up crystallization trials for *T*-p53C-G245S, Dr. Frank Boeckler for help generating Fig. 4 *B* and *C*; and the staff at the Synchrotron Radiation Source for helpful advice and assistance in data collection. This work was supported by the Agency for Science, Technology, and Research of Singapore (H.C.A.), Cancer Research UK, the Medical Research Council, and European Community FP6 funding.

1. Vogelstein B, Lane D, Levine AJ (2000) *Nature* 408:307–310.
2. Vousden KH, Lu X (2002) *Nat Rev Cancer* 2:594–604.
3. Olivier M, Eeles R, Hollstein M, Khan MA, Harris CC, Hainaut P (2002) *Hum Mutat* 19:607–614.
4. Beroud C, Soussi T (2003) *Hum Mutat* 21:176–181.
5. Cho Y, Gorina S, Jeffrey PD, Pavletich NP (1994) *Science* 265:346–355.
6. Clore GM, Ernst J, Clubb R, Omichinski JG, Kennedy WM, Sakaguchi K, Appella E, Gronenborn AM (1995) *Nat Struct Biol* 2:321–333.
7. Jeffrey PD, Gorina S, Pavletich NP (1995) *Science* 267:1498–1502.
8. Bell S, Klein C, Muller L, Hansen S, Buchner J (2002) *J Mol Biol* 322:917–927.
9. Veprintsev DB, Freund SM, Andreeva A, Rutledge SE, Tidow H, Canadillas JM, Blair CM, Fersht AR (2006) *Proc Natl Acad Sci USA* 103:2115–2119.
10. Kitayner M, Rozenberg H, Kessler N, Rabinovich D, Shaulov L, Haran TE, Shakked Z (2006) *Mol Cell* 22:741–753.
11. Ho WC, Fitzgerald MX, Marmorstein R (2006) *J Biol Chem* 281:20494–20502.
12. Canadillas JM, Tidow H, Freund SM, Rutherford TJ, Ang HC, Fersht AR (2006) *Proc Natl Acad Sci USA* 103:2109–2114.
13. Bullock AN, Fersht AR (2001) *Nat Rev Cancer* 1:68–76.
14. Bullock AN, Henckel J, Fersht AR (2000) *Oncogene* 19:1245–1256.
15. Wong KB, DeDecker BS, Freund SM, Proctor MR, Bycroft M, Fersht AR (1999) *Proc Natl Acad Sci USA* 96:8438–8442.
16. Joerger AC, Ang HC, Veprintsev DB, Blair CM, Fersht AR (2005) *J Biol Chem* 280:16030–16037.
17. Joerger AC, Allen MD, Fersht AR (2004) *J Biol Chem* 279:1291–1296.
18. Ang HC, Joerger AC, Mayer S, Fersht AR (2006) *J Biol Chem* 281:21934–21941.
19. Pan Y, Ma B, Levine AJ, Nussinov R (2006) *Biochemistry* 45:3925–3933.
20. Rippin TM, Freund SM, Veprintsev DB, Fersht AR (2002) *J Mol Biol* 319:351–358.
21. Klein C, Planker E, Diercks T, Kessler H, Kunkele KP, Lang K, Hansen S, Schwaiger M (2001) *J Biol Chem* 276:49020–49027.
22. Tidow H, Veprintsev DB, Freund SM, Fersht AR (2006) *J Biol Chem*, 10.1074/jbc.M604725200.
23. Zupnick AE, Prives C (2006) *J Biol Chem* 281:20464–20473.
24. Zhang W, Guo XY, Hu GY, Liu WB, Shay JW, Deisseroth AB (1994) *EMBO J* 13:2535–2544.
25. Di Como CJ, Prives C (1998) *Oncogene* 16:2527–2539.
26. Xu J, Baase WA, Baldwin E, Matthews BW (1998) *Protein Sci* 7:158–177.
27. Buckle AM, Henrick K, Fersht AR (1993) *J Mol Biol* 234:847–860.
28. Buckle AM, Cramer P, Fersht AR (1996) *Biochemistry* 35:4298–4305.
29. Shiraishi K, Kato S, Han SY, Liu W, Otsuka K, Sakayori M, Ishida T, Takeda M, Kanamaru R, Ohuchi N, *et al.* (2004) *J Biol Chem* 279:348–355.
30. Derbyshire DJ, Basu BP, Serpell LC, Joo WS, Date T, Iwabuchi K, Doherty AJ (2002) *EMBO J* 21:3863–3872.
31. Joo WS, Jeffrey PD, Cantor SB, Finnin MS, Livingston DM, Pavletich NP (2002) *Genes Dev* 16:583–593.
32. Gorina S, Pavletich NP (1996) *Science* 274:1001–1005.
33. Friedler A, Veprintsev DB, Rutherford T, von Glos KI, Fersht AR (2005) *J Biol Chem* 280:8051–8059.
34. Bykov VJ, Selivanova G, Wiman KG (2003) *Eur J Cancer* 39:1828–1834.
35. Wiman KG (2006) *Cell Death Differ* 13:921–926.
36. Brachmann RK, Yu K, Eby Y, Pavletich NP, Boeke JD (1998) *EMBO J* 17:1847–1859.
37. Nikolova PV, Wong KB, DeDecker B, Henckel J, Fersht AR (2000) *EMBO J* 19:370–378.
38. Sunyaev S, Ramensky V, Koch I, Lathe W, III, Kondrashov AS, Bork P (2001) *Hum Mol Genet* 10:591–597.
39. Huyen Y, Jeffrey PD, Derry WB, Rothman JH, Pavletich NP, Stavridi ES, Halazonetis TD (2004) *Structure (London)* 12:1237–1243.
40. Tang KS, Guralnick BJ, Wang WK, Fersht AR, Itzhaki LS (1999) *J Mol Biol* 285:1869–1886.
41. Resnick MA, Inga A (2003) *Proc Natl Acad Sci USA* 100:9934–9939.
42. Menendez D, Inga A, Resnick MA (2006) *Mol Cell Biol* 26:2297–2308.
43. Olivier M, Langerod A, Carrieri P, Bergh J, Klaar S, Eyfjord J, Theillet C, Rodriguez C, Lidereau R, Bieche I, *et al.* (2006) *Clin Cancer Res* 12:1157–1167.
44. Leslie AGW (1992) *Joint CCP4 and ESF-EACMB Newsletter on Protein Crystallography* (Daresbury Lab, Warrington, UK), Vol 26.
45. Collaborative Computational Project, Number 4 (1994) *Acta Crystallogr D* 50:760–763.
46. Brünger AT, Adams PD, Clore GM, DeLano WL, Gros P, Grosse-Kunstleve RW, Jiang J-S, Kuszewski J, Nilges M, Pannu NS, *et al.* (1998) *Acta Crystallogr D* 54:905–921.
47. Turk D (1992) PhD thesis (Technische Universität, Munich).
48. Laskowski RA, MacArthur MW, Moss DS, Thornton JM (1993) *J Appl Crystallogr* 26:283–291.
49. Kleywegt GJ, Jones TA (1994) *Acta Crystallogr D* 50:178–185.
50. Kraulis PJ (1991) *J Appl Crystallogr* 24:946–950.
51. Merritt EA, Bacon DJ (1997) *Methods Enzymol* 277:505–524.

One of the mutants, Y220C, is a perfect paradigm as a target for stabilization, having a deep, 'druggable', mutationally induced cleft far away from the functionally active regions of the protein[214]. This paper opened a whole new world for me in developing novel anti-cancer drugs. There was a reprise of my jumping up and down with joy in the corridors of Stanford when, at the end of a lecture in a summer school I was organising on the Greek Island of Spetses, David Lane phoned me that one of our lead compounds was active in a cancer cell line with Y220C. It was another eureka moment when students saw me jumping up and down with joy.

18.6 2009: Andreas Joerger and Joel Kaar.

Targeted rescue of a destabilized mutant of p53 by an *in silico* screened drug

Frank M. Boeckler*†, Andreas C. Joerger*, Gaurav Jaggi, Trevor J. Rutherford, Dmitry B. Veprintsev, and Alan R. Fersht‡

Centre for Protein Engineering, Medical Research Council Centre, Hills Road, Cambridge CB2 0QH, United Kingdom

Contributed by Alan R. Fersht, University of Cambridge, Cambridge, United Kingdom, June 1, 2008 (sent for review May 8, 2008)

The tumor suppressor p53 is mutationally inactivated in ≈50% of human cancers. Approximately one-third of the mutations lower the melting temperature of the protein, leading to its rapid denaturation. Small molecules that bind to those mutants and stabilize them could be effective anticancer drugs. The mutation Y220C, which occurs in ≈75,000 new cancer cases per annum, creates a surface cavity that destabilizes the protein by 4 kcal/mol, at a site that is not functional. We have designed a series of binding molecules from an *in silico* analysis of the crystal structure using virtual screening and rational drug design. One of them, a carbazole derivative (PhiKan083), binds to the cavity with a dissociation constant of ≈150 μM. It raises the melting temperature of the mutant and slows down its rate of denaturation. We have solved the crystal structure of the protein–PhiKan083 complex at 1.5-Å resolution. The structure implicates key interactions between the protein and ligand and conformational changes that occur on binding, which will provide a basis for lead optimization. The Y220C mutant is an excellent "druggable" target for developing and testing novel anticancer drugs based on protein stabilization. We point out some general principles in relationships between binding constants, raising of melting temperatures, and increase of protein half-lives by stabilizing ligands.

NMR screen | oncogenic mutant | protein stabilization | virtual drug design | crystal structure

The tumor suppressor p53 is a key protein in the cell's defense against cancer. If p53 and its associated cell-cycle pathways are active, then p53 will arrest the cell cycle of a potentially cancerous cell and induce apoptosis (1, 2). It is such a potent tumor suppressor that it or its pathways must be inactivated by mutation for cancer to proceed. p53 is inactivated directly by mutation in ≈50% of human cancers. It has a complex structure, being composed of several domains (reviewed in ref. 3). Nearly all of the oncogenic mutations occur in its core or DNA-binding domain, contained within the sequence of residues 94-292 (4, 5). The core domain of wild-type protein is rather unstable, with a melting temperature of ≈44°C and a short half-life of ≈9 min at body temperature (6–8). Many of the oncogenic mutants are inactivated simply because their stability is lowered so that the protein denatures very rapidly and is either too unstable to function at body temperature or rapidly depleted by denaturation and aggregation (9).

In principle, destabilized mutants of p53 can be stabilized by the binding of other molecules, as shown by the binding of a specific double-stranded DNA (6), heparin (10), or a designed peptide (11). Those molecules are targeted against wild-type p53 core domain and should bind generically to most mutants. Other molecules have been proposed to stabilize the folded state of the core domain of p53 [e.g., CP-31398 (12)], but, unlike the aforementioned examples, do not bind reversibly to the core domain (10, 13, 14) and promote anticancer effects *in vivo* via routes other than thermal stabilization of p53 (15).

We have proposed an alternative strategy to stabilizing p53 in a generic mode: the targeting of specific lesions in the protein that are induced by mutation so a drug may be designed that binds tightly to a mutation-induced binding site but weakly to wild-type p53 (5, 16). Y220C is the ninth most frequent p53 cancer mutant and accounts for an estimated 75,000 new cancer cases per annum worldwide based on cancer incidence statistics by the World Health Organization and reported mutation frequencies (www-p53.iarc.fr) (17). The mutation creates a surface crevice, and the protein is highly destabilized as a result (16). Importantly, this crevice is distant from the surface regions that are known to be involved in DNA recognition or protein–protein interactions, making it a particularly attractive target site for stabilizing small-molecule drugs. Here, we have: discovered a family of molecules that bind to the cavity of the oncogenic mutant Y220C using *in silico* screening based on the crystal structure and NMR screening within selected compounds; measured the binding of a representative compound, PhiKan083, to the target cavity using NMR spectroscopy; confirmed the dissociation constant by various biophysical methods; and shown that it raises the apparent melting temperature of the Y220C mutant and slows down the rate of thermal denaturation. We have solved the crystal structure of PhiKan083 bound to the mutant Y220C, thus identifying the residues of the protein that are in contact with the small molecule, which will provide a template for the design of drugs that may be therapeutically important.

Results and Discussion

Experiments were performed on the core domain *T*-p53C-Y220C (16), which has the mutation Y220C in a stabilized framework of human p53 that contains four mutations (18, 19).

***In Silico* Screening.** We applied a structure-based *in silico* screening approach to the crystal structure of *T*-p53C-Y220C [Protein Data Bank (PDB) ID code 2J1X] (16). Starting from the ZINC database (release 5) (20), a virtual collection of commercially available screening compounds, we chose a subset of 2,066,906 compounds according to the Lipinski "rule-of-five" (21). We explicitly considered multiple tautomers and protonation states when we created our initial database of 2,529,908 structures. Adding up to 10 alternative low-energy conformations per structure by the high-throughput conformational sampling module of MOE (22), we generated a total pool of >24.8 million conformers. We applied a series of increasingly more sophisticated filters including: (*i*) structure-based pharmacophore mod-

Author contributions: F.M.B., A.C.J., G.J., T.J.R., and A.R.F. designed research; F.M.B., A.C.J., G.J., T.J.R., and D.B.V. performed research; A.R.F. contributed new reagents/analytic tools; F.M.B., A.C.J., T.J.R., D.B.V., and A.R.F. analyzed data; and F.M.B., A.C.J., and A.R.F. wrote the paper.

The authors declare no conflict of interest.

Data deposition: The atomic coordinates and structure factors have been deposited in the Protein Data Bank, www.pdb.org (PDB ID code 2VUK).

*F.M.B. and A.C.J. contributed equally to this work.

†Present address: Department of Pharmacy, Center for Drug Research, Ludwig Maximilians-University Munich, Butenandtstrasse 7, D-81377 Munich, Germany.

‡To whom correspondence should be addressed. E-mail: arf25@cam.ac.uk.

This article contains supporting information online at www.pnas.org/cgi/content/full/0805326105/DCSupplemental.

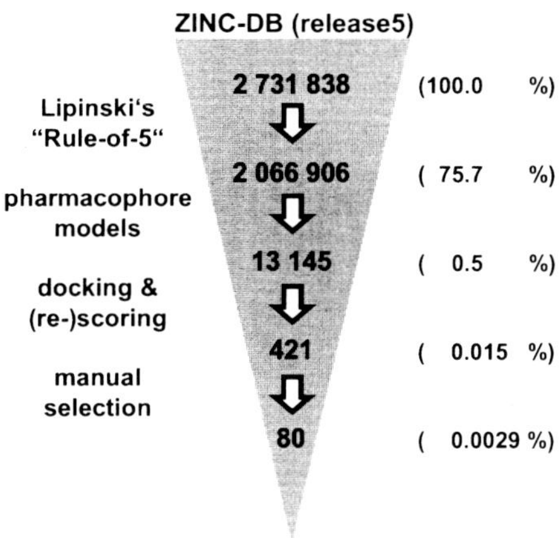

Fig. 1. *In silico* screening overview.

els [supporting information (SI) Fig. S1] in MOE (22), (*ii*) ligand docking into the crystal structure with GOLD (23, 24) and rescoring of the best poses to form a consensus score (Fig. S2), and (*iii*) manual selection from the consensus hit lists according to criteria of medicinal chemistry and crystallography (Fig. 1).

Primary *in Vitro* Screening Using $^{1}H/^{15}N$-HSQC NMR Spectroscopy. We screened 80 compounds from the final list of selected molecules in cocktails of four ligands using $^{1}H/^{15}N$-HSQC (HSQC, heteronuclear single-quantum coherence) NMR spectroscopy. Four compounds each at 2.5 mM concentration in d6-DMSO were added to ^{15}N-labeled *T*-p53C-Y220C to give a final concentration of 70 μM of protein, 114 μM of each compound, and 4.6% d6-DMSO. The compounds from any pot that caused significant changes in chemical shifts were then reexamined in four separate experiments. PhiKan059 (Fig. 2) was discovered to bind to the mutation-induced cleft of *T*-p53C-Y220C but not to *T*-p53C. We performed a series of $^{1}H/^{15}N$-HSQC experiments

9H carbazole

2-(9-ethyl-9*H*-carbazol-3-ylamino)-2-oxoethyl 5-oxopyrrolidine-2-carboxylate

PhiKan059

1-(9-ethyl-9*H*-carbazol-3-yl)-*N*-methylmethanamine

PhiKan083

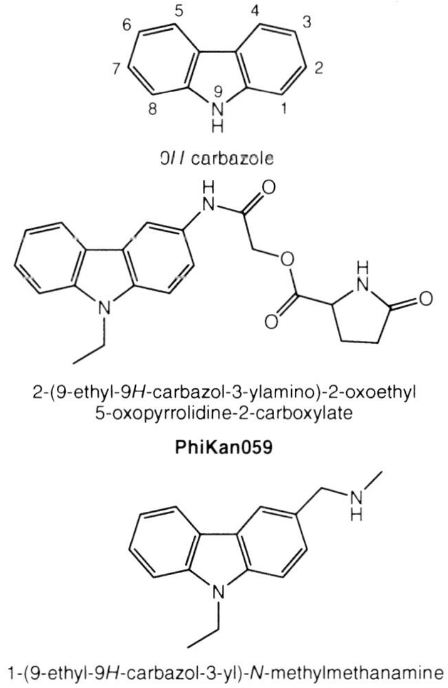

Fig. 2. Compounds and numbering.

with varying concentrations of PhiKan059 and used 15 characteristic chemical-shift differences to derive an average K_d of 213 μM for the binding of PhiKan059 to *T*-p53C-Y220C at 20°C (data not shown).

Extension of the Compound Library and Further Biophysical Characterization. We screened a secondary library of PhiKan059 analogues based on the carbazole scaffold, which yielded PhiKan083 (Fig. 2) among others with improved binding affinity to *T*-p53C-Y220C, with a K_d measured by NMR of 167 $\pm$ 12 μM for PhiKan083 (Figs. 3*A* and S3). Analytical ultracentrifugation (25) gave K_ds of 300 and 170 μM for PhiKan059 and PhiKan083, respectively, at 10°C (data not shown). Isothermal titration of PhiKan083 with *T*-p53C-Y220C gave 1:1 stoichiometry and a K_d of 125 $\pm$ 10 μM (Fig. 4).

Thermal Stabilization and Kinetics of Denaturation. We found initially from differential scanning calorimetry that PhiKan083 stabilized *T*-p53C-Y220C in a concentration-dependent manner. *T*-p53C-Y220C denatures irreversibly, and its apparent T_m varies with heating rate as does the denaturation of any protein where reversible and irreversible denaturation compete. At very fast heating, the measured T_m approximates to its true value, because the irreversible process is slower than equilibration, which in turn is still fast compared with the heating rate. The T_m is raised nearly 2°C from 316 K by 2.5 mM PhiKan083, and the data fit the equation expected for stabilization by simple binding with an approximate K_d of 140 $\pm$ 73 μM at 316–318 K (Fig. 3*B*).

The kinetics of denaturation of *T*-p53C-Y220C at 310 K (37°C) was fitted to a simple binding model for PhiKan083 (Fig. 3*C*). In the absence of ligand, the protein had a half-life of 3.8 min. This increased to 15.7 min at saturating concentrations of PhiKan083.

Crystal Structure of PhiKan083 Bound to *T*-p53C-Y220C and Implications for Drug Design. We solved the crystal structure of the *T*-p53C-Y220C:PhiKan083 complex at 1.5-Å resolution by soaking the small molecule into crystals of the mutant (Fig. 5*A*, Table 1). One of the two molecules in the asymmetric unit (chain B) showed unambiguous electron density for a PhiKan083 molecule bound to the mutation-induced surface cleft (Fig. 5*C*), whereas the pocket was partly occupied in chain A (Fig. S4). The central carbazole moiety is largely buried in the cleft, with the 9-ethyl group occupying the deepest part of the hydrophobic pocket (Fig. 5*B*). Binding would appear to have an important contribution from hydrophobic packing interactions. The ethyl group is in close contact to the sulfhydryl group of the mutated residue Cys-220, which adopts two alternative conformations, and a number of hydrophobic side chains (Phe-109, Leu-145, Val-147, and Leu-257), thus anchoring the ligand to the pocket. The planar carbazole ring system is sandwiched between the hydrophobic side chains of Pro-222 and Pro-223 on one side, and Val-147 and Pro-151 on the other side of the binding cleft. The ring nitrogen sits close to the position of the hydroxyl group of the tyrosine residue in the wild-type structure (1.0-Å distance) (Fig. 5*E*). The *N*-methylmethanamine moiety forms a hydrogen bond with the main-chain carbonyl of Asp-228 (2.7-Å distance). Only very small structural shifts occur upon ligand binding to the mutant. The residues that are within 5 Å of PhiKan083 (residues 109, 145–147, 150, 151, 220–223, 228–230, and 257) superimpose with a rmsd of 0.3 Å (all atoms). The most significant shift is observed for the side chain of Thr-150, which is displaced by up to 1.4 Å upon binding, thus widening the entrance of the pocket (Fig. 5*D*).

The occupancy of PhiKan083 in the binding pocket upon soaking is significantly different in the two molecules in the asymmetric unit, even though the architecture of the crystal lattice would suggest a similar accessibility of both sites. In

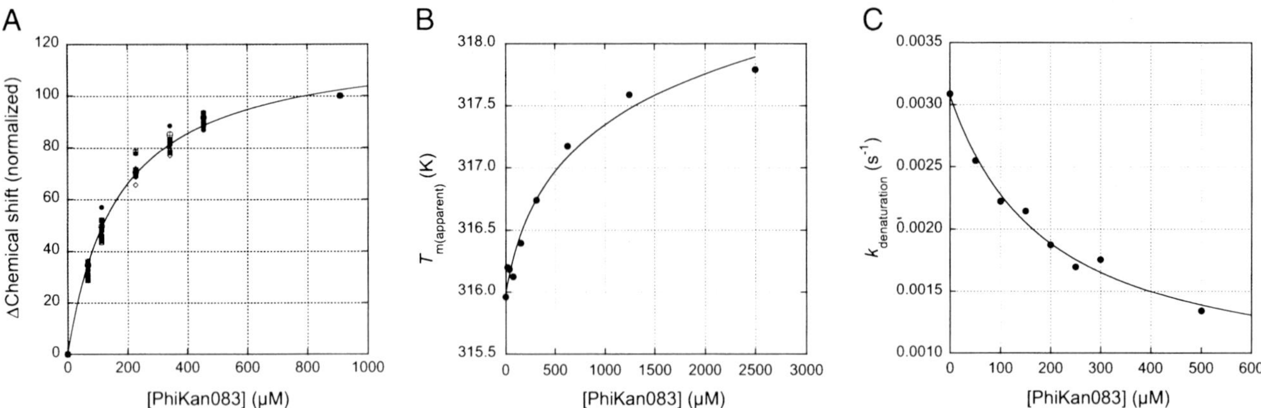

Fig. 3. Effects of PhiKan083 on *T*-p53C-Y220C. (*A*) Changes in chemical shifts (normalized) vs. concentration for 15 resonances of *T*-p53C-Y220C in the presence of PhiKan083 at 20°C. The data are fitted to a single-site binding model. (*B*) Thermal denaturation of *T*-p53C-Y220C (10 μM) in the presence of PhiKan083. Denaturation is irreversible. However, at the very high heating rate of 270 K/h, the measured T_m is close to the reversible value. The data are fitted to the equation: $T = T_m/(1 - (R/\Delta S_{D\text{-}N(Tm)})\ln(1 + [L]/K_d))$, where T is the observed melting temperature, T_m that in the absence of ligand L, K_d its dissociation constant, and $\Delta S_{D\text{-}N(Tm)}$ the entropy of denaturation at T_m (the derivation is in the legend to Fig. S5). (*C*) Effect of PhiKan083 on kinetics of thermal denaturation at 37°C.

subunit A, however, parts of the S7/S8 loop (residues 225 and 226) interact with a neighboring molecule, whereas this region is not engaged in packing interactions in molecule B. In this packing arrangement, the structural plasticity required for efficient binding may be reduced in chain A. Moreover, in the ligand-free structure of the mutant (PDB ID code 2J1X), the side chain of Thr-230 inside the binding pocket adopts two alternative conformations in molecule A (but not in molecule B), which may hamper effective binding to molecule A upon soaking of the crystals, because one of these conformations narrows the bottom of the binding pocket (Fig. 5*F*).

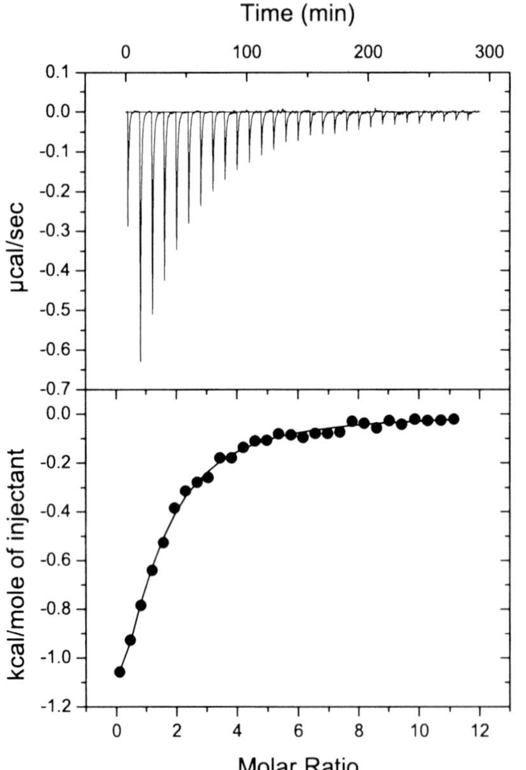

Fig. 4. Isothermal titration calorimetry of PhiKan083 binding at 20°C showing raw data (*Upper*) and fit after integration (*Lower*).

Comparison of modeled and observed binding modes nicely shows how relatively small changes in the protein environment that occur upon ligand binding can be crucial for the accuracy of binding-mode predictions. If the ligand is docked to chain A of the free structure with the Cγ-methyl group of the Thr-230 side chain occupying part of the binding site (purple structure in Fig. 5*F*), the anchoring ethyl group is positioned at ≈1.5 Å from its actual binding position (rmsd of the whole ligand, 6.0 Å), and the carbazole is flipped by 180°, placing the amine group at the opposite end of the binding pocket. Furthermore, this altered binding mode corresponds to a dramatic decrease in predicted affinity, as represented by its DrugScore^{CSD} value (26) (Table 2). In contrast, if the ligand is docked into the structure of the complex (after removing the coordinates of the bound ligand), i.e., taking the observed small induced-fit movements of the protein residues into account, the modeled binding mode (yellow molecule in Fig. 5*F*) almost perfectly matches the observed binding mode (green molecule in Fig. 5*F*). Small induced-fit structural changes bedevil *in silico* screening methods, and so it is important to have solved the structure of the complex. Even small changes in the structures of substrates and ligands have been known from early studies to cause radical changes in modes of binding (27).

Inspection reveals contacts, where the central, hydrophobic carbazole scaffold contacts a polar atom. The hydroxyl of Thr-150 at the entrance of the cleft, for example, is in close contact to C5 of the bound carbazole (3.5 Å). Probably more crucially, the main-chain oxygen of Leu-145 at the bottom of the cleft is only 3.3 Å away from C1 of the carbazole ring system. In the unligated structure (PDB ID code 2J1X), this carbonyl group forms a hydrogen bond with a water molecule that is displaced upon ligand binding (Fig. 5*D*). These observations indicate that a substantial increase in binding energy, and hence a lower dissociation constant, should be achieved through structure-guided modification of the central scaffold, e.g., by providing a hydrogen-bond donor for the carbonyl of Leu-145 or by replacing part of the aromatic structure by suitable nonaromatic bioisosteres (28–30). Moreover, the part of the crevice that preexists in the wild type is not addressed by the binding mode of PhiKan083, which offers the potential for substantial extension of the ligand to create additional stabilizing interactions. Further hit optimi-

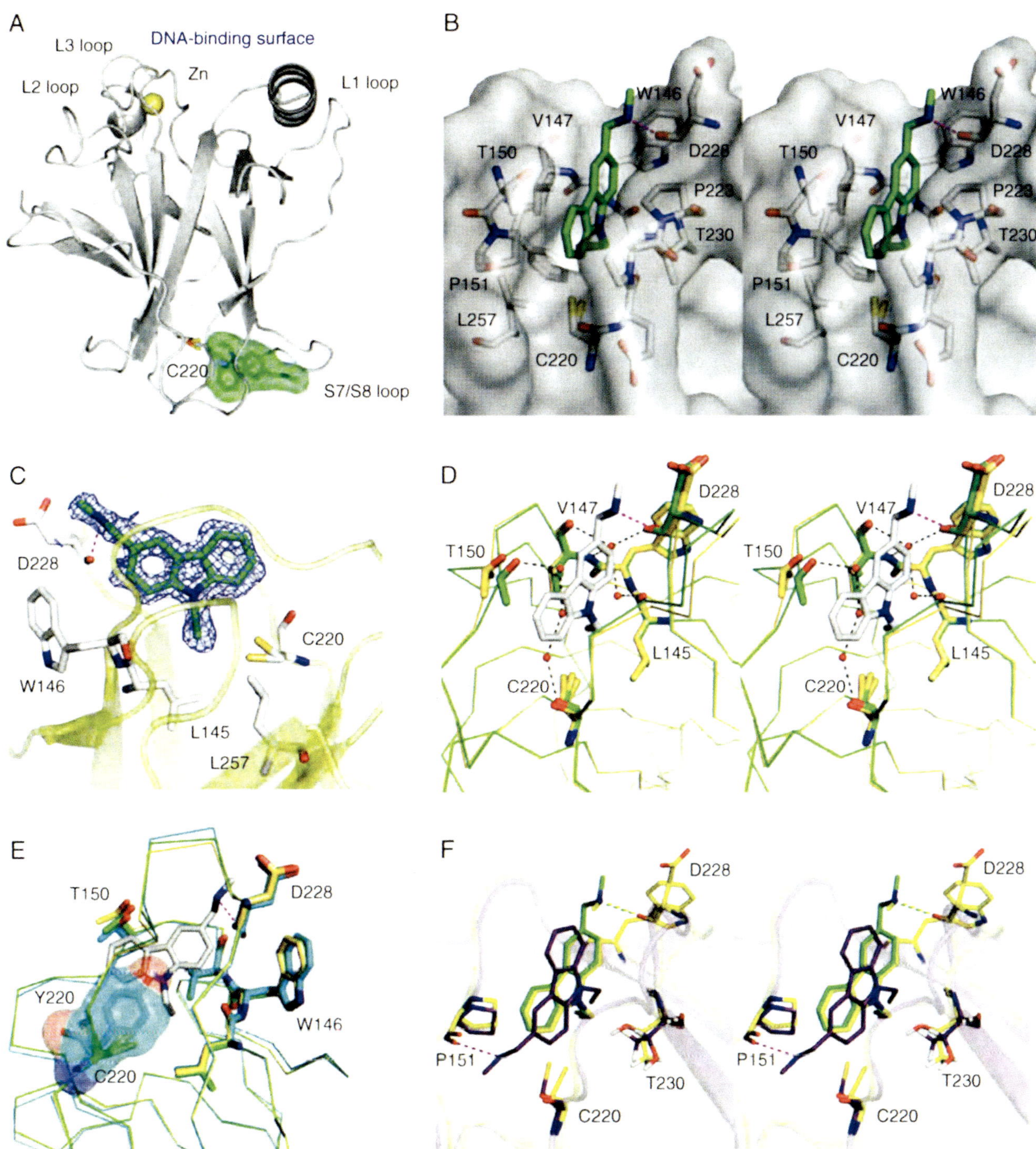

Fig. 5. Crystal structure of *T*-p53C-Y220C in complex with PhiKan083. (*A*) Ribbon representation of the overall structure of *T*-p53C-Y220C in complex with PhiKan083 (PDB ID code 2VUK, chain B). PhiKan083 is shown in green as a stick model with its molecular surface. It binds to the mutation-induced cleft on the protein surface that is distant from the known functional interfaces of the protein. The side chain of Cys-220 at the mutation site, which adopts two alternative conformations, is highlighted in orange. (*B*) Stereoview of the PhiKan083-binding site. p53 residues within a 5-Å distance of the ligand are shown as gray stick models. The protein surface is highlighted in semitransparent gray. (*C*) $|F_o - F_c|$ simulated-annealing omit map of PhiKan083 bound to chain B of *T*-p53C-Y220C contoured at 3.0 σ. (*D*) Superposition of *T*-p53C-Y220C in its free (PDB ID code 2J1X chain B; green) (16) and PhiKan083-bound form (yellow), indicating small structural shifts upon ligand binding. PhiKan083 is depicted as a gray stick model. The small red spheres represent water molecules in the ligand-free structure that are displaced upon ligand binding. (*E*) In wild-type p53, Tyr-220 blocks part of the Phikan083-binding pocket, as shown for the structure of wild-type core domain (PDB code 2AC0, chain B; cyan) (42) superimposed onto Phikan083-bound *T*-p53C-Y220C (yellow protein chain and gray PhiKan083 molecule) and free *T*-p53C-Y220C (green). (*F*) Docking of Phikan083 to the structure of ligand-free *T*-p53C-Y220C (PDB ID code 2J1X, chain A, Thr-230 rotamer A; purple) and to the protein chain of the complex structure (yellow) compared with its actual binding mode in the crystal structure of the complex (green). All images were prepared with PYMOL (http://pymol.sourceforge.net).

Table 1. Data collection and refinement statistics

Data collection	
Space group	$P2_12_12_1$
Cell, *a*, *b*, *c*, Å	65.09, 71.23, 105.21
Molecules per AU	2
Resolution, Å*	65.1 – 1.50 (1.58 – 1.50)
Unique reflections	76,025
Completeness, %*	96.6 (83.4)
Multiplicity*	5.6 (4.6)
R_{merge}, %*†	6.6 (22.9)
$<I/\sigma_I>$*	17.7 (5.6)
Wilson *B* value, Å^2	13.8
Refinement	
Number of atoms	
Protein‡	3,119
Water	434
Zinc	2
PhiKan083	18
R_{cryst}, %§	18.6
R_{free}, %§	20.8
rmsd bonds, Å	0.009
rmsd angles, °	1.5
Mean *B* value, Å^2	15.9
Ramachandran plot statistics¶	
Most favored/additional allowed, %	91.0/9.0
Generously allowed/disallowed, %	0/0

*Values in parentheses are for the highest-resolution shell.
†$R_{merge} = \Sigma(I_{h,i} - \langle I_h \rangle)/\Sigma I_{h,i}$.
‡Number includes alternative conformations.
§R_{cryst} and $R_{free} = \Sigma||F_{obs}| - |F_{calc}||/\Sigma|F_{obs}|$, where R_{free} was calculated over 5% of the amplitudes chosen at random and not used in the refinement.
¶Calculated with PROCHECK (43).

Table 2. Consequences of different crystal structures on *in silico* prediction of the binding mode

Structure	Binding mode	Structure in Fig. 5*F*	rmsd of PhiKan083*, Å	DrugScore value†
2VUK (chain B)	Experimental	Green	0	−187,720
2VUK (chain B)	Docked	Yellow	0.56	−175,488
2J1X (chain A)	Docked	Purple	6.01	−26,247

*rmsd vs. coordinates of experimental structure.
†Calculated with DrugScore online (www.agklebe.de) (26). Higher negative values indicate a higher-affinity prediction.

zation could, therefore, combine ligand-based SAR data with structure-based design (31, 32).

Conclusions

We have demonstrated that the site of mutation in the oncogenic Y220C mutant is a druggable target. PhiKan083 binds to it with reasonable affinity. The crystal structure of the complex will be a starting point for further rounds of drug design and refinement. Because the site of mutation does not appear to be in a region of the protein that is functionally important, it is an excellent target for drug stabilization therapy. The degree of stabilization given by a drug is related to:

$T = T_m/(1 - (R/\Delta S_{D\text{-}N(Tm)})\ln(1 + [L]/K_d))$, where T is the observed melting temperature, T_m that in the absence of ligand (drug) L, K_d its dissociation constant, and $\Delta S_{D\text{-}N(Tm)}$ the entropy of denaturation at T_m (see legend to Fig. S5). For $\Delta S_{D\text{-}N(Tm)}$ for a protein such as p53, increases in T_m of 3.7, 4.8, 6.2 and 7.3°C are expected for the ratios of $[L]/[K_d]$ of, 10, 20, 50, and 100 respectively; see Fig. S5. Further, the half life of the mutant protein is significantly increased by the drug, and the increase will be greater for future generations of drugs that have more intrinsic binding energy. There is every possibility that an anticancer drug could be developed for the Y220C mutant using compounds that bind more tightly than the current lead.

Methods

Materials. *T*-p53C-Y220C for protein crystallography was purified as described (16, 33). For biophysical measurements and NMR spectroscopy, we used a pET24a-HLTV-derived vector (courtesy of M. D. Allen, Medical Research Council Centre), containing an N-terminal fusion of a 6xHis/lipoamyl domain/TEV protease cleavage site sequence. Expression and purification followed published protocols (34). The N-terminally fused 6xHis/lipoamyl domain sequence was removed by TEV protease digestion. PhiKan059 and PhiKan083 were obtained from Enamine, Ltd.

Biophysical Methods. ***Isothermal titration calorimetry.*** Binding was measured with an isothermal titration calorimeter (VP-ITC, MicroCal) (35) at 20°C in a 25 mM sodium phosphate buffer (pH 7.2), 150 mM NaCl, and 1 mM DTT. For binding of PhiKan083 to *T*-p53C-Y220C, 5 mM PhiKan083 in 5% d6-DMSO in buffer was titrated into the sample cell containing 100 μM protein in 5% d6-DMSO in buffer. The DMSO content was matched using a high-precision Mettler Toledo balance. Injection steps were 10 μl (first injection, 3 μl) with 600-s spacing. Further data evaluation was done by using the MicroCal Origin program.

NMR spectroscopy. The low-molecular-weight compounds were dissolved in d6-DMSO to make 10 mM stock solutions. To screen compound mixtures by chemical-shift mapping, 10 μl each of four different compounds were mixed together, and 25 μl of this mixture was added to 25 μl of D_2O and 500 μl of 70 μM *T*-p53C-Y220C (in 25 mM sodium phosphate, 150 mM NaCl, and 5 mM DTT, pH 7.2). The final concentration for each compound was 114 μM at a concentration of 4.5% (vol/vol) d6-DMSO. NMR samples were freshly prepared and kept sealed under argon after degassing. $^1H/^{15}N$ HSQC correlation spectra were acquired at 20°C on Bruker AvanceII+ 700 and Avance 800 spectrometers using a $^1H/^{13}C/^{15}N$ triple resonance inverse, cryogenic 5-mm probe (Bruker), with the following parameters: 16 scans, 128 complex points in t1, recycle time of 0.95 sec, and 1,024 total points in t2. Using Bruker's TopSpin 2.0 software, the number of complex points in t1 was doubled by forward-complex linear prediction, and shifted squared sine bell window functions were applied to both dimensions before zero filling and Fourier transformation. A digital resolution of 2.0 Hz per point in the ^1H frequency dimension and 4.7 Hz per point in the ^{15}N frequency dimension was used. Spectra were analyzed by using Sparky 3.113 (36).

Thermal stability. Thermal unfolding was followed either by differential scanning calorimetry as described in ref. 37 at a scan rate of 250 K/h or by monitoring unfolding by the binding of the dye SYPRO Orange (5×) using a Rotor-gene 6000 (Corbett Life Science) at 270 K/h in 25 mM sodium phosphate, 150 mM NaCl, and 5 mM DTT, pH 7.2, with a protein concentration of 10 μM.

Time-dependent fluorescence studies. Unfolding kinetics was performed as described by Friedler *et al.* (11) at 37°C in 50 mM Hepes, pH 7.2, 1 mM Tris-2-carboxyethylphosphine, by following the emission of tryptophan at 340 nm on excitation at 280 nm (6), using a Cary Eclipse fluorescence spectrophotometer controlled by the supplied Cary software. Reactions were followed for 10,000 s. Data were fitted to a single exponential followed by a linear-drift term.

X-Ray Crystallography Methods. Crystals of *T*-p53C-Y220C in space group $P2_12_12_1$ with two molecules in the asymmetric unit were grown at 21°C by sitting-drop vapor diffusion under the conditions described in ref. 16. PhiKan083 was soaked into crystals of *T*-p53C-Y220C by stepwise addition of cryo buffer (19% polyethylene glycol 4,000, 20% glycerol, 10 mM sodium phosphate, pH 7.2, 100 mM Hepes, pH 7.2, 150 mM KCl) with increasing concentration of PhiKan083 over a period of 2 h. After reaching the final concentration of 10 mM, soaking was continued for another 30 min before the crystals were flash frozen in liquid nitrogen. An x-ray dataset to 1.5-Å resolution was collected at 100 K on beamline I04 at the Diamond Light Source. Data processing was performed by using Mosflm (38) and Scala (39). Structure solution and refinement were performed with CNS (40). After an initial round of rigid body refinement using the structure of free *T*-p53C-Y220C (PDB ID code 2J1X) as starting model, the structure of the complex was refined by iterative cycles of refinement with CNS and manual model building with MAIN (41). Water molecules were added to the structure by using the water-pick option implemented within CNS and manual model building. At this stage of the refinement, PhiKan083 was built into the model of chain B, and the structure was further refined, including incorporation of alternative conformations for selected side chains. For the cavity in chain A, we observed significant difference density having contributions from PhiKan083 in the same binding mode as in chain B but bound with a low occupancy and a network of water molecules as observed in the unbound state (Fig. S4), but we did not include the ligand in the final model of chain A. The data collection and refinement statistics are shown in Table 1.

ACKNOWLEDGMENTS. We thank Caroline Blair for protein purification for crystallographic studies, Dr. Chris Johnson for advice and helpful discussions on DSC and ITC, Dr. Wolfgang Utz (Friedrich-Alexander University, Erlangen, Germany) for help with part of the rescoring procedure, and the staff at beamline I04 of the Diamond Light Source for technical support in data collection. F.M.B. was supported by German Research Foundation (Deutsche Forschungsgemeinschaft) Fellowship BO 3029/1-1 and subsequently by Marie-Curie Intra-European Fellowship EU MEIF-CT-2006-039961. This work was supported by Cancer Research U.K., the Medical Research Council, and by EC FP6 funding.

1. Vogelstein B, Lane D, Levine AJ (2000) Surfing the p53 network. *Nature* 408:307–310.
2. Vousden KH, Lu X (2002) Live or let die: The cell's response to p53. *Nat Rev Cancer* 2:594–604.
3. Joerger AC, Fersht AR (2008) Structural Biology of the tumor suppressor p53. *Annu Rev Biochem* 77:557–582.
4. Olivier M, *et al.* (2002) The IARC TP53 database: New online mutation analysis and recommendations to users. *Hum Mutat* 19:607–614.
5. Joerger AC, Fersht AR (2007) Structure-function-rescue: The diverse nature of common p53 cancer mutants. *Oncogene* 26:2226–2242.
6. Bullock AN, *et al.* (1997) Thermodynamic stability of wild-type and mutant p53 core domain. *Proc Natl Acad Sci USA* 94:14338–14342.
7. Ang HC, Joerger AC, Mayer S, Fersht AR (2006) Effects of common cancer mutations on stability and DNA binding of full-length p53 compared with isolated core domains. *J Biol Chem* 281:21934–21941.
8. Friedler A, Veprintsev DB, Hansson LO, Fersht AR (2003) Kinetic instability of p53 core domain mutants: implications for rescue by small molecules. *J Biol Chem* 278:24108–24112.
9. Bullock AN, Henckel J, Fersht AR (2000) Quantitative analysis of residual folding and DNA binding in mutant p53 core domain: Definition of mutant states for rescue in cancer therapy. *Oncogene* 19:1245–1256.
10. Rippin TM, *et al.* (2002) Characterization of the p53-rescue drug CP-31398 *in vitro* and in living cells. *Oncogene* 21:2119–2129.
11. Friedler A, *et al.* (2002) A peptide that binds and stabilizes p53 core domain: Chaperone strategy for rescue of oncogenic mutants. *Proc Natl Acad Sci USA* 99:937–942.
12. Foster BA, Coffey HA, Morin MJ, Rastinejad F (1999) Pharmacological rescue of mutant p53 conformation and function. *Science* 286:2507–2510.
13. Issaeva N, *et al.* (2003) Rescue of mutants of the tumor suppressor p53 in cancer cells by a designed peptide. *Proc Natl Acad Sci USA* 100:13303–13307.
14. Tanner S, Barberis A (2004) CP-31398, a putative p53-stabilizing molecule tested in mammalian cells and in yeast for its effects on p53 transcriptional activity. *J Negat Results Biomed* 3:5.
15. Tang X, *et al.* (2007) CP-31398 restores mutant p53 tumor suppressor function and inhibits UVB-induced skin carcinogenesis in mice. *J Clin Invest* 117:3753–3764.
16. Joerger AC, Ang HC, Fersht AR (2006) Structural basis for understanding oncogenic p53 mutations and designing rescue drugs. *Proc Natl Acad Sci USA* 103:15056–15061.
17. Petitjean A, *et al.* (2007) Impact of mutant p53 functional properties on TP53 mutation patterns and tumor phenotype: lessons from recent developments in the IARC TP53 database. *Hum Mutat* 28:622–629.
18. Nikolova PV, Henckel J, Lane DP, Fersht AR (1998) Semirational design of active tumor suppressor p53 DNA binding domain with enhanced stability. *Proc Natl Acad Sci USA* 95:14675–14680.
19. Joerger AC, Allen MD, Fersht AR (2004) Crystal structure of a superstable mutant of human p53 core domain. Insights into the mechanism of rescuing oncogenic mutations. *J Biol Chem* 279:1291–1296.
20. Irwin JJ, Shoichet BK (2005) ZINC–a free database of commercially available compounds for virtual screening. *J Chem Inf Model* 45:177–182.
21. Lipinski CA, Lombardo F, Dominy BW, Feeney PJ (2001) Experimental and computational approaches to estimate solubility and permeability in drug discovery and development settings. *Adv Drug Deliv Rev* 46:3–26.
22. Chemical Computing Group (2005) MOE (Chemical Computing Group Inc, Montreal, Canada), Ver. 2005.06.
23. Jones G, Willett P, Glen RC (1995) Molecular recognition of receptor sites using a genetic algorithm with a description of desolvation. *J Mol Biol* 245:43–53.
24. Jones G, Willett P, Glen RC, Leach AR, Taylor R (1997) Development and validation of a genetic algorithm for flexible docking. *J Mol Biol* 267:727–748.
25. Weinberg RL, Veprintsev DB, Fersht AR (2004) Cooperative binding of tetrameric p53 to DNA. *J Mol Biol* 341:1145–1159.
26. Velec HFG, Gohlke H, Klebe G (2005) DrugScoreCSD-knowledge-based scoring function derived from small molecule crystal data with superior recognition rate of near-native ligand poses and better affinity prediction. *J Med Chem* 48:6296–6303.
27. Fastrez J, Fersht AR (1973) Mechanism of chymotrypsin, structure, reactivity, and nonproductive binding relationships. *Biochemistry* 12:1067–1074.
28. Lenz C, Boeckler F, Hubner H, Gmeiner P (2004) Analogues of FAUC 73 revealing new insights into the structural requirements of nonaromatic dopamine D3 receptor agonists. *Bioorg Med Chem* 12:113–117.
29. Lenz C, Haubmann C, Hubner H, Boeckler F, Gmeiner P (2005) Fancy bioisosteres: Synthesis and dopaminergic properties of the endiyne FAUC 88 as a novel non-aromatic D3 agonist. *Bioorg Med Chem* 13:185–191.
30. Lenz C, Boeckler F, Hubner H, Gmeiner P (2005) Fancy bioisosteres: Synthesis, SAR, and pharmacological investigations of novel nonaromatic dopamine D3 receptor ligands. *Bioorg Med Chem* 13:4434–4442.
31. Boeckler F, Lanig H, Gmeiner P (2005) Modeling the similarity and divergence of dopamine D2-like receptors and identification of validated ligand-receptor complexes. *J Med Chem* 48:694–709.
32. Boeckler F, *et al.* (2005) CoMFA and CoMSIA investigations revealing novel insights into the binding modes of dopamine D3 receptor agonists. *J Med Chem* 48:2493–2508.
33. Joerger AC, Ang HC, Veprintsev DB, Blair CM, Fersht AR (2005) Structures of p53 cancer mutants and mechanism of rescue by second-site suppressor mutations. *J Biol Chem* 280:16030–16037.
34. Canadillas JM, *et al.* (2006) Solution structure of p53 core domain: structural basis for its instability. *Proc Natl Acad Sci USA* 103:2109–2114.
35. Tidow H, Andreeva A, Rutherford TJ, Fersht AR (2007) Solution structure of ASPP2 N-terminal domain (N-ASPP2) reveals a ubiquitin-like fold. *J Mol Biol* 371:948–958.
36. Goddard TD, Kneller DG (2006) *SPARKY* 3 (University of California, San Francisco).
37. Mayer S, Rudiger S, Ang HC, Joerger AC, Fersht AR (2007) Correlation of levels of folded recombinant p53 in *Escherichia coli* with thermodynamic stability *in vitro*. *J Mol Biol* 372:268–276.
38. Leslie AGW (1992) Recent Changes to the MOSFLM package for processing film and image plate data. *Joint CCP4 and ESF-EACMB Newsletter on Protein Crystallography* (Daresbury Laboratory, Warrington, UK), Vol. 26.
39. Collaborative Computational Project N (1994) The CCP4 suite: Programs for protein crystallography. *Acta Crystallogr D* 50:760–763.
40. Brünger AT, *et al.* (1998) Crystallography and NMR system: A new software suite for macromolecular structure determination. *Acta Crystallogr D* 54: 905–921.
41. Turk D (1992) PhD thesis (Technische Universität München, Munich, Germany).
42. Kitayner M, *et al.* (2006) Structural basis of DNA recognition by p53 tetramers. *Mol Cell* 22:741–753.
43. Laskowski RA, MacArthur MW, Moss DS, Thornton JM (1993) PROCHECK: A program to check the stereochemical quality of protein structures. *J Appl Crystallogr* 26:283–291.

Chapter 19

p53 Structural Studies

'And, you know, it'll take time to restore chaos and order — order out of chaos. But we will.'

George W. Bush

p53 is a representative of an emerging class of proteins that have large disordered regions. Their structures are not amenable to solution by the conventional methods of X-ray crystallography and NMR used on globular proteins, but require application of a batch of methods from small-angle X-ray diffraction and NMR in solution to electron microscopy on immobilised samples. Our superstable variant of full-length p53 is one of the first such proteins to be solved in such a manner[215,216]. Our paper is the prelude to studies in which I, and no doubt other groups, will build up the structures of the protein complexes involving p53 and other cell-cycle proteins that cannot be crystallised. Such studies require extensive teamwork because so many different techniques are involved. I am very proud to have led a team that has produced the structure of full-length p53.

19.1 2009: Henning Tidow at Spetses Summer School.

Quaternary structures of tumor suppressor p53 and a specific p53–DNA complex

Henning Tidow*, Roberto Melero†, Efstratios Mylonas‡, Stefan M. V. Freund*, J. Guenter Grossmann§, José María Carazo†, Dmitri I. Svergun‡¶‖, Mikel Valle†‖**, and Alan R. Fersht*‖

*Medical Research Council Centre for Protein Engineering, Hills Road, Cambridge CB2 0QH, United Kingdom; †Centro Nacional de Biotecnología, Darwin 3, Cantoblanco 28049 Madrid, Spain; **CIC-bioGUNE, Parque Tecnológico de Bizkaia, 48160 Derio, Spain; ‡European Molecular Biology Laboratory, Hamburg Outstation, Notkestrasse 85, 22603 Hamburg, Germany; ¶Institute of Crystallography, Russian Academy of Sciences, Leninsky pr. 59, 117333 Moscow, Russia; and §Molecular Biophysics Group, Council for the Central Laboratory of the Research Councils (CCLRC) Daresbury Laboratory, Warrington, Cheshire WA4 4AD, United Kingdom

Contributed by Alan R. Fersht, May 30, 2007 (sent for review May 8, 2007)

The homotetrameric tumor suppressor p53 consists of folded core and tetramerization domains, linked and flanked by intrinsically disordered segments that impede structure analysis by x-ray crystallography and NMR. Here, we solved the quaternary structure of human p53 in solution by a combination of small-angle x-ray scattering, which defined its shape, and NMR, which identified the core domain interfaces and showed that the folded domains had the same structure in the intact protein as in fragments. We combined the solution data with electron microscopy on immobilized samples that provided medium resolution 3D maps. *Ab initio* and rigid body modeling of scattering data revealed an elongated cross-shaped structure with a pair of loosely coupled core domain dimers at the ends, which are accessible for binding to DNA and partner proteins. The core domains in that open conformation closed around a specific DNA response element to form a compact complex whose structure was independently determined by electron microscopy. The structure of the DNA complex is consistent with that of the complex of four separate core domains and response element fragments solved by x-ray crystallography and contacts identified by NMR. Electron microscopy on the conformationally mobile, unbound p53 selected a minor compact conformation, which resembled the closed conformation, from the ensemble of predominantly open conformations. A multipronged structural approach could be generally useful for the structural characterization of the rapidly growing number of multidomain proteins with intrinsically disordered regions.

DNA binding | intrinsically unfolded | modular | natively disordered | protein

The tumor suppressor p53 is a tetrameric, multidomain transcription factor that plays a central role in the cell cycle and maintaining genomic integrity (1, 2). It binds to specific DNA response elements, is integrated in various signaling networks by a multitude of protein–protein interactions, and is controlled by extensive posttranslational modifications (1, 3, 4). p53 protein is a homotetramer of 4 × 393 residues. Each chain consists of two folded domains [the core, or DNA-binding, domain (94–294) and the tetramerization domain (323–360)] that are linked by an intrinsically disordered sequence. The transactivation domain (1–67) (5), proline-rich region (67–94), nuclear localization signal (NLS)-containing region (303–323) (6), and C-terminal negative regulatory domain (360–393) are also intrinsically disordered (7–9) (see refs. 10 and 11 for reviews). The DNA-binding core domain (residues 94–294) binds to sequence-specific response elements associated with p53 target gene promoters (12–14). The structures of the core domain complexes with DNA have been solved by crystallography (15–17), and in solution in the absence of DNA by NMR (18). The structure of the tetramerization domain has been solved by both NMR and x-ray crystallography (19–21).

Structural studies on full-length p53 have been impeded both by its intrinsic instability and the presence of disordered regions (8, 18, 22, 23). There is an increasing number of proteins being discovered to have globular domains linked by intrinsically disordered regions (24), and so the determination of such structures will be a recurring problem. Recently, the first structure of full-length murine p53, obtained by cryoelectron microscopy, was reported (25). It gave a radically new concept of the molecular organization of p53. We have solved the instability problem of human p53 by designing a biologically active mutant with a superstable core domain (26, 27) that is suitable for NMR studies. NMR studies have given a picture of the p53 tetramer as a dimer of loosely tethered core dimers of appropriate symmetry to be poised to bind target DNA with well resolved transverse relaxation optimized spectroscopy (TROSY) NMR spectra (23). But, on addition of DNA, the protein rigidifies with poorly resolved spectra (23).

Here, we used small-angle x-ray scattering (SAXS), electron microscopy (EM), and NMR spectroscopy in a multitechnique approach to determine the structures of tetrameric full-length human p53 and various truncation mutants, both unligated and as a complex with a DNA response element.

Results and Discussion

Quaternary Structure of p53 from SAXS. SAXS experiments were performed on solutions of three p53 constructs: p53 core plus tetramerization domain (94–360) (CTetD), p53 core plus tetramerization plus C-terminal domain (94–393) (CTetCD), and full-length p53 (1–393) (flp53). All had four stabilizing mutations in the core domain (M133L/V203A/N239Y/N268D) (27) that do not perturb the structure (26). The processed experimental patterns (scattering intensity I versus momentum transfer $s = 4\pi\sin\theta/\lambda$, where 2θ is the scattering angle and $\lambda = 1.5$ Å is the wavelength) are displayed in Fig. 1*a*, and the overall parameters are summarized in Table 1. The solutions are monodisperse [supporting information (SI) Fig. 6], and the molecular weights (M_r) of the solutes determined from the extrapolated zero angle intensity $I(0)$ consistently point to a tetrameric assembly for each construct. The radius of gyration (R_g) increases from CTetD (52.2 Å) to CTetCD (54.0 Å) to flp53 (68.2 Å) (Table 1). The

Author contributions: H.T. and R.M. contributed equally to this work; H.T., R.M., E.M., S.M.V.F., J.M.C., M.V., and A.R.F. designed research; H.T., R.M., E.M., S.M.V.F., J.G.G., and M.V. performed research; H.T., R.M., E.M., S.M.V.F., J.M.C., D.I.S., M.V., and A.R.F. analyzed data; and H.T., D.I.S., M.V., and A.R.F. wrote the paper.

The authors declare no conflict of interest.

Abbreviations: CTetD, p53 core and tetramerization domains; CTetCD, p53 core, tetramerization and C-terminal domain; flp53, full-length p53; SAXS, small-angle X-ray scattering; TROSY, transverse relaxation optimized spectroscopy.

See Commentary on page 12231.

‖To whom correspondence may be addressed. E-mail: arf25@cam.ac.uk, mvalle@cicbiogune.es, or svergun@embl-hamburg.de.

This article contains supporting information online at www.pnas.org/cgi/content/full/0705069104/DC1.

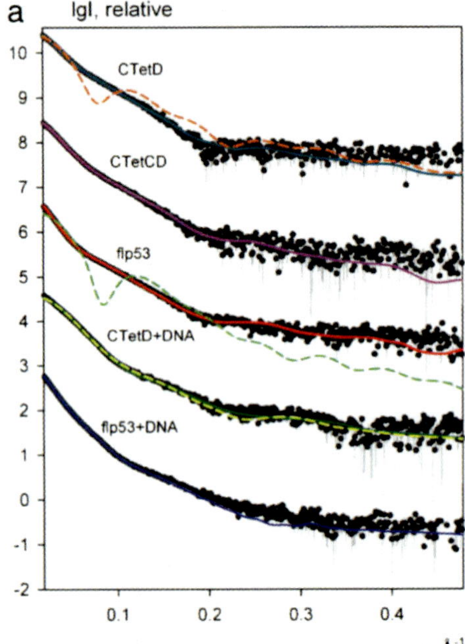

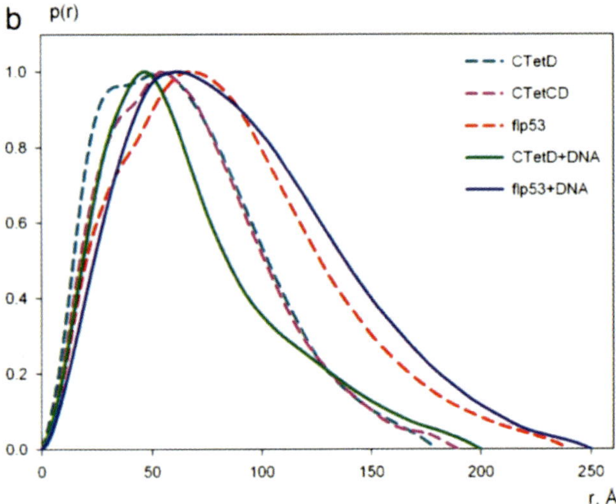

Fig. 1. SAXS analysis of p53 constructs. (*a*) Experimental intensities (dots with error bars) and fits computed from the structural models by rigid body modeling: CTetD (cyan), CTetCD (magenta), flp53 (red), CTetD+DNA (dark green). For the flp53–DNA complex the fit from the *ab initio* model (SI Fig. 9) is displayed in blue. The scattering profiles are displaced along the ordinate for better visualization. The fit to the CTetD data from the model proposed by Okorokov *et al.* (25) and the fit to flp53 data from the EM-map by Okorokov *et al.* (25) are shown in dashed orange and dashed light green, respectively. The fit to our EM-based model is shown in dashed yellow. (*b*) The distance distribution plots computed from the experimental data and normalized to the maximum value of unity.

distance distribution functions of all of the constructs display peaks or shoulders at shorter distances (20–50 Å), indicating the presence of separated subdomains (Fig. 1*b*). CTetCD reveals only small expansion of the structure compared with CTetD, whereas flp53 demonstrates a significant increase in the overall dimensions (Table 1). This increase indicates that the N-terminal domain is not folded back onto the core protein but instead is largely extended and potentially flexible (although, as evidenced by the so-called Kratky plot in SI Fig. 7, not completely unfolded). This finding agrees with NMR results showing that the N-terminal region within the full-length protein is flexible (S.M.V.F. and A.R.F., unpublished results).

Table 1. Overall parameters of p53 constructs determined by SAXS

Construct	R_g, Å	D_{max}, Å	Mass, kDa*
p53 CTetD	52.2 ± 0.4	180 ± 10	105 ± 10 (120)
p53 CTetCD	54.0 ± 0.3	190 ± 15	118 ± 12 (134)
Full-length p53	68.2 ± 0.3	240 ± 30	199 ± 20 (175)
p53 CTetD + DNA	52.8 ± 0.5	200 ± 20	140 ± 15 (135)
flp53 + DNA	72.3 ± 0.8	250 ± 30	290 ± 40 (190)

R_g, radius of gyration; D_{max}, maximum size of particle.

*Mass values were evaluated by normalization against the BSA solution accounting for the contrast reduction due to 5% glycerol and (for the DNA-containing constructs) also for the higher DNA contrast compared with that of the protein. In parentheses, the expected masses of tetrameric constructs are given.

Low-resolution models of the constructs generated *ab initio* by the program DAMMIN (28) revealed elongated cross-shaped structures (a typical model is displayed in SI Fig. 8). There are only small changes in chemical shifts of the core and tetramerization domains in flp53 compared with separate domains in solution (23). Accordingly, the known high-resolution structures of p53 core (PDB ID code: 1TUP, 2AC0) and tetramerization domains (PDB ID code: 1C26) were used to build models of p53 constructs by using the programs SASREF and BUNCH (29). These programs perform rigid-body refinement against single or multiple data sets but also allow one to model the missing linkers as dummy residue chains. The configuration of the tetramerization domain was fixed as in the crystal structure, and the linkers and core domains were added to fit the experimental data assuming the P222 symmetry of the constructs. Independently determined models of all constructs consistently display extended cross-like structures, their overall shape being compatible with the *ab initio* reconstructions (SI Fig. 8). Importantly, the tetramerization domain in the middle of these models is surrounded by two pairs of associated core domains, which is consistent with previous NMR results on the mobility of core domains (23).

To constrain the modeling further, we fitted the three scattering patterns from CTetD, CTetCD, and flp53 simultaneously and added information about the interacting surface of the core dimers. The residues likely to be involved in domain interfaces are found from comparing chemical shifts in the tetrameric constructs with those in isolated core domains (23). The interface found previously (23) has now been reassigned to that of the interaction of core domain in flp53 with its N-terminal extension (S.M.V.F., C. Baloglu, and A.R.F., unpublished results). The interface between the core domains in the crystal structure of the complex of p53 monomers with DNA has self-complementarity of residues 177, 178, 180, 181, 243, and 244 (17). These residues are also implicated in the NMR spectra of dimeric constructs of two core domains with half-site DNA (30, 31). Signals for residues 181, 243, and 244 are in clear isolated positions in the 800 and 900 MHz ^{1}H-^{15}N TROSY spectrum of core domain (Fig. 2). Although there are insignificant changes in chemical shifts of nearly all of the remaining 147 assigned residues of the core domain in CTetD relative to core, the signals from 181, 243, and 244 are missing. We attribute the absence of those signals to line broadening from chemical exchange as each pair of core domains associates and dissociates, lowering the signals below the threshold for detection. Further, the normalized changes in ^{1}H-^{15}N chemical shifts for 178 and 179 are among the two greatest in the spectrum at 0.034 and 0.059 ppm, respectively. These data are compatible with a transient inter-

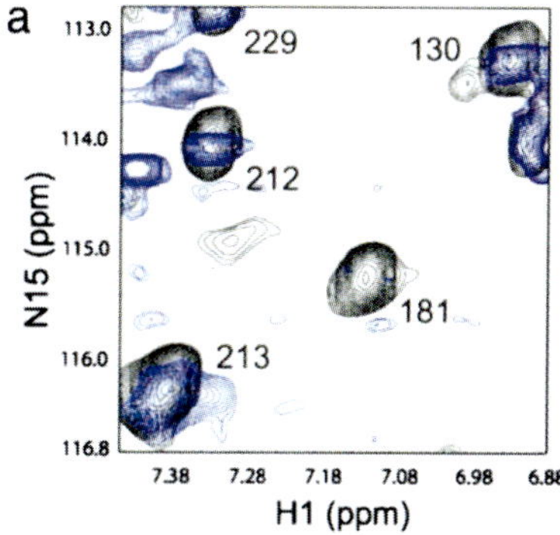

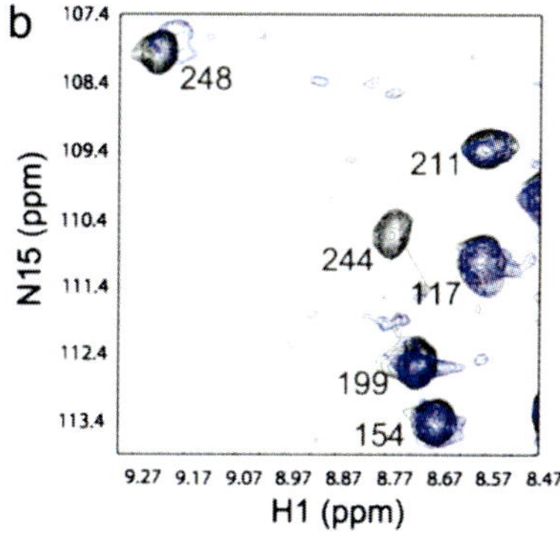

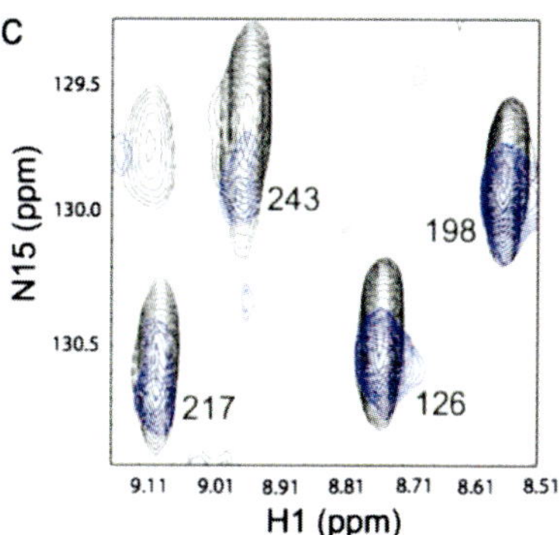

Fig. 2. Overlay of TROSY spectra acquired for p53 core domain (black) and p53 CTetD (blue) of 100-μM samples. Chemical shift deviations for the p53 CTetD spectra were estimated from the overlay and partly confirmed with TROSY-based triple resonance spectra (data not shown). Most of the core resonances showed insignificant chemical shift deviations between the two spectra. The expansions *a*, *b*, and *c* show signals that are strong and well isolated in the spectrum of isolated core domains but are not seen in the spectrum of CTetD most likely because of substantial line broadening associated with residues in a protein–protein interface. In *a* 181 and in *b* 244 are lost, and in *c* 243 is severely diminished. Spectra were plotted close to the noise level; additional signals in *a* are from the linker region between core and tetramerization domain.

action between core domains involving the same interface observed in the crystal structure of the core domain–DNA complex (17).

We included the interface detected by NMR constraints, using the crystal structure coordinates (17). A typical constrained model displayed in Fig. 3 fits simultaneously the scattering patterns from CTetD, CTetCD, and flp53 with discrepancy χ = 0.82, 0.83, and 1.24, respectively (Fig. 3). The overall shape of this model is compatible with those of the unconstrained model and with the *ab initio* model (SI Fig. 8), but the constrained model displays a somewhat different orientation of the core domains.

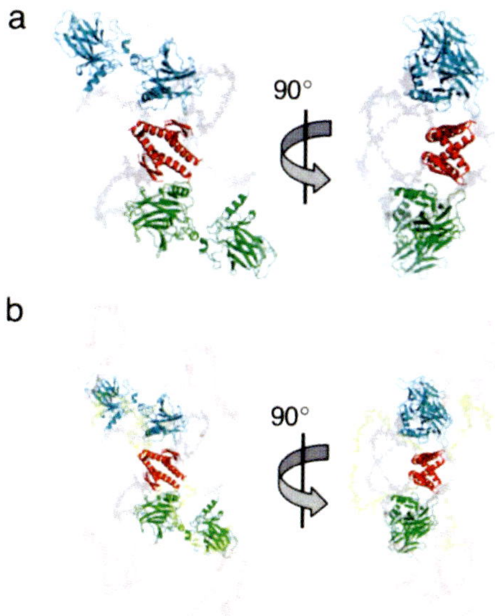

Fig. 3. SAXS models of free p53 in solution from rigid body analysis and addition of missing fragments. (*a*) CTetD portion of flp53 shown in *b*. Core domains and tetramerization domain are displayed in cartoon representation, connecting linkers (gray), N termini (salmon), and C termini (yellow) in semitransparent spacefill mode. Both models are presented in two orthogonal views. The model of flp53 and its CTetD and CTetCD portions fit simultaneously the experimental SAXS data from appropriate constructs in Fig. 1*a*.

The C termini are located in the central part surrounding the tetramerization domain, whereas the N-terminal domains are rather extended and point toward the periphery of the molecule.

It cannot be excluded that p53 has a considerable degree of conformational flexibility in solution and that the SAXS-generated models represent an average over such an ensemble. The conformation of potentially flexible linkers and termini, in particular in Fig. 3*b*, should be taken only as an indication of the volume occupied by these chains. On the other hand, the configuration of the tetramerization and core domains in CTetD appears rather rigid and reproducibly reconstructed in the independent modeling of different constructs. Summarizing, the model in Fig. 3 is compatible with a large body of SAXS and NMR evidence. In contrast, the calculated scattering from the recently published EM structure of p53 (25) does not agree with experimental SAXS data, yielding a poor fit with χ = 9.2 (Fig. 1*a*).

Structure of p53–DNA Complex from SAXS. Tetrameric flp53 and CTetD bind 24-residue *gadd45* or *p21* response element DNA with K_d in the low nanomolar range (32, 33). The scattering profiles from the DNA complexes are displayed in Fig. 1*a* and the overall parameters are summarized in Table 1. The CTetD–DNA complex is clearly compatible with tetrameric CTetD and a single DNA molecule, whereas the M_r of the flp53–DNA complex somewhat exceeds the expected value, which may point to some aggregation or high flexibility. The distance distributions of DNA complexes display no "fine-structure" at small distances observed for free p53 constructs, suggesting that the internal structure of the complexes is more compact than that of free protein (Fig. 1*b*). This compaction is further confirmed by comparison of the Kratky plots of the free and DNA-bound p53 constructs (SI Fig. 7).

We generated the structural model for the CTetD–DNA complex by rigid body modeling using the core domains, tetramerization domain, and DNA. This procedure yielded two

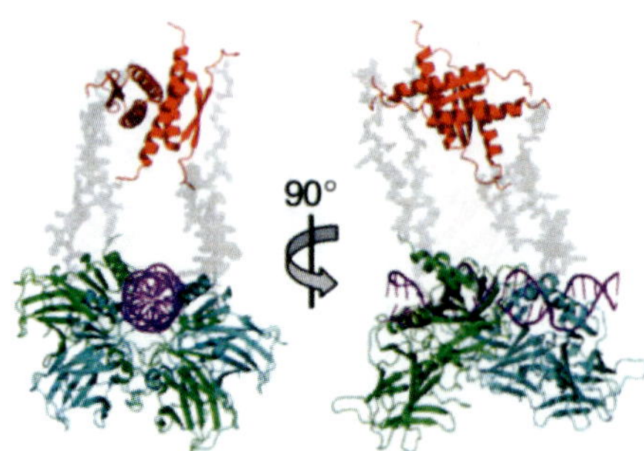

Fig. 4. Rigid body model of a p53–DNA complex from SAXS data. Core domains, tetramerization domain and DNA are shown in cartoon representation, connecting linkers in semitransparent spacefill mode. The model is displayed in two orthogonal views.

types of alternative models both showing four core domains binding to DNA and a tetramerization domain in its tetrameric form. These two structures fit the experimental data equally well and differ only by the positioning of the tetramerization domain with respect to the core domains–DNA complex (data not shown). Because one SAXS model was very similar to our EM model (see *Structure of p53–DNA Complex from EM*), we used this latter model as a starting point to add missing linkers and termini that could not be visualized in the EM map. The final model of the CTetD–DNA complex fitting the experimental data with χ = 1.48 (Fig. 1*a*) reveals two superdomains: four core domains bound to DNA as described by Kitayner *et al.* (17), and the tetramerization domain (Fig. 4). These superdomains are separated by 40–120 Å and connected by linkers. The full-length p53–DNA complex appears much more extended than the CTetD–DNA complex (Table 1), suggesting that the N terminus is extremely extended. This extended structure agrees with the TROSY-NMR spectra on these complexes, which show that the N-terminal resonances remain unchanged (ref. 23). Attempts to add the missing termini yielded poor fits to the scattering pattern from the flp53–DNA complex, suggesting that the N termini may be even more flexible than those in the free-full length p53 (as indicated above, aggregation could also not be excluded). A tentative model of the flp53–DNA complex constructed *ab initio* by DAMMIN accommodates the rigid-body model of the CTet-D–DNA complex and leaves ample space for the missing termini (SI Fig. 9).

Structure of p53 from EM. The EM images for negative-stained flp53 were heterogeneous in shape and size, and so the global set was unsuitable for 3D averaging techniques. A fraction, <20%, of the total particles adsorbed to the carbon support layer, however, was compact, with uniform structures of a size that was compatible with 2D projections of flp53 tetramers. We calculated the 3D map for this group following a reference-free scheme and imposing an overall C2 symmetry (see *SI Materials and Methods* and SI Fig. 10). At the defined density threshold, the map reveals a structure with two main masses linked by two faint connecting regions (SI Fig. 10 *a* and *b*). Despite the limitations of the negative staining approach and the moderate resolution of the 3D map (estimated in the range of 25–30 Å), the boundaries allowed us to fit available atomic coordinates for p53 core (15, 17) and tetramerization (19, 20) domains. The 3D map (SI Fig. 10 *d–f*) resembles and accommodates both structures with no need of major rearrangements, and the C termini of the core domains point toward the tetramerization domains for their connection. This structure is not the major one in solution that was found from the SAXS studies, and the computed scattering pattern does not match the experimental profile from CTetD (χ = 4.37). It may well be that this is a genuine low occupancy conformation that has to open to allow access of DNA to the binding site.

Structure of p53–DNA Complex from EM. EM images of stained flp53 in a complex with a 60-bp dsDNA probe containing a sequence-specific binding site DNA had a more uniform size distribution than the unligated protein. The volume, rendered from the calculated 3D map (Fig. 5 *a–c*), does not show the dsDNA. This invisibility is a typical drawback of negative stained protein-nucleic acid samples, where nucleic acids (thin and charged strings) are rarely visible and sometimes positively stained. The EM map, however, contains structural features that confirm the presence of the DNA probe, and we independently confirmed the presence of the flp53–DNA complex by EMSA (data not shown).

As with the EM structure of free flp53, there were two regions of density, at the top and bottom, ≈40 Å apart, connected by two linkers (Fig. 5 *a* and *b*). Importantly, the connecting densities in the map for p53–DNA complex were more obvious than for the free flp53 (see SI Fig. 10 *a* and *b* and Fig. 5 *a* and *b*) and were retained at high cut-off density thresholds. Those densities clearly delineate a see-through channel with the shape of a circle in one of the side views (Fig. 5*a*). The crystal structure of four core domains bound to dsDNA (17) fits very well around the channel of missing density of DNA (Fig. 5*d*). The EM map is understood as of a stained flp53 tetramer wrapped around unstained dsDNA.

The fitted atomic structures for the core (17) and tetramerization domains (19) are consistent with the EM envelope for the p53–DNA complex (Fig. 5 *d–f*). Importantly, the SAXS pattern computed from the complete EM-based model, including the core and tetramerization domains and the DNA, is fully consistent with the experimentally measured scattering data from CTetD+DNA construct (Fig. 1*a*, dashed yellow curve; the fit neatly is graphically indistinguishable from the dark green curve). A local disagreement is seen for two of the core domains that protrude outside the 3D map exposing their C-terminal region (apparent for the right cyan monomer in Fig. 5*d*). We have used the atomic coordinates as rigid bodies, and no alternative arrangement is presented. Some differences between our p53–DNA complex and the crystallographic structure of the core domain tetramer bound to DNA (17) could explain this discrepancy. The x-ray structure was determined for the isolated core domain tetramer and in complex with two DNA molecules, whereas our p53–DNA sample covers the entire protein bound to a unique DNA probe. However, we cannot rule out either a partial distortion of the p53-DNA structure because of the aggressive nature of the negative staining technique.

Kitayner *et al.* (17) modeled the structure of flp53 bound to DNA based on their crystal structure of four core domains bound to two segments of DNA and the positions of where the linker sections to the tetramerization domain would exit the truncated structure. The relative positions of core and tetramerization domains are in excellent agreement with the structures determined here by both SAXS and EM.

General Implications for Structure Determination. p53 is typical of the class of proteins containing a mixture of ordered and disordered domains. The strategy used here of fitting the high resolution structures of individual domains, solved by x-ray crystallography or NMR, into low resolution SAXS and EM data, guided by TROSY NMR constraints on interfaces, may well be a useful general approach to solving their structures. SAXS, like conventional NMR spectroscopy, has the advantage of analyzing structures in solution, obviating any artifacts associated with freezing of samples or immobilizing them on a grid, as used in EM. However, it does not generate unique models. SAXS optimizes the spatial arrangement of subunits by direct

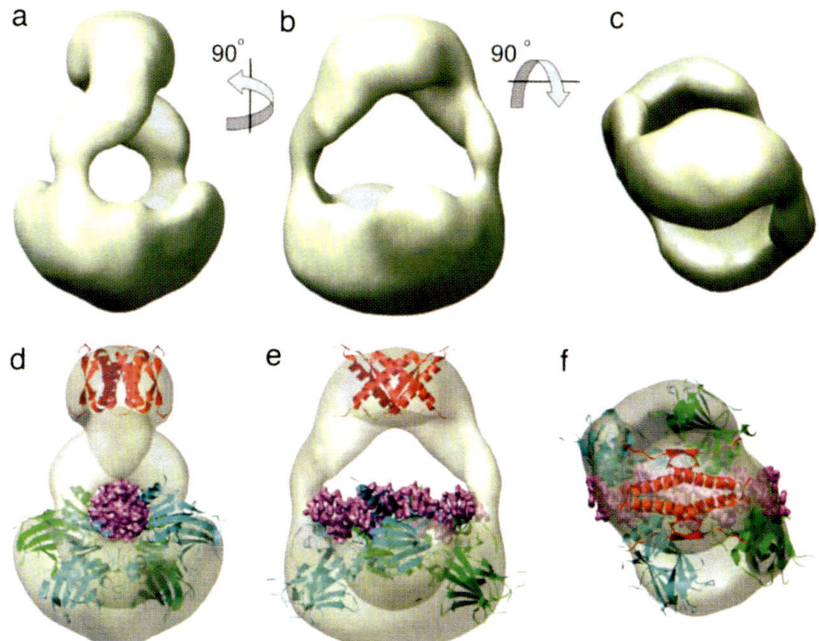

Fig. 5. 3D-EM map for flp53–DNA complex. The 3D map for the p53–DNA complex is rendered with a volume that covers 100% of the expected volume for a p53 tetramer in solid (*a–c*) and semitransparent (*d–f*) views. The fitted coordinates place the representation for the dsDNA in the see-through channel of one of the side views (see *a* and *d*).

fitting of experimental scattering data on docking high-resolution structures into shapes of macromolecular complexes, but there may be more than one solution (29). EM of negative stained samples uses an envelope obtained after 3D image reconstruction for docking of high-resolution models of subunits, and the overall shape should not be ambiguous if the high-resolution structures of domains are assigned correctly in the envelope. However, immobilization on the carbon support and differential staining may lead to selection of a particular conformation. SAXS data are particularly useful for discriminating between independently determined models by comparing calculated and observed scattering profiles because SAXS can eliminate incorrect models (34), and differences between SAXS solution models and EM-generated models have been reported (35). The differences between the structures of unligated p53 determined in solution and by EM show that a multimethod approach is necessary.

It is striking, here, that both EM and SAXS generate a very similar model for a p53–DNA complex, which is a relatively rigid structure. For the free p53, however, the solution studies using SAXS and NMR indicate a structure of loosely tethered pairs of core domains, with the possibilities of facile conformational rearrangement. Our EM studies, in contrast, show a structure that is preformed for binding the DNA response element. However, the calculated scattering profile is inconsistent with that found from the overall population in solution. It is likely that the experimental conditions for EM have selected a minor closed conformation that has to open for DNA to have access to the binding site. The model proposed for murine p53 (25) may also be a minor closed conformation. Key features of the murine model are not consistent with our data: for example, the murine model has the tetramerization domain dissociated and the N and C termini in contact. We find from SAXS that unligated p53 and its truncation variants all have the same quaternary arrangements of core and tetramerization domains, showing that the structure does not depend on the interaction of the N and C termini. Further, the dissociation constants of flp53 and CTetD lacking the C and N termini are quite similar (23). For the conformationally flexible free p53, the structures from SAXS, with supplementary information from NMR, must have precedence over the structures purely from EM.

Biological Relevance of the Quaternary Structure. The presence of an extended N-terminal domain in our model is consonant with its biological function. The N-terminal domain of p53 is subject to extensive posttranslational modifications that potentially regulate transactivational activity (36, 37). It also binds to a variety of proteins that further modulate p53 activity, like MDM2 (38), p300 (39), several transcription activators of the TFIID complex (40–42), and RPA (43). Upon binding to other proteins, small regions within the intrinsically unfolded p53 N terminus form helical structures (38, 43), a phenomenon also observed for other transcription factors (44–46). Further, our model of loosely coupled core domains assembled in an elongated cross-shaped structure allows binding of regulatory protein like Bcl-xL (47), Bcl2 (48), or 53BP2 (49, 50) to p53 core domain.

The open arrangement in flp53 of two separate pairs of core domains (Fig. 3) allows access of DNA to the binding site: first, one pair of core domains binds to two of the four binding site on a response element; then, the flexible linkers between the core and tetramerization domains allow the second pair of core domains to bind to the remaining two sites, thus burying the DNA within the protein.

Methods

Human p53 CTetD, CTetCD, and full-length p53 were expressed in *Escherichia coli* and purified as described (23). SAXS data were collected at the X33 beamline at European Molecular Biology Laboratory/Deutsches Elektronen Synchrotron (EMBL/DESY), Hamburg and at station 2.1 of the Daresbury Synchrotron Radiation Source (SRS) following standard procedures. Electron microscopy using negative staining involved manual selection of particles and reference-free classification (see SI Fig. 11). TROSY NMR spectra were acquired on 800

MHz and 900 MHz spectrometers by using perdeuterated, isotopically ^{2}H, ^{15}N-labeled samples. Details of sample preparation, SAXS measurements and modeling, electron microscopy, image processing and domain fitting, and NMR spectroscopy are given in *SI Materials and Methods*.

We thank Dr. A. Joerger for helpful discussions, C. Blair (MRC Centre for Protein Engineering, Cambridge, U.K.) for TEV protease and protein preparation, Dr. E. Orlova for communication of coordinates of murine p53 and kind comments, Dr. J. Milner for kind comments, and Dr. R. Nuñez for help using EMAN. M.V. is supported by Spanish Ministry of Education and Science Grant BFU2006-09648, by Etortek Research Programs 2005/2006 (The Department of Industry, Tourism and Trade of the Government of the Autonomous Community of the Basque Country), and by the Innovation Technology Department of the Bizkaia County. H.T. is supported by a fellowship from the Boehringer Ingelheim Fonds. We acknowledge support from European Community–Research Infrastructure Action under the FP6 (RII3/CT/2004/5060008) for access to EMBL/DESY, Hamburg, and access to the European Large Scale Facility for NMR Spectroscopy in Utrecht, The Netherlands. D.I.S. and J.G.G. acknowledge support from the European Union FP6 Design Study Small-Angle X-Ray Scattering Initiative for Europe (SAXIER), RIDS 011934.

1. Hainaut P, Wiman KG, eds (2005) *25 Years of p53 Research* (Springer, New York).
2. Vogelstein B, Lane D, Levine AJ (2000) *Nature* 408:307–310.
3. Hupp TR, Lane DP (1994) *Cold Spring Harbor Symp Quant Biol* 59:195–206.
4. Hupp TR, Sparks A, Lane DP (1995) *Cell* 83:237–245.
5. Dawson R, Muller L, Dehner A, Klein C, Kessler H, Buchner J (2003) *J Mol Biol* 332:1131–1141.
6. Shaulsky G, Goldfinger N, Ben-Ze'ev A, Rotter V (1990) *Mol Cell Biol* 10:6565–6577.
7. Ayed A, Mulder FA, Yi GS, Lu Y, Kay LE, Arrowsmith CH (2001) *Nat Struct Biol* 8:756–760.
8. Bell S, Klein C, Muller L, Hansen S, Buchner J (2002) *J Mol Biol* 322:917–927.
9. Lee H, Mok KH, Muhandiram R, Park KH, Suk JE, Kim DH, Chang J, Sung YC, Choi KY, Han KH (2000) *J Biol Chem* 275:29426–29432.
10. Joerger AC, Fersht AR (2007) *Oncogene* 26:2226–2242.
11. Roemer L, Klein C, Dehner A, Kessler H, Buchner J (2006) *Angew Chem Int Ed Engl* 45:6440–6460.
12. el-Deiry WS, Kern SE, Pietenpol JA, Kinzler KW, Vogelstein B (1992) *Nat Genet* 1:45–49.
13. Kern SE, Kinzler KW, Bruskin A, Jarosz D, Friedman P, Prives C, Vogelstein B (1991) *Science* 252:1708–1711.
14. Wei CL, Wu Q, Vega VB, Chiu KP, Ng P, Zhang T, Shahab A, Yong HC, Fu Y, Weng Z, *et al.* (2006) *Cell* 124:207–219.
15. Cho Y, Gorina S, Jeffrey PD, Pavletich NP (1994) *Science* 265:346–355.
16. Ho WC, Fitzgerald MX, Marmorstein R (2006) *J Biol Chem* 281:20494–20502.
17. Kitayner M, Rozenberg H, Kessler N, Rabinovich D, Shaulov L, Haran TE, Shakked Z (2006) *Mol Cell* 22:741–753.
18. Canadillas JM, Tidow H, Freund SM, Rutherford TJ, Ang HC, Fersht AR (2006) *Proc Natl Acad Sci USA* 103:2109–2114.
19. Clore GM, Omichinski JG, Sakaguchi K, Zambrano N, Sakamoto H, Appella E, Gronenborn AM (1995) *Science* 267:1515–1516.
20. Jeffrey PD, Gorina S, Pavletich NP (1995) *Science* 267:1498–1502.
21. Lee W, Harvey TS, Yin Y, Yau P, Litchfield D, Arrowsmith CH (1994) *Nat Struct Biol* 1:877–890.
22. Bullock AN, Henckel J, DeDecker BS, Johnson CM, Nikolova PV, Proctor MR, Lane DP, Fersht AR (1997) *Proc Natl Acad Sci USA* 94:14338–14342.
23. Veprintsev DB, Freund SM, Andreeva A, Rutledge SE, Tidow H, Canadillas JM, Blair CM, Fersht AR (2006) *Proc Natl Acad Sci USA* 103:2115–2119.
24. Dyson HJ, Wright PE (2002) *Curr Opin Struct Biol* 12:54–60.
25. Okorokov AL, Sherman MB, Plisson C, Grinkevich V, Sigmundsson K, Selivanova G, Milner J, Orlova EV (2006) *EMBO J* 25:5191–5200.
26. Joerger AC, Allen MD, Fersht AR (2004) *J Biol Chem* 279:1291–1296.
27. Nikolova PV, Henckel J, Lane DP, Fersht AR (1998) *Proc Natl Acad Sci USA* 95:14675–14680.
28. Svergun DI (1999) *Biophys J* 76:2879–2886.
29. Petoukhov MV, Svergun DI (2005) *Biophys J* 89:1237–1250.
30. Klein C, Planker E, Diercks T, Kessler H, Kunkele KP, Lang K, Hansen S, Schwaiger M (2001) *J Biol Chem* 276:49020–49027.
31. Rippin TM, Freund SM, Veprintsev DB, Fersht AR (2002) *J Mol Biol* 319:351–358.
32. Ang HC, Joerger AC, Mayer S, Fersht AR (2006) *J Biol Chem* 281:21934–21941.
33. Weinberg RL, Veprintsev DB, Fersht AR (2004) *J Mol Biol* 341:1145–1159.
34. Vestergaard B, Sanyal S, Roessle M, Mora L, Buckingham RH, Kastrup JS, Gajhede M, Svergun DI, Ehrenberg M (2005) *Mol Cell* 20:929–938.
35. Pioletti M, Findeisen F, Hura GL, Minor DL, Jr (2006) *Nat Struct Mol Biol* 13:987–995.
36. Banin S, Moyal L, Shieh S, Taya Y, Anderson CW, Chessa L, Smorodinsky NI, Prives C, Reiss Y, Shiloh Y, Ziv Y (1998) *Science* 281:1674–1677.
37. Shieh SY, Ikeda M, Taya Y, Prives C (1997) *Cell* 91:325–334.
38. Kussie PH, Gorina S, Marechal V, Elenbaas B, Moreau J, Levine AJ, Pavletich NP (1996) *Science* 274:948–953.
39. Grossman SR (2001) *Eur J Biochem* 268:2773–2778.
40. Lu H, Levine AJ (1995) *Proc Natl Acad Sci USA* 92:5154–5158.
41. Naar AM, Lemon BD, Tjian R (2001) *Annu Rev Biochem* 70:475–501.
42. Thut CJ, Chen JL, Klemm R, Tjian R (1995) *Science* 267:100–104.
43. Bochkareva E, Kaustov L, Ayed A, Yi GS, Lu Y, Pineda-Lucena A, Liao JC, Okorokov AL, Milner J, Arrowsmith CH, Bochkarev A (2005) *Proc Natl Acad Sci USA* 102:15412–15417.
44. Bowers PM, Schaufler LE, Klevit RE (1999) *Nat Struct Biol* 6:478–485.
45. Graham TA, Weaver C, Mao F, Kimelman D, Xu W (2000) *Cell* 103:885–896.
46. Spolar RS, Record MT, Jr (1994) *Science* 263:777–784.
47. Chipuk JE, Bouchier-Hayes L, Kuwana T, Newmeyer DD, Green DR (2005) *Science* 309:1732–1735.
48. Tomita Y, Marchenko N, Erster S, Nemajerova A, Dehner A, Klein C, Pan H, Kessler H, Pancoska P, Moll UM (2006) *J Biol Chem* 281:8600–8606.
49. Gorina S, Pavletich NP (1996) *Science* 274:1001–1005.
50. Tidow H, Veprintsev DB, Freund SM, Fersht AR (2006) *J Biol Chem* 281:32526–32533.

Chapter 20

Some Reflections

'Everyone sits in the prison of his own ideas.'

Albert Einstein

All scientists who innovate suffer problems of acceptance of their work, and so I end with a paper I was commissioned to write by *Nature Reviews Molecular and Cell Biology* for the 50th anniversary of the determination of the first crystal structure, in which I took as its theme the battle of excitement versus scepticism over ground-breaking innovation[217].

I must thank all of the research assistants, students, post-doctorals, colleagues and administrators who made my work possible. Only a few have been mentioned in the text. My work in the last twenty years has required considerable technical backup, supplied by the skilled staff of the CPE. The work on folding has had excellent support from Chris Johnson with expertise in thermodynamics, folding and building instruments. Our superb NMR centre has been run by Stefan Freund and Trevor Rutherford in recent years, and by Andrea Hownslow and Mark Howard in the early days of the CPE. Fortunately, the LMB has realised the importance of the CPE infrastructure and recruited Stefan, Trevor and Chris. Dima (Dmitry) Veprintsev has also transferred from CPE to LMB. Dima, aside from doing his own work, was able to automate much of our equipment as well as write software. Their contributions are well documented in the papers they have co-authored with me. Alexei Murzin, Tony Andreeva and David Howarth who ran the SCOP database from CPE have also joined suit. It is more than gratifying that the pioneering, interdisciplinary set up of the CPE will live on in LMB. There has been a succession of research assistants, from Yolanda Requena to Caroline Blair, who have been essential to my experimental work. Some of them, Meg Kaethner (Meg Ralph), Colin Dingwall and Tammy Gray, went on to obtain a Ph.D. and become independent scientists. My twenty plus years in Cambridge were made so much easier by two long serving PAs, Heather Thomson and Paula Murphy, who were able to put up with me. An experimental scientist requires a well-funded laboratory. I have been most fortunate that the Medical Research Council has funded me magnificently throughout my career and surrounded me with inspiring mentors and colleagues. In return, the Centre for Protein Engineering has not only produced science worthy of their investment, but thanks to Sir Gregory Winter, has repaid it in multiples in royalties from patents in protein engineering.

Chess players progress through three phases. When they are young, they are often 'combination' players, looking for brilliant combinations of moves that will destroy an opponent in a vicious onslaught. They mature into 'positional' players who have more strategy and slowly build up winning positions. Then, as they pass their prime, they rely heavily on 'technique', using the experience of a lifetime. I will have spent 45 years as a scientist in a laboratory when I reach the mandatory retirement age in 2010. Fortunately, I can revert to my youth because LMB has invited me to join them as a group leader. But, scientists mature like chess players. I will not be that young man of 40 years ago straining constantly for novel thoughts and clever new experiments. Instead, I will be putting my experience into practice to design those novel anti-cancer drugs and use state-of-the-art technologies and collaborations in structural biology and capitalise on my five decades of scientific experience. It is a far cry from my work and ideas when starting out, but it is so exciting still because it is being applied to uncharted areas where I can make discoveries and novel contributions. And it is so rewarding to continue to be invited to workshops and conferences and present work that is innovative and at the forefront of interesting research and not just recapitulation of past successes.

20.1 1999: Heather Thomson.

20.2 2010: Paula Murphy.

Alan has been fortunate to have had the two best secretaries ever!

PERSPECTIVES

TIMELINE

From the first protein structures to our current knowledge of protein folding: delights and scepticisms

Alan R. Fersht

Abstract | Every breakthrough that opens new vistas also removes the ground from under the feet of other scientists. The scientific joy of those who have seen the new light is accompanied by the dismay of those whose way of life has been changed for ever. The publication of the first structures of proteins at atomic resolution 50 years ago astounded and inspired scientists in every field, but caused others to flee or scoff. That advance and every subsequent paradigm-shifting breakthrough in protein science have met with some resistance before universal acceptance. I relate these events and their impact on the field of protein folding.

Well-funded laboratories can today run the gamut of protein studies; from molecular biology, biophysics and computation to X-ray crystallography and nuclear magnetic resonance (NMR). We take it for granted that protein science is underpinned by a multitude of complementary disciplines and methods. But it is easy to forget that each of the underpinning technologies represented a major breakthrough and each had to overcome initial, and sometimes continuing, scepticism to be universally adopted. The progress from the initial determination of protein structures to our present understanding of protein folding typifies this process of simultaneous delight and scepticism about new methods before their eventual acceptance. In this article, I do not track all of the developments in protein science, just some key advances, mainly technological and experimental, that have led to our present understanding of protein folding.

The first 3D structures

For half a century, the three-dimensional structures of proteins were studied indirectly. Details of a mechanism and its relationship to structure could only be inferred from the properties of the protein. In 1958, fifty years ago this year, a landmark paper[1] on the structure of myoglobin by John Kendrew and co-workers showed the first 3D structure of a protein (FIG. 1; TIMELINE). Its resolution was too low to show the atomic details of what seemed to be a multiply bent sausage, and the crude map gave only a hint of the revolution to come. The subsequent low-resolution structure of haemoglobin, which resembled four myoglobin molecules stacked together (FIG. 2), was reported by Max Perutz and colleagues in 1960 (REF. 2) and gave the first intimation of protein families.

> We take it for granted that protein science is underpinned by a multitude of complementary disciplines and methods.

The floodgates burst open after the publication of the high-resolution structure of hen egg-white lysozyme by David Phillips and his colleagues[3]. The detailed molecular structure at atomic resolution inspired the proposal of a chemical mechanism for the enzymatic reaction that this protein catalysed, parts of which have stood the test of time. Soon afterwards, the structures of α-chymotrypsin[4], ribonuclease[5], carboxypeptidase[6] and *Staphylococcus aureus* nuclease[7] followed, each of which gave new information on the protein structure and its function. Molecular biologists (as structural biologists called themselves then) were convinced that function and mechanism would always spring obviously from structure. It was therefore a "disappointment" to Francis Crick that this was not generally true (F. Crick, personal communication). Regardless, the rules had changed: from then on, mechanistic work on proteins had to be based on atomic-level-resolution structures.

The beauty of the structures, and especially the proposed atomic-level mechanism for lysozyme, inspired a new generation of experimental scientists to test proposals for mechanisms of action that were based on the protein structure with ingenious new experiments. Theoreticians worked in parallel to calculate the interactions within proteins and test ideas, such as whether an enzyme could distort a substrate towards the structure of its transition state. Perhaps the most illuminating work of all came from Perutz, whose subsequent studies on the conformational changes of haemoglobin showed much of the full power of X-ray crystallography combined with both protein chemistry and work on natural mutants, which had been isolated by Herman Lehmann[8].

A wave of enthusiasm engulfed the world of protein science, but not all of the older guard were thus inspired. Fine work had been done over the years on indirectly inferring the structures of active sites of enzymes through experiments on their activity with different substrates and by identifying functional groups from their pH titrations and chemical modification. The combination of protein chemistry and kinetics had laid the foundation stones of modern protein science. Some traditional enzymologists and protein chemists focused their attention on proteins that they thought would never be crystallized in order to avoid the new technocrats, who they thought would put them out of business but who were to become, in fact, indispensable partners. Others complained that crystalline proteins had unrepresentative structures and that protein crystals were unnatural, which, in a small minority of cases, is true. To make matters worse, some crystallographers kept their structure coordinates as closely guarded secrets and regarded the protein chemists as vultures. So, some enzymologists were indeed frozen out and had little choice but to flee, and it took three decades for every journal to insist on depositing atomic coordinates as a prerequisite for publication.

From structure to protein folding

Solving the structures of proteins at high resolution uncovered a new problem and initiated a novel field of research — that of protein folding. It was known from seminal experiments by Christian Anfinsen that small proteins could spontaneously refold from their denatured states, and so the primary structure (sequence) of a protein dictates its tertiary structure[9]. The 'protein folding problem' consists of two parts: first, the tertiary structures of proteins need to be predicted from their primary sequences, and second, the pathway of folding and unfolding must be predicted.

Cyrus Levinthal and others, such as Michael Levitt and Oleg Ptitsyn, wanted to predict tertiary structures by predicting their folding pathways. Levinthal famously pointed out in almost offhand remarks during a meeting in 1969 (REF. 10) that it seemed impossible that an unfolded protein could fold spontaneously by a random process on a biological time scale. Mechanisms were proposed that could overcome the 'Levinthal paradox' by simplifying the folding process and breaking it down into subprocesses that could occur stepwise. Ptitsyn suggested that folding could occur in a hierarchical process with the initial rapid formation of secondary structures, such as α-helices or β-sheets. At each step, the formation of a new layer of structure stabilized the previous one[11]. The framework mechanism was tested by Robert Baldwin[12] and formulated analytically and computationally in the diffusion–collision model of Karplus and Weaver[13]. Ptitsyn further proposed that proteins could undergo rapid hydrophobic collapse to form a 'molten globule', in which native tertiary interactions were rapidly formed that directed the subsequent formation of native secondary structure[14]. Levinthal had suggested a nucleation-growth mechanism in which the slow formation of a nucleus of secondary structure was followed by its rapid growth. Nucleation mechanisms went out of favour because the first proteins to be studied folded through kinetically observable intermediates. This is consistent with the framework and molten globule mechanisms, but not the nucleation mechanism, which implies only high-energy, kinetically silent intermediates.

How do we investigate protein folding mechanisms that involve the formation of a myriad of non-covalent interactions that give states that cannot be isolated and studied directly? One early approach, pioneered by Tom Creighton, was to study proteins that contained disulphide bonds and trap the covalent intermediates that contain both correctly and incorrectly formed disulphide bonds[15]. But modern mechanistic studies have to achieve atomic-resolution information and so it was necessary to develop methods that could analyse folding at the level of individual residues and atoms. It was also essential to extract and interpret the information encoded in a large number of proteins. The solution, albeit partial, of both parts of the folding problem required the introduction of new technologies, including recombinant DNA technology, protein engineering, advanced computer simulation and bioinformatics.

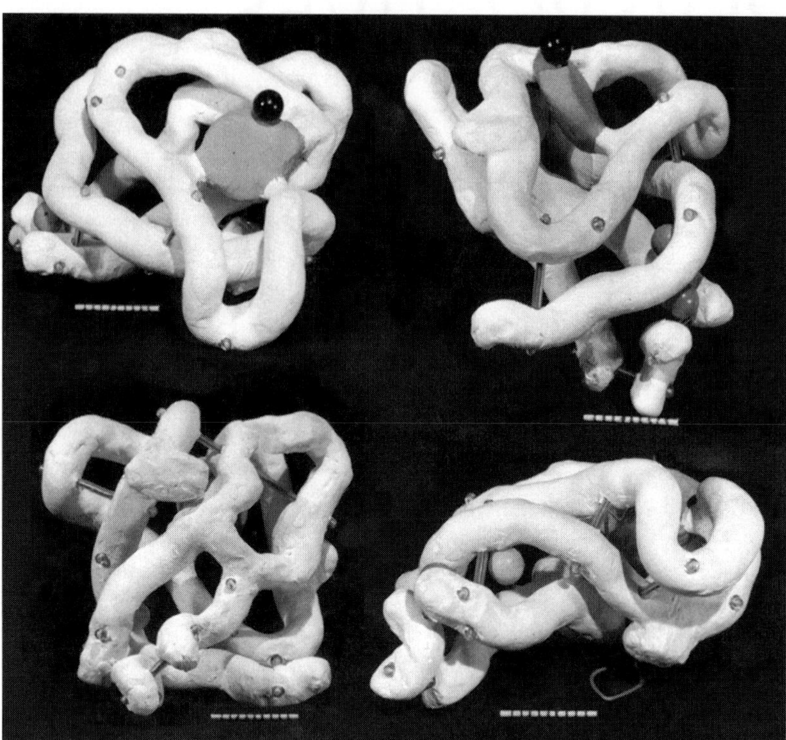

Figure 1 | **Three-dimensional structure of myoglobin.** The low-resolution structure of myoglobin that was published by John Kendrew and colleagues in 1958 (REF. 1). This figure in the *Nature* paper was reconstructed by the author using the original figures in the archives of the Medical Research Council Laboratory of Molecular Biology, Cambridge, UK. Polypeptide chains are in white and the grey disc represents the haem group. The three spheres show positions at which heavy atoms were attached to the molecule (black, Hg of *p*-chloro-mercuri-benzene-sulphonate; dark grey, Hg of mercury diammine; light grey, Au of auri-chloride). The marks on the scale are 1 Å apart.

Recombinant DNA technology. The march of X-ray crystallography lost some momentum in the 1970s because studies were restricted to proteins that could be isolated from natural sources in large amounts. The development of recombinant DNA technology reinvigorated the field by providing the resources for making large quantities of previously rare and unknown proteins[16]. However, even Perutz was heard to comment when so many colleagues immediately took up DNA cloning that "they must have been bored with what they were doing to drop it so readily", and the Medical Research Council Laboratory of Molecular Biology in Cambridge, UK — the epicentre of the structural biology revolution where not only Perutz and Kendrew had been working but also Sydney Brenner, Crick, Aaron Klug, Cesar Milstein and Fred Sanger — was slow in adopting the technology. Recombinant DNA technology proved essential for protein folding studies as it provided a source of experimentally tractable proteins[17–19], which replaced an older generation of readily available proteins. The new proteins lacked disulphide bridges, the formation of which dominated the folding pathways of the former generation. The recombinant proteins were amenable to the all-important structure–function studies through protein engineering.

PERSPECTIVES

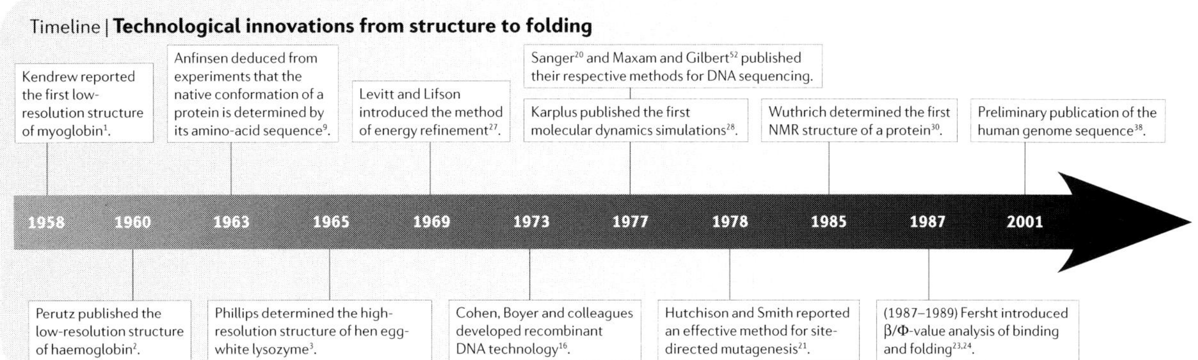

Protein engineering. DNA sequencing, in particular the method that was developed by Sanger[20] and its consequent refinements, profoundly altered the course of biology and protein science, without a murmur of dissent. Until the introduction of site-directed mutagenesis[21], and subsequently protein engineering[22] by the manipulation of genes, protein scientists were just observers and users of proteins. But by changing first individual amino-acid residues and then whole segments of proteins, precise structure–function–activity experiments could be designed and new functional proteins made.

There was a time lag of 5 years between the introduction of the technology for site-directed mutagenesis[21] and it being put into practice, initially to analyse enzyme catalysis[22]. The delay occurred, perhaps, because people were not sure how to use the technology rather than because they objected to it. Even so, there were initial murmurings that point mutations would radically alter the structure and function of proteins, which ill-designed mutations might, of course, do.

Protein engineering proved indispensable to protein folding studies because it allowed structure–activity studies that monitored the perturbation of the mechanism and kinetics of folding by making tiny changes in the protein structure. In particular, Φ-value analysis (a way of estimating the extent of non-covalent bonding during folding by examining the changes in folding kinetics and equilibria following targeted mutation[23], a method that was first implemented as β-value analysis[23,24]) has given near-atomic-level descriptions of protein folding transition states. Φ-value analysis took more than a decade for its near-universal adoption, and even the concept of transition states in protein folding met with initial scepticism. Nevertheless, protein engineering studies led to the discovery of a widespread and fundamental mechanism of protein folding — nucleation condensation — whereby the protein seems to collapse around a diffuse nucleus with most of the final native interactions being partly formed[25]. The framework and molten-globule mechanisms are extreme manifestations of this basic mechanism[25,26].

Figure 2 | **Max Perutz and John Kendrew.** Max Perutz (left) holding a balsawood model of the structure of haemoglobin solved at 6-Å resolution, and John Kendrew (right) holding a wire model of myoglobin solved at 1.4-Å resolution, which was determined remarkably soon after the low-resolution structure was published in 1958 (REF 1). Photo, which was taken in 1962, courtesy of the Medical Research Council Laboratory of Molecular Biology, Cambridge, UK.

Simulation and computer graphics. The first atomic representations of protein structures were built from pieces of screwed-together wire and inter-atomic distances were measured with rulers. In fact, one could buy plastic ball-and-sticks modelling kits. The first attempts at computer graphics were met with some wry amusement before such programs rapidly, and initially very expensively, displaced the old, crude mechanical models.

Computer calculations of energetics within proteins were used originally to refine the structures that were derived from fitting side chains to electron-density maps, by optimising the non-covalent interactions[27]. Molecular dynamics simulation then burst on to the scene[28]. The first attempts at simulation were severely limited by computational power and by approximations in energy functions, which are still imperfect today. Initially disparaged, these methods are now indispensable for understanding the mechanisms and folding of proteins as well as their intricate details because atomistic simulation, benchmarked by experimentation, is the only way of analysing a complete folding pathway and calculating the folding energetics[29].

NMR methods. NMR spectroscopy is such an important experimental procedure for determining structure in both solution and the solid state that it is difficult to conceive the battle that it took to establish its importance following the first determination of the structure of a protein[30]. Part of the problem was that NMR spectroscopy was seen as a rival, rather than complementary, technology to X-ray crystallography. NMR is, however, the only procedure available for studying the structures of disordered, denatured and partly folded states at atomic-level resolution[31]. This is done in solution, in which the proteins display their natural dynamics, which can also be analysed[32].

Stable folding intermediates have been engineered so that their structures can be solved directly by NMR methods[33]. High-energy folding intermediates that are only fractionally populated and are otherwise undetectable can be analysed structurally and their rates of interconversion measured[34]. The use of hydrogen–deuterium exchange between backbone NH groups and solvent water, introduced by Linderstrøm-Lang in

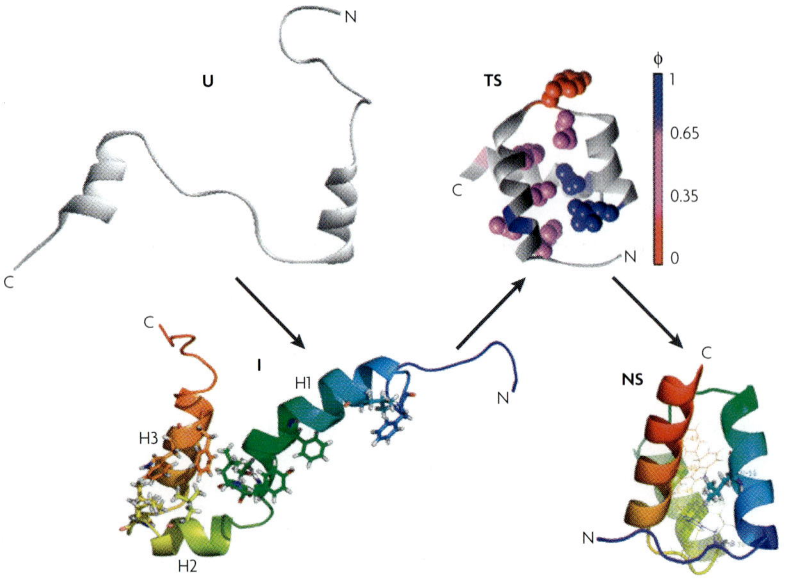

Figure 3 | **The pathway of folding of the Engrailed homeodomain of *Drosophila melanogaster*.** From right to left: the structure of the native state (NS) was solved by nuclear magnetic resonance (NMR) and X-ray crystallography; the transition state (TS) by Φ analysis (colour-coded from red, meaning unstructured, to blue, meaning highly native-like); the folding intermediate (I) was generated as a stable entity using protein engineering and its structure was solved by NMR; the structure of the denatured state (U), under conditions that favour folding, was simulated using molecular dynamics; and the entire unfolding pathway was simulated by molecular dynamics. Nearly 50 years of technological advances were needed to proceed from the structural resolution of Figure 1 to the dynamics and structures in Figure 3. H1, H2 and H3 represent helices 1, 2 and 3, respectively.

the 1950s (REF. 35), is exquisitely exploited by NMR, especially by S. Walter Englander[36], as a residue-specific probe for structural organization in folding intermediates. NMR has, as a result, become indispensable for protein folding studies.

Bioinformatics and structural genomics. In the 1970s, Levitt and Cyrus Chothia showed that proteins could be classified according to their secondary structure[37], which inexorably led to modern bioinformatics studies whereby protein structures are analysed by homology. Such studies require the knowledge of protein sequences. However, establishing one of the finest initiatives of the twentieth century — the determination of the DNA sequence of a human genome[38] — was highly controversial. Here, the arguments centred on 'small' science versus 'big' science, the possible diversion of funds from individual projects and the usefulness of the whole initiative. The huge amount of data that stemmed from the Human Genome Project and the technology that was introduced for rapid sequencing has been invaluable for areas as diverse as bioinformatics and personalized medicine. The two major repositories of the information that has been obtained are the National Center for Biotechnology Information (NCBI), and the European Bioinformatics Institute (EBI; see Further information). The databases are used to look for structural homology and so aid in the prediction of protein structure by analogy, and also provide libraries of local structural elements to be used as analogues or for calculating empirical energy functions.

“Several protein folding pathways are now known in detail at atomic resolution[29] thanks to the application of a combination of technologies...”

The arguments regarding the Human Genome Project have been repeated for the Protein Structure Initiative. This multinational structural genomics project aims to determine the 3D structures of all proteins by organizing known protein sequences into families, solving the structures of representative examples by X-ray crystallography or NMR spectroscopy and building models of the other proteins by homology. It is hoped that just 2,000 structures are required to be able to model every protein. Some of the arguments against the initiative almost reprise those that opposed the relevance of crystal structures 40–50 years ago, and both sides are worth reading for the discussion of the issues involved[39,40]. Even if 2,000 structures turn out to be insufficient, the structural information will be invaluable.

Current status of protein folding

Despite the huge advances in computational power, it is still not possible, in general, to predict the structures of proteins *de novo*. What has been possible is to harness bioinformatics and use databases of known structural elements to compute the structures of small proteins with high precision and give good models of multidomain proteins. The procedures rely, fundamentally, on the accumulated experimental information that has stemmed from structural biology, using local and global homology modelling and empirical equations that describe the non-covalent interactions between atoms. The progress in prediction methods is assessed every two years (see REF. 41 for the latest assessment).

The pathway of protein unfolding can be calculated by molecular dynamics simulation from the known 3D structure[42]. Folding simulations are more difficult, but have been aided by the discovery of ultra-fast-folding proteins, which fold a million times faster than those that prompted the Levinthal paradox 40 years ago — these proteins fold within a few microseconds on the time scale that is accessible to full atomistic simulation. Such pathways are benchmarked by experimentation, with Φ-value analysis of transition states and NMR spectroscopic structural determination of intermediates and analysis of denatured states. The pathway of folding and unfolding of the Engrailed homeodomain from *Drosophila melanogaster*, for example, has been solved at atomic resolution using this whole gamut of techniques[33,43] (FIG. 3). Simulations using Φ values as constraints can be used to construct transition states[44]. In addition, we can identify from databases structurally homologous proteins of vastly different amino-acid sequence to study the change of folding mechanism with structure and derive more general principles about the mechanism[45]. Several protein folding pathways are now known in detail at atomic resolution[29] thanks to the application of a combination of technologies, all of which initially met with some scepticism.

PERSPECTIVES

Challenges ahead

After half a century of structural studies on beautifully folded globular proteins, it is perhaps a shock to discover that up to some 40% of the proteins in the human proteome are estimated to be intrinsically disordered and become fully or partly structured on binding to binding partners in the cell[46]. Analysing the structures of disordered or partly disordered proteins, especially in the context of the cell, is a real challenge and requires a combination of structural approaches[47]. The concept of functional disorder is only slowly percolating through the scientific community. The role of protein instability, misfolding and aggregation, which might lead to novel structures and disease pathogenesis, is a further challenge for therapy[48] and structural studies.

The ability to design a functionally active protein *de novo* is still in its infancy despite 40 years of atomic-level structural studies and 25 years of protein engineering. Our most effective strategy is still to weave themes on what nature has already provided, such as producing humanized antibodies[49], which are now used in the clinic. We can routinely increase the thermostability of proteins, not only for biotechnology applications but also to facilitate structural studies. We can even change specificity, although the design of catalytic function is still extremely difficult. The first enzymes to be designed *de novo* catalyse reactions that require only minimal catalysis[50,51], but this achievement remains very impressive. Protein design and the structural analysis of proteins within the context of the cell are two of the great future challenges. Given the staying power of scientists in the face of difficulty and opposition, we can be optimistic that those challenges will be met.

Alan Fersht is the Herchel Smith Professor of Organic Chemistry at the Department of Chemistry, Lensfield Road, Cambridge, CB2 1EW, UK, and is the Director of the MRC Centre for Protein Engineering, MRC Centre, Hills Road, Cambridge, CB2 0QH, UK.
e-mail: arf25@cam.ac.uk
doi: 10.1038/nrm2446
Published online 25 June 2008

1. Kendrew, J. C. *et al.* A three-dimensional model of the myoglobin molecule obtained by X-ray analysis. *Nature* **181**, 662–666 (1958).
2. Perutz, M. F. *et al.* A three-dimensional fourier synthesis at 5.5-Å resolution, obtained by X-ray analysis. *Nature* **185**, 416–422 (1960).
3. Blake, C. C. *et al.* Structure of hen egg-white lysozyme. A three-dimensional Fourier synthesis at 2 Å resolution. *Nature* **206**, 757–761 (1965).
4. Matthews, B. W., Sigler, P. B., Henderson, R. & Blow, D. M. Three-dimensional structure of tosyl-α-chymotrypsin. *Nature* **214**, 652–656 (1967).
5. Wyckoff, H. W. *et al.* The structure of ribonuclease-S at 3.5 Å resolution. *J. Biol. Chem.* **242**, 3984–3988 (1967).
6. Lipscomb, W. N., Hartsuck, J. A., Quiocho, F. A. & Reeke, G. N. Jr. The structure of carboxypeptidase A. IX. The x-ray diffraction results in the light of the chemical sequence. *Proc. Natl Acad. Sci. USA* **64**, 28–35 (1969).
7. Arnone, A. *et al.* A high resolution structure of an inhibitor complex of the extracellular nuclease of *Staphylococcus aureus*. I. Experimental procedures and chain tracing. *J. Biol. Chem.* **246**, 2302–2316 (1971).
8. Perutz, M. F. Stereochemistry of cooperative effects in haemoglobin. *Nature* **228**, 726–739 (1970).
9. Epstein, C. J., Goldberger, R. F. & Anfinsen, C. B. The genetic control of tertiary protein structure. Model systems. *Cold Spring Harb. Symp. Quant. Biol.* **28**, 439–449 (1963).
10. Levinthal, C. in *Mossbauer Spectroscopy in Biological Systems* (eds Debrunner, P. *et al.*) 22–24 (University of Illinois Press, Urbana, Illinois, 1969).
11. Ptitsyn, O. B. Stages in the mechanism of self-organization of protein molecules. *Dokl. Akad. Nauk SSSR* **210**, 1213–1215 (1973) (in Russian).
12. Kim, P. S. & Baldwin, R. L. Specific intermediates in the folding reactions of small proteins and the mechanism of protein folding. *Annu. Rev. Biochem.* **51**, 459–489 (1982).
13. Karplus, M. & Weaver, D. L. Diffusion-collision model for protein folding. *Biopolymers* **18**, 1421–1437 (1979).
14. Dolgikh, D. A. *et al.* α-Lactalbumin: compact state with fluctuating tertiary structure? *FEBS Lett.* **136**, 311–315 (1981).
15. Creighton, T. E. The two-disulphide intermediates and the folding pathway of reduced pancreatic trypsin inhibitor. *J. Mol. Biol.* **95**, 167–199 (1975).
16. Cohen, S. N., Chang, A. C., Boyer, H. W. & Helling, R. B. Construction of biologically functional bacterial plasmids *in vitro*. *Proc. Natl Acad. Sci. USA* **70**, 3240–3244 (1973).
17. Shortle, D. A genetic system for analysis of staphylococcal nuclease. *Gene* **22**, 181–189 (1983).
18. Perry, L. J., Heyneker, H. L. & Wetzel, R. Non-toxic expression in *Escherichia coli* of a plasmid-encoded gene for phage T4 lysozyme. *Gene* **38**, 259–264 (1985).
19. Hartley, R. W. Barnase and barstar. Expression of its cloned inhibitor permits expression of a cloned ribonuclease. *J. Mol. Biol.* **202**, 913–915 (1988).
20. Sanger, F. *et al.* Nucleotide sequence of bacteriophage φX174 DNA. *Nature* **265**, 687–695 (1977).
21. Hutchison, C. A., *et al.* Mutagenesis at a specific position in a DNA sequence. *J. Biol. Chem.* **253**, 6551–6560 (1978).
22. Winter, G., Fersht, A. R., Wilkinson, A. J., Zoller, M. & Smith, M. Redesigning enzyme structure by site-directed mutagenesis: tyrosyl tRNA synthetase and ATP binding. *Nature* **299**, 756–758 (1982).
23. Matouschek, A., Kellis, J. T. Jr., Serrano, L. & Fersht, A. R. Mapping the transition state and pathway of protein folding by protein engineering. *Nature* **340**, 122–126 (1989).
24. Fersht, A. R., Leatherbarrow, R. J. & Wells, T. N. Structure–activity relationships in engineered proteins: analysis of use of binding energy by linear free energy relationships. *Biochemistry* **26**, 6030–6038 (1987).
25. Itzhaki, L. S., Otzen, D. E. & Fersht, A. R. The structure of the transition state for folding of chymotrypsin inhibitor 2 analysed by protein engineering methods: evidence for a nucleation-condensation mechanism for protein folding. *J. Mol. Biol.* **254**, 260–288 (1995).
26. Daggett, V. & Fersht, A. R. Is there a unifying mechanism for protein folding? *Trends Biochem. Sci.* **28**, 18–25 (2003).
27. Levitt, M. & Lifson, S. Refinement of protein conformations using a macromolecular energy minimization procedure. *J. Mol. Biol.* **46**, 269–279 (1969).
28. McCammon, J. A., Gelin, B. R. & Karplus, M. Dynamics of folded proteins. *Nature* **267**, 585–590 (1977).
29. Daggett, V. & Fersht, A. The present view of the mechanism of protein folding. *Nature Rev. Mol. Cell Biol.* **4**, 497–502 (2003).
30. Williamson, M. P., Havel, T. F. & Wuthrich, K. Solution conformation of proteinase inhibitor IIA from bull seminal plasma by ^{1}H nuclear magnetic resonance and distance geometry. *J. Mol. Biol.* **182**, 295–315 (1985).
31. Baum, J., Dobson, C. M., Evans, P. A. & Hanley, C. Characterization of a partly folded protein by NMR methods: studies on the molten globule state of guinea pig α-lactalbumin. *Biochemistry* **28**, 7–13 (1989).
32. Bax, A. & Grishaev, A. Weak alignment NMR: a hawk-eyed view of biomolecular structure. *Curr. Opin. Struct. Biol.* **15**, 563–570 (2005).
33. Religa, T. L., Markson, J. S., Mayor, U., Freund, S. M. & Fersht, A. R. Solution structure of a protein denatured state and folding intermediate. *Nature* **437**, 1053–1056 (2005).
34. Korzhnev, D. M. & Kay, L. E. Probing invisible, low-populated states of protein molecules by relaxation dispersion NMR spectroscopy: an application to protein folding. *Acc. Chem. Res.* **41**, 442–451 (2008).
35. Hvidt, A. & Linderstrøm-Lang, K. Exchange of hydrogen atoms in insulin with deuterium atoms in aqueous solutions. *Biochim. Biophys. Acta* **14**, 574–575 (1954).
36. Englander, S. W., Mayne, L., Bai, Y. & Sosnick, T. R. Hydrogen exchange: the modern legacy of Linderstrøm-Lang. *Protein Sci.* **6**, 1101–1109 (1997).
37. Levitt, M. & Chothia, C. Structural patterns in globular proteins. *Nature* **261**, 552–558 (1976).
38. Lander, E. S. *et al.* Initial sequencing and analysis of the human genome. *Nature* **409**, 860–921 (2001).
39. Petsko, G. A. An idea whose time has gone. *Genome Biol.* **8**, 107 (2007).
40. Banci, L. *et al.* An idea whose time has come. *Genome Biol.* **8**, 408 (2007).
41. Moult, J. *et al.* Critical assessment of methods of protein structure prediction — Round VII. *Proteins* **69** (Suppl. 8), 3–9 (2007).
42. Daggett, V. & Levitt, M. Protein unfolding pathways explored through molecular dynamics simulations. *J. Mol. Biol.* **232**, 600–619 (1993).
43. Mayor, U. *et al.* The complete folding pathway of a protein from nanoseconds to microseconds. *Nature* **421**, 863–867 (2003).
44. Vendruscolo, M., Paci, E., Dobson, C. M. & Karplus, M. Three key residues form a critical contact network in a protein folding transition state. *Nature* **409**, 641–645 (2001).
45. Gianni, S. *et al.* Unifying features in protein-folding mechanisms. *Proc. Natl Acad. Sci. USA* **100**, 13286–13291 (2003).
46. Fink, A. L. Natively unfolded proteins. *Curr. Opin. Struct. Biol.* **15**, 35–41 (2005).
47. Wells, M. *et al.* Structure of tumor suppressor p53 and its intrinsically disordered N-terminal transactivation domain. *Proc. Natl Acad. Sci. USA* **105**, 5762–5767 (2008).
48. Chiti, F. & Dobson, C. M. Protein misfolding, functional amyloid, and human disease. *Annu. Rev. Biochem.* **75**, 333–366 (2006).
49. Jones, P. T., Dear, P. H., Foote, J., Neuberger, M. S. & Winter, G. Replacing the complementarity-determining regions in a human antibody with those from a mouse. *Nature* **321**, 522–525 (1986).
50. Jiang, L. *et al.* *De novo* computational design of retro-aldol enzymes. *Science* **319**, 1387–1391 (2008).
51. Rothlisberger, D. *et al.* Kemp elimination catalysts by computational enzyme design. *Nature* **453**, 190–195 (2008).
52. Maxam, A. M. & Gilbert, W. A new method for sequencing DNA. *Proc. Natl Acad. Sci. USA* **74**, 560–564 (1977).

Acknowledgements
I thank the Medical Research Council for 40 years of funding, without which this article would not have been written.

DATABASES
UniProtKB: http://ca.expasy.org/sprot
haemoglobin | hen egg-white lysozyme | myoglobin | nuclease

FURTHER INFORMATION
Alan Fersht's homepage:
http://www.ch.cam.ac.uk/staff/arf.html
European Bioinformatics Institute: http://www.ebi.ac.uk
National Center for Biotechnology Information:
http://www.ncbi.nlm.nih.gov
Protein Structure Initiative:
http://www.structuralgenomics.org
ALL LINKS ARE ACTIVE IN THE ONLINE PDF

Group members 1999: *From top row to bottom*: Cara Vaughan, Doug Axe, Chris Johnson and Julia Henckel. Jose-Luis Neira, Kit Tang, Karoly von Glos and Geoff Symonds. Laura Itzhaki, Myriam Altamirano, Mark Bycroft and Naoki Tanaka. Nick Foster, Qinghua Wang, Rivka Isaacson and Ugo Mayor, Teikichi Ikura, Thierry de Lumley, Yu Wai Chen and Wooi Koon Wang. Alex Bullock, Kelvin Stott, Ssrah Perrett, and Luis Briseño-Roa. Alison Meekhof, Alex Buchberger, Sue Ellis and Heather Thomson.

Group members 2003: *From top row to bottom*: Zoryana Oliynik, Charlotte Dodson (Rusby), Dima Veprintsev and Tim Sharpe. Miriana Petrovich, Pamela Schartau, Grace Yu and Hwee Ching Ang. Per Jemth, Stefano Gianni, Assaf Friedler and Andreas Joerger. Faizah Khan, Satoshi Sato, Ugo Mayor and Neil Ferguson. Tomek Religa, Joe Markson, Pia Sondergeld and Roberto Canales. Luis Briseño-Roa, Nuria Sanchez Puig, Caroline Blair and Tobias Janowitz.

Group members 2010: *From top row to bottom*: Tuck Seng Wong, Robert Sade, Eyal Arbely, Anita Rea, Jenifer Lum, Hannes Neuweiler, Trevor Rutherford and Tobias Brandt. Matt Biancalana, Dima Veprintsev, Tony Andreeva, Grace Yu, Miriana Petrovich, Marina Vaysburd, Caroline Blair and Gianni Settanni. Dan Teufel, Chris Johnson, Sarah Burge and Jan van Dieck. Cetin Baloglu, Grace Yu and Alessando Vezzoli, Christine Barrie and Joel Kaar. Sridharan Rajagopalan, Eviatar Natan, Kian Hoe Khoo and Wiktor Banachewicz. Gaurav Jaggi, Fiona Townsley and Sarah Burge, Karoly von Glos and Andrew Northrop, and Andreas Joerger and Tony Andreeva.

T.L. Religa F. Khan J. Jacobsen F. Khan R. Golbik A. Sathyamurthy M.J.E. Sternberg R.J. Leatherbarrow P.D. Barker C. Frisch D.E. Mossakowksa D. Lipovsek G. Schreiber T.E. Gray A.P. Jasanoff N.F. elMasry A.J. Wilkinson

A. Horovitz A.G. Day D.M. Lowe D.M. Jones T.R. Killick S.M.V. Freund C.M. Santiveri T. Janowitz Z. Oliynyk R.L. Weinberg P.I. Sondergeld G. de Prat Gay J. Eder J. Avis A. Buchberger S. Rudiger E. Cota-Segura O. Schon S. Cotterill

A. Tissot M. Allen A. Campbell N. Brand W. Tsui J. Shindler J. Fastrez C. Vaughan P. Dalby S. Hamill V. Arcus A. Kippen D. Otzen B. Davis J. Matthews M. Petrovich U. Mayor P. Jemth S. Sato M. Bycroft N. Foster X. Lou

N. Sanchez Puig S. Mayer T. Sharpe J. Tran P. Schartau T. Bentin L. Fourage S. Arnould J.D. Beggs P. Nikolova A. Friedler S. Evans S. Freeman R. Isaacson T. Rutherford T. Hubbard L. Briseno-Roa C. Johnson

A. Bullock H.C. Ang A. Joerger H. Peto C. Blair P. Taylor C. Fernandez W.G. Yu J. Sancho G. Stenberg L. Hansson M. Moracci G. Baker D.N. Jones Y.W. Chen C-P.B. Yu J. Kellis C. Longstaff K. Brown R. Canales S. Gianni

B. DeDecker J. Henckel D. Teufel A. Murzin K. von Glos N. Ferguson L. Itzhaki L. Serrano A. Matouschek V. Daggett A. Kirby G. Winter A. Fersht B. Hartley C. Dobson J. Clarke S. Meiering S. Jackson M. Oliveberg P. Harryson D. Lane J. Sanders A. Ladurner M. Howard D. Veprintsev

Attendees of 60th Birthday Symposium for Alan Fersht.

Audience and lecturers at 60th Birthday Symposium for Alan Fersht.

Top: audience in Chemistry Department lecture theatre. *Second row*: Luis Serrano, Greg Winter, Chris Dobson and Amnon Horovitz. *Third row*: Elizabeth Meiering, Mark Bycroft, Jane Clarke and Andreas Matouschek. *Fourth row*: Alan Fersht, Valerie Daggett, Tony Wilkinson and Mikael Oliveberg.

65th Birthday Symposium and Dinner for Alan Fersht in Shanghai.

Top row: Qinghua Wang, Hualiang Jiang and Jianpeng Ma. *Second row*: p53 group members and Alan cutting cake. *Third row*: Ada Yonath, Hualiang Jiang, Jianpeng Ma and Sarah Perrett. *Fourth row*: Amnon Horovitz, Qinghua Wang and p53 group members; *Bottom row*: Qinghua Wang, Marilyn Fersht, Qinghua Wang, Jianpeng Ma and Albert Ma.

References

1. Fersht, A. R. & Kirby, A. J. (1967). Structure and mechanism in intramolecular catalysis. The hydrolysis of substituted aspirins. *J Am Chem Soc* **89**, 4853–7.
2. Fersht, A. R. & Kirby, A. J. (1967). The hydrolysis of aspirin. Intramolecular general base catalysis of ester hydrolysis. *J Am Chem Soc* **89**, 4857–63.
3. Fersht, A. R. & Kirby, A. J. (1967). Structure and mechanism in intramolecular catalysis. The hydrolysis of substituted aspirins. *J Am Chem Soc* **89**, 4853–6.
4. Fersht, A. R. & Kirby, A. J. (1968). Intramolecular nucleophilic catalysis in the hydrolysis of substituted aspirin acids. *J Am Chem Soc* **90**, 5826–32.
5. Fersht, A. R. & Kirby, A. J. (1968). Intramolecular nucleophilic catalysis of ester hydrolysis by the ionized carboxyl group. The hydrolysis of 3,5-dinitroaspirin anion. *J Am Chem Soc* **90**, 5818–26.
6. Fersht, A. R. & Jencks, W. P. (1969). The acetylpyridinium ion intermediate in pyridine-catalyzed acyl transfer. *J Am Chem Soc* **91**, 2125.
7. Fersht, A. R. & Jencks, W. P. (1970). Acetylpyridinium ion intermediate in pyridine-catalyzed hydrolysis and acyl transfer reactions of acetic anhydride — observation, kinetics, structure–reactivity correlations, and effects of concentrated salt solutions. *J Am Chem Soc* **92**, 5432–42.
8. Fersht, A. R. & Jencks, W. P. (1970). Reactions of nucleophilic reagents with acylating agents of extreme reactivity and unreactivity. Correlation of beta values for attacking and leaving group variation. *J Am Chem Soc* **92**, 5442–52.
9. Fersht, A. R. (1971). Acyl-transfer reactions of amides and esters with alcohols and thiols. A reference system for the serine and cysteine proteinases. Concerning the N protonation of amides and amide-imidate equilibria. *J Am Chem Soc* **93**, 3504–15.
10. Fersht, A. R. & Requena, Y. (1971). Free energies of hydrolysis of amides and peptides in aqueous solution at 25 degrees. *J Am Chem Soc* **93**, 3499–504.
11. Fersht, A. R. & Kirby, A. J. (1967). Intramolecular nucleophilic catalysis of ester hydrolysis by the carboxylate group. *J Am Chem Soc* **89**, 5960.
12. Fersht, A. R. & Kirby, A. J. (1967). Intramolecular general acid catalysis of ester hydrolysis by the carboxylic acid group. *J Am Chem Soc* **89**, 5961.
13. Fersht, A. R. & Kirby, A. J. (1968). Series nucleophilic catalysis in hydrolysis of 3-Aacetoxyphthalate. Intramolecular catalysis of ester hydrolysis by the carboxyl group once removed. *J Am Chem Soc* **90**, 5833–8.
14. Fersht, A. R. & Requena, Y. (1971). Equilibrium and rate constants for the interconversion of two conformations of α-chymotrypsin. The existence of a catalytically inactive conformation at neutral pH. *J Mol Biol* **60**, 279–90.

15. Fersht, A. R. (1972). Conformational equilibria in α-and δ-chymotrypsin. The energetics and importance of the salt bridge. *J Mol Biol* **64**, 497–509.
16. Fastrez, J. & Fersht, A. R. (1973). Demonstration of the acyl-enzyme mechanism for the hydrolysis of peptides and anilides by chymotrypsin. *Biochemistry* **12**, 2025–34.
17. Fersht, A. R. & Requena, Y. (1971). Mechanism of the α-chymotrypsin-catalyzed hydrolysis of amides. pH dependence of kc and Km. Kinetic detection of an intermediate. *J Am Chem Soc* **93**, 7079–87.
18. Fersht, A. R. (1972). Mechanism of the α-chymotrypsin-catalyzed hydrolysis of specific amide substrates. *J Am Chem Soc* **94**, 293–5.
19. Fersht, A. R. (1972). Conformational equilibria and the salt bridge in chymotrypsin. *Cold Spring Harb Symp Quant Biol* **36**, 71–3.
20. Fastrez, J. & Fersht, A. R. (1973). Mechanism of chymotrypsin. Structure, reactivity, and nonproductive binding relationships. *Biochemistry* **12**, 1067–74.
21. Fersht, A. R. (1973). The catalytic activity of the inactive conformation of delta-chymotrypsin. *FEBS Lett* **29**, 283–285.
22. Fersht, A. R., Blow, D. M. & Fastrez, J. (1973). Leaving group specificity in the chymotrypsin-catalyzed hydrolysis of peptides. A stereochemical interpretation. *Biochemistry* **12**, 2035–41.
23. Fersht, A. R. & Sperling, J. (1973). The charge relay system in chymotrypsin and chymotrypsinogen. *J Mol Biol* **74**, 137–49.
24. Renard, M. & Fersht, A. R. (1973). Anomalous pH dependence of k_{cat}/K_M in enzyme reactions. Rate constants for the association of chymotrypsin with substrates. *Biochemistry* **12**, 4713–8.
25. Fersht, A. R. & Renard, M. (1974). pH dependence of chymotrypsin catalysis. Appendix: substrate binding to dimeric α-chymotrypsin studied by X-ray diffraction and the equilibrium method. *Biochemistry* **13**, 1416–26.
26. Fersht, A. R. (1974). Catalysis, binding and enzyme–substrate complementarity. *Proc R Soc Lond B Biol Sci* **187**, 397–407.
27. Brocklehurst, K. & Cornish-Bowden, A. (1976). The pre-eminence of k(cat) in the manifestation of optimal enzymic activity delineated by using the Briggs-Haldane two-step irreversible kinetic model. *Biochem J* **159**, 165–6.
28. Cornish-Bowden, A. (1976). The effect of natural selection on enzymic catalysis. *J Mol Biol* **101**, 1–9.
29. Fersht, A. R. (1975). Demonstration of two active sites on a monomeric aminoacyl-tRNA synthetase. Possible roles of negative cooperativity and half-of-the-sites reactivity in oligomeric enzymes. *Biochemistry* **14**, 5–12.
30. Hutton, R. L. & Boyer, P. D. (1979). Subunit interaction during catalysis. Alternating site cooperativity of mitochondrial adenosine triphosphatase. *J Biol Chem* **254**, 9990–3.
31. Fersht, A. R., Ashford, J. S., Bruton, C. J., Jakes, R., Koch, G. L. & Hartley, B. S. (1975). Active site titration and aminoacyl adenylate binding stoichiometry of aminoacyl-tRNA synthetases. *Biochemistry* **14**, 1–4.

32. Fersht, A. R. & Jakes, R. (1975). Demonstration of two reaction pathways for the aminoacylation of tRNA. Application of the pulsed quenched flow technique. *Biochemistry* **14**, 3350–6.
33. Fersht, A. R. (1977). Editing mechanisms in protein synthesis. Rejection of valine by the isoleucyl-tRNA synthetase. *Biochemistry* **16**, 1025–30.
34. Fersht, A. R. & Dingwall, C. (1979). Evidence for the double-sieve editing mechanism in protein synthesis. Steric exclusion of isoleucine by valyl-tRNA synthetases. *Biochemistry* **18**, 2627–31.
35. Fersht, A. R. & Dingwall, C. (1979). An editing mechanism for the methionyl-tRNA synthetase in the selection of amino acids in protein synthesis. *Biochemistry* **18**, 1250–6.
36. Fersht, A. R. & Dingwall, C. (1979). Cysteinyl-tRNA synthetase from Escherichia coli does not need an editing mechanism to reject serine and alanine. High binding energy of small groups in specific molecular interactions. *Biochemistry* **18**, 1245–9.
37. Fersht, A. R., Shindler, J. S. & Tsui, W. C. (1980). Probing the limits of protein-amino acid side chain recognition with the aminoacyl-tRNA synthetases. Discrimination against phenylalanine by tyrosyl-tRNA synthetases. *Biochemistry* **19**, 5520–4.
38. Tsui, W. C. & Fersht, A. R. (1981). Probing the principles of amino acid selection using the alanyl-tRNA synthetase from Escherichia coli. *Nucleic Acids Res* **9**, 4627–37.
39. Kellis, J. T., Jr., Nyberg, K. & Fersht, A. R. (1989). Energetics of complementary side-chain packing in a protein hydrophobic core. *Biochemistry* **28**, 4914–22.
40. Kellis, J. T., Jr., Nyberg, K., Sali, D. & Fersht, A. R. (1988). Contribution of hydrophobic interactions to protein stability. *Nature* **333**, 784–6.
41. Baldwin, A. N. & Berg, P. (1966). Transfer Ribonucleic acid-induced hydrolysis of valyladenylate bound to isoleucyl ribonucelic acid synthetase. *J. Biol Chem* **241**, 839.
42. Fersht, A. R. & Kaethner, M. M. (1976). Enzyme hyperspecificity. Rejection of threonine by the valyl-tRNA synthetase by misacylation and hydrolytic editing. *Biochemistry* **15**, 3342–6.
43. Fersht, A. R. (1981). Enzymic editing mechanisms and the genetic code. *Proc R Soc Lond B Biol Sci* **212**, 351–79.
44. Fersht, A. R. (1998). Sieves in sequence. *Science* **280**, 541.
45. Fersht, A. R. (1979). Fidelity of replication of phage phi X174 DNA by DNA polymerase III holoenzyme: Spontaneous mutation by misincorporation. *Proc Natl Acad Sci USA* **76**, 4946–50.
46. Fersht, A. R. & Knill-Jones, J. W. (1981). DNA polymerase accuracy and spontaneous mutation rates: Frequencies of purine.purine, purine.pyrimidine, and pyrimidine.pyrimidine mismatches during DNA replication. *Proc Natl Acad Sci USA* **78**, 4251–5.
47. Fersht, A. R., Knill-Jones, J. W. & Tsui, W. C. (1982). Kinetic basis of spontaneous mutation. Misinsertion frequencies, proofreading specificities and cost of proofreading by DNA polymerases of *Escherichia coli*. *J Mol Biol* **156**, 37–51.

48. Fersht, A. R. & Knill-Jones, J. W. (1983). Contribution of 3'→5' exonuclease activity of DNA polymerase III holoenzyme from *Escherichia coli* to specificity. *J Mol Biol* **165**, 669–82.
49. Fersht, A. R. & Knill-Jones, J. W. (1983). Fidelity of replication of bacteriophage ϕX174 DNA *in vitro* and *in vivo*. *J Mol Biol* **165**, 633–54.
50. Fersht, A. R., Shi, J. P. & Tsui, W. C. (1983). Kinetics of base misinsertion by DNA polymerase I of *Escherichia coli*. *J Mol Biol* **165**, 655–67.
51. Grosse, F., Krauss, G., Knill-Jones, J. W. & Fersht, A. R. (1983). Accuracy of DNA polymerase-α in copying natural DNA. *EMBO J* **2**, 1515–9.
52. Fersht, A. R. (1984). Fidelity of DNA replication *in vitro*. *Adv Exp Med Biol* **179**, 525–33.
53. Grosse, F., Krauss, G., Knill-Jones, J. W. & Fersht, A. R. (1984). Replication of ϕX174 DNA by calf thymus DNA polymerase-α: Measurement of error rates at the amber-16 codon. *Adv Exp Med Biol* **179**, 535–40.
54. Shi, J. P. & Fersht, A. R. (1984). Fidelity of DNA replication under conditions used for oligodeoxynucleotide-directed mutagenesis. *J Mol Biol* **177**, 269–78.
55. Hutchison, C. A., 3rd & Edgell, M. H. (1971). Genetic assay for small fragments of bacteriophage ϕX174 deoxyribonucleic acid. *J Virol* **8**, 181–9.
56. Middleton, J. H., Edgell, M. H. & Hutchison, C. A., 3rd. (1972). Specific fragments of ϕX174 deoxyribonucleic acid produced by a restriction enzyme from *Haemophilus aegyptius*, endonuclease Z. *J Virol* **10**, 42–50.
57. Hutchison, C. A., 3rd, Phillips, S., Edgell, M. H., Gillam, S., Jahnke, P. & Smith, M. (1978). Mutagenesis at a specific position in a DNA sequence. *J Biol Chem* **253**, 6551–60.
58. Winter, G., Fersht, A. R., Wilkinson, A. J., Zoller, M. & Smith, M. (1982). Redesigning enzyme structure by site-directed mutagenesis: Tyrosyl tRNA synthetase and ATP binding. *Nature* **299**, 756–8.
59. Dalbadie-Mcfarland, G., Cohen, L. W., Riggs, A. D., Morin, C., Itakura, K. & Richards, J. H. (1982). Oligonucleotide-directed mutagenesis as a general and powerful method for studies of protein function. *Proc Natl Acad Sci USA* **79**, 6409–13.
60. Sigal, I. S., Harwood, B. G. & Arentzen, R. (1982). Thiol-beta-lactamase - Replacement of the active-site serine of RTEM beta-lactamase by a cysteine residue. *Proc Natl Acad Sci USA* **79**, 7157–60.
61. Carter, P. J., Winter, G., Wilkinson, A. J. & Fersht, A. R. (1984). The use of double mutants to detect structural changes in the active site of the tyrosyl-tRNA synthetase (*Bacillus stearothermophilus*). *Cell* **38**, 835–40.
62. Wilkinson, A. J., Fersht, A. R., Blow, D. M., Carter, P. & Winter, G. (1984). A large increase in enzyme–substrate affinity by protein engineering. *Nature* **307**, 187–8.
63. Fersht, A. R., Shi, J. P., Knill-Jones, J., Lowe, D. M., Wilkinson, A. J., Blow, D. M., Brick, P., Carter, P., Waye, M. M. & Winter, G. (1985). Hydrogen bonding and biological specificity analysed by protein engineering. *Nature* **314**, 235–8.

64. Fersht, A. R. (1988). Relationships between apparent binding energies measured in site-directed mutagenesis experiments and energetics of binding and catalysis. *Biochemistry* **27**, 1577–80.
65. Fersht, A. R. (1987). The hydrogen-bond in molecular recognition. *Trends Biochem Sci* **12**, 301–4.
66. Wells, T. N. C. & Fersht, A. R. (1985). Hydrogen bonding in enzymatic catalysis analysed by protein engineering. *Nature* **316**, 656–7.
67. Leatherbarrow, R. J., Fersht, A. R. & Winter, G. (1985). Transition-state stabilization in the mechanism of tyrosyl-tRNA synthetase revealed by protein engineering. *Proc Natl Acad Sci USA* **82**, 7840–4.
68. Wells, T. N. C. & Fersht, A. R. (1986). Use of binding energy in catalysis analyzed by mutagenesis of the tyrosyl-tRNA synthetase. *Biochemistry* **25**, 1881–6.
69. Thomas, P. G., Russell, A. J. & Fersht, A. R. (1985). Tailoring the pH-dependence of enzyme catalysis using protein engineering. *Nature* **318**, 375–6.
70. Russell, A. J. & Fersht, A. R. (1987). Rational modification of enzyme catalysis by engineering surface charge. *Nature* **328**, 496–500.
71. Sternberg, M. J., Hayes, F. R., Russell, A. J., Thomas, P. G. & Fersht, A. R. (1987). Prediction of electrostatic effects of engineering of protein charges. *Nature* **330**, 86–8.
72. Fersht, A. R., Leatherbarrow, R. J. & Wells, T. N. C. (1986). Quantitative analysis of structure–activity relationships in engineered proteins by linear free-energy relationships. *Nature* **322**, 284–6.
73. Fersht, A. R., Leatherbarrow, R. J. & Wells, T. N. C. (1987). Structure–activity relationships in engineered proteins: Analysis of use of binding energy by linear free energy relationships. *Biochemistry* **26**, 6030–8.
74. Straub, J. E. & Karplus, M. (1990). The interpretation of site-directed mutagenesis experiments by linear free energy relations. *Protein Eng* **3**, 673–5.
75. Fersht, A. R. & Wells, T. N. (1991). Linear free energy relationships in enzyme binding interactions studied by protein engineering. *Protein Eng* **4**, 229–31.
76. Warshel, A., Hwang, J. K. & Aqvist, J. (1992). Computer simulations of enzymatic reactions: Examination of linear free-energy relationships and quantum-mechanical corrections in the initial proton-transfer step of carbonic anhydrase. *Faraday Discuss*, 225–38.
77. Bycroft, M. & Fersht, A. R. (1988). Assignment of histidine resonances in the ^{1}H NMR (500 MHz) spectrum of subtilisin BPN′ using site-directed mutagenesis. *Biochemistry* **27**, 7390–4.
78. Sali, D., Bycroft, M. & Fersht, A. R. (1988). Stabilization of protein structure by interaction of alpha-helix dipole with a charged side chain. *Nature* **335**, 740–3.
79. Bycroft, M., Ludvigsen, S., Fersht, A. R. & Poulsen, F. M. (1991). Determination of the three-dimensional solution structure of barnase using nuclear magnetic resonance spectroscopy. *Biochemistry* **30**, 8697–701.
80. Creighton, T. E. (1988). The protein folding problem. *Science* **240**, 267, 344.

81. Matsumura, M., Becktel, W. J. & Matthews, B. W. (1988). Hydrophobic stabilization in T4 lysozyme determined directly by multiple substitutions of Ile 3. *Nature* **334**, 406–10.
82. Serrano, L., Kellis, J. T., Jr., Cann, P., Matouschek, A. & Fersht, A. R. (1992). The folding of an enzyme. II. Substructure of barnase and the contribution of different interactions to protein stability. *J Mol Biol* **224**, 783–804.
83. Serrano, L. & Fersht, A. R. (1989). Capping and alpha-helix stability. *Nature* **342**, 296–9.
84. Serrano, L., Neira, J.-L., Sancho, J. & Fersht, A. R. (1992). Effect of alanine versus glycine in alpha-helices on protein stability. *Nature* **356**, 453–5.
85. Nicholson, H., Becktel, W. J. & Matthews, B. W. (1988). Enhanced protein thermostability from designed mutations that interact with alpha-helix dipoles. *Nature* **336**, 651–6.
86. Sali, D., Bycroft, M. & Fersht, A. R. (1991). Surface electrostatic interactions contribute little to stability of barnase. *J Mol Biol* **220**, 779–88.
87. Dao-pin, S., Soderlind, E., Baase, W. A., Wozniak, J. A., Sauer, U. & Matthews, B. W. (1991). Cumulative site-directed charge-charge replacements in bacteriophage T4 lysozyme suggest that long-range electrostatic interactions contribute little to protein stability. *J Mol Biol* **221**, 873–87.
88. Serrano, L., Sancho, J., Hirshberg, M. & Fersht, A. R. (1992). α-helix stability in proteins. I. Empirical correlations concerning substitution of side-chains at the N and C-caps and the replacement of alanine by glycine or serine at solvent-exposed surfaces. *J Mol Biol* **227**, 544–59.
89. Horovitz, A., Matthews, J. M. & Fersht, A. R. (1992). α-helix stability in proteins. II. Factors that influence stability at an internal position. *J Mol Biol* **227**, 560–8.
90. Blaber, M., Zhang, X. J. & Matthews, B. W. (1993). Structural basis of amino-acid α-helix propensity. *Science* **260**, 1637–40.
91. Scott, K. A., Alonso, D. O., Sato, S., Fersht, A. R. & Daggett, V. (2007). Conformational entropy of alanine versus glycine in protein denatured states. *Proc Natl Acad Sci USA* **104**, 2661–6.
92. Horovitz, A. & Fersht, A. R. (1990). Strategy for analysing the co-operativity of intramolecular interactions in peptides and proteins. *J Mol Biol* **214**, 613–7.
93. Serrano, L., Horovitz, A., Avron, B., Bycroft, M. & Fersht, A. R. (1990). Estimating the contribution of engineered surface electrostatic interactions to protein stability by using double-mutant cycles. *Biochemistry* **29**, 9343–52.
94. Serrano, L., Bycroft, M. & Fersht, A. R. (1991). Aromatic-aromatic interactions and protein stability. Investigation by double-mutant cycles. *J Mol Biol* **218**, 465–75.
95. Loewenthal, R., Sancho, J. & Fersht, A. R. (1992). Histidine-aromatic interactions in barnase. Elevation of histidine pKa and contribution to protein stability. *J Mol Biol* **224**, 759–70.
96. Otzen, D. E., Rheinnecker, M. & Fersht, A. R. (1995). Structural factors contributing to the hydrophobic effect: The partly exposed hydrophobic minicore in chymotrypsin inhibitor 2. *Biochemistry* **34**, 13051–8.

97. Tissot, A. C., Vuilleumier, S. & Fersht, A. R. (1996). Importance of two buried salt bridges in the stability and folding pathway of barnase. *Biochemistry* **35**, 6786–94.
98. Ladurner, A. G., Itzhaki, L. S. & Fersht, A. R. (1997). Strain in the folding nucleus of chymotrypsin inhibitor 2. *Fold Des* **2**, 363–8.
99. Otzen, D. E. & Fersht, A. R. (1999). Analysis of protein-protein interactions by mutagenesis: direct versus indirect effects. *Protein Eng* **12**, 41–5.
100. Horovitz, A., Serrano, L. & Fersht, A. R. (1991). COSMIC analysis of the major alpha-helix of barnase during folding. *J Mol Biol* **219**, 5–9.
101. Vaughan, C. K., Harryson, P., Buckle, A. M. & Fersht, A. R. (2002). A structural double-mutant cycle: Estimating the strength of a buried salt bridge in barnase. *Acta Crystallogr D Biol Crystallogr* **58**, 591–600.
102. Meiering, E. M., Serrano, L. & Fersht, A. R. (1992). Effect of active site residues in barnase on activity and stability. *J Mol Biol* **225**, 585–9.
103. Serrano, L., Day, A. G. & Fersht, A. R. (1993). Step-wise mutation of barnase to binase. A procedure for engineering increased stability of proteins and an experimental analysis of the evolution of protein stability. *J Mol Biol* **233**, 305–12.
104. Matouschek, A., Kellis, J. T., Jr., Serrano, L. & Fersht, A. R. (1989). Mapping the transition state and pathway of protein folding by protein engineering. *Nature* **340**, 122–6.
105. Buchner, J. & Kiefhaber, T. (1990). Folding pathway enigma. *Nature* **343**, 601–2.
106. Matouschek, A., Kellis, J. T., Jr., Serrano, L., Bycroft, M. & Fersht, A. R. (1990). Transient folding intermediates characterized by protein engineering. *Nature* **346**, 440–5.
107. Fersht, A. R. (2000). A kinetically significant intermediate in the folding of barnase. *Proc Natl Acad Sci USA* **97**, 14121–6.
108. Khan, F., Chuang, J. I., Gianni, S. & Fersht, A. R. (2003). The kinetic pathway of folding of barnase. *J Mol Biol* **333**, 169–86.
109. Sancho, J. & Fersht, A. R. (1992). Dissection of an enzyme by protein engineering. The N and C-terminal fragments of barnase form a native-like complex with restored enzymic activity. *J Mol Biol* **224**, 741–7.
110. Sancho, J., Neira, J. L. & Fersht, A. R. (1992). An N-terminal fragment of barnase has residual helical structure similar to that in a refolding intermediate. *J Mol Biol* **224**, 749–58.
111. Horovitz, A. & Fersht, A. R. (1992). Co-operative interactions during protein folding. *J Mol Biol* **224**, 733–40.
112. Fersht, A. R., Matouschek, A. & Serrano, L. (1992). The folding of an enzyme. I. Theory of protein engineering analysis of stability and pathway of protein folding. *J Mol Biol* **224**, 771–82.
113. Matouschek, A., Serrano, L. & Fersht, A. R. (1992). The folding of an enzyme. IV. Structure of an intermediate in the refolding of barnase analysed by a protein engineering procedure. *J Mol Biol* **224**, 819–35.

114. Matouschek, A., Serrano, L., Meiering, E. M., Bycroft, M. & Fersht, A. R. (1992). The folding of an enzyme. V. $^1H/^2H$ exchange-nuclear magnetic resonance studies on the folding pathway of barnase: complementarity to and agreement with protein engineering studies. *J Mol Biol* **224**, 837–45.
115. Serrano, L., Matouschek, A. & Fersht, A. R. (1992). The folding of an enzyme. VI. The folding pathway of barnase: comparison with theoretical models. *J Mol Biol* **224**, 847–59.
116. Serrano, L., Matouschek, A. & Fersht, A. R. (1992). The folding of an enzyme. III. Structure of the transition state for unfolding of barnase analysed by a protein engineering procedure. *J Mol Biol* **224**, 805–18.
117. Fersht, A. R. (1993). The sixth Datta Lecture. Protein folding and stability: The pathway of folding of barnase. *FEBS Lett* **325**, 5–16.
118. Fersht, A. R. (1995). Characterizing transition states in protein folding: An essential step in the puzzle. *Curr Opin Struct Biol* **5**, 79–84.
119. Fersht, A. R. & Sato, S. (2004). Φ-value analysis and the nature of protein-folding transition states. *Proc Natl Acad Sci USA* **101**, 7976–81.
120. Jackson, S. E. & Fersht, A. R. (1991). Folding of chymotrypsin inhibitor 2. 1. Evidence for a two-state transition. *Biochemistry* **30**, 10428–35.
121. Otzen, D. E., Itzhaki, L. S., elMasry, N. F., Jackson, S. E. & Fersht, A. R. (1994). Structure of the transition state for the folding/unfolding of the barley chymotrypsin inhibitor 2 and its implications for mechanisms of protein folding. *Proc Natl Acad Sci USA* **91**, 10422–5.
122. Itzhaki, L. S., Otzen, D. E. & Fersht, A. R. (1995). The structure of the transition state for folding of chymotrypsin inhibitor 2 analysed by protein engineering methods: Evidence for a nucleation-condensation mechanism for protein folding. *J Mol Biol* **254**, 260–88.
123. Fersht, A. R. (1995). Optimization of rates of protein folding: The nucleation-condensation mechanism and its implications. *Proc Natl Acad Sci USA* **92**, 10869–73.
124. Fersht, A. R. (2000). Transition-state structure as a unifying basis in protein-folding mechanisms: Contact order, chain topology, stability, and the extended nucleus mechanism. *Proc Natl Acad Sci USA* **97**, 1525–9.
125. Matouschek, A. & Fersht, A. R. (1993). Application of physical organic chemistry to engineered mutants of proteins: Hammond postulate behavior in the transition state of protein folding. *Proc Natl Acad Sci USA* **90**, 7814–8.
126. Matouschek, A., Otzen, D. E., Itzhaki, L. S., Jackson, S. E. & Fersht, A. R. (1995). Movement of the position of the transition state in protein folding. *Biochemistry* **34**, 13656–62.
127. Matthews, J. M. & Fersht, A. R. (1995). Exploring the energy surface of protein folding by structure–reactivity relationships and engineered proteins: Observation of Hammond behavior for the gross structure of the transition state and anti-Hammond behavior for structural elements for unfolding/folding of barnase. *Biochemistry* **34**, 6805–14.

128. Daggett, V., Li, A. J. & Fersht, A. R. (1998). Combined molecular dynamics and Phi-value analysis of structure–reactivity relationships in the transition state and unfolding pathway of barnase: Structural basis of Hammond and anti-Hammond effects. *J Am Chem Soc* **120**, 12740–54.
129. Oliveberg, M., Vuilleumier, S. & Fersht, A. R. (1994). Thermodynamic study of the acid denaturation of barnase and its dependence on ionic strength: Evidence for residual electrostatic interactions in the acid/thermally denatured state. *Biochemistry* **33**, 8826–32.
130. Oliveberg, M., Arcus, V. L. & Fersht, A. R. (1995). pK_A values of carboxyl groups in the native and denatured states of barnase: The pK_A values of the denatured state are on average 0.4 units lower than those of model compounds. *Biochemistry* **34**, 9424–33.
131. Oliveberg, M., Tan, Y. J. & Fersht, A. R. (1995). Negative activation enthalpies in the kinetics of protein folding. *Proc Natl Acad Sci USA* **92**, 8926–9.
132. Tan, Y. J., Oliveberg, M., Davis, B. & Fersht, A. R. (1995). Perturbed pK_A-values in the denatured states of proteins. *J Mol Biol* **254**, 980–92.
133. Oliveberg, M. & Fersht, A. R. (1996). New approach to the study of transient protein conformations: The formation of a semiburied salt link in the folding pathway of barnase. *Biochemistry* **35**, 6795–805.
134. Oliveberg, M. & Fersht, A. R. (1996). Thermodynamics of transient conformations in the folding pathway of barnase: reorganization of the folding intermediate at low pH. *Biochemistry* **35**, 2738–49.
135. Oliveberg, M. & Fersht, A. R. (1996). Formation of electrostatic interactions on the protein-folding pathway. *Biochemistry* **35**, 2726–37.
136. Tan, Y. J., Oliveberg, M. & Fersht, A. R. (1996). Titration properties and thermodynamics of the transition state for folding: comparison of two-state and multi-state folding pathways. *J Mol Biol* **264**, 377–89.
137. Johnson, C. M., Oliveberg, M., Clarke, J. & Fersht, A. R. (1997). Thermodynamics of denaturation of mutants of barnase with disulfide crosslinks. *J Mol Biol* **268**, 198–208.
138. Tan, Y. J., Oliveberg, M., Otzen, D. E. & Fersht, A. R. (1997). The rate of isomerisation of peptidyl-proline bonds as a probe for interactions in the physiological denatured state of chymotrypsin inhibitor 2. *J Mol Biol* **269**, 611–22.
139. Dalby, P. A., Oliveberg, M. & Fersht, A. R. (1998). Folding intermediates of wild-type and mutants of barnase. I. Use of Φ-value analysis and m-values to probe the cooperative nature of the folding pre-equilibrium. *J Mol Biol* **276**, 625–46.
140. Dalby, P. A., Oliveberg, M. & Fersht, A. R. (1998). Movement of the intermediate and rate determining transition state of barnase on the energy landscape with changing temperature. *Biochemistry* **37**, 4674–9.
141. Oliveberg, M., Tan, Y. J., Silow, M. & Fersht, A. R. (1998). The changing nature of the protein folding transition state: Implications for the shape of the free-energy profile for folding. *J Mol Biol* **277**, 933–43.

142. Silow, M., Tan, Y. J., Fersht, A. R. & Oliveberg, M. (1999). Formation of short-lived protein aggregates directly from the coil in two-state folding. *Biochemistry* **38**, 13006–12.
143. Daggett, V., Li, A., Itzhaki, L. S., Otzen, D. E. & Fersht, A. R. (1996). Structure of the transition state for folding of a protein derived from experiment and simulation. *J Mol Biol* **257**, 430–40.
144. Bond, C. J., Wong, K. B., Clarke, J., Fersht, A. R. & Daggett, V. (1997). Characterization of residual structure in the thermally denatured state of barnase by simulation and experiment: Description of the folding pathway. *Proc Natl Acad Sci USA* **94**, 13409–13.
145. Ladurner, A. G., Itzhaki, L. S., Daggett, V. & Fersht, A. R. (1998). Synergy between simulation and experiment in describing the energy landscape of protein folding. *Proc Natl Acad Sci USA* **95**, 8473–8.
146. Clarke, J., Hounslow, A. M., Bond, C. J., Fersht, A. R. & Daggett, V. (2000). The effects of disulfide bonds on the denatured state of barnase. *Protein Sci* **9**, 2394–404.
147. Mayor, U., Johnson, C. M., Daggett, V. & Fersht, A. R. (2000). Protein folding and unfolding in microseconds to nanoseconds by experiment and simulation. *Proc Natl Acad Sci USA* **97**, 13518–22.
148. Wong, K. B., Clarke, J., Bond, C. J., Neira, J. L., Freund, S. M., Fersht, A. R. & Daggett, V. (2000). Towards a complete description of the structural and dynamic properties of the denatured state of barnase and the role of residual structure in folding. *J Mol Biol* **296**, 1257–82.
149. Ferguson, N., Pires, J. R., Toepert, F., Johnson, C. M., Pan, Y. P., Volkmer-Engert, R., Schneider-Mergener, J., Daggett, V., Oschkinat, H. & Fersht, A. (2001). Using flexible loop mimetics to extend Φ-value analysis to secondary structure interactions. *Proc Natl Acad Sci USA* **98**, 13008–13.
150. Kazmirski, S. L., Wong, K. B., Freund, S. M., Tan, Y. J., Fersht, A. R. & Daggett, V. (2001). Protein folding from a highly disordered denatured state: The folding pathway of chymotrypsin inhibitor 2 at atomic resolution. *Proc Natl Acad Sci USA* **98**, 4349–54.
151. Fersht, A. R. & Daggett, V. (2002). Protein folding and unfolding at atomic resolution. *Cell* **108**, 573–82.
152. Kazmirski, S. L., Isaacson, R. L., An, C., Buckle, A., Johnson, C. M., Daggett, V. & Fersht, A. R. (2002). Loss of a metal-binding site in gelsolin leads to familial amyloidosis-Finnish type. *Nat Struct Biol* **9**, 112–6.
153. Daggett, V. & Fersht, A. (2003). The present view of the mechanism of protein folding. *Nat Rev Mol Cell Biol* **4**, 497–502.
154. Daggett, V. & Fersht, A. R. (2003). Is there a unifying mechanism for protein folding? *Trends Biochem Sci* **28**, 18–25.
155. Gianni, S., Guydosh, N. R., Khan, F., Caldas, T. D., Mayor, U., White, G. W., DeMarco, M. L., Daggett, V. & Fersht, A. R. (2003). Unifying features in protein-folding mechanisms. *Proc Natl Acad Sci USA* **100**, 13286–91.

156. Mayor, U., Guydosh, N. R., Johnson, C. M., Grossmann, J. G., Sato, S., Jas, G. S., Freund, S. M., Alonso, D. O., Daggett, V. & Fersht, A. R. (2003). The complete folding pathway of a protein from nanoseconds to microseconds. *Nature* **421**, 863–7.
157. Jemth, P., Gianni, S., Day, R., Li, B., Johnson, C. M., Daggett, V. & Fersht, A. R. (2004). Demonstration of a low-energy on-pathway intermediate in a fast-folding protein by kinetics, protein engineering, and simulation. *Proc Natl Acad Sci USA* **101**, 6450–5.
158. Sato, S., Religa, T. L., Daggett, V. & Fersht, A. R. (2004). Testing protein-folding simulations by experiment: B domain of protein A. *Proc Natl Acad Sci USA* **101**, 6952–6.
159. Ferguson, N., Day, R., Johnson, C. M., Allen, M. D., Daggett, V. & Fersht, A. R. (2005). Simulation and experiment at high temperatures: Ultrafast folding of a thermophilic protein by nucleation-condensation. *J Mol Biol* **347**, 855–70.
160. Jemth, P., Day, R., Gianni, S., Khan, F., Allen, M., Daggett, V. & Fersht, A. R. (2005). The structure of the major transition state for folding of an FF domain from experiment and simulation. *J Mol Biol* **350**, 363–78.
161. White, G. W., Gianni, S., Grossmann, J. G., Jemth, P., Fersht, A. R. & Daggett, V. (2005). Simulation and experiment conspire to reveal cryptic intermediates and a slide from the nucleation-condensation to framework mechanism of folding. *J Mol Biol* **350**, 757–75.
162. Petrovich, M., Jonsson, A. L., Ferguson, N., Daggett, V. & Fersht, A. R. (2006). Phi-analysis at the experimental limits: mechanism of beta-hairpin formation. *J Mol Biol* **360**, 865–81.
163. Sharpe, T., Jonsson, A. L., Rutherford, T. J., Daggett, V. & Fersht, A. R. (2007). The role of the turn in β-hairpin formation during WW domain folding. *Protein Sci* **16**, 2233–9.
164. Schaeffer, R. D., Fersht, A. & Daggett, V. (2008). Combining experiment and simulation in protein folding: Closing the gap for small model systems. *Curr Opin Struct Biol* **18**, 4–9.
165. Nolting, B., Golbik, R. & Fersht, A. R. (1995). Submillisecond events in protein folding. *Proc Natl Acad Sci USA* **92**, 10668–72.
166. Nolting, B., Golbik, R., Neira, J. L., Soler-Gonzalez, A. S., Schreiber, G. & Fersht, A. R. (1997). The folding pathway of a protein at high resolution from microseconds to seconds. *Proc Natl Acad Sci USA* **94**, 826–30.
167. Religa, T. L., Markson, J. S., Mayor, U., Freund, S. M. & Fersht, A. R. (2005). Solution structure of a protein denatured state and folding intermediate. *Nature* **437**, 1053–6.
168. Huang, F., Sato, S., Sharpe, T. D., Ying, L. & Fersht, A. R. (2007). Distinguishing between cooperative and unimodal downhill protein folding. *Proc Natl Acad Sci USA* **104**, 123–7.
169. Huang, F., Ying, L. & Fersht, A. R. (2009). Direct observation of barrier limited folding of BBL by single-molecule fluorescence resonance energy transfer. *Proc Natl Acad Sci USA* **106**, 16239–44.

170. Ferguson, N., Schartau, P. J., Sharpe, T. D., Sato, S. & Fersht, A. R. (2004). One-state downhill versus conventional protein folding. *J Mol Biol* **344**, 295–301.
171. Ferguson, N., Sharpe, T. D., Johnson, C. M. & Fersht, A. R. (2006). The transition state for folding of a peripheral subunit-binding domain contains robust and ionic-strength dependent characteristics. *J Mol Biol* **356**, 1237–47.
172. Ferguson, N., Sharpe, T. D., Johnson, C. M., Schartau, P. J. & Fersht, A. R. (2007). Structural biology: Analysis of 'downhill' protein folding. *Nature* **445**, E14–5; discussion E17–8.
173. Ferguson, N., Sharpe, T. D., Schartau, P. J., Sato, S., Allen, M. D., Johnson, C. M., Rutherford, T. J. & Fersht, A. R. (2005). Ultra-fast barrier-limited folding in the peripheral subunit-binding domain family. *J Mol Biol* **353**, 427–46.
174. Neuweiler, H., Sharpe, T. D., Rutherford, T. J., Johnson, C. M., Allen, M. D., Ferguson, N. & Fersht, A. R. (2009). The folding mechanism of BBL: Plasticity of transition-state structure observed within an ultrafast folding protein family. *J Mol Biol* **390**, 1060–73.
175. Settanni, G. & Fersht, A. R. (2009). Downhill versus barrier-limited folding of BBL 3. Heterogeneity of the native state of the BBL peripheral subunit binding domain and its implications for folding mechanisms. *J Mol Biol* **387**, 993–1001.
176. Arbely, E., Rutherford, T. J., Sharpe, T. D., Ferguson, N. & Fersht, A. R. (2009). Downhill versus barrier-limited folding of BBL 1: Energetic and structural perturbation effects upon protonation of a histidine of unusually low pKa. *J Mol Biol* **387**, 986–92.
177. Neuweiler, H., Johnson, C. M. & Fersht, A. R. (2009). Direct observation of ultra-fast folding and denatured state dynamics in single protein molecules. *Proc Natl Acad Sci USA*.
178. Gray, T. E. & Fersht, A. R. (1991). Cooperativity in ATP hydrolysis by GroEL is increased by GroES. *FEBS Lett* **292**, 254–8.
179. Gray, T. E., Eder, J., Bycroft, M., Day, A. G. & Fersht, A. R. (1993). Refolding of barnase mutants and pro-barnase in the presence and absence of GroEL. *EMBO J* **12**, 4145–50.
180. Gray, T. E. & Fersht, A. R. (1993). Refolding of barnase in the presence of GroE. *J Mol Biol* **232**, 1197–207.
181. Buckle, A. M., Zahn, R. & Fersht, A. R. (1997). A structural model for GroEL–polypeptide recognition. *Proc Natl Acad Sci USA* **94**, 3571–5.
182. Kobayashi, N., Freund, S. M., Chatellier, J., Zahn, R. & Fersht, A. R. (1999). NMR analysis of the binding of a rhodanese peptide to a minichaperone in solution. *J Mol Biol* **292**, 181–90.
183. Chatellier, J., Buckle, A. M. & Fersht, A. R. (1999). GroEL recognises sequential and non-sequential linear structural motifs compatible with extended beta-strands and α-helices. *J Mol Biol* **292**, 163–72.
184. Wang, Q., Buckle, A. M. & Fersht, A. R. (2000). From minichaperone to GroEL 1: information on GroEL–polypeptide interactions from crystal packing of minichaperones. *J Mol Biol* **304**, 873–81.

185. Zahn, R., Buckle, A. M., Perrett, S., Johnson, C. M., Corrales, F. J., Golbik, R. & Fersht, A. R. (1996). Chaperone activity and structure of monomeric polypeptide binding domains of GroEL. *Proc Natl Acad Sci USA* **93**, 15024–9.
186. Chatellier, J., Hill, F., Lund, P. A. & Fersht, A. R. (1998). In vivo activities of GroEL minichaperones. *Proc Natl Acad Sci USA* **95**, 9861–6.
187. Chatellier, J., Hill, F., Foster, N. W., Goloubinoff, P. & Fersht, A. R. (2000). From minichaperone to GroEL 3: Properties of an active single-ring mutant of GroEL. *J Mol Biol* **304**, 897–910.
188. Altamirano, M. M., Golbik, R., Zahn, R., Buckle, A. M. & Fersht, A. R. (1997). Refolding chromatography with immobilized mini-chaperones. *Proc Natl Acad Sci USA* **94**, 3576–8.
189. Altamirano, M. M., Garcia, C., Possani, L. D. & Fersht, A. R. (1999). Oxidative refolding chromatography: folding of the scorpion toxin Cn5. *Nat Biotechnol* **17**, 187–91.
190. Altamirano, M. M., Woolfson, A., Donda, A., Shamshiev, A., Briseno-Roa, L., Foster, N. W., Veprintsev, D. B., De Libero, G., Fersht, A. R. & Milstein, C. (2001). Ligand-independent assembly of recombinant human CD1 by using oxidative refolding chromatography. *Proc Natl Acad Sci USA* **98**, 3288–93.
191. Karadimitris, A., Gadola, S., Altamirano, M., Brown, D., Woolfson, A., Klenerman, P., Chen, J. L., Koezuka, Y., Roberts, I. A., Price, D. A., Dusheiko, G., Milstein, C., Fersht, A., Luzzatto, L. & Cerundolo, V. (2001). Human CD1d-glycolipid tetramers generated by in vitro oxidative refolding chromatography. *Proc Natl Acad Sci USA* **98**, 3294–8.
192. Batuwangala, T., Shepherd, D., Gadola, S. D., Gibson, K. J., Zaccai, N. R., Fersht, A. R., Besra, G. S., Cerundolo, V. & Jones, E. Y. (2004). The crystal structure of human CD1b with a bound bacterial glycolipid. *J Immunol* **172**, 2382–8.
193. Koch, M., Stronge, V. S., Shepherd, D., Gadola, S. D., Mathew, B., Ritter, G., Fersht, A. R., Besra, G. S., Schmidt, R. R., Jones, E. Y. & Cerundolo, V. (2005). The crystal structure of human CD1d with and without alpha-galactosylceramide. *Nat Immunol* **6**, 819–26.
194. Buckle, A. M., Schreiber, G. & Fersht, A. R. (1994). Protein-protein recognition: Crystal structural analysis of a barnase-barstar complex at 2.0-Å resolution. *Biochemistry* **33**, 8878–89.
195. Frisch, C., Fersht, A. R. & Schreiber, G. (2001). Experimental assignment of the structure of the transition state for the association of barnase and barstar. *J Mol Biol* **308**, 69–77.
196. Frisch, C., Schreiber, G., Johnson, C. M. & Fersht, A. R. (1997). Thermodynamics of the interaction of barnase and barstar: Changes in free energy versus changes in enthalpy on mutation. *J Mol Biol* **267**, 696–706.
197. Schreiber, G., Buckle, A. M. & Fersht, A. R. (1994). Stability and function: Two constraints in the evolution of barstar and other proteins. *Structure* **2**, 945–51.
198. Schreiber, G. & Fersht, A. R. (1995). Energetics of protein–protein interactions: Analysis of the barnase–barstar interface by single mutations and double mutant cycles. *J Mol Biol* **248**, 478–86.

199. Schreiber, G. & Fersht, A. R. (1996). Rapid, electrostatically assisted association of proteins. *Nat Struct Biol* **3**, 427–31.
200. Vijayakumar, M., Wong, K. Y., Schreiber, G., Fersht, A. R., Szabo, A. & Zhou, H. X. (1998). Electrostatic enhancement of diffusion-controlled protein–protein association: Comparison of theory and experiment on barnase and barstar. *J Mol Biol* **278**, 1015–24.
201. Bullock, A. N., Henckel, J., DeDecker, B. S., Johnson, C. M., Nikolova, P. V., Proctor, M. R., Lane, D. P. & Fersht, A. R. (1997). Thermodynamic stability of wild-type and mutant p53 core domain. *Proc Natl Acad Sci USA* **94**, 14338–42.
202. Bullock, A. N., Henckel, J. & Fersht, A. R. (2000). Quantitative analysis of residual folding and DNA binding in mutant p53 core domain: Definition of mutant states for rescue in cancer therapy. *Oncogene* **19**, 1245–56.
203. Bullock, A. N. & Fersht, A. R. (2001). Rescuing the function of mutant p53. *Nat Rev Cancer* **1**, 68–76.
204. Friedler, A., Hansson, L. O., Veprintsev, D. B., Freund, S. M., Rippin, T. M., Nikolova, P. V., Proctor, M. R., Rudiger, S. & Fersht, A. R. (2002). A peptide that binds and stabilizes p53 core domain: Chaperone strategy for rescue of oncogenic mutants. *Proc Natl Acad Sci USA* **99**, 937–42.
205. Issaeva, N., Friedler, A., Bozko, P., Wiman, K. G., Fersht, A. R. & Selivanova, G. (2003). Rescue of mutants of the tumor suppressor p53 in cancer cells by a designed peptide. *Proc Natl Acad Sci USA* **100**, 13303–7.
206. Nikolova, P. V., Henckel, J., Lane, D. P. & Fersht, A. R. (1998). Semirational design of active tumor suppressor p53 DNA binding domain with enhanced stability. *Proc Natl Acad Sci USA* **95**, 14675–80.
207. Joerger, A. C., Allen, M. D. & Fersht, A. R. (2004). Crystal structure of a superstable mutant of human p53 core domain. Insights into the mechanism of rescuing oncogenic mutations. *J Biol Chem* **279**, 1291–6.
208. Joerger, A. C., Ang, H. C., Veprintsev, D. B., Blair, C. M. & Fersht, A. R. (2005). Structures of p53 cancer mutants and mechanism of rescue by second-site suppressor mutations. *J Biol Chem* **280**, 16030–7.
209. Ang, H. C., Joerger, A. C., Mayer, S. & Fersht, A. R. (2006). Effects of common cancer mutations on stability and DNA binding of full-length p53 compared with isolated core domains. *J Biol Chem* **281**, 21934–41.
210. Joerger, A. C., Ang, H. C. & Fersht, A. R. (2006). Structural basis for understanding oncogenic p53 mutations and designing rescue drugs. *Proc Natl Acad Sci USA* **103**, 15056–61.
211. Joerger, A. C. & Fersht, A. R. (2007). Structural biology of the tumor suppressor p53 and cancer-associated mutants. *Adv Cancer Res* **97**, 1–23.
212. Joerger, A. C. & Fersht, A. R. (2007). Structure-function-rescue: The diverse nature of common p53 cancer mutants. *Oncogene* **26**, 2226–42.
213. Joerger, A. C. & Fersht, A. R. (2008). Structural biology of the tumor suppressor p53. *Annu Rev Biochem* **77**, 557–82.

214. Boeckler, F. M., Joerger, A. C., Jaggi, G., Rutherford, T. J., Veprintsev, D. B. & Fersht, A. R. (2008). Targeted rescue of a destabilized mutant of p53 by an *in silico* screened drug. *Proc Natl Acad Sci USA* **105**, 10360–5.
215. Tidow, H., Melero, R., Mylonas, E., Freund, S. M., Grossmann, J. G., Carazo, J. M., Svergun, D. I., Valle, M. & Fersht, A. R. (2007). Quaternary structures of tumor suppressor p53 and a specific p53–DNA complex. *Proc Natl Acad Sci USA* **104**, 12324–9.
216. Wells, M., Tidow, H., Rutherford, T. J., Markwick, P., Jensen, M. R., Mylonas, E., Svergun, D. I., Blackledge, M. & Fersht, A. R. (2008). Structure of tumor suppressor p53 and its intrinsically disordered N-terminal transactivation domain. *Proc Natl Acad Sci USA* **105**, 5762–7.
217. Fersht, A. R. (2008). From the first protein structures to our current knowledge of protein folding: Delights and scepticisms. *Nat Rev Mol Cell Biol* **9**, 650–4.

214. Boeckler, F. M., Joerger, A. C., Jaggi, G., Rutherford, T. J., Veprintsev, D. B. & Fersht, A. R. (2008). Targeted rescue of a destabilized mutant of p53 by an in silico screened drug. Proc. Natl Acad Sci USA 105, 10360–5.
215. Tidow, H., Melero, R., Mylonas, E., Freund, S. M. V., Grossmann, J. G., Carazo, J. M., Svergun, D. I., Valle, M. & Fersht, A. R. (200 [illegible]

[illegible]

[illegible]